nature

The Living Record of Science

《自然》百年科学经典

（英汉对照版）

第一卷

总顾问：李政道（Tsung-Dao Lee）

英方主编：Sir John Maddox
Sir Philip Campbell

中方主编：路甬祥

I

1869-1930

外语教学与研究出版社 · 麦克米伦教育 · 自然科研

FOREIGN LANGUAGE TEACHING AND RESEARCH PRESS · MACMILLAN EDUCATION · NATURE RESEARCH

北京 BEIJING

京权图字：01-2006-7364

图书在版编目（CIP）数据

《自然》百年科学经典．第1卷，1869～1930：英汉对照／（英）马多克斯（Maddox, J.），（英）坎贝尔（Campbell, P.），路甬祥主编．— 北京：外语教学与研究出版社，2009.10（2020.5 重印）
ISBN 978-7-5600-9091-7

Ⅰ．自… Ⅱ．①马… ②坎… ③路… Ⅲ．自然科学－文集－英、汉 Ⅳ．N53

中国版本图书馆 CIP 数据核字（2009）第 186327 号

出 版 人　徐建忠
项目统筹　章思英　Charlotte Liu（加拿大）
项目负责　刘晓楠　黄小斌　Mary-Jane Newton（德）
责任编辑　刘　明　黄小斌　张梦璇
装帧设计　孙莉明
出版发行　外语教学与研究出版社
社　　址　北京市西三环北路 19 号（100089）
网　　址　http://www.fltrp.com
印　　刷　北京华联印刷有限公司
开　　本　787×1092　1/16
印　　张　58.5
版　　次　2009 年 11 月第 1 版　2020 年 5 月第 3 次印刷
书　　号　ISBN 978-7-5600-9091-7
定　　价　568.00 元

购书咨询：（010）88819926　电子邮箱：club@fltrp.com
外研书店：https://waiyants.tmall.com
凡印刷、装订质量问题，请联系我社印制部
联系电话：（010）61207896　电子邮箱：zhijian@fltrp.com
凡侵权、盗版书籍线索，请联系我社法律事务部
举报电话：（010）88817519　电子邮箱：banquan@fltrp.com
物料号：190910101

《自然》百年科学经典（英汉对照版）

编译委员会

Foreword by Lu Yongxiang

Since the birth of modern science, and in particular throughout the 20th century, we have continuously deepened our understanding of Nature, and developed more means and methods to make use of natural resources. Technological innovation and industrial progress have become decisive factors in promoting unprecedented development of productive forces and the progress of society, and have greatly improved the mode of production and the way we live.

The 20th century witnessed many revolutions in science. The establishment and development of quantum theory and the theory of relativity have changed our concept of time and space, and have given us a unified understanding of matter and energy. They served as a theoretical foundation upon which a series of major scientific discoveries and technological inventions were made. The discovery of the structure of DNA transformed our understanding of heredity and helped to unify our vision of the biological world. As a corner-stone in biology, DNA research has exerted a far-reaching influence on modern agriculture and medicine. The development of information science has provided a theoretical basis for computer science, communication technology, intelligent manufacturing, understanding of human cognition, and even economic and social studies. The theory of continental drift and plate tectonics has had important implications for seismology, geology of ore deposits, palaeontology, and palaeoclimatology. New understandings about the cosmos have enabled us to know in general terms, and also in many details, how elementary particles and chemical elements were formed, and how this led to the formation of molecules and the appearance of life, and even the origin and evolution of the entire universe.

The 20th century also witnessed revolutions in technology. Breakthroughs in fundamental research, coupled to the stimulus of market forces, have led to unparalleled technological achievements. Energy, materials, information, aviation and aeronautics, and biological medicine have undergone dramatic changes. Specifically, new energy technologies have helped to promote social development; new materials technologies promote the growth of manufacturing and industrial prosperity; information technology has ushered in the Internet and the pervasive role of computing; aviation and aeronautical technology has broadened our vision and mobility, and has ultimately led to the exploration of the universe beyond our planet; and improvements in medical and biological technology have enabled people to live much better, healthier lives.

Outstanding achievements in science and technology made in China during its long history have contributed to the survival, development and continuation of the Chinese nation. The country remained ahead of Europe for several hundred years before the 15th century. As Joseph Needham's studies demonstrated, a great many discoveries and innovations in understanding or practical capability—from the shape of snowflakes to the art of cartography, the circulation of the blood, the invention of paper and sericulture

路甬祥序

自近代科学诞生以来，特别是20世纪以来，随着人类对自然的认识不断加深，随着人类利用自然资源的手段与方法不断丰富，技术创新、产业进步已成为推动生产力空前发展和人类社会进步的决定性因素，极大地改变了人类的生产与生活方式，使人类社会发生了显著的变化。

20世纪是科学革命的世纪。量子理论和相对论的创立与发展，改变了人类的时空观和对物质与能量统一性的认识，成为了20世纪一系列重大科学发现和技术发明的理论基石；DNA双螺旋结构模型的建立，标志着人类在揭示生命遗传奥秘方面迈出了具有里程碑意义的一步，奠定了生物技术的基础，对现代农业和医学的发展产生了深远影响；信息科学的发展为计算机科学、通信技术、智能制造提供了知识源泉，并为人类认知、经济学和社会学研究等提供了理论基础；大陆漂移学说和板块构造理论，对地震学、矿床学、古生物地质学、古气候学具有重要的指导作用；新的宇宙演化观念的建立为人们勾画出了基本粒子和化学元素的产生、分子的形成和生命的出现，乃至整个宇宙的起源和演化的图景。

20世纪也是技术革命的世纪。基础研究的重大突破和市场的强劲拉动，使人类在技术领域获得了前所未有的成就，能源、材料、信息、航空航天、生物医学等领域发生了全新变化。新能源技术为人类社会发展提供了多元化的动力；新材料技术为人类生活和科技进步提供了丰富的物质材料基础，推动了制造业的发展和工业的繁荣；信息技术使人类迈入了信息和网络时代；航空航天技术拓展了人类的活动空间和视野；医学与生物技术的进展极大地提高了人类的生活质量和健康水平。

历史上，中国曾经创造出辉煌的科学技术，支撑了中华民族的生存、发展和延续。在15世纪之前的数百年里，中国的科技水平曾遥遥领先于欧洲。李约瑟博士曾经指出，从雪花的形状到绘图的艺术、血液循环、造纸、养蚕，包括更有名的指南针和

and, most famously, of compasses and gunpowder—were first made in China. The Four Great Inventions in ancient China have influenced the development process of the world. Ancient Chinese astronomical records are still used today by astronomers seeking to understand astrophysical phenomena. Thus Chinese as well as other long-standing civilizations in the world deserve to be credited as important sources of modern science and technology.

Scientific and technological revolutions in 17th and 18th century Europe, the First and Second Industrial Revolutions in the 18th and 19th centuries, and the spread of modern science education and knowledge sped up the modernization process of the West. During these centuries, China lagged behind.

Defeat in the Opium War (1840-1842) served as strong warning to the ancient Chinese empire. Around and after the time of the launch of *Nature* in 1869, elite intellectuals in China had come to see the importance that science and technology had towards the country's development. Many scholars went to study in Western higher education and research institutions, and some made outstanding contributions to science. Many students who had completed their studies and research in the West returned to China, and their work, together with that of home colleagues, laid the foundation for the development of modern science and technology in the country.

In the six decades since the founding of the People's Republic of China, the country has made a series of achievements in science and technology. Chinese scientists independently developed the atomic bomb, the hydrogen bomb and artificial satellite within a short period of time. The continental oil generation theory led to the discovery of the Daqing oil field in the northeast. Chinese scientists also succeeded in synthesizing bovine insulin, the first protein to be made by synthetic chemical methods. The development and popularization of hybrid rice strains have significantly increased the yields from rice cultivation, benefiting hundreds of millions of people across the world. Breakthroughs in many other fields, such as materials science, aeronautics and life science, all represent China's progress in modern science and technology.

As the Chinese economy continues to enjoy rapid growth, scientific research is also producing increasing results. Many of these important results have been published in first-class international science journals such as *Nature*. This has expanded the influence of Chinese science research, and promoted exchange and cooperation between Chinese scientists with colleagues in other countries. All these indicate that China has become a significant global force in science and technology and that greater progress is expected in the future.

Science journals, which developed alongside modern science, play an essential role in faithfully recording the path of science, as well as spreading and promoting modern science. Such journals report academic development in a timely manner, provide a platform for scientists to exchange ideas and methods, explore the future direction of science, stimulate academic debates, promote academic prosperity, and help the public

火药,都是首先由中国人发现或发明的。中国的“四大发明”影响了世界的发展进程，古代中国的天文记录至今仍为天文学家在研究天体物理现象时所使用。中华文明同其他悠久的人类文明一样，成为了近代科学技术的重要源泉。

但我们也要清醒地看到，发生在17~18世纪欧洲的科学革命、18~19世纪的第一次和第二次工业革命,以及现代科学教育与知识的传播,加快了西方现代化的进程,同时也拉大了中国与西方的差距。

鸦片战争的失败给古老的中华帝国敲响了警钟。就在《自然》创刊前后，中国的一批精英分子看到了科学技术对于国家发展的重要性，一批批中国学子到西方高校及研究机构学习，其中一些人在科学领域作出了杰出的贡献。同时，一大批留学生回国，同国内的知识分子一道，为现代科学技术在中国的发展奠定了基础。

新中国成立60年来,中国在科学技术方面取得了一系列成就。在很短的时间里,独立自主地研制出“两弹一星”；在陆相生油理论指导下，发现了大庆油田；成功合成了牛胰岛素，这是世界上第一个通过化学方法人工合成的蛋白质；杂交水稻研发及其品种的普及，显著提高了水稻产量，造福了全世界几亿人。中国人在材料科学、航天、生命科学等许多领域，也取得了一批重要成果。这些都展现了中国在现代科技领域所取得的巨大进步。

当前，中国经济持续快速增长，科研产出日益增加，中国的许多重要成果已经发表在像《自然》这样的世界一流的科技期刊上，扩大了中国科学研究的影响，推动了中国科学家和国外同行的交流与合作。现在，中国已成为世界重要的科技力量。可以预见，在未来，中国将在科学和技术方面取得更大的进步。

伴随着现代科学产生的科技期刊，忠实地记录了科学发展的轨迹，在传播和促进现代科学的发展方面发挥了重要的作用。科技期刊及时地报道学术进展，交流科学思想和方法,探讨未来发展方向,以带动学术争鸣与繁荣,促进公众对科学的理解。中国在推动科技进步的同时，应更加重视科技期刊的发展，学习包括《自然》在内

to better understand science. While promoting science and technology, China should place greater emphasis on the betterment of science journals. We should draw on the philosophies and methods of leading science journals such as *Nature*, improve the standards of digital access, and enable some of our own science journals to extend their impact beyond China in the not too distant future so that they can serve as an advanced platform for the development of science and technology in our country.

In the 20th century, *Nature* published many remarkable discoveries in disciplines such as biology, geoscience, environmental science, materials science, and physics. The selection and publication of the best of the more than 100,000 articles in *Nature* over the past 150 years or so in English-Chinese bilingual format is a highly meaningful joint undertaking by the Foreign Language Teaching and Research Press, Macmillan Publishers Limited and the Nature Publishing Group. I believe that *Nature: the Living Record of Science* will help bridge cultural differences, promote international cooperation in science and technology, prove to be high-standard readings for its intended large audience, and play a positive role in improving scientific and technological research in our country. I fully endorse and support the project.

The volumes offer a picture of the course of science for nearly 150 years, from which we can explore how science develops, draw inspiration for new ideas and wisdom, and learn from the unremitting spirit of scientists in research. Reading these articles is like vicariously experiencing the great discoveries by scientific giants in the past, which will enable us to see wider, think deeper, work better, and aim higher. I believe this collection will also help interested readers from other walks of life to gain a better understanding of and care more about science, thus increasing their respect for and confidence in science.

I should like to take this opportunity to express my appreciation for the vision and joint efforts of Foreign Language Teaching and Research Press, Macmillan Publishers Limited and the Nature Publishing Group in bringing forth this monumental work, and my thanks to all the translators, reviewers and editors for their exertions in maintaining its high quality.

President of Chinese Academy of Sciences

的世界先进科技期刊的办刊理念和方法，提高期刊的数字化水平，使中国的一些科技期刊早日具备世界影响力，为中国科学技术的发展创建高水平的平台。

20 世纪的生物学、地球科学、环境科学、材料科学和物理学等领域的许多重大发现，都被记录在《自然》上。外语教学与研究出版社、麦克米伦出版集团和自然出版集团携手合作，从《自然》创刊近一百五十年来发表过的十万余篇论文中撷取精华，并译成中文，以双语的形式呈现，纂为《〈自然〉百年科学经典》丛书。我认为这是一项很有意义的工作，并相信本套丛书的出版将跨越不同的文化，促进国际间的科技交流，向广大中国读者提供高水平的科学技术知识文献，为提升我国科学技术研发水平发挥积极的作用。我赞成并积极支持此项工作。

丛书将带领我们回顾近一百五十年来科学的发展历程，从中探索科学发展的规律，寻求思想和智慧的启迪，感受科学家们百折不挠的钻研精神。阅读这套丛书，读者可以重温科学史上一些科学巨匠作出重大科学发现的历程，拓宽视野，拓展思路，提升科研能力，提高科学道德。我相信，这套丛书一定能成为社会各界的良师益友，增强他们对科学的了解与热情，加深他们对科学的尊重与信心。

借此机会向外语教学与研究出版社、麦克米伦出版集团、自然出版集团策划出版本丛书的眼光和魄力表示赞赏，对翻译者、审校者和编辑者为保证丛书质量付出的辛苦劳动表示感谢。

是为序。

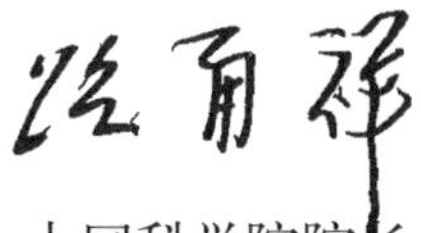

中国科学院院长

Foreword by Tsung Dao Lee

We can appreciate the significance of natural science to human life in two aspects. Materially, natural science has achieved many breakthroughs, particularly in the past hundred years or so, which have brought about revolutionary changes to human life. At the same time, the spirit of science has taken an ever-deepening root in the hearts of the people. Instead of alleging that science is omnipotent, the spirit of science emphasizes down-to-earth and scrupulous research, and critical and creative courage. More importantly, it stands for the dedication to working for the wellbeing of humankind. This is perhaps more meaningful than scientific and technological achievements themselves, which may be closely related to specific backgrounds of the times. The spirit of science, on the other hand, constitutes a most valuable and constant component of humankind's spiritual civilization.

In this sense, *Nature: the Living Record of Science* presents not only the historical paths of the various fields of natural science for almost a century and a half, but also the unremitting spirit of numerous scientists in their pursuit of truth. One of the most influential science journals in the whole world, *Nature*, reflects a general picture of different branches of science in different stages of development. It has also reported many of the most important discoveries in modern science. The collection of papers in this series includes breakthroughs such as the special theory of relativity, the maturing of quantum mechanics and the mapping of the human genome sequence. In addition, the editors have not shunned papers which were proved to be wrong after publication. Included also are the academic debates over the relevant topics. This speaks volumes of their vision and broadmindedness. Arduous is the road of science; behind any success are countless failures unknown to outsiders. But such failures have laid the foundation for success in later times and thus should not be forgotten. The comprehensive and thoughtful coverage of these volumes will enable readers to gain a better understanding of the achievements that have tremendously promoted the progress of science and technology, the evolution of key and cutting-edge issues of the relevant fields, the inspiration brought about by academic controversies, the efforts and hardships behind these achievements, and the true meaning of the spirit of science.

China now enjoys unprecedented opportunities for the development of science and technology. At the policy level, the state has created a fine environment for scientific research by formulating medium- and long-term development programs. As for science and technology, development in the past decades has built up a solid foundation of research and a rich pool of talent. Some major topics at present include how to introduce the cream of academic research from abroad, to promote Sino-foreign exchange in science and technology, to further promote the spirit of science, and to raise China's development in this respect to the advanced international level. The co-publication of *Nature: the Living Record of Science* by the Foreign Language Teaching and Research

李政道序

如何认识自然科学对人类生活的意义，可以从两个方面来分析：一是物质层面，尤其是近百年来，自然科学取得了很多跨越性的发展，给人类生活带来了许多革命性的变化；二是精神层面，科学精神日益深入人心，这种科学精神并不是认为科学万能、科学可以解决一切问题，它应该是一种老老实实、严谨缜密、又勇于批判和创造的精神，更重要的是，它具有一种坚持为人类福祉而奋斗的信念。这种科学精神可能比物质意义上的科技成就更重要，因为技术进步的影响可能与时代具体的背景有密切关系，但科学精神却永远是人类精神文明中最可宝贵的一部分。

从这个意义上，这套《〈自然〉百年科学经典》丛书的出版，不仅为读者呈现了一个多世纪以来自然科学各个领域发展的历史轨迹，更重要的是，它展现了无数科学家在追求真理的过程中艰难求索、百折不回的精神世界。《自然》作为全世界最有影响力的科学期刊之一，反映了各个学科在不同发展阶段的概貌，报道了现代科学中最重要的发现。这套丛书的可贵之处在于，它不仅汇聚了狭义相对论的提出、量子理论的成熟、人类基因组测序完成这些具有开创性和突破性的大事件、大成就，还将一些后来被证明是错误的文章囊括进来，并展现了围绕同一论题进行的学术争鸣，这是一种难得的眼光和胸怀。科学之路是艰辛的，成功背后有更多不为人知的失败，前人的失败是我们今日成功的基石，这些努力不应该被忘记。因此，《〈自然〉百年科学经典》这套丛书不但能让读者了解对人类科技进步有着巨大贡献的科学成果，以及科学中的焦点和前沿问题的演变轨迹，更能使有志于科学研究的人感受到思想激辩带来的火花和收获背后的艰苦努力，帮助他们理解科学精神的真意。

当前，中国科学技术的发展面临着历史上前所未有的机遇，国家已经制定了中长期科学和技术发展纲要，为科学研究创造了良好的制度环境，同时中国的科学技术经过多年的积累也已经具备了很好的理论和人才基础。如何进一步引进国外的学术精华，促进中外科技交流，使科学精神深入人心，使中国的科技水平迅速提升至世界前列就成为这一阶段的重要课题。因此，外语教学与研究出版社和麦克米伦出

Press, Macmillan Publishers Limited and the Nature Publishing Group will prove to be a huge contribution to the country's relevant endeavors. I sincerely wish for its success.

Science is a cause that does not have a finishing line, which is exactly the eternal charm of science and the source of inspiration for scientists to explore new frontiers. It is a cause worthy of our uttermost exertion.

T. D. Lee

版集团合作出版这套《〈自然〉百年科学经典》丛书，对中国的科技发展可谓贡献巨大，我衷心希望这套丛书的出版获得极大成功，促进全民族的科技振兴。

科学的事业永无止境。这是科学的永恒魅力所在，也是我们砥砺自身、不断求索的动力所在。这样的事业，值得我们全力以赴。

李政道

Preface

Nature is the world's most influential science journal. It has published some of the most important discoveries in modern science, and has carried contributions from leading scientists, ranging from Charles Darwin and Albert Einstein to James Watson, Francis Crick and Stephen Hawking. Since its earliest days it has reported on all areas of science, from the study of human origins to the structure of the universe, from genetics to nuclear physics.

So it is surprising that no substantial overview of *Nature*'s publication history has been attempted until now. And while *Nature* is known globally, access to its full archive has been rather less easy outside of Western countries (although the full archive is now available online). That is why this collection, titled *Nature: the Living Record of Science*, will provide an indispensable resource. It supplies an unparalleled view of how the preoccupations and priorities of science have changed over the last century and more, often in a way that reflects currents in the broader social and political landscape. The collected papers—more than 840 selected from over 100,000 published in the journal over the past century and a half—offer a vision of what society wants from science, and what science has given society.

The evolution of *Nature*

Nature is almost unique in publishing leading research in every area of science. The journal was begun in 1869 by the enterprising English astronomer J. Norman Lockyer. Its aim, announced (for reasons now forgotten) only in the second issue of 11 November, was:

> "First, to place before the general public the grand results of Scientific Work and Scientific Discovery; and to urge the claims of Science to a more general recognition in Education and in Daily Life;
>
> And, secondly, to aid Scientific men themselves, by giving early information of all advances made in any branch of Natural knowledge throughout the world, and by affording them an opportunity of discussing the various Scientific questions which arise from time to time."

That is a fair statement of *Nature*'s goal today. In the first issue, the reader could find Lockyer's description of a recent total solar eclipse in America, Thomas Henry Huxley's analysis of some newly discovered dinosaur fossil bones from the Triassic period, some observations of the absorption and radiation of heat by the German physicist Heinrich Gustav Magnus, and an obituary of the Scottish chemist Thomas Graham, the father of colloid chemistry. Such breadth of subject matter has been characteristic of the journal ever since.

前言

《自然》是全世界最具影响力的科学期刊。它报道过现代科学中一些最重要的发现，并刊登过如查尔斯·达尔文、阿尔伯特·爱因斯坦、詹姆斯·沃森、弗朗西斯·克里克、斯蒂芬·霍金等顶尖级科学家的文章。从创立初期，《自然》就涵盖了所有科学领域，从人类的起源到宇宙的结构、从遗传学到核物理学。

然而，令人颇为诧异的是，此前居然没有任何关于《自然》出版历史的有分量的概述。虽然《自然》是全球发行的，但在西方国家之外查阅其全文并不容易（当然，现在可以通过网络进行全文检索）。因此，《〈自然〉百年科学经典》这套选集可能会成为相关研究的第一手资料。通过这套选集提供的独特视角，读者可以了解过去一个多世纪中科学的前沿和热点经历了怎样的变迁，这通常能够折射出当时更为广阔的社会和政治生活的变化趋势。从《自然》在过去近一个半世纪里发表过的十万多篇论文中精选出来的这八百四十余篇文章，展现了社会对科学的冀求和科学对社会的贡献。

《自然》的变迁

《自然》几乎是独一无二的发表所有科学领域中开创性研究成果的杂志，由非常有魄力的英国天文学家约瑟夫·诺曼·洛克耶于 1869 年创立。其办刊宗旨发表在当年 11 月 11 日的《自然》第 2 期上（为什么没有发表在第 1 期上原因不详），内容如下：

首先，将科学研究和科学发现的重大成果呈现给公众，并促使科学理念在教育和日常生活中得到更为普遍的认可。

其次，帮助科学家自己，为他们提供自然科学各个分支在世界范围内取得的所有进展的最新信息，为他们探讨不时出现的各种科学问题提供交流平台。

上述内容也是对《自然》现在的办刊宗旨的一个恰当诠释。在《自然》第 1 期中，读者们可以读到洛克耶对当时的美洲日全食的描述，托马斯·亨利·赫胥黎对最新发现的一些三叠纪恐龙化石的分析，德国物理学家海因里希·古斯塔夫·马格努斯对热辐射和热吸收的一些观察数据，还有胶体化学之父、苏格兰化学家托马斯·格雷姆的讣告。自那时起，如此广泛的学科覆盖就一直是《自然》的特色。

The first issue of *Nature* appeared at a time when periodical publishing was booming and science was increasingly seen as an integral part of daily life. There was a general consensus that scientists deserved greater respect, social distinction and financial support. At the same time, there was a calling for scientific education to be expanded and interest in science to be encouraged. From its inception, *Nature* has been produced by the British publishers Macmillan & Co., although there is now no record of how that arrangement came about.

Lockyer was an astronomer and civil servant who had been elected as a Fellow of the Royal Society six months before *Nature* first appeared in print. He was well connected in the scientific community, counting the biologist Thomas Henry Huxley and the physicist John Tyndall among his circle of associates. He called on the services of both men in the early days of *Nature*, helping to establish its authoritative status. As editor, Lockyer displayed from the outset some of the characteristics that *Nature* went on to display in later times. He was unashamed to parade his own enthusiasms, making the journal particularly welcoming to research on the physics of the sun. He gave it an international flavour, including reports on meetings in such places as St. Petersburg, Vienna and Philadelphia. He was willing, indeed eager, to include news and gossip from within the scientific community, and happy to court controversy and to report it plainly: some of the arguments that rage in the early pages have an acerbic tone that reveals a hands-off editorial touch. And he was ready to offer robust opinions on public affairs and matters of state that might seem only tangentially relevant to science.

Yet despite *Nature*'s mission statement to "place before the general public the grand results of Scientific Work and Scientific Discovery", the journal made few concessions to the non-scientist. It was not until well into the following century that *Nature* underwent its metamorphosis. John Maddox became editor in 1966 and set about making the journal less scholarly and more engaging to readers, while in no way compromising its academic stature. It was Maddox's hope that everyone would be able to read and understand reports of new discoveries in any area of science.

That was an ambitious goal. Even now, the research papers in *Nature* are not easy reading for the lay person with no scientific education, and with the increasing specialization of science it is often difficult even for scientists to understand papers outside their own field. On the other hand, the expansion of *Nature* into a publishing group with many "sister" journals, such as *Nature Genetics*, *Nature Geoscience* and *Nature Materials*, as well as the advent of new media for communicating and providing content such as news, has given *Nature* further tools and avenues for reaching new, broader audiences. It remains the most highly cited interdisciplinary science journal.

Probably the most famous, and arguably the most influential, paper to appear in *Nature* was that written by Francis Crick and James Watson, published in 1953, describing the

《自然》第 1 期的面世恰逢期刊出版蓬勃发展之时，科学也日益成为日常生活中不可或缺的一部分。当时的普遍共识是科学家们理应得到更多的尊重、社会荣誉和资金支持。同时，人们呼吁普及科学教育并倡导对科学的兴趣。《自然》从创刊至今一直由英国麦克米伦公司出版，至于缘何这样安排，并没有任何记录。

洛克耶是天文学家，同时也是公职人员，在《自然》首期刊印半年以前，他当选为英国皇家学会会员。他与科学界保持着良好的关系，生物学家托马斯·亨利·赫胥黎和物理学家约翰·廷德尔都与他有来往。在《自然》创刊之初，洛克耶就邀请上述两位为《自然》供稿以确立《自然》的权威地位。作为主编，洛克耶从一开始就展示了《自然》后来一直坚持的一些特色。他不回避个人兴趣，从而使《自然》特别欢迎太阳物理学方面的研究。他赋予《自然》国际化的特色，刊登对各地（诸如圣彼得堡、维也纳和费城等）举行的学术会议的报道。他愿意甚至热衷于报道科学界的新闻和杂谈，也乐于引发争论并坦率地进行报道：早期刊载的一些激烈论战，就显示出未经编辑的尖锐语调。洛克耶还不吝对公众事务和国家大事发表立场鲜明的观点，这些看上去都与科学没有太大关联。

尽管《自然》的宗旨是“将科学研究和科学发现的重大成果呈现给公众”，但它并没有为一般读者降低专业水平，这种状况直到进入之后的一个世纪才有所改观。约翰·马多克斯于 1966 年成为《自然》的主编，他决定降低这本杂志的专业性以增加它对普通读者的吸引力，同时保证其学术水准。马多克斯希望每一个人都能够阅读和理解关于任何科学领域中的新发现的报道。

这是一个雄心勃勃的目标。即使是现在，《自然》刊登的研究论文对于未受过科学教育的普通人来说也是很有难度的；随着科学专业化程度的提高，对于科学家来说，理解自己专业领域以外的论文常常也是很困难的。另一方面，《自然》已经成为了一个拥有众多“姊妹”刊物的出版物集群，诸如《自然-遗传学》、《自然-地球科学》、《自然-材料学》等，这种扩展与传播和提供新闻之类的内容的新媒体的出现一起，使得《自然》有了更多的工具和渠道来发展新的、更宽泛的读者群。迄今为止，《自然》仍然是被引用得最多的跨学科科学期刊。

发表在《自然》上最有名的、也可能最具影响力的文章当数弗朗西斯·克里克

structure of DNA, the molecule that carries every organism's genes. It was quickly seen as a major breakthrough in understanding how heredity works: the molecule's structure, in which genetic information is encoded in the sequence of chemical building blocks as a four-letter code, immediately suggested how this information might be copied and passed on from one generation to the next. This was the missing link in Darwin's theory of evolution, showing how genes work at the molecular scale. The understanding that flowed from Crick and Watson's epoch-defining paper paved the way to developments such as the decoding of the entire human genome and the cloning of Dolly the sheep, also both reported first in *Nature*, in papers included in this collection.

Several other disciplines have been transformed by discoveries that *Nature* has reported. The theory of plate tectonics, advanced in the 1930s, was verified by the discovery of seafloor spreading by Fred Vine and Drummond Matthews in 1963. The birth of nuclear physics, which profoundly affected not only physical science but society and international relations, was traced in contributions to *Nature* from the likes of Ernest Rutherford, Niels Bohr, Otto Hahn, Irène Joliot-Curie and Otto Frisch in the 1920s and 30s. The quantum-mechanical nature of matter was revealed by the discovery in 1927 by Clinton Davisson and Lester Germer of wave-like properties in beams of electrons. This collection also records the birth of key developments in atmospheric and environmental sciences, from the discovery in 1985 of the destruction of ozone in the stratospheric layer above Antarctica to the gradual recognition of human-made global warming and the interactions between the biosphere, oceans, geological earth and ice sheets in bringing about the climate fluctuations of the past.

Tribute to John Maddox, Co-Editor-in-Chief

Sir John Maddox died, aged 83, while this project was in its final stages. He served as chief editor of *Nature* from 1966 to 1973, and then again from 1980 to 1995. Under his leadership, *Nature* was transformed from a rather austere periodical to a publication that combined the strengths of a major scientific journal of record and a magazine that made the latest advances in science understandable to a wider audience. This character reflected that of Maddox himself, who trained as a physicist at King's College London before, in 1955, becoming one of the leading science journalists of his day at the *Manchester Guardian* (now the *Guardian*).

It was his journalistic instinct that enabled Maddox to restore *Nature*'s position as an influential voice in the scientific community. In the tradition of Lockyer, he was willing to make the journal vigorously outspoken on matters that demanded it, as for example when *Nature* campaigned against the dangerous and damaging claims made by the *Sunday Times* newspaper in the early 1990s that HIV was not the cause of AIDS. This forthright attitude earned Maddox something of a reputation as a maverick, and he himself adopted a contrarian stance against some aspects of scientific orthodoxy—he remained sceptical about the Big Bang (which he worried was "too neat"), and for many years expressed

和詹姆斯·沃森对 DNA 这一携带有机体基因的分子的结构的描述，该文章发表于1953 年。它很快被视为理解遗传机制的重大突破：这种将遗传信息存储在由特定的化学组成单元构成的四码系统中的分子结构，立即揭示了这种信息被复制并遗传到下一代的可能方式。这是达尔文的进化论中缺失的环节，它显示了在分子尺度上基因是如何工作的。正是基于对沃森和克里克的这篇划时代文章的理解，才有了诸如解码人类基因组和克隆多莉羊等一系列进展，这些结果都由《自然》首先发表，并被收录在本套选集中。

其他的一些学科也曾因为《自然》刊载过的发现而发生彻底的改变。形成于20 世纪 30 年代的板块构造学说被 1963 年弗雷德里克·瓦因和卓门·马修斯发现的海底扩张所证实。核物理学的诞生不仅深深地影响了物理科学，也对社会甚至国际关系产生了深远的影响，而这正是基于 20 世纪二三十年代欧内斯特·卢瑟福、尼尔斯·玻尔、奥托·哈恩、伊雷娜·约里奥–居里和奥托·弗里施等人在《自然》上发表的文章。物质的量子力学本质被克林顿·戴维森和莱斯特·革末在 1927 年进行的电子束的波动性实验所证实。本套选集还见证了大气和环境科学中出现的许多重大进展，从 1985 年发现南极上空平流层的臭氧被破坏到逐渐意识到人类活动导致的全球变暖，以及生物圈、海洋、地壳和冰盖之间相互作用导致的过去的一系列气候波动。

悼念主编约翰·马多克斯

在这个项目的最后阶段，约翰·马多克斯与世长辞了，享年 83 岁。1966~1973 年和 1980~1995 年，他两度担任《自然》的主编。在他的领导下，《自然》从一个完全针对专业人士的期刊转型成为一个集记录科学与将科学的最新进展以易于理解的方式呈现给读者为一体的出版物。这一特质恰恰与马多克斯本人的特点相符，他在伦敦的国王学院接受学习并被培养成为专业的物理学家，而到 1955 年他成了当时《曼彻斯特卫报》（现称《卫报》）的顶尖级科学新闻记者。

记者的本能使马多克斯将《自然》重新变为科学界很有影响力的声音。他继承了洛克耶的传统，愿意就重大事件在杂志上直言不讳地发表看法，比如，《自然》就曾对《星期日泰晤士报》在 20 世纪 90 年代初提出的 HIV 不是艾滋病的起因这一危险而且危害极大的观点展开论战。这一直率的态度为马多克斯赢得了特立独行的评

doubts about the notion that human activity was causing global warming. (Maddox later changed his mind on this.) Most famously, perhaps, Maddox published a paper from French scientists that appeared to lend scientific support to homeopathy. He argued that "there are good and particular reasons why prudent people should, for the time being, suspend judgment" about the experiments, and soon after he led an equally controversial investigation into the claims.

Nonetheless, Maddox was a tireless opponent of what he viewed as anti-scientific and anti-rational views. A lifelong atheist, his funeral service was conducted in a tent just outside the church grounds in which he was buried.

Maddox was also a strong advocate of international research, and acutely aware of the social impacts of scientific advance—he was a member of the Pugwash Group that campaigns for nuclear non-proliferation. In 1995 he was awarded a knighthood, which he said he accepted for the sake of *Nature* rather than himself, and in 2000 he was the first person ever to be elected an Honorary Fellow of the Royal Society, an accolade almost unheard of in recent times for anyone whose career has not been primarily in scientific research. A scientific polymath, it was sometimes said that John Maddox was one of the few people capable of reading *Nature* from cover to cover and understanding more or less all of it. Certainly, it is hard to think of anyone else with the breadth of knowledge needed to oversee this collection.

Selection of papers for *Nature: the Living Record of Science*

The well-known papers mentioned above were obvious candidates for inclusion, and are among the 840 odd papers which make up this collection. The initial selection was made by John Maddox, who chose over 2,000 of *Nature*'s most prestigious articles. This expansive list was then condensed and supplemented by Philip Ball, then a consultant editor for *Nature* and a freelance writer on all areas of science, who became the Managing Editor of the project in October 2007.

It was clearly necessary for all the papers to have been "important" in some sense, but the process to assess the worthiness of each paper was by no means straightforward. For one thing, most of the truly important currents in science are more aptly seen as tidal swells than tsunamis: they arrive by steady accumulation, not in one sudden inundation. It was sometimes necessary to arrive at a compromise between the need to do justice to the important themes in scientific research, as they were reflected in the pages of *Nature*, and the absence of any single contribution that was transformative to a given theme.

There are several ways in which a paper can be deemed important. Some may have helped to establish a particular field of enquiry—for example, the search for gravity waves in space, or for planets around other stars—even if the actual findings reported in the paper itself have not stood the test of time. Such papers are not necessarily of only

价，他对科学界有些所谓的正统观点采取不同立场，生前一直对宇宙大爆炸持怀疑态度（他认为这一理论“过于完美”），他还在相当长的时间里对人类活动造成全球变暖的观点表示怀疑（但他后来改变了看法）。最使马多克斯名噪一时的，可能是他刊登了一篇几位法国科学家关于顺势疗法似乎有科学依据的文章。他认为，对这些实验而言，“当时有很多令人满意的、详尽的理由去解释为什么谨慎的人应该终止他们的偏见”，在那之后不久他又领导了一次同样有争议的顺势疗法调查。

马多克斯一生都抵制反科学、反理性的观点。作为终生的无神论者，他的追悼会是在教堂陵地外的一个帐篷中举行的，而他就葬在那片陵地中。

马多克斯也大力倡导研究的国际化，对科学进步产生的社会影响有着敏锐的洞察力——他是致力于核不扩散的帕格沃什组织的成员。1995 年他获得了爵士头衔，他强调能获此殊荣是因为《自然》而不是因为他自己。2000 年，他当选为英国皇家学会名誉会员，之前几乎从未听说有非专业从事科学研究的人士获此荣誉。作为一个博学者，马多克斯被认为是少数几位能够从头至尾阅读《自然》并且大致理解全部内容的人士之一。很难想象还有其他人能够具备足够宽的知识面来筹划这一选集。

文章的遴选

上面提到的那些著名的文章很明显应该被收录在这套选集之中，然而整套选集由八百四十余篇文章组成，遴选的难度相当大。最初的选篇工作由约翰·马多克斯完成，他选出了大约两千篇《自然》上最有影响的文章。这些篇目后来由菲利普·鲍尔进行了浓缩和补充，他当时是《自然》的顾问编辑和针对科学各领域进行写作的自由撰稿人，他于 2007 年 10 月成为本选集的项目执行编辑。

很显然，所有曾经在某种程度上“重要”的文章都有被收录的必要，但操作起来并不像看上去那么容易。首先，科学中真正重要的发现更像潮水涨潮而非海啸：它们是经过稳步积累而逐渐显露出来的，而不会如洪水般乍现。有时候必须在两种情况之间折衷：一是需要对科学研究中的重点论题予以公正的评判，像《自然》的字里行间体现出来的那样，二是不能忽略任何为某一学科带来重大变革的工作。

一篇文章被视为重要的原因有多种。其中一些可能是由于为某一特定的探索领域的确立提供了帮助，例如，寻找空间中的重力波，或者探寻地外行星，尽管文章

historical interest. They might, for example, illuminate what was happening in other areas of science, and shed light on how the ideas of the time were received. This seems true of the many discussions in *Nature*, for at least 50 years after the publication of Charles Darwin's *Descent of Man*, of eugenics: efforts to engineer the genetics of human populations by selective breeding. These ideas are now discredited, but it would seem not only dishonest in historical terms to ignore them but also a distortion of how biologists at that time were interpreting Darwinian theory.

One underlying principle of the selection process was to maintain an international perspective on the importance of each paper, rather than a bias towards scientific discoveries related to China, or of discoveries made by Chinese scientists. Within this context, the important findings of Chung-yao Chao, Pei-ji Chen and Zhi-ming Dong are included in the collection purely for their substantial contributions to the global scientific community.

Some papers may look, from today's perspective, to be of limited relevance—often simply because their findings have been so thoroughly assimilated. For example, some of the early studies in the chemistry of vitamins, or attempts to understand the nature of the atomic nucleus and its constituent particles, might not seem like "classics" in their own right, but at the time they were key stepping-stones in the journey towards a deeper understanding of the phenomena in question.

A few papers in the collection are not just "wrong", but notoriously so. The "polywater" affair of the late 1960s and 1970s, in which Soviet scientists claimed to have discovered a new, ultra-viscous form of water, proved groundless, most probably being an artefact of experimental contamination. But both it and the putative "memory of water" reported in *Nature* in 1988 not only illustrate how science is done (and how it may be done badly) but also highlight ongoing themes in the investigation of water's molecular structure. The papers included here that investigate the alleged "cold fusion" of hydrogen not only serve to mark this prominent scientific controversy of the late 1980s but also played an important role in showing it to be the result of faulty experimentation. The same is true of the papers that uncover the palaeontological fraud of "Piltdown Man".

Some reviews of the state of play of a discipline by leading scientists have been included, even though they do not in themselves report new discoveries, in part because such contributions serve as landmarks that, with the imprimatur of *Nature*, established the work they describe as now a solid part of the body of science and helped to gain its recognition from a wide audience. That is true, for example, of the sole contribution in the collection by Albert Einstein, a 1921 survey of the historical development of his theory of relativity. By this time, general relativity was widely accepted, especially after the demonstration in 1919 by Arthur Eddington that light from distant stars is bent by gravity as the theory predicts. Yet a paper in *Nature* was still a powerful signal to readers

中的实际发现并没有经受住时间的检验。这样的文章并不只是具有历史价值，它们可能对科学的其他领域发生的事情有启示作用，并说明了当时的概念是如何被接受的。例如，就《自然》中的许多讨论而言上述说法看上去都是正确的，在查尔斯·达尔文的《人类起源》发表至少 50 年之后，《自然》发表了许多关于优生学的讨论：通过选择性生育来调控人类的遗传。这些想法在现代社会已被摒弃，但是忽略掉它们不仅是对历史的不忠实，也会使读者曲解当时的生物学家对于达尔文理论的解释。

选篇的重要原则之一是从国际化的视角考虑每篇文章的重要性，并不偏向与中国相关的或中国科学家的发现。因此，中国科学家们（赵忠尧，陈培基，董枝明等）的研究成果的入选，纯粹是基于其对全球科学界的重大贡献。

还有一些文章从今天的视角来看价值有限——往往是因为对它们的发现已经有了透彻的理解。比如，一些关于维生素化学的早期研究，或者试图理解原子核的本质及其构成粒子的早期研究，这些研究看上去并不是那么“经典”，但它们当时是深入理解相关现象过程中的关键基石。

选集中也有少数文章不仅仅是“错误的”，而且简直算得上是臭名昭著。20 世纪 60 年代末和 20 世纪 70 年代的“聚合水”事件中，前苏联科学家声称发现了一种新的超黏滞性的水，这一结果被证明是毫无根据的，很可能是实验中的人为因素造成的。但是，它和 1988 年《自然》发表的推测“水的记忆”的文章一起，不仅揭示了科学研究是如何进行的（以及可能是如何错误地进行的），也凸显了当时进行的对水分子结构研究的主题。本选集还收录了一些调查所谓氢的“冷核聚变”的文章，它们不仅在今天再现了 20 世纪 80 年代末那场尖锐的科学争论，而且在当时揭示那是一个错误实验的过程中也发挥了重要作用。那些揭露古生物学中“皮尔当人”骗局的文章也具有同样的意义。

我们还选入了一些顶级科学家对某一学科进展的综述，尽管它们本身没有报道新的发现。这在某种程度上是因为这些文献是科学发展中的里程碑，借助于《自然》的认可，它们确立了所报道的工作后来成为今天科学重要组成部分的地位，并在当时使其得到广大读者的认可。阿尔伯特·爱因斯坦 1921 年对他的相对论的历史回顾的文章入选，就是基于上述原因。那时，广义相对论已经被广泛接受，尤其是 1919 年阿瑟·爱丁顿证实了遥远星体发出的光会被引力弯曲，正如广义相对论所预言的那

in the English-speaking world that general relativity was here to stay.

Such review-type papers were an important component of what *Nature* published in the pre-war era. Whereas in recent decades there has been a rather clear demarcation between original research papers, reviews, and commentary on developments reported elsewhere, in earlier times the distinctions were more fluid, and strictly new research was less prevalent in the journal's pages. Without these "overview" papers, much of what was happening in science at that time would have been missed. This is increasingly true the further back one goes, so that in its first several decades *Nature* seems more a kind of "scientific newspaper", reporting at least as much gossip, rumour, curiosities and accounts of learned meetings as any novel research. Perhaps this is a reflection of how the reporting of science has become increasingly professionalised and standardised—with both advantages and drawbacks.

In general, however, the selected papers in this collection are primary scientific contributions, not comments on work reported elsewhere, or editorials, book reviews, obituaries and so forth. That is a reflection of the intended aims of this collection, although it means that it provides only a partial view of the "flavour" of the journal itself. Some of the liveliest writing has appeared in these "excluded" genres, and the reader will need to look elsewhere for a more general sense of the journal's full character—most notably, it can be sampled in *A Bedside Nature* (W. H. Freeman, 1997), edited by Walter Gratzer (who also assisted the current selection process) and providing a miscellany of contributions between 1869 and 1953.

Introductions to the papers

Each of the selected papers is accompanied by a brief introduction that explains the key findings and places them in the context of their times. These introductions, which were written by John Maddox and several members of *Nature*'s current and former editorial staff, strive to sketch (they can do no more, given space constraints) the broader picture: why the work was important, how it changed or contributed to thinking at the time and how the conclusions have been vindicated, modified or rendered obsolete by later work. Occasionally they provide a glimpse of the paper's author(s), who are often figures extremely eminent in their time (if perhaps less known today)—or people who were later to become so. One of the most rewarding aspects of the process of editing *Nature: the Living Record of Science* was that it served as a reminder of what an extraordinary range and depth of research has been announced in *Nature*, and by what a remarkable collection of scientists. Inevitably, the collection can do only partial justice to that—but the introductory pieces go some way to compensating for what could not be included, by filling in some of the gaps and forging links that are not immediately evident from the papers themselves. In some cases, it is only with hindsight that the significance of a particular discovery becomes apparent, and we have occasionally needed to point out implications that were not even clear to a paper's author(s).

样。尽管如此，刊登在《自然》上的这篇文章仍然给予英语世界的读者一个强有力的信号——广义相对论确立了。

战前这些综述性的文章是《自然》的重要组成部分。最近几十年，原创的研究论文、综述和对别处发表的科学进展的评论之间有明晰的界限，而早些时候三者之间的区别并不明确，而且当时杂志所报道的严格意义上的新研究不像后来那么多。没有这些“综述”文章，当时科学界的许多事件都会无人知晓。越久远的时期越是这样，所以在《自然》初创的前几十年里，它更像是一份“科学新闻”，在报道新颖研究和学术会议的同时，也报道了至少同样多的杂谈、传闻和奇闻轶事。也许这反映了科学报道越来越专业化和标准化的过程，这一变化有利也有弊。

从总体上看，本选集收录的文章都是重要的科学论文，而不是对发表于别处的研究工作的评论、社论、书评、讣告或诸如此类的文章。这是本选集宗旨的体现，尽管这样只能够展示《自然》的部分“特色”。一些最活泼的论述就包含在那些“被排除”的文体中，读者需要在其他地方寻找对《自然》所有特点的记述——显而易见，*A Bedside Nature*（弗里曼著，1997年出版）就是一个这样的尝试，这本书由沃尔特·格拉泽（他也参与了本选集的选篇工作）编辑，收录了1869~1953年的各类文章。

每篇文章的导读

本选集收录的每篇文章都配有一篇简短的导读，介绍该文主要的发现并介绍当时的研究背景。这些导读由约翰·马多克斯和《自然》的几位现任和前任编辑撰写，力图勾勒出（受篇幅限制，他们只能点到为止）一幅更为广阔的图景：为什么这项研究很重要？它如何改变或者深化当时的思想？以及这个结论如何被证明、改进或者如何被后来的研究结论替代？有些导读会简略介绍一下文章的作者，他们或者在当时特别杰出（也许现在少有人知），或者后来变得非常杰出。遴选和编辑这套选集的过程中最有价值的方面之一，就是让我们意识到《自然》曾经报道过这么宽广而有深度的研究，以及众多非凡的科学家。不可避免的是，本选集做不到面面俱到，但是导读填补了一些空白并建立起了文章之间并不明显的联系，从某种程度上弥补了未能收录某些文章而造成的遗憾。有些情况下，只有事后才能明白一个发现的重要性，因此我们偶尔需要指出某项研究可能的影响，而那些文章的作者当时可能并不清楚。

Scientific publishing in China

Publishing and the media have played a significant role in spreading and promoting modern science. In late 19th century China, for example, translations of Western natural and social science works added tremendous momentum to the country's movement from a feudal society into the modern age. One of the most influential translations was that of Thomas Henry Huxley's *Evolution and Ethics* by the scholar Yan Fu (1854-1921), who studied at the Naval Academy in Greenwich, England, in 1877-1879. It immediately became essential reading for progressive young thinkers, and the ideas of natural selection that it described influenced several generations of intellectuals. Huxley was one of the key figures in the development of *Nature*, and made many contributions to early issues. Charles Darwin, whose ideas Huxley expounded, was also a reader and correspondent, and one of his contributions features in the first volume of this collection. The implications of his evolutionary theory stimulated many discussions in *Nature*'s pages, and indeed continue to do so today, as illustrated in the special issue of *Nature* in 11 February 2009 to celebrate the 200th anniversary of Darwin's birth and the 150th anniversary of the publication of his *On the Origin of Species*.

With the rapid growth of its economy and an increasing output of scientific research in the past three decades, China is now regarded as a major scientific force in the Asia-Pacific region and in the world at large. The number of active scientists and researchers in China is now second only to that of the United States, and the number of research papers they generate overtook those from the United Kingdom and Japan in 2008. These developments make it ever more important that high-quality scientific research papers, both past and present, be readily accessible within China. Most of the scientific literature is not available in the Chinese language; this collection takes the first step to change that situation, with the original articles represented as they first appeared in *Nature* and the translations strictly reviewed for faithfulness. The articles are arranged in chronological order, and an index of classification is provided for quick reference.

An unprecedented collection

Nature: the Living Record of Science is unprecedented in that it represents the first ever large collection of science articles of any sort written for students and researchers worldwide—regardless of their nationality. The collection covers a wide range of disciplines, from biology to the earth and environmental sciences, materials science and physics. It is translated into the Chinese language and published in bilingual format so as to bridge different cultures in a fashion that can stimulate international research.

We wish to take this opportunity to thank all the translators, reviewers and editors for their painstaking efforts in making the project a success. We also look forward to closer

中国的科学出版

出版和传媒在传播和促进现代科学的发展方面发挥着重要的作用。比如，在 19 世纪末的中国，西方自然科学和社会科学著作的翻译作品极大地推动了中国从封建社会进步到现代社会的步伐。最有影响力的翻译作品之一就是学者严复（1854~1921 年）翻译的托马斯·亨利·赫胥黎的作品《天演论》（严复于 1877~1879 年在英国格林尼治海军学院学习）。这一作品立即成为当时进步的年轻学者们的必读书，书中自然选择的思想影响了几代知识分子。赫胥黎是《自然》发展过程中的重要角色，在杂志的早期发表过很多文章。查尔斯·达尔文也是《自然》的读者和作者，赫胥黎传播的就是他的思想，在选集的第 1 卷中就收录有他的文章。关于达尔文的进化论的讨论占据了《自然》相当大的篇幅，而且时至今日依然如此，正如 2009 年 2 月 11 日《自然》庆祝达尔文诞辰 200 周年暨《物种起源》发表 150 周年的专刊所示。

随着最近三十年经济的快速增长和科研产出的日益增加，中国已成为亚太地区和全世界重要的科技力量。中国的一线科研研究人员的数量仅次于美国，他们发表的研究论文的数量在 2008 年已经超过了英国和日本。这也使得能够在中国方便地获得过去及现在的高水平研究论文变得前所未有的重要。大部分科学文献都没有中文版本，这套选集第一次尝试改变这种状况，我们对原文未作任何改动，对译文进行了严格的审订以确保其准确性。所有文章都按发表时间顺序排列，每卷附有按学科分类的文章索引以便快速查找。

一部开创性的选集

《〈自然〉百年科学经典》的开创性在于，此前从未有过如此鸿篇巨制的科学论文选集，而且针对的是全世界的学生与研究人员。它涵盖广泛的学科领域，从生物学到地球与环境科学、材料科学及物理学等等。我们以英汉双语形式出版的目的是为了促进不同文化之间的沟通并激发国际化的研究。

我们希望借此机会感谢所有的译者、审稿专家和编辑，是他们的精心努力成就了本套选集。我们也希望外语教学与研究出版社、麦克米伦出版集团和自然出版集

cooperation between the Foreign Language Teaching and Research Press, the Macmillan Publishing Group and Nature Publishing Group in providing more high-quality science works for scientists and students, and other interested readers in China. Such cooperation, we believe, will effectively promote the exchange of ideas and sharing of achievements between Chinese and Western scientists.

Philip Campbell

Editor-in-Chief of *Nature*

团能够更加紧密地合作，向中国的科学家、学生和科技爱好者提供更多高水准的科学作品。我们相信，这样的合作一定会有效地促进中国科学家和西方科学家之间的思想交流与成果共享。

菲利普·坎贝尔

《自然》主编

Contents
目录

Volume I
(1869-1930)

Nature's Aims

Editor's Note

The only existing published statement of *Nature*'s aims appeared here among the advertisements in the second issue of the journal. The illustration at the top was used until the late 1950s. It seems likely that printed copies of this announcement were used as publicity material before the journal appeared. The two lines of poetry at the top are by William Wordsworth, one of the foremost English poets of the early nineteenth century, and they hint at the associations then current between Romantic poets and early scientists such as Humphry Davy. This statement of intent shows how *Nature* intended from the outset to address both working scientists and the general public—an aim still observed today.

"To the solid ground
Of Nature trusts the mind that builds for aye."
—Wordsworth

THE object which it is proposed to attain by this periodical may be broadly stated as follows. It is intended,

First, to place before the general public the grand results of Scientific Work and Scientific Discovery; and to urge the claims of Science to a more general recognition in Education and in Daily Life;

And, secondly, to aid Scientific men themselves, by giving early information of all advances made in any branch of Natural knowledge throughout the world, and by affording them an opportunity of discussing the various Scientific questions which arise from time to time.

To accomplish this twofold object, the following plan will be followed as closely as possible:

NATURE

《自然》的宗旨

编者按

这是唯一现存的公开阐述《自然》的宗旨的文章，发表在《自然》第2期的公告部分，其示意图一直沿用到20世纪50年代。在《自然》问世之前，这篇公告可能已经被印刷出来用作宣传材料了。开篇的两行诗引自19世纪早期英国杰出的诗人之一威廉·华兹华斯，它暗示了当时浪漫主义诗人与早期的科学家（如汉弗莱·戴维）之间的种种联系。这一意向性声明显示，《自然》从一开始就打算既面向一线科学家，又面向普通大众。这个宗旨一直遵循至今。

“思想常新者
以自然为其可靠之依据。”
——华兹华斯

这份期刊的办刊宗旨可以概述如下：

首先，将科学研究和科学发现的重大成果呈现给公众，并促使科学理念在教育和日常生活中得到更为普遍的认可。

其次，帮助科学家自己，为他们提供自然科学各个分支在世界范围内取得的所有进展的最新信息，为他们探讨不时出现的各种科学问题提供交流平台。

为实现这一双重目标，本刊将尽可能严格地遵循下列计划：

Those portions of the Paper more especially devoted to the discussion of matters interesting to the public at large will contain:

I. Articles written by men eminent in Science on subjects connected with the various points of contact of Natural knowledge with practical affairs, the public health, and material progress; and on the advancement of Science, and its educational and civilizing functions.

II. Full accounts, illustrated when necessary, of Scientific Discoveries of general interest.

III. Records of all efforts made for the encouragement of Natural knowledge in our Colleges and Schools, and notices of aids to Science-teaching.

IV. Full Reviews of Scientific Works, especially directed to the exact Scientific ground gone over, and the contributions to knowledge, whether in the shape of new facts, maps, illustrations, tables, and the like, which they may contain.

In those portions of "Nature" more especially interesting to Scientific men will be given:

V. Abstracts of important Papers communicated to British, American, and Continental Scientific societies and periodicals.

VI. Reports of the Meetings of Scientific bodies at home and abroad.

In addition to the above, there will be columns devoted to Correspondence.

Many eminent scientific men* are among those who have already promised to contribute Articles, or to otherwise aid in a work which it is believed may, if rightly conducted, materially assist the development of Scientific thought and work in this country.

(**1**, 66-67; 1869)

* The list of names continues on a second page in the original journal. The notable figures among them are Darwin, Huxley, Joule, Kékulé, Tyndall, Wallace and Wöhler.

本杂志将刊载一些公众普遍感兴趣的论文，内容包括：

一、由科学界杰出人士撰写的文章，这些文章将涉及与实际事务、公共健康以及物质文明相关的自然知识的各个领域，以及科学进展及其教育与教化功能等方面。

二、对公众普遍关注的科学发现所作的详尽叙述，必要时会辅以图示。

三、对大中院校倡导自然知识所取得的成果的报道，以及旨在为科学教育提供各种援助的公告。

四、对科研成果的全面综述，特别是对前人科研成果的总结回顾，以及对知识进步有所贡献的文章，形式不限，包括文章内的新论据、地图、插图、表格等。

本杂志还会刊载一些更受科学界人士关注的文章，这些将包括：

五、提交给英、美以及欧洲大陆各科学学会和期刊的重要论文的摘要。

六、对国内外科学机构召开会议的报道。

除此之外，本杂志还将开辟通信专栏。

许多杰出的科学家*已承诺向本刊供稿，或者以其他方式向本刊提供帮助。如果处理得当，所有这些可能会对我国科学认识和研究工作水平的进步有所帮助。

（陈丹 翻译；王式仁 审稿）

* 在当初的杂志上，紧随其后的一页是一份名单。名单中有许多非常知名的人物，如达尔文、赫胥黎、焦耳、凯库勒、廷德尔、华莱士和维勒等。

On the Fertilisation of Winter-Flowering Plants

A. W. Bennett

Editor's Note

If many plants rely on insects for pollination, how is this accomplished in those plants that habitually flower in winter, when flying insects are few? Such was the question posed in the very first research paper to be published in *Nature*. The answer, according to Alfred Bennett, a London publisher and bookseller, is that many winter-flowering plants are self-fertile, the pollen discharging from the anthers while the flower is yet a bud. This is obviously not possible in those plants in which male and female organs are carried in separate flowers. In those cases, such as the hazel (*Corylus*), pollen is shed very liberally from male flowers close to female ones, which "favours the scattering of the pollen by the least breath of wind".

THAT the stamens are the male organ of the flower, forming unitedly what the older writers called the "androecium", is a fact familiar not only to the scientific man, but to the ordinary observer. The earlier botanists formed the natural conclusion that the stamens and pistil in a flower are intended mutually to play the part of male and female organs to one another. Sprengel was the first to point out, about the year 1790, that in many plants the arrangement of the organs is such, that this mutual interchange of offices in the same flower is impossible; and more recently, Hildebrand in Germany, and Darwin in England, have investigated the very important part played by insects in the fertilisation of the pistil of one individual by the stamens of another individual of the same species. It is now generally admitted by botanists that cross-fertilisation is the rule rather than the exception. The various contrivances for ensuring it, to which Mr. Darwin has especially called the attention of botanists, are most beautiful and interesting; and the field thus opened out is one which, from its extent, importance, and interest, will amply repay the investigation of future observers. For this cross-fertilisation to take place, however, some foreign agency like that of insects is evidently necessary, for conveying the pollen from one flower to another. The question naturally occurs, How then is fertilisation accomplished in those plants which flower habitually in the winter, when the number of insects that can assist in it is at all events very small? I venture to offer the following notes as a sequel to Mr. Darwin's observations, and as illustrating a point which has not been elucidated by any investigations that have yet been recorded. I do not here refer to those flowers of which, in mild seasons, stray half-starved specimens may be found in December or January, and of which we are favoured with lists every year in the corners of newspapers, as evidence of "the extraordinary mildness of the season." I wish to call attention exclusively to those plants, of which we have a few in this country, whose normal time of flowering is almost the depth of winter, like the hazel-nut *Corylus avellana*, the butcher's broom *Ruscus aculeatus*, and the gorse *Ulex europoeus*; and to that more numerous class which flower and fructify all through the year, almost regardless of season or temperature; among which

论冬季开花型植物的受精作用

贝内特

编者按

如果说植物的传粉过程都依赖于昆虫，那么那些通常在冬季开花的植物是如何完成传粉的呢？冬季可是很少有昆虫飞行的。这正是《自然》上发表的第一篇研究论文提出的问题。根据阿尔弗雷德·贝内特（伦敦的一位出版商和书商）的观点，许多冬季开花型植物是自花受精的，在花朵还没绽放的时候花粉就从花药中释放出来了。对于那些雄性器官和雌性器官分别位于不同花朵上的植物来说，这显然是不可能的。而在诸如榛树之类的一些个例中，雄花中的花粉能够自由地释放到邻近的雌花中，这种情况“有利于花粉在最微弱风力作用下成功散播”。

雄蕊是花的雄性器官，它们聚集在一起就形成了年长的学者所称的“雄蕊群”，这是科学工作者和普通观察者都很熟悉的事实。早期的植物学家自然而然地得出了如下结论：一朵花的雄蕊和雌蕊倾向于对其他花朵发挥雄性器官和雌性器官的作用。大概在 1790 年，施普伦格尔就首先指出许多植物中器官的安排都是如此，这种职责的互换是不可能在同一朵花里实现的。最近，德国的希尔德布兰德和英国的达尔文对昆虫在同种植物不同个体的雄蕊与雌蕊间的受精中所起的重要作用进行了研究。异花受精是自然界的规律而非例外这一观点，现在已经得到了植物学家的普遍认可。植物有很多种确保异花受精的策略，达尔文认为这些策略是非常美妙有趣的，他特别呼吁植物学家能对此给予关注。于是一个新的研究领域出现了，从广度、重要性和利益等方面来看，这一领域都将给未来从事此领域研究的人员带来充分的回报。然而，一些外界媒介对异花受精的发生显然也是必要的，例如昆虫可以将花粉从一朵花传到另一朵花。于是我们便有了如下问题：对于那些总是在冬天开花的植物来说，那时能够帮助它们传粉的昆虫数量特别少，在这种情况下，这些花是如何完成受精的呢？本人斗胆给出如下摘记，作为对达尔文先生的观察研究的延续，同时也是对一个在已有记载的研究中从未被阐明过的问题进行的说明。有些植物在气候温和的年份里开的花可能会一直保持到 12 月或 1 月，但这只是些零散出现而且濒临死亡的花朵；还有些植物每年都被列在报纸的边边角角处，作为“今年气候异常温和”的证据。本人在这里并不想谈这些植物，而是想专门关注那些在本国数量不多、正常开花时间几乎在深冬的植物，例如欧洲榛、假叶树和荆豆，同时也会关注那些几乎不受季节或温度影响、一年四季都可以开花结果的植物，这样的植物种类相对更多，其中包括短柄野芝麻、紫花野芝麻、阿拉伯婆婆纳、雏菊、蒲公英、千里光、

may be mentioned the white and red dead-nettles *Lamium album* and *purpureum*, the *Veronica Buxbaumii*, the daisy, dandelion, and groundsel, the common spurge *Euphorbia peplus*, the shepherd's purse, and some others.

During the winter of 1868–1869, I had the opportunity of making some observations on this class of plants; the result being that I found that, as a general rule, fertilisation, or at all events the discharge of the pollen by the anthers, takes place in the bud before the flower is opened, thus ensuring *self-fertilisation* under the most favourable circumstances, with complete protection from the weather, assisted, no doubt, by that rise of temperature which is known to take place in certain plants at the time of flowering. The dissection of a flower of *Lamium album* (Fig. A) gathered the last week in December, showed the stamens completely curved down and brought almost into contact with the bifid stigma, the pollen being at that time freely discharged from the anthers. A more complete contrivance for self-fertilisation than is here presented would be impossible. The same phenomena were observed in *Veronica Buxbaumii*, where the anthers are almost in contact with the stigma before the opening of the flower, which occurs but seldom, *V. agrestis* and *polita*, the larger periwinkle *Vinca major*, the gorse, dandelion, groundsel, daisy, shepherd's purse, in which the four longer stamens appear to discharge their pollen in the bud, the two shorter ones not till a later period, *Lamium purpureum, Cardamine hirsuta*, and the chickweed *Stellaria media*, in which the flowers open only under the influence of bright sunshine. In nearly all these cases, abundance of fully-formed, seed-bearing capsules were observed in the specimens examined, all the observations being made between the 28th of December and the 20th of January.

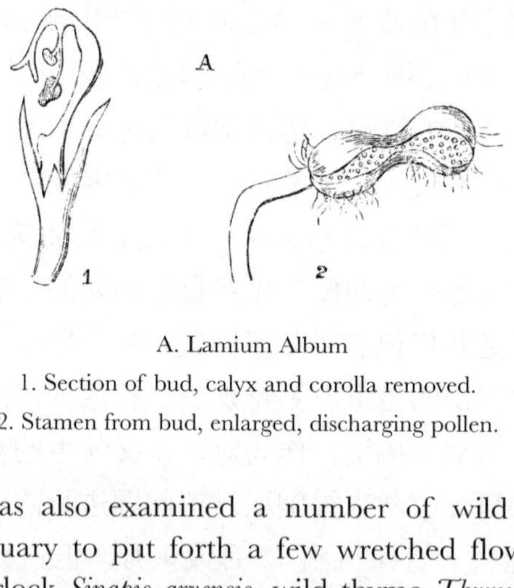

A. Lamium Album

1. Section of bud, calyx and corolla removed.

2. Stamen from bud, enlarged, discharging pollen.

In contrast with these was also examined a number of wild plants which had been tempted by the mind January to put forth a few wretched flowers at a very abnormal season, including the charlock *Sinapis arvensis*, wild thyme *Thymus serpyllum*, and fumitory *Fumaria officinalis*; in all of which instances was there not only no pollen discharged before the opening of the flower, but no seed was observed to be formed. An untimely specimen of the common garden bean *Faba vulgaris*, presented altogether different phenomena from its relative the gorse, the anthers not discharging their pollen till after the opening of the flower; and the same was observed in the case of the *Lamium Galeobdolon* or yellow archangel (Fig. B) gathered in April, notwithstanding its consanguinity to the dead-nettle.

南欧大戟、荠菜等。

在 1868~1869 年间的冬天，本人有幸对这类植物进行了一些观察。结果发现：受精作用是在开花之前发生于花蕾中的，这是一般规律，或者说在花药释放花粉的所有情况下都是如此。这样就保证了**自花受精**能在最有利的环境下进行而完全不受天气的影响，而且我们知道某些植物是在温度升高时开花的，这就更加加强了环境的有利性。从采集于 12 月最后一周的短柄野芝麻花的解剖图（图 A）可以看出，雄蕊完全弯曲下垂，几乎与二裂柱头相接触，此时花粉便从花药中自由散出。对于自花受精而言，不可能存在比这更完善的设计了。同样的现象在阿拉伯婆婆纳中也有发现，即开花前花药几乎与柱头相接触。在直立婆婆纳、双生婆婆纳、大蔓长春花、荆豆、蒲公英、千里光、雏菊和荠菜中，这种现象并不多见，而是 4 枚较长的雄蕊将花粉释放到花蕾上，2 枚较短的雄蕊则需再经过一段时间才能成熟。紫花野芝麻、碎米荠和繁缕是否开花只受日照影响。在检测过的样本中，几乎所有我们观察到的植物都产生了大量完全成形的结有种子的蒴果。上述所有观察是在 12 月 28 日到 1 月 20 日之间进行的。

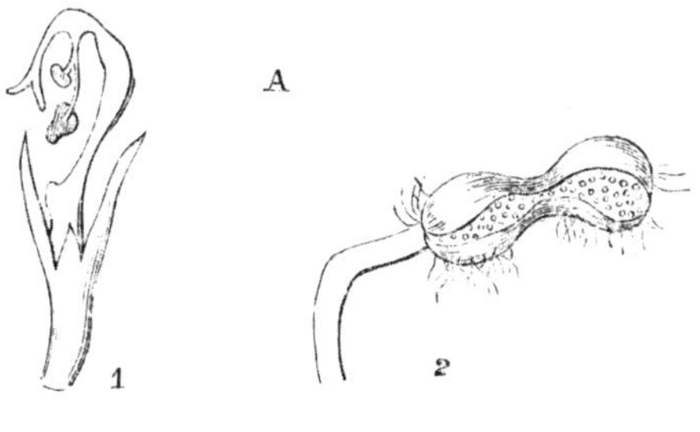

A. 短柄野芝麻

1．去除了花萼和花冠后的花蕾部分。

2．花蕾上生出的膨大雄蕊，花粉正从中散出。

作为对照，我们也研究了另外一些野生植物，这些植物在 1 月里受温暖天气的诱导反季节地开出了几朵可怜的小花，这包括野芥、铺地百里香和球果紫堇。对于所有这些植物，不仅没有发现开花前花粉的释放，而且也没有观察到种子的形成。一个不太合常理的例子是蚕豆。蚕豆的花药是在开花之后才释放花粉，这与其亲缘植物荆豆的表现完全不同。无独有偶，我们于 4 月采集的花叶野芝麻（图 B）也是这样，尽管它是野芝麻的近亲，但其花药也是在开花后才释放花粉。

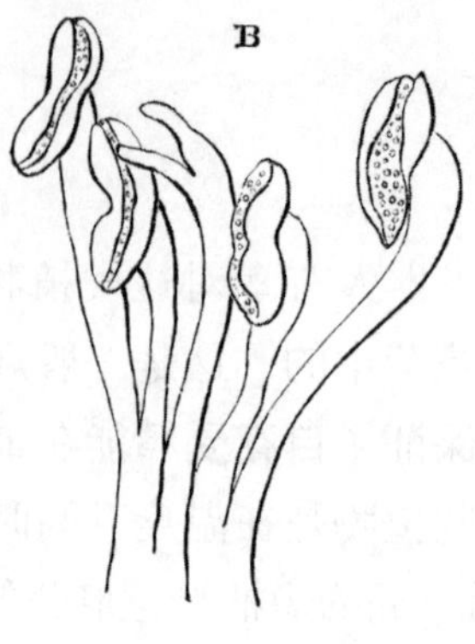

B. Lamium Galeobdolon—Pistil and stamens from open flower; the latter discharging pollen.

Another beautiful contrast to this arrangement is afforded by those plants which, though natives of warmer climates, continue to flower in our gardens in the depth of winter. An example of this class is furnished by the common yellow jasmine, *Jasminium nudiflorum*, from China, which does not discharge its pollen till considerably after the opening of the flower, and which never fructifies in this country. But a more striking instance is found in the "allspice tree", the *Chimonanthus fragrans*, or *Calycanthus praecox* of gardeners, a native of Japan, which, flowering soon after Christmas, has yet the most perfect contrivance to prevent self-fertilisation (Fig. C). In a manner very similar to that which has been described in the case of *Parnassia palustris**, the stamens, at first nearly horizontal, afterwards lengthen out, and rising up perpendicularly, completely cover up the pistil, and then discharge their pollen outwardly, so that none can possibly fall on the stigma. As a necessary consequence, fruit is never produced in this country; but may we not conjecture that in its native climate the *Chimonanthus* is abundantly cross-fertilised by the agency of insects, attracted by its delicious scent, in a similar manner to our Grass of Parnassus?

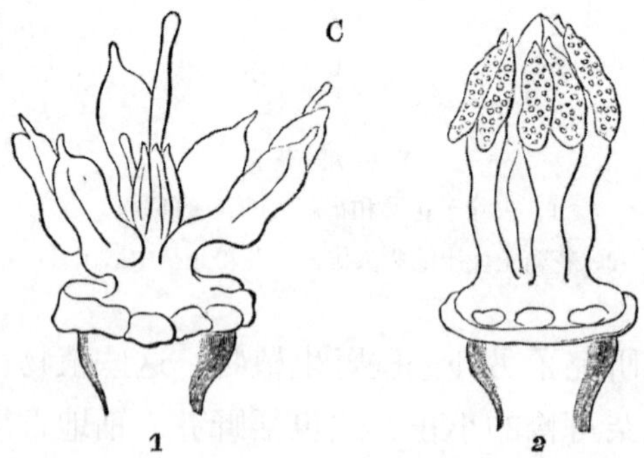

C. Chimonanthus Fragrans

1. Early stage of flower, calyx and corolla removed.

2. Later stage, stamens surrounding the pistil, and discharging their pollen outwardly.

* *Journal of the Linnaean Society* for 1868–1869, Botany, p.24.

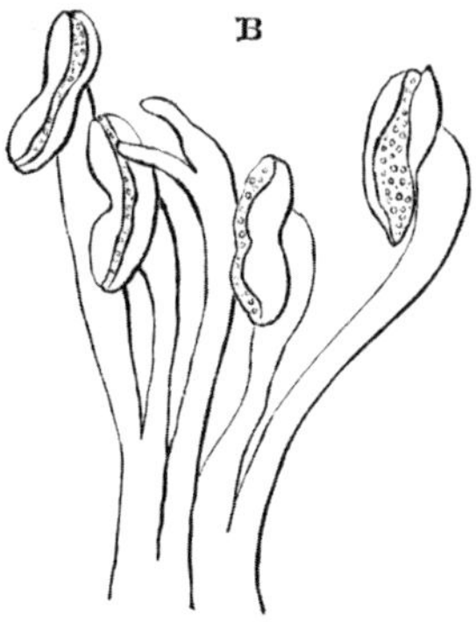

B. 花叶野芝麻——盛开花朵中的雌蕊和雄蕊；雄蕊正在释放花粉。

另一个比较典型的对照是我们花园里的植物，尽管它们的原产地是气候较为温暖的地方，但在我们的花园里它们却一直开花到深冬。这类植物的一个例子是原产于中国的迎春花，其花粉在开花后相当长一段时间才散出，而且这种植物在我国从不结果。另一个更具代表性的例子是原产于日本的腊梅，这种植物在圣诞后不久就开花，它有最完美的阻止自花受精的机制（图 C）。与梅花草采取的方式相似*，腊梅的雄蕊最初几乎是水平的，随后不断延长并垂直耸立，完全包裹住雌蕊，然后才向外散出花粉，这样就使得花粉不可能落到柱头上。一个必然的结果就是这些花在我国永远不会结出果实。但是，难道我们不能猜想腊梅在其原产地的气候下会像我国的帕纳色斯草一样，通过花朵的诱人香气来吸引昆虫作为媒介而完成异花受精吗？

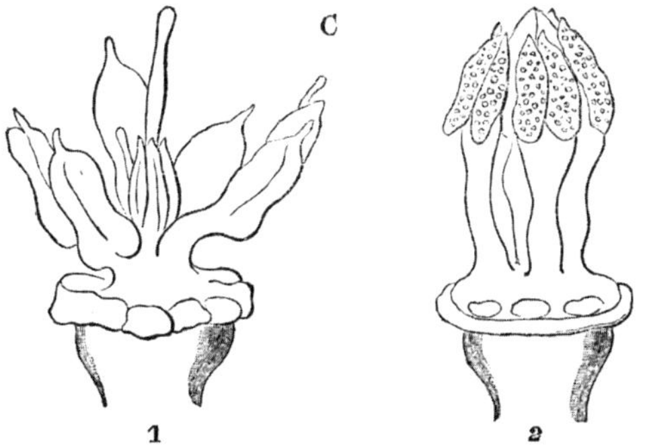

C. 腊梅
1. 去除花萼和花冠后的早期花朵。
2. 晚期花朵，雄蕊包围着雌蕊并在向外散发花粉。

* 《林奈学会会刊》，1868~1869年，植物学，第24页。

The description detailed above cannot of course apply to those winter-flowering plants in which the male and female organs are produced on different flowers; but here we find commonly another provision for ensuring fertilisation. In the case of the hazel-nut the female flowers number from two to eight or ten in a bunch, each flower containing only a single ovule destined to ripen. To each bunch of female flowers belongs at least one catkin (often two or three) of male flowers, consisting of from 90 to 120 flowers, and each flower containing from three to eight anthers. The pollen is not discharged till the stigmas are fully developed, and the number of pollen-grains must be many thousand times in excess of what would be required were each grain to take effect. The arrangement in catkins also favours the scattering of the pollen by the least breath of wind, the reason probably why so many of the timber-trees in temperate climates, many of them flowering very early in the season, have their male inflorescence in this form.

The *Euphorbias* or spurges have flowers structurally unisexual, but which, for physiological purposes, may be regarded as bisexual, a single female being enclosed along with a large number of male flowers in a common envelope of involucral glands. Two species are commonly found flowering in the winter, and producing abundance of capsules, *E. peplus* and *helioscopia*. In both these species the pistil makes its appearance above the involucral glands considerably earlier than the bulk of the stamens (Fig. D).

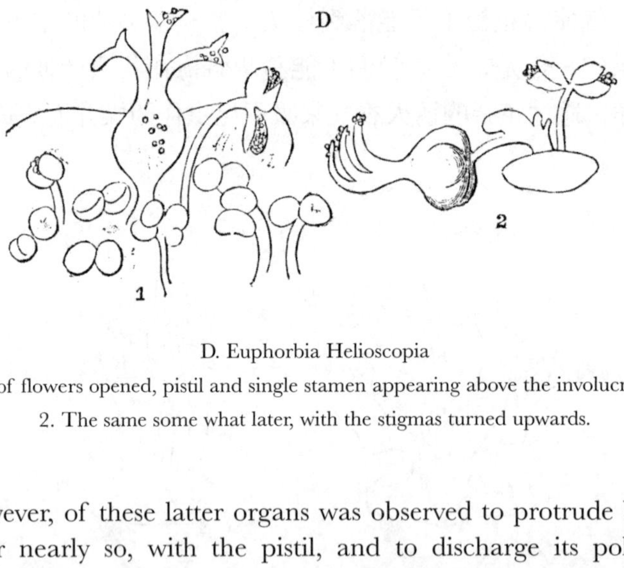

D. Euphorbia Helioscopia

1. Head of flowers opened, pistil and single stamen appearing above the involucral glands.
2. The same some what later, with the stigmas turned upwards.

A single one, however, of these latter organs was observed to protrude beyond the glands simultaneously, or nearly so, with the pistil, and to discharge its pollen freely on the stigmas, thus illustrating a kind of quasi-self-fertilisation. The remaining stamens do not discharge their pollen till a considerably later period, after the capsule belonging to the same set has attained a considerable size. In *E. helioscopia* the capsules are always entirely included within the cup-shaped bracts, and the stigmas are turned up at the extremity so as to receive the pollen freely from their own stamens. Now contrast with this the structure of *E. amygdaloides*, which does not flower before April (Fig. E). The heads of flowers which first open are entirely male, containing no female flower; in the hermaphrodite heads, which open subsequently, the stigmas are completely exposed beyond the involucral

上述描述当然并不适用于那些雄性器官和雌性器官分别生长于不同花朵上的冬季开花型植物。但是我们发现了另一种保证受精的常见方式。以欧洲榛为例，一簇花中雌花的数目是2~8或2~10不等，每朵花只含有一个可以成熟的胚珠。每簇雌花至少对应有一个（通常是2个或3个）雄花序，每个雄花序由90~120朵花（每朵花有3~8个花药）组成。在柱头充分发育后花粉才会散发出来。如果每个花粉粒真的都能发挥作用，那么实际花粉粒的数量就会超出所需数量数千倍。雄花序中的这种安排也有利于花粉在最微弱风力作用下成功散播，这很可能就是温带气候下有许多成材木的原因，因为其中许多在刚进入冬季时开花的植物都具有这种形式的雄花序。

大戟属的花从结构上来说是单性花，但是从生理学功能上来说又可以将其看作是双性的，因为一朵雌花与许多雄花被共同包裹在总苞腺体中。该属的南欧大戟和泽漆两种植物都是在冬季开花的，并且能产生大量蒴果。在这两种植物中，雌蕊出现在总苞腺体之上的时间比大部分雄蕊都早许多（图D）。

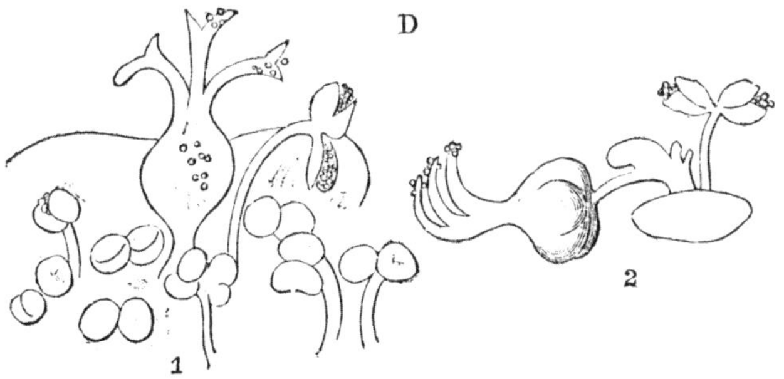

D. 泽漆

1．盛开花朵的头部，雌蕊和单枚雄蕊位于总苞腺体之上。

2．同一朵花较晚期时的头部，这时柱头已经转弯向上。

然而，我们观察到只有一枚雄蕊是和雌蕊同时或者接近同时显露在总苞腺体外的，这枚雄蕊会自由地将花粉散发到柱头上，这种受精方式被称为准自花受精。其余的雄蕊则是经过很长一段时间直到同组的蒴果长到一定大小的时候才会释放花粉。泽漆的蒴果自始至终都被完全包裹在杯状的苞叶之中，而柱头的末端会转而向上以便接受自身雄蕊自由释放的花粉。与此形成对比的是在4月之前并不开花的扁桃状大戟（图E）的结构。最先开放的一批花朵，其头部是完全雄性的，不含任何雌花。随后开放的花朵头部则是雌雄同体的，在雄蕊从总苞腺体中伸出之前，柱头就已经完全暴露在总苞腺体之外了。因此，就会发生完全的异花受精，即首先开放的雄花

glands long before any stamens protrude from the same glands. Here, therefore, complete cross-fertilisation takes place, the pollen from the first-opened male heads no doubt fertilising the stigma from the next-opened hermaphrodite heads, and so on. In this species the bracts are not cup-shaped, but nearly flat; the stigmas hang out very much farther than in *E. helioscopia;* and the styles are perfectly straight.

E. Euphorbia Amygdaloides—Head of flower, pistil appearing above the involucral glands, all the stamens still undischarged.

The above observations are very imperfect as a series, and I can only offer them as a contribution towards an investigation of the laws which govern the cross-fertilisation or self-fertilisation of winter-flowering plants. On communicating some of them to Mr. Darwin, he suggested that the self-fertilised flowers of *Lamium album*, and other similar plants, may possibly correspond to the well-known imperfect self-fertilised flowers of *Oxalis* and *Viola*; and that the flowers produced in the summer are cross-fertilised; a suggestion which I believe will be found correct.

In conclusion, I may make two observations. The time of flowering of our common plants given in our textbooks is lamentably inexact; for the hazel, March and April for instance! and for the white dead-nettle, May and June! according to Babington. Great care also should be taken to examine the flowers the moment they are brought in-doors; as the heat of the room will often cause the anthers to discharge their pollen in an incredibly short space of time. This is especially the case with the grasses.

(**1**, 11-13; 1869)

头部释放的花粉必然会使随后开放的雌雄同体花头部的柱头受精，以此类推。这种植物的苞叶不是杯状的，而是几乎扁平的，其柱头也比泽漆的远得多，并且是完全直立的。

E

E. 扁桃状大戟——花朵头部，雌蕊位于总苞腺体之上，所有雄蕊尚未开始释放花粉。

上述观察研究作为一个系列还很不完善，所以我只希望这些结果可以为研究冬季开花型植物的自花受精或异花受精的调控规律提供一些参考。与达尔文先生进行相关交流时，他提出短柄野芝麻和其他类似植物的自花受精方式可能与酢浆草和堇菜属植物那著名的不完全自花受精方式相似，而夏天开的花是异花受精的。我相信这种说法终将会被证明是正确的。

综上，通过观察研究我可以提出如下两点。首先，我们的教科书中给出的普通植物的开花时间是非常不准确的。例如，巴宾顿的书中竟然说榛树是在 3 月和 4 月开花，短柄野芝麻是在 5 月和 6 月开花！另外，我们也应该特别关注花被搬进室内时的反应，因为室内的热量通常会使花药在很短的时间内就释放出花粉，稻科植物尤其如此。

（刘皓芳 翻译；刘京国 审稿）

Fertilisation of Winter-Flowering Plants

A. W. Bennett

Editor's Note

Alfred Bennett's article in *Nature*'s first issue on fertilisation in winter-flowering plants made much mention of work on plant fertilisation by Charles Darwin. Following the publication of that paper, Darwin drew Bennett's attention to various small errors and clarifications. Thus Bennett got a letter in *Nature*'s second issue to address Darwin's points, notably that *Vinca* (periwinkle) absolutely requires the attention of insects for pollination. Bennett went back to his notes and found that Darwin was, of course, right—apparently self-fertile periwinkles do not set fruit in the winter.

MR. Darwin has done me the honour of calling my attention to one or two points in my paper, published in your last number, "On the Fertilisation of Winter-Flowering Plants". He thinks there must be some error in my including *Vinca major* among the plants of which the pollen is discharged in the bud, as he "knows from experiment that some species of *Vinca* absolutely require insect aid for fertilisation." On referring to my notes, I find them perfectly clear with respect to the time at which the pollen is discharged. My observation, however, so far agrees with Mr. Darwin's, that I find no record of any fruit being produced in January; it was, in fact, the absence of capsules on the *Vinca* which induced me to qualify the sentence on this subject, and to say "in nearly *all* these cases, abundance of fully formed seed-bearing capsules were observed." It is worthy of remark, that the *Vinca* is the only species in my list of apparently bud-fertilised plants not indigenous to this country. The second point relates to the white dead-nettle, with respect to which Mr. Darwin says, "I covered up *Lamium album* early in June, and the plants produce no seed, although surrounding plants produced plenty." This again would agree with my conjecture that it is only the flowers produced in winter that are self-fertilised. I may, however, be permitted to suggest that the test of covering up a plant with a bell-glass is not conclusive on the point of cross-fertilisation, as it is quite probable that with plants that are ordinarily self-fertilised, the mere fact of a complete stoppage of a free circulation of air may prevent the impregnation taking place. Has the experiment ever been tried with grasses, which, according to the French observer, M. Bidard are necessarily self-fertilised?

(**1**, 58; 1869)

Alfred W. Bennett: 3, Park Village East, Nov. 8, 1869.

冬季开花型植物的受精作用

贝内特

编者按

阿尔弗雷德·贝内特在《自然》的第1期上发表了一篇关于冬季开花型植物受精作用的论文，该论文中大量提到了查尔斯·达尔文关于植物受精作用的研究工作。在贝内特的论文发表后，达尔文提请贝内特注意文章中存在的多处小错误，并请他作一些说明。于是，在这篇发表于《自然》第2期的快报文章中，贝内特公布了达尔文的观点，即长春花的传粉完全需要昆虫的参与。贝内特在重新查阅了他的记录后发现，达尔文的观点确实是正确的，很显然自花受精型的长春花在冬天是不结果的。

在贵刊的上一期中有一篇我发表的论文——《论冬季开花型植物的受精作用》。我很荣幸达尔文先生给了我一些提醒，让我注意到该文中的一两个问题。我在论文中提到，大蔓长春花是一种在花蕾中就有花粉散落出来的植物，他认为这种说法肯定存在某些错误，因为他“由实验得知，长春花属的某些种只有在昆虫的辅助下才能完成受精。”于是我再次查阅了我的记录，发现关于花粉释放时间的记录是非常清楚的。不过，到目前为止，我的观察结果与达尔文先生的是一致的，因为我并没有发现长春花属的植物在1月结果的任何记录。事实上，正是长春花属的植物没有结出蒴果这一现象让我仔细考虑了该如何用文字来表述这个问题，后来我写下了“几乎**所有**我们观察到的植物都产生了大量完全成形的结有种子的蒴果”这句话。需要说明的是，长春花属是我观察到的花蕾受精植物中唯一一种非本土的植物。第二点是与短柄野芝麻有关的问题。对于这方面，达尔文先生说：“我在6月初将一株短柄野芝麻罩住，当它周围的植物都硕果累累的时候，它却并没有结出任何种子。”这再一次与我提出的只有冬季开花型植物才能自花受精的推测相吻合。不过，请允许我提出如下建议：用钟形玻璃罩罩住植物以检验异花受精的方法并不完全可信，因为这样对于本来是自花受精的植物来说，很可能会由于完全中断了空气的自由流通而阻止受精作用的发生。这不就是当初法国观察者比达尔自以为是地认为肯定发生了自花受精所依据的那个用稻科植物进行的实验吗？

（刘皓芳 翻译；刘京国 审稿）

The Fertilisation of Winter-Flowering Plants

C. Darwin

Editor's Note

The correspondence on fertilisation in winter-flowering plants, begun by Alfred Bennett in the journal's first issue, continues here with a letter from Charles Darwin himself, who cautions against using bell-jars to isolate flowers from their environment (and thus any external pollination agency) because such a practice "is injurious from the moisture of the contained air". Darwin recommends "what is called by ladies, 'net' ", a practice which he had followed for twenty years and was able to observe fertilisation in thousands of plants. As regards Bennett's observation in *Vinca*, Darwin cautions that the observation of pollen falling on a stigma, and the formation of pollen tubes, is in itself "a most fallacious indication of self-fertilisation".

WILL you permit me to add a few words to Mr. Bennett's letter, published at p. 58 of your last number? I did not cover up the *Lamium* with a bell-glass, but with what is called by ladies, "net". During the last twenty years I have followed this plan, and have fertilised thousands of flowers thus covered up, but have never perceived that their fertility was in the least injured. I make this statement in case anyone should be induced to use a bell-glass, which I believe to be injurious from the moisture of the contained air. Nevertheless, I have occasionally placed flowers, which grew high up, within small wide-mouthed bottles, and have obtained good seed from them. With respect to the *Vinca*, I suppose that Mr. Bennett intended to express that pollen had actually fallen, without the aid of insects, on the stigmatic surface, and had emitted tubes. As far as the mere opening of the anthers in the bud is concerned, I feel convinced from repeated observations that this is a most fallacious indication of self-fertilisation. As Mr. Bennett asks about the fertilisation of Grasses, I may add that Signor Delpino, of Florence, will soon publish some novel and very curious observations on this subject, of which he has given me an account in a letter, and which I am glad to say are far from being opposed to the very general law that distinct individual plants must be occasionally crossed.

(**1**, 85; 1869)

Charles Darwin: Down, Beckenham, Kent, Nov. 13.

冬季开花型植物的受精作用

达尔文

编者按

关于冬季开花型植物受精作用这一主题的讨论，始于阿尔弗雷德·贝内特在《自然》第1期中发表的文章，随后就有了这篇来自查尔斯·达尔文的快报文章。达尔文提出不要使用钟形广口瓶将植物从其周围环境（和任何外部媒介）中隔离开来，因为这种操作"不利于保持容器内空气的湿度"。达尔文建议使用一种"被女士们称为'网'的工具"，他持续使用这种工具有20年，并观察到数千种植物的受精作用。至于贝内特对长春花的观察研究，达尔文提出，把花粉掉落到柱头上的现象以及花粉管的形成"看作是自花受精的象征是非常荒谬的"。

请允许我对贝内特先生发表于贵刊上一期第58页的文章补充几句。我并没有用钟形玻璃罩罩住野芝麻，而是用了一种被女士们称为"网"的工具。在过去的20年中，我一直用这种方法，并且已经使成数千种被这样遮罩的花成功受精，但从未观察到因为这样的处理而使它们的生殖能力受到丝毫损伤的例子。我作此声明以防有人可能会被误导而使用钟型玻璃罩，我认为钟型玻璃罩不利于保持容器内空气的湿度。不过，我曾经偶然地将一些已经长得很高的花放到一个小广口瓶中，最后这些花也结出了很好的种子。关于长春花属的植物，我猜想贝内特先生想表达的意思是：在没有昆虫帮助的情况下，花粉实际上是落到了柱头表面并萌发出花粉管。至于文中提到的在花蕾中出现的花药释放，通过反复观察我确信，把这种现象看作是自花受精的象征是非常荒谬的。此外，既然贝内特先生说到了稻科植物的受精作用，我想补充的是，佛罗伦萨的德尔皮诺先生不久就会发表一些针对此问题的观察结果，那将是非常新颖、非常奇妙的。他已经给我来信说明了这些结果，我非常高兴地告诉大家，他的观察结果与完全不同的植物个体间肯定会偶尔发生杂交这一基本规律并不矛盾。

（刘皓芳 翻译；刘京国 审稿）

The Recent Total Eclipse of the Sun

J. N. Lockyer

Editor's Note

The first Editor of *Nature*, J. Norman Lockyer, had a broad interest in astronomy but particularly in the constitution of the Sun. His chosen technique was that of spectroscopy, by which means he aimed to identify the chemical constituents of the Sun. His distinctive achievement was to have identified spectroscopic lines corresponding to a then unknown atom which eventually turned out to be helium. In the year of *Nature*'s founding, he was concerned to gather evidence about the nature of the Sun's corona—the Sun's plasma "atmosphere", extending millions of kilometres beyond its surface, visible only during a total solar eclipse. His article on page 14 of the first issue of *Nature* advertises his belief that the Sun's corona is not a part of the sun at all but a phenomenon caused either by the atmosphere of the Earth or that of the Moon (which at the time was supposed to be capable of retaining an atmosphere similar to that of the Earth).

IF our American cousins in general hesitate to visit our little island, lest, as some of them have put it, they should fall over the edge; those more astronomically inclined may very fairly decline, on the ground that it is a spot where the sun steadily refuses to be eclipsed. This is the more tantalising, because the Americans have just observed their third eclipse this century, and already I have been invited to another, which will be visible in Colorado, four days' journey from Boston (I suppose I am right in reckoning from Boston?) on July 29, 1878.

Thanks to the accounts in *Silliman's Journal* and the *Philosophical Magazine*, and to the kindness of Professors Winlock and Morton, who have sent me some exquisite photographs, I have a sufficient idea of the observations of this third eclipse, which happened on the 7th of August last, to make me anxious to know very much more about them—an idea sufficient also, I think, to justify some remarks here on what we already know.

A few words are necessary to show the work that had to be done.

An eclipse of the sun, so beautiful and yet so terrible to the mass of mankind, is of especial value to the astronomer, because at such times the dark body of the moon, far outside our atmosphere, cuts off the sun's light from it, and round the place occupied by the moon and moon-eclipsed sun there is therefore none of the glare which we usually see—a glare caused by the reflection of the sun's light by our atmosphere. If, then, there were anything surrounding the sun ordinarily hidden from us by this glare, we ought to see it during eclipses.

近期的日全食

洛克耶

编者按

诺曼·洛克耶是《自然》的第一任主编，他对天文学的诸多领域都有浓厚的兴趣，尤其关注太阳的组成。他通过光谱技术研究太阳以确定太阳的化学组成。他最突出的成就是分辨出了一些谱线，这些谱线对应于一种当时人们还不知道的原子，这种原子后来被确认为氦。在《自然》创刊的这一年，他关心的是收集有关日冕（仅在日全食发生时可见的太阳的等离子体“大气”，它从太阳的表面向上延伸达数百万公里）本质的证据。在这篇发表于《自然》第1期第14页的文章中，他提出了这样的观点：日冕并不是太阳自身的一部分，而是由地球大气或月球大气（当时人们认为月球可以像地球一样保持住一个大气层）引起的一种特殊现象。

如果说我们的美国表兄弟因为担心会从边缘掉到海里（就像他们中的某些人说的那样）而不愿意造访我们这座小岛，那么那些对天文学有强烈兴趣的人可能就完全不愿意来了，因为在我们不列颠岛这样的小地方基本上就看不到日食。这实在是很让人干着急的事情，美国人刚刚已经观测了他们本世纪的第3次日食，而我们在英格兰还一次都没有看到。好在我已经被邀请去观察另一次日食了，这次日食将于1878年7月29日在美国科罗拉多州发生，那里距离波士顿有4天的路程（我想我对从波士顿到那里的行程估算应该是正确的吧？）。

多亏《斯灵曼杂志》和《哲学杂志》的报道，以及温洛克教授和莫顿教授慷慨相送的一些精美照片，我对去年8月7日发生的第3次日食的观测结果已经有了比较充分的了解，这使我渴望知道更多观测信息；此外，我觉得我也有足够的准备在这里评述我们已经知道的一些说法。

首先有必要用一点篇幅来介绍一下已经做过的工作。

日食是一种非常美丽但也令广大民众感到十分恐慌的现象，然而它对于天文学家有着特殊的价值，因为在这些时候，不能发光而又远离地球大气层的月球切断了来自太阳的光线，从而在月球周围以及被月亮蚀掉的太阳的周围都不再有我们平常能看到的那种由于地球大气层对太阳光的反射而产生的炫目的光芒。所以，太阳周围如果有什么东西在平时会因为这种光芒而被掩盖的话，那么在日食的时候我们就应该能够看到。

In point of fact, strange things are seen. There is a strange halo of pearly light visible, called the corona, and there are strange red things, which have been called red flames or red prominences, visible nearer the edge of the moon—or of the sun which lies behind it.

Now, although we might, as I have pointed out, have these things revealed to us during eclipses if they belonged to the sun, it does not follow that they belong to the sun because we see them. Halley, a century and a half ago, was, I believe, the first person to insist that they were appearances due to the moon's atmosphere, and it is only within the last decade that modern science has shown to everybody's satisfaction—by photographing them, and showing that they were eclipsed as the sun was eclipsed, and did not travel with the moon—that the red prominences really do belong to the sun.

The evidence, with regard to the corona, was not quite so clear; but I do not think I shall be contradicted when I say, that prior to the Indian eclipse last year the general notion was that the corona was nothing more nor less than the atmosphere of the sun, and that the prominences were things floating in that atmosphere.

While astronomers had thus been slowly feeling their way, the labours of Wollaston, Herschel, Fox Talbot, Wheatstone, Kirchhoff, and Bunsen, were providing them with an instrument of tremendous power, which was to expand their knowledge with a suddenness almost startling, and give them previously undreamt-of powers of research. I allude to the spectroscope, which was first successfully used to examine the red flames during the eclipse of last year. That the red flames were composed of hydrogen, and that the spectroscope enabled us to study them day by day, were facts acquired to science independently by two observers many thousand miles apart.

The red flames were "settled", then, to a certain extent; but what about the corona?

After I had been at work for some time on the new method of observing the red flames, and after Dr. Frankland and myself had very carefully studied the hydrogen spectrum under previously untried conditions, we came to the conclusion that the spectroscopic evidence brought forward, both in the observatory and in the laboratory, was against any such extensive atmosphere as the corona had been imagined to indicate; and we communicated our conclusion to the Royal Society. Since that time, I confess, the conviction that the corona is nothing else than an effect due to the passage of sunlight through our own atmosphere near the moon's place has been growing stronger and stronger; but there was always this consideration to be borne in mind, namely, that as the spectroscopic evidence depends mainly upon the brilliancy of the lines, that evidence was in a certain sense negative only, as the glare might defeat the spectroscope with an un-eclipsed sun in the coronal regions, where the temperature and pressure are lower than in the red-flame region.

The great point to be settled then, in America, was, what is the corona? And there were many less ones. For instance, by sweeping round the sun with the spectroscope, both before

事实上，人们的确发现了一些奇怪的事情。在月球的边缘处，也可能是在月球后面的太阳的边缘处，可以看到有一个珍珠色的发光晕和一些奇特的红色物质，这个发光晕被称作日冕，而那些红色物质则被称作红色火焰或者日珥。

就像我已经说过的，如果这些东西属于太阳，那么在日食时我们就有可能看到它们，但并不能因为我们能看到它们而得出它们属于太阳的结论。我相信，一个半世纪以前的哈雷是坚持认为这些现象都是由月球大气造成的第一人。而直到近十年，现代科学才给出了令人满意的答案：日珥确确实实属于太阳。这是因为，通过摄影技术，人们发现，在日食发生时日珥也被蚀去了，它并不随着月球的移动而移动。

虽然关于日冕的证据还不是十分清楚，但我想应该不会有人质疑下面这个说法：在去年的印度日食之前，人们一般认为日冕就是太阳的大气，而日珥则是在这个大气上漂浮着的物质。

在天文学家缓慢地摸索前行之时，沃拉斯顿、赫歇尔、福克斯·塔尔博特、惠斯通、基尔霍夫和本生所做的工作给他们提供了一个强有力的仪器，它以惊人的速度拓宽天文学家的知识并给他们以前难以想象的研究能力，这就是光谱仪。在去年日食期间，天文学家第一次成功地将光谱仪用在对日珥的研究上。两个相隔数千英里的观测者独立得到了这样的科学事实：日珥是由氢组成的；我们每天都可以利用光谱仪对日珥进行研究。

因此，从某种程度上来说，日珥的问题已经“得到解决”了，但是日冕又如何呢？

我用观测日珥的这种新方法工作了一段时间，另外我还和弗兰克兰在各种以前未曾尝试过的条件下对氢的光谱进行了非常仔细的研究，在这之后，我们得出了这样的结论：尽管此前人们总是设想存在日冕这样一层非常厚的大气，但是我们在实验室和天文台得到的光谱学证据都反对这一点。我们也向皇家学会提交了这一结论。从那时开始，我越来越确信日冕只是太阳光在通过靠近月球一侧的地球大气层时产生的一种效应。不过也有一个问题一直困扰着我：既然光谱学证据主要依赖于谱线的亮度，那么我们的证据就只是在一定程度上否定了以前的设想，因为强烈的太阳光可能会影响光谱仪对没有发生日蚀的日冕区（温度和压强都低于日珥区）的检测。

对于下一次美国的日食，需要解决的最大问题是，什么是日冕？除此之外，还有许多比较小的问题。比如，通过在日食前和日食后用光谱仪绕太阳扫描一周，或

and after the eclipse, and observing the prominences with the telescope merely during the eclipse, we should get a sort of key to the strange cypher band called the spectrum, which might prove of inestimable value, not only in the future, but in a proper understanding of all the telescopic observations of the past. We should, in fact, be thus able to translate the language of the spectroscope. Again, by observing the spectrum of the same prominence both before and during, or during and after the eclipse, the effect of the glare on the visibility of the lines could be determined—but I confess I should not like to be the observer charged with such a task.

What, then, is the evidence furnished by the American observers on the nature of the corona? It is *bizarre* and puzzling to the last degree! The most definite statement on the subject is, that it is nothing more nor less than a *permanent solar aurora*! the announcement being founded on the fact, that three bright lines remained visible after the image of a prominence had been moved away from the slit, and that one (if not all) of these lines is coincident with a line (or lines) noticed in the spectrum of the aurora borealis by Professor Winloch.

Now it so happens that among the lines which I have observed up to the present time—some forty in number—this line is among those which I have most frequently recorded: it is, in fact, the first iron line which makes its appearance in the part of the spectrum I generally study when the iron vapour is thrown into the chromosphere. Hence I think that I should always see it if the corona were a permanent solar aurora, and gave out this as its brightest line; and on this ground alone I should hesitate to regard the question as settled, were the new hypothesis less startling than it is. The position of the line is approximately shown in the woodcut (Fig. 1) near E, together with the other lines more frequently seen.

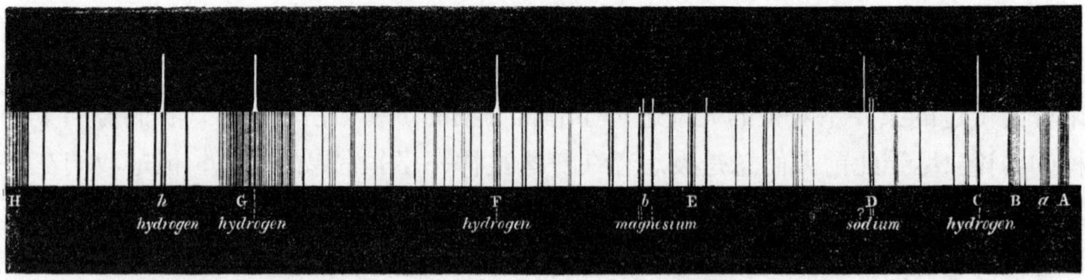

Fig. 1. Showing the solar spectrum, with the principal Fraunhofer lines, and above it the bright-line spectrum of a prominence containing magnesium, sodium, and iron vapour at its base.

It is only fair, however, to Professor Young, to whom is due this important observation, to add that Professor Harkness also declares for one bright line in the spectrum of the corona, but at the same time he, Professor Pickering, and indeed others, state its spectrum to be also continuous, a remark hard to understand unless we suppose the slit to have been wide, and the light faint, in either of which cases final conclusions can hardly be drawn either way.

者只是在日食时用望远镜来观测日珥，我们都可以得到一类解开太阳的奇怪光谱带的办法。应该说，这对于正确地理解过去用望远镜观测得到的结果和将来进行的研究都具有不可估量的价值。事实上，这样我们应该就能够翻译光谱仪的语言。另外，通过在日食前和日食中，或者日食中和日食后观测同一日珥的光谱，我们可能就可以确定强烈的太阳光对谱线可见性的影响。不过，我觉得自己并不是能够胜任这一工作的观测者。

那么美国的观测者们关于日冕的本质提供了什么样的证据呢？实际情况真是极其**荒诞**而又莫名其妙！关于这个问题最明确的说法是：日冕是一种**永久性的太阳极光**！得出这个结论的事实基础是：将来自日珥的光线从狭缝中移走后，仍然能看到3条明亮的谱线，而且其中一条谱线（如果不是全部的话）与温洛克教授在北极光的光谱中观察到的一条（或几条）谱线相吻合。

现在看来，在迄今为止我观察过的所有谱线（大约40条）中，这条谱线是我最常记录到的谱线之一。实际上，这是我通常研究的部分光谱中的第一条铁线，它是在铁蒸气被抛入色球时产生的。因此我认为，如果日冕是永久性的太阳极光，并且这条铁线是它发出的最亮谱线的话，那么我应该总能够看到这条谱线。如果不是后来出现了更令人惊奇的新假说，仅仅依靠这一点我想我是不会认为这个问题已经得到解决的。这条铁线的位置大约在E点附近的刻线（图1）处，图中还显示了其他一些更为常见的谱线。

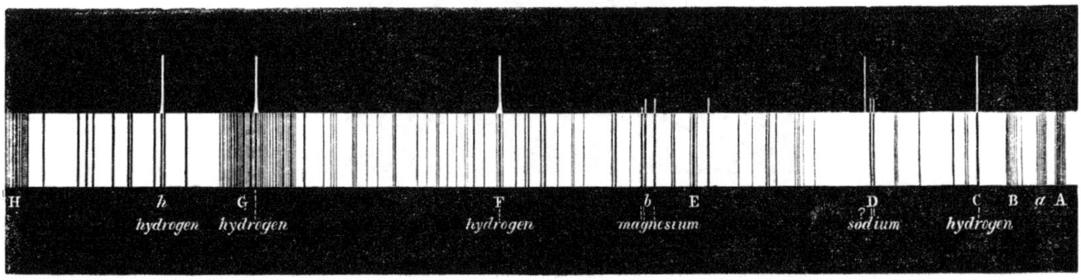

图1. 包括主要的夫琅和费谱线的太阳光谱，其上方是日珥所包含的镁和钠的亮线光谱，其底部则是铁蒸气的光谱。

另外还有一点，不过应该由发现上述重要观测结果的扬教授来补充才是公平的，这就是：哈克尼斯教授也曾宣称在日冕的光谱中发现了一条明亮的谱线，但与此同时，他、皮克林教授以及其他一些人又声称，日冕的光谱也是连续的。这实在是一个令人难以理解的论断，除非我们假设狭缝比较宽或者光线比较微弱，但无论是哪种情况，我们都无法得出上述的最终结论。

So much, then, for the spectroscopic evidence with which we are at present acquainted on the most important point. The results of the other attacks on the same point are equally curious and perplexing. Formerly, a favourite argument has been that because the light of the corona is polarised; therefore it is solar. The American observers state that the light is *not* polarised—a conclusion, as M. Faye has well put it, "very embarrassing for Science". Further,—stranger still if possible,—it is stated that another line of inquiry goes to show that, after all, Halley may be right, and that the corona may really be due to a lunar atmosphere.

I think I have said enough to show that the question of the corona is by no means settled, and that the new method has by no means superseded the necessity of carefully studying eclipses; in fact, their observation has become of much greater importance than before; and I earnestly hope that all future eclipses in the civilised area in the old world will be observed with as great earnestness as the last one was in the new. Certainly, never before was an eclipsed sun so thoroughly tortured with all the instruments of Science. Several hundred photographs were taken, with a perfection of finish which may be gathered from the accompanying reproduction of one of them.

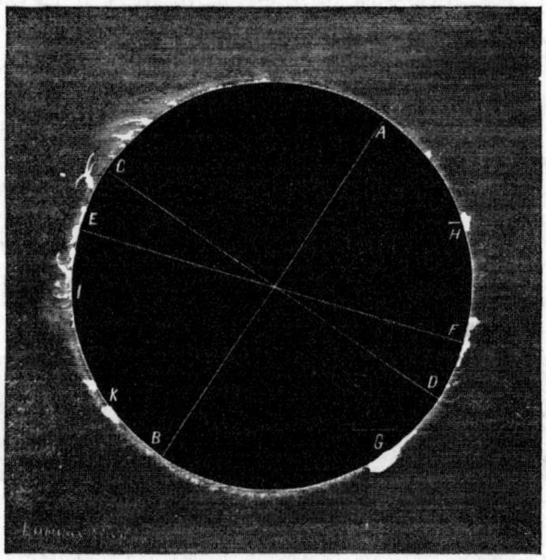

Fig. 2. Copy of a photograph of the Eclipse of August 7, obtained by Professor Mortor's party

The Government, the Railway and other companies, and private persons threw themselves into the work with marvellous earnestness and skill; and the result was that the line of totality was almost one continuous observatory, from the Pacific to the Atlantic. We read in *Silliman's Journal*, "There seems to have been scarcely a town of any considerable magnitude along the entire line, which was not garrisoned by observers, having some special astronomical problem in view." This was as it should have been, and the American Government and men of science must be congratulated on the noble example they have shown to us, and the food for future thought and work they have accumulated.

以上就是我们目前在最重要的问题上获得的光谱学证据。关于这一问题的其他一些结果也同样古怪而又令人困惑。之前有个广受欢迎的说法认为：日冕的光是偏振的，所以它就是太阳光。但美国的观测者们声称这些光**不是**偏振的。就像费伊所说，这实在是一个“让科学界异常尴尬”的结论。另外更奇怪的是（如果是真的话，就更奇怪了），据称有调查表明可能哈雷终究是对的，日冕可能真的是由月球的大气产生的。

我想我说了这么多已经足以表明，日冕的问题还没有得到解决，并且新方法也完全没有取代对日食进行仔细观测的必要性。事实上，天文学家的观测变得比以前更加重要。我非常真诚地希望，对于此后将在旧大陆的文明地区发生的所有日食，我们也能像对最近这次发生在新大陆上的日食一样进行非常认真的观测。无疑，以前人们从来没有像这次这样用所有的科学仪器来彻底地拷问日食。在这次日食中，人们拍摄了几百张照片，我复制了其中一张附在文中，从这张图可以看到拍照工作做得十分完美。

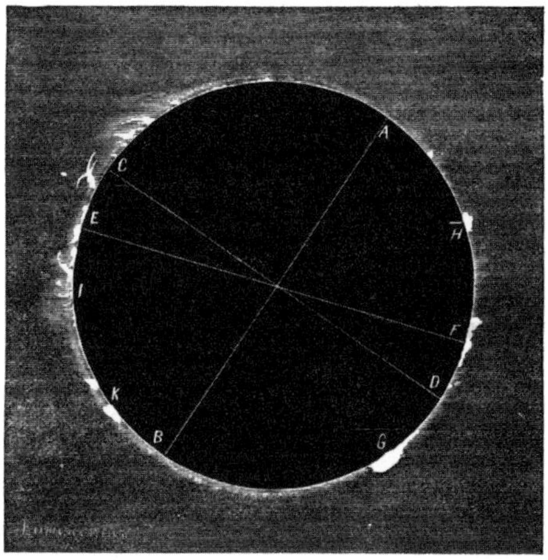

图 2. 莫顿教授的观测队拍摄的一张照片（8月7日的日食）的拷贝

政府、铁路和其他公司，以及许多个人都以极大的热情和娴熟的技巧投入到了这项工作中。结果是从太平洋到大西洋的整个观测线几乎成了一个连续的天文台。我们在《斯灵曼杂志》上读到，“在整条观测线上，几乎所有有点名头的城镇中都驻扎着一些正在思索某些特殊天文学问题的观测者。”这一切好像本来就该如此。美国政府和科学工作者们给我们树立了一个光辉的榜样，也积累了一些有助于进一步研究的证据和思路，应该为此而向他们致敬。

* * *

Since writing the above, I find the following independent testimony in favour of Dr. Frankland's and my own notion of the corona in the *Astronomische Nachrichten*, from the pen of Dr. Gould. He says:—"Its form varied continually, and I obtained drawings for three epochs at intervals of one minute. It was very irregular in form, and in no apparent relation with the protuberances on the sun, or the position of the moon. Indeed, there were many phenomena which would almost lead to the belief that it was an atmospheric rather than a cosmical phenomenon. One of the beams was at least 30′ long."

(**1**, 14-15; 1869)

*　*　*

写完以上这些后，我在《天文通报》上看到了谷德博士独立发现的一些能够支持弗兰克兰博士以及我自己关于日冕观点的证据。谷德博士称："它的形状在不断变化，我得到了间隔一分钟的 3 个不同时刻的图像。可以看到，它的形状很不规则，与太阳表面的突起或月球上的某些位置都没有明显的联系。事实上，有很多现象都能够推出这是一种大气现象而非宇宙现象的结论。其中一条光束至少有 30 角分长。"

（汪浩 翻译；蒋世仰 审稿）

The Suez Canal

T. Login

Editor's Note

British engineer Thomas Login, who had recently worked on the Ganges Canal, here writes to *Nature* to suggest a way to prevent the new Suez Canal from silting up. Operating the canal might require heavy dredging, he notes, to remove sediments from the nearby Nile River deposited at its mouth by waters flowing past in the Mediterranean. But dredging might be avoided by constructing two long barriers, the first to block water with suspended material from sweeping past the canal mouth, and the second to direct clean water from further out to sea to flow in toward the mouth. If this were done, he suggests, the canal will be a success.

THE all-engrossing topic of the day is the Suez Canal, about which some diversities of opinion still exist. As for many years back I have had my attention particularly drawn to some of the chief matters in dispute, having been engaged on the largest irrigation works in India, I venture to trouble you with the following remarks.

Engineering science and indomitable energy have, in the case of the Suez Canal, overcome difficulties which at one time were considered insurmountable; but even up to the present moment doubts still exist, and some fear that the whole scheme may yet prove a failure, owing to the débris of the Nile travelling eastward transported by the currents of air and water. That we can overcome the former is, in my opinion, beyond all doubt; for it is found that whenever an irrigation channel is run out from the Jumna Canal into the great desert of Northern India, rich vegetation takes the place of arid sand. And so in Egypt will irrigation force back the desert; so the only question is, can irrigation be carried out on an extensive scale? And of this also I have no doubt, for the enormous volume of water which now flows into the sea and is lost, is quite sufficient to reclaim the whole of the desert.

It may be asked, can the water be made to flow over the desert? And of this I hold that there can also be no doubt. The very name of the Timsa Lake proves, I think, that the Nile, or at least a branch of it, flowed eastward, for the word Timsa signifies crocodile, showing that the water must at one time have been brackish or fresh, for these creatures could not have existed in this lake had it been salt as at present. If, therefore, a portion of the Nile water at one time flowed eastward, there can be no great engineering difficulty to make it do so again; and I am almost inclined to think that it would have been better to have made the canal a fresh-water one, for it is only by vegetation, the produce of irrigation, that the desert can be kept under control. Other advantages may be cited, such as cleaning the bottom of ships by bringing them into fresh water, and the prevention of any of the disturbed and very muddy waters along the Mediterranean coast getting

苏伊士运河

洛金

编者按

近期一直致力于恒河运河工程的英国工程师托马斯·洛金给《自然》来稿，提出了一种防止新通航的苏伊士运河出现淤塞的办法。他认为，要经营好运河，可能需要通过繁重的疏浚工作来清除尼罗河水流经地中海沿岸地区后在河口处沉积产生的泥沙。不过，如果能修建两座长堤可能就不用疏浚了，第一座长堤用以阻挡混有沉积物的河水流入运河口，第二座长堤用以从远处引导干净的河水流向河口。洛金认为，如果能这么做的话，苏伊士运河将会取得成功。

苏伊士运河是当今一个引人入胜的话题，对此一直存在多种不同的意见。多年来我一直特别关注争论中的一些主要议题，而且也曾参与建设印度最大的水利工程，因此在这里我想冒昧地提请大家注意下列意见。

在苏伊士运河开凿过程中，工程科学和不屈不挠的精神克服了那些曾经被认为不可克服的困难。但是，至今仍然有一些疑问。有些人担心，气流和水流会使尼罗河的泥沙向东转移扩散，从而使整个计划失败。我认为，这一点毫无疑问是不能成立的，因为从朱木拿运河到北印度大沙漠的水道开通后，我们看到干旱的沙漠中出现了丰富的植被。同样，在埃及，通过水道灌溉也必将迫使沙漠后退，那么唯一的问题就是灌溉能否覆盖一个很广泛的范围？对这一点我也没有疑问，因为现在由于注入海洋而流失的水量足够浇灌整个沙漠。

有人可能会问，水能流经沙漠吗？我认为这也没有问题。在我看来，提姆萨湖这个名字就表明尼罗河（或至少它的一条支流）曾经是向东流的，因为提姆萨这个词是鳄鱼的意思，这说明湖水曾经是淡水或者微咸的水，因为鳄鱼在现在这样的盐水湖里不能生存。如果部分尼罗河水曾经是向东流的，那么要让它再次东流，工程上应该没有多大困难。另外，我倾向于认为，如果开挖一条淡水运河将会更好，因为只有淡水灌溉产生的植被才能使沙漠得到控制。淡水运河还有别的一些优点，例如，船舶驶入淡水运河后船底可以得到清洗，运河中的淡水可以阻止地中海沿岸泥

admission into the canal; for by keeping the water in the canal at a higher level than that of the sea at both ends there could only be an outflow. So all the water wasted would be expended on lockage.

It may be objected that the fresh-water canal would get silted up by the muddy waters of the Nile; but could not this Timsa Lake be used as a silt-trap? I do not mean to say, that the present canal will be a failure because it has not been made a fresh-water one; but what I do think is, that possibly in the end a fresh-water canal would have been best and perhaps cheapest, as the dredging of the canal might have been much reduced*, as the water could have been kept at a higher level in the canal.

The great difficulty, however, to contend against, appears to me to be to keep a deep-water channel at the Mediterranean end of the Canal; and what drew my attention to this more than a dozen years ago, was the fact that the harbour of Alexandria does not get silted up. Some have supposed that the subsidence of the delta accounts for this, and that the small advance of the land on the sea in this direction is owing to a constant sinking of the land. In my opinion a very different cause can be assigned: Nature here is working by a very different agency, namely, the current in the Mediterranean which flows eastward all along the African coast, and transports the débris of the nile, depositing it all along the western portion of the Mediterranean. The fact of the Timsa Lake being at one time fresh or brackish, goes to support this view; so the only question is, will the cost of continuous dredging be so excessive that the Canal will become a financial failure? On this point I cannot venture to give an opinion, as I have no data, but I think this difficulty may be met by forcing this easterly current to aid in keeping the mouth of the Canal clear of silt deposits.

What aids this current to transport the earthy matter is the beat of the sea always stirring the mud and sand up on the coast, and enabling the water to hold a large proportion of matter in suspension, and even to transport heavy matter.†

The proportion of earthy matter a short distance out to sea is comparatively little, so the great object appears to me to prevent the agitated water traveling as it does at present, and this can be done by arranging the breakwaters somewhat as shown in this diagram.

---

* I observe that, in a discussion at the Civil Engineers Institution, the total excavation of the Suez Canal is stated to be 70,000,000 cubic metres. The excavation of the Ganges Canal was 2,547,000,000 cubic feet, or a little over 70,000,000 metres; but this latter does not include some 3,000 miles of distribution channels.

† At Felixstowe, last March, during a gale of wind, I watched a mass of brickwork, some eighteen inches square and about six inches thick, moved along the coast by the action of the waves, which were in an oblique direction to the coast, and no doubt the same takes place along the mouths of the Nile. By a sample I took of this agitated water, I found it contained 0.70375 percent of its weight of small pebbles, sand, and mud. This sample was taken at a height of nearly ten feet above the sea, and was got by catching the spray of the sea as it was falling.

沙含量很高的水受扰动后流入，因为只要保持运河中的水位高于两端海洋的水位，水就只会向外流。这样，通过船闸系统就能将所有多余的水利用起来。

也许有人会这样反对，说淡水运河会被尼罗河的泥沙淤塞。难道就不能用提姆萨湖来沉积淤泥吗？我的意思并不是说现在的苏伊士运河将会因为不是淡水运河而失败，而是说淡水运河很可能是最好的而且最便宜的选择，因为如果是淡水运河的话，我们可以使其保持一个较高的水位而大大降低疏浚运河的耗费*。

不过，对我来说反对者提出的最具挑战的困难是如何保持运河水道在地中海一端的水深。实际上，亚历山大港并未被淤塞的事实让我在十多年前就开始思考这个问题。有人认为这可以用三角洲的下沉来解释，海边的陆地向海中延伸的速度非常缓慢是因为这些陆地一直在下沉。在我看来，这是由另一种完全不同的原因造成的，在这里自然界中的另一种力量起了作用，那就是沿着非洲海岸向东流动的地中海洋流，它转移了尼罗河的泥沙并使其沉积在地中海西部。提姆萨湖的湖水曾经是淡水或微咸的水这一事实支持了这种观点。那么唯一的问题就是，是否会因连续不断的疏浚造成成本太高而导致运河成为一个财务上失败的工程呢？我没有这方面的资料，因此不敢冒昧地就此发表意见，但是我想如果可以利用向东的洋流的帮助而使运河河口免除淤泥沉积的话，这一难题就能得到解决。

海水的律动使地中海洋流能够转移泥沙。正是这种作用使泥沙经常被卷到岸上，使海水可以携带大量的悬浮物并且甚至可以运送重物。†

从运河向海延伸一小段距离后，泥沙的含量相对就比较少了，所以我认为主要目标应该是防止涌动的水以现在的方式流动，这可以通过建造如图所示的防浪堤来实现。

---

* 在土木工程研究院的一次讨论中，我得知苏伊士运河的总挖土量是 70,000,000 立方米。恒河运河的总挖土量是 2,547,000,000 立方英尺，稍多于 70,000,000 立方米，但后者并未包括 3,000 英里分渠的挖掘。

† 在去年 3 月费利克斯托的一次大风期间，我注意到，在斜冲向岸边的海浪的作用下，许多碎石泥沙沿着河岸移动，其中有些甚至是 18 英寸见方、6 英寸厚的大块，毫无疑问同样的事情在尼罗河河口也会发生。我采集了一份这里的涌动海水的水样，通过检测我发现其中 0.7375%（质量百分比）是小鹅卵石、沙子和泥。这份水样是我在海浪落下时采集的，其采集位置大约在海平面以上 10 英尺。

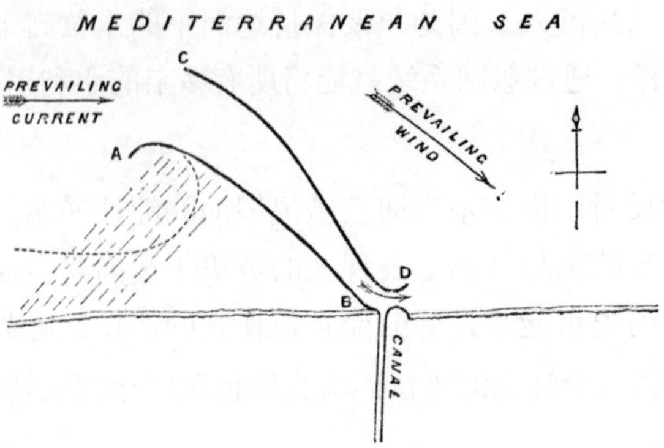

The breakwater AB is intended to prevent the very muddy water traveling along the coast, and the point A should extend well out into deep water. The breakwater CD is to direct the comparatively pure water where the sea is deep to pass across the mouth of the canal; and by the funnel-mouthed shape thus given, the velocity at D will be increased, and thus keep deep water at the head of the canal. Some may say that the expense will be enormous, and that it will have to be year after year extended. But, in reply to this, I say that deltas do not extend out into the sea at so rapid a rate as some suppose; and that the formation of a delta takes several thousands of years to accomplish, so that in this very delta, the advance is hardly perceptible; and that a sinking of the land has been brought forward, to account for the very slow progress made; while, in fact, Nature has at present a power at work which is quite sufficient to explain the reason why so little advance is made on the sea during the historic period (see my paper on the Delta of the Irrawaddy, read before the Royal Society of Edinburgh in 1857).

In conclusion, I have no doubt this Suez Canal will have many ready to abuse it and say it is a total failure, as has been said of the Ganges Canal; but like the latter work, which last year saved some three million human beings from starvation, so will this canal, I have little doubt, outlive the abuse, and become one of the greatest blessings to the civilized world.

(**1**, 24; 1869)

T. Login: C. E., late of the Ganges Canal.

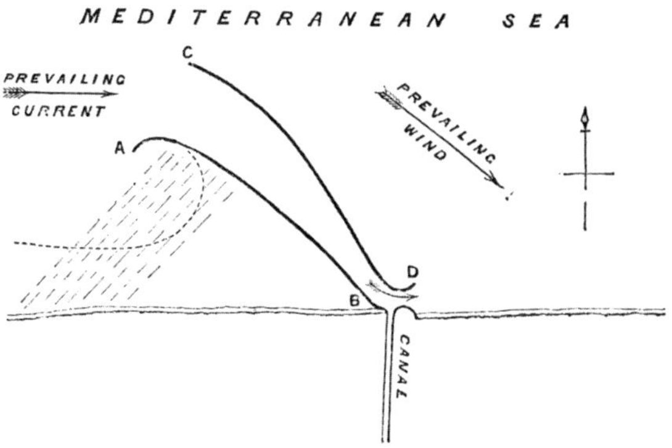

防浪堤AB用以防止泥沙含量很大的水沿岸移行，A点应伸到深水区中。防浪堤CD用以引导深海中相对纯净的海水流经运河河口。通过建造这样两条漏斗口状的防浪堤，D处水流速度将会增大，从而就能使运河河口处总是深海海水。也许有人会说这样做花费巨大，而且以后还要不时地延展这些防浪堤。对此，我的答复是，河口三角洲向海中延伸的速度并没有某些人想象的那样快。三角洲的形成要经过数千年，因此人们几乎不会感觉到这处三角洲向海中的推进，而且还有人曾在解释三角洲推进缓慢时提出过陆地沉降的观点，尽管自然界中的一种确定存在的力量实际上已经足以解释在历史上陆地向海洋延伸得很慢的原因（见1857年我就伊洛瓦底三角洲这个主题向爱丁堡皇家学会宣读的论文）。

总之，我肯定苏伊士运河将会饱受责难并被称为一个完全失败的项目，正像现在人们评论恒河运河那样。但正如去年就使大约 300 万人免受饥饿的恒河运河一样，我坚信苏伊士运河也将经受住人们的指责并成为文明世界的最大幸事之一。

（孙惠南 翻译；刘盛和 审稿）

The Suez Canal

Editor's Note

As the Suez Canal officially opened, this editorial discusses the historic nature of the event. The canal, linking the Mediterranean Sea with the Gulf of Suez in the Red Sea, provided a marine link between Europe and Asia that did not require circumnavigation of Africa. Earlier efforts to build such a route had discovered remnants of ancient canals in the same region. But a British study of the outflow of the Nile showed the great difficulty of maintaining such a canal, as immense quantities of sediment would be constantly deposited near its mouth. Tremendous advances in engineering had finally allowed the construction of a canal able to carry large vessels, but further problems in maintaining it were anticipated.

IF all went well, and we hope it did, yesterday witnessed a grand gathering on the sandy shores of a dreary bay in the Midland Sea—that sea around which so much of history has been enacted, and in whose annals the gathering in question will not be the least noteworthy incident. The Suez Canal—that problem of many centuries—is to be opened in presence of emperors, kings, princes, and potentates; of eminent engineers, famous warriors, and distinguished *savants* invited from the East and from the West; and while the ceremonial lasts the very dreariest of the dreary wastes that here and there border the blue waters of the Mediterranean will be animated by a brilliant throng and the sound of music; and speeches will be made and healths will be drunk, and all present will join in wishing success to the memorable enterprise, which, for a time, is to furnish to Arab story-tellers and Frankish newsmongers a topic to talk about.

Dreary as the region is, it has a history. There marched with invading armies the kings whose names are recorded in Scripture; there Artaxerxes was stayed in his victorious advance by the siege of Pelusium; there are yet to be seen relics of cities and towns named in the "Itinerary" of Antoninus; there Titus marched to the siege of Jerusalem; there Baldwin and his Crusaders took the city of Pharamia: the actors in these and other exploits never dreaming that the sands of the desert, drifted by the winds and by the stream of the Nile, would so bury and alter the surface of the land, that after generations should be puzzled to identify its historical localities.

The question of a canal dates from a very early period. In high floods the waters of the Nile spread to within two or three miles of the Red Sea, which would suggest the idea of a permanent communication between the river and the great Arabian Gulf. This communication was actually established, as is said, under Ptolemy Philadelphus; but of course it fell into neglect, and was buried under the drifting sands, until one of the caliphs had it cleared out, after which there was a navigable canal between the Nile and the Red

苏伊士运河

编者按

在苏伊士运河正式通航之际，这篇评论文章讨论了运河开凿的历史。苏伊士运河连接了地中海与红海中的苏伊士海湾，为欧洲和亚洲之间提供了一条不必绕道非洲的海上通道。早先，人们在试图建设这样一条通道时在该地区发现了古代运河的遗迹。不过，英国人对尼罗河河口进行的一项研究表明，维护这样一条运河有相当大的困难，因为在它的河口附近将会沉积大量的泥沙。后来，工程学的巨大进步使人们终于能够建造这样一条可供大型船舰航行的运河，不过人们也预料到要维护好运河还会有其他困难。

昨天有一个盛大的集会在地中海的一个寂静海湾的沙滩上举行，我们希望一切顺利。地中海周围曾发生过很多历史事件，而这次盛会在地中海的历史记录中也算得上是值得注意的事件。在被邀请的来自东方和西方的皇帝、国王、王子、当权者，杰出的工程师、著名的勇士、杰出的专家的见证下，苏伊士运河这一困扰了人们几个世纪的困难工程终于开通了。开通仪式进行时，在这片处处与地中海蓝色海水相连的最寂静的荒地上，精英云集，乐声激荡，还有演讲和祝酒。所有来宾都希望这个有纪念意义的事业取得成功，这个计划曾一度是阿拉伯的讲故事能手们和法兰克的新闻传播人员谈论的中心话题。

虽然这个地区是荒凉的，但也有不少历史记载。圣经中载有名字的国王们曾经率领入侵军队在那里行军；阿尔塔薛西斯在那里取得了培琉喜阿姆围攻的胜利；安东尼在征战途中命名的城市和村镇的遗迹在那里尚能见到；提图斯曾经在那里进军耶路撒冷围城；鲍德温和他的十字军战士在那里攻下了法拉米亚。这些著名人物和其他丰功伟业的建立者们做梦也不会想到：风和尼罗河水带来的沙漠里的沙子就这样掩埋并改变了土地的表面，几代之后人们就很难辨认出它的历史位置了。

关于修建运河的问题可以追溯到很久以前。在大水期间尼罗河水淹没了红海附近 2~3 英里范围内的土地，这可能使人们产生了把尼罗河与广阔的阿拉伯湾永久联系在一起的想法。据说这个联系实际上在托勒密·费拉德尔甫斯时代就已建立起来了，但毫无疑问它已被人们遗忘，而且被流沙掩埋，直到有一个哈里发下令将它清理出来，从此这里才出现了一条尼罗河到红海之间的航道。一百多年后它又消失了，

Sea for more than a hundred years. Then it was again lost, and so completely that its ever having existed became matter of doubt and dispute.

But the main project was a ship canal across the isthmus. There is some tradition that Alexander consulted with his engineer officers as to its feasibility, and that they reported against it on account of the difficulty in preventing the mouth of the canal from silting up. In a later age Sultan Selim, who had been baffled in his scheme for a canal to connect the Don and Volga, resolved on cutting one from Pelusium to Suez; and he took an important step towards accomplishing his purpose, for he conquered the country all across, and made his name a terror to the Arabs. But he did not live to cut the canal. The first Napoleon revived the project, and ordered a survey, during which the long-buried remains of the canal above-mentioned were discovered, and the question as to its having existed was settled. From that time the question of a ship-canal became a standing topic, enlisting divers opinions, among which were some to the effect that the project was simply impossible, because, as the level of the Red Sea was so much higher than that of the Mediterranean, the swift current in one direction would prevent navigation.

During this time of debate, Captain Spratt of the Royal Navy was sent, with the ship *Medina*, to make a survey along the shores of Egypt and of the Isthmus, of which an account was published by the Admiralty in 1859, entitled, "An Investigation of the Effect of the Prevailing Wave Influence on the Nile's Deposits"; and this was followed by "A Dissertation on the True Position of Pelusium and Farama". Beginning at the western extremity of the Egyptian coast, Captain Spratt found that the Nile there exerted no influence, but that, owing to the prevalent north-westerly and westerly winds, the deposits brought down by the Nile were drifted to the eastward in prodigious quantity, even to the shores of Syria. This was no hasty conclusion: by a careful series of soundings and dredgings, Captain Spratt determined the identity of the sand along the sea bottom, within a given distance of the shore, with that of the deserts through which the Nile flows. Farther out to sea the sand was coralline, and of an entirely different character, while the Nile drift is made up of quartzose sand, with fine mud and particles of mica. The verifications in this particular were too numerous and too exact to leave room for doubt. "By this means," writes Captain Spratt, "I was enabled to trace the extent of the Nile's influence both directly off the coast and along it, as well as to ascertain the large quantity of sand—pure silicious sand—it must annually bring to the sea; and to an amount which far exceeded my expectations and experience in respect to other rivers, particularly that of the Danube, which, in comparison, brings a very much less proportion of sand to mud. The Danube sand, also, is of the finest quality. The Nile sand, on the contrary, is much coarser generally, and forms sandbanks off the coast that are composed of quartzose sand nearly as large as mustard seed."

The quantity of solid matter brought down by the Nile when in flood is prodigious, and precisely at this season—that is, for three or four months—the north-west winds blow strongest. Indeed, if the wind did not blow with the violence of a monsoon it would be impossible for sailing-vessels to navigate the river during the time of its rise. The suspended

而且连它究竟是否存在过也成了被怀疑和争论的问题。

但主要工程是开通一条穿过地峡的能行船的运河。依照惯例，亚历山大向工程官员们咨询了开凿运河的可行性，他们因为难以克服运河河口被淤塞而持反对态度。后来，苏丹塞利姆二世计划开通一条运河以连接顿河与伏尔加河的方案受阻后，他转而决定开凿一条从培琉喜阿姆到苏伊士的运河。他为完成这个目标迈出了重要的一步，因为他征服了横跨这一区域的国家，阿拉伯人都畏惧他的名字。但他没能活到运河开通。拿破仑一世重提了这个方案，并下令进行勘察，这期间才发现了前面提到的已沉埋多年的运河遗址，由此运河是否确实存在过的问题也得到了解决。自那以后开挖一条能行船的运河成了一个时常被考虑的问题。为此，征集过种种意见，其中有些人认为该方案是完全不可能的，因为红海水位比地中海水位高出很多，一个方向的湍急水流将阻碍航行。

争论期间，皇家海军的斯普拉特上校被派去进行调查，他乘**梅迪纳号**沿埃及和苏伊士地峡的海岸进行了调查。1859 年，英国海军发布了题为《关于盛行海浪对尼罗河沉积物的影响的调查结果》的报告，随后又发布了《关于培琉喜阿姆和法拉玛真实位置的论证》。斯普拉特上校发现，从埃及海岸的最西端开始，尼罗河并未造成什么影响，但是由于盛行西北风和西风，尼罗河带来的沉积物会大量向东漂移，甚至可以到达叙利亚海岸。这绝不是草率的论断：经过一系列的探测和挖掘，斯普拉特上校确定了沿岸一定距离内海底的沙子就是尼罗河流经沙漠时带来的沙子。在海洋深处，沙子是珊瑚色的，而且完全是另外一种性质，而尼罗河的泥沙是由石英砂组成的，伴随着细泥和云母颗粒。这方面的证据很充分，很准确，不容置疑。“用这种方法，”斯普拉特上校写道，“我可以去追踪调查尼罗河对海岸外侧和海岸沿线地区的影响，还可以断定每年都有大量纯硅质的沙子被带到海里，其数量远远超过我的预期，也远远超过在其他河流中见到过的量，特别是多瑙河，相比之下，多瑙河所带的沙泥少很多。而且，多瑙河的沙也是最细的。相反地，尼罗河携带的沙通常粗得多，且能在海岸外侧形成一个由芥菜籽般大小的石英砂组成的沙洲。”

在洪水期，尼罗河带来的固体物质数量惊人，就在这个季节——有三四个月时间——西北风最强劲。实际上，如果此时没有强劲的季风，那么帆船在丰水期就无法航行。悬浮物不断沿海岸向东漂移，随后在岸边堆积成沙丘，每一阵狂风过后这

matter is consequently driven to the eastward along the coast, and there accumulating forms dunes or sandhills, which shift their position with every gale, "burying at times the huts of the coastguard men". The hollows between the dunes are cultivated by the Arabs, but the plots must be protected by screens of reeds, against which the sand accumulates by repetition, until in some instances the hill is a hundred feet in height. Captain Spratt here remarks: "The best efforts of a population of several thousand Arabs, who inhabit the villages along this strip of land, fail in permanently fixing these dunes. For as the sea continually reaccumulates the sand upon the beach, onward it moves, in spite of those efforts, and the rate of progress may be imagined when I state that a mosque near Brulos has in about twelve months been nearly buried in one of the dunes" advancing from the westward. "And as the coarse sand of which these hills are composed is not distinguishable in differing from the sands of the desert near the Pyramids, or that on the route to Suez, they must undoubtedly be all the gifts of the Nile."

Besides coarse sand the Nile carries down fragments of brick, pottery, and other heavy substances, which are also drifted along the coast by the combined action of wind and current. When the wind blows its strongest the coastguard men say they cannot walk against it. To test these facts, Captain Spratt one day landed eleven bags of ashes and clinkers, five of the bags containing pure clinkers, the largest of which weighed from four to five pounds. The whole were laid in a heap just above the water's edge, and left to the care of wind and sea. Twelve days later, when the party returned, not a vestige of the heap, which had weighed nearly two tons, was to be seen. The shore was examined towards the quarter from which the wind blew, but without result; while in the other direction, that of the prevailing wave movement, clinkers weighing about two ounces were found dispersed to a distance of fully 1,500 yards, one of 3.5 pounds was picked up at 240 yards, and others from 4 to 8 ounces at from 600 to 700 yards. The greater portion had, however, been buried by the movement of the sand. "Thus this evidence," writes Captain Spratt, "of the movement of the beach in only twelve days, in the month of May, during which there was but one strong westerly breeze and several fresh easterly breezes, is a positive evidence of the great easterly movement of the shore and littoral shallows along this coast, but which, during a succession of winter gales, and during the prevailing north-west breezes at the period of high Nile, must cause a continuous progression of an immense quantity of the sands and matter carried out by the turbid river."

We quote another passage bearing on this point. The captain was walking along the coast for the purpose of observation, from the beacon marking the site of Port Said, to the head of the bay of Tineh, when he found a great quantity of broken pottery, broken jars, ancient and modern, and broken bricks scattered on the shore, at the highest and lowest surf margin. "On discovering them in such quantity," he continues, "I was naturally anxious to trace out their origin, thinking they must have come from some adjacent ruin. But I found eventually that they had come wholly from the mouths of the Nile, and that they were the positive *débris* from the towns situated on the banks of the river, and brought out by the strength of the current at high Nile, but then dispersed along the coast to the eastward by its littoral currents and prevailing ground swell."

些沙丘还会移动位置，“有时会埋没海岸警卫员的小屋”。沙丘间的洼地则由阿拉伯人耕种，但那些小块土地必须用芦苇屏障围起来，以阻挡不断堆积的沙子，某些情况下，沙丘可高达 100 英尺。斯普拉特上校写道：“在这个狭长地带居住着的数千名阿拉伯人尽了最大的努力，但仍未能固定那些沙丘。因为不管人们怎么努力，大海都不断地把沙子堆到海滩上，并向前推动，如果我用布鲁罗斯附近一个清真寺在大约 12 个月的时间就几乎被沙丘埋掉来说明，你就能想象这些沙丘向前推进的速度了”，这些沙丘是从西边移过来的。“因为组成这些沙丘的沙子是粗沙，与金字塔附近或去苏伊士路上的沙漠里的沙子没有什么区别，它们无疑都是尼罗河送来的礼物。”

除了粗沙以外，尼罗河还带来了砖头、陶器以及其他重物的碎片，它们在风和水流的共同作用下沿海岸漂移。海岸警卫人员说，当风力最大的时候，他们无法迎风走路。为检验这些事实，有一天斯普拉特上校运来了 11 袋灰和砖块，其中 5 个袋子只装砖块，最大的砖块有 4~5 磅重。这些袋子全部被堆成一堆放在水边，任由风和海水搬运。12 天以后船队回来时，这一堆接近 2 吨重的灰和砖块全部消失了。他们沿着风吹过来的方向在海岸边搜索，但毫无结果。而在另一个方向，也就是盛行海浪涌动的方向，他们找到了重约 2 盎司的砖块，散落在至少 1,500 码远的地方，一个 3.5 磅重的砖块在 240 码处被发现，而另外 4~8 盎司重的砖块在 600~700 码之间的地方被找到。然而更多的砖块则已被流沙掩埋。斯普拉特上校写道，“因此，这证明仅在这 12 天中，在只有一次强西风和几次东风的 5 月，海岸和岸边的浅滩沿着海岸大幅度地向东推进，而当冬季连续刮大风时，以及尼罗河水高涨且西北风盛行的时候，大量沙子和其他物质必然随着浑浊的河水不断向前移动。”

我们再引用一段与这一观点有联系的文字。为了便于观察，上校曾沿岸步行，从标志塞得港的灯塔到泰尼赫海湾的顶端，他发现在低潮位和高潮位的地方，大量破陶器、古代的和现代的破坛子和碎砖块散布在岸边。“在发现了这么多碎片以后，”他继续写道，“我自然想要探寻它们的源头。我以为它们可能来自附近倒塌的建筑，但我最后发现，它们全都来自尼罗河河口，而且确确实实是来自尼罗河两岸村镇的破旧物品。它们在尼罗河涨水时由水流带来，然后在沿岸海流和盛行海浪的作用下，沿着海岸向东边散去。”

It would be easy to multiply facts, if further evidence were wanted, that the Nile is no exception in the great transforming powers of Nature, washing down the dry land into the sea, and forming there beneath and on the margin of the waves new continents and islands. The Mississippi, the Ganges, the Yang-tse-Kiang, and other rivers of the great continents, carry down millions of tons of solid matter every year. The North Sea is gradually being silted up by the rivers of Belgium, Holland, and the British islands. At the mouth of the Ebro, on the northern side of the Mediterranean, the deposits brought down by the river are in course of reclamation by an eminent English engineer. Hence we need not feel surprise that the Nile—one of the greatest of rivers—has during long ages wrought great changes on the southern shores of the same sea. In the face of facts such as are above adduced, a government or a nation might well be justified in believing the project of a harbour and canal on the Bay of Pelusium to be, if not impossible of execution, at least unprofitable. Places which were on the shore when Strabo wrote are now from four to six miles inland, as is shown on the accompanying map, reduced from that published with Captain Sprat's report; and this modifying action is still going on.

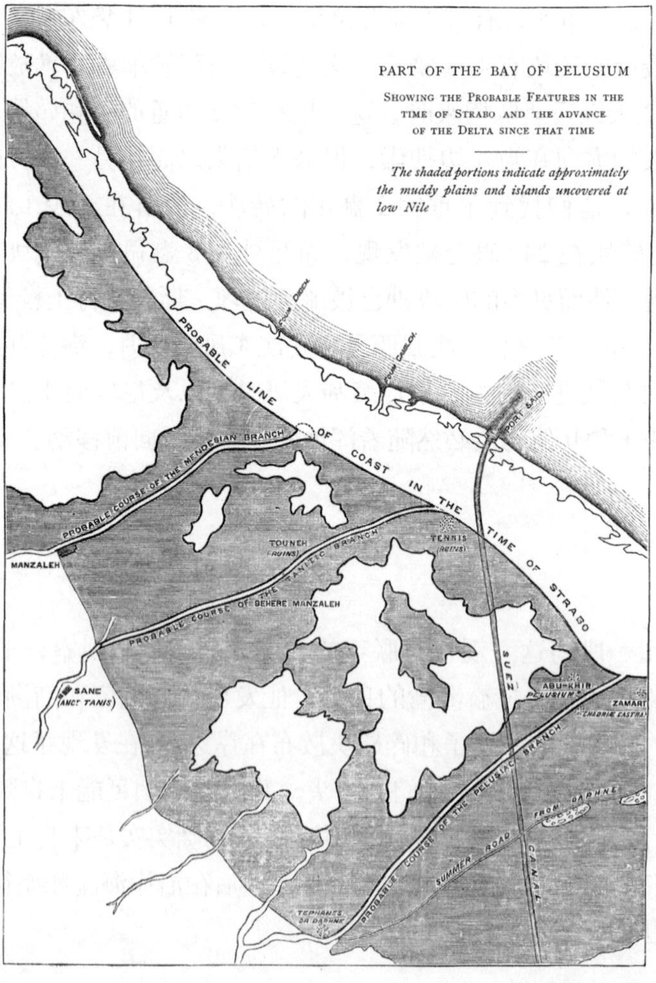

如果需要更多的证据，我可以很容易地列举出更多的事实。毫无例外，尼罗河也是自然界强大的改造力量之一，它把干旱的土地冲刷到海里，在海边和海上形成新的陆地和岛屿。密西西比河、恒河、长江和各大陆上的其他河流每年都要携带数百万吨的固体物质。北海正逐渐被比利时、荷兰和大不列颠诸岛的河流带来的泥沙淤塞。一位著名的英国工程师正在治理地中海北侧埃布罗河河口的沉积物。因此我们无需对尼罗河——世界上最大的河流之一——在漫长的岁月中使地中海南岸发生的巨大变化感到惊讶。面对前面所引证的事实，一个政府或者一个国家可能就有理由相信，培琉喜阿姆海湾的港口和运河工程，即使可行，至少也是无利可图的。斯特拉博作记录时还在岸边的地方，现在在斯普拉特上校公布的报告中的图上已经向内陆方向推进了 4~6 英里（如附图所示），而且这种改变仍在继续。

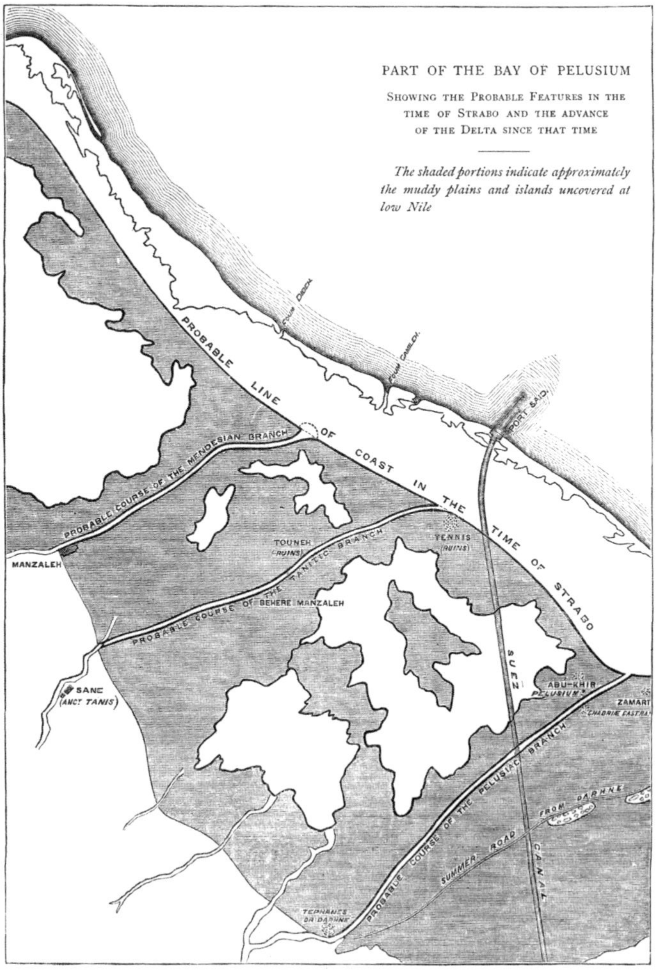

Since the Suez Canal was first projected engineering science has advanced; and though the sands will accumulate at Port Said as from of old, the piers and breakwaters will be periodically lengthened, made to stretch further and further into the sea, while powerful steam-dredges will scrape away the sand from the mouth of the harbour. Whether heavy gales will effect any more serious choking of the approaches, or drift tons of blowing sand into the canal itself remains to be seen. But while the world is greeting, and worthily greeting, the great work as a triumph of engineering skill, it may be well, at the same time, to bestow a little thought on the facts and conclusions here brought under notice, which in the pre-scientific age rendered man's contests against the works of the winds and sea perfectly hopeless.

(**1**, 81-83; 1869)

苏伊士运河的破土动工推动了工程学的发展；虽然沙子将一如既往地堆积在塞得港，人们将周期性地延长码头和防浪堤，使其不断向海中延伸，而强有力的挖泥船会挖走港口出口处的沙子。强风是否会带来更严重的堵塞或者把成吨的流沙吹向运河尚未可知。但是当全世界都在庆贺这一值得庆贺的、作为工程技术重要成就的伟大工程时，也应当同时稍微留意一下这里引起人们注意的事实，即在前科学时代，人类要与风和海斗争是完全无望的。

（孙惠南 翻译；郭华东 审稿）

The Atomic Controversy

Editor's Note

In 1869, the atomic hypothesis of the constitution of matter, as proposed 60 years earlier by the Englishman John Dalton, had yet to receive definitive confirmation. This short essay reviews arguments for and against the atomic hypothesis which, as Dalton had pointed out, lends order to many otherwise puzzling phenomena. It can explain the law of definite proportion: the fact that chemical compounds always involve elements in simple rational proportions. And the hypothesis had been shown to explain why atomic elements in the gaseous state all had similar heat capacities (dependence of temperature on heat input). Even so, the essay concludes that such arguments for atomism remained circumstantial.

IT is one of the most remarkable circumstances in the history of men, that they should in all times have sought the solution of human problems in the heavens rather than upon the earth. Sixty years ago a memorable instance of this truth occurred when Dalton borrowed from the stars an explanation of the fundamental phenomena of chemical combination. Carbon and oxygen unite in a certain proportion to form "carbonic acid"; and this proportion is found to be invariable, no matter from what source the compound may have been prepared. But carbon and oxygen form one other combination, namely, "carbonic oxide"—the gas whose delicate blue flame we often see in our fires. Carbonic oxide may be obtained from many sources; but, like carbonic acid, its composition is always exactly the same. These two bodies, then, illustrate the law of *Definite* Proportions. But Dalton went a step further. He found that, for the same weight of carbon, the amount of oxygen in "carbonic acid" was *double* that which exists in carbonic oxide. Several similar instances were found of two elements forming compounds in which, while the weight of the one remained constant, the other doubled, trebled, or quadrupled itself. Hence the law of *Multiple* Proportions. The question was—in fact, the question is—how to account for these laws. Dalton soon persuaded himself that matter was made up of very small particles or *minima naturae*, not by any possibility to be reduced to a smaller magnitude. Matter could not be divisible without limit; there must be a barrier somewhere. No doubt, as a chemist, he would have rejected the famous couplet—

Big fleas have little fleas, upon their backs, to bite 'em;
And little fleas have smaller fleas, and so *ad infinitum*.

"Let the divisions be ever so minute," he said, "the number of particles must be finite; just as in a given space of the universe, the number of stars and planets cannot be infinite. We might as well attempt to introduce a new planet into the solar system, or to annihilate one already in existence, as to create or destroy a particle of hydrogen." All substances, then, are composed of atoms; and these attract each other, but at the same time keep their distance, just as is the case with the heavenly bodies. The atoms of one compound

原子论战

编者按

1869 年，关于物质组成的原子理论仍旧没有得到确定性的验证，尽管这一理论已经由英国人约翰·道尔顿提出达 60 年之久。这篇短文综述了人们对原子理论的各种支持意见和反对意见。正如道尔顿当初说的，原子理论揭示了许多分散的、令人困惑的现象背后的规律。原子理论可以解释定比定律（自然界所有化合物中的各种组成元素之间总是存在简单的比例），通过原子理论也可以解释为什么各种元素在气态时都具有相似的比热容（温度变化对于供热的依赖程度）。不过，本文最后还是认为，关于原子理论的这些论述都只是描述性的，尚未得到证实。

人类历史上最值得注意的情况就是，我们在任何时候都应该从天空而不是从大地去探求人类问题的解决之道。60 年前发生了一件证实这个真理的令人难忘的事例，道尔顿先生藉由恒星获得灵感，解释了化合物组成的基本现象。碳和氧以一定的比例结合形成“碳酸”，不管是用什么原料来制备碳酸这种化合物，其中碳和氧的比例都是恒定不变的。但是碳和氧还能形成另一种叫作“一氧化碳”的化合物，这种气体的火焰是蓝色的，就像我们经常在火中见到的那样。尽管一氧化碳来源广泛，但和碳酸一样，不同来源的一氧化碳在组成上总是完全相同的。这两个例子说明了什么是**定比**定律。但是道尔顿的研究更进了一步，他发现对于一定质量的碳，“碳酸”中氧的量是一氧化碳中的**两倍**。还有几个类似的例子，由两种元素组成的多种化合物，当其中一种元素的质量一定时，另外一种元素的量可能相差一倍、两倍甚至三倍，于是道尔顿提出了**倍比**定律。实际上，问题在于如何解释这些定律。他很快作出猜测：物质是由很小的颗粒或者**不可再分的微粒**组成的，这些颗粒不能再被分解成更小尺度的颗粒。物质不能被无限分割，肯定存在某种界限。作为一名化学家，他应该会批判下面这首著名的小诗：

大蚤生小蚤，蝇蝇啮其背；
小蚤复微蚤，夫何有止归。

他说：“即使分割得非常小，微粒的数目也一定是有限的，就像在一定的宇宙空间内行星和恒星的数目是有限的一样。我们可以试图向太阳系中引入一颗新的行星，或者毁掉其中一颗已存在的行星，同样地，我们也可以创造或剔除一个氢粒子。”所有的物质都是由原子构成的，这些原子就像天体一样相互吸引而又保持一定的距离。不同化合物中原子的质量、大小以及相互间的引力是不同的。但因为原子是不可再

do not resemble those of another in weight, or size, or mutually gravitating power. But as they are indivisible, it is between them that we must conceive all chemical action to take place; and an atom of any particular kind must always have the same weight. The atom of carbon weighs 5; the atom of oxygen weighs 7. Carbonic oxide, containing one of each must therefore be invariably constituted of 5 carbon, and 7 oxygen: carbonic acid must in like manner contain 5 carbon, and 14 oxygen. Here, then, Dalton not only states that he has accounted for the two laws we have mentioned by making a single assumption; but he evidently intends his theory to be used as a criterion or control in all future analytical results, and already views it as the birth-place of chemical enterprise.

Such, and so great, was the atomic theory of Dalton; founded, certainly, on erroneous numbers, but containing in itself the germ of their correction; aspiring to the command in innumerable conquests; and setting itself for the rise or fall of the chemical spirit.

It is hardly necessary to make any detailed review of the history of the atomic theory. Berzelius made it a starting-point for researches which, on the whole, have been unsurpassed in their practical importance, and engrafted upon it his celebrated electrical doctrine. Davy and Faraday refused to admit it; Laurent and Gerhardt accepted it doubtfully, or in a much modified form. Henry declared that it did not rest on an inductive basis. There can be no doubt, however, that the atomic theory has been accepted by the majority of chemists, as may be seen on even a cursory inspection of the current literature of their science. Our present intention is to give such a summary of the atomic question as may be serviceable to those who take an interest in the discussion at the Chemical Society on Thursday last.

The modern supporters of the atomic theory agree with Dalton in the fundamental suppositions we have given above; but assert that they have a much stronger case. The phenomena of gaseous combination and specific heat have indeed changed the numerical aspect of the theory, but not its substance. The simplicity of all the results we have accumulated with respect to combining proportions is itself a great argument for the existence of atoms. They all, for example, have the same capacity for heat; they all, when in the gaseous state, have a volume which is an even multiple of that of one part by weight of hydrogen. But bodies in the free or uncombined state—such, in fact, as we *see* them—more commonly consist of many clusters of atoms (*molecules*) than of simple atoms. These molecules are determined by the fact that when in the gaseous state they all have the same volume. Again, select a series of chemical equations, in which water is formed, and eliminate between them so as to obtain the smallest proportion of water, taking part in the transformations they represent. It will be found that the number is 18; which necessarily involves the supposition that the oxygen (16) in water (18) is an indivisible quantity. To put this last point another way: hydrochloric acid, if treated with soda, no matter in what amount, only forms one compound (common salt). Now we know that the action in this case consists in the exchange of hydrogen for sodium. But if hydrogen were infinitely divisible, we ought to be able to effect an inexhaustible number of such exchanges, and produce an interminable variety of compounds of hydrogen, sodium, and chlorine;

分的，所以我们必须设想所有的化学反应发生在原子之间，并且一种特定原子的质量是不变的。假设碳原子的质量是 5，氧原子的质量是 7。一氧化碳分子包含一个碳原子和一个氧原子，因此，它的质量肯定恒定为一个碳（5）与一个氧（7）之和。同理，碳酸分子的质量一定为一个碳（5）与两个氧（14）之和。由此，道尔顿通过一个简单的假设解释了前面提到的两个定律，同时，他显然还希望他的理论被当作将来所有分析结果的标准或参照，并且他已把这视为化学工业的起源。

这就是道尔顿伟大的原子论。虽然它建立在错误的数据之上，但其中包含了正确理论的萌芽。它立志征服无数的困难，引领化学精神的荣衰。

我们就不去回顾原子论的历史细节了。伯齐利厄斯以原子论为起点进行了研究，总体上来说，他的这些研究的应用价值还未被超越，他还在原子论的基础上衍生出了著名的电子学说。戴维和法拉第拒绝接受原子论。洛朗和凯哈德则半信半疑，确切地说是认可经过了很大修正后的原子论。亨利则声称原子论并非建立在归纳学的基础之上。然而，毫无疑问原子论已被大部分科学家接受，这只要大致浏览一下目前这方面的科学文献就知道了。这里我们简单总结一下关于原子论的探讨，以供那些对上周四化学学会举行的讨论感兴趣的读者参考。

原子论的现代支持者同意前面所述的道尔顿的基本假设，同时他们声称拥有说服力更强的证据。气体化合现象和比热方面的研究确实改变了原子论的数值形式，但并没有改变其本质。我们积累的有关化合比例的所有结果都非常简单，这本身就是原子存在的强力论据。例如，它们都具有同样的热容；处于气态时它们的体积都是单位质量氢气体积的偶数倍。但实际上，就像我们**看到**的，游离态或非化合态的物质在大多数情况下是由许多原子团（**分子**）而非单个原子组成。实验表明，处于气态时它们具有相同的体积，这就证明了分子的存在。再者，我们挑选一些有水生成的化学方程式，通过比较、消约可以推算出水的最小组分，我们发现其分子量是 18。根据这一结果，很自然地可以推测：氧（16）在水（18）中是个不可分割的量。另外一个例子也能够说明这个观点：如果向盐酸中加入苏打，无论苏打的用量是多少，反应都只能产生一种化合物（食盐）。现在我们知道这个例子中发生的反应就是氢被钠替换了。但如果假设氢是无限可分的话，那么通过控制用量应该能够产生多种多样的交换，从而产生无数种由氢、氯、钠形成的化合物，盐酸只是反应这一侧的终点，食盐（氯化钠）是另一侧的终点。可是这种现象从来都没有发生过。可以肯定，物质要么是无限可分的要么是有限可分的，既然现在已经证明不是前者，那

hydrochloric acid being the limit on the one side, and common salt (sodic chloride) terminating the other. No such phenomenon occurs; and, since matter must be infinitely or finitely divisible, and has been thus proved not to be the former, it must be the latter. Atoms, therefore, really exist; and chemical combination is inconsistent with any other supposition. Those who hold the contrary opinion are bound to produce an alternative theory, which shall explain the facts in some better way.

Now let us hear the plaintiff in reply.

The atomic theory has undoubtedly been of great service to science, since the laws of definite and multiple proportions would probably not have received the attention they deserve, but for being stated in terms of that theory. Yet we must discriminate between these laws, which are the simple expression of experimental facts, and the assumption of atoms, which preceded them historically, and therefore has no necessary connection with them. For it was the Greek atomic theory which Dalton revived. Nor has any substance yet been produced by the atomists, which we cannot find means to divide. If, moreover, we have no alternative but to admit the infinite divisibility of matter, even that is consistent with the simple ratios in which bodies combine; for two or more infinites may have a finite ratio. Therefore, the observed simplicity, if used as an argument, cuts both ways. Possibly we are mistaken in connecting the ideas of matter and division at all; at any rate, the connection has never been justified by the opposite side. Again, admitting the argument based on the formation of common salt, the atomic theory does not tell us why only one third of the hydrogen in tartaric acid can be exchanged for sodium; why, indeed, only a fraction of the hydrogen in most organic substances can be so exchanged. Yet, the explanation of the one fact, when discovered, will evidently include that of the other. On the whole, it appears that the atomic theory demands from us a belief in the existence of a limit to division. No such limit has been exhibited to our senses; and the facts themselves do not raise the idea of a limit, which Dalton really borrowed from philosophy. The apparent simplicity of chemical union we do not profess to explain, but to be waiting for any experimental interpretation that may arise. The atomists, in bringing forward their theory, are bound to establish it, and with them lies the *onus probandi*.

The above are a few broad outlines of the existing aspect of the atomic controversy, and may somewhat assist in forming an estimate of it. The general theoretical tone of the discussion last Thursday must have surprised most who were present. Our own position is necessarily an impartial one; but it will probably be agreed that between the contending parties there is a gulf, deeper and wider than at first appears, and perhaps unprovided with a bridge.

(**1**, 44-45; 1869)

么必然只能是后者了。因此，可以肯定原子是确实存在的，并且化合反应的结果与其他任何假设都是不相吻合的。那些持有相反意见的人，必须提出另一种可以更好地解释这些现象的理论。

现在让我们听听反对者的意见。

毫无疑问原子论对科学是极有帮助的，因为如果不是用原子论的形式表述出来，倍比定律和定比定律的法则恐怕就不会受到应得的重视。但是，我们必须将这两大法则与原子假设加以区分，前者是对实验事实的简单描述，而后者在历史上先于两大法则出现，因此它们之间并无必然联系。道尔顿接受的只是古希腊的原子论而已。原子论者从来都没有得到过无法分割的物质。再者，如果我们别无选择而只能承认物质的无限可分性的话，那么这与物质化合中的简单比例也是相符合的，因为两种或两种以上的无限可分物之间也可以有一个有限的确定的比例。因此，就化合过程中观察到的简单比例来说，两种理论都能解释。很可能，我们把物质和分割这两个概念联系起来本身就是完全错误的。至少从来没有证明这种联系的正确性。另外，原子论虽然能够解释食盐形成的现象，但它并未告诉我们为什么酒石酸中只有三分之一的氢可以与钠交换，还有为什么大多数有机物中只有一部分氢可以这样交换。作为一种正确的理论，不但要能解释已经发现的某种现象，同样也要能够解释其他的现象。总体来说，原子论似乎就是要求我们坚信存在分割的极限，但我们从来没有感到过这种极限的存在，事实本身也并没有使人们想到极限这一概念。可以说，道尔顿只是从哲学那里借来了分割的极限这个想法。我们并未声称已经解释了化学结合过程中明显的简单比例，这要等待可能出现的实验性的解释。原子论者自从提出他们理论的那天起就背负着提供证据的责任，他们必须证实这一理论。

上面概述了目前原子论战的各方的观点，可能会有助于人们形成自己对原子论的评价。上周四的讨论中众人的观点可能使一些与会人员受惊了。我们的立场必须保持中立，但是有一点大家达成了共识，就是在论战双方之间存在着一条鸿沟，它比刚出现时更深、更宽，或许根本无法架通。

（高如丽 翻译；汪长征 审稿）

Dr. Livingstone's Explorations

F.R.G.S.

Editor's Note

David Livingstone's expedition in search of the source of the River Nile, begun in 1866, supplied Great Britain with an unfolding narrative of adventure and discovery to rival any serialized fiction, informed more by rumour than solid evidence. Some of the porters who deserted him claimed that he was dead; but this article reports letters from Livingstone dating from July 1868 that had recently reached the Royal Geographical Society, describing him as being "in good health and spirits". Livingstone's remarks enabled his route to be reconstructed, but the correspondent "F.R.G.S." here is rather sceptical that Livingstone was close to achieving his goals, and criticizes his general sense of geography. Indeed Livingstone, in poor health by 1869, never did succeed.

THE letters from Dr. Livingstone lately read at the Royal Geographical Society, give the grateful assurance, not only that he was in good health and spirits in July 1868, but also that he was under no apprehension of ill-treatment from the Cazembe. Visiting that chief without a numerous escort, he created no alarm. He has, in truth, notwithstanding seeming difficulties, been singularly fortunate; for his rumoured death and expected captivity have created a sensation of much greater value to him than the discovery of the Nile's sources. Dr. Livingstone's account of his journey northwards from the Aroangoa is in general reconcilable with those given by the Portuguese expeditions, with some difference, however, arising from difference of route. He seems to have crossed that river much further to the west than Monteiro, whose line of march was ten or twelve miles more west than that of Lacerda. He saw mountains, he tells us, and the Portuguese saw none. Herein he is greatly mistaken: Monteiro's expedition crossed over the flanks of a wondrous mountain, supposed to be a Portuguese league (about 20,000 feet) high, with trees, population, but no snow on its broad summit. The account of this mountain, called by mistake Muchingue (the glen or defile), given by a writer in the *Journal of the Royal Geographical Society* (vol. XXVI), improves the original by a precise statement of longitude and latitude, and by a description of the panoramic view from the summit to a distance of 200 miles.

The high land which culminates towards the east in Muchingue was ascended on leaving the valley of the Aroangoa. The traveller then came in lat. 10° 34′ S., to the river Chambezi, called by Lacerda the New Zambezi, flowing from east to west, and rarely fordable. He remarks that it resembles the Zambezi, not in name only, but also in the abundance of food found in the stream or on its banks. He forgets that the critic who denied his explanation of the name Zambezi (river *par excellence*), showed that in all its forms, Liambegi, Chambezi, Yabengi, &c., it means simply (river) "of meat" or animal food. The Chambezi abounds in oysters, but we know nothing of their flavour. This river,

利文斯通博士的探险

F.R.G.S.

编者按

利文斯通旨在寻找尼罗河源头的探险活动始于1866年，他撰写了一系列记叙性的文章逐渐展开式地讲述他的奇遇和发现，这些文章与其他的系列小说一样受到当时英国读者的喜爱。从他的队伍中逃离的一些搬运工人说利文斯通已经去世了。不过，这篇文章报道了皇家地理学会最近刚刚收到的利文斯通的来信，在这封1868年7月发出的信件中，利文斯通说他“身体和精神都很好”。来自利文斯通的这些消息使他的行程再次受到人们的重视，不过，通讯员“F.R.G.S.”在这篇文章中对利文斯通快要实现其目标表示非常怀疑，而且对利文斯通的地理学常识提出了批评。实际上，1869年时利文斯通身体已经很不好了，他最终也没能成功找到尼罗河的源头。

刚刚在皇家地理学会宣读的利文斯通博士的来信表明，1868年7月时他不仅身体和精神都很好，而且一点也不惧怕卡仁比族的苛待。他拜访酋长时只带了几个陪同，所以也没有引起任何惊慌。尽管看上去他遇到了很多困难，但是实际上他是异常幸运的，因为，关于他已死亡的传闻和可能被囚禁的猜测早已引起了轰动，而这种轰动为他创造的价值，比他发现尼罗河源头带来的价值大得多。利文斯通博士记载的关于他从阿罗安瓜出发北上的旅行与葡萄牙探险队的描述基本一致，只是在路线上存在一些差别。其中，他渡河后向西走得比蒙泰罗远，而蒙泰罗向西行进得又比拉塞尔达远10~12英里。并且，利文斯通还告诉我们他看到了一些山脉，而葡萄牙人则一座未见。不过在这里他是大错特错了：蒙泰罗探险队曾从一座奇特山脉的侧面穿过，据推测该山脉高达1葡萄牙里格（约20,000英尺），其山顶宽阔，丛林密布，经常有动物出没，不过未见积雪。这段记录来自《皇家地理学会会刊》（第26卷）的一位作者，虽然这座山脉被误认为是姆沁盖（峡谷或山隘），但他通过对经纬度的精确描述以及对山顶以下200英里内全景的描写，使原作的价值得到了提升。

离开阿罗安瓜山谷，爬过姆沁盖东部最高点的高地以后，旅行家利文斯通来到了南纬10° 34′附近的谦比西河，拉塞尔达称之为新赞比西河，该河自东向西流，很少有人涉水通过。他认为，这条河不仅名称与赞比西河很相似，而且在河流内部或两岸也发现了丰富的食物。但是他忘了，不同意他对赞比西这一名称的解释（**最卓越的**河）的那些批评家们曾指出，该河的各种名称，诸如Liambegi、Chambezi、Yabengi等，都是“鲜肉”或动物性食物（之河）的意思。谦比西河还盛产牡蛎，

according to Dr. Livingstone, forms in the west the great Lake Bengweolo, from which it again issues to the north under the name of Luapula; but we believe it would be more correct to say that it joins the Luapula, a much larger river, the great marsh Pampage, which is, doubtless, often overflowed and converted into a lake, lying in the angle between the two rivers. Then we are told—"The Luapula flows down north past the town of Cazembe, and twelve miles below it enters the Lake Moero." From this it might be concluded that the river flows by the chief's town, and that twelve miles lower down, or further north, it enters the lake, but this cannot possibly be the traveller's meaning. Lake Moero forms a remarkable feature in Dr. Livingstone's latest discoveries, but his account of it is singularly perplexed and obscure. We know that the Luapula flows to the north or N.N.E., some eight or ten miles west of the Cazembe. Lake Moero, by our traveller's account, is fifty miles long, and from 30 to 60 miles wide. "Passing down," he says, "the eastern side of Moero, we came to the Cazembe;" and again he states that "the Cazembe's town stands on the north-east bank of the lakelet Mofwe, two or three miles broad and four long, totally unconnected with Lake Moero." In endeavouring to reconcile these statements it is necessary to beware of rash conclusions and inaccurate expressions. It is a hazardous thing to pronounce upon the length, breadth, and boundaries of lakes without surveying them. The Portuguese officers in 1831 obtained leave to examine Lake Mofo or Mofwe, and for that purpose went four and a half leagues N.N.E. along its shore, till they came to the Lounde, a river, as they called it, two miles wide, where they expected to find boats. These, however, had been purposely removed, so that the explorers were brought to a stand. They had proceeded far enough, however, to perceive that the lake turned to the north-west. They did not see the end of it, but state distinctly that it communicates with other large lakes. Dr. Livingstone, describing the flooded state of the country, tells his experience of two rivers which flow into the north end of Moero: the Luo, which was crossed by the Portuguese, thirty miles south of the Cazembe's town; and the Chungu, near which Lacerda died, about ten miles south of that place. From these particulars we cannot help concluding that the Moero of our traveller, who has found the country in a state of flood, is the Carucuige of the Portuguese, or at least that these names apply to parts of the same great marsh or lagoon. At the eastern side of it, visited by Dr. Livingstone, is the Fumo Moiro, whose title is probably taken from his district. Manoel Gaetano Pereira, who first visited the Cazembe, related, that near the chief's town he spent a whole day wading breast-deep through a lagoon. It was subsequently found that the lagoon in question was Carucuige. The strength of the Cazembe's position lies in the difficulty of approaching it through a labyrinth of swamps, lakes, and wide drains. The Portuguese spent some hours in crossing a river, as they called it, two miles wide, on matted vegetation which sank under their feet. This and the Lounde above mentioned were probably the connecting arms of large lakes.

Our traveller informs us, that "the Luapula, leaving Moero at its northern end by a rent in the mountains of Rua, takes the name of Lualaba, and, passing on N.N.W., forms Ulenge in the country west of Tanganyika." He saw the Luapula only at this gap in the mountains. Ulenge is a lake with many islands, or a separation of the river into many branches. "These branches," he goes on to say, "are all gathered up by the Lufira...

不过我们对其味道一无所知。根据利文斯通博士的说法，该河流在西部形成了广阔的班韦乌卢湖，而从班韦乌卢湖流出后，该河继续向北流，此时被称为卢阿普拉河。不过说它汇入了卢阿普拉河应该更确切一些，因为卢阿普拉河更大。在两条河流之间是彭佩西大沼泽，毫无疑问，由于河流经常泛滥，它变成了湖泊。接下来，利文斯通告诉我们“卢阿普拉河向北流去，穿过卡仁比城，在其下游 12 英里处注入姆韦鲁湖。”从这里我们可能会得出结论认为：这条河流过了酋长所在的城镇，并在下游约 12 英里或者更北的地方注入湖泊。不过这可能并不是他要表达的意思。在利文斯通最新的发现中，姆韦鲁湖呈现出不寻常的特征，然而他写的有关记录却不可思议地被复杂化、模糊化了。我们知道卢阿普拉河流经卡仁比城西约 8~10 英里处，是向北或东北偏北方向流的。而根据我们的旅行家利文斯通的记录，姆韦鲁湖长 50 英里，宽 30~60 英里。他还说“穿过姆韦鲁湖东部，我们来到了卡仁比”，然后他又叙述到“卡仁比城位于孟弗苇湖东北岸，孟弗苇湖宽 2~3 英里，长 4 英里，与姆韦鲁湖完全不相连。”在竭力理解这些描述时，我们应该保持警惕，不可轻率地下结论或作出错误的解释。要知道，不进行调查就对湖泊的长、宽以及边界加以评论是件可怕的事情。葡萄牙官员曾在 1831 年出发前考察了莫佛湖（或称孟弗苇湖），并为此沿岸边向东北偏北方向走了 4.5 里格的路程，最后来到了他们称之为卢恩德的宽约 2 英里的一条河边。他们希望能在此找到渡船（这说明水很深）。但是，船早就被故意移走了，所以探险者们只好停了下来。不过他们已经走得够远了，足以发现那个湖是转向西北去的。虽然他们并没有看到湖的尽头，但已清楚地认识到它是与其他大湖相连的。另外，利文斯通博士不仅描述了这个国家洪水泛滥的情况，还讲述了他在探索注入姆韦鲁湖北端的两条河流期间的经历。这两条河流分别是：葡萄牙人曾经穿过的、位于卡仁比城南部 30 英里处的罗河和距离拉塞尔达死去的地方往南约 10 英里处的淳古河。从这些细节来看，我们不难得到如下结论：发现这个国家正值洪水泛滥的旅行家眼中的姆韦鲁湖就是葡萄牙人所说的卡鲁奎哥湖，至少它们应该是同一个大沼泽或泻湖的不同部分。湖的东部就是利文斯通博士曾到过的弗莫–姆韦鲁，这个名字很可能是从他所在的教区得来的。第一个到达卡仁比的马诺埃尔 · 加埃塔诺 · 佩雷拉曾讲过，他曾为涉水通过酋长城附近一个水深及胸的泻湖而花了整整一天的时间。人们随后发现他所讲的泻湖正是卡鲁奎哥湖。确定卡仁比城确切位置的困难在于要到达那里必须穿过一系列错综复杂的沼泽、湖泊和深沟。葡萄牙人曾花了几个小时才穿过一条河，据他们说，这条河宽约 2 英里，他们经过时脚下都是纠缠在一起的水草。这条河以及前面提到的卢恩德河很可能就是各个大湖之间的连接通道。

我们的旅行家利文斯通告诉我们“卢阿普拉河通过卢阿山脉的间隙从姆韦鲁湖的北端流出，之后更名为卢阿拉巴河，然后向西北偏北方向流去，进而在坦噶尼喀西部的地区形成了乌莱西湖。”也就是说，他仅在山脉中间的缺口处看到了卢阿普拉

I have not seen the Lufira; but, pointed out west of 11° S., it is there asserted always to require canoes.…This is purely native information." Now it is quite possible that the traveller totally misunderstood his native informants. They spoke of the waters to the S.W., and he understood them to speak of the N. or N.E. The great river Luviri, called by the Arabs Lufira, flows into the Luapula from the west, about 100 miles S.W. or S.S.W. from the Cazembe. The Lualaba, the sacred river of the Alunda, whence their forefathers emigrated, still farther west (a month's journey), falls into the Lulua, and so joins the Zaire. The great salt marshes, which chiefly supply the interior of Africa, are situated on its banks at its southern bend; these may be the Ulenze above described, if it be not a marshy tract lying between the sources of the two rivers. The native information here given cannot be received as perfectly pure. When our author speaks of the Luviri entering Tanganyika at Uvira, he evidently casts the dimly discerned views of the natives into his own preconceived mould, and clothes them in his own language.

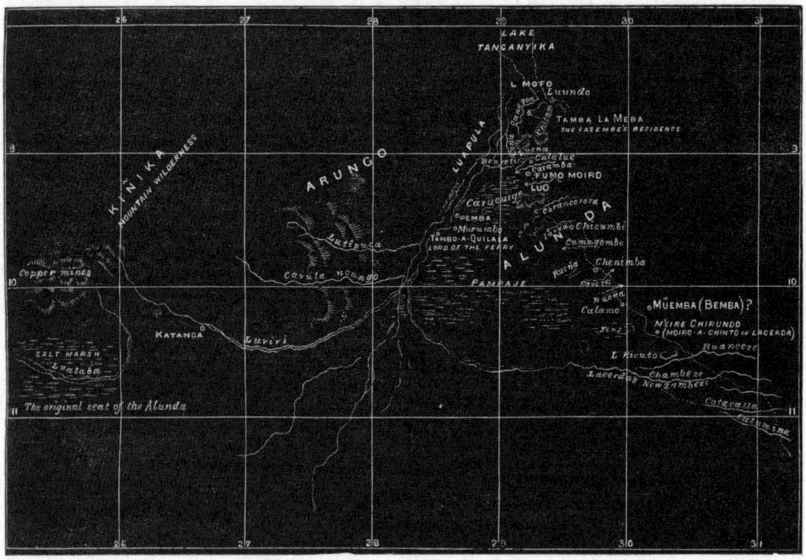

Respecting the language of this country, our author informs us that "the people are known by the initial Ba instead of the initial Lo or U for the country." This is not merely confused, but also, we believe, wholly erroneous. The initial U never forms the name of a country, but the collective name of a nation, chief, and people. The Portuguese, who on this point are the best authorities, use the names Alunda, Arungo, Acumbe, not Balunda, &c. They tell us that the Alunda never pronounce the letter r, and that in writing the names Arungo, Moiro, &c., in which that letter occurs, they have adhered to the Maravi dialect. We thence conclude that for the names Rua, Moero, Lufira, &c., and perhaps for the initial Ba above alluded to, Dr. Livingstone is probably indebted to his Arab friends, who rest satisfied with a jargon, in some degree intelligible everywhere, and nowhere perfect.

河。乌莱西湖则是一个拥有众多岛屿的大湖，或者说它是卢阿普拉河众多支流的交界处。他还说“这些支流都汇集到卢菲拉河……我并没有见到卢菲拉河，不过它应该位于南纬 11° 西侧的某个地方，据说那里需要独木舟才能通过……这完全是从土著人那里得到的信息。”现在我们知道他很可能完全误解了为他提供信息的土著人。当地土著人说的是西南的河流，而利文斯通则理解成了北或东北。被阿拉伯人称为卢菲拉河的罗非河，从位于卡仁比城西南或西南偏南方向约 100 英里处，自西流入卢阿普拉河。被阿隆达人尊为圣河的卢阿拉巴河（阿隆达人的祖先就是从那里移民而来的），还在更偏西的地方（大约需要一个月的行程），它先注入了卢卢阿河，后汇入扎伊尔河。作为非洲内陆主要补给地的大盐沼都位于扎伊尔河南部转弯处的堤岸上；如果这不是一个位于两条河流源头之间的沼泽地，那么它们就很有可能是上面所说的乌莱西湖。这里给出的从土著人那里得到的信息我们不能全盘接受。当我们的作者提到罗非河在乌维拉注入坦噶尼喀湖时，很明显他是将土著人朦胧认识到的观点纳入到自己预定的模式中，并且冠以自己的语言。

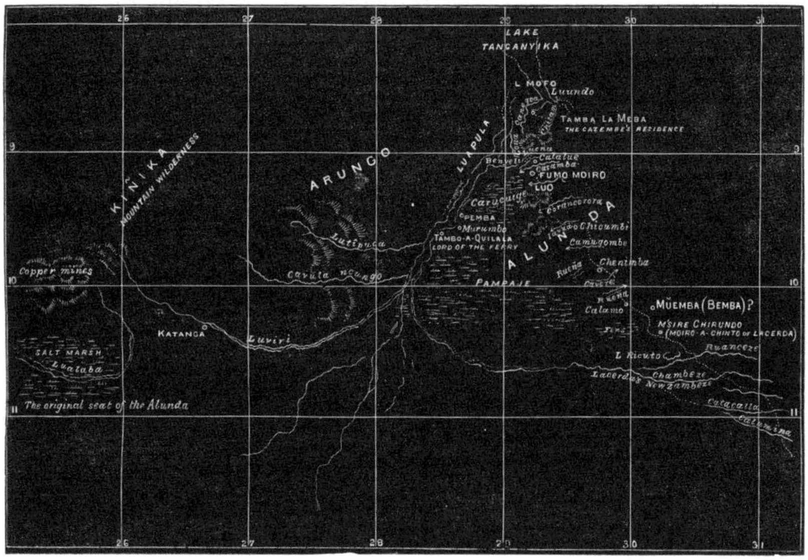

谈及该地区的方言，作者告诉我们：“人们的名字通常以 Ba 开头，而不是像地名那样以 Lo 或 U 开头。”但我们认为，这不但令人感到困惑不解，而且根本就是完全错误的。以 U 开头的单词并不是指一个地名，而是指一个部落、酋长以及族人集体的名称。在这一点上，葡萄牙人是最具权威的，他们用 Alunda、Arungo、Acumbe 这些名字，而不是 Balunda 等。他们告诉我们，阿隆达人从来不发字母 r 的音，而且在写 Arungo、Moiro 等出现该字母的名字时，他们会采用马拉维方言。所以，我们可以得出如下结论：就 Rua、Moero、Lufira 等名字以及上面提到的以 Ba 开头的一些名字而言，很可能是利文斯通博士受了他的阿拉伯朋友们的影响。我们知道阿拉伯人很满意自己对方言的了解，他们对各地方言都能理解一些，但对每一种方言都不精通。

Dr. Livingstone seems to be elated with the discovery that "the chief sources of the Nile arise between 10° and 12° S. lat. or nearly in the position assigned to them by Ptolemy, whose river Rhapta (?) is probably the Rovuma." Here two different problems are attempted to be solved at once—one touching the Sources of the White Nile, and the other, those of Ptolemy's Nile. With respect to these latter, it will be enough to observe that Ptolemy's Lakes of the Nile, two in number, 8 degrees asunder, are placed by him respectively in lats. 6° and 7° S., but his graduation being defective, through an imperfect estimate of the length of a degree, the positions thus assigned to the lakes fall under true graduation, to 11″ N., and 40″ S. of the equator. Ptolemy's Lakes, therefore, have not been reached by the zealous traveller.

With respect to the sources of the Bahr el Abyad, they may of course be traced to the head waters of the Luapula, provided that the results of Capt. Burton's observations on the altitude of Nyanza and the character of its northern end are completely thrown aside. With a greater elevation, and an outlet through Speke's Mountains of the Moon, the waters of the lake may reach Egypt.

It is to be regretted that Dr. Livingstone missed the opportunity of viewing the highest mountain in this part of the world, now known only by a ridiculously exaggerated description; and also a most interesting point in the centre of Africa. The great town, Katanga, as described by the Arabs, is near the copper mines, where 75 lbs. of copper may be bought for 4 cubits of American sheeting. The town is larger than Roonda (the Cazembe's town), and has good bazaars; it stands on the Rafira (Luvira) which joins the Ruapura (Luapula). The people are peaceable, and kind to strangers. The people from Zanzibar learned the language almost immediately.

[We give a map of the region recently traversed by Livingstone, showing its connection with the known points in this part of Africa. We owe this map to the courtesy of the officers of the Royal Geographical Society. —ED.]

(**1**, 72-74; 1869)

利文斯通博士发现“尼罗河的主要源头位于南纬 10°~12°之间，或是托勒密所说的位置附近，其中托勒密指的拉普塔河（?）很可能就是鲁伍马河。”他似乎为此而兴高采烈。在这里需要同时解决两个问题——其一是白尼罗河的源头问题，另外一个则是托勒密所指的尼罗河的源头问题。关于后者，我们已经知道，托勒密所说的尼罗河的源头湖泊有两个，相隔 8°，分别位于南纬 6°和 7°，但他的分度结果存在缺陷，因为他对纬度 1°代表的长度估计得不准确，所以，湖泊所在的正确位置应该分别是赤道两侧北纬 11″和南纬 40″处。因此，执着的旅行家利文斯通并没有到过托勒密所说的湖泊那里。

至于白尼罗河的源头问题，假设完全抛开伯顿上校在尼亚萨湖所在的海拔高度上观测到的结果以及关于其北端特征的描述，那么当然可以将其追溯为卢阿普拉河的源头。考虑到尼亚萨湖的海拔较高，又有一个出口可以穿过斯皮克所称的月亮山系，所以湖水应该能够到达埃及。

非常遗憾，利文斯通博士错过了欣赏地球这一端的最高山脉的机会，这是非洲中部最令人感兴趣的一个地方，现在仅留下一段荒谬夸张的描述。据阿拉伯人记述，伟大的加丹加城位于铜矿附近，据说在那里用 4 腕尺美式被单料子就可以换取 75 磅铜。加丹加城比卢恩达（卡仁比族的一个城镇）大，位于卢阿普拉河的支流卢菲拉河边上，那里还有不错的集市。那里的居民和平安宁，对陌生人也非常友好。来自桑给巴尔的人很快就学会了当地的语言。

[我们提供了一幅利文斯通刚刚穿过的这个地区的地图，地图中给出了他穿越的地区与非洲的这部分区域中已知地点之间的关系。本地图的使用得到了皇家地理学会的允许。——编辑注]

（齐红艳 翻译；郭华东 审稿）

Dr. Livingstone's Explorations

C. Beke

Editor's Note

The renowned English traveller Charles Tilstone Beke here defends David Livingstone, then in Africa in search of the source of the River Nile, against the somewhat sceptical remarks made by a correspondent "F.R.G.S." in the light of recent letters from Livingstone received in London. Beke asserts that Livingstone has made some solid contributions to the knowledge of African geography, particularly of its networks of rivers and lakes in the centre and northwest of the continent. Moreover, Beke argues that F.R.G.S. is wrong to dismiss out of hand Livingstone's claim to have discovered the Nile's sources—although significantly, Beke says that claim supports "the views I have long entertained". Yet Livingstone was, in this respect, indeed mistaken.

IT certainly is to be regretted that the information received from Dr. Livingstone should be so imperfect. Still, though insufficient in itself, perhaps, to warrant our arriving at any positive conclusion respecting his claim to have discovered the chief sources of the Nile, the information furnished by him affords material aid towards the solution of that great problem of African geography, and is generally of much greater value, in my estimation, than it would appear to be in that of your learned correspondent "F.R.G.S."

Before adverting to the main subject, I desire to point out, in the first place, that Dr. Livingstone has definitively settled that the Chambeze—the New Zambesi of some of our maps—is not an affluent of the well-known river Zambesi, which flows eastward into the Indian Ocean, but is a distinct stream, of which the course is to the west and north-west. On this point it is due to Mr. Cooley to say, that, although he was mistaken respecting the upper course of the Zambesi itself, he has long contended for the separate existence of the "New Zambesi", or Chambeze.

Secondly, Dr. Livingstone has ascertained that the Chambeze, in its lower course beyond the capital of the Cazembe, is joined by another large river, the Lufira, coming from the south and southwest, which drains the western side of the country south of Tanganyika, as the Chambeze drains the east side. The Lufira was not seen by the traveller; but when he was at some place, not named by him, in 11° S. lat., that river was pointed out to him as being at some distance west of that spot, and was described as being so large there as always to require canoes; for so I read his words:—"I have not seen the Lufira, but, pointed out west of 11° S., it is there asserted always to require canoes;"—which shows that it must come from a considerable distance south of that parallel.

In the next place, Dr. Livingstone informs us that the Chambeze enters Lake Bangweolo,

利文斯通博士的探险

比克

编者按

最近伦敦方面收到了一些来自当时正在非洲寻找尼罗河源头的利文斯通的来信，一位名为 F.R.G.S. 的通讯员对此提出了一些怀疑，在这篇文章中，英国著名的旅行家查尔斯·蒂尔斯通·比克针对这些怀疑为利文斯通进行了辩护。比克指出，利文斯通对于有关非洲地理的知识有许多实质性的贡献，特别是关于非洲大陆中心地区和西北部地区河流与湖泊的分布网络。另外，比克认为，F.R.G.S. 不假思索地驳斥利文斯通已经发现尼罗河源头的声明，这是错误的行为。不过值得注意的是，比克认为该声明肯定了“我长期以来持有的观点”。然而就这一点而言，利文斯通确实是错的。

从利文斯通博士那里收到的信息有非常多的缺陷实在非常令人遗憾。然而，尽管它本身存在一些不足，但或许对于保证我们得出与他声称的发现了尼罗河主要源头有关的正确结论仍起到了重要作用。他提供的材料对解决非洲地理中的这个重大难题有实质性的帮助，而且，我认为，其价值远远大于贵刊那位有学问的通讯员 F.R.G.S. 所认为的那样。

在转入正题之前，我想要指出的是：首先，利文斯通博士明确地澄清了谦比西河——在我们的一些地图中是新赞比西河——并不是著名的赞比西河的一条支流。赞比西河向东流入印度洋，而谦比西河则是向西和西北方向流的另一条截然不同的河流。在这一点上，库利先生也曾说过，尽管在赞比西河的上游路径问题上他弄错了，但他一直坚持认为“新赞比西河”或者谦比西河独立存在。

其次，利文斯通博士探知在谦比西河远离卡仁比首都的下游处有另一条大河——卢菲拉河汇入。卢菲拉河发源于南部和西南部，流经坦噶尼喀湖南部的那个国家的西侧，而谦比西河则从东侧流过。旅行家利文斯通并没有看到卢菲拉河，但当他到达南纬 11°附近一个他没有说出具体名字的地方时，有人告诉他那条河在他所在位置西边不远处，并说那条河很大以至于要有独木舟才能通过。关于这一点我读到了他的这样一段文字：“我并没有见到卢菲拉河，不过它应该位于南纬 11°西侧的某个地方，据说那里要有独木舟才能通过。”——这表明它一定发源于距离与它平行的谦比西河很远的南部地方。

第三，利文斯通博士告诉我们，谦比西河注入了班韦乌卢湖，然后更名为卢阿

and then changes its name to Luapula; that this river flowing north enters Moero Lake, and "on leaving Moero at its northern end by a rent in the mountains of Rua it takes the name of Lualaba, and passing on N.N.W. forms Ulenge in the country west of Tanganyika." This, it must be remarked, is not native information, but the result of the traveller's own personal observation on the spot. His letters are dated "near Lake Bangweolo"; and in speaking of the Lualaba he says, "I have seen it only where it leaves Moero, and where it comes out of the crack in the mountains of Rua."

To make it more certain that he is speaking of the Lua*laba*, and not of the Lua*pula*, the traveller expresses his intention "to follow down the Lua*laba* and see whether, as the natives assert, it passes Tanganyika to the west, or enters it and finds an exit by the river called Locunda [or Loanda] into Lake Chowambe;" which lake, he says, "I conjecture to be that discovered by Mr. Baker;"—adding, "I shall not follow Lua*laba* by canoes", &c.

Nothing could well be more explicit than this. And yet your correspondent represents Dr. Livingstone as saying that "he saw the Lua*pula* only at this gap in the mountains", and describes the Lua*laba* as being a month's journey further west, and as falling into the Lulua and so joining the Zaire, or great river of Congo, on the west coast of Africa. There must clearly be some mistake here.

I think, too, there must be some misapprehension respecting "the great salt marshes, which chiefly supply the interior of Africa", described by "F.R.G.S." as situated on the banks of the Lualaba, a great running stream of fresh water. Is it not more likely that those salt marshes lie in some extensive depression in the interior of the continent, having no outlet, but in which the rivers that may flow into it are absorbed and lost?

Further, according to Dr. Livingstone, the Lualaba, after leaving Moero beyond the town of the Cazembe to the north, forms Ulenge, either a lake with many islands or a division into several branches, which are taken up by the Lufira. This I understand to mean, that the junction of the Lualaba and the Lufira is in Ulenge, *north* of the Cazembe's residence. "F.R.G.S." says, on the contrary, that the Lufira "flows into the Luapula from the west about 100 miles S.W., or S.S.W., from the Cazembe." How are these two statements to be reconciled?

Then "F.R.G.S." says, "When our author speaks of the Luviri (Lufira) entering Tanganyika at Uvira, he evidently casts the dimly discerned views of the natives into his own preconceived mould, and clothes them in his own language." But Dr. Livingstone could scarcely have had any "preconceived" notions on the subject, unless he took with him Mr. Cooley's map of 1852, in which the Chambeze, under the name of the New Zambesi, is laid down as joining the Luviri and then, under the name of Luapula, falling into the lake of "Zangañika" on its west side in about 8° S. lat. And this opinion Mr. Cooley would seem to regard still as the correct one; for in a letter which appeared in the *Daily Telegraph* of the 27th August last, with his initials "W.D.C.", he expressly states that "the drainage of the Cazembe's country is all into the Nyanza on the east." Though

普拉河，而卢阿普拉河则向北流入了姆韦鲁湖，并且“通过卢阿山脉的间隙从姆韦鲁湖的北端流出，之后更名为卢阿拉巴河，然后向西北偏北方向流去，进而在坦噶尼喀湖西部的地区形成了乌莱西湖。”值得注意的是，这并不是来自土著人的信息，而是旅行家利文斯通在这个地方亲自观察获得的结果。他在信中曾注明是“在班韦乌卢湖附近”，并且在提及卢阿拉巴河时，他说，“我仅在它流出姆韦鲁湖，从卢阿山脉的缺口处流出来时看到过它。”

为了进一步确认他提到的是卢阿拉巴河而不是卢阿普拉河，旅行家利文斯通阐明了他的意图，即“沿卢阿拉巴河顺流而下，去看看到底是不是像土著人所说的那样，河流穿过坦噶尼喀湖向西流去，还是进入坦噶尼喀湖以后又在被称为罗村达［或者罗安达］的河边找到出口注入了查旺比湖。”对于查旺比湖，他说，“我推测是由比克先生发现的”，并且“我不会乘独木舟沿卢阿拉巴河继续前行”等等。

没有比这更清晰的了。然而，贵刊的通讯员却对利文斯通博士提出质疑，说“他仅在山脉中间的缺口处看到了卢阿普拉河”，并且认为卢阿拉巴河位于向西约有一个月路程的地方，并流入卢卢阿河，进而汇入了非洲西海岸的扎伊尔河，也就是刚果河。很明显这里一定存在一些问题。

我还觉得，F.R.G.S. 所说的认为“作为非洲内陆主要供给地的大盐沼”位于卢阿拉巴河（一条很大的淡水河）两岸也有一些问题。那些盐沼难道不是更有可能位于大陆内部一些广阔的洼地上吗？洼地没有出口，河流流入洼地则被吸收而消失。

再者，利文斯通博士认为，卢阿拉巴河离开卡仁比城远处的姆韦鲁湖向北流去，形成了乌莱西湖——可能是一个拥有众多岛屿的湖泊，也可能是形成卢菲拉河若干支流的分流区。我认为，这说明卢阿拉巴河与卢菲拉河是在卡仁比族聚居地**北部**的乌莱西湖汇合的。相反，F.R.G.S. 却说卢菲拉河“从位于卡仁比城西南或西南偏南方向约 100 英里处，自西流入卢阿普拉河。”如何才能调和这两种不同的陈述呢？

F.R.G.S. 又说，“当我们的作者提到罗非河（卢菲拉河）在乌维拉注入坦噶尼喀湖时，很明显他是将土著人朦胧认识到的观点纳入到自己预定的模式中，并且冠以自己的语言。”但在这个问题上，利文斯通博士几乎不可能有任何预定的想法，除非他随身带着库利先生 1852 年绘制的地图。在这幅地图中，谦比西河被称为新赞比西河，汇入罗非河后消失，之后被称为卢阿普拉河，在南纬约 8° 处从“坦噶尼喀”湖西侧汇入该湖。库利先生似乎还认为这一观点是正确的，因为在去年 8 月 27 日的《每日电讯报》中刊载的以他的姓名首字母“W.D.C.”署名的一封信中，他清楚地写道，“卡仁比国所有的河流都流入了东部的尼亚萨湖。”不过不知为什么会给坦噶尼喀湖取这样一个名字。我们知道斯皮克称之为“维多利亚–尼亚萨湖”，贝克称

why this name should be applied to the Lake of Tanganyika is not patent. We know the "Victoria Nyanza" of Speke, the "Albert Nyanza" of Baker, the "Lake Tanganyika" of Burton, and the "Lake Nyassa" of Livingstone. We also know that in Mr. Cooley's maps of 1845 and 1852, Tanganyika and the more southerly Lake Nyassa are made to form one continuous body of water under the name of "Nyassa, or the Sea". But the present seems to be the first time that the designation of "Nyanza" has been applied, without any qualification, to the separate Lake Tanganyika. I perceive that "F.R.G.S." associates Captain Burton with this "Nyanza"; but such a name was never given to it by its discoverer, neither is it generally known by any other designation than that of "Lake Tanganyika": whether or not it should properly be called the "Lake *of* Tanganyika" is of no moment.

I come now to the consideration of Dr. Livingstone's claim to the discovery of the sources of the Nile, which will be best given in his own words: "I think that I may safely assert that the chief sources of the Nile rise between 10° and 12° south latitude, or nearly in the position attributed to them by Ptolemy, whose river Rhaptus is probably the Rovuma." On this "F.R.G.S." acutely remarks: "Here two different problems are attempted to be solved at once—the one touching the sources of the White Nile, and the other those of Ptolemy's Nile;" in which remark he is no doubt substantially correct. Into the question of Ptolemy's sources of the Nile, on which subject "F.R.G.S." and I differ widely, I need not now enter: what I have here to do with, is the question of the chief sources of the Nile. And in order to decide whether Dr. Livingstone has really discovered these sources, it is, in the first instance, requisite to define the limits of the basin of the Nile, so as absolutely to determine where the sources of the river can or can not be situated. As those limits were approximatively determined in a paper "On the Nile and its Tributaries", communicated to the Royal Geographical Society in 1846, and published in the seventeenth volume of the Society's Journal, I cannot do better than reproduce the portion of it relating to this particular subject.

After describing the physical character of the table-land of Eastern Africa, of which Abyssinia forms the northern extremity, and its rivers as far as they were then known,—on which subject I need not dilate, as the substantial correctness of my views is now established,—I proceeded in these terms:—

"All the streams of the plateau or western counter-slope of the Abyssinian chain are affluents of the Nile, and their easternmost branches take their rise at the extreme eastern edge of the table-land, which is the limit of the basin of the Nile, and the watershed between its tributaries and the rivers flowing E. and S.E. towards the Indian Ocean. On the seaward side of this watershed, the declivity being much more abrupt and its extent much more limited, the rivers must necessarily be of secondary importance. Thus, proceeding from the N., we do not meet with a stream deserving of name until we come to the Hawash; and even that river is, near Aussa, lost in Lake Abhebbad before reaching the ocean. The river Haines of Lieutenant Christopher, which is the next in succession, appears, in like manner, not to have sufficient power to reach the sea, at least not at all

之为“艾伯特–尼亚萨湖”，伯顿称之为“坦噶尼喀湖”，利文斯通称之为“尼亚萨湖”。我们还知道在库利先生 1845 年和 1852 年版的地图中，坦噶尼喀湖和更南部的尼亚萨湖共同构成了一片连续的水体，被称为“尼亚萨湖（或者海）”。但这似乎是第一次将“尼亚萨”这个名称不加任何限定地用于单独的坦噶尼喀湖。我发现 F.R.G.S. 将伯顿上校与“尼亚萨”联系在了一起，可是，这个名字以及其他那些我们通常所知的“坦噶尼喀湖”以外的名称都绝不是由其发现者取的，况且将其称为“坦噶尼喀湖”是否合适并不重要。

现在我来说一下利文斯通博士宣称发现了尼罗河源头的问题，这最好用他自己的话来说明——“我认为我可以有把握地断言尼罗河的主要源头位于南纬 10°~12° 之间，或是托勒密所说的位置附近，其中托勒密指的拉普塔河很可能是鲁伍马河。”关于这一点，F.R.G.S. 曾尖锐地评论道，“在这里需要同时解决两个问题——其一是白尼罗河的源头问题，另外一个则是托勒密所指的尼罗河的源头问题。”毫无疑问，他的上述评论是完全正确的。但在托勒密所指的尼罗河的源头问题上，我和 F.R.G.S. 有不同的看法，我觉得现在没必要介入，在这里我必须解决的是尼罗河的主要源头问题。为了确定利文斯通博士是否确实发现了这些源头，首先必须确定尼罗河流域的界限，以便能够完全确定河流的源头在或不在某个地方。由于在 1846 年写给皇家地理学会的一篇题为《尼罗河及其支流》的文章已经大致给出了尼罗河流域的界限，且该文已在《皇家地理学会会刊》第 17 卷中发表，所以在这里我只重述一下其中与此主题有关的部分。

在描述了以阿比西尼亚为北端界限的东非高原及其已知的一些河流的自然特征以后——关于该问题我已没必要详述，因为我的观点已被充分证明是正确的——我从如下文字着手继续阐述：

“高原或阿比西尼亚山系西部反坡上的所有河流都是尼罗河的支流，其最东端的支流都发源于高原东端边缘处，而这里正是尼罗河流域的边界，也是尼罗河各支流与流向东和东南注入印度洋的诸河之间的分水岭。分水岭向海的一侧非常陡峭且范围有限，所以这里的河流只能是一些次级河流。因此，从北侧前行时，我们没有遇到一条值得一提的河流，直到我们来到了哈瓦希河；但这条河也位于豪萨附近，在流入大洋之前汇入阿拜巴德湖之中了。接下来看到的是克里斯托弗上尉所说的海因斯河，它同样没有足够的能量到达大海，至少不是全年都可以。再向南我们见到了科维河（也就是瓦比–吉维娜河）或称朱卜河，它有大洋河流的独立特征，不过在

times of the year. Further to the S. we find the river Gowin (*i.e.* Wabbi-Giweyna) or Jubb, possessing a substantive character as an ocean stream; but this river, during the dry season, has at its mouth a depth of only two feet. At a short distance to the S. of the equator is the Ozay, which river, though said to be of great extent, has very little water at the entrance. Further S. the same law appears to prevail, as is exemplified in the Lufiji or Kwavi (Quavi), the Livuma [Rovuma] and the Kwama (Quama) or Kilimane (Quilimane), which rivers rise on the eastern edge of the elevated plain in which Lake Zambre or N'yassi is situate, and flow into the Indian Ocean. *Here, however, the southern extremity of the basin of the Nile having been passed*, the larger streams of the counter-slope no longer join that river, but take their course westwards into the Atlantic, belonging in fact to a distinct hydrographical basin."

What I thus wrote three-and-twenty years ago requires now but little modification. The erroneous identification of Lake Zambre with N'yassi was simply adopted from Mr. Cooley's learned and valuable paper in the fifteenth volume of the Society's Journal, which was then our only authority on the subject. I also followed him in his alteration of the spelling of the name "Zambre", which in my paper was printed "Zambeze", with the explanatory note, "This name is usually printed Zembere, Zembre, or Zambre. It is the Lake *Maravi* of the maps." Though even this was wrong; for Nyassa is properly Lake Maravi, and Tanganyika is the Great Lake, or Zambre. The blending of the two together by Delille and D'Anville was the primary cause of the long-existing misapprehension of the subject.

In my paper from which the foregoing extract is taken, when speaking of the lakes and swamps of the Upper Nile as then known, I added in a note, "May not Lake Zambre ('Zambeze'), or Nyassi, be the continuation of this series of lakes? In this case it would be simply the upper course of the Nile."

Acting on this suggestion, Professor Berghaus, in 1850, laid down Mr. Cooley's "Nyassi, or the Sea" as the head of the Nile; but, as I pointed out to him, he had under any circumstances carried the river too far south, because the Chevalier Bunsen and I had in the previous year come to the positive conclusion, on the reports of the Church missionaries at Mombas, that Zambre (now Tanganyika) and Nyassa were two separate lakes, a conclusion which every fresh discovery only tended to confirm.

The Cuama and Quilimane mentioned by me were all that we then well knew of the Zambesi, the great western extent of which river only became revealed to us through the former explorations of Livingstone. He thus absolutely closed the basin of the Nile in that direction; though the fact of his having done so was not then demonstrable. When he wrote to Lord Clarendon in February 1867, as he says in his present letter, he "had the impression that he was then on the watershed between the Zambesi and either the Congo or the Nile." His present determination of the want of connection between the Chambeze and the Zambesi, and of the western and northwestern course of the former river, has proved the soundness of his impression of February 1867.

枯季时，其河口处水深只有 2 英尺。在赤道以南不远处就是奥溡河。据说这条河很长，但在河口处也只有很少的水量。再往南还是跟前面一样的规律，正如卢斐济河或卡维河（库维河），里伍玛河［鲁伍马河］和卡玛河（库玛河）或称基里曼河（库里曼河）所表现的那样，这些河流发源于赞比西湖或尼亚萨湖所在高原的东缘，都流入了印度洋。**不过，这里已经超出尼罗河流域最南端了**，反坡上较大的河流已不再汇入尼罗河，而是向西流入大西洋。这里实际上属于一个截然不同的水文地理流域。”

这是我在 23 年前写下的东西，现在只需要进行很少的修正。将赞比西湖修正为尼亚萨只不过是采用了库利先生发表于《皇家地理学会会刊》第 15 卷中那篇颇有学术价值的文章中的提法，要知道《皇家地理学会会刊》可是当时关于该问题的唯一权威。我还参照了他对该名称所用的“Zambre”这种特别的拼写，在我原文中拼写的“Zambeze”后面加了一个注解，“这个名字通常还被写成 Zembere，Zembre，或者 Zambre，也就是地图上的**马拉维**湖。”即使这样仍然不对，因为尼亚萨才是真正的马拉维湖，而坦噶尼喀湖才是大湖或赞比西湖。德利尔和德安维尔对两者的混淆是导致该问题长期存在误解的主要原因。

前述摘录出自我的一篇文章，在这篇文章中，谈及当时已知的尼罗河上游湖泊和沼泽时，我加了一个注释，“赞比西湖或尼亚萨湖或许并不是这一系列湖泊的延续？如果的确如此，它可能就是尼罗河的上游。”

根据上述思路，1850 年贝格豪斯教授否定了库利先生的“尼亚萨湖（或者海）”是尼罗河源头的说法。但正如我对他指出的那样，无论怎么看，他都把尼罗河搬得太靠南了，因为在前一年我和希瓦利埃 · 邦森在蒙巴萨的传教士教堂作的报告中就已经得出了明确的结论，即赞比西湖（也就是现在的坦噶尼喀湖）与尼亚萨湖是两个独立的湖，并且这一结论正越来越被最新的发现证实。

我提到的库玛河和基里曼河正是那时我们已非常了解的赞比西河，但通过利文斯通早先的探索，我们才真正看到了它西部广阔地区的真面目。所以，利文斯通绝对是从那个方向接近尼罗河流域的，只不过那时不能证明他已做到了这一点而已。当他于 1867 年 2 月份写信给克拉伦登伯爵时，正如他在这封信中说的那样，他“已经感觉到他当时就处在赞比西河与刚果河或者尼罗河之间的分水岭上。”他现在对以往谦比西河与赞比西河之间缺失的联系和前者在西侧和西北侧部分的流经路线的确定，已经证明了他在 1867 年 2 月的直觉是正确的。

The question is therefore now narrowed to this:—Do the united streams of the Chambeze and the Lufira, under the name first of Luapula, and then of Lualaba, flow into the Nile or into the Congo? I am of opinion that they join the former river, and that the explorations of Dr. Livingstone have established the correctness of the views I have long entertained, and especially those enunciated in the *Athenaeum* No. 1,969, of July 22, 1865, on the first announcement by Sir Samuel Baker of his (unconscious) discovery of the main stream of the Nile under the name of "Albert Nyanza", and consequently I believe we are at length enabled to strip the veil from the Nile Mystery.

(**1**, 240-241; 1869)

Charles Beke: Bekesbourne, December 1.

因此，现在只剩一个问题，那就是，先叫卢阿普拉河后又被称为卢阿拉巴河的谦比西河与卢菲拉河的联合体是否流入了尼罗河或刚果河？我认为它们汇入了尼罗河，而且利文斯通博士的探查也说明我长期以来持有的观点是正确的，尤其是我在1865年7月22日第1,969期的《科学协会》上就塞缪尔·贝克爵士的第一份声明阐述的那些观点。那篇声明是塞缪尔·贝克爵士关于他（无意中）发现尼罗河干流并命名为“艾伯特–尼亚萨”的声明。因此，我相信，我们最后一定可以解开尼罗河之谜的面纱。

（齐红艳 翻译；郭华东 审稿）

The Origin of Species Controversy

A. R. Wallace

Editor's Note

The British naturalist Alfred Russel Wallace, whose theory of evolution paralleled Darwin's, here reviews a new book of scientific essays by Joseph John Murphy. Murphy had argued that evolution by natural selection could not account for the observed species if all variations were small, for the time required would be longer than earth had apparently been habitable. Wallace argued that Murphy was mistaken, suggesting that the difficulty might be resolved with further understanding of geological history. Murphy also touched on the nature of intelligence, both unconscious and conscious, the latter which he proposed must be linked to human instinct. Wallace (who developed mystical and spiritualist enthusiasms in later life) felt more favourably towards these speculations.

Habit and Intelligence, in their Connection with the Laws of Matter and Force.
A Series of Scientific Essays. By Joseph John Murphy. (Macmillan and Co., 1869)

I

THE flood of light that has been thrown on the obscurest and most recondite of the forces and forms of Nature by the researches of the last few years, has led many acute and speculative intellects to believe that the time has arrived when the hitherto insoluble problems of the origin of life and of mind may receive a possible and intelligible, if not a demonstrable, solution. The grand doctrine of the conservation of energy, the all-embracing theory of evolution, a more accurate conception of the relation of matter to force, the vast powers of spectrum analysis on one side, showing us as it does the minute anatomy of the universe, and the increased efficiency of the modern microscope on the other, which enables us to determine with confidence the structure, or absence of structure, in the minutest and lowest forms of life, furnish us with a converging battery of scientific weapons which we may well think no mystery of Nature can long withstand. Our literature accordingly teems with essays of more or less pretension on the development of living forms, the nature and origin of life, the unity of all force, physical and mental, and analogous subjects.

The work of which I now propose to give some account, is a favourable specimen of the class of essays alluded to, for although it does not seem to be in any degree founded on original research, its author has studied with great care, and has, in most cases, thoroughly understood, the best writers on the various subjects he treats of, and has brought to the task a considerable amount of original thought and ingenious criticism. He thus effectually raises the character of his book above that of a mere compilation, which, in less able hands, it might have assumed.

物种起源论战

华莱士

编者按

英国博物学家阿尔弗雷德·拉塞尔·华莱士关于进化的观点与达尔文的观点很类似，这里他评论了约瑟夫·约翰·墨菲新出的一本科学论文集。墨菲早先提出，如果所有的变异都很微小的话，那么依靠自然选择的进化将无法解释可见的物种，因为进化所需的时间将比地球适宜生物生存的时间更漫长。华莱士认为墨菲是错误的，他指出，在深入了解地质历史之后这一问题将迎刃而解。墨菲还提到了智能的本质，包括无意识的和有意识的，他认为有意识的智能一定与人类的本能有关联。对于这些猜想，华莱士（他晚年对神秘主义与唯灵论产生了兴趣）则比较赞成。

《习性与智能，及其与物质和力的法则的关系》
约瑟夫·约翰·墨菲的一本科学论文集（麦克米伦公司，1869 年）

I

通过最近几年的研究，人们已经揭示了关于自然界中力和组成结构的许多最为晦涩且极其深奥的问题，这使很多性急又有投机心理的知识分子们以为，对于到目前为止还未解决的关于生命和思维起源的问题，即使还不能得到一个可论证的答案，也应该是时候给出一个合理而又清楚的解释了。一方面，极其重要的能量守恒学说、囊括一切的进化论、关于物质和力关系的更准确的概念以及具有巨大威力的光谱分析，给我们展示了宇宙万物结构和机理的本来面目；另一方面，现代显微镜技术性能的提升，使我们可以很有把握地确定最微小和最低等的生命形式的结构或者确认它们没有结构；这些都给我们提供了强大的科学武器，使我们相信自然界没有永久的秘密。因此，这部书里的文章或多或少都对生命形式的发展、生命的起源和本质、各种力的统一、肉体和精神以及诸如此类的问题给出了自信的答案。

我将要介绍的这本书，是人们曾经提到的这类著作中的一个很好的范例。虽然这本书不是建立在原始研究的基础之上，但作者进行了细致谨慎的研究，而且在大多数情况下都能深刻地理解他所涉及的各个科目中最优秀的作者，此外，还精心引入了大量原创性的思想和独创性的批评。这样，作者就有效地提高了这本书的品质，使之超过了能力稍显逊色的人可能炮制出的纯粹的合集。

The introductory chapter treats of the characteristics of modern scientific thought, and endeavours to show, "that the chief and most distinctive intellectual characteristic of this age consists in the prominence given to historical and genetic methods of research, which have made history scientific, and science historical; whence has arisen the conviction that we cannot really understand anything unless we know its origin; and whence also we have learned a more appreciative style of criticism, a deeper distrust, dislike, and dread, of revolutionary methods, and a more intelligent and profound love of both mental and political freedom." The first six chapters are devoted to a careful sketch of the great motive powers of the universe, of the laws of motion, and of the conservation of energy. The author here suggests the introduction of a useful word, *radiance*, to express the light, radiant heat, and actinism of the sun, which are evidently modifications of the same form of energy,—and a more precise definition of the words *force* and *strength*, the former for forces which are capable of producing motion, the latter for mere resistances like cohesion.

He enumerates the primary forces of Nature as, gravity, capillary attraction, and chemical affinity, and notices as an important generalisation "that *all primary forces are attractive*; there is no such thing in Nature as a primary repulsive force" (p. 43). Now here there seem to be two errors. Cohesion, which is entirely unnoticed, is surely as much a primary force as capillary attraction, and, in fact, is probably the more general force, of which the other is only a particular case; and elasticity is the effect of a primary repulsive force. In fact, at p. 26, we find the author arguing that *all matter is perfectly elastic*, for, when two balls strike together, the lost energy due to imperfect elasticity of the mass is transferred to the molecules, and becomes heat. But this surely implies repulsion of the molecules; and Mr. Bayma has shown, in his "Molecular Mechanics", that repulsion is as necessary a property of matter as attraction.

The eighth chapter discusses the phenomena of crystallisation; and the next two, the chemistry and dynamics of life. The reality of a "vital principle" is maintained as "the unknown and undiscoverable something which the properties of mere matter will not account for, and which constitutes the differentia of living beings." Besides the formation of organic compounds, we have the functions of organisation, instinct, feeling, and thought, which could not conceivably be resultants from the ordinary properties of matter. At the same time it is admitted that conceivableness is not a test of truth, and that all questions concerning the origin of life are questions of fact, and must be solved, not by reasoning, but by observation and experiment; but it is maintained that the facts render it most probable that "life, like matter and energy, had its origin in no secondary cause, but in the direct action of creative power." Chapters X to XIV treat of organisation and development, and give a summary of the most recent views on these subjects, concluding with the following tabular statement of organic functions:—

Formative or Vegetative Functions, essentially consisting in the Transformation of Matter

Chemical	Formation of organic compounds
Structural	Formation of tissue
	Formation of organs

前言中提到了现代科学思想的特点，并力图展示“这一时期最重要也是最独具一格的思维特点是人们大多习惯于采用历史学和遗传学的研究方法，这些方法使历史科学化，也使科学历史化，由此我们可以确信：除非我们知道起源，否则我们不能够真正地认识任何事物；从这里我们也学到了带有更多赞赏眼光的批判风格，带有更深的怀疑、不赞成和担忧的创新方法，以及对精神和政治自由更理智更深刻的爱。”前 6 章主要致力于详细描述宇宙巨大的动力、运动的规律和能量守恒。作者在这里引入了**辐射**这个有用的词语，用来表示光、热辐射以及太阳的光化作用，这些实际上是同一种形式的能量的不同表现。此外，作者还更精确地定义了 *force* 和 *strength* 这两个词，前者代表能够产生运动的力，而后者只代表阻力，比如黏滞力。

作者列举了自然界中几种基本的力，如重力、毛细引力和化学亲和力，然后他总结出了一个重要的结论：“**所有基本的力都是引力**，自然界中没有任何一种基本的力是排斥力”（第 43 页）。这里似乎有两个错误。首先，和毛细引力一样，黏滞力显然也是一种基本的力，而作者则完全忽视了这种力。事实上，黏滞力可能是更普遍存在的，相比而言，毛细引力倒要算是特殊的力了。另外，弹力就是一种基本的排斥力。实际上，在该书第 26 页，我们发现作者对**所有物质都是完全弹性的**这一观点提出了质疑，他认为，当两个球撞击时，损失的能量转移给了分子并最终变成热，而这部分能量的损失应归于物质的不完全弹性。但是这显然暗示了分子排斥的存在。就像拜马先生在他的《分子力学》中所写的那样，排斥力就像引力一样是物质必需的属性。

第 8 章讨论了结晶现象，紧接着的两章讲了生命化学和生命动力学。“生命法则”存在于“未知的和未发现的事物”这些实体当中，“这些事物不能用纯粹的物质属性来解释，它们造成了生物体的差异性。”除了有机化合物的形成外，还有组织功能、本能、感觉和思想，这些都不能用普通的物质属性来解释。同时，我们也承认猜测绝不是对真理的检验，关于生命起源的所有问题都是事实的问题，必须通过观察和实验来解决，而不是只靠思考与推理。但是书中声称，事实最可能是“就像物质和能量一样，生命的起源也是第一创造力的直接作用结果，而没有任何别的原因。”第 10~14 章提到了组织和发育，总结了这些领域的最新观点，并以如下的表格形式就有机物的功能给出了最终的结论：

构形功能或营养功能，本质上来说都是通过物质转化而形成的

化学的	有机化合物的形成
结构的	组织的形成
	器官的形成

Animal Functions, consisting essentially in the Transformation of Energy

Motor	Spontaneous
	Reflex
	Consensual
	Voluntary
Sensory	Sensation
	Mind

In the fifteenth chapter we first come to one of the author's special subjects;—the Laws of Habit. He defines habit as follows: "The definition of habit and its primary law, is that all vital actions tend to repeat themselves; or, if they are not such as can repeat themselves, they tend to become easier on repetition." All habits are more or less hereditary, are somewhat changeable by circumstances, and are subject to spontaneous variations. The *prominence* of a habit depends upon its having been recently exercised; its *tenacity* on the length of time (millions of generations it may be) during which it has been exercised. The habits of the species or genus are most tenacious, those of the individual often the most prominent. The latter may be quickly lost, the former may appear to be lost, but are often latent, and are liable to reappear, as in cases of reversion. The fact that active habits are strengthened, while passive impressions are weakened, by repetition, is due in both cases to the law of habit; for, in the latter, the organism acquires the habit of not responding to the impression. As an example, two men hear the same loud bell in the morning; it calls the one to work, as he is accustomed to listen to it, and so it always wakes him; the other has to rise an hour later, he is accustomed to disregard it, and so it soon ceases to have any effect upon him. Habit has produced in these two cases exactly opposite results. Habits are capable of any amount of change, but only a slight change is possible in a short time; and in close relation with this law are the following laws of variation.

Changes of external circumstances are beneficial to organisms if they are slight; but injurious if they are great, unless made gradually.

Changes of external circumstances are agreeable when slight, but disagreeable when great.

Mixture of different races is beneficial to the vigour of the offspring if the races mixed are but slightly different; while very different races will produce either weak offspring, or infertile offspring, or none at all. Even the great law of sexuality, requiring the union of slightly different individuals to continue the race, seems to stand in close connection with the preceding laws.

The next seven chapters treat of the laws of variation, distribution, morphology, embryology, and classification, as all pointing to the origin of species by development; and we then come to the causes of development, in which the author explains his views as

动物性机能，本质上来说是通过能量转化而形成的

动作	自发性的
	反射性的
	交感性的
	自主性的
感官	感觉
	思维

在第 15 章，我们首次看到了作者一个独特的观点——习性法则。他是这样定义习性的："习性的定义及基本法则是，所有的生命活动都有自我重复的趋势，或者如果不能进行自我重复，那么它们倾向于使重复变得更加容易。"所有的习性几乎都是遗传性的，随着环境变化会有少许改变，并且受到自发变异的影响。一个习性是否有**明显的影响力**取决于它最近是否得到运用，它的**强度**取决于运用时间的长短（可能是几百万代）。种或属的习性是非常稳定不太变化的，而个体的习性经常是差异较大而十分凸显的。个体习性可能会很快消失，种属习性可能也会消失，但通常是很难察觉到的，而且种属习性还容易再次出现，例如返祖现象。通过不断重复，主动习性得到强化，而被动受到的影响会减弱，这两个事实也可以根据习性法则推断出来。因为，在后一种情况下，生物体获得了对被动受到的影响不作出反应的习性。比如，有两个人在早上听到相同的钟声。对于其中一个人，钟声会叫醒他去上班，而他也习惯于去听钟声，因此钟声总能叫醒他。而对另一个人，在钟声响起一小时后他才起床，他习惯于忽略钟声，这样不久之后钟声对他就没有任何作用了。在这两个例子中习性产生了两种完全相反的结果。习性可以发生任何程度的改变，但在短期内它只能发生微小的改变。与习性法则密切相关的是接下来要谈及的变异法则。

外界环境的轻微改变对生物体是有利的，但是，外界环境的较大改变对生物体却是有害的，除非这种改变是逐渐发生的。

当外界环境发生轻微的改变时，生物体是可以适应的，而当外界环境变化较大时，生物体就不能适应了。

如果混居在一起的各种族之间只有微小的差异，那么种族混居是有利于提高后代素质的。但是，差别较大的种族却会产生虚弱的后代或者不育的后代，甚至根本无法产生后代。伟大的性法则告诉我们，要延续种族就需要群体中有稍微不同的个体。这样看来，性法则与我们前面提到的法则之间似乎是有紧密联系的。

接下来的 7 章谈到了变异法则、分布、形态学、胚胎学以及分类法则，然后谈到了发育的原因。这样，以生物体的发育为纽带，所有问题都指向了物种的起源。

follows:—

> These two causes, self-adaptation and natural selection, are the only *purely physical* causes that have been assigned, or that appear assignable, for the origin of organic structure and form. But I believe they will account for only part of the facts, and that no solution of the questions of the origin of organization, and the origin of organic species, can be adequate, which does not recognise an Organising Intelligence, over and above the common laws of matter....But we must begin the inquiry by considering *how much* of the facts of organic structure and vital function may be accounted for by the two laws of self-adaptation and natural selection, before we assert that any of those facts can only be accounted for by supposing an Organising Intelligence.

Again:

> Life does not suspend the action of the ordinary forces of matter, but works through them. I believe that wherever there is life there is intelligence, and the intelligence is at work in every vital process whatever, but most discernibly in the highest...Nutrition, circulation, and respiration are in a great degree to be explained as results of physical and chemical laws;—but sensation, perception, and thought cannot be so explained. They belong exclusively to life; and similarly the organs of those functions—the nerves, the brain, the eye, and the ear—can have originated, I believe, solely by the action of an Organising Intelligence.

Admitting Mr. Herbert Spencer's theory of the origin of the vascular system, and possibly of the muscular, by self-adaptation, he denies that any such merely physical theory will account for the origin of the special complexities of the visual apparatus:

> Neither the action of light on the eye, nor the actions of the eye itself, can have the slightest tendency to produce the wondrous complex histological structure of the retina; nor to form the transparent humours of the eye into lenses; nor to produce the deposit of black pigment that absorbs the stray rays that would otherwise hinder clear vision; nor to produce the iris, and endow it with its power of closing under a strong light, so as to protect the retina, and expanding again when the light is withdrawn; nor to give the iris its two nervous connections, one of which has its root in the sympathetic ganglia, and causes expansion, while the other has its root in the brain and causes contraction.

Nor will he allow that Natural Selection (which he admits may produce any simple organ, such as a bat's wing) is applicable to this case; and he makes use of two arguments which have considerable weight. One is that of Mr. Herbert Spencer, who shows that in all the higher animals natural selection must be aided by self-adaptation, because an alteration in any part of a complex organ necessitates concomitant alterations in many other parts, and these cannot be supposed to occur by spontaneous variation. But in the case of the eye he shows that self-adaptation cannot occur, whence he conceives it may be proved to be almost an infinity of chances to one against the simultaneous variations necessary to produce an eye ever having occurred. The other argument is, that well-developed eyes occur in the higher orders of the three great groups, Annulosa,

在书中，作者是这样阐释他的观点的：

对于生物体结构和形态的起源，在人们明确给出或者暗示性地提出的原因中，只有两个是**完全只与自然界有关的**原因，这两个原因是：自适应和自然选择。但是我认为这两个原因只能解释部分事实。而且，对于组织起源和物种起源这些问题，这两个原因给出的解释并不比根据普通的物质法则作出的解释更充分，并且根本没有注意到组织智能……但是，在我们断言只要加入组织智能的假定就能解释所有这些事实之前，我们还必须弄清楚，自适应和自然选择这两个法则到底能够解释**多少**有关有机体结构和功能的问题。

此外，作者还写道：

生命并没有中止物质通常的力的作用，而是通过它们来发挥作用。我相信，所有的生命都是有智能的。智能在所有的生命过程中都发挥作用，只是在最高等的生命过程中最容易被觉察到……营养、循环和呼吸在很大程度上是可以用物理和化学规律解释的。但是，感觉、感知和思考却不能这样解释。这些过程属于生命所专有。类似地，对于具有这些功能的器官，即神经、脑、眼睛和耳朵，我相信它们应该都是起源于组织智能的活动。

作者赞同赫伯特·斯宾塞先生的理论，即血管系统起源于自适应，肌肉系统也很可能起源于自适应。由此，作者认为单靠物理学理论并不能解释异常复杂的视觉器官的起源。他写道：

眼睛的结构和功能是很神奇的：视网膜的组织构造非常巧妙而复杂；眼睛里的透明体液形成了晶状体；眼睛中黑色素的沉淀能够吸收杂散光线从而使视觉清晰不受影响；眼睛中的虹膜具有在强光下关闭以保护视网膜而在光线转弱时又再次张开的能力；虹膜中有两类神经连接，一类是根部位于交感神经节中的神经连接，它能引起扩张，另一类是根部位于脑中的神经连接，它能引起收缩。所有这些结构与功能的产生，应该都不会受到光对眼睛的作用和眼睛自身活动的丝毫影响。

作者也不认为自然选择适用于这个例子（他承认，自然选择可以产生一些简单的器官，比如蝙蝠的翅膀）。他提到了两个相当有分量的观点。其中一个是来自赫伯特·斯宾塞先生的观点，即，在所有高等动物中自然选择必须借助于自适应，因为一个复杂器官任何部位的改变必然伴随着其他部位的许多改变，而这些改变是不可能通过自发变异产生的。但是作者指出，在眼睛这个例子中自适应并不会发生，由此，他觉得这个例子可能给那些认为同步变异并不是产生眼睛的必要条件的人提供了一个屡用不爽的例证（事实上这些人此前确实用过这个例子）。另一个观点是，在环节动物、软体动物和脊椎动物这三类高等群体中才出现了结构和功能发展得很好的眼睛，而在较低等的群体中，眼睛还是很初等的或者根本就没有眼睛。因此，完美

Mollusca, and Vertebrata, while the lower orders of each have rudimentary eyes or none; so that the variations requisite to produce this wonderfully complicated organ must have occurred three times over independently of each other. In the first of these objections, he assumes that many variations must occur simultaneously, and on this assumption his whole argument rests. He notices Mr. Darwin's illustration of the greyhound having been brought to its present high state of perfection by breeders selecting for one point at a time, but does not think it possible "that any apparatus, consisting of lenses, can be improved by any method whatever, unless the alterations in the density and the curvature are perfectly simultaneous." This is an entire misconception. If a lens has too short or too long a focus, it may be amended either by an alteration of curvature, or an alteration of density; if the curvature be irregular, and the rays do not converge to a point, then any increased regularity of curvature will be an improvement. So the contraction of the iris and the muscular movements of the eye are neither of them essential to vision, but only improvements which might have been added and perfected at any stage of the construction of the instrument. Thus it does not seem at all impossible for spontaneous variations to have produced all the delicate adjustments of the eye, once given the rudiments of it, in nerves exquisitely sensitive to light and colour; but it does seem certain that it could only be effected with extreme slowness; and the fact that in all three of the primary groups, Mollusca, Annulosa, and Vertebrata, species with well-developed eyes occur so early as in the Silurian period, is certainly a difficulty in view of the strict limits physicists now place to the age of the solar system.

(**1**, 105-107; 1869)

II

In his chapter on "The Rate of Variation", Mr. Murphy adopts the view (rejected after careful examination by Darwin) that in many cases species have been formed at once by considerable variations, sometimes amounting to the formation of distinct genera and he brings forward the cases of the Ancon sheep, and of remarkable forms of poppy and of *Datura tatula* appearing suddenly, and being readily propagated. He thinks this view necessary to get over the difficulty of the slow rate of change by natural selection among minute spontaneous variations; by which process such an enormous time would be required for the development of all the forms of life, as is inconsistent with the period during which the earth can have been habitable. But to get over a difficulty it will not do to introduce an untenable hypothesis; and this one of the rapid formation of species by single variations can be shown to be untenable, by arguments which Mr. Murphy will admit to be valid. The first is, that none of these considerable variations can possibly survive in nature, and so form new species, unless they are *useful* to the species. Now, such large variations are admittedly very rare compared with ordinary spontaneous variability, and as they have usually a character of "monstrosity" about them, the chances are very great against any particular variation being useful. Another consideration pointing in the same direction is, that as a species only exists in virtue of its being tolerably well adapted

的具有复杂结构的眼睛的产生，必然经过了三次以上的独立变异过程。作者对这个观点的第一条反对意见是，他认为大多数变异都是同时发生的。他的所有论述都是基于这个假设的。达尔文在关于灰色猎犬的阐述中说到，饲养者通过每次只筛选某一个特征可以使猎犬达到目前的高级状态。虽然作者也注意到了这些，但是他认为下面的论述是不可能的："除非透镜材料的密度和曲率同时得到完美的改变，否则，任何方法都不可能改进任何由透镜组成的仪器的性能。"这个观点是完全错误的。如果一个透镜的焦距太长或者太短，那么我们就可以通过改变曲率或者改变密度使透镜的性能得到改善。如果曲率不规则导致光线无法汇聚到一点，那么任何能使曲率变得规则的方法都可以改善透镜的性能。因此，对于视力来说，虹膜的收缩和眼部肌肉的运动都不重要，只有那些可能在眼睛形成过程的某些阶段已经发生而且完善化了的改进才是至关重要的。这样看来，一旦有了眼睛的雏形，那么在那些对光和色彩具有敏锐感知能力的神经中发生的多个自发变异完全有可能都使眼睛的微观结构发生有利改变。但是应该可以肯定的是，眼睛受到这种影响的速度是异常缓慢的。不过，考虑到已经被严格界定了的志留纪的确切时限，那么三个主要类群（软体动物、环节动物和脊椎动物）中具备完美进化的眼睛的物种早在志留纪就已出现这一事实无疑就很难解释了。

（郑建全 翻译；陈平富 审稿）

II

在《变异速率》一章中，墨菲先生采纳的观点（被达尔文在认真检验之后驳斥了的）是，在许多情况下，物种是通过显著的变异即刻形成的，有时可以积累变异而形成不同的属。他列举了一些案例，如安康羊，多种类型的罂粟，以及突然出现变异并迅速繁殖的紫花曼陀罗。他认为有必要凭借这一观点来克服微小的自发变异中自然选择面临变化速度太慢的难题；任何形式的生命通过这一过程发育的话，都将需要无比漫长的时间，而这与地球上生命出现的时期相矛盾。但是为了克服这种困难，引入一个站不住脚的假说也毫无意义；而通过墨菲一厢情愿地认为有效的论据来看，这个通过单一变异就能快速形成物种的论点确实是站不住脚的。首先，这些显著的变异都不可能在自然状态下幸存，更别提形成新物种了，除非他们对物种是**有用的**。现在，人们认为大变异与普通的自发变异相比少之又少，而通常这种变异的特征很“畸形”，于是出现任何有用的特定变异的机率非常小。另外，通常要思考的另一点是，由于物种只是因为具有良好的耐性而与环境相适应才生存下来，同时环境只是**缓慢地**变化，所以物种更需要小的变化而非大的变化来维持适应性。但是即使环境迅速发生了巨大的变化，比如一些新物种的侵入或者由于几英尺的下陷

to its environment, and as that environment only changes *slowly*, small rather than large changes are what are required to keep up the adaptation. But even if great changes of conditions may sometimes occur rapidly, as by the irruption of some new enemy, or by a few feet of subsidence causing a low plain to become flooded, what are the chances that among the many thousands of *possible* large variations the one exactly adapted to meet the changed conditions should occur at the right time? To meet a change of conditions this year, the right large variation *might* possibly occur a thousand years hence.

The second argument is a still stronger one. Mr. Murphy fully adopts Mr. Herbert Spencer's view, that a variation, however slight, absolutely requires, to ensure its permanence, a number of concomitant variations, which can only be produced by the slow process of self-adaptation; and he uses this argument as conclusive against the formation of complex organs by natural selection in all cases where there is no tendency for action to produce self-adaptation; *à fortiori*, therefore, must a sudden large variation in any one part require numerous concomitant variations; it is still more improbable that they can accidentally occur together; it is impossible that the slow process of self-adaptation can produce them in time to be of any use; so that we are driven to the conclusion, that any large single variation, unsupported as it must be by the necessary concomitant variations, can hardly be other than hurtful to the individuals in which it occurs, and thus lead in a state of nature to its almost immediate extinction. The question, therefore, is not, as Mr. Murphy seems to think, whether such large variations occur in a state of nature, but whether, having occurred, they could possibly maintain themselves and increase. A calculation is made by which the more rapid mode of variation is shown to be necessary. It is supposed that the greyhound has been changed from its wolf-like ancestor in 500 years; but it is argued that variation is much slower under nature than under domestication, so that with wild animals it would take ten times as long for the same amount of variation to occur. It is also said that there is ten times less chance of favourable variations being preserved, owing to the free intermixture that takes place in a wild state; so that for nature to produce a greyhound from a wolf would have required 50,000 years. Sir W. Thomson calculates that life on the earth must be limited to some such period as one hundred million years, so that only two thousand times the time required to produce a well-marked specific change has, on this theory, produced all the change from the protozoon to the elephant and man.

Although many of the data used in the above calculation are quite incorrect, the result is probably not far from the truth; for it is curious that the most recent geological researches point to a somewhat similar period as that required to change the specific form of mammalia. The question of geological time is, however, so large and important that we must leave it for a separate article.

The second volume of Mr. Murphy's work is almost wholly psychological, and can be but briefly noticed. It consists to a great extent of a summary of the teachings of Bain, Mill, Spencer, and Carpenter, combined with much freshness of thought and often submitted to acute criticism. The special novelty in the work is the theory as to the "intelligence"

导致低地平原洪水泛滥，那么在成千上万种**可能的**大变异中，恰好适应变化之后的环境的变异能在恰当的时间发生吗？为了适应今年的一种环境变化，相应的大变异**也许**需要一千年才会发生。

第二个论据更充分一些。墨菲先生全盘接受了赫伯特·斯宾塞先生的观点，即，认为一种变异无论多么微不足道都一定需要许多伴生的变异来保证它的持久性，这些伴生的变异只能通过缓慢的自我适应过程来产生；他认为这一论据无可置疑，并借此反对在没有产生自我适应的行为趋势的所有情况中复杂器官都是通过自然选择形成的；因此，更不容置疑的是突然发生在任何部位的一种大变异肯定需要大量伴生变异；这些变异偶然同时出现是非常不可能的；自我适应的缓慢过程不可能及时产生具有任何作用的伴生变异；因此我们不可避免地得到如下结论：任何单一的大变异如果缺乏必需的伴生变异，那它们除了对发生这些变异的个体产生伤害之外几乎不会起别的作用，因此在自然状态下，它们几乎可以导致物种的迅速灭绝。所以问题并不如墨菲先生臆想的那样，也就是说，问题并不是这种大变异能否在自然状态下发生，而是已经发生的变异能否维持并加强。计算结果表明必然有更快的变异模式存在。人们猜测灰狗在 500 年内从它的类狼祖先变化而来；但是有争论说，变异在自然条件下比在驯养条件下发生得更加缓慢，因此野生动物发生同样程度的变异花费的时间将是驯养动物变异时间的 10 倍。也有人说，由于野生状态下会发生自由交配，保留有利变异的机会将少 10 倍；因此在自然界中狼演化为灰狗将需要 50,000 年时间。经计算，汤姆孙爵士认为地球上的生命肯定是在一亿年这样有限的时段内产生的，因此根据这一理论，只需要 2,000 倍于产生可以明确识别的特异性变化的时间，就可以形成从原生动物到大象和人所需的特异性变化。

尽管上述计算中使用的许多数据都相当不准确，但是结果可能距离真相并不遥远；因为最近的地质学研究指出了一个与哺乳动物发生特定形式的变化所需要的时间有点类似的时期，这点很令人好奇。然而，地质年代这个问题非常庞大也非常重要，所以我们不得不在另外一篇文章里单独进行介绍。

墨菲先生著作的第 2 卷几乎全是心理学方面的内容，这里只能简要地提一下。这一卷主要概述了贝恩、米尔、斯宾塞和卡彭特所倡导的学说，并结合了很多引发激烈批评的新鲜思想。这篇著作的特别新颖之处在于提出“智能”是通过组织结构和

manifested in organisation and mental phenomena, and this is so difficult a conception that it must be presented in the author's own words:—

> "I believe the unconscious intelligence that directs the formation of the bodily structures is the same intelligence that becomes conscious in the mind. The two are generally believed to be fundamentally distinct: conscious mental intelligence is believed to be human, and formative intelligence is believed to be Divine. This view, making the two to be totally unlike, leaves no room for the middle region of instinct; and hence the marvellous character with which instinct is generally invested. But if we admit that all the intelligence manifested in the organic creation is fundamentally the same, it will appear natural, and what might be expected, that there should be such a gradation as we actually find, from perfectly unconscious to perfectly conscious intelligence; the intermediate region being occupied by intelligent though unconscious motor actions—in a word, by instinct.... The intelligence which forms the lenses of the eye is the same intelligence which in the mind of man understands the theory of the lens; the intelligence that hollows out the bones and the wing-feathers of the bird, in order to combine lightness with strength, and places the feathery fringes where they are needed, is the same intelligence which in the mind of the engineer has devised the construction of iron pillars hollowed out like those bones and feathers.... It will probably be said that this identification of formative, instinctive, and mental intelligence is Pantheistic.... I am not a Pantheist: on the contrary, I believe in a Divine Power and Wisdom, infinitely transcending all manifestations of power and intelligence that are or can be known to us in our present state of being.... Energy or force is an effect of Divine power; but there is not a fresh exercise of Divine power whenever a stone falls or a fire burns. So with intelligence. All intelligence is a result of Divine Wisdom, but there is not a fresh determination of Divine thought needed for every new adaptation in organic structure, or for every new thought in the brain of man. Every Theist will admit that there is not a fresh act of creation when a new living individual is born. I go a little further, and say that I do not believe in a fresh act of creation for a new species. I believe that the Creator has not separately organised every structure, but has endowed vitalised matter with intelligence, under the guidance of which it organised itself; and I think there is no more Pantheism in this than in believing that the Creator does not separately cause every stone to fall and every fire to burn, but has endowed matter with energy, and has given energy the power of transposing itself."

I am not myself able to conceive this impersonal and unconscious intelligence coming in exactly when required to direct the forces of matter to special ends, and it is certainly quite incapable of demonstration. On the other hand, the theory that there are various grades of conscious and personal intelligences at work in nature, guiding the forces of matter and mind for their purposes as man guides them for his, is both easily conceivable and is not necessarily incapable of proof. If therefore there are in nature phenomena which, as Mr. Murphy believes, the laws of matter and of life will not suffice to explain, would it not be better to adopt the simpler and more conceivable solution, till further evidence can be obtained?

The only other portion of the work on which my space will allow me to touch, is the chapter on the Classification of the Sciences, in which a scheme is propounded of great simplicity and merit. Mr. Murphy does not appear to be acquainted with Mr. Herbert

精神现象彰显出来的这一理论。这种概念太难理解，因此这里必须奉上作者的原话：

“我相信指导身体结构形成的无意识的智能，与能在头脑中变成意识的智能是相同的。人们通常认为这两种智能具有本质的不同：有意识的精神智能只为人类所具有，而造型智能则是神所拥有的。这种观点，使得两种智能毫不相干，没有为中间领域的本能所赋予的奇异性留下任何余地。但是如果我们承认生物创造中反映的所有智能在本质上都是相同的话，我们就会自然而然地想到并作出预期，即现实中应存在一种智能的等级，如同我们也确实发现的一样，从完全无意识的智能到完全有意识的智能之间的这种等级；而这些中间区域由无意识但有智能的动机行为所占据——用一个词形容就是本能……形成眼睛晶状体的智能与人类的头脑理解透镜理论的智能是一样的；使鸟类的骨头和翅膀羽毛成为空心的以将轻巧和力量结合起来的智能以及在需要的地方长出羽毛状边缘的智能，与工程师的头脑里将建筑物的铁柱设计得像那些空心的骨头和羽毛的智能没有差别……有人可能会说，这种对造型的、本能的、精神的智能的鉴别是泛神论的……我不是一个泛神论者：相反，我相信神的支配力量和智慧，它们远远超越了我们现存生命状态中所知道的或可能知道的所有力量和智能的体现……能量或者力量是神圣支配力量的结果；但是无论何时，石头掉落或大火燃烧都不是神圣支配力量的一种冒失操练。智能也是这样。所有智能都是神的智慧的结果，但是并不存在一种神圣意志为需要它的生物结构的每种新的适应，或者人类头脑中的每种新思想给出冒失判决。每位有神论者都会承认当一个新生个体降生时，不会有创造的冒失行为，我将这点引申一下，就是说我不相信存在一种创造新物种的冒失行为。我相信创世主并没有分别组建每种结构，而是赋予有生命的物质以智能，使其在这一智能的指导下自我组建；我认为泛神论是不会相信创世主没有令每块石头分别掉落、没有令每场火分开燃烧，只是赋予物质以能量，并且给予能量变换自身的能力。”

我自己不能设想出这样的情况：非人的、无意识的智能在需要指导物质力量达到特定目标时恰好出现，而且这显然是无法论证的。另一方面，自然界中处于运作中的有意识的、个体的智能所具有的不同等级，按照它们的目的引导物质和精神的力量，就像人类为自己的目的来引导它们一样，上述理论既容易想到又必然能被证明。因此如果自然界存在如墨菲先生相信的那种物质和生命法则所不能解释的现象的话，在得到进一步的证据之前，采用简单的、更容易想到的解决方法不是更好么？

在该著作其余部分中我具有发言权的只有谈论科学的分类那一章，这部分提出了一种非常简单又有价值的方案。墨菲先生好像对赫伯特·斯宾塞先生关于这一主题的文章并不熟悉，而且值得注意的是，他得到了非常相似的结果，尽管这一结

Spencer's essay on this subject, and it is somewhat remarkable that he has arrived at so very similar a result, although less ideal and less exhaustively worked out. In one point his plan seems an improvement on all preceding ones. He arranges the sciences in two series, which we may term primary and secondary. A primary science is one which treats of a definite group of *natural laws*, and these are capable of being arranged (as Comte proposed) in a regular series, each one being more or less dependent on those which precede it, while it is altogether independent of those which follow it. A secondary science, on the other hand, is one which treats of a group of *natural phenomena*, and makes use of the primary sciences to explain those phenomena; and these can also be arranged in a series of decreasing generality and independence of those which follow them, although the series is less complete and symmetrical than in the case of the primary sciences. The two series somewhat condensed are:—

Primary Series	Secondary Series
1. Logic	1. Astronomy
2. Mathematics	2. Terrestrial Magnetism
3. Dynamics	3. Meteorology
4. Sound, Heat, Electricity, &c.	4. Geography
5. Chemistry	5. Geology
6. Physiology	6. Mineralogy
7. Psychology	7. Palaeontology
8. Sociology	8. Descriptive Biology

Taking the first in the list of secondary or compound sciences, Astronomy, we may define it as the application of the first five primary sciences to acquiring a knowledge of the heavenly bodies, and we can hardly say that any one of these sciences is more essential to it than any other. We are, perhaps, too apt to consider, as Comte did, that the application of the higher mathematics through the law of gravitation to the calculation of the planetary motions, is so much the essential feature of modern astronomy as to render every other part of it comparatively insignificant. It will be well, therefore, to consider for a moment what would be the position of the science at this day had the law of gravitation remained still undiscovered. Our vastly multiplied observations and delicate instruments would have enabled us to determine so many empirical laws of planetary motion and their secular variations, that the positions of all the planets and their satellites would have been calculable for a moderate period in advance, and with very considerable accuracy. All the great facts of size and distance in planetary and stellar astronomy, would be determined with great precision. All the knowledge derived from our modern telescopes, and from spectrum analysis, would be just as complete as it is now. Neptune, it is true, would not have been discovered except by chance; the nautical almanack would not be published four years in advance; longitude would not be determined by lunar distances, and we should not have that sense of mental power which we derive from the knowledge of Newton's grand law;—but all the marvels of the nebulae, of solar, lunar, and planetary structure, of the results of spectrum analysis, of the velocity of light, and of the vast

果不够完美也没有完全解决该问题，不过他的计划看上去对前人的各种研究成果有所改进。他将科学划分成两个系列，我们可以将它们称为初级的和次级的。初级科学是探讨一组确切的**自然法则**的科学领域，这些科学领域能够按一规则序列罗列下来（正如孔德提议的），每一领域都或多或少依赖于在它之前产生的那些领域，而总的来说，它又独立于随后产生的那些领域。另一方面，次级科学是探索一系列**自然现象**的科学领域，这些科学领域利用初级科学来解释这些现象；这些领域也可以通过一系列逐渐减弱的普遍性和随后产生的领域的独立性加以罗列，但是该系列与初级科学的系列相比，它们不够完整也不够系统。这两个压缩后的系列如下：

初级系列	次级系列
1. 逻辑学	1. 天文学
2. 数学	2. 地磁学
3. 动力学	3. 气象学
4. 声学、热学、电学等	4. 地理学
5. 化学	5. 地质学
6. 生理学	6. 矿物学
7. 心理学	7. 古生物学
8. 社会学	8. 描述生物学

以次级科学或者复合科学目录中的第一个（即天文学）为例来说明，我们可以将其定义为使用前 5 种初级科学来获取关于天体的知识的学科，我们几乎不能说这些科学中的哪一个比另外某个更重要。也许我们会有与孔德一样的倾向，认为将高等数学运用到地心引力法则来计算行星运动是现代天文学的基本特征，相比之下天文学的其他部分都无关紧要了。因此，不妨考虑一下如果地心引力法则尚未发现，那么今天这一科学会被置于何地？我们大大增加的观测资料和精密的仪器使我们能够确定许多行星运动及其长期变化的经验法则，从而可以提前一段时间精确地将所有行星及其卫星的位置计算出来。所有这些关于行星天文学和恒星天文学的大小和距离的事实都将被非常精确地确定，所有这些出自现代望远镜以及光谱分析的知识都会像现在一样完备。如若不然，那么除非靠运气，不然就不会发现海王星；航海天文年历也不会提前 4 年就出版；我们也不能通过月球距离确定出经度，我们也不会知晓由牛顿伟大定律的知识而获得的那种智能；但关于星云、太阳、月亮和行星结构，光谱分析结果，光速以及行星和恒星空间的广阔维度的所有奇迹都会像现在一样全部为我们所熟知，这些知识将形成一门天文学，而这门科学在尊严、庄重和强烈的趣味性方面都不会逊色于我们现在拥有的天文学。

dimensions of planetary and stellar spaces, would be as completely known to us as they now are, and would form a science of astronomy hardly inferior in dignity, grandeur, and intense interest, to that which we now possess.

Mr. Murphy guards us against supposing that the series of sciences he has sketched out includes all that is capable of being known by man. He professes to have kept himself in this work to what may be called positive science, but he believes equally in metaphysics and in theology, and proposes to treat of their relation to positive science in a separate work, which from the author's great originality and thoughtfulness will no doubt be well worthy of perusal.

(**1**, 132-133; 1869)

墨菲先生反对我们把他概括出来的科学系列假定为囊括所有能够为人类所知的领域。他声称已经全身心投入到这一可以被称为实证科学的工作之中了，但是他同样相信形而上学和神学，准备在另一篇著作中单独讨论它们与实证科学的关系，从作者的伟大原创性和思虑的慎重性方面考虑，这本著作无疑具备精读的价值。

（刘皓芳 翻译；陈平富 审稿）

The Isthmian Way to India

Editor's Note

The Suez Canal has been opened, *Nature* here reports, and a flotilla of boats of all kinds is now streaming through it from the Mediterranean to the Gulf of Arabia. The distance by water to India has now been reduced to a mere 8,000 miles, just over half of the previous 15,000 on the old route around the Cape of Good Hope. But one must wonder about the future of the canal, the author notes, given the weight of sediment delivered at its mouth by currents drifting eastward from the outflow of the Nile River. Of course, it may be cleared with constant dredging; but would the canal's French operators be so diligent once the excitement has died away?

THE Canal has been opened. The flotilla, with its noble, royal, imperial, and scientific freight, has progressed along the new-made way from sea to sea. From Port Saïd, that new town between the sea and the wilderness, with its ten thousand inhabitants, and acres of workshops and building-yards, and busy steam-engines, the naval train floated through sandy wastes, across lakes of sludge and lakes of water filled from the Salt Sea; past levels where a few palm-trees adorn the scorched landscape; past hill-slopes on which the tamarisk waves its thready arms; past swamps where flocks of flamingoes, pelicans, and spoonbills, disturbed by the unwonted spectacle, sent up discordant cries; through deep excavations of hard sand or rock; across the low flat of the Suez lagoons, where Biblical topographers have searched for the track of the children of Israel; and so to the "red" waters of the great Gulf of Arabia. The flotilla has done its work: the Canal has been opened; and the distance by water to India is now 8,000 miles, instead of the 15,000 miles by the old route round the Cape of Good Hope.

It is a great achievement. So great, that we need not wonder that the capital of 8,000,000 *l.* sterling with which it was commenced in 1859 was all expended, and as much more required, before the work was half accomplished. And perhaps we ought not to be too much overcome with pity for the 20,000 unlucky Egyptians—natives of the house of bondage—pressed every month up to the year 1863 by their paternal Government to labour, wherever required, along the line of excavations. How persistent are Oriental customs! Here we have in modern days—the days of power-looms, of steam printing-presses, and under-sea telegraphs—a touch of the old tyranny, the taskmasters and the groanings, associated in our memories with the very earliest of Egyptian history.

The length of the Canal is one hundred miles, and the depth, as the French engineers inform us, is to be everywhere twenty-eight feet, so as to admit of the passage of large vessels. It must not be supposed that an excavation of the depth above mentioned has been dug all across the Isthmus, for the level of the country is, for the most part, below that of the Mediterranean; consequently, miles of banks have been thrown up across the

通向印度的运河之路

编者按

根据《自然》上这篇文章的报道，苏伊士运河已经通航了，而且有一支由多种类型的船只组成的舰队正沿着运河行驶在从地中海到阿拉伯海湾的航道上。苏伊士运河开通后，通往印度的水路距离减少到只有 8,000 英里，这与以前绕道好望角的 15,000 英里的水路距离相比，只是一半多一点。在这篇文章中作者提出，考虑到尼罗河向东流时携带到河口处的沉积物的堆积量，我们不得不担心运河的未来。当然，这些泥沙可以通过持续的疏浚来清除，但是，在令人激动的通航纪念过去之后，经营运河的法国人会勤快地疏浚运河么？

运河通航了！满载着来自贵族、王室、帝国和科学界的货物的小型船队在这条连接地中海与红海的新运河上开始航行了。船队从塞得港出发，这是位于地中海和荒漠地区之间的一座新建城镇，它拥有上万居民、大片工场和建筑工地以及繁忙的蒸汽机。一路上，船队经过了沙荒地、淤泥湖和盐水湖，经过了有零星的棕榈树点缀其间的焦灼平原，经过了有柽柳树在摇动细枝的山坡，还经过了聚集着火烈鸟、鹈鹕和蓖鹭的沼泽（船队经过时，鸟儿被这罕见的奇观惊得发出了不安的叫声）。之后，船队又通过了在坚硬的沙或岩石中开凿出来的隧道，穿过了苏伊士礁湖浅滩（这里正是圣经地形学家寻觅希伯来人踪迹的地方）。就这样，船队来到了阿拉伯湾这片“红色”的水域。船队也完成了它的任务：运河已经可以通航了；经水路到印度的距离由原来绕经好望角的 15,000 英里缩短到了 8,000 英里。

这是一项伟大的成就，伟大到工程在进行不到一半时就花光了 1859 年起投入使用的 800 万英镑，而且还需要更多的资金，但我们无需为此感到惊讶。或许我们也不应过于怜悯那 20,000 名不幸的埃及本地劳工，他们在 1863 年之前的每个月里，只要工程需要，就要在当地政府的强迫下开凿运河。东方大众是多么坚持不懈啊！在拥有电力纺织机、蒸汽印刷机和海底电缆等现代文明成果的今天，我们在这片土地上看到的却是旧时的暴政统治、严厉的监工和痛苦的呻吟，这使我们想起了古老的埃及历史。

运河长 100 英里。法国的工程师告诉我们，全程的深度都达到了 28 英尺，这样就可以使大型船舶通过。但并不是横跨苏伊士地峡的所有地方都被挖得这么深，因为这里大部分地区都低于地中海。因此，在地势低洼的广阔地域上数英里的浅滩已经被挖开以形成水道。仔细研究一下从塞得港到苏伊士全程中的一段——从海到

lowest tracts to form a channel for the water. In looking at a section of the whole route from Saïd to Suez—seventy-five miles in a direct line from sea to sea—the great extent of depression is well seen. In Lake Timsah it is about eighteen feet; in the Bitter Lakes, which stretch to a length of twenty-five miles, it is in places twenty-six feet. On the other hand, the elevations, though comparatively few, are somewhat formidable of aspect, particularly at El Guier and at Chalouf. The more this section is studied, the more forcible becomes the impression on the mind that a strait thickly studded with islands, as Behring's Strait, once separated Asia and Africa, and that by the drift from the Nile and the desert the sea has been filled up around the islands, with the exception of the lake depressions, until the present Isthmus was formed. Hence the difference of soil. The islands rising boldly up: El Guier, ten miles long, layers of sand and hard clay; Serapeum, three miles long, a kind of shelly limestone; and Chalouf, six miles long, composed of hard clay, sandstone rock, and conglomerate, the severest part of the excavation. Geologists have remarked upon the fact that the fossils found in the Chalouf ridge are identical with those of the London basin and the hill of Montmartre, whereby we learn that parts of Egypt, France, and England are of the same age.

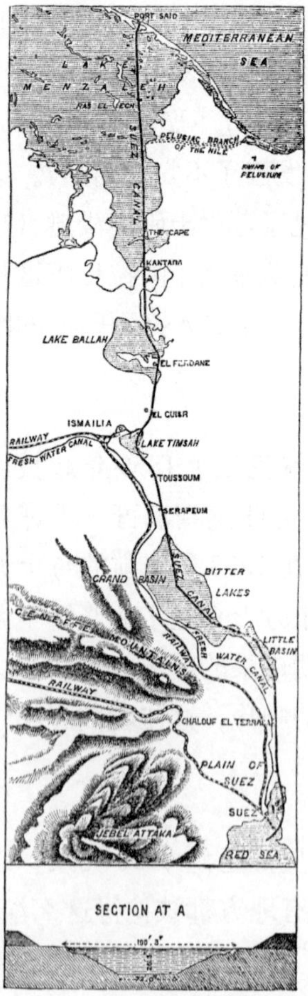

Map and Section of the Suez Canal

The mountains of Abyssinia are every year diminished in size and height by the enormous periodical rains which wash down millions of cubic feet of mud and clay into the Nile. Vast clouds of sand are blown into the great river in its long course through the deserts; and these transported matters, caught by the strong current setting in from the Straits of Gibraltar, have been drifted to the eastward during immemorial ages, with consequences which are well known to those who have studied the geography and geology of the Isthmus. Such a transformation will be recognised as one of the ordinary operations of nature, when we remember that in 4,150 years the valley of the Nile has been raised eleven feet by deposits from the periodical floods, and that the land of Egypt is supposed to have been at one time a gulf stretching from the Mediterranean towards the Mountains of the Moon, but which became silted up by slow accumulations.

We may now form a clear notion of the region through which the Canal has been cut. A low, sandy shore is generally washed by a shallow sea. At Southend the pier extends for a mile and a quarter into the sea before meeting depth enough for an ordinary steamer; and the long piers at Lowestoft and other places on our eastern coast present themselves as illustrations in point. So shallow is the sea off Port Saïd, that the mouth of the harbour had to be commenced two miles from the shore, for there only did

海的直线距离 75 英里——可以看到一片广阔洼地。提姆萨湖区深约 18 英尺，而跨度达 25 英里的比特湖区的某些地方深度甚至达到 26 英尺。不过，高地虽然相对较少但都有令人敬畏的一面，尤其是在厄吉欧地区和查洛夫地区。对这一区域研究得越多，我们就会更加强烈地认识到，如同白令海峡一样，苏伊士海峡岛屿密布，曾经将亚洲与非洲分开。除了湖区低洼处，这些小岛周围的海域都被来自尼罗河和沙漠的漂流物填充，最终形成了现在的苏伊士地峡。这也使得地峡中具有多种不同的土壤。这些岛屿大幅抬升：厄吉欧地区长 10 英里，具有沙质地层和硬质黏土地层；塞拉比尤姆地区长 3 英里，具有一种贝壳灰岩的地层；查洛夫地区长 6 英里，由硬质黏土、砂岩和砾岩组成，这是运河开凿中最艰难的部分。地质学家认为，在查洛夫山脊发现的化石与在伦敦盆地和蒙马特山发现的化石完全一样。据此我们可以认定，埃及、法国和英国的部分地区处于同一地质年代。

由于大量季节性降雨将数百万立方英尺的泥浆和泥土冲入到尼罗河中，阿比西尼亚山脉的山体和山高每年都在减少。在尼罗河流经沙漠的途中大量的沙尘落入河里。远古时代，这些泥沙被直布罗陀海峡强劲的海潮截获，向东漂移。只要研究过苏伊士地峡的地理和地质的人，都会知道这些造成了什么样的后果。不过，当我们回忆起尼罗河流域在过去的 4,150 年间由于周期性的洪水泛滥而使河床抬高了 11 英尺，以及埃及大陆被认为曾经是从地中海向月亮山系延伸的一个海湾，经过常年累积现在已被淤泥充塞时，只能把上述的泥沙转移看成是多种普通的自然作用中的一种而已。

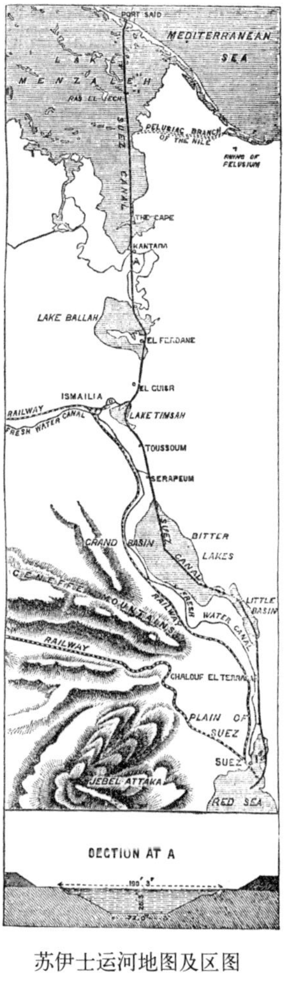

苏伊士运河地图及区图

现在，我们对截断运河的这个地区应该有了清楚的认识。这是一个被浅海冲刷的沙质浅滩。在运河南端，为了使水深足够普通轮船的停靠，码头向海中延伸了 1.25 英里。在运河东岸，洛斯托夫特及其他地方的长码头本身也都恰恰说明了这一点。而在塞得港，由于水太浅，港口只好设在距离海岸 2 英里处，那里的水深刚刚达到 26 英尺，如果水深比这还小，大型轮船就不能停靠。另外，

the required depth of twenty-six feet of water begin. Less than this will keep out vessels of the largest class. The western pier, the one against which the whole weight of the powerful current falls, projects more than two miles into the sea; the one to the east is half a mile shorter. These breakwaters have been built up of concrete blocks weighing twenty tons each, made on the spot from the sand dredged out of the harbour mixed with hydraulic lime brought from Marseilles. Spaces were left between the blocks to be filled up by the seadrift; but though there have been great deposits of sand and mud outside the breakwater, the filling up of the gaps has not been so speedy as was anticipated, and heaps of sand which drifted through have had to be dredged out again. Of course, while money for payment is forthcoming, any number of dredging-machines may be employed; but can that process be depended on when enthusiasm shall have evaporated, and there is nothing but the prosaic work of letting ships in or out to animate the promoters? Will it always be possible to prevent the formation of such soft banks as that on which the "Prince Consort" and the "Royal Oak" grounded on their arrival to take part in the opening of the Canal?

There is something instructive in the operations which have so diligently been carried on at the mouth of the Tyne, where a passage through the bar is essential. To maintain this passage, eighteen feet deep only, more than four million tons of sand must be dredged out every year. This has been going on for ten years or more, and the channel is not yet secure.

Not only the harbour of Port Saïd, but the greater part of the Canal itself, has been formed by dredging; and this, in soft ground or through the sludge of Lake Menzaleh, was comparatively easy work. The mud raised from the bottom was spread along each margin of the newly-scooped-out channel; but it would not stay there, and for a time the prospect of maintaining an open channel seemed as hopeless as George Stephenson's first attempt to carry the Liverpool and Manchester railway across Chat Moss. No sooner was the Menzaleh mud deposited in its new position, than it either slipped back into its former bed, or squeezed the soft soil on which it lay into the channel. The dredgers were in despair over a task in which no progress could be made, until one day one of the labourers showed that if, instead of great heaps, a thin layer of the mud were spread and left to harden in the sun, it would not slip back. So layer by layer the mud was spread, the banks were built up, and a way for the Canal was opened through such slime as was used in ancient days for the making of bricks.

In the hard ground the "bondagers" dug with pick and spade, and carried away the loosened soil in baskets. But when they were supplemented by European labourers, powerful excavating machinery was employed, and the line of works presented as busy a spectacle as an English railway in course of construction, or the main drainage works in their progress towards Barking Creek across the Essex marshes. The slopes of the cuttings were alive with labourers and machines, by which the excavated earth was lifted and run off to a distance. The power of the digging-machines may be judged of from the fact that some of them could dig out 80,000 cubic metres of soil every month, and that on one occasion the quantity was 120,000. A dozen or two of machines working at this rate would soon make a big gap through the high grounds before them.

西部码头由于要承受洋流的强力冲击，因此向海中延伸的距离超过了 2 英里，而在它东边的一个码头向海中延伸的距离则比它短 0.5 英里。它们附近的防浪堤是用很多 20 吨重的混凝土块构筑成的，混凝土块是在现场用港口挖出来的沙子与从马赛运来的水硬石灰混合制成。混凝土块之间留出了缝隙以供海水中的漂浮物填充。尽管防浪堤外沉积了很多泥沙，但是那些缝隙并没有像预期的那样很快被填满，因此不得不对沉积的沙堆再次进行清理。当然，为此要支付的资金很快可以到位，许多挖掘机也都可以租来，但是在大家已经丧失工作热情，除了指挥船舶进进出出的乏味工作外就无所事事的当下，还能期望清理工作能进行下去吗？另外，是否能够防止防浪堤变得像以前“亲王号”和“皇家橡树号”驶来参加运河开航仪式时停靠过的堤坝那样松垮也是个问题。

泰恩河河口是一处航道要冲，在这里常年进行的一些工作对于苏伊士运河的维护是有意义的。为了确保泰恩河正常通航，即便只是保证航道有 18 英尺深，每年也必须挖出超过 400 万吨的泥沙。这样的挖掘工作已经持续进行了十多年，但是河道依然不够安全。

不仅仅是建设塞得港需要挖掘泥沙，实际上运河大部分河道都要通过挖掘泥沙来开通。在土质松软的地区或是在门扎勒湖的淤泥区，挖掘工作相对比较容易。从底部挖出的泥浆被堆放在新挖开河道的两边，但它不会总停留在那里，因此，使运河保持畅通的期望曾经一度就像当年乔治 · 斯蒂芬森最初试图跨越查特沼泽建造从利物浦到曼彻斯特的铁路一样完全落空了。这些门扎勒湖的泥浆刚刚被堆放到一个新的地方后，马上又会滑回原来的河床，或者将附近的软泥挤入运河。挖泥沙的工人们因为工作没有进展而近乎绝望，直到后来有一天一位工人提出，如果把挖出的软泥摊成薄层而不是堆成堆，并且在太阳下硬化，它就不会滑回去了。就这样，泥浆被一层层地摊开，河岸终于建成，夹在这种古代用来造砖的软泥中的运河航道终于开通了。

在土质坚硬的地区，“奴隶工”们用镐和铲挖掘，再将挖出的土用筐搬走。但当欧洲来的工人们补充进来时，大马力的开凿机就开始投入使用了。施工现场一片繁忙，就像是在英国修建铁路或者横跨埃塞克斯沼泽向巴金克里克方向推进主排水工程的工作场景。开凿运河的斜坡上到处是工人和机器，通过这些大型机器，挖出来的泥土被吊起来转运到了远处。挖掘机的功率非常强大，有的挖掘机每个月可以挖 80,000 立方米的泥土，有时甚至可以挖到 120,000 立方米。如果都以这种速度工作，那么一二十个挖掘机很快就可以在高地中挖出一条大沟。

The lakes of water on the Isthmus may be regarded as Nature's contribution towards the success of the Canal; for in them the only labour required is to dredge a channel which will give a depth of twenty-eight feet. Moreover, they may be used as ports. This is especially the case with Lake Timsah, on the shore of which stands the newly-built town of Ismaïlia, the half-way stopping place for travellers on the Canal. Here anchored the flotilla during the progress of the opening, and the dark-skinned children of the Prophet were seen mingled with throngs of fair-complexioned Giaours in friendly rejoicings.

Ismaïlia is an important place, for it is the pumping-station of the fresh-water canal which was first made in order to supply the thousands of labourers with drink, and water for their works. On this pumping-station all the country between Lake Timsah and Port Saïd depends for it supplies of the precious element.

The hollow of the Bitter Lakes, six miles wide in the widest part, is believed to have been at one time connected with the Red Sea. The level of the water in these lakes has been brought up to that of the sea by a re-opening of the connection. In March of the present year, all preparations being complete, the water was admitted, and a great stream, pouring in from the Mediterranean and from the Red Sea, gradually rose upon the arid saline slopes of the deep and desolate basin. For some weeks the flow went on, until, as was estimated, two thousand million cubic metres of water had flowed in, and the level was established. The area of the lakes will be largely increased by this contribution from the two seas; and it will be interesting to watch whether in connection with the two canals—the salt-water and fresh-water—any modification of the climate of the Isthmus may be produced. Much has been said, too, about the loss that will take place by evaporation under the sun of Egypt: the amount is so great as to be almost incredible. This loss will have to be provided for; as also the effect of blowing sands, which will accelerate the tendency of the bottom to grow towards the surface, always observable in canals.

Up to the last moment predictions from various quarters have been heard that no big ships would ever effect the passage of the canal. But while we write these lines, telegrams from the East inform us that *L'Aigle*, the French yacht, with her Majesty the Empress on board, had got through, and were anchored in the Red Sea. From the same source we learn that the Peninsular and Oriental steamer *Delta*, drawing 15.5 feet of water, had arrived at Ismaïlia from Port Saïd, but had touched ground a few times on the way. The Egyptian vessel *Lattif* attempted the passage, but for want of sufficient depth had to return; difficulties occurred with other vessels, and the banks of the Canal were much damaged.

But the Khédive has invested M. de Lesseps with the Grand Cross of the Order of the Osmanli, and the Emperor Napoleon has appointed him to the rank of Grand Cross of the Legion of Honour. We may therefore hope for the best in all that appertains to the Suez Canal, and that foreigners will believe that Englishmen are too ready to admire good work to feel jealous of the energetic hearts by whom it has been accomplished.

(**1**, 110-112; 1869)

分布在苏伊士地峡中的多处湖泊真可以看作是大自然对运河的捐赐，因为在这些湖区唯一要做的就是挖掘水道使之达到 28 英尺深，而且这些湖泊还可以被辟为港口。提姆萨湖就是个典型的例子，新建的伊斯梅利亚镇就矗立在其岸边，这里成了运河上的旅行者中途停歇的地方。运河开通时小型船队就停泊在此，先知穆罕默德的深肤色子民和非伊斯兰教的白种人友好地聚在一起欢庆通航。

伊斯梅利亚镇是一个重要的地方，因为当初为了给数千劳工供应饮用水和工作用水而修建的淡水输运管道的泵站就在这里。提姆萨湖和塞得港之间的所有村镇都要靠这个泵站才能获得宝贵的淡水。

比特湖区洼地最宽处达 6 英里，被认为曾经一度与红海相连。运河的开通使其水位提升到了红海的水位。今年 3 月一切准备就绪后，海水便被引入到洼地之中，来自地中海和红海的海水大量涌入，原本纵深、荒凉的盆地之中的干旱贫瘠的盐渍斜坡就逐渐被海水淹没了。海水的涌入持续了几个星期，直到估计入水总量达到 20 亿立方米后水位才稳定下来。由于有大量的海水从两个大洋涌入，湖区的面积将大幅增加。在这个咸水渠和淡水渠的连接处，苏伊士的气候是否会发生变化成了一个有趣的话题。另一个人们谈论很多的话题是，埃及境内强烈的日照会使水大量蒸发，其总量大得令人难以置信。对此一定要做好准备，同样，加速运河河床上升趋势的扬沙在运河沿线也会随处可见。

最近听到来自各种渠道的预报说运河已经不能使大船通过了。但就在我们写这篇文章的时候，来自东方的电报报道说，法国女皇乘坐着的**拉艾格尔号**游轮已经通过运河并停靠在红海。我们还从那里了解到，吃水深度达 15.5 英尺的半岛东方**戴尔塔号**已从塞得港驶到伊斯梅利亚，不过在途中船底碰到河床好几次。埃及的**拉提夫号**试图通过运河，但因水深不够只得返航。其他一些大型船舰也遇到了同样的困难，运河的堤坝也遭受了严重的损害。

不过，赫迪夫已经向德雷塞布授予了土耳其大十字勋章，皇帝拿破仑也向他颁发了法国荣誉军团勋章大十字勋位。因此，我们可以对所有与苏伊士运河有关的事情报以最乐观的期望。相信英国人也乐于赞美这项伟大的工程，而不会嫉妒达成这一壮举的精力充沛的法国人。

（孙惠南 翻译；郭华东 审稿）

On the Dinosauria of the Trias, with Observations on the Classification of the Dinosauria

Editor's Note

As a news magazine, *Nature* reported on the activities of various scientific societies. This report of a meeting of the Geological Society held on 24 November 1869 describes the reading of a paper by the President of the Society, Thomas Henry Huxley, entitled "On the Dinosaurs of the Trias, with Observations on the Classification of the Dinosauria". The anonymous third-person account notes the discussion, in which Harry Seeley (an expert on Pterosaurs) and others disputed various points with the President. Huxley's paper was followed by one on biogeography as elucidated by fossil corals. This and many other detailed accounts give a flavour of the daily scientific life of London: from today's perspective, they read like the business of titans.

GEOLOGICAL Society, November 24.—"On the Dinosauria of the Trias, with observations on the Classification of the Dinosauria", by Prof. T. H. Huxley, LL.D., F.R.S., President. The author commenced by referring to the bibliographical history of the Dinosauria, which were first recognised as a distinct group by Hermann von Meyer in 1830. He then indicated the general characters of the group, which he proposed to divide into three families, viz.: —

I. The Megalosauridae, with the genera *Teratosaurus, Palaeosaurus, Megalosaurus, Poikilopleuron, Laelaps,* and probably *Euskelosaurus;*

II. The Scelidosauridae, with the genera *Thecodontosaurus, Hylaeosaurus, Pholacanthus,* and *Acanthopholis;* and

III. The Iguanodontidae, with the genera *Cetiosaurus, Iguanodon, Hypsilophodon, Hadrosaurus,* and probably *Stenopelys.*

Compsognathus was said to have many points of affinity with the Dinosauria, especially in the ornithic character of its hind limbs, but at the same time to differ from them in several important particulars. Hence the author proposed to regard *Compsognathus* as the representative of a group (*Compsognatha*) equivalent to the true Dinosauria, and forming, with them, an order to which he gave the name of Ornithoscelida. The author then treated of the relations of the Ornithoscelida to other Reptiles. He indicated certain peculiarities in the structure of the vertebrae which serve to characterise four great groups of Reptiles, and showed that his Ornithoscelida belong to a group in which, as in existing Crocodiles, the thoracic vertebrae have distinct capitular and tubercular processes springing from the arch of the vertebra. This group was said to include also the Crocodilia, the Anomodontia, and the Pterosauria, to the second of which the author was inclined to approximate the Ornithoscelida. As a near ally of these reptiles, the author cited the Permian *Parasaurus*, the structure of which he discussed, and stated that it seemed to be a terrestrial reptile, leading back to some older and less specialised reptilian form.

关于三叠纪恐龙以及恐龙分类的研究

编者按

作为新闻杂志，《自然》一直在报道各个科学学会的活动。这篇关于 1869 年 11 月 24 日地质学会举行的一次会议的通讯，记述了学会主席托马斯·亨利·赫胥黎在会上所作的报告，报告的题目是《关于三叠纪恐龙以及恐龙分类的研究》。匿名的第三方提到了哈里·西利（翼龙方面的专家）等人与赫胥黎主席在许多观点上的争论。在赫胥黎的这篇文章之后，是一篇生物地理学方面的阐述珊瑚化石的文章。从这篇文章以及其他许多详细的报道，可以看到当时伦敦日常学术活动的情况。以现在的眼光来看，他们作学术报告就像是现在的商界巨亨们发表演说一样。

11 月 24 日，在地质学会的会议上宣读了《关于三叠纪恐龙以及恐龙分类的研究》，报告人是法学博士、英国皇家学会会员、地质学会主席赫胥黎教授。作者在文章开篇首先介绍了有关恐龙的文献历史，然后又指出了这一类群的基本特征。1830 年，赫尔曼·冯·迈尔首次将恐龙划定为一个独立的类群。作者提议将恐龙这个类群分为 3 个科，即：

I. 巨齿龙科，包括怪晰龙属、远古龙属、巨齿龙属、杂肋龙属、暴风龙属，此外很可能还包括优肢龙属。

II. 腿龙科，包括槽齿龙属、林龙属、多刺甲龙属和棘甲龙属。

III. 禽龙科，包括鲸龙属、禽龙属、棱齿龙属、鸭嘴龙属和狭盘龙属。

有人曾指出，有很多特征可以说明美颌龙与恐龙具有较近的亲缘关系，特别是它的后肢具有鸟类的特征，但同时，二者在一些重要细节上又存在差异。因此作者提出将以美颌龙为代表的一个类（美颌龙类）与真正的恐龙类视为同一等级，并共同构成一个目，命名为鸟臀目。之后作者论述了鸟臀目与其他爬行动物的关系。他列举了爬行动物 4 个主要类群的脊椎结构特征，并证明鸟臀目属于其中的一类。这一类群的胸椎有一个独特的从椎骨弓形区发出的小头突，与现存的鳄鱼相似。这一类群还包括鳄目，缺齿亚目和翼龙目，而作者认为缺齿亚目与鸟臀目亲缘关系更近。由于这些爬行动物关系较近，作者引述了二叠纪的鸡冠龙属，讨论了它的结构特征，认为它属于陆生爬行动物，是较古老的、非特化的爬行动物类群。关于鸟臀目与鸟类的关系问题，作者说，他所知的所有特征都表明，鸟类在结构上和爬行动物并没有区别，而爬行动物的结构在鸟臀目中已有先兆。他还简要讨论了翼手龙和鸟类

With regard to the relation of the Ornithoscelida to birds, the author stated that he knew of no character by which the structure of birds as a class differs from that of reptiles which is not foreshadowed in the Ornithoscelida, and he briefly discussed the question of the relationship of Pterodactyles to birds. He did not consider that the majority of the Dinosauria stood so habitually upon their hind feet as to account for the resemblance of their hind limbs to those of birds, by simple similarity of function. The author then proceeded to notice the Dinosauria of the Trias, commencing with an historical account of our knowledge of the occurrence of such reptilian forms in beds of that age. He identified the following Triassic reptilian-forms as belonging to the Dinosauria:—*Teratosaurus*, *Plateosaurus*, and *Zanclodon* from the German trias; *Thecodontosaurus* and *Palaeosaurus* from the Bristol conglomerate (the second of these genera he restricted to *P. cylindrodon* of Riley and Stutchbury, their *P. platyodon* being referred to *Thecodontosaurus*); *Cladyodon* from Warwickshire; *Deuterosaurus* from the Ural; *Ankistrodon* from Central India; *Clepsysaurus* and *Bathygnathus* from North America; and probably the South African *Pristerosaurus*.—Sir Roderick Murchison, who had taken the chair, inquired as to the lowest formation in which the bird-like character of Dinosaurians was apparent, and was informed that it was to be recognised as low as the Trias, if not lower.—Mr. Seeley insisted on the necessity for defining the common plan both of the Reptilia and of the ordinal groups before they could be treated of in classification. He had come to somewhat different conclusions as to the grouping and classification of Saurians from those adopted by the President. This would be evident, in so far as concerned Pterodactyles, from a work on Ornithosauria which he had just completed, and which would be published in a few days.—Mr. Etheridge stated that the dolomitic conglomerate, in which the Thecodont remains occurred near Bristol, was distinctly at the base of the Keuper of the Bristol area, being beneath the sandstones and marls which underlie the Rhaetic series. There were no Permian beds in the area. He regarded the conglomerates as probably equivalent to the Muschelkalk. It was only at one point near Clifton that the Thecodont remains had been found.—Prof. Huxley was pleased to find that there was such a diversity of opinion between Mr. Seeley and himself, as it was by discussion of opposite views that the truth was to be attained. He accepted Mr. Etheridge's statement as to the age of the Bristol beds.

(**1**, 146-147; 1869)

的关系。作者认为，虽然大部分恐龙类也习惯后脚站立，但不能仅凭这种功能上的简单相似就认为恐龙类的后肢和鸟类的后肢相似。之后，作者开始关注三叠纪恐龙，首先阐述了那个时期出现的爬行类相关的历史发展知识。他鉴别出以下几个属于恐龙目的三叠纪爬行类：德国的三叠纪巨齿龙属、远古龙属和镰齿龙属；英国布里斯托尔砾岩中的槽齿龙属和远古龙属（其中远古龙属的代表是由赖利和斯塔奇伯里命名的柱齿龙，而槽齿龙属的代表则是板齿龙）；英格兰沃里克郡的独巨齿龙属；乌拉尔地区的亚次龙属；印度中部的钩齿龙属；北美的伏龙属和深颚龙属；还可能包括来自南非的原始龙属。曾任学会主席的罗德里克·麦奇生爵士就恐龙类最早是在哪一个地层出现了与鸟类相似的特征提出问题，他得到了如下的答案：如果在更早期的地层中没有发现的话，那就是出现在三叠纪。西利先生坚持认为，无论是对于爬行类，还是对于有序排列的各个类群，在进行分类之前都必须先明确基本的排序方法。对于蜥臀类的分组和分类，他得到了一些与学会主席所持观点不太相同的结论。根据他的一部关于鸟蜥亚纲的著作，就翼手龙而言这一点是非常明显的。西利先生的这部著作刚刚完成，不久后将会发表。埃瑟里奇先生则提到，在布里斯托尔附近，发现过槽齿类遗骸的白云石砾岩地层明显是处在考依波地层的底部，其上方是瑞替期的砂岩和泥灰岩。这个地区没有二叠纪的地层。他认为白云石砾岩地层和壳灰岩地层很可能是等同的。在克利夫顿附近，只在一处发现有槽齿类的遗骸。赫胥黎教授表示，很高兴听到西利先生提出不同的观点，因为真理是从不同意见的碰撞和冲突中产生的。对于埃瑟里奇先生关于布里斯托尔地区地层年代的论述，他则表示认可。

（刘冉冉 翻译；徐星 审稿）

Spectroscopic Observations of the Sun

Editor's Note

This paper reports some of the first detailed observations of solar flares, then called "solar protuberances". These had been discovered just ten years before, and were associated with sunspots. American astronomer C. A. Young describes an immense cloudy mass on the Sun's edge, which he estimates to be 22,500 miles high (three times the Earth's diameter) and 1,350,000 miles wide. His spectroscopic observations of the feature show it to be predominantly hydrogen and provide information about velocities of this material. The Doppler shift of the hydrogen emission line implies that the material was moving at roughly 55 miles per second. The feature persisted for five minutes before dissolving into several lumps.

PROFESSOR C. A. Young, of Dartmouth, U.S., has communicated to the October number of the *Journal of the Franklin Institute* the following important observations of solar protuberances, which entirely endorse the work done by Mr. Lockyer in this country. We are enabled to place them thus early before our readers by the kindness of Professor Morton.

September 4th, 1869.—Prominences were noted on the sun's limb at 3 p.m. today in the following positions, angles reckoned from North point to the East:—

1. +70° to +100°, very straggling, not very bright.

2. −10°, large and diffuse.

3. −90°, small, but pretty bright.

September 13th, 1869.—The following protuberances were noted today.

1. Between +80° and +110°, a long straggling range of protuberances, whose form was as in Fig. 1. I dare not profess any very extreme accuracy in the drawings, not being a practised draughtsman, but the sketch gives a very fair idea of the number, form, and arrangement of the immense cloudy mass, whose height was about 50″ and its length 330″ (22,500 miles by 1,350,000). The points *a* and *b* were very bright.

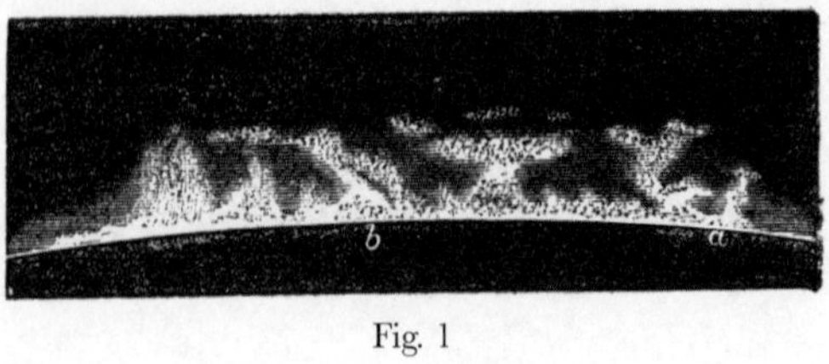

Fig. 1

观测太阳光谱

编者按

这篇文章报道了一些关于太阳耀斑的最早的详细研究。这种当时还被称为“太阳突出物”的现象是10年前才发现的，它与太阳黑子相关联。美国天文学家扬在本文中描述了在太阳边缘出现的巨大云块状物质，他估计这些物质高达22,500英里（地球直径的3倍），宽达1,350,000英里。他对这一现象进行的光谱学研究表明，该物质中的主要成分应该是氢，他还给出了这种物质运动速度的信息。氢发射谱线的多普勒频移意味着，这种物质移动的速度大约是每秒55英里。这种特征持续存在5分钟后，这个巨大云块状物质逐渐分解成了几个团块。

来自美国达特茅斯大学的扬教授已经在10月份的《富兰克林研究院院刊》上发表了关于太阳突出物的一些重要观测结果，这些重要发现是对本国的洛克耶先生所做工作的充分肯定。多亏莫顿教授的热心相助，我们才可以早早地把这些工作呈现给读者。

1869年9月4日，当天下午3点，我们在太阳边缘的下列位置发现了日珥，角度由北向东，分别是：

1. +70°～+100°，非常发散并且不太明亮。

2. –10°，大且弥散。

3. –90°，小但是相当明亮。

1869年9月13日，当天发现了下面的日珥：

1. +80°～+110°，有一长条发散的日珥，形状如图1所示。由于不是经验丰富的绘图员，我不敢说图中的描绘绝对精确，但从草图中完全可以分辨出巨大云块的数量、形状和排列。这个不透明的巨大云块高约50角秒，长330角秒（22,500英里 × 1,350,000英里），并且 *a* 点和 *b* 点非常明亮。

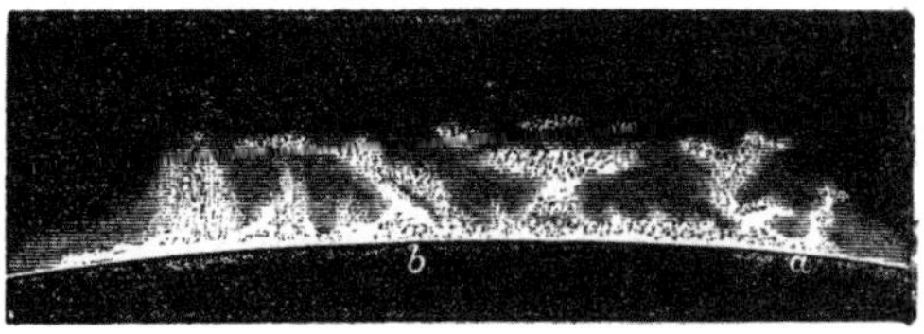

图1

2. +135° small, but very bright at the base, of this form (Fig. 2).

Fig. 2

3. −85° of this form (Fig. 3).

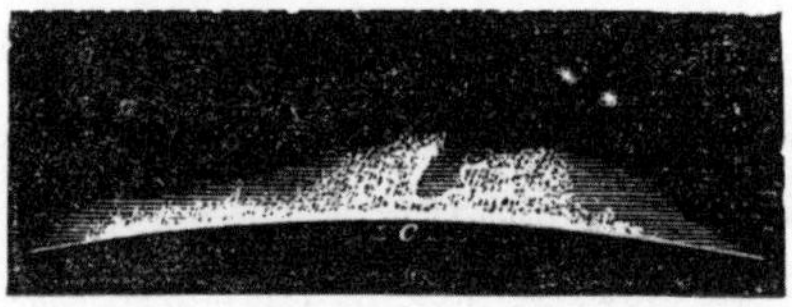

Fig. 3

The dark spot, marked *c*, was very curious, reminding one strongly of the so-called fish-mouth in the nebula of Orion. I saw no change in it for 20 minutes. On the other hand, the first series mentioned were changing rapidly, so that at five o'clock the sketch which was drawn at two was quite inapplicable, only the general features remaining unaltered.

4. −128°, about 20″ high, forked, as in Fig. 4.

Fig. 4

The structure was *cirrus* in every one but No.3, which seemed more like a mass of cumulus.

Today, for the first time, I saw b_1 reversed in the chromosphere when the slit was tangent to disc; 1,474 was easy; the new line at 2,602 cannot be detected as yet.

At 2.25, while examining the spectrum of a large group of spots near the sun's western limb, my attention was drawn to a peculiar double *knobbishness* of the F line (on the sun's disc, not at the edge), represented by Fig. 5, *a*, at the point *e*. In a very few moments a brilliant spot replaced the knobs, not merely interrupting and reversing the dark line, but blazing like a star near the horizon, only with blue instead of red light; it remained for about two minutes, disappearing, unfortunately, while I was examining the sun's image

2. +135°，小，但底部非常明亮，形状如图 2 所示。

图 2

3. –85°，形状如图 3 所示。

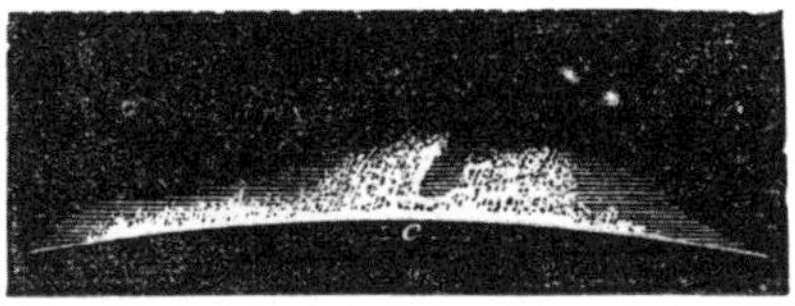

图 3

标注为 c 的暗黑子非常奇妙，很容易让人想起猎户座大星云中被称作鱼嘴的形状。我观察了 20 分钟，没有看到什么变化。另一方面，前述一系列现象却变化得非常迅速，所以在 2 点钟画的草图到 5 点钟时已经不适用了，但总的特征仍然没有改变。

4. –128°，高约 20 角秒，叉状，如图 4 所示。

图 4

除了第 3 个外，其他日珥的结构都是“卷云”，而第 3 个则更像是一块积云。

今天是我第一次看到当狭缝与日面相切时 b_1 在色层球中发生反转，检测 1,474 线相对容易，而在 2,602 的新线到现在还不能被检测出来。

在 2 点 25 分时，当观测靠近太阳西部边缘附近的一大片黑子的光谱时，我注意到了在 F 线处（在日面上，而不是在边缘）有一个奇特的双**球柄结构**，由图 5 中 a 上的 e 点表示。没过多久，一个明亮的亮点取代了球柄结构的突出物，不但打断和反转了那条暗线，而且亮得像一颗在地平线附近闪耀的恒星，但是闪着蓝色而不是

upon the graduated screen at the slit, in order to fix its position, which was at −82.5, about 43″ from the edge of the limb, about 15″ inside of the inner edge of the spot-cluster. I do not know, therefore, whether it disappeared instantaneously or gradually, but presume the latter. Fig. 5, *b*, attempts to give an idea of the appearance. When I returned to the eye-piece, I saw what is represented at Fig. 5, *c*, &c. On the upper (more refrangible) edge of F there seemed to hang a little black mote, making a *barb*, whose point reached nearly to the faint iron line just above F. As given on Ångström's atlas, the wavelength of F is 486.07, while that of the iron line referred to is 485.92 (the units being millionths of a millimetre). This shows an absolute change of 0.15 in the wavelength, or a fraction of its whole amount, represented by the decimal 0.00030, and would indicate an advancing velocity of about 55.5 miles per second in the mass of hydrogen whose absorption produced this barbed displacement.

Fig. 5

The barb continued visible for about five minutes, gradually resolving itself into three small lumps, one on the upper, and two on the lower line, Fig. 5, *d*. In about ten minutes more, the F line resumed its usual appearance. I did not examine the C line, as I did not wish to disturb the adjustments and risk losing some of the curious changes going on under my eye.

After the close of this strange phenomenon, I examined, with our large telescope of 6-inch aperture, the neighbourhood in which this took place, and found a very small spot exceedingly close to, if not actually *at*, the place. This was at 2.45. At 5.30 it had grown considerably.

Undoubtedly, the phenomenon seen was the same referred to by Mr. Lockyer when he speaks of often seeing the bright lines of the prominences not only at the sun's limb but on his disc. It is the only time I have had the good fortune to see it as yet.

(**1**, 172-173; 1869)

红色的光芒。这种现象只持续了约 2 分钟就消失了，遗憾的是当时我正在查看狭缝处刻度屏上的太阳的像，想把它的位置固定住，这时位置是 –82.5，大约离突出物的边缘约 43 角秒，在黑子云内缘内部约 15 角秒处。因此我不知道它到底是瞬间消失的还是逐渐消失的，但我认为是后者。图 5 中的 *b* 试图给出这一现象的概貌。当我再次用目镜观察的时候，我看到了图 5 中的 *c* 等所示的图像。在 F 折射较大的上端，似乎挂着一小块黑的微粒，构成一个**鱼钩状**，钩的顶端接近 F 上方那条微弱的铁谱线。在波长表中查到，F 的波长为 486.07，而铁谱线的波长为 485.92（单位是百万分之一毫米）。这说明波长的确改变了 0.15，或者用变化量占原始值的比例表示，即小数 0.00030，这也表明其吸收谱线产生鱼钩状位移的这部分氢的速度提高到了每秒约 55.5 英里。

图 5

这个鱼钩持续可见约 5 分钟，之后逐渐分为 3 个小块，一块在上面，另两块相对靠下，如图 5 的 *d* 所示。大约 10 分钟之后，F 线恢复到它之前的样子。我没有检查 C 线，因为我不想冒着失掉正发生在眼皮底下的一些奇特变化的风险去改变调好的位置。

在这个奇特的现象结束之后，我用直径为 6 英寸的大望远镜观测了此现象发生处的周边，发现了一个非常小的点，即使不是恰好**在此现象发生处**，也是在离得特别近的位置。这是发生在 2 点 45 分的事，到 5 点 30 分它已经变得相当大了。

毫无疑问，上述观测到的现象与洛克耶先生提及的不仅在太阳边缘，而且在日面上都能经常观测到的日珥的亮线是一回事。但到目前为止，只有这一次我有幸看到了这种现象。

（刘东亮 翻译；蒋世仰 审稿）

Spectroscopic Observations of the Sun

Editor's Note

This editorial reports on several papers read recently at the Royal Society, pertaining to studies probing the constitution and dynamics of the Sun with the new technique of spectroscopy. *Nature*'s founder and editor J. Norman Lockyer notes that velocities of around 100 miles per second had been detected for matter in the chromosphere. He and others argued that the photosphere and chromosphere together constitute the atmosphere of the Sun, contrary to some alternative proposals. The papers discuss detailed spectroscopic clues about structure and composition obtained from a range of elements, and reflect the great and sudden observational progress on a topic not long before considered inaccessible to empirical science.

ROYAL Society, December 9.—Dr. W. A. Miller, V.P., in the chair. The following papers were read: —

"Spectroscopic Observation of the Sun" —No. V. By J. Norman Lockyer, F.R.S.

The author first referred to several new facts of importance as follows:

"I. The extreme rates of movement in the chromosphere observed up to the present time are—

Vertical movement	40 miles a second
Horizontal or cyclonic movement	120 miles a second

"II. I have carefully observed the chromosphere when spots have been near the limb. The spots have sometimes been accompanied by prominences, at other times they have not been so accompanied. Such observations show that we may have spots visible without prominences in the same region, and a prominences without spots; but I do not say that a spot is not accompanied by a prominence, *at some stage of its life*, or that it does not result from some action which, in the majority of cases, is accompanied by a prominence.

"III. At times, when a prominence is seen bright on the sun itself, the bright F line varies considerably, both in thickness and brilliancy, within the thickness of the dark line. The appearances presented are exactly as if we were looking at the prominences through a grating.

"IV. Bright prominences, when seen above spots on the disk, if built up of other substances

观测太阳光谱

编者按

这篇评论报道了最近在皇家学会宣读的几篇论文，这些论文涉及的都是人们利用新的光谱技术对太阳组成及其动力学进行的探索研究。《自然》的创始人及主编诺曼·洛克耶指出，探测结果表明色球层中物质的运动速度大约是每秒 100 英里。他和其他一些科学家都认为光球层和色球层一起组成了太阳的大气层，这与另外一些说法相悖。这些文章讨论了从一系列元素得出的一些关于太阳结构和组成的详细光谱线索，同时也反映了在不久之前还被认为是无法通过实证科学解决的问题在最近已经取得的重大而迅速的观测进展。

皇家学会，12 月 9 日——副会长米勒博士主持会议，会议中宣读了下面的文章：

《观测太阳光谱》——第 5 篇，作者是诺曼·洛克耶（皇家学会会员）。

作者第一次提到了如下一些重要的新发现：

"I. 到目前为止观测到的色球层中物质运动的最大速率是：

纵向运动	每秒40英里
横向运动或回旋运动	每秒120英里

"II. 当黑子出现在日面边缘时，我很仔细地观测了色球层。黑子有时候与日珥相伴出现，另一些时候又不是。这项观测表明，在同一区域内我们可能只看到黑子而没有看到日珥，或者只看到日珥而没有看到黑子。但是我并没有说黑子**在它生命周期中的某个阶段**不会伴随着日珥的出现，也不意味着，黑子的产生不是由某个在大多数情况下与日珥相伴的过程造成的。

"III. 有时候，当在太阳上看到一个明亮的日珥时，明亮的 F 线将在宽度和色度上发生显著的变化，但其宽度不会超过暗线宽度。其呈现出的恰似我们透过光栅观察日珥时的表现。

"IV. 在日面上看到黑子群时出现的明亮日珥，如果包括除氢以外的其他物质，

besides hydrogen, are indicated by the bright lines of those substances in addition to the lines of hydrogen. The bright lines are then seen very thin, situated centrally (or nearly so) on the broad absorption-bands caused by the underlying less-luminous vapours of the same substances.

"V. I have at last detected an absorption-line corresponding to the orange line in the chromosphere. Father Secchi states* that there is a line corresponding to it much brighter than the rest of the spectrum. My observation would seem to indicate that he has observed a bright line less refrangible than the one in question, which bright line is at times excessively brilliant. It requires absolutely perfect atmospheric conditions to see it in the ordinary solar spectrum. It is best seen in a spot-spectrum when the spot is partially covered by a bright prominence.

"VI. In the neighbourhood of spots the F bright line is sometimes observed considerably widened out in several places, as if the spectroscope were analysing injections of hydrogen at great pressure in very limited regions into the chromosphere.

"VII. The brilliancy of the bright lines visible in the ordinary solar spectrum is extremely variable. One of them, at 1871.5, and another, at 1529.5 of Kirchhoff's scale, I have detected in the chromosphere at the same time that they were brilliant in the ordinary solar spectrum.

"VIII. Alterations of wavelength have been detected in the sodium-, magnesium-, and iron-lines in a spot-spectrum. In the case of the last substance, the lines in which the alteration was detected were *not* those observed when iron (if we accept them to be due to iron alone) is injected into the chromosphere.

"IX. When the chromosphere is observed with a tangential slit, the F bright line close to the sun's limb shows traces of absorption, which gradually diminish as the higher strata of the chromosphere are brought on to the slit, until the absorption-line finally thins out and entirely disappears. The lines of other substances thus observed do not show this absorption.

"X. During the most recent observations, I have been able to detect traces of magnesium and iron in nearly all solar latitudes in the chromosphere. If this be not merely the result of the good definition lately, it would indicate an increased general photospheric disturbance as the maximum sunspot period is approached. Moreover, I suspect that the chromosphere has lost somewhat of its height."

The author appends a list of the bright lines, the position of which in the chromosphere have been determined absolutely, with the dates of discovery, remarking that in the case of C and F his observations were anticipated by M. Janssen:—

* *Comptes Rendus*, 1869, I sem, p. 358.

则谱线中除了氢线之外还应当有其他物质产生的亮线。这时看到的亮线都非常细，位于宽吸收谱带区域的中心（或接近中心），这是由该物质底层一些低亮度的蒸气发射出来的。

“V. 我终于发现了一条与色球层上橘黄色谱线相对应的吸收谱线。塞奇神父说过*，有一条比光谱中其他谱线亮很多的谱线与之相对应。我的观测似乎表明，他观测到的那条亮线，折射率比有争议的那条谱线小，这就是为什么亮线有时候会显得格外明亮的原因。必须在非常好的大气条件下才能在普通的太阳光谱中看到它，尤其是当一个黑子被亮的日珥遮挡住一部分的时候，在黑子光谱中看得更清楚。

“VI. 在黑子周围观察到 F 亮线有时候在一些地方明显加宽，就好像分析氢气在很大压力下注入到色球层上很小的区域时的光谱一样。

“VII. 可视亮线的亮度在普通太阳光谱中的变化非常大。在普通太阳光谱中有两条明亮的谱线，我在色球层中也同时观察到了，按照基尔霍夫标度，其中一条在 1871.5，另一条在 1529.5。

“VIII. 在一个黑子光谱中，我发现钠线、镁线和铁线的波长发生了改变。对于铁线，检测到波长发生改变的谱线并**不是**当铁（如果我们认为它们就是单一物质铁发出的）射入色球层时观测到的谱线。

“IX. 当用一个切向夹缝观测色球层时，在日面边缘附近的 F 亮线显示出吸收的迹象，而当夹缝对准色球层的较高位置时，F 亮线逐渐减弱，直到吸收线越来越窄，然后全部消失。其他物质的吸收谱线并没有表现出这种现象。

“X. 在最近的观测活动中，我在太阳色球层的几乎所有纬度都检测到了少量的镁和铁。如果这不仅仅是因为最近透镜清晰度较好的缘故，那就说明当太阳黑子极大期到来的时候，光球层受到了更大的扰动。另外，我猜测色球层的高度也有所降低。”

作者附了一个亮线的明细表，这些亮线在色球层里的位置已完全确定，其中还列出了发现它们的日期，并指出让森曾预测到了他观测到的 C 线和 F 线：

---

* 《法国科学院院刊》，1869年，第1期，第358页。

	Hydrogen	
C.	October 20, 1868	
F.	October 20, 1868	
near D.	October 20, 1868*	
	[* *Hydrogen?*—G. G. S.]	
near G.	December 22, 1868	
h.	March 14, 1869	
	Sodium	
D.	February 28, 1869	
	Barium	
1989.5†	March 14, 1869	
2031.2	July 5, 1869	
	Magnesium and included line	
b^1, b^2, b^3, b^4	February 21, 1869	
	Other Lines	
Iron	1,474	June 6, 1869
?	1515.5	June 6, 1869
Bright line	1529.5	July 5, 1869
?	1567.5	March 6, 1869
?	1613.8	June 6
Iron	1867.0	June 26
Bright line	1871.5	,,
Iron	2001.5	,,
?	2003.4	,,
? band or line near black line, very delicate ...	2054.0	July 5

Other lines besides these have been seen at different times; but their positions have not been determined absolutely.

The author points out that taking iron as an instance, and assuming that the iron-lines mapped by Ångström and Kirchhoff are due to iron only, he has only been able, up to the present time, to detect three lines out of the total number (460) in the spectrum of the lower regions of the chromosphere,—a fact full of promise as regards the possible results of future laboratory work. The same remark applies to magnesium and barium.

The paper then proceeded as follows:—

* *Comptes Rendus*, 1869, I sem, p. 358.

† This reference is to Kirchhoff's scale.

	氢	
C.		10月20日，1868年
F.		10月20日，1868年
接近D.		10月20日，1868年*
	[*氢？—G. G. S.]	
接近G.		12月22日，1868年
h.		3月14日，1869年
	钠	
D.		2月28日，1869年
	钡	
1989.5†		3月14日，1869年
2031.2		7月5日，1869年
	镁及相关谱线	
b^1 b^2 b^3 b^4		2月21日，1869年
	其他谱线	
铁线	1,474	6月6日，1869年
?	1515.5	6月6日，1869年
亮线	1529.5	7月5日，1869年
?	1567.5	3月6日，1869年
?	1613.8	6月6日
铁线	1867.0	6月26日
亮线	1871.5	6月26日
铁线	2001.5	6月26日
?	2003.4	6月26日
? 靠近暗线的谱带或谱线，非常细	2054.0	7月5日

除上述谱线外，其他谱线也被观测了许多次，但是它们的位置仍然没有被严格确定。

作者指出，以铁为例，如果埃斯特朗和基尔霍夫图谱中标注的铁线是由铁这一种元素的吸收造成的，那么到目前为止，他只能在色球层较低的区域中检测到全部 460 条谱线中的 3 条——也许未来的观测工作中可能会观测到全部。镁和钡也如此。

文章接着写道：

* 《法国科学院院刊》，1869年，第1期，第358页。

† 参照基尔霍夫标度。

"Dr. Frankland and myself have determined that the widening out of the sodium-line in the spectrum of a spot which I pointed out in 1866, and then stated to be possibly an evidence of greater absorption, indicates a greater absorption due to greater pressure.

"The continuous widening out of the sodium-line in a spot must therefore be regarded as furnishing an additional argument (if one were now needed) in favour of the theory of the physical constitution of the sun first put forward by Dr. Frankland and myself—namely, that the chromosphere and the photosphere form the true atmosphere of the sun, and that under ordinary circumstances the absorption is continuous from the top of the chromosphere to the bottom of the photosphere, at whatever depth from the bottom of the spot that bottom may be assumed to be.

"This theory was based upon all our observations made from 1866 up to the time at which it was communicated to the Royal Society and the Paris Academy of Sciences, and has been strengthened by all our subsequent work; but several announcements made by Father Secchi to the Paris Academy of Sciences and other learned bodies are so opposed to it, and differ so much from my own observations, that it is necessary that I should refer to them, and give my reasons for still thinking that the theory above referred to is not in disaccord with facts.

"Father Secchi states that the chromosphere is often separated from the photosphere, and that between the chromosphere and the photosphere there exists a stratum giving a continuous spectrum, which he considers to be the base of the solar atmosphere, and in which he thinks that the inversion of the spectrum takes place.

"With regard to the first assertion, I may first state that all the observations I have made have led me to a contrary conclusion. Secondly, in an instrument of comparatively small dispersive power, such as that employed by Father Secchi, in which the widening out of the F line at the base of the chromosphere is not clearly indicated, it is almost impossible to determine, by means of the spectroscope, whether the chromosphere rests on the sun or not, as the chromosphere is an envelope and we are not dealing merely with a section. But an instrument of great dispersive power can at once settle the question; for since the F line widens out with pressure, and as the pressure increases as the sun is approached, the continuous curvature of the F line must indicate really the spectrum of a section; and if the chromosphere were suspended merely at a certain height above the photosphere, we should not get a widening due to pressure; but we always do get such a widening.

"With regard to the second assertion, I would remark that if such a continuous-spectrum-giving envelope existed, I entirely fail to see how it could be regarded as a region of selective absorption. Secondly, my observations have indicated no such stratum, although injections of sodium, magnesium, &c. into the chromosphere not exceeding the limit of the sun's limb by 2″ have been regularly observed for several months past. Today I have even detected a low level of barium in the chromosphere not 1″ high. This indicates, I think, that my instrument is not lacking in delicacy; and as I have never seen

“弗兰克兰博士和我已经确定了我在1866年提出的黑子光谱中钠线的展宽效应，并称这可能是存在更强吸收作用的证据，说明较大的压力导致强的吸收。

“在一个黑子光谱中钠线的持续展宽可以被看作是一个附加的证据（如果现在需要一个证据的话），这支持由弗兰克兰博士和我首先提出的有关太阳物质组成的理论，即色球层和光球层组成了太阳的大气，而且在通常情况下，从色球层顶部到光球层底部距离黑子底部（这一底部可以是假定的）任意深度处，吸收作用是连续进行的。

“这个理论基于我们从1866年到现在的所有观测结果，曾报告给皇家学会和巴黎科学院，并已经得到我们所有后续观测结果的进一步验证。但是塞奇神父在巴黎科学院和其他学术机构公布的一些言论极力反对这个理论，而且他的观点与我们的观测结果有非常大的差异，因此我有必要提及这些，并给出我一直认为上述理论与事实相符的理由。

“塞奇神父宣称色球层经常与光球层分离，且在色球层和光球层之间存在一个能发出连续光谱的层，他认为这个层是太阳大气的底部，并且认为在这一层中光谱会发生反转。

“对于他的第一个论断，我首先要说的是，我所有的观测结果都把我引向了一个相反的结论。其次，用色散率较小的仪器，例如塞奇神父使用的那种，不能清楚地看到色球层底部F线的展宽效应，用光谱仪几乎不可能分辨出色球层到底是不是依附在太阳上，因为色球层是一个包层，而且我们要考虑的不仅仅是一个截面。大色散率的仪器可以马上解决这一问题，因为既然F线由于压力作用而展宽，越接近太阳内部的地方压力越大，所以F线的连续弯曲说明这是一个截面上的光谱，而如果色球层只是悬浮在光球层上方的某一特定高度处，我们就不会看到由于压力而导致的展宽效应，但事实上我们总是能观察到这样的展宽。

“对于第二个论断，我想说的是，如果这样一个产生连续光谱的包层确实存在，那么我实在看不出来它怎么能被看作是具有选择吸收的区域。其次，我的观测结果表明没有这样的一层，尽管在前几个月里经常观测到钠、镁等射入色球层的深度没有超过日面边缘2角秒。今天我在色球层中不到1角秒的高度处检测到了少量的钡。我觉得这表明我的仪器精度很高，而且在我的仪器状态良好的条件下，我从未看到过哪怕是一个近似连续的光谱，我倾向于将观测结果归结为仪器误差。这个现象可

anything approaching to a continuous spectrum when my instrument has been in perfect adjustment, I am inclined to attribute the observation to some instrumental error. Such a phenomenon might arise from a local injection of solid or liquid particles into the chromosphere, if such injection were possible. But I have never seen such an injection. If such an occurrence could be observed, it would at once settle that part of Dr. Frankland's and my own theory, which regards the chromosphere as the last layer of the solar atmosphere; and if it were possible to accept Father Secchi's observation, the point would be settled in our favour.

"The sodium experiments to which I have referred, however, and the widening out of the lines in the spot-spectra, clearly indicate, I think, that the base of the atmosphere is below the spot and not above it. I therefore cannot accept Father Secchi's statement as being final against another part of the theory to which I have referred—a conclusion which Father Secchi himself seems to accept in other communications.

"Father Secchi remarks also that the F line is produced by the absorption of other bodies besides hydrogen, because it never disappears. This conclusion is also negatived by my observations; for it has very often been observed to disappear altogether and to be replaced by a bright line. At times, as I pointed out to the Royal Society some months ago, when a violent storm is going on accompanied by rapid elevations and depressions of the prominences, there is a black line on the less-refrangible side of the bright one; but this is a phenomenon due to a change of wavelength caused by a rapid motion of the hydrogen.

"With regard to the observation of spot-spectra, I find that every increase of dispersive power renders the phenomenon much more clear, and at the same time more simple. The selective absorption I discovered in 1866 comes out in its most intense form, but without any of the more complicated accompaniments described by Father Secchi. I find, however, that by using three prisms this complexity vanishes to a great extent. We get portions of the spectrum here and there abnormally bright, which have given rise doubtless to some of the statements of the distinguished Roman observer; but the bright lines, properly so-called, are as variable as they are in any other part of the disk, but not much more so. I quite agree that the 'interpretation' of sun-spot phenomena to which Father Secchi has referred, which ascribes the appearances to anything but selective plus general absorption, is erroneous. But as I was not aware that it had ever been propounded, I can only refer to my own prior papers in support of my assertion which were communicated to the Royal Society some three years ago."

"Researches on Gaseous Spectra in relation to the Physical Constitution of the Sun, Stars, and Nebulae."—Third Note. By E. Frankland, F.R.S., and J. Norman Lockyer, F.R.S.

The authors remark that it has been pointed out by Mr. Lockyer that the vapours of magnesium, iron, &c., are sometimes injected into the sun's chromosphere, and are then rendered sensible by their bright spectral lines. (*Proc. Roy. Soc.*, vol. XVII, p. 351)

能是由局部区域的固体或者液体微粒注入到色球层引起的，如果这种注入过程可能的话。但我从未见到过这种注入过程。当然如果这个过程能够被观测到的话，它将立刻证明弗兰克兰博士和我的理论是正确的，即认为色球层是太阳大气的最后一层，而即使认为塞奇神父的观察结果是可以接受的，那也将印证我们的观点。

“但是，对于以前提到的有关钠的实验和黑子光谱中出现的谱线展宽，我认为都明确地说明了大气层的底部在黑子之下而不是在黑子之上，因此我不认为塞奇神父的报告是对我提出的理论中的另一部分观点的彻底否定——塞奇神父自己似乎在其他报告中也是表示同意这一点的。

“塞奇神父还指出，由于 F 线从不消失，所以它是氢以外的其他物质的吸收产生的。这个结论也被我的观测结果否定了；因为我经常观察到它完全消失，并被一条亮线取代。正如我几个月前在皇家学会上所说的，当强烈的太阳风暴来临的时候，伴随着日珥快速的升降，在亮线折射率较小的一侧有时会有一条暗线，但这种现象是由于氢的快速运动导致的波长改变造成的。

“在观察黑子光谱的时候，我发现色散率的提高每次都能使这个现象更加明显，同时也更简单。我在 1866 年发现的选择性吸收谱线达到了它的最强状态，但没有产生塞奇神父所说的其他更复杂的现象。然而，我发现使用三棱镜后，这种复杂的现象基本上就会消失。我们得到了一些有的地方异常明亮的光谱，这使得我们对这位杰出的罗马观测者的一些论断深信不疑。但是那些亮线，严格地说是所谓的亮线，和它们在日面上任何其他位置时一样变化。我认为塞奇神父对黑子现象的‘解释’是不正确的，即认为这些现象绝不是选择性吸收和一般吸收共同作用的结果。但由于我不知道这个观点是否曾经被提出过，我只能参考我自己以前的文章来支持我 3 年前向皇家学会提出的论点。”

《关于太阳、恒星以及星云物质组成的气态光谱研究》——第 3 篇，作者是弗兰克兰（皇家学会会员）和诺曼·洛克耶（皇家学会会员）。

作者谈到，洛克耶先生曾经指出：镁、铁等的蒸气有时候会注入到太阳的色球层，从而产生明亮的谱线而被人察觉到。（《皇家学会学报》，第 17 卷，第 351 页）

2. It has also been shown (1) that these vapours, for the most part, attain only a very low elevation in the chromosphere, and (2) that on rare occasions the magnesium vapour is observed like a cloud separated from the photosphere.

3. It was further established on the 14th of March, 1869, and a drawing was sent to the Royal Society indicating, that when the magnesium vapour is thus injected, the spectral lines do not all attain the same height.

Thus, of the *b* lines, b^1 and b^2 are of nearly equal height, but b^4 is much shorter.

4. It has since been discovered that of the 450 iron lines observed by Ångström, only a very few are indicated in the spectrum of the chromosphere when iron vapour is injected into it.

5. The authors' experiments on hydrogen and nitrogen enabled them at once to connect these phenomena, always assuming that the great bulk of the absorption to which the Fraunhofer lines are due takes place in the photosphere itself.

It was only necessary, in fact, to assume that, as in the case of hydrogen and nitrogen, the spectrum became simpler where the density and temperature were less, to account at once for the reduction in the number of lines visible in those regions where, on the authors' theory, the pressure and temperature of the absorbing vapours of the sun are at their minimum.

6. It became important, therefore, to test the truth of this assumption by some laboratory experiments, the preliminary results of which are communicated in this note.

The spark was taken in air between two magnesium poles, so separated that the magnesium spectrum did not extend from pole to pole, but was visible only for a little distance, indicated by the atmosphere of magnesium vapour round each pole.

The disappearance of the *b* lines was then examined, and it was *found that they behaved exactly as they do on the sun.* Of the three lines, the most refrangible was the shortest; and shorter than this were other lines, *which Mr. Lockyer has not detected in the spectrum of the chromosphere.*

This preliminary experiment, therefore, quite justified the assumption, and must be regarded as strengthening the theory on which the assumption was based, namely, that the bulk of the absorption takes place in the photosphere, and that it and the chromosphere form the true atmosphere of the sun. In fact, had the experiment been made in hydrogen instead of in air, the phenomena indicated by the telescope would have been almost perfectly reproduced; for each increase in the temperature of the spark caused the magnesium vapour to extend further from the pole, and where the lines disappeared a band was observed surmounting them, which is possibly connected with one which at times is observed in the spectrum of the chromosphere itself when the magnesium lines are not visible.

(**1**, 195-197; 1869)

2．这也表明：（1）这些气体中的绝大部分都处于色球层的底部；（2）在极少数情况下，镁蒸气看上去像是一块与光球层分离的云。

3．1869 年 3 月 14 日向皇家学会提交了一幅草图，进一步说明当镁蒸气注入的时候，并不是所有谱线都达到了同一高度。

比如，对于 *b* 线而言，b^1 和 b^2 的高度近似相等，但 b^4 短很多。

4．在埃斯特朗先前发现的 450 条铁线中，只有极少数会在铁蒸气注入色球层时出现在色球层的光谱中。

5．作者关于氢和氮的实验可以立刻与这些现象联系起来，这通常要假设大部分产生夫琅和费线的吸收发生于光球层内部。

实际上，就像在氢和氮的实例中一样，只需要假定在密度和温度较低的地方光谱会变得更简单，就可以马上用作者的理论解释为什么可见谱线的数量会在太阳内部吸收气体的压力和温度最小的地方下降。

6．因此，用实验来验证这一假设的真实性就显得尤为重要，初步的实验结果通报如下：

当在两个镁电极之间的空气中产生电火花时，镁谱线并非从一极延伸到另一极，而是只在一个很短的距离内可见，说明镁蒸气只位于两个电极附近。

接下来检测到了 *b* 线的消失，**发现它们的表现与它们在太阳上的表现完全相同**。在这 3 条线中，折射率最大的也是最短的，而比这条更短的是**洛克耶先生在色球层光谱中没有探测到的一些其他的谱线**。

因此这个初步的实验完全证明了这个假设，而且强化了该假设所基于的理论，即大部分吸收发生于光球层，且光球层和色球层组成了太阳真正的大气。实际上，如果这个实验是在氢气中而不是空气中进行，望远镜中观察到的现象就会几乎完美地重现。因为火花温度的每一次增加，都会使镁蒸气扩展到离电极更远的地方，并且在谱线消失的地方可以看到其上有一个谱带。这可能与镁线不可见的时候在色球层光谱中看到的现象有关联。

（刘东亮 翻译；邓祖淦 审稿）

Darwinism and National Life

H.

Editor's Note

This comment by a writer identified only as "H." (perhaps Thomas Huxley?) shows how Darwinian evolutionary theory was from its inception linked to social engineering and eugenics. H. implies that natural selection should be able to explain all national characteristics (which here are really just stereotypes), such as the "self-reliance" of the inhabitants of the United States. Little account is taken of the tremendous timescales that Darwin assumed to be necessary for significant evolutionary change (although Darwin himself echoed many of the assumptions and presumptions made here in "The Descent of Man" (1877)). Nonetheless, the idea that evolutionary forces can explain why one culture thrives and another fails is still current, thanks to researchers who are probing the links between environment and history.

THE Darwinian theory has a practical side of infinite importance, which has not, I think, been sufficiently considered. The process of natural selection among wild animals is of necessity extremely slow. Starting with the assumption (now no longer a mere assumption) that the creature best adapted to its local conditions must prevail over others in the struggle for existence, the final establishment of the superior type is dependent at each step upon three accidents—first, the accident of an individual sort or variety better adapted to the surrounding conditions than the then prevailing type; secondly, the accident that this superior animal escapes destruction before it has had time to transmit its qualities; and, thirdly, the accident that it breeds with another specimen good enough not to neutralise the superior qualities of its mate. In the case of domesticated animals the progress is incomparably more rapid, because it is practicable, first, to modify the conditions of life, so as to encourage the appearance of an improved specimen; next, to cherish and protect it against disaster; and, lastly, to give it a consort not altogether unworthy of the honour of reproducing its qualities. The case of man is intermediate in rapidity of progress to the other two. The development of improved qualities cannot be insured by judicious mating, because as a rule human beings are capricious enough to marry without first laying a case for opinion before Mr. Darwin. Neither would it be easy, nor, perhaps, even allowable, to extend any special protection by law or custom to those who may be physically and intellectually the finest examples of our race. Still, two things may be done: we may vary the circumstance of life by judicious legislation, and still more easily by judicious non-legislation, so as to multiply the conditions favourable to the development of a higher type; and by the same means we may also encourage, or at least abstain from discouraging, the perpetuation of the species by the most exalted individuals for the time being to be found. Parliament, being an assembly about as devoid of any scientific insight as a body of educated men could possibly be, has not as yet consciously legislated with a view to the improvement of the English type of character. Without knowing it, however, the legislature has sometimes stumbled on the right course, though

达尔文学说与国民生活

H.

编者按

通过这篇作者仅署名为“H.”（也许是托马斯·赫胥黎？）的评论文章我们可以看到，从一开始达尔文的进化论就与社会工程学和优生学紧密相连。作者暗示，应该可以用自然选择来解释所有民族的特征（文章中论述的实际上只是些模式化的见解），例如美国土著居民“自力更生”的特点。文章几乎没有考虑漫长的时间尺度这一在达尔文看来对明显的进化改变来说十分必要的因素（尽管达尔文本人在文中反复提到了1877年的《人类的由来》一书中的多项假设）。不过，根据一直在探索环境与历史之间关联的研究人员的工作，进化动力的概念可以解释为什么一种文化繁盛而另一种文化衰落的现象持续存在。

达尔文的进化理论有极其重要的实际应用，我认为人们并没有充分认识到这一点。在野生动物中，自然选择的过程必然是十分缓慢的。我们从一个假设（现在已经不仅仅是一个假设了）开始论述这个问题，假设最适应当地环境的生物必定在生存斗争中胜过其他生物，那么，一种高等生物最终确立的每一步都是依赖于以下三个事件：第一，出现了一个新种类的个体或变异体，比当时占优势的物种更适应周围环境；第二，这个更高等的个体在死亡之前有足够的时间能够产生后代以传递自己的特性；第三，这个个体与其他足够优良的个体交配来繁殖后代，以保证它的优等生物性状不会被中和掉。对于家养动物来说，进化的速度无比之快，因为在实际应用中为了促进改良品种的产生首先会改变动物的生活环境；其次，会照料并保护出现的改良品种使其免受灾害；最后，还会给它指定配偶以保证其优良性状的延续。人类的进化速度则介于两者之间。人类中出现的优良性状不能通过明智的交配保证其延续，因为在达尔文提出进化论的观点之前，人类本能地在婚配问题上表现得十分任性，通常不会提出任何婚配的理由。企图通过法律或者惯例给那些在体能和智力两方面都是我们种族的精英的个体提供特别的保护既不容易，甚至也不被允许。尽管如此，我们仍然可以做这样两件事情：我们也许可以通过明智的立法来改变生活的环境，不过可能通过明智的非立法手段会更容易些，这样一来就能提供更多适合于更高等个体出现和发展所需要的环境；我们也可以通过同样的办法来促进，或者至少不去妨碍由那些在当时能找到的最好的个体所组成的物种的绵延不绝。国会尚未有意识地从提高英国人素质的角度来进行立法，因为国会就是这样的一个群体，它缺乏一批有素养的人可能具备的科学的洞察力。然而，由于不理解这一观点，立法机构有时会在正确的问题上犹豫，而更经常的则是犯下愚蠢的错误。我们的自由

it has more often blundered into the wrong. Our free trade policy has furnished special scope and special advantages to the energetic enterprising character, and so far has tended to perpetuate and intensify the type which has given to little England her wonderful prominence in the world. On the other hand, the steady refusal to make a career for scientific men has drained away most of our highest intellect from its proper field, and has subjected the rest to an amount of discouragement by no means favourable to increase and improvement. Our laws and customs practically check the growth of the scientific mind as much as they tend to develop the speculative and energetic commercial character.

We do not expect for a long time to hear an orator in the House of Commons commence his speech by announcing, (as a distinguished member of the Austrian Reichsrath recently did, in a debate on the relation of the different nationalities in the empire), that the whole question is whether we are prepared to accept and act upon the Darwinian theory. But even an average English M.P. may be brought to see that it may be possible, indirectly, to influence the character and prosperity of our descendants by present legislation, and none will deny that, if this is practicable, a higher duty could not be cast upon those who guide the destinies of a nation.

A glance at the operation of Darwinism in the past, will best show how potent it may be made in the future. Look at English progress and English character, and consider from this point of view to what we owe it. There were originally some natural conditions favourable to the growth of our commercial and manufacturing energy. We had an extensive coast and numerous harbours. We had also abundance of iron-stone in convenient proximity to workable coal. Other nations either wanted these advantages or were ignorant that they possessed them. These favourable conditions developed in many individuals a special adaptability to commercial pursuits. The type was rapidly reproduced and continually improved until England stood, in the field of commerce, almost alone among the nations of the world. And what is there now to sustain our pre-eminence? Nothing, or next to nothing, except the type of national character, which has been thus produced. Steam, by land and sea, has largely diminished the superiority which we derived from the nature of our coast; and coal and iron are now found and worked in a multitude of countries other than our own. Our strength in commerce, like our weakness in art, now rests almost exclusively on the national character which our history has evolved.

Take another example of the character of a people produced partly by natural conditions of existence, but far more by the artificial conditions to which evil legislation has exposed it. What has made the typical Irishman what he now is? The Darwinian theory supplies the answer. Ireland is mainly an agricultural country, with supplies of mineral wealth altogether inferior to those of England, though by no means contemptible if they were but developed. This is her one natural disadvantage, and it is trifling compared with those which we in our perversity created. For a long period we ruled Ireland on the principles of persecution and bigotry, and left only two great forces at work to form the character of the people. All that there was of meanness and selfishness and falsehood was tempted to servility and apostacy, and flourished and perpetuated itself accordingly. All that there was

贸易政策为那些富有强烈进取心的人们提供了特殊的机会和特别有利的条件，迄今为止，此政策也使得已经让不大的英格兰在世界范围内获得荣耀的那种特质类型得以存续和加强。另一方面，由于从事科学研究的人们在成就一番事业上还存在很大的阻力，使得我们这些智力最高的人中的大多数无法在合适的领域从事相关的工作，而其余的人也由于没有良好的发展环境而备受挫折。我们的法律和习俗在很大程度上发展了富有活力的商业投机特质，但反过来实际上也同样多地限制了科学思维的发展。

长期以来，我们并不期望听见一个演说家在下议院开始他的演讲时就宣称（就像奥地利议会的一位著名成员最近在关于帝国中不同民族之间关系的辩论中所做的那样），所有的问题就是我们是否准备接受并且按照达尔文的理论行事。但是，即使是一名普通的英国下议院议员也应该明白，通过现有的立法是有可能间接地对我们后代的特质和富足产生影响的，而且如果这是切实可行的话，也不会有人否认这才是赋予那些引导民族发展方向的人的最为重大的责任。

回顾一下进化论在过去所起的作用会让我们对它在未来将有多大的潜力看得更加清楚。看看英国人取得的进步和英国人的特质，并从这个角度来考虑一下我们应该把这一切归因于什么。这里本来就有适合我们的商业和制造业发展的一些自然条件。我们拥有广阔的海岸线和大量的港口。我们也拥有丰富的与可经营煤矿相邻的铁矿。其他国家要么没有这样的便利，要么还不知道他们拥有这样的便利。这些有利条件造就了许多个体在商业活动方面的特殊适应能力。这些特质迅速地扩散并且持续提高，直至英格兰在商业领域几乎是独一无二地屹立于世界民族之林。如今，又要靠什么来维持我们的这种卓越呢？除了这种历史造就的民族特质，我们没有或者说几乎没有别的任何东西可以依赖。无论是在内陆还是沿海，蒸汽的应用都已经大大地降低了我们从海岸线得到的优越性，煤和钢铁如今在其他许多国家也被大量发现和使用。我们在商业领域的优势，就好比我们在艺术领域的弱势一样，现在几乎只能完全依赖于我们从历史演化中得来的民族特质了。

再举另外一个例子，其中一部分人们的特质是由于自然条件造成的，而更多的则是遭受了那些有害的立法的影响。是什么造就了现在典型的爱尔兰人呢？达尔文的进化理论提供了答案。爱尔兰很大程度上是一个农业国家，矿藏资源总量比英格兰少，即便他们真的发展起来也还是如此，这绝不是轻视他们的看法。这对他们来说是一个天然的不足，而这个不足与我们自以为是地创造出的东西相比就显得微不足道了。长期以来，我们带着迫害和偏见的原则统治着爱尔兰，结果只留下了两种巨大的力量影响着这里人民特质的形成。一方面，吝啬和自私到了如此地步，为了达到使他们接受奴役并放弃信仰的目的，谎言得到鼓励和使用，以至于谎言自身竟然因

of nobleness and heroic determination was drawn into a separate circle, where the only qualities that throve and grew were irreconcilable hatred of the oppressor and resolute but not contented endurance. The two types rapidly reproduced themselves, and as long as the external conditions remained unaltered, they absorbed year by year more and more of the people's life; as, if Darwinism is true, they could not but do. And what is the result now? A great part of a century has elapsed since we abandoned the wretched penal laws, and yet none can fail to see in Ireland the two prevailing types of character which our ancestors artificially produced, the only change being that the two types have become, to a certain extent, amalgamated in a cross which reflects the peculiarities of each. Whether future legislation may so far modify the conditions of Irish existence as to work a gradual change in the national character, is a question of much interest, but too large to be discussed just now. In any case we can scarcely expect the results of centuries upon a national type to be reversed in less than a succession of generations.

Still confining myself to the past, let me point again to the very marked qualities which the conditions of their existence have produced in the people of the United States. They started with a large element of English energy already ingrained into them; they have been reinforced by millions of emigrants presumably of more than the average energy of the various races which have contributed to swell the tide. Added to this, the Americans have enjoyed the natural stimulus of a practically unlimited field for colonisation. Only the resolute, self-reliant settler could hope to prosper in the early days of their national existence; and self-reliance approaching to audacity is the special type of character which on the Darwinian hypothesis we should expect to see developed, transmitted, and increased. How far this accords with actual experience, no one can be at a loss to say. There is probably not a nation in the world whose peculiarities might not be traced with equal ease to the operation of the same universal principle. And the moral of the investigation is this: Whenever a law is sufficiently ascertained to supply a full explanation of all past phenomena falling within its scope, it may be safely used to forecast the future; and if so, then to guide our present action with a view to the interest and well-being of our immediate and remote descendants. Read by the light of Darwinism, our past history ought to solve a multitude of perplexing questions as to the probable supremacy of this or that nation in times to come in the field of commerce, as to the effects of emigration and immigration on the ultimate type likely to be developed in the country that loses and in that which gains the new element of national life, and many another problem of no less interest to ourselves and to humanity.

The subject I have thus slightly indicated seems to me to deserve a closer investigation than it has yet received: and, strange as it will sound to the ears of politicians, I cannot doubt that, in this and other ways, statesmen, if they could open their eyes, might derive abundant aid from the investigations of science, which they almost uniformly neglect and despise.

(**1**, 183-184; 1869)

此而大行其道。另一方面，高尚和英勇却被引入另一个不同的循环，在这里唯一得到繁荣和发展的特性就是对于压迫者无可调和的憎恨以及坚定而无尽的忍耐。这两种类型的特质迅速地自我复制着，并且只要外部环境保持不变，它们就会年复一年地同化越来越多的人；如果达尔文的进化论是正确的，那么他们本不该如此，但却确实如此。那么，现在的结果是什么呢？从我们废除了那些肮脏的刑律算起，已经过去大半个世纪了，然而在爱尔兰，没有人不会发现由我们的祖先人为导致的这两种主流特质。唯一的变化是，这两种特质在一定程度上混杂在一起，交叉体现着各自的特性。是否未来的立法能在一定程度上改变爱尔兰的生存环境，从而使得这个民族的特质慢慢有所变化，这个问题是十分有意义的，但它太大了，不是现在就能讨论的。不管怎样，我们都难以期待短短几代人的时间就能使几个世纪以来形成的民族特质得到改变。

我还是把眼光放在过去吧，让我再来说一下由美国人自身的生存环境造就的美国人的显著的特质。一开始，他们身上就携带着大量的英国人所具有的根深蒂固的特质，而数百万超出各个种族平均水平的移民可能使这种特质得到了进一步加强，促进了这种趋势的增强。另外，因殖民而占据的几乎无限的土地极大地激励着美国人。也只有坚定的、自力更生的移民者有希望能在他们民族生存的早期获得成功；自力更生以至无畏进取，这种特殊的特质，正是我们根据达尔文的理论预期会得到发展、传承和提高的。这与实际的经历究竟有多一致，大概没有人会说不出来。可能对于世界上任何一个民族的特质，都能够按照相同的普遍原则轻易地追溯其来源。研究的意义应当是这样的：如果一个法则能充分确定地为过去在这个范围内的现象提供充分解释的话，那么用它来预测未来应该就是可行的。如果确实如此，那么为我们子孙后代的利益和安康着想，就应该用它来引导我们目前的行为。从达尔文的进化论来看，通过研究我们过去的历史应该可以解决很多复杂的问题，如关于将来哪个民族在商业领域可能占据至高无上的地位的问题，关于出入境移民致使民族特质元素丧失或新增，从而影响民族特质最终发展方向的问题，以及其他许多对我们自身和人类的利益来说都相当重要的问题。

我在这里极其简略地提出的这个主题，在我看来实在是值得进行比以往更深入的调查研究，这对政治家来说可能还是一个陌生的东西，但我不会怀疑，无论通过什么样的方法，如果政治家们打开视野，他们将能够从科学研究中得到充分的帮助，而这些他们几乎都无一例外地忽略和轻视了。

（刘晓辉 翻译；江丕栋 审稿）

A Deduction from Darwin's Theory

W. S. Jevons

Editor's Note

This comment on Darwin's theory of evolution, from British economist W. Stanley Jevons, is typical of the racial chauvinism of its time, from which Darwin himself was by no mean immune. It argues that temperate climates, like those of Jevons' own country, are best suited to nurturing the "highest forms of civilization". Jevons takes it for granted that Europeans have a "superior degree of energy and intellect" than black Africans. Despite Darwin's humanitarian principles and opposition to slavery, he and most of his supporters shared this notion of racial hierarchy, which is very evident in Darwin's "The Descent of Man", published two years after this letter.

THERE is one important consequence deducible from Darwin's profound theory which has not yet been noticed so far as I am aware. The theory is capable under certain reasonable conditions of accounting for the fact that the highest forms of civilisation have appeared in temperate climates.

Although some apparent exceptions might be adduced, it is no doubt true that man displays his utmost vigour and perfection, both of mind and body, in the regions intermediate between extreme heat and extreme cold, allowance being made for the reduced temperature of elevated mountain districts. The explanations hitherto given of this fact are of a purely hypothetical and shallow character. It is said, for instance, that the prolific character of the tropical climate too easily furnishes man with subsistence, so that his powers are never properly called into action. On the other hand in the Arctic regions nature is too sterile and no exertions can lead to the accumulation of much wealth. This explanation obviously involves the gratuitous hypothesis that man has been created with powers exactly suited to be called forth by just that degree of difficulty experienced in a temperate climate. There are those even who maintain our peculiar British climate to be the very best possible, because it taxes our powers of endurance to the last point which they can bear, and thus calls forth the greatest amount of energy. But here again is the assumption that the British people and the British climate were specially created to suit each other.

The theory of natural selection, on the other hand, represents that great method by which infinitely numerous adaptations will always be produced throughout time. Whatever happens in this material world must happen in consequence of the properties originally impressed upon matter, and our notions of the wisdom embodied in the Creation must be infinitely raised when we understand, however imperfectly, its true method. The continual resort to special inventions and adaptations must surely be below the greatness of a Power which could so design and create matter from the first that it must go on thenceforth

达尔文理论的一个推论

杰文斯

编者按

这篇文章是英国经济学家斯坦利·杰文斯对达尔文进化论的评论，可以说颇具代表性地反映了当时的种族沙文主义，达尔文本人也不是完全没有受到影响。文章声称，包括杰文斯自己的祖国在内的许多地方所具有的温带气候最适宜于孕育“最高级别的文明”。杰文斯据此就想当然地认为，与非洲黑人相比，欧洲人拥有“更高等的体力和智力”。与达尔文提倡人道主义精神并反对奴隶制不同，杰文斯和他的许多支持者都持这种种族等级观念，这一点从本文发表两年后出版的达尔文的著作《人类的由来》一书中可以很明显地看到。

据我所知，至今还没有人注意到，达尔文理论有一个重要的推论，即在一定的理想条件下，根据达尔文理论可以证明，最高等形式的文明出现在温带气候中。

尽管也可能存在一些明显的例外，但无疑，生活在介于极寒冷和极炎热地区之间的中间气候带中的人类，无论是身体方面还是心智方面都是最富有活力、最完美的，这里也考虑到了因海拔升高而温度降低的山区。至今为止，对这一现象的解释还都只是一些单纯的假说或粗浅的说明。例如，有些人认为热带地区自然物产十分丰富，可以很容易地满足人类的物质需要，因此在热带地区居住的人类的能力永远无需全面施展从而也难以发展。另一方面，极地地区的自然条件则太恶劣，以至于任何努力都不会带来大量财富的积累。这种说法中显然包含着一个毫无根据的假设：人类被创造出来时具备的能力恰好与温带气候下会遭遇的困难程度相一致。一些人坚持认为我们不列颠特殊的气候是最佳的可能，因为这样的气候使我们的忍耐力达到我们所能承受的极限程度，因此最大限度地挖掘出了我们的潜力。这里又是一个假设：不列颠的人和气候是特别创造出来的，彼此相适应。

另一个方面，自然选择的理论表明：伴随着时间的推移，会有多种多样的适应现象不断产生。物质世界发生的任何事情必然反馈到原本强加于物质的特性上，尽管这看起来不那么完美，但当我们真正理解了这一点时，可以肯定，我们对蕴涵在创造过程中的智慧的认知也必将不断得到提升。持续不断的创造和适应一定是受某种伟大力量支配的，就好像从一开始就是这种力量按照某种始终如一的法则设计并

inventing and adapting forms of life without apparent limit, in pursuance of one uniform principle.

I conceive it to be the essential consequence of Darwin's views that no form of life is to be regarded as a fixed form; but that all living beings, including man, are in a continual process of adjustment to the conditions in which they live. If this be so, it will of necessity follow that the longer any race dwells in given circumstances, the more perfectly will it become adapted to those circumstances. A migratory race, on the contrary, will always be liable to enter climates unsuited to it, and less favourable to the development of the greatest amount of energy. Negroes can bear a tropical heat simply because the race has grown more accustomed to it than Europeans, who bring with them indeed a superior degree of energy and intellect, but soon sicken and fail to reproduce themselves in equal perfection.

The intellect of man renders him far more migratory than most other animals, and when we look over long periods of time we must regard him as in a constant state of oscillation between the equator and the borders of perpetual snow. It will of necessity follow that the race, as a whole, will be better adapted to a medium than to an extreme climate. Not only may the same race have passed alternately through colder and hotter climates, but it is obvious that the tribes which intermix and intermarry in temperate regions will have come, some from a hotter and some from a colder region. The amalgamated race will therefore be precisely adapted to a medium climate. The inhabitants of the Arctic regions, on the contrary, must have come entirely from a warmer climate, and those of a tropical region from a colder climate, so that ages must pass before either re-adapts itself perfectly to its new circumstances.

It is hardly to be expected that history can afford complete corroboration of this theory; but I do not think that historical facts can be adduced in serious opposition to it. The progress of archaeological and linguistic inquiry shows more and more clearly that the civilised parts of the earth have been inhabited by a succession of different races. A really aboriginal and indigenous people, growing upon a single island or spot of ground without kinship with other races, is not known to exist; and it is almost certain that all races have descended from a few stocks, if not from a single one. The evidences of extensive and frequent migrations are thus most complete, even if we had not distinct historical facts concerning the rapid and extensive movements of the Goths, Huns, Moors, Scandinavians, and many other races.

If the historical evidence disagrees with the theory in any point, it is that the migrations from temperate to extreme climates greatly over-balance any opposite movement. It would hardly, perhaps, be too much to represent the temperate regions of the Old World as the birthplace of successive races, which have diverged and died away more or less rapidly in distant and extreme climates. But if such be the conclusion from historical periods, it would only indicate that the human race had already acquired, in prehistoric

创造了各种物质，之后又不受明显约束地继续创造了各种生命形式并使它们不断适应。

我认为达尔文观点的一个必然结论是，没有任何一种生命形式是固定不变的；所有的生物，包括人类，都是处于不断调整改变自己以适应生存环境的过程之中。如果的确如此，那么一个必然的结论是，任何一个种族在给定的环境中生活的时间越长，他们就越适应这种环境。相反，一个种族迁移后，很有可能迁入他们不适应的气候环境，因而不利于他们最大能力的发展。例如，黑人能忍受热带的高温，只不过是因为他们在热带生活了更长时间而更加习惯热带的环境，相反，欧洲人虽然在身体和智力上的确都更有优势，但他们并不习惯热带的环境，因而很快就会生病，并且无法繁衍和他们一样优良的后代。

人类的智慧使他们比其他大部分动物都更具迁移性，当我们纵观历史时，我们可以看出人类一直在赤道和终年积雪的区域之间不断徘徊着。这样就可以得出一个必然的结论，即人类作为一个整体应该更适应于温和的气候而不是极端的气候。不但同一个种族可能交替地经历较冷和较热的气候，而且很明显温带地区将出现混居或通婚的部落，一部分是来自较热地区，而另一部分可能来自较寒冷地区。融合后的种族因此也正好可以适应温和的气候。相反，如果北极地区的居民一定是全部来自较温暖气候的，而热带地区的居民全部是来自较寒冷气候带，那么许多年后他们才能重新适应他们的新环境。

我们很难期待历史能提供事实来完全证实这一理论，但我也相信不会有与这一理论强烈相悖的历史事实。考古学和语言学的研究成果越来越清晰地告诉我们，地球上出现过文明的地方都曾相继居住过不同的种族。至今还没有发现那种生活在一个孤岛或某一片孤立的土地上，和其他外族没有任何亲缘关系的纯粹的土著种族；另外也几乎可以确定，任何一个种族如果不是一个血统的话则肯定是来源于几个血统。就算没发现像哥特人、匈奴人、摩尔人、斯堪的纳维亚人以及其他一些种族的快速而广泛的迁徙这样清楚的历史记录，我们关于广泛而频繁迁移的证据也是充分的。

如果说有历史证据在某些方面与该理论不相吻合，那便是人类从温带地区向极端气候区的迁移要比相反方向的迁移多很多。可能我们可以毫不夸张地认为，旧大陆温带地区就是那些在偏远且气候极端的地区中很快分散并渐渐消失了的后继种族的发祥地。但是，如果通过历史学的研究发现果真如此的话，那只能认为在温带地区的人类在史前就已经获得一种非凡的体质，从而展示出他们强大的生命力。无疑，

times, a constitution displaying its greatest vitality in temperate regions. There can be no doubt that, were the rest of the world uninhabited by man, a very inferior race, such as the negroes of tropical Africa, would gradually re-people it; but they cannot do so in the present state of things, because they come into conflict with races of superior intellect and energy.

I would add in conclusion that the utmost result of speculations of this kind, supposing them to be valid, would consist in establishing a *general tendency*, so that the probabilities will be in favour of a great display of civilisation occurring in temperate climates rather than elsewhere. I do not for a moment suppose that any common physical cause, such as soil, climate, mineral wealth, or geographical position, or any combination of such causes, can alone account for the rise and growth of civilisation in Assyria, Egypt, Greece, Italy, or England. Material resources are nothing without the mind which knows how to use them. No physiology of protoplasm, no science that yet has a name, or perhaps ever will have a name, can account for the evolution of intellect in all its endless developments. The vanity of the Comtists leads them to suppose that their philosophy can compass the bounds of existence and account for the evolution of history; but the scientific man remembers that however complicated the facts which he reduces under the grasp of his laws, yet beyond all doubt there remain other groups of facts of surpassing complication. Science may ever advance, but, like an improved telescope in the hands of an astronomer, it only discloses the unsuspected extent and difficulty of the phenomena yet unreduced to law.

(**1**, 231-232; 1869)

世界上可能还有一些无人居住的地方，那么某些劣等的种族，比如非洲热带地区的黑人，将逐渐移居进去；但是在目前的状况下，他们就无法做到了，因为他们正在与更具智慧和力量的优等种族发生冲突。

最后，我还要再补充一点，假设前面的论述都是正确的话，那么最可能的推论结果将体现为一种**普遍趋势**的建立，那就是来自温带而不是其他地区的文明将有更大的机会得到展示。我一直认为，单靠各种常见的物质因素，如土壤、气候、矿藏或地理位置，或是这些因素的各种组合，是不能解释亚述、埃及、希腊、意大利或英格兰等地区文明的出现和发展的。如果没有利用物质资源的意识，物质本身就是完全无用的。如果没有原生质基础上的生理机能，现在或将来都不会有一门能够对智力进化的整个无穷的发展过程作出说明的科学学科。实证主义哲学家们自夸地认为他们的哲学可以解释世间万物，可以解释历史的演变。但是从事科学的人们清醒地认识到，尽管复杂的事物可以简化到已有的科学规律中，但毋庸置疑，总有更复杂的事物没法简化处理。科学会不断进步，但正如天文学家手中不断改进的望远镜一样，它揭示的只是那些未知的问题和尚未简化成科学规律的疑难现象。

（刘冉冉 翻译；陈平富 审稿）

The Veined Structure of Glaciers

E. Whymper

Editor's Note

Edward Whymper was a renowned British explorer, and the first person to ascend the Matterhorn in the Alps. He commonly made geological observations on his expeditions, and here he weighs in on a debate about glaciers that had involved some leading scientists of the age, including Louis Agassiz, who first proposed the ice ages, and John Tyndall. The point at issue was why some glaciers have vertical or steeply inclined veins. Agassiz regards them as akin to geological strata, caused by the deposition process itself, while Tyndall says they are created subsequently by pressure. Whymper argues that these veins may have more than one cause, and adds the "healing" of crevasses to the possible mechanisms.

I think there is no one point in connection with glaciers more interesting than their veined structure, or one upon which so much has been written that remains equally unsettled. The differences of opinion about it between the authors who have published most upon the subject are not less remarkable than the phenomenon itself: no two are agreed, except in considering it as a constitutional feature.

Professor Agassiz maintains (*Atlantic Monthly*, Dec, 1863) that the horizontal layers of pure ice which are formed between the beds of snow from which a glacier is born, constitute many of the identical veins or plates of pure ice which pervade the glacier when it is in full life and activity; and attributes the inclination which they make, in the latter case, to their former horizontal position, to the contortion, bending, or folding, to which they have been subjected on their downward course; but, at the same time, he distinguishes between these veins—the result of stratification, and others which he terms bands of infiltration, and which he believes to have been formed by the infiltration and freezing of water.

The late Principal J. D. Forbes maintained ("Occasional Papers on the Theory of Glaciers", 13th letter) that the veins of stratification were annihilated at a certain point, and that at precisely the same time other veins, approximately at right angles to the former ones, were formed. These effects he referred to intense pressure.

Professor Tyndall ("Glaciers of the Alps", pp. 380, 425–426), agrees with Professor Forbes "in ascribing to the structure a different origin from stratification", and, if I understand him rightly, does not believe that any portion of the (approximately) vertical veins have such an origin. He divides the veins into marginal, transverse, and longitudinal structure, and asserts that all are produced by pressure, which causes partial liquefaction of the ice, and that the water is refrozen when the pressure is relieved.

冰川的纹理构造

温珀

编者按

爱德华·温珀是英国著名的探险家，也是第一个登上阿尔卑斯山脉马特洪峰的人。在探险过程中，他通常会进行一些地质观测。本文是他在一场关于冰川的辩论中提出的自己的观点。许多杰出的科学家参与了那次辩论，其中包括约翰·廷德尔和第一个提出冰期概念的路易斯·阿加西斯。争论的焦点是为什么一些冰川发育出垂直或者陡倾的纹理。阿加西斯认为这些纹理类似于由沉积过程形成的地层，但廷德尔却提出，纹理的形成与后期压力作用有关。温珀认为产生这些纹理的原因可能不止一个，裂隙的“愈合”可能是这些纹理形成的另一个原因。

我认为，在有关冰川的问题中，冰川的纹理构造是最令人感兴趣的，虽然对冰川纹理进行了大量的研究，但其成因仍没有确定。纹理现象引人注目，不同作者之间的观点分歧也非常明显：除了都认为纹理构造是冰川的本质特征外，没有两个人的意见是一致的。

阿加西斯教授坚持认为，纯冰的水平层形成于冰川的雪层之间，在整个冰川活动期内，它们构成了冰川内部许多类似的纯冰纹理或板条（《大西洋月刊》，1863 年 12 月）。他提出纯冰板条倾斜是由于原来呈水平状的板条在下移过程中发生了变形、弯曲和褶皱。阿加西斯教授还区分出两种冰川纹理，即分层纹理和渗透条带，后者被认为是由于水的渗透和凝结而形成的。

已故的福布斯校长认为：分层纹理在某一时刻会消失，但与分层纹理近于垂直的其他纹理便立刻形成。他认为这是由于（冰川）高压作用造成的（《冰川理论部分研究论文集》，第 13 篇）。

廷德尔教授同意福布斯教授关于“垂直纹理构造不是由分层作用形成”的观点（《阿尔卑斯山脉的冰川》，第 380、425~426 页）。如果我没有理解错的话，福布斯教授不相信任何（近于）垂直的纹理是分层作用的结果。他将纹理分为边缘、横向和纵向三种构造，认为它们都与压力有关，压力可造成冰的部分液化，当压力消失后，水将再次结冰。

If any one cause produced the whole of the veins of pure ice that are found in the imperfect ice of glaciers (which all are agreed are a constitutional feature of those bodies), it is obvious that that cause would have to be equally generally distributed. It is indisputable that all the veins are not veins of stratification, because examples have been frequently observed crossing (cutting) the strata lines at a larger or smaller angle. But although such observations prove conclusively that all the veins must not be attributed to stratification, they do not prove any more. I believe, with Professor Agassiz, for reasons advanced elsewhere*, it can be demonstrated, equally conclusively, that many of the veins which are seen in the lower courses of glaciers in the Alps are veins originally produced by stratification, and dissent entirely from the "annihilation" of Principal Forbes. But as it is proved that some have a different origin, we must look to other causes for an explanation. It is probable that the theories quoted above offer a practical solution of the difficulty, although they are unfortified by direct proofs. But I have seen examples which it was difficult to explain by either one or the other.

There is one means by which the veins might be produced, which, if not overlooked, is at least not generally advanced. All glaciers have crevasses; a glacier is known by its crevasses. The sides of all crevasses become more or less weathered and coated with a glaze of pure ice. When they close up again, when the sides join by virtue of regelation, does this leave no trace? Can it be annihilated? Or, do the two coalesced films leave their mark as a vein of pure ice throughout the generally whitish mass of the glacier? I consider a large number of the veins of ice which constitute the "veined structure" of glaciers as nothing more than the *scars of healed crevasses*.

It is not easy to say whether this was the meaning of the following passage, taken from p. 201 of Forbes's "Occasional Papers": "Most evidently, also, the icy structure is first induced near the sides of the glacier where the pressure and working of the interior of the ice, accompanied with intense friction, comes into play, and the multitudinous incipient fissures occasioned by the intense strain are reunited by the simple effects of time and cohesion." Judged by his preceding pages, it is not, and I am unaware that it has been, advanced in any other place. Some of your readers may perhaps be able to throw some light upon the subject.

(**1**, 266-267; 1870)

* British Association, 1866 (Nottingham).

如果存在一种机制，它在纯冰中产生的所有纹理同样出现在冰川不纯的冰中（普遍认为这种纹理属于冰体的结构特征），那么这个机制应该普遍存在。毋庸置疑，并非所有纹理都是分层纹理，因为经常可以观察到一些纹理以不同角度切割地层线的现象。这种观察虽然明确指示并非所有纹理都是由分层作用引起的，但却不能说明更多的问题。根据其他文献*，我和阿加西斯教授同样可以证明阿尔卑斯山许多冰川前部的纹理最初都是由分层作用造成的。这与福布斯教授“纹理消失”的理论完全相悖。有证据表明部分纹理有别的起因，我们必须寻找其他原因来解释。上面所引用的理论可能有助于解释上述问题，但尚缺乏直接证据，还存在一些实例，不论哪种理论都难以对它们进行解释。

冰川纹理还可以通过另一种方式形成，但这种方式即使没有被忽略，至少也是很少被提及。所有冰川都有冰隙，一个冰川可以通过它的冰隙来辨别。所有冰隙的边缘都会发生不同程度的风化，并且被纯冰薄膜覆盖。当冰隙重新闭合，以及当冰隙边缘被重新凝结的冰连接起来时，难道不会留下痕迹吗？这些痕迹会消失吗？两个冰膜层结合后能在近乎白色的冰川体中留下纯冰的纹理吗？我认为构成冰川“纹理构造”的大量纯冰纹理正是**冰隙愈合后留下的痕迹**。

下面是摘自福布斯的《冰川理论部分研究论文集》第 201 页的一段文字，很难说其中所表述的意思与我的想法一致：“很显然，这种冰结构首先出现在冰川边缘附近，因为冰川内部的压力以及伴随的巨大摩擦力在那里产生效应。由强烈应变产生的大量初始裂痕随时间很容易重新黏合。”从他论文前面几页的论述判断，他表述的意思与我的观点不同。我不知道他是否在其他地方还有论述。一些读者也许能对冰川纹理问题提出自己的看法。

（孙惠南 翻译；孟庆任 审稿）

* 英国科学促进会，1866年（诺丁汉）。

Are Any of the Nebulae Star-Systems?

R. A. Proctor

Editor's Note

In 1870, the nature of many extended astronomical objects generally referred to as "nebulae", apparently devoid of stars, remained mysterious. Some astronomers, notably William and John Herschel, had argued that many of these objects were probably extragalactic star systems: galaxies much like our own, but too far away for individual stars to be resolved. But Richard Proctor here argues that the balance of evidence pointed instead to nebulae being extended objects within our own galaxy. Today we know that these "nebulae" include a wide range of astrophysical objects, including some star clusters in our own galaxy, but also other galaxies, galactic clusters, interstellar clouds of dust and gas, and remnants of supernovae.

THIS may seem a bold question, for it is commonly believed that Sir William and Sir John Herschel—the Ajax and the Achilles of the astronomical host—have long since proved that many of the nebulae are star-systems. If we inquire, however, into what the Herschels have done and said, we shall find that not only have they not proved this point, but that the younger Herschel, at any rate, has expressed an opinion rather unfavourable than otherwise to the theory that the nebulae are galaxies in any sense resembling our own sidereal system.

Sir William Herschel, by his noble plan of star-gauging, proved that the stars aggregate along a certain zone, which in one direction is double. He argued, therefore, that presuming a general equality to exist among the stars and among the distances separating them from each other, the figure of the sidereal system resembles that of a cloven disc. And as the only system from which he could form a probable judgment—I mean the planetary system—presented to him a number of bodies, widely separated from each other and each a globe of considerable importance, he reasoned from analogy that similar relations exist in the sidereal spaces. This being so, his cloven disc theory of the sidereal system seemed satisfactorily established.

Then, of course, those nebulae which exhibit a multitude of minute points of light very close together, and those other nebulae which, while not thus resolvable into minute points, yet in other respects resemble those which are, came naturally to be looked upon as distinct from the sidereal system. The analogy of this system, in fact, pointed to them as external star-systems, resembling it in all important respects.

Then there were certain other objects, which seemed to present no analogy either to the sidereal system or to separate stars. These objects Sir Wm. Herschel considered to belong to our sidereal system; for he could not put them outside its range without looking on

所有的星云都是恒星系统吗？

普洛克特

编者按

1870 年，许多延展型天体的本质仍然是个谜。这些天体通常被称为“星云”，其中似乎看不见恒星。包括著名的威廉·赫歇尔和约翰·赫歇尔在内的许多天文学家们认为，这些天体中大部分可能是类似银河系的河外恒星系统，不过因为过于遥远而无法分辨单个恒星。不过，理查德·普洛克特在这篇文章中指出，各方面的证据综合起来反而更支持星云是银河系内的延展型天体。现在我们知道，这些“星云”所涵盖的天体类型非常广泛，不仅包括一些银河系中的星团，还包括其他一些星系、星系团、由气体和尘埃组成的星际星云，以及超新星遗迹。

这个问题提得也许有些鲁莽，因为威廉·赫歇尔爵士和约翰·赫歇尔爵士——天文学界的埃阿斯和阿喀琉斯——在很久以前就已经证明了很多星云都是恒星系统。然而如果我们深入研究赫歇尔父子说过的话和他们做过的事情，我们就会发现他们俩不但没有证实这个观点，并且小赫歇尔至少还表达了一些不赞同的看法，而没有说明星云就是与我们所在恒星系统类似的星系这一观点。

威廉·赫歇尔爵士凭借他闻名于世的恒星标定法证明了恒星会在一定区域内聚集，并在某个方向上数目加倍。因此他认为，假设恒星和恒星间相隔的距离是均匀分布的，则恒星系统的外形类似于一个裂开的盘。由于他只能根据一个系统——我指的是行星系统——作出可能的判断，这唯一的系统展示在他面前的是许多相距甚远的行星，每一个行星的地位都很重要，于是他由类推法得出相似的关系也存在于恒星之间。这样，他关于恒星系统是一个裂开的盘的理论也就似乎圆满地建立起来了。

另外，有些星云看上去是由许多相距很近的小亮点组成的，而在另一些星云中则无法分辨出微小的光点，但在其他方面与前一种星云相似，后者自然就被看作是有别于恒星系统。实际上，这种星云与恒星系统在所有的重要特征上都很相似，这种相似性向他们表明这种星云是银河系外的恒星系统。

还有某些其他的天体，它们看起来既不像恒星系统也不像一个单独的恒星。威廉·赫歇尔爵士认为，这种天体属于我们的恒星系统，因为，他不能把它们置于

them as objects *sui generis*, which would have been to abandon the argument from analogy. In order to explain their appearance, he suggested that they might be gaseous bodies, by whose condensation stars would one day be formed.

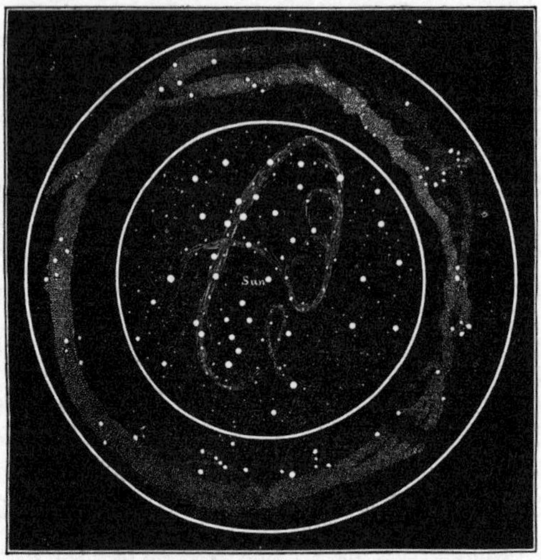

The value of Sir Wm. Herschel's work is not in the least affected even if science have to reject every one of these opinions. He himself held them with a light hand; he had once held other opinions; and he was gradually modifying these. Had he seen one sound reason for rejecting any or all of them he would have done so instantly. For it belonged to the strength of his character that he was never fettered by his own opinions, as weak men commonly are.

Sir John Herschel did for the southern heavens what his father had done for the northern. He completely surveyed and gauged them. It is commonly believed that the results of his labours fully confirmed the opinions which his father had looked upon as probable.

Let us see if this is so.

Sir W. Herschel thought the Milky Way indicated that the sidereal system has the figure of a cloven disc; Sir John Herschel judges rather that the sidereal system has the figure of a flattened ring. Sir Wm. Herschel thought the stellar nebulae are probably external galaxies; Sir John gives reasons for believing that they lie within our system, and Whewell considered that these reasons amount to absolute proof.

It has been further believed and stated that the researches of the elder Struve go far to confirm the opinions put forward by Sir W. Herschel as probable.

Let us inquire how far this is true.

我们的恒星系统范围之外，除非把它们视作自成一类的天体系统，但这样又必须抛弃关于相似性的原则。为了解释这种天体系统的外貌，他认为这些天体系统可能是一些气体，总有一天这些气体会凝聚成恒星。

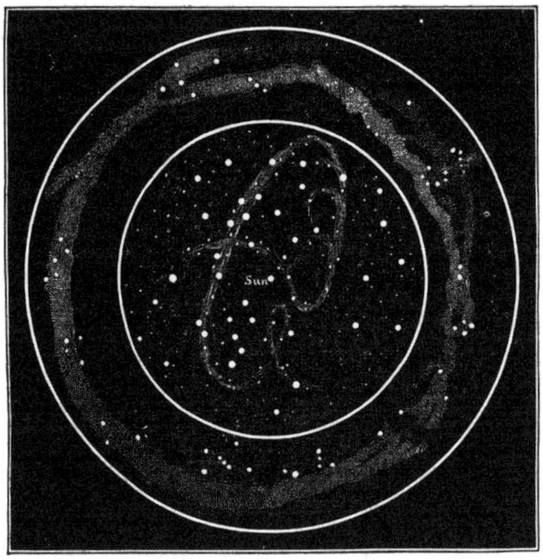

威廉·赫歇尔爵士工作的价值绝对不会因为科学否定了他在这方面的所有观点而受影响。在这些问题上，他的态度灵活多变，他也曾有过其他观点，并逐渐地对这些观点进行修正。假如有可靠的理由让他相信应该放弃他的某些或是全部观点，他就会立即那样做。因为他从来都不会像那些软弱的人们一样被自己的观点束缚，这就是他人格中的闪光之处。

约翰·赫歇尔爵士对南天所做的工作与他的父亲对北天所做的工作一样。他全面地观测并标定了南天的恒星。人们普遍认为，他的研究成果充分肯定了他父亲认为很有可能的观点。

让我们来看一下事实是否如此。

威廉·赫歇尔爵士认为，银河系是一个裂开的盘状恒星系统。约翰·赫歇尔爵士判断，这个恒星系统更像是一个被压平的环。威廉·赫歇尔爵士认为，恒星星云可能是外面的星系。约翰·赫歇尔爵士则给出了一些理由，认为它们处于我们星系内部，休厄尔认为这些理由已经构成了确凿的证据。

人们已经确信，老斯特鲁韦的研究工作在很大程度上证明威廉·赫歇尔爵士提出的观点很可能是正确的。

那就让我们来研究一下他的论断在多大程度上是正确的。

Struve found that the numbers of stars of given magnitudes exhibit nearly the same proportion in different directions. Thus supposing that in a given direction there are three times as many stars of a certain magnitude as there are of the next highest magnitude, then in other directions, also, the same relation is observed. This is a very striking law; but to make it serve as a proof of the opinion which Sir William Herschel had put forward as probable, it would be necessary that another law should be exhibited. For clearly, if that opinion were just, it would be easy to calculate what the relation should be between stars of different magnitudes. Had Struve been able to show that the numbers actually seen corresponded to the relations thus calculated, he would have gone far to render that view certain which Herschel always spoke of as merely an assumption.

But Struve found no such law of stellar distribution. On the contrary, he found a law so different, that in order to force the facts into agreement with Sir William Herschel's views about the sidereal system, he had to invent his famous theory of the extinction of light in traversing space. Now, according to this theory, we cannot see to the limits of our sidereal system, even though we could increase the powers of our telescopes a million-fold; so that if the theory is true, the question which heads this paper is at once disposed of. Obviously, we cannot see galaxies beyond the sidereal system if we cannot see to the limits of that system. And I may note in passing that (independently of Struve's theory) the most powerful telescopes cannot render visible the most distant stars of our sidereal scheme; so that if the nebulae are really external galaxies, the stars we see in them must be enormously greater than those in our galaxies, supposing Herschel was right in thinking these tolerably uniform in magnitude.

Before proceeding to exhibit the evidence which has led me to the conviction that the nebulae belong to our sidereal system, I may mention some reasons for believing that if Sir William Herschel's labours in the sidereal heavens were to be begun now, not only would he not have been led to adopt as probable the view on which he formed his opinions, but he would have rejected it as opposed to known analogies.

He had argued that because the planetary system exhibits a definite number of bodies separated by wide distances, therefore analogy should lead us to regard the sidereal system as similarly constituted, though on a much larger scale. This was perfectly just. Despite the various differences which no one recognised more clearly than he did, this view was the only one he could safely adopt for his guidance, ninety years ago.

But would not he have been the first to reject that view if he had known what we now know of the solar system? If he had known that besides the primary planets, there are hundreds of minute bodies forming a zone between the orbits of Mars and Jupiter; that the rings of Saturn are formed of a multitude of minute satellites; that innumerable meteor-systems circle in orbits of every conceivable degree of eccentricity; that near the sun these systems grow denser and denser; that the comets of the solar system must be counted by millions on millions; that, in fine, every conceivable form of matter, every conceivable degree of aggregation, and every conceivable variety of size, exists within the

斯特鲁韦发现，一定星等的恒星数目在不同方向上所占的比例大致相同。因此，如果假定在一个给定方向上某个星等的恒星数目是比该星等亮一等的恒星数目的3倍，那么在其他方向上，这个关系也会成立。这是一个非常值得注意的定律，但是把它作为证明威廉·赫歇尔爵士提出的论点的证据是不够的，还需要另一条定律。显然，如果那个论点是正确的，将很容易推算出不同星等的恒星之间是什么样的关系。如果斯特鲁韦能够说明实际观测到的数据符合计算得出的关系，那他也就可以确定那个赫歇尔常常只是作为一项假设而提及的论点。

但是斯特鲁韦并没有发现这样一个恒星分布定律。相反地，他发现了一个完全不同的定律，为了强行使实际情况能与威廉·赫歇尔爵士提出的恒星系统的观点保持一致，他不得不发明那个著名的光穿越星际空间时的消光理论。根据这个理论，即便我们把望远镜的放大倍率提高一百万倍，我们也不可能看见自己星系的边界，所以如果这个理论是对的，那么这篇文章标题中提出的那个问题就可以立即迎刃而解了。显然，如果我们无法看见自己星系的边界，也就不能看见外面的星系了。或许我可以顺便提一下（与斯特鲁韦的理论无关），用目前放大倍率最高的望远镜也看不到我们所在星系中离我们最远的恒星，所以如果星云真的是银河系以外的星系，假设赫歇尔认为它们具有差不多相同的星等的观点是对的，那么我们在星云中看见的恒星肯定比我们自己星系中的恒星大许多。

在向你们展示那些使我确信星云属于我们所在星系的证据之前，我先陈述一些理由来说明，如果威廉·赫歇尔爵士现在才开始他对恒星系统的工作的话，那么他不但不可能接受他自己提出的那个观点，还会因为它与现在已知的类推观点相悖而抛弃那个观点。

他曾经争辩说，行星系统是由彼此之间分隔较远的确定数目的星体组成的，因此我们可以类推出恒星系统也具有相似的结构，只不过是在一个更大的尺度上罢了。这是非常合理的。在90年前这是他唯一可以放心采用的观点，尽管对于这两种系统的各种差别没有人比他认识得更清楚。

但是如果他可以了解到我们现在所掌握的关于太阳系的知识，他会不会第一个抛弃他那个观点呢？如果他知道，除了几个主要的大行星之外，在火星轨道和木星轨道之间有一个由数百颗小星体构成的带；土星环是由许多微小的卫星聚集而成；无数的流星群可以在任意可想象到的偏心率轨道上环绕运行；越靠近太阳，星体的密度越大；太阳系中的彗星数以百万计；总之，任何可以想象到的物态、任何可以想象到的凝聚程度、任何可以想象到的尺度的物质在太阳系范围内都存在，难道他能以此类推出恒星系统中只有离散的恒星和正在形成恒星的物质吗？

limits of the solar system,—would he, then, have been led by analogy to recognise in the sidereal system only discrete stars and masses forming into stars?

From a careful study of all that Sir William Herschel has written, I feel certain, that in the case I have imagined, he would have been prepared, even before commencing his labours, to expect precisely that variety of matter, size, and aggregation, which modern observations, rightly understood, prove actually to exist within the range of the sidereal system.

The Herschels, father and son, discovered about 4,500 nebulae. Other observers have brought up the number to about 5,400. When these are divided into classes, it appears that some 4,500 must be looked on as irresolvable into discrete points of light. But of these the greater proportion so far resemble resolvable nebulae as to lead to the belief that increase of optical power alone is wanting to resolve them.

Taking these irresolvable nebulae, however, as we find them, and marking down their places over the celestial sphere, we recognise certain peculiarities in their arrangement. In the northern heavens they gather into a clustering group as far as possible from the Milky Way. In the southern heavens they form into streams, which run out from a region nearly opposite the northern cluster of nebulae; but the *extremities* of the streams are the region where nebulae are most closely crowded. The Milky Way is almost clear of nebulae.

This withdrawal of the nebulae from the Milky Way has been accepted by many as clearly indicating that there is no association between them and the sidereal system. The opinion of the Herschels, if they had been led to pronounce definitively on this point, would have been different, however; for the younger Herschel quotes (as agreeing with it) a remark of his father's to the effect that the peculiar position of the northern nebular group is not accidental. If not accidental, it can only be due to some association between the nebular group and the galaxy. Every other conceivable explanation will be found to make the relation merely apparent—that is, accidental, which neither of the Herschels admit.

But yet stronger evidence of association exists; evidence which I do not hesitate to speak of as incontrovertible. Space will only permit me to treat it very briefly.

There is a certain well-marked stream of nebulae in the southern heavens leading to a well-marked cluster of nebulae. There is an equally well-marked stream of stars leading to an equally well-marked cluster of stars. The nebular stream agrees in position with the star-stream, and the probability is small that this coincidence is accidental. The nebular cluster agrees in position with the star-cluster, and the probability is still smaller that this second coincidence is accidental. Such are the separate chances. It will be seen at once, therefore, how small the chance is that both coincidences are accidental.

The cluster here referred to is the greater of the celebrated Magellanic Clouds. When it is added that the evidence is repeated point for point in the case of the lesser Magellanic

在仔细研究了威廉·赫歇尔爵士的所有著作之后，我确实感到，如果情况正如我已经设想过的那样，那么他甚至可能会在开始他的工作之前就已经做好明确地预期各种不同物质、尺度和凝聚度的存在的准备，这些已被现代观测手段证实确实存在于恒星系统之中。

赫歇尔父子发现了大约4,500个星云。其他观测者又把这个数字提升到了5,400。当对这些星云进行分类时，大约有4,500个星云是不能分辨出离散光点的。但是这些星云中的大多数都非常类似于那些可以从中分辨出离散光点的星云，这使人们确信只要提高望远镜的放大倍率就可以分辨它们。

然而，当挑出这些不能从中分辨出光点的星云，并把发现它们的位置标在天球上时，我们发现这些星云的排布有些奇特。在北天，它们在尽可能远离银河的地方聚成一团。在南天，它们形成许多星云流，发端于与北方星云团正对的位置附近，但是在这些星云流的**尽头**则是星云最密集之处。银河区域内几乎没有星云。

很多人认为，银河区域没有星云已经清楚地说明星云和我们所在的恒星系统没有关联。然而如果要求赫歇尔父子在这个问题上明确表态，那他们的观点就会与此不同；因为小赫歇尔提到过（以赞同的口吻），他父亲曾评论说北天区星云团的特殊位置并非出于偶然。如果不是偶然现象，那只能是由于星云团与我们的星系之间存在相关性。所有能找到的其他解释只不过是让它们之间的关系更明显而已——即赫歇尔父子都不能接受偶然性的存在。

但是有更强有力的证据表明相关性的存在。我可以毫不犹豫地说这个证据是不容置疑的。但是由于版面所限，我只能对此作简要的介绍。

在南天有一个十分明显的星云流，一直通向一个十分明显的星云团。同样存在一个十分明显的恒星流一直通向同样十分明显的恒星团。星云流的位置与恒星流的位置相符，这种一致性源于巧合的可能性是很小的。星云团在位置上也与恒星团相符，这第二种一致性源于巧合的可能性更小。这两个事件是相对独立的。于是我们立刻可以想到，这两个事件都源于巧合的几率会是多么小。

这里提到的星团就是著名的大麦哲伦星云。当我们把小麦哲伦星云的情况也考虑进来的时候，该现象又一次重复出现了，这说明这种相关性的存在是毋庸置疑的。

Cloud, the indications of association appear overwhelmingly convincing. If the nebulae really are associated in this manner with fixed stars, the question which heads this paper is disposed of at once.

But there is yet further evidence.

The nebulae pass by insensible gradations from clusters less and less easily resolvable, to nebulae properly so called, but still resolvable, and so to irresolvable nebulae. Now clusters are found not only to aggregate in a general manner near the Milky Way, but in some cases (on which Sir John Herschel has dwelt with particular force) to exhibit the clearest possible signs of belonging to that zone. If they then belong to the Milky Way, can any good reason be given for believing that the various other classes of nebulae are not associated with the sidereal scheme? Where should the line be drawn?

Again, some of the nebulae are gaseous, and all the gaseous nebulae exhibit the same spectrum. Now, two classes of gaseous nebulae, the planetary and the irregular nebulae, exhibit a marked preference for the Milky Way, and therefore we must admit the probability that they, at any rate, belong to the sidereal scheme. But then a large proportion of the irresolvable nebulae are also gaseous, and as they are formed of the same gases, we see good reason for believing that they also must belong to our galaxy. This, however, brings in all the nebulae, since the recent detection by Lieut. Herschel of the same bright lines in or rather on the continuous spectrum of a star-cluster, shows the great probability which exists that with more powerful spectroscopes all the nebulae may be found to exhibit these bright lines, that is, to contain these particular gases. I pass over the facts, that many nebulae are found to be closely associated with stars, and that if any doubt could remain as to the association being real and not apparent, it would be removed by a picture of the nebula M 17, as seen in Mr. Lassell's great reflector at Malta. The reader will be more interested by the following quotation, which I extract (by permission) from a letter of Sir John Herschel's:—

"A remark which the structure of Magellanic Clouds has often suggested to me has been strongly recalled by what you say of the inclusion of every variety of nebulous or clustering forms within the galaxy, viz., that if such be the case—*i.e.* if these forms belong to, and form part and parcel of the Galactic system—then that system includes *within itself miniatures of itself* on an almost infinitely reduced scale; and what evidence, then, have we that there exists a universe beyond—unless a sort of argument from analogy, that the Galaxy with all its contents may be but one of these miniatures of a more vast universe, and that there may, in that universe of other systems on a scale as vast as our galaxy, be the analogues of those other nebulous and clustering forms which are not miniatures of our galaxy?"

It will be seen that, while Sir John Herschel is quite ready (should the evidence require it) to adopt altogether new views about the nebulae, he is not ready to forego the grandeur

如果星云真的以这种方式与固定的恒星相联系，那么这篇文章标题中的问题就会马上迎刃而解。

但是这里还有更进一步的证据。

通过难以察觉的渐变，星云从非常不容易分辨的星团，逐渐过渡到严格来说可被称为星云但仍然可以分辨的阶段，最终成为不可分辨的星云。现在发现，星团不仅以普通的方式在银河系附近聚集，而且在一些情况下（约翰·赫歇尔爵士在研究这个问题时曾认为存在一种特殊的作用力）还很明白地显示出可能属于那个区域的迹象。如果它们是属于银河系的，那么有什么好的理由可以使我们相信其他类型的星云与我们的恒星系统没有关系呢？这个界线应该划在哪里呢？

此外，有一些星云是气态的，并且所有的气态星云都具有相同的光谱。有两种气态星云，即行星状星云和不规则星云，具有明显的与银河系相似的谱线特征，因此我们必须承认这两种星云有可能属于我们所在的这个恒星系统。但是很大一部分不可分辨的星云也是气态的，由于它们也是由同样的气体组成的，因此我们就有充分的理由相信这些星云也属于我们所在的星系。因为赫歇尔中尉最近在一个恒星团的连续光谱中，或者更确切地说是在连续光谱上，发现了相同的亮线，这表明使用分辨率更高的分光镜很可能会让我们在所有星云的光谱中都能找到这些亮线，也就是说这些星云中都含有同样的气体。我在此略去了一些论据，诸如人们发现很多星云与恒星具有密切的相关性，如果有人对星云与恒星间关系的真实性和明确性表示怀疑，那么拉塞尔先生在马耳他使用大型反射式望远镜对 M17 星云进行观测所得到的一幅照片应该可以消除这方面的疑问。读者可能会对下面我从约翰·赫歇尔爵士的一封信中摘录出来（已获同意）的片段更感兴趣：

“你所提出的银河系中每一种星云和团状结构中包含的物质总使我回忆起麦哲伦星云的结构给我的启发，即，如果事实确实如此——也就是说，假如这些星云和团状结构属于或者组成了银河系的一部分或一块——那么这个系统内部就包含了许多在尺度上无限缩小的**它自身的微结构**。但是，如果不是通过类推法，我们又有什么证据能够证明在星系外部的宇宙的存在，银河系连同它内部所包含的一切物质只不过是更大宇宙中众多微结构中的一个，以及在其他与银河系同样大小的星系中也许会有与我们的星系类似但又绝非与我们星系的微结构一样的星云和团状物呢？”

由此可见，虽然约翰·赫歇尔爵士十分乐意（需要一些证据）全盘接受关于星云的新观点，但他不愿破坏他与他父亲建立的神圣的宇宙观，正是这个宇宙观使他

of those noble views of the universe which he and his father have established, thereby earning the well-deserved gratitude of every lover of astronomy.

And then with regard to the actual form of our galaxy or Milky Way, the figure introduced shows that its apparent one as projected on the heavens may really be due to an arrangement differing both from the cloven disc or flattened ring, a point to which I shall return in a subsequent article.

(**1**, 331-333; 1870)

们理所当然地赢得了每一位天文爱好者的感激。

有关我们所在星系或银河系的形状，文中插入了一张银河系在天球上的投影图，图中显示的排列确实既不像是裂开的盘，也不是被压平的环。我会在以后的文章中讲述这个问题。

（史春晖 翻译；邓祖淦 审稿）

Where Are the Nebulae?

H. Spencer

Editor's Note

Stimulated by Proctor's arguments on the nature of nebulae, the polymath Herbert Spencer notes here that he too had previously questioned the popular view that nebulae were distant galaxies like our own. He points out that some astronomers had noted correlations between the apparent locations of some nebulae and stars in our local galaxy, which should not exist if nebulae were much more distant. Spencer also argues that the relationship between the apparent luminosity of astronomical objects and their distance also works against the idea. In general, larger nebulae should be closer, and therefore easier to resolve into independent stars, whereas in practice, on the contrary, smaller nebulae were more easily resolvable.

MR. Proctor's interesting paper in your last number reminded me of an essay on "The Nebular Hypothesis", originally published in 1858, and re-published, along with others, in a volume in 1863 ("Essays: Scientific, Political, and Speculative". Second Series), in which I had occasion to discuss the question he raises. In that essay I ventured to call in question the inference drawn from the revelations of Lord Rosse's telescope, that nebulae are remote sidereal systems—an inference at that time generally accepted in the scientific world. On referring back to this essay, I find that, besides sundry of the reasons enumerated by Mr. Proctor for rejecting this inference, I have pointed out one which he has omitted.

Here are some of the passages:—

"'The spaces which precede or which follow simple nebulae,' says Arago, 'and, *à fortiori*, groups of nebulae, contain generally few stars. Herschel found this rule to be invariable. Thus, every time that, during a short interval, no star approached, in virtue of the diurnal motion, to place itself in the field of his motionless telescope, he was accustomed to say to the secretary who assisted him, "Prepare to write; nebulae are about to arrive."' How does this fact consist with the hypothesis that nebulae are remote galaxies? If there were but one nebula, it would be a curious coincidence were this one nebula so placed in the distant regions of space as to agree in direction with a starless spot in our own sidereal system? If there were but two nebulae, and both were so placed, the coincidence would be excessively strange. What, then, shall we say on finding that they are habitually so placed? (the last five words replace some that are possibly a little too strong) . . . When to the fact that the general mass of nebulae are antithetical in position to the general mass of stars, we add the fact that local regions of nebulae are regions where stars are scarce, and the further fact that single nebulae are habitually found in comparatively starless spots, does not the proof of a physical connection become overwhelming?"

星云在哪里?

斯宾塞

编者按

在普洛克特关于星云本质的论述的启发下，博学的赫伯特·斯宾塞发表了这篇文章，文中他提到，先前他也质疑过星云是类似于银河系的遥远星系这一流行的观点。斯宾塞指出，有些天文学家已经注意到一些星云的视位置与我们所在的银河系中某些恒星位置之间的相关性，如果星云是非常遥远的类似银河系的星系，那么这种现象就不该存在。斯宾塞还指出，天体视光度与其距离之间的关系也不支持流行的观点。简言之，越大的星云应该越近，从而应该越容易分辨出独立的恒星，但是实际情况与此完全相反，倒是越小的星云中越容易分辨出恒星。

普洛克特先生在上一期《自然》中发表了一篇关于星云的饶有趣味的文章，这使我想起我的一篇题为《星云假说》的论文，这篇关于星云的论文最初发表于1858年，后来又与其他文章一起再次发表于1863年的某部书上（《论文集：科学、政治与推理》，第2部），我在该书中也讨论了相同的问题。当时科学界普遍认为星云是遥远的恒星系统，这一观点是根据罗斯勋爵用望远镜观测到的新结果推出的，我在自己的论文中对这一推论提出了异议。再次回顾普洛克特先生的文章，我发现除了普洛克特先生在他的文章中列举的各种反驳以上推论的理由之外，我曾经指出过的一条原因被他遗漏了。

以下几个段落是从我的文章中节选出来的：

“阿拉戈说：‘在单个星云或更准确地说在星云群前后的星际空间，一般只包含很少的恒星。赫歇尔发现这一规则是永恒不变的。因此，每当由于周日运动，在一个比较短的时间间隔内没有恒星进入他的固定望远镜的视场的时候，他总是习惯对助手说：“准备记录，星云马上就要来了。”’这一现象怎么能够和星云是遥远的星系这一假说一致呢？如果只存在一个星云，为什么它在遥远宇宙中的位置恰好分布于我们所在恒星系统中没有恒星的方向上呢？如果只存在两个星云，而二者的位置都是如此，这岂不是更令人不可思议吗？那么，我们将如何解释星云为什么总处于这样的特殊位置呢？（最后5个词替换掉了以前可能有点偏激的措词）……考虑到大部分星云处在与绝大多数恒星相反的位置上这一事实，再加上恒星的数量在星云所在区域相当稀少，以及单个星云通常位于几乎没有恒星的地方这些事实，难道这些证明它们之间具有物理关联的证据还不够充分吗？”

The reasonings of Humboldt and others proceeded upon the tacit assumption that differences of apparent magnitude among the stars result mainly from differences of distance. The necessary corollaries from this assumption I compared with the hypothesis that the nebulae are remote sidereal systems in the following passage:—

"In round numbers, the distance of Sirius from the earth is a million times the distance of the earth from the sun; and according to the hypothesis, the distance of a nebula is something like a million times the distance of Sirius. Now, our own 'starry island, or nebula', as Humboldt calls it, 'forms a lens-shaped, flattened, and everywhere-detached stratum, whose major axis is estimated at seven or eight hundred, and its minor axis at a hundred and fifty times the distance of Sirius from the earth.' And since it is concluded that our solar system is near the centre of this aggregation, it follows that our distance from the remotest parts of it is about four hundred distances of Sirius. But the stars forming these remotest parts are not individually visible, even through telescopes of the highest power. How, then, can such telescopes make individually visible the stars of a nebula which is a million times the distance of Sirius? The implication is, that a star rendered invisible by distance becomes visible if taken two thousand five hundred times further off!"

This startling incongruity being deducible if the argument proceeds on the assumption that differences of apparent magnitude among the stars result mainly from differences of distance, I have gone on to consider what must be inferred if this assumption is not true; observing that "awkwardly enough, its truth and its untruth are alike fatal to the conclusions of those who argue after the manner of Humboldt. Note the alternatives":—

"On the one hand, what follows from the untruth of the assumption? If apparent largeness of stars is not due to comparative nearness, and their successively smaller sizes to their greater and greater degrees of remoteness, what becomes of the inferences respecting the dimensions of our sidereal system and the distances of the nebulae? If, as has lately been shown, the almost invisible star, 61 Cygni, has a greater parallax than α Cygni, though, according to an estimate based on Sir W. Herschel's assumption, it should be about twelve times more distant—if, as it turns out, there exist telescopic stars which are nearer to us than Sirius, of what worth is the conclusion that the nebulae are very remote, because their component luminous masses are made visible only by high telescopic powers? . . . On the other hand, what follows if the truth of the assumption be granted? The arguments used to justify this assumption in the case of the stars, equally justify it in the case of the nebulae. It cannot be contended that, on the average, the *apparent* sizes of the stars indicate their distances, without its being admitted that, on the average, the *apparent* sizes of the nebulae indicate their distances—that, generally speaking, the larger are the nearer, and the smaller are the more distant. Mark, now, the necessary inference respecting their resolvability. The largest or nearest nebulae will be most easily resolved into stars; the successively smaller will be successively more difficult of resolution; and the irresolvable ones will be the smaller ones. This, however, is exactly the reverse of the fact. The largest nebulae are either wholly irresolvable, or but partially resolvable under the highest telescopic powers; while a great proportion of quite small nebulae are easily resolvable by far less powerful telescopes."

洪堡和其他研究者的推理基于以下默认的假设：各个恒星视星等的不同主要源于距离的差异。我对这一假设和认为星云是遥远的恒星系统的假说进行过比较。下面的段落给出了上述假设的一些必然的推论：

“如果以整数计算，天狼星与地球之间的距离是地球与太阳距离的100万倍；根据星云假说，一个星云离我们的距离大约是我们与天狼星之间距离的100万倍。这样，洪堡所谓的‘恒星岛或星云’就会‘形成扁平的、具有透镜形状的、处处分离的层，其长轴大约是天狼星与地球之间距离的700或800倍，其短轴则大约是天狼星与地球之间距离的150倍。’既然认为我们的太阳系靠近这一聚集体的中心，那么我们与聚集体最远端的距离大约是我们与天狼星的距离的400倍。然而即使借助目前威力最大的望远镜，也不能逐个分辨出位于聚集体最远端的恒星。那么，用这样的望远镜如何才能分辨出这些100万倍于天狼星距离的星云中的单个恒星呢？这意味着，一颗由于距离过远而无法观测到的恒星，在将其距离拉远2,500倍后，反而可以观测到了！”

如果承认恒星之间视星等的区别主要源于距离的差异，那么从以上结果得到的推论会存在惊人的矛盾。因此，我继续思考，如果该假设不成立，那么究竟可以推出什么。我也注意到“情况非常尴尬，无论是否正确，由洪堡的方法得出的结论都是同样致命的。只能考虑其他可能的选择”：

“一方面，如果这一假设不成立我们会得到什么结论呢？如果一颗看上去体积巨大的恒星不是由于离我们很近，而看上去尺寸较小的恒星也不是由于它们非常非常遥远，那么我们根据太阳系的尺寸和星云的距离又能得出什么推论呢？就像最近已经被指出的，如果天鹅座61这颗几乎不可见的恒星比天鹅座α星的视差更大，虽然按照赫歇尔爵士的假设推算，天鹅座61离我们的距离约为后者的12倍——但是如果事实证明存在可以用望远镜观测到的比天狼星离我们更近的恒星，那么因为只有通过高倍望远镜才能观测到星云中的发光物质而认为星云非常遥远又有什么价值呢？……另一方面，如果假设成立又能得到什么结论呢？用于证明该假设在恒星情况下成立的论据，同样也可以用于星云。既然认为恒星的**视**尺寸一般可以表明它们与我们的距离的远近，那么就必须承认，星云的**视**尺寸也同样对应于它们的距离——即一般而言，尺寸越大表明距离越近，尺寸越小表明距离越远。下面考虑由星云的可分辨性得到的推论。最大或者最近的星云将最容易从中分辨出恒星，而越小的星云则越难从中分辨出各个恒星，那些无法分辨的星云将会是尺寸较小的星云。然而，这与事实刚好相反。即使用最高放大倍率的望远镜观测，那些最大的星云不是完全不可分辨就是只能部分分辨，而大部分尺寸很小的星云，反而可以借助倍率低得多的望远镜轻易地分辨出来。”

At the time when these passages were written, spectrum-analysis had not yielded the conclusive proof which we now possess, that many nebulae consist of matter in a diffused form. But quite apart from the evidence yielded by spectrum-analysis, it seems to me that the incongruities and contradictions which may be evolved from the hypothesis that nebulae are remote sidereal systems, amply suffice to show that hypothesis to be untenable.

(**1**, 359-360; 1870)

Herbert Spencer: 37, Queen's Gardens, Jan. 31.

在写下以上段落的时候，光谱分析尚未得出结论性的证据，即我们现在知道的许多星云由弥散物质组成。但是除了光谱分析的结果可以作为证据以外，我认为由星云是遥远的恒星系统这一假说推导出来的自相矛盾的结果就足以表明该假说是站不住脚的了。

（金世超 翻译；邓祖淦 审稿）

The Measurement of Geological Time

A. R. Wallace

Editor's Note

The measurement of geological sedimentation—together with biological evolution—made it clear that the Earth was much older than conventional wisdom had assumed. But how much older? When Alfred Russel Wallace turned his attention to the problem, radiometric dating still lay far in the future, and there was no clear way of estimating the Earth's antiquity. Wallace found some solace in "a series of admirable papers in the *Philosophical Magazine*" in which Scottish geologist James Croll (1821–1890) advanced his ideas of climate change based on cyclic changes in the eccentricity of the Earth's orbit and the precession of the equinoxes. But could this idea, applied to the relatively recent ice ages, be thrown back into the far past? In 1870, nobody knew.

MODERN geological research has rendered it almost certain, that the same causes which produced the various formations with their imbedded fossils, have continued to act down to the present day. It has therefore become possible that, by means of changes which are known to have occurred in a given number of years, some measurement of the time represented by the whole series of geological formations might be obtained. It is true, that changes in the earth's surface, the records of which constitute the materials for geological research, occur very slowly, yet not so slowly as to be quite imperceptible in historical time. Land has risen or sunk beneath the sea, rivers have deepened their channels and have brought down sediment which has converted water into land, cliffs have been eaten away and the surface of the earth has been, in many ways, perceptibly and measurably altered during an ascertained number of centuries. But it is found that these changes are too minute, too limited and too uncertain, to afford the basis of even an approximate measurement of the time required for those grand mutations of sea and land, those contortions of rocky strata many thousands of feet thick, those upheavals of mountain-chains and that elaborate modelling of the surface into countless hills and valleys, with long inland escarpments and deep rock-bound gorges, which form the most prominent and most universal characteristics of the earth's superficial structure. Another deficiency in this mode of measurement arises from the fact, now universally admitted, that the record of past changes is excessively imperfect, so that even if we could estimate with tolerable accuracy the time required to deposit and upheave the series of strata of which we have any knowledge; still that estimate would only represent an unknown proportion, perhaps a minute fraction of the whole time which has elapsed since the strata began to be formed.

But there is another class of geological phenomena which enable us to measure those very gaps in the record of which we have just spoken, and it is now generally admitted that the continual change of the forms of animal and vegetable life which each succeeding

地质时间的测量

华莱士

编者按

通过对地质沉积和生物进化的测定，人们了解到地球比先前猜想的古老得多。但到底有多古老呢？在阿尔弗雷德·拉塞尔·华莱士关注这个问题的年代，放射性定年法还属于很遥远的未来，当时没有任何能够确切估计地球年龄的方法。在此之前，苏格兰地质学家詹姆斯·克罗尔（1821~1890）根据地球轨道偏心率和岁差的周期性变化提出了气候变迁的观点，并就这方面的成果"在《哲学杂志》上发表了一系列令人钦佩的论文"。华莱士正是从这些论文中得到了一些启发。但是，那些应用在相对较晚的冰川时代的观点能用在更早的时代吗？1870年时，没人知道答案。

现代地质学研究已经可以确定，形成埋有不同化石的各种地层的活动至今仍在发挥作用。因此有可能通过在一定年代中发生的已知变化来确定代表一系列地质过程发生的时间。地球表层的变化的确非常缓慢，但并没有慢到在历史时期内无法感知的地步，记录地球表层变化也是地质学的研究范畴。陆地抬升或者下沉到海底；河流冲刷下切了河床，并用带来的沉积物把曾经有水的地方变成陆地；悬崖峭壁被侵蚀；地球表面在若干世纪里以各种形式发生着变化，这些变化都是可以被人感知和测量的。但是，想要确定下面一些过程的时间，这些变化就显得太渺小、太局限、太不确定了，因此不能作为提供以下现象哪怕是估算时间的依据，诸如海陆的巨大变迁，几千英尺厚的岩层扭曲，山脉的隆起，以及将内陆高峭的陡崖和巉岩深谷塑造成不计其数的山丘和溪沟，而后者正是地球表面结构最显著和最普遍的特征。这种测量模式的另外一个不足在于：现在普遍认为，对远古变化的记录太不完整，因此即使我们能够用可接受的精度去估算迄今已知的一系列地层沉积和抬升需要的时间，估算时间所占的比例也是未知的，也许只是地层开始形成以来经历的全部时间中的一小部分。

但是有一类地质现象可以让我们度量上述地质记录中的那些空缺时期，我们已知的动物和植物的生命演化是目前人们普遍接受的能够胜任估算地质年代相对长度

formation presents to us, affords the best means of estimating the proportionate length of geological epochs. Though we have no reason to think that this change was at all times effected by a uniform and regular process; yet believing, as we now do, that it was due to the action of a vast number and variety of natural causes acting and reacting on each other, according to fixed general laws, it seems probable that, with much local and temporary irregularity, there has been on the whole a considerable degree of uniformity in the rate at which organic forms have become modified. It may indeed be the case that this rate of variation has continually increased or diminished from the first appearance of life upon the earth until the present day, or has been subject to temporary changes; but so long as we have no proof that such was the case, we shall be safer in considering that the change has been tolerably uniform.

To measure geological time, therefore, all we require is a trustworthy unit of measurement for the change of species; but this is exactly what we have not yet been able to get; for the whole length of the historical period has not produced the slightest perceptible change in any living thing in a state of nature. Moreover, though, the much longer time that has elapsed since the Neolithic or Newer Stone age, has been sufficient for some changes of physical geography and has, to some extent, altered the distribution of animals and plants, it has not effected any alteration in their form. It is only when we get back to the Palaeolithic or Older Stone age, when men used chipped flints for weapons and Europe was, probably, either just emerging from the severity of the glacial epoch, or in some of the intercalated milder periods, that we meet with a decided change in the forms of life. Elephants and rhinoceroses, bears, lions and hyenas then inhabited Europe; but they were nearly all of species slightly different from any now existing, while the reindeer, the musk-sheep, the lemming and some other animals, were the same as those that still live in the Arctic regions: All the mollusca, however, were identical with living species. In the newer Pliocene Crag, on the other hand, which seems to have been deposited just as the glacial epoch was coming on, there are 11 percent of extinct species of shells and about 55 percent of extinct mammalia. What we want, therefore, is to be able to estimate, by means of the physical changes before alluded to the time since the beginning or the end of the glacial epoch. Then we should have the unit we require for measuring geological time by the repeated changes in the forms of life as we go further and further back into the past; but before showing how this may perhaps be done, something must be said about physical and astronomical determinations of the age of our globe.

A few years ago, Sir W. Thomson startled geologists by placing a limit to the time at their disposal, which they had been in the habit of regarding as practically infinite. He showed, from the known laws of heat and the conservation of energy, that there are determinable limits to the age of the sun. Then, applying the same principles to the earth, he showed that, from the known increase of heat towards its interior and from experiments on the rate of cooling of various rocks, it cannot have existed in a habitable state for more than about one hundred million years. It is within that time, therefore, that the whole series of geological changes, the origin and development of all forms of life, must be comprised. But, geologists had been accustomed to demand a much vaster period than this for

的最好方法。虽然我们没有理由认为这类变化在所有时间里都在匀速地和有规律地进行，但我们相信，正如我们现在所做的那样，由于大量不同类别的自然因素之间存在作用和反作用，按照一般规律，很可能出现局部的和暂时的不规则性，但生物体的进化速度在整体上仍是相当均匀的。从生命形式最初在地球上出现一直到今天，其演化的速度可能确实曾持续增加或减少过，或者曾经历过一些短期的变化，但是直到现在我们还没有这方面的实例证据。我们可以有把握地认为，生命的演变基本上是匀速进行的。

因此，为了测定地质年代，我们只需要为物种变化确定一个可靠的测量单位，但这恰恰是我们至今还没有解决的问题，因为在人类有记载的全部历史时期中，自然界的生物体本身还没能产生人类能够感觉到的哪怕是最轻微的变化。况且，虽然从新石器时代至今这段相当漫长的时间里，自然地理已发生了某些变化，而且在一定程度上改变了动植物的分布，但是并没有对生物形态的演化产生影响。只有当我们回到旧石器时代，当人类使用打制的石器作为武器，而欧洲可能或是刚刚度过严酷的冰川期，或是处在某个比较温暖的过渡时期的时候，这些生物形态才有了决定性的变化。后来，大象、犀牛、狗熊、狮子和鬣狗来到欧洲定居，但几乎所有这些物种都与现存的物种只有少许的差别，而驯鹿、麝牛、旅鼠和其他一些动物与现在仍然生活在北极的同种动物是一样的，所有的软体动物也都和现存物种完全相同。另一方面，在似乎是冰川期即将到来时沉积的上新世晚期的沙质泥灰岩中，有 11% 的贝壳类灭绝，有 55% 的哺乳动物灭绝。因此，我们希望能够利用自然界的上述变化去估算冰川期开始或结束的时间。然后，当进一步地追溯到更古老的时代时，我们应该有一个用生命形态反复变化折算出的单位去测量地质年代；但是在表述我们可能会怎么做以前，必须先说一说物理学界和天文学界对地球年龄的测定。

几年前，汤姆孙爵士提出的时间有限性震惊了地质学界，长期以来地质学家们都习惯于认为时间实际上是无限的。他指出，根据已知的热力学定律和能量守恒定律可以确定太阳的年龄。那么，将同一原理应用于地球，他认为，根据已知的地表向地球内部转移的热量值和有关各种岩石冷却速度的实验结果，地球上适宜居住的时间不会超过一亿年，因此所有的地质变化，各种形式生命的起源和发展，都应该包含在这个时间段里。但是地质学家们已习惯于认为在地壳中形成含有化石的沉积物需要比这更长的时间；而达尔文先生的研究几乎可以肯定，虽然从志留纪和寒武

the production of the series of fossiliferous deposits in the crust of the earth; while the researches of Mr. Darwin render it almost certain that, however vast the time since the Silurian and Cambrian epochs, yet anterior to these, at least an equal, and probably a much longer, series of ages must have elapsed since life first appeared upon the earth, in order to allow for the slow development of the varied and highly organised forms which we find in existence at those early epochs. Sir Charles Lyell is not disposed to admit the accuracy of these calculations, and Professor Huxley has criticised them in detail, with a view of showing that they are, in many respects, unsound; while Mr. Croll as strenuously maintains that they are sound in principle and accurate within certain limits.

We have now to consider the bearing of Astronomy upon the problem. In a series of admirable papers in the *Philosophical Magazine*, Mr. Croll has fully discussed the question, how far variations in the excentricity of the earth's orbit, together with the precession of the equinoxes, have produced variations of climate in past ages. He has endeavoured to show that the date of the last glacial epoch and those preceding, may be determined by such considerations. With this view he has laboriously calculated tables showing the amount of excentricity for a period of three million years, at intervals of 10,000 years for a large portion of that time, and 50,000 for the remainder. These tables show that the amount of excentricity is alternately great and small at intervals of 50,000 or 100,000 years, as represented with sufficient accuracy in the diagram, which I have constructed by means of his figures. Owing to the precession of the equinoxes, combined with the revolution of the apsides, either pole will be presented towards the sun (constituting summer in that hemisphere and winter in the opposite one) at a different point in the earth's orbit on each succeeding year, the motion being such as to cause a complete revolution in 21,000 years. If, therefore, at any one period, winter in the northern hemisphere occurs when the earth is nearest the sun or in *perihelion* (as is the case now), in 10,500 years it will occur in *aphelion*; at the one period the winters will be shorter and warmer, at the other longer and colder. When the excentricity is great (say two, three, or four times what it is now), Mr. Croll shows that, from the known laws of heat in reference to air and water, winter in *aphelion* will lead to an accumulation of snow, in the polar regions, which the summer will not be able to melt. This will go on increasing for many thousand years, till winter occurs near the *perihelion*, when the snow will be melted and transferred to the opposite pole. When the excentricity was very great a glacial epoch would occur in each hemisphere for more or less than 10,500 years, the other portion of the period of 21,000 years being occupied by an almost perpetual spring, with two transition periods from that to the glacial epoch. By examining the diagram of excentricity, we see that during the last three million years there have been more than twelve periods of great excentricity, each long enough to admit two or three, and several of them eight or ten, complete revolutions of the equinoctial points, thus sufficing for the production of not less than fifty or sixty glacial epochs in each hemisphere, with intervening phases of perpetual spring or summer.

纪以来经历了很长时间，但在此之前至少需要同样长的时间，或许还要更长一些，才能使地球上的生命从萌芽状态缓慢地进化到我们在较早的年代中发现的各种高等的生物组织形式。查尔斯·赖尔爵士不认为这种估算是精确的。赫胥黎教授则在具体细节上批驳了这一观点，认为它在很多方面都是不可靠的；而克罗尔先生则全力地支持这些观点，认为它们在原理上是可靠的，且在一定条件下是精确的。

现在我们来考虑一下天文学家对这个问题的看法。克罗尔先生在《哲学杂志》上的一系列优秀文章中充分讨论了在过去的年代里地球轨道偏心率的改变和岁差的变化如何引起气候变迁的问题。他力图表明，用这种推理方式可以把最近一次以及那些更早期的冰川期的日期估算出来。为了达到这一目的，他对记载了 300 万年内地球轨道偏心率数据的表格进行了深入细致的研究，大部分时间以 10,000 年为一个区间，其余时间以 50,000 年为一个区间。这些表格表明，偏心率的大小是交替变化的，周期是 50,000 年或 100,000 年，正如我利用他的数据，以足够精度制作的这张曲线图所示。由于岁差和拱点的周期变化，每一年南、北极点都会在地球轨道上的不同位置朝向太阳（面向太阳的半球是夏天，另一个半球是冬天），这种运动的一个完整周期是 21,000 年。因此，假如在某个时期，北半球的冬天降临时地球在靠太阳最近的地方，或称**近日点**（现在即如此），那么 10,500 年后北半球的冬天降临时地球将在**远日点**。在一个时期内北半球的冬天将较短和较暖，而在另一个时期内则会较长和较冷。克罗尔先生认为，当偏心率很大（比如说是现在的 2 倍、3 倍甚至 4 倍）时，根据空气和水的热力学性质，冬天地球位于**远日点**时极地地区将有积雪，到夏天也不能融化。积雪在几千年间将不断积累，直到该半球的冬天发生在**近日点**附近，积雪才能融化，并转向另一个极端。当偏心率很大时，在两个半球都将出现冰川期，历时 10,500 年左右。在 21,000 年周期的其余时间里，几乎是四季常春，并有两个向冰川期转变的过渡期。通过研究偏心率图，我们发现，在最近的 300 万年中，有超过 12 个偏心率较大的时期，每一次的时间长度足以使二分点发生 2~3 次，有时甚至是 8~10 次的周期变化，这足以使每个半球出现不少于 50 个或 60 个冰川期，中间伴随有常年春天或常年夏天的间隔期。

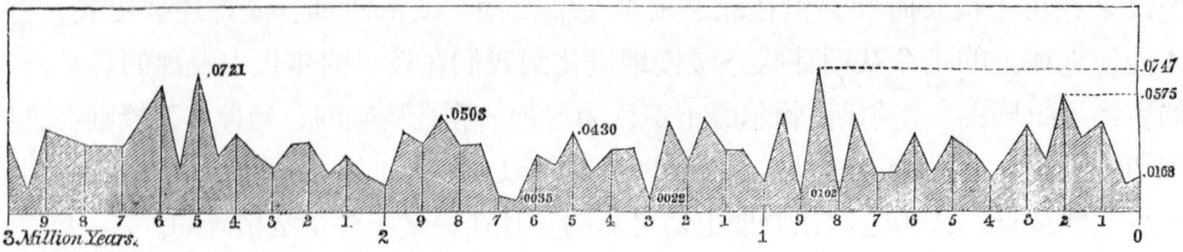

The diagram also shows us (and this is of very great importance) that the present amount of excentricity is exceptionally small. During the last three million years there have only been five occasions, always of very short duration, when it has been less than it is now, while periods of high excentricity have often lasted for two hundred thousand years at a time. This period of three million years probably includes a large portion of the tertiary period, which therefore should have mainly consisted of alternations of warm and cold climates in each hemisphere, the latter generally forming true glacial epochs. This seems the legitimate deduction from Mr. Croll's reasoning and from the tables of excentricity with which he has furnished us; but, as he very justly argues, we cannot expect to find geological evidence of all these changes of climate. The warm and temperate periods will naturally leave the best records, while the cold epochs will generally be characterised only by an absence of organic remains. Besides, we must consider 10,500 years as a very small fragment of time in geology and we have good reason for thinking that several such periods might pass away without the occurrence of those exceptional conditions which Mr. Darwin and Sir C. Lyell have shown to be necessary for the preservation of any geological record. As to physical proofs of ice-action, very few could survive the repeated denudations, upheavals and subsidencies, which the surface must have undergone since any of the earlier glacial epochs; so that it may be fairly argued that these repeated changes of climate may have occurred and yet have left no distinct record by which the geologist could interpret their history.

Throughout the whole of his argument, Mr. Croll considers astronomical causes to be the most important and effective agents in modifying climate, while Sir Charles Lyell maintains that the distribution of land and water, with their action on each other by influencing marine and aerial currents, are of prepondering importance. He has certainly shown that these causes have an immense influence at the present time. The effects which, on Mr. Croll's theory, ought to be produced by the existing phase of precession combined with even the small amount of excentricity that now exists, is not only neutralised, but actually reversed by terrestrial causes. Dove has shown that the whole earth is really warmer when it is furthest from the sun in June, than when it is nearest in December, a fact which is to be explained by the northern hemisphere (turned toward the sun in June) having so much more land than the southern. So, the northern hemisphere being three millions of miles nearer the sun in winter than in summer, while the southern hemisphere is the reverse, the northern winter ought to be warmer and the northern summer cooler than the southern; but this, too, is the opposite of the fact, for the southern summer is more than 11 °F cooler than ours, while its winter is nearly 5 °F warmer. The immense differences of temperature of places in the same latitude, sometimes amounting to nearly

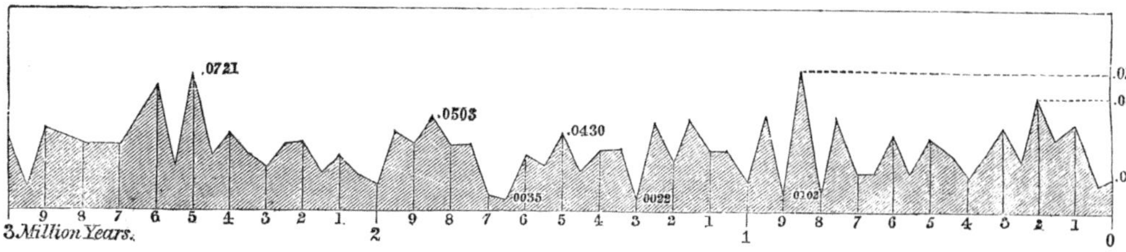

这张图还告诉我们（这一点很重要）：地球目前的偏心率格外低。在最近的300万年中只遇到过5次偏心率比现在低的情况，并且持续时间都非常短，而偏心率较大的时期几乎每次都要延续20万年。最近的300万年可能包含了第三纪的大部分时间，因此在这段时间里主要是两个半球中温暖和寒冷气候的交替变化，后者通常能形成真正的冰川期。这似乎是从克罗尔先生的推理和他提供给我们的偏心率表格中得出的合理推论，但正如他论证的那样，我们无法找到这些气候变化的地质学方面的证据。气候温暖的时期自然会留下最好的纪录，而寒冷时期则只能是以生物遗骸的缺失为特征。此外，我们必须考虑到10,500年在地质学中只是一个很小的时间段，我们有理由认为，可能有几个这样的时期并没有发生那些达尔文先生和赖尔爵士认为是保存地质记录所必需的异常情况。至于冰川活动的自然证据，很少能在反复的剥蚀、抬升和沉降作用后保存下来，这也是自以前任何一个冰川期以来地球表面必然要经受的过程。因此，可以合理地论证，气候的反复变化可能发生过，但并没有留下足以使地质学家辨认其历史的清晰记录。

在整个论证过程中，克罗尔先生一直认为天文方面的原因在气候变化中是最重要和最有效的因素。而查尔斯·赖尔爵士坚持认为：陆海分布以及它们之间通过洋流和大气环流形成的相互作用特别重要。他明确指出，这些因素对目前也有极大的影响。而在克罗尔先生看来，气候变化应该是现阶段岁差和很小量的偏心率联合作用的结果，但会被地球上的因素抵消甚至逆转。达夫指出，整个地球实际上在离太阳最远的6月比离太阳最近的12月更温暖，原因是北半球（在6月份朝向太阳）比南半球的陆地面积大很多。北半球在冬天时比在夏天时距太阳近300英里，而南半球则相反，因此，北半球的冬天应该较南半球暖，而夏天则应比南半球冷，但这也与事实相反，因为南半球夏天的气温比北半球低11°F，而冬天却高5°F。在同一纬度上各地的温差之大有时竟然可达30°F，这些个例几乎都可以归因于海陆分布以及风和洋流的分布。查尔斯·赖尔爵士进一步论证说，现在的陆地分布是非常不规则的——在极地附近陆地过多，而在赤道和南温带又太少——这使得无论过去已经出现了什么样的差异，

30 °F, can also be traced, in almost every instance, to the distribution of land and water and of winds and currents. Sir Charles Lyell further argues that the existing distribution of land is so extremely irregular—such an undue proportion being near the poles, while there is such a deficiency at the equator and in the south temperate zone—that whatever differences may have occurred in past time, they can hardly fail to have often been such as to cause a more uniform climate. Therefore he believes that if the poles were tolerably free from land, so as to admit of the uninterrupted circulation of the warmer equatorial waters and to afford no lodgment for great accumulations of snow and ice, a glacial epoch would be impossible even during the most extreme phases of excentricity.

We have now much evidence to show that three distinct modifications in physical geography occurred just before or during the Glacial epoch, which would each tend to lower the temperature. The first is the submergence of the Sahara, which would have caused the southerly winds to be charged with aqueous vapour, condensing on the Alps into snow instead of being, as now, dry and heated and acting powerfully to melt the glaciers. The second is the submergence of Lapland, which would have admitted the cold iceberg-laden waters of the Arctic Sea into the very heart of Europe. The third is the probable submergence of part of Central America, causing the Gulf Stream to be diverted into the Pacific. The only proof of this is the fact that one-third of the known species of marine fishes are absolutely identical on the two sides of the isthmus of Panama; but it is impossible to conceive any means by which such an amount of identity could have been brought about except by a recent, if only a temporary, communication. A subsidence and elevation no greater than what occurred in Wales about the same time—as proved by Arctic shells of existing species in drift 1,300 feet above the sea—would have effected the communication by a broad and deep channel. Now if any two of these changes of physical geography occurred together, we may be sure that a very small increase of excentricity would have led to a more severe glacial epoch than would be possible, under existing conditions, with a much larger excentricity. We must keep this in mind when attempting to fix the most probable date for the last glacial epoch.

(**1**, 399-401; 1870)

这些差异几乎都无法使得全球形成较为均一的气候。因此，他相信，如果两极没有那么多陆地，赤道的暖流就不会被打断，极地也就不会积累那么多冰雪，即使在地球轨道偏心率最大的时候，冰川期也不可能出现。

我们有很多证据表明，在最近一次冰川期之前或之中，曾发生过 3 次自然地理的显著变化，每次变化都使温度有所降低。第一次淹没了撒哈拉沙漠，使南风携带的水汽在阿尔卑斯山上凝结成雪，而现在刮的风又干又热，猛烈得足以融化冰川；第二次淹没了拉普兰，它使北极海域浸泡着冰山的冷水流向欧洲的中心地区；第三次可能淹没了中美洲的部分地区，它使墨西哥湾流转向流入太平洋。第三次变化的唯一证据是：在巴拿马地峡的两侧，有 1/3 已知海洋鱼种是完全一样的。但是，几乎不可能想出一种方法能够在如此短的时间里带来如此大的一致性，除非是现代交通工具。一次地壳升降过程（程度与大约同时在威尔士发生的地壳运动差不多）开通了一条宽广深邃的水道，这一点可以用在海拔 1,300 英尺的冰碛中发现了现存种的北冰洋贝类来证明。现在，如果任何两种上述的自然地理过程一起发生，我们可以确信，只要偏心率略有增加，就会导致更为严酷的冰川期，也许比现有条件下让偏心率增大很多对冰川期的影响更大。当我们试图确定最后一次冰川期最可能出现的时间时，我们必须记住这一点。

（孙惠南 翻译；张忠杰 审稿）

The Solution of the Nile Problem

C. Beke

Editor's Note

Here Charles Beke continues the debate about the watersheds of northwest Africa that stemmed from recent letters of David Livingstone claiming to have found the sources of the Nile. (Livingstone was shortly forced to curtail his expedition through ill health.) Central to Livingstone's claim was the idea that the Chambeze river and its lakes flow into the Nile, not the Congo—a notion Beke had advanced earlier. Despite Beke's confidence, it was wrong. But Beke argues that his previous discoveries point to a great parting of waterways in Africa: to the Nile in the north, the Congo to the west and the Zambesi to the east. The discussion is ample reminder of how poorly Africa was mapped out in the late nineteenth century.

I have read with much pleasure Mr. Keith Johnston's remarks in your impression of the 27th ult. on the subject of Dr. Livingstone's explorations, not only because they manifest an intimate acquaintance with the general physical features of the field of inquiry and a proper estimate of the merits of the question; but because they help to establish the correctness of my opinion, that the Chambeze and its lakes belong to the Nile system, and not to that of the Congo. I have only to explain that, in my letter of December 1st (*Nature*, No. 9), I did not "give the opinion that the river which forms the main part of the great traveller's latest discoveries is *the head stream* of the Nile", but merely said that it "joins" it.

On the question of levels your correspondent is substantially correct, and if he will look to the *Illustrated Travels* of the 1st inst., he will see how far I agree with him. From Dr. Livingstone's statements it appears that the general drainage level of the basin of the Chambeze does not exceed 3,000 feet; and it is not improbable that in the passage of the waters northwards on the west side of Tanganyika, they fall two or even three hundred feet lower, so as to descend nearly if not entirely to the level of the Albert Nyanza. But even if this be the case, I fail to see how the difference in height, however small, "could not give a sufficient lowness to the latter lake (Albert Nyanza) to allow this river (Chambeze) to flow down to it through the five degrees of latitude which separate them." The levels of the Lakes Liemba, Tanganyika, and Albert Nyanza—of which the first is in about 10° S. lat. and the last has its northern end in about 3° N. lat.—are respectively *circa* 2,800, 2,844, and 2,720 feet; and as the continuity of these three bodies of water is assumed by Mr. Keith Johnston, it follows that there is here a virtual dead level extending over not five, but *thirteen* degrees of latitude, or 780 geographical miles! If then it is possible for the waters of Lake Liemba, the head of Livingstone's "eastern line of drainage", to flow into the Albert Nyanza, it is equally possible for those of the Chambeze and its lakes, forming that traveller's central line of drainage, to do so.

尼罗河问题的答案

比克

编者按

戴维·利文斯通在最近的信中宣布已经发现了尼罗河的源头（不久之后利文斯通就因为身体不好而被迫缩短了考察行程），由此引起了关于非洲西北部地区流域的争论，查尔斯·比克在这篇文章中继续围绕这一主题进行了争辩。利文斯通的声明的核心意思是，谦比西河及其湖泊流入了尼罗河，并不是像此前比克提出的那样流入刚果河。尽管比克充满自信，但他是错误的。不过，比克辩解到，他先前的发现揭示了非洲水道大尺度的分界：北流的尼罗河，西流的刚果河以及东流的赞比西河。这篇文章充分显示了在19世纪后期人们对非洲的勘察还非常不足。

读了贵刊上月第27期刊载的基思·约翰斯顿先生关于利文斯通博士的探险的评论，我非常高兴，因为这些评论不仅深刻地揭示了所探讨地区的基本自然风貌，并恰当地评估了该问题的价值，而且，还有助于证实我的看法的正确性，即谦比西河及其湖泊归属于尼罗河水系而不是刚果河水系。只是有一点我要解释一下，在12月1日的那篇快报（《自然》，第9期）中，我并没有“认为构成伟大的旅行家最新发现的主要部分的那条河就是尼罗河的**源头**”，而仅仅是说它“汇入”了尼罗河。

在水位问题上，贵刊的通讯员是完全正确的，并且，如果他看一下本月的《图说旅行》第1期，就会知道我与他的观点有多一致。从利文斯通博士的陈述来看，谦比西河流域的一般水位不超过3,000英尺，并且，河流在坦噶尼喀湖西侧向北流去的过程中，水位下跌了200英尺甚至300英尺，进而接近（即使不能完全达到）艾伯特–尼亚萨湖的水位也不无可能。然而，即便如此，我也不明白高度上的差异，尽管很小，为什么还是“不能使艾伯特–尼亚萨湖的水位低到足以使谦比西河跨越它们之间5个纬度的距离而向下流入其中。”列姆巴湖、坦噶尼喀湖和艾伯特–尼亚萨湖的水位分别约为2,800英尺、2,844英尺和2,720英尺，其中列姆巴湖位于南纬10° 附近，而艾伯特–尼亚萨湖的北端约在北纬3°。由于基思·约翰斯顿先生已经假定这三片水体是相连的，那么它们之间的实际距离应该跨越**13**个纬度而非5个，或者说约为780地理英里！因此，如果列姆巴湖的湖水，也就是利文斯通所说的“流域东部水系”的源头的水能够进入艾伯特–尼亚萨湖，那么同样地，构成旅行家所在流域中央水系的谦比西河及其附属湖泊也可以到达那里。

In his last letter from Ujiji, Dr. Livingstone says that "the western and central lines of drainage converge into an unvisited lake west or south-west of this"—that is to say, situated in the unexplored regions west of Tanganyika, in the north-north-west direction in which he saw the Lualaba (as he calls the lower course of the Chambeze) flowing, after it had emerged from the crack in the mountains of Rua, north of Lake Moero. This "unvisited lake" is evidently the Lake Chowambe of the traveller's former communications, which by his now calling Baker's Albert Nyanza by the name of "Nyigi Chowambe", he would seem to identify with it. But this is quite consistent with Baker's own statement, that, to the south of about 1° 30′ S. lat., the Albert Nyanza "turns suddenly to the west, in which direction its extent is unknown."

"Albert Nyanza", "Nyigi Chowambe", and this "unvisited lake west or south-west of Ujiji", are, therefore, one continuous body of water, which, being on the lowest level of all, must form not merely the "western line of drainage", but the *main* drainage of the upper Nile Basin; and as, on its eastern side, it is the recipient of the waters of the lakes Victoria Nyanza and Tanganyika, so, on its western side, it receives those of the great lake discovered by Signor Piaggia, with an elevation (as I believe) of four or five thousand feet.

This is entirely in accordance with the opinion I have always entertained that the water-parting between the basin of the Nile and those of the rivers flowing into the Atlantic—the Ogowai, the Kuango (Congo), the Kwanza, and the Kunene—is on about the twentieth meridian of east longitude, as it is, in fact, marked on my maps of "The Basin of the Nile" of 1849, 1859, and 1864. The Mossamba range of mountains, situate to the east of the Portuguese colony of Benguela, on the west coast of Africa, forms the southern extremity of this water-parting, and it is in these mountains that I find the head of the great river, which with the Lufira forms Livingstone's "western line of drainage", or, as it should be more correctly designated, the main stream of the Nile. This river is the Kasái, Kassávi, or Loke, whose sources are in the forests of Quibokoe or Kibokoe, on these Mossamba Mountains, within 300 miles of the Atlantic Ocean; which river was crossed by Dr. Livingstone within 160 or 170 miles of its head, on February 27th, 1854, in his adventurous journey across the African continent, and is described by him in page 332 of his "Missionary Travels", and the lower course of which river was followed down by the Hungarian traveller, Ladislaus Magyar, in 1850, as far as about 6° 30′ S. lat., *where he heard that it flowed eastward into Lake "Nhanja"*—a statement strikingly in accordance with Mr. Cooley's assertion, adverted to in my former communication, that "the drainage of the Cazembe's country is all into the Nyanza on the east."

The Nile of Egypt, in thus having its source at the opposite side of the continent of Africa, within a short distance of that ocean into which it does *not* flow, only follows an almost general law of Nature. In the *Athenaeum* of July 22nd, 1865, when commenting on Sir Samuel Baker's announcement of his discovery of the Albert Nyanza, I compared the Nile and its Lakes with the Po and its Lakes, pointing out how the two rivers have some of their sources in *snowy* mountains, not at the extremity but at the side of their respective basins. Dr. Livingstone's present discoveries seem to establish the fitness of this comparison, and

在上一封从乌吉吉发来的信中，利文斯通博士说："流域的西部水系和中央水系均汇聚于西部或西南部的一个人们未曾到过的湖中"——也就是说，位于坦噶尼喀湖西部未经勘察的区域内，他在西北偏北的方向上看到了卢阿拉巴河（他将其称为谦比西河的下游）从姆韦鲁湖北端卢阿山脉的裂隙中流出。很显然，这个"人们未曾到过的湖"就是他之前提到的查旺比湖，即他现在用"Nyigi Chowambe"这个名字表示的贝克称之为艾伯特–尼亚萨的湖，他好像把它们看作同一个了。不过这与贝克自己的叙述还是非常一致的，即，向南在约南纬 1°30′处，艾伯特–尼亚萨湖"突然转而向西，在这个方向上它的延伸范围不明。"

因此，"艾伯特–尼亚萨湖"、"查旺比湖"以及这里的"乌吉吉西部或西南部的一个人们未曾到过的湖"都是同一片连续的水体。由于位于海拔最低处，它们不仅构成了"流域的西部水系"，还是尼罗河流域上游的**主要**水系；而且，在东侧它接收了维多利亚–尼亚萨湖和坦噶尼喀湖的湖水，在西侧，它同样接收了皮亚贾先生发现的大湖的湖水，我认为这个湖的海拔约为 4,000~5,000 英尺。

这与我一直坚持的观点完全一致，即，尼罗河流域与其他流入大西洋的河流——奥戈瓦河、刚果河、宽扎河、库内内河——之间的分水岭大约位于东经 20°左右，实际上和我分别于 1849 年、1859 年和 1864 年出版的《尼罗河流域》地图中标出的一样。位于非洲西海岸葡萄牙殖民地本格拉市东部的莫桑巴山系是该分水岭的最南端，而我就是在这些山系中发现了一条大河的源头，这条大河与卢菲拉河一起构成了利文斯通所称的"流域西部水系"，或者，更确切一点，应该称之为尼罗河的干流。这条河就是开赛河，或称卡萨维河、洛克河，其源头位于莫桑巴山系的奎博克森林中，距大西洋不到 300 英里。利文斯通博士在穿越非洲大陆的冒险旅行中，曾于 1854 年 2 月 27 日在距源头 160~170 英里的地方穿过了这条河流，且在他自己写的《传教旅行》一书的第 332 页曾对此作过描述。匈牙利的旅行家拉迪斯劳斯·毛焦尔也曾于 1850 年沿其下游顺流而下到达了南纬约 6° 30′之处，**并在那里听说该河向东流入了"尼亚萨"湖**。我以前信件中谈到的库利先生的说法，即"卡仁比国的河流全都向东流入了尼亚萨湖"，与这一点惊人地一致。

埃及的尼罗河，发源于非洲大陆的另一侧，它并**没有**流入与它相距很近的大洋，这仅仅是遵循了大自然的一般规律。在 1865 年 7 月 22 日的《科学协会》中，我在评论塞缪尔·贝克爵士发现艾伯特–尼亚萨湖的声明时，曾将尼罗河及其湖泊与波河及其湖泊进行了比较，并指出这两条河流的一些源头都在**雪山**之中，但这些源头不是在山的顶端，而是位于这些雪山的山坳里。利文斯通博士如今的发现似乎也证明了这一比较是恰当的，而且作了补充。波河流入亚得里亚海，而其源头则在距热那

to extend it. For as the Po, whose exit is in the Adriatic, has its head sources in the Cottian and Maritime Alps, within a few miles of the Gulf of Genoa; so, in like manner, the Nile, which flows into the Mediterranean, has its head on the Mossamba Mountains, within 300 miles of the Atlantic Ocean.

The spot which I have thus discovered to contain the hitherto hidden Source of the Nile, and so to reveal is the most remarkable culminating point and water-parting of the African Continent, if not of the whole world; for, within the space of a degree east and west (between 18° and 19° E. long.) and half as much north and south (between 11° 30′ and 12° S. lat.) it includes not only the head of the mighty Nile, which runs northwards over one-eighth of the entire circumference of the globe, but likewise those of the Kuango (Congo), the Kuanza and the Kunene flowing westwards; those of the Kuivi and the Kubango running to the south; and that of the Lungebungo having its course eastward and forming the head stream of the Zambesi. It is, in fact, what I have been endeavouring to determine since 1846, "the great *hydrophylacium* of the continent of Africa, the central point of division between the waters flowing to the Mediterranean, to the Atlantic, and to the Indian Ocean" (*Journal of the Royal Geographical Society*, vol. XVII, p. 82), as likewise to Lake Nyami, or some other depression in the interior of the continent.

(**1**, 405-406; 1870)

Charles Beke: Bekesbourne, Feb. 2.

亚湾仅几英里的科蒂安山脉和滨海阿尔卑斯山脉中。与此类似，流入地中海的尼罗河，其源头却是在距大西洋不到 300 英里的莫桑巴山脉。

我已经发现的包含此前一直隐藏着的尼罗河源头的那个地方就算不是全世界的极点和分水岭，至少也是非洲大陆上最显著的极点和分水岭。因为在东西距离不到 1°（位于东经 18°~19°之间），南北距离仅为 0.5°（南纬 11° 30′~12°）的范围内，发源于此的河流不仅包括流向北方的全长超过地球周长 1/8 的尼罗河，还有那些向西流去的刚果河、宽扎河和库内内河，以及向南流去的奎维河和库班河，另外，这里还是兰热邦戈河（流向东方并构成赞比西河源头）的发源地。实际上，它就是自 1846 年以来我一直努力要确定的“非洲大陆的巨型水库，是分别流入地中海、大西洋、印度洋的各河流的中心分界处”（《皇家地理学会会刊》，第 17 卷，第 82 页），同样也是流向尼亚萨湖和其他内陆低地的各河流的中心分界点。

（齐红艳 翻译；刘盛和 审稿）

The Velocity of Thought

M. Foster

Editor's Note

The author here is Michael Foster, a pioneer of physiology at Cambridge. He describes experiments reminiscent of Luigi Galvani's famous studies of muscle action in a dead frog's leg induced by electrical currents. Foster aims to measure how fast nerve signals travel when inducing muscle contraction. He uses an ingenious apparatus of levers, rotating cylinders and tuning forks to measure the very short time intervals involved, and finds a resulting speed—about 28 metres per second—similar to that measured by the German physiologist Hermann von Helmholtz. But he points out that mental processes involve some processing time too. His measurements of reaction times lead him to assert that it takes about 1/26 of a second "to think".

"As quick as thought" is a common proverb, and probably not a few persons feel inclined to regard the speed of mental operations as beyond our powers of measurement. Apart, however, from those minds which take their owners so long in making up because they are so great, rough experience clearly shows that ordinary thinking does take time; and as soon as mental processes were brought to work in connection with delicate instruments and exact calculations, it became obvious that the time they consumed was a matter for serious consideration. A well-known instance of this is the "personal equation" of the astronomers. When a person watching the movement of a star, makes a signal the instant he sees it, or the instant it seems to him to cross a certain line, it is found that a definite fraction of a second always elapses between the actual falling of the image of the star on the observer's eye, and the making of the signal—a fraction, moreover, varying somewhat with different observers, and with the same observer under differing mental conditions. Of late years considerable progress has been made towards an accurate knowledge of this mental time.

A typical bodily action, involving mental effort, may be regarded as made up of three terms; of sensations travelling towards the brain, of processes thereby set up within the brain, and of resultant motor impulses travelling from the brain towards the muscles which are about to be used. Our first task is to ascertain how much time is consumed in each of these terms; we may afterwards try to measure the velocity of the various stages and parts into which each term may be further subdivided.

The velocity of motor impulses is by far the simplest case of the three, and has already been made out pretty satisfactorily. We can assert, for instance, that in frogs a motor impulse, the message of the will to the muscle, travels at about the rate of 28 metres a second, while in man it moves at about 33 metres. The method by which this result is obtained may be described in its simplest form somewhat as follows: —

思考的速度

福斯特

编者按

这篇文章的作者迈克尔·福斯特是剑桥生理学领域的先驱。他描述的实验很容易让我们想到路易吉·伽伐尼用电流引起死蛙腿肌肉运动的著名研究。福斯特的目标是测量神经信号在诱导肌肉收缩时传递得有多快。他使用了一个精巧的杠杆装置，利用该装置可以通过旋转圆筒和调整叉子来测量过程中所涉及的极短的时间间隔。通过测量他得到的速度是大约每秒 28 米，这与德国生理学家赫尔曼·冯·亥姆霍兹测得的结果很接近。不过，福斯特指出精神活动同样需要一些处理时间。通过对反应时间的测量，福斯特提出人的"思考"大约需要 1/26 秒。

人们常说"像思考一样快"，而且可能不少人都倾向于认为思考的速度已经快到无法测量的程度。然而，人们除了作重大决定时需要花很长时间思考以外，粗略的经验清楚地告诉我们，普通的思考也是需要花费时间的。而且，一旦将思考过程与精密仪器和精确计算联系起来以后，思考所花的时间就变得不容忽视了。天文学家的"测者误差"就是一个著名的例子。当观测者在观察星体运动时，会在看到星体或其特定的运动轨迹后立即记录，人们发现观测者作出记录的时间总是比星体实际在他眼中成像的时间滞后零点几秒，并且这个滞后时间的长短因人而异，就算是同一个观测者，其精神状态不同，滞后时间也不同。近年来，人们在对这一思考时间的准确认识方面已经取得了相当大的进展。

一个典型的身体动作需要大脑的帮助，可能包括三个步骤：首先感觉信息向大脑传输，然后在大脑中建立动作过程，最后大脑再将产生的运动脉冲传递给相关的肌肉。我们的首要任务是确定每个步骤花费的时间，然后测定每个步骤可能再被细分成的不同阶段和部分的速率。

目前，运动脉冲的传输速率是三个速率中最容易测定的，现在已经得到了满意的结果。例如，我们可以确定，运动脉冲（即由大脑向肌肉传输的信息）的传输速率对于蛙来说是每秒 28 米，而对于人则是每秒 33 米。现将获得上述速率的测定方法以最简单的方式描述如下：

The muscle which in the frog corresponds to the calf of the leg, may be prepared with about two inches of its proper nerve still attached to it. If a galvanic current be brought to bear on the nerve close to the muscle, a motor impulse is set up in the nerve, and a contraction of the muscle follows. Between the exact moment when the current breaks into the nerve, and the exact moment when the muscle begins to contract, a certain time elapses. This time is measured in this way: —A blackened glass cylinder, made to revolve very rapidly, is fitted with two delicate levers, the points of which just touch the blackened surface at some little distance apart from each other. So long as the levers remain perfectly motionless, they trace on the revolving cylinder two parallel, horizontal, unbroken lines; and any movement of either is indicated at once by an upward (or downward) deviation from the horizontal line. These levers further are so arranged (as may readily be done) that the one lever is moved by the entrance of the very galvanic current which gives rise to the motor impulse in the nerve, and thus marks the beginning of that motor impulse; while the other is moved by the muscle directly this begins to contract, and thus marks the beginning of the muscular contraction. Taking note of the direction in which the cylinder is revolving, it is found that the mark of the setting-up of the motor impulse is always some little distance ahead of the mark of the muscular contraction; it only remains to be ascertained to what interval of time that distance of space on the cylinder corresponds. Did we know the actual rate at which the cylinder revolves this might be calculated, but an easier method is to bring a vibrating tuning-fork, of known pitch, to bear very lightly sideways on the cylinder, above or between the two levers. As the cylinder revolves, and the tuning fork vibrates, the latter will mark on the former a horizontal line, made up of minute, uniform waves corresponding to the vibrations. In any given distance, as for instance in the distance between the two marks made by the levers, we may count the number of waves. These will give us the number of vibrations made by the tuning-fork in the interval; and knowing how many vibrations the tuning-fork makes in a second, we can easily tell to what fraction of a second the number of vibrations counted corresponds. Thus, if the turning-fork vibrates 100 times a second, and in the interval between the marks of the two levers we count ten waves, we can tell that the time between the two marks, *i.e.* the time between the setting-up of the motor impulse and the beginning of the muscular contraction, was 1/10 of a second.

Having ascertained this, the next step is to repeat the experiment exactly in the same way, except that the galvanic current is brought to bear upon the nerve, not close to the muscle, but as far off as possible at the furthest point of the two inches of nerve. The motor impulse has then to travel along the two inches of nerve before it reaches the point at which, in the former experiment, it was first set up.

On examination, it is found that the interval of time elapsing between the setting up of the motor impulse and the commencement of the muscular contraction is greater in this case than in the preceding. Suppose it is 2/10 of a second—we infer from this that it took the motor impulse 1/10 of a second to travel along the two inches of nerve: that is to say, the rate at which it travelled was one inch in 1/20 of a second.

将蛙小腿的肌肉制成带有大约 2 英寸长固有神经的样品。如果给临近肌肉的神经通电，神经内部就会产生运动脉冲，进而导致随后的肌肉收缩。从给神经通电到肌肉开始收缩需要经过一段时间。而这段时间是这样测定的：将两根细杆连到一个快速旋转的涂黑的玻璃圆柱上，让它们的末端彼此分开一段微小的距离，并分别与圆柱黑色表面接触。只要两根细杆保持完全静止，那么当圆柱体旋转时，它们就会在圆柱黑色表面上画出两条平行的水平实线。而如果任何一根细杆有一丁点的运动，水平线就会立即发生向上（或向下）的偏移。进一步对这些细杆作出如下安排（实施起来很容易）：一根细杆的运动由在神经中引起运动脉冲的特定电流控制，以此来标记运动脉冲的产生；另一根细杆的运动由产生收缩的肌肉控制，以此来标记肌肉收缩的开始。记录下玻璃圆柱的旋转方向，我们发现运动脉冲产生的标记总是比肌肉收缩开始的标记靠前一小段距离。现在我们只需要确定圆柱上这段距离所对应的时间间隔。如果知道圆柱旋转的准确速度，那么就可以计算出这段时间的间隔，然而我们还有更简单的方法：在两根细杆之间或上面放置一个已知音调的振动音叉，轻轻贴在圆柱两侧。随着圆柱的旋转和音叉的振动，音叉会在圆柱表面标记出一条水平线，这条水平线由与振动相对应的微小而又均一的波组成。对于任何给定的距离，比如细杆画出的两个标记之间的距离，我们都能数出其中的波数。这就给出了在上述时间间隔内音叉的振动次数。我们知道了一秒内音叉振动的次数，就能很容易地算出相应音叉振动次数对应的时间间隔。因此，假设音叉在一秒内振动 100 次，在两个细杆画出的标记之间我们数出了 10 个波，那么我们就可以算出这两个标记之间的时间间隔，也就是说，从产生运动脉冲到肌肉开始收缩的时间间隔是 1/10 秒。

确定了上述时间间隔，下一步就是以几乎完全相同的方式重复上述实验，唯一不同的是，不是给临近肌肉的神经通电，而是在 2 英寸长的神经上尽可能远离肌肉的末端通电。在到达临近肌肉的末端之前，运动脉冲沿着这 2 英寸神经传递，而在之前的实验中，运动脉冲则是在这一临近肌肉的末端产生的。

检测后发现，在这种实验条件下，从产生运动脉冲到肌肉开始收缩的时间间隔比之前的实验中的时间间隔更长。假设从产生运动脉冲到肌肉开始收缩的时间间隔是 2/10 秒，那么我们就可以由此推测出运动脉冲在 2 英寸神经中传递的时间是 1/10 秒，也就是说，运动脉冲的传递速度是每 1/20 秒 1 英寸。

By observations of this kind it has been firmly established that motor impulses travel along the nerves of a frog at the rate of 28 metres a second, and by a very ingenious application of the same method to the arm of a living man, Helmholtz and Baxt have ascertained that the velocity of our own motor impulses is about 33 metres a second.* Speaking roughly this may be put down as about 100 feet in a second, a speed which is surpassed by many birds on the wing, which is nearly reached by the running of fleet quadrupeds, and even by man in the movements of his arm, and which is infinitely slower than the passage of a galvanic current. This is what we might expect from what we know of the complex nature of nervous action. When a nervous impulse, set up by the act of volition, or by any other means, travels along a nerve, at each step there are many molecular changes, not only electrical, but chemical, and the analogy of the transit is not so much with that of a simple galvanic current, as with that of a telegraphic message carried along a line almost made up of repeating stations. It has been found, moreover, that the velocity of the impulse depends, to some extent, on its intensity. Weak impulses, set up by slight causes of excitement, travel more slowly than strong ones.

The contraction of a muscle offers us an excellent objective sign of the motor impulse having arrived at its destination; and, all muscles behaving pretty much the same towards their exciting motor impulses, the results obtained by different observers show a remarkable agreement. With regard to the velocity of sensations or sensory impulses, the case is very different; here we have no objective sign of the sensation having reached the brain, and are consequently driven to roundabout methods of research. We may attack the problem in this way. Suppose that, say by a galvanic shock, an impression is made on the skin of the brow, and the person feeling it at once makes a signal by making or breaking a galvanic current. It is very easy to bring both currents into connection with a revolving cylinder and levers, so that we can estimate by means of a tuning-fork, as before, the time which elapses between the shock being given to the brow and the making of the signal. We shall then get the whole "physiological time", as it is called (a very bad name), taken up by the passage of the sensation from the brow to the brain, by the resulting cerebral action, including the starting of a volitional impulse, and by the passage of the impulse along the nerve of the arm and hand, together with the muscular contractions which make the signal. We may then repeat exactly as before, with the exception that the shock is applied to the foot, for instance, instead of the brow. When this is done, it is found that the whole physiological time is greater in the second case than in the first; but the chief difference to account for the longer time is, that in the first case the sensation of the shock travels along a short tract of nerve (from the brow to the brain), and in the second case through a longer tract (from the foot to the brain). We may conclude, then, that the excess of time is taken up by the transit of the sensation through the distance by which the sensory nerves of the foot exceed in length those of the brow. And from this we can calculate the rate at which the sensation moves.

* Quite recently M. Place has determined the rate to be 53 metres per second. This discordance is too great to be allowed to remain long unexplained, and we are very glad to hear that Helmholtz has repeated his experiments, employing a new method of experiment, the results of which we hope will soon be published.

通过这样的研究已经确定了蛙神经中运动脉冲传递的速率是每秒 28 米。亥姆霍兹和巴克斯特将该方法非常巧妙地应用在活人的胳膊上，测定出我们人类神经中运动脉冲的传递速度是每秒 33 米。* 粗略来说这个速度大约为每秒 100 英尺，这超过了许多鸟的飞行速度，几乎接近四足动物的奔跑速度，甚至接近人挥动手臂的速度，但远比电流的传输速度慢得多。这可能就是神经行为的复杂性所在。在由动作意识或其他方式激发的神经脉冲沿神经传递的过程中，每一步都发生许多分子变化，不仅有电学的还有化学的变化。神经脉冲的传递，与其说是类似于简单的电流传递，不如说是类似于含有多个重复信号站的电话线中电信号的传递。此外，研究发现，神经脉冲的传递速度，在某种程度上取决于其强度。轻度兴奋导致的弱神经脉冲比强神经脉冲传递得要慢一些。

肌肉收缩是运动脉冲到达目的地的客观标志。所有肌肉对兴奋性运动脉冲的反应完全相同，而且由不同观测者得到的观测结果相当一致。感觉或感觉脉冲的传递速度，却有着不同的情况。我们没有找到感觉到达大脑的客观指标，因此只能采取间接的研究方法。我们或许可以这样做：假设当一个人的额头皮肤受到电击后，人在感知的同时立即通过形成或切断电流而产生一个信号。我们可以很容易地通过旋转的圆柱和细杆将这两种电流联系起来，这样我们就能用上文中描述的音叉法测定出从额头遭受电击到产生信号的时间间隔。这样我们就可以获得整个的“生理时间”，正如它的名字（一个很糟糕的名字）所表示的一样，在这段时间里感觉从额头传递到大脑，然后大脑皮层作出反应，形成意识脉冲，意识脉冲沿着胳膊和手的神经传递，到达目的地后引起肌肉收缩从而产生信号。然后我们又进行了类似的测量，不过这次电击的是脚部而不是额头。结果发现脚部的整个生理时间比额头的长，主要原因是刺激感觉从脚部到大脑的神经传递路径比从额头到大脑长得多。由此，我们可以得出结论，多出的时间花费在感觉的传递上了，传递的距离就是从脚部到大脑的神经长度超过从额头到大脑的神经长的那段距离。这样我们就可以算出感觉的传递速度了。

* 最近，普莱斯测定出这一速度是每秒 53 米。这么大的差异，不能任其长期得不到解释。我们很高兴地听说，亥姆霍兹已经用一种新的实验方法重复了他的实验，我们期望能尽快看到该结果的公布。

Unfortunately, however, the results obtained by this method are by no means accordant; they vary as much as from 26 to 94 metres per second. Upon reflection, this is not to be wondered at. The skin is not equally sentient in all places, and the same shock might produce a weak shock (travelling more slowly) in one place, and a stronger one (travelling more quickly) in another.

Then, again, the mental actions involved in the making the signal may take place more readily in connection with sensations from certain parts of the body than from others. In fact, there are so many variables in the data for calculation that though the observations hitherto made seem to show that sensory impressions travel more rapidly than motor impulses (44 metres per second), we shall not greatly err if we consider the matter as yet undecided.

By a similar method of observation certain other conclusions have been arrived at, though the analysis of the particulars is not yet within our reach. Thus nearly all observers are agreed about the comparative amount of physiological time required for the sensations of sight, hearing, and touch. If, for instance, the impression to be signalled be an object seen, a sound heard, or a galvanic shock felt on the brow, while the same signal is made in all three cases, it is found that the physiological time is longest in the case of sight, shorter in the case of hearing, shortest of all in the case of touch. Between the appearance of the object seen (for instance, an electric spark) and the making of the signal, about 1/6; between the sound and the signal, 1/5; between the touch and signal, 1/7 of a second, is found to intervene.

This general fact seems quite clear and settled; but if we ask ourselves the question, why is it so? Where, in the case of light, for instance, does the delay take place? We meet at once with difficulties. The differences certainly cannot be accounted for by differences in length between the optic, auditory, and brow nerves. The retardation in the case of sight as compared with touch may take place in the retina during the conversion of the waves of light into visual impressions, or may be due to a specifically lower rate of conduction in the optic nerve, or may arise in the nervous centre itself through the sensations of light being imperfectly connected with the volitional mechanism in the brain put to work in the making of the signal. One observer (Wittich) has attempted to settle the first of these questions by stimulating the optic nerve, not by light, but directly by a galvanic current, and has found that the physiological time was thereby decidedly lessened; while conversely, by substituting a prick or pressure on the skin for a galvanic shock, the physiological time of touch was lengthened. But there is one element, that of intensity (which we have every reason to think makes itself felt in sensory impressions, and especially in cerebral actions even more than in motor impulses), that disturbs all these calculations, and thus causes the matter to be left in considerable uncertainty. How can we, for instance, compare the intensity of vision with that either of hearing or of touch?

The sensory term, therefore, of a complete mental action is far less clearly understood than the motor term; and we may naturally conclude that the middle cerebral term is

然而，不幸的是，用这种方法得出的结果彼此之间很不一致，传输速度从每秒26米到每秒94米不等。细想之下，有这样的结果也并不奇怪。因为各个部分皮肤的敏感度是不一样的，同样的刺激在一个地方引起较弱的脉冲（传递得较慢），而在另一个地方则引起较强的脉冲（传递得较快）。

同样地，人体某些部位的感觉比其他部位的感觉更容易引起这种产生信号的神经行为。事实上，这些用于计算的数据变化太大，以至于尽管已有许多测定结果表明感觉的传递快于运动脉冲的传递（每秒44米），我们却无法精确计算。如果我们能将未知因素考虑在内，就不会犯太大的错误。

对那些特例我们鞭长莫及，不过用类似的方法我们还是获得了另外一些结果。几乎所有的观测者在视觉、听觉及触觉的生理时间比较上都获得了一致的结论。例如，研究发现，如果信号的获得方式是看到一个物体，听到一个声音，或者额头感觉到电击，尽管这三种情况都产生同样的信号，但是视觉的生理时间最长，其次是听觉，而触觉的生理时间最短。不过也出现了不同于上述结论的研究结果：从看到物体（如电火花）到产生信号的时间是1/6秒；从听到声音到产生信号的时间是1/5秒；而从接触物体到产生信号的时间是1/7秒，发现这一结果造成了干扰。

看起来大体上事实问题似乎已经很清楚并已经得到了确定，不过如果我们自问，为什么会是这样呢？比如光，是在哪一步发生了延迟呢？我们立刻遇到了问题。很显然，生理时间长短的不同不是由视神经、听觉神经和额头的神经在长度上的差异引起的。与触觉相比，视觉延迟可能发生在视网膜将光波转化为视觉印象的过程中，也可能与视神经中特殊的低传导速度有关，或者还可能发生在中枢神经中，对光的感觉并没有与产生信号的大脑意识机制很好地协调起来。一位观测者（威蒂克）已经尝试解决上述问题中的首要问题，他用电替代光去刺激视神经，发现生理时间明显减少了；然而相反地，用尖刺或按压替代电击去刺激皮肤，结果触觉的生理时间延长了。不过有一个因素——强度（我们有理由相信，我们可以感受到感觉印象的强度，而与感受运动脉冲的强度相比，我们能更强烈地感受到大脑行为的强度），扰乱了所有这些计算结果，使得问题具有相当大的不确定性。例如，我们如何对视觉的强度与听觉或触觉的强度进行比较呢？

因此，对于感觉这样一种完全精神性的活动，人们对它的理解远没有对运动的理解那么清晰。于是我们可能会很自然地认为，人们对处于中间地位的大脑仍然

still less known. Nevertheless, here too it is possible to arrive at general results. We can, for instance, estimate the time required for the mental operation of deciding between two or more events, and of willing to act in accordance with the decision. Thus, if a galvanic shock be given to one foot, and the signal be made with the hand of the same side, a certain physiological time is consumed in the act. But if the apparatus be so arranged that the shock may be given to either foot, and it be required that the person experimenting, not knowing beforehand to which foot the shock is coming, must give the signal with the hand of the same side as the foot which receives the shock, a distinctly longer physiological time is found to be necessary. The difference between the two cases, which, according to Donders, amounts to 66/1,000, or about 1/15 of a second, gives the time taken up in the mental act of recognising the side affected and choosing the side for the signal.

A similar method may be employed in reference to light. Thus we know the physiological time required for any one to make a signal on seeing a light. But Donders found that when matters were arranged so that a red light was to be signalled with the left hand and a white with the right, the observer not knowing which colour was about to be shown, an extension of the physiological time by 154/1,000 of a second was required for the additional mental labour. This of course was after a correction (amounting to 9/1,000 of a second) had been made for the greater facility in using the right hand.

The time thus taken up in recognising and willing, was reduced in some further observations of Donders, by the use of a more appropriate signal. The object looked for was a letter illuminated suddenly by an electric spark, and the observer had to call out the name of the letter, his cry being registered by a phonautograph, the revolving cylinder of which was also marked by the current giving rise to the electric spark.

When the observer had to choose between two letters, the physiological time was rather shorter than when the signal was made by the hand; but when a choice of five letters was presented, the time was lengthened, the duration of the mental act amounting in this case to 170/1,000 of a second.

When the exciting cause was a sound answered by a sound, the increase of the physiological time was much shortened. Thus, the choice between two sounds and the determination to answer required about 50/1,000 of a second; while, when the choice lay between five different sounds, 88/1,000 of a second was required. In these observations two persons sat before the phonautograph, one answering the other, while the voices of both were registered on the same revolving cylinder.

These observations may be regarded as the beginnings of a new line of inquiry, and it is obvious that by a proper combination of changes various mental factors may be eliminated and their duration ascertained. For instance, when one person utters a sound, the nature of which has been previously arranged, the time elapsing before the answer is given corresponds to the time required for simple recognition and volition. When, however, the first person has leave to utter any one, say of five, given sounds, and the second person

了解甚少。然而，事实并非如此。例如，我们能估计出大脑在两三件事之间作出抉择以及将决定付诸行动所需的时间。因此，如果一只脚遭到电击后，用同一侧的手作出反应，这一动作过程需要一定的生理时间。如果受试者事先并不知道哪只脚会遭到电击，并被要求必须用与被电击的那只脚同侧的手作出反应的话，那么需要的生理时间会明显变长。东德斯得出，这两种情况的生理时间差是 66/1,000 秒，大约 1/15 秒，这就是精神活动用于识别哪一侧受刺激（识别哪只脚被电击）和决定用哪一侧作出反应（决定用哪只手作出反应）所需的时间。

类似的方法也可以用于光刺激的实验中，这样我们就可以知道人看见一束光后作出反应所需的生理时间。不过东德斯发现，当受试者被要求看到红光举左手，看到白光举右手，并且事先同样不知道将会出现什么颜色的光时，生理时间延长了 154/1,000 秒，这是额外的精神活动需要的时间。当然，这是针对人们使用右手更为熟练这种情况进行校正（9/1,000 秒）后的结果。

东德斯的进一步研究发现，如果使用更合适的信号，大脑进行识别和作出决定所需的时间将会缩短。受试者需要去寻找突然被电火花照亮的字母，看到后要大喊出字母名称，喊声用声波记振仪记录，记振仪的旋转圆柱也用产生电火花的电流进行了标记。

受试者在 2 个字母之间进行选择需要的生理时间比前述的用手作出反应所需的生理时间短很多，而当被选择的字母增加到 5 个时，生理时间变长，这一过程中思考的时间是 170/1,000 秒。

当用声音回应声音刺激时，延长的生理时间将大大缩短。在 2 种声音之间作出选择并决定回应所需的时间大约是 50/1,000 秒。然而，当在 5 种声音之间作出选择时，所需的生理时间为 88/1,000 秒。在上述观测研究中，2 名受试者坐在声波记振仪前，他们的声音被记录在同一个旋转圆柱上。

这些观测结果也许可以看作是新一轮研究的起点，很明显，适当地综合运用多种方法，可以减少精神因素的影响，并确定它们的生理时间。例如，人发出声音的过程是预先安排好的，在给出回答之前消耗的时间与简单的识别和决定所需的时间相对应。然而，当第一个人可以发出 5 种音中的任一种，而第二个人需在听到后发出同样的声音回应前者时，这一思维过程要复杂很多。在这一过程中，首先要对声

to make answer by the same sound to any and every one of the five which he thus may hear, the mental process is much more complex. There is in this case first the perception and recognition of sound, then the bare volition towards an answer, and finally the choice and combination of certain motor impulses which are to be set going, in order that the appropriate sound may be made in answer. All this latter part of the cerebral labour may, however, be reduced to a minimum by arranging that though any one of five sounds may be given out, answer shall be made to a particular one only. The respondent then puts certain parts of his brain in communication with the origin of certain out going nerves; he assumes the attitude, physical and mental, of one about to utter the expected sound. To use a metaphor, all the trains are laid, and there is only need for the match to be applied. When he hears any of the four sounds other than the one he has to answer, he has only to remain quiet. The mental labour actually employed when the sound at last is heard is limited almost to a recognition of the sound, and the rise of what we may venture to call a bare volitional impulse. When this is done, the time is very considerably shortened. In this way Donders found, as a mean of numerous observations that the second of these cases required 75/1,000 of a second, and the third only 39/1,000 over and above the first. That is to say, while the complex act of recognition, rise of volitional impulse, and inauguration of an actual volition, with the setting free of coordinated motor impulses, took 75/1,000 of a second, the simple recognition and rise of volitional impulse took 39/1,000 only. We infer, therefore, that the full inauguration of the volition took (75 – 39)/1,000 = 36/1,000. In rough language, it took 1/20 of a second to think, and rather less to will.

We may fairly expect interesting and curious results from a continuation of these researches. Two sources of error have, however, to be guarded against. One, and that most readily appreciated and cared for, refers to exactitude in the instruments employed; the other, far more dangerous and less readily borne in mind, is the danger of getting wrong in drawing averages from number of exceedingly small and variable differences.

(**2**, 2-4; 1870)

音进行感知和识别，然后形成初步的回答意识，最后对某些运动脉冲进行选择和组合，以便选用合适的声音进行回答。然而，所有这些后来的脑力活动是可以通过如下安排而减到最少的：尽管发出声音时可以选择 5 种声音中的任何一种，但是回答时只能选择特定的一种。然后，应答者将大脑中的某些部分与某些外周神经联系起来，身体上和精神上都采取一种准备发出预期声音的姿态。打个比方，所有的火车都停在那，只能开那辆最合适的。当应答者听到其他 4 种声音中的一种，而不是那个他必须回答的声音时，他就得保持沉默。他最后听到声音时消耗的脑力劳动几乎只是识别声音并唤起意识脉冲。完成了这些脑力劳动，生理时间就大大缩短了。用这样的方式，东德斯发现，综合大量观察结果，第二种情况所花的平均时间是 75/1,000 秒，而第三种情况只需 39/1,000 秒，优于第一种情况。也就是说，识别、唤起意识脉冲和作出决定，在不产生与之相协调的运动脉冲的情况下，需要的时间是 75/1,000 秒，而识别和唤起意识冲动的时间只有 39/1,000 秒。因此，我们推出，作出决定的时间是 (75 – 39)/1,000 = 36/1,000 秒。粗略地说，思考只需要 1/20 秒，作出决定的时间则更短。

我们很期待后续的实验研究能进一步获得有趣而奇妙的结果。不过，应该注意避免两种错误来源：首先，对于所使用仪器的准确性要多加注意，研究者在这一点上是最小心谨慎的；其次，对只有细微差别的数据取平均值时，要避免出现错误，研究者经常在这个问题上犯错并且很容易忽视它。

（高如丽 翻译；刘力 审稿）

Why is the Sky Blue?

H.A.N.

Editor's Note

In 1870, no one knew why the sky is blue. Here "H.A.N." offers some speculations on the matter. Perhaps, he suggests, as the colour of sunlight is orange (a common belief at the time), regions of the sky from which we receive no sunlight naturally appear as complementary blue. If so, hypothetical beings living on planets of the stars Sirius and Vega, which are white, might see black skies, while reddish Betelgeuse would induce a green sky. In passing, the author also notes that blue might simply be the intrinsic colour of the sky, as suggested by the 1859 experiments of John Tyndall on light scattering. We now know that light scattering from air molecules predominantly causes the colour.

CAN any of your readers inform me why the sky is blue? Is it that the predominant colour of sunlight being orange, the regions devoid of sunlight appear of the complementary colour? If so, the planets of Sirius and Vega would have a black sky, those of Betelgeux a green sky, while those of the double stars would have different coloured skies at different times, according to their position with respect to their two luminaries. Or again, is the blueness merely the colour of our atmosphere, as Prof. Tyndall's experiments have led some to believe? In favour of the former explanation, is the fact that the maximum intensity of the light of the solar spectrum is in the orange, and indeed that the sun looks orange, and if we close our eyes after gazing a moment at him when high up in the sky, we see a blue image. When the sun is low, his colour changes from orange to red, and this would explain the green tints so often seen in the cloudless parts of the sky at sunset. Possibly Mr. Glaisher, who has seen the sky through a thinner stratum of air than most of us, could help us to a solution.

(**2**, 7; 1870)

H.A.N.: Hampstead, April 24.

为什么天空是蓝色的？

H.A.N.

编者按

1870 年时，没有人知道为什么天空是蓝色的。在本文中，“H.A.N.”针对这一现象提出了一些假设。他说，也许因为太阳光是橙色的（当时大家都这么认为），所以在天空中阳光照不到的地方自然就会显现出橙色的互补色——蓝色。如果事实果真如此，那么假如人类生活在天狼星或织女星（它们都是白色的）的行星上就会看到黑色的天空，而在参宿四（红色的）的行星上则会看到绿色的天空。作者还顺便提到，蓝色也许就是天空本来的颜色，就像 1859 年约翰·廷德尔在光散射实验中指出的那样。现在我们都知道空气中各种分子对光的散射是造成天空显现出蓝色的主要原因。

《自然》的读者中有没有人能告诉我为什么天空是蓝色的？是因为太阳光的主导色是橙色，而那些没有太阳光的地方就会显示出其互补色吗？如果是这样的话，那在天狼星和织女星的行星上看到的天空就会是黑色的，在参宿四的行星上看到的天空就会是绿色的，而在双星的行星上看到的天空则会随着两颗星相对位置的变动而在不同时刻呈现出不同的颜色。又或者，如同廷德尔教授的实验已经使一些人相信的那样，天空的蓝色仅仅是因为我们的大气层是蓝色的？实际上，太阳光谱的最大强度出现在橙色位置，而且太阳看起来就是橙色的，但当它高挂在空中时，我们盯着它看一会儿后再闭上眼睛，会看到蓝色的图像，这些事实则更有利于前一种解释。当太阳比较低的时候，它的颜色会从橙色转变为红色，这可能就能解释为什么日落时在晴朗的天空我们经常能看到绿色。至于天空到底为什么是蓝色的，也许格莱舍先生能帮我们找到答案吧，与我们中的大多数人相比，他曾透过更加稀薄的空气观察过天空。

（刘明 翻译；邓祖淦 审稿）

The Physical Constitution of the Sun

Editor's Note

What gives the Sun its heat? Hermann von Helmholtz and William Thomson had recently speculated on the physical constitution of the Sun, and this editorial explores one of their ideas concerning its energy source. Their so-called "contraction theory" offered a conceivable explanation for the output of both solar light and heat. If the sun were gradually contracting, they supposed, while maintaining its temperature, then it would lose energy each year. Helmholtz had shown that a contraction of as little as 0.1 percent in diameter per year would liberate more than 20,000 times the observed solar energy. The hypothesis seemed to give plausible estimates of energy release for a wide range of materials that might make up the Sun.

DR. Gould has addressed an important letter on the above subject to the *Journal of the Frankland Institute*. In the first part he refers to the new light recently thrown on the sun's physical constitution by the observations of Mr. Lockyer, and agrees with him and Dr. Frankland, both as to the absorption taking place in the chromosphere and photosphere itself, and also as to the possible telluric origin of the corona.

He then proceeds with regard to the probable age of the sun:—

"The researches of Helmholtz and Thomson regarding the age of the sun as a source of cosmical heat have shown us limits within which, in the absence of more decisive evidence, we must restrict our theories as to the length of time during which he has warmed the earth. The contraction-theory has been most ably discussed by these eminent physicists, and seems to afford the only satisfactory mode of accounting for the solar light and heat, now that we know both that the meteors generally revolve in cometary orbits, and that the habitability of the earth, as well as the apparent unchanged mutual attraction of the planets, bears testimony to the incorrectness of the meteoric theory. From Pouillet's data (derived from experiments which ought to be repeated in some year when the solar spots are at a minimum) Helmholtz has shown that, even were the sun's density uniform, a contraction of 1/10 percent in his diameter would evolve 20,000 times the present annual supply of solar heat. But when the sun was hotter the same proportional contraction would have evolved yet more heat; so that we must consider the above estimate as a minimum.

"The expansibility of hydrogen gas for 100 °C is 0.3661. No gas appears to have so small a coefficient as 0.360, which would correspond to a linear expansion of 0.108. The expansibility of glass, the smallest known, I believe, even for a solid, is about 1/150 part as great; say 0.00244 in volume, or 0.00081 linear. Therefore for glass even, a contraction of 1 percent in diameter would imply a fall of temperature by 1,230 °C, and a mean specific heat of 218. This seems certainly a minimum value.

关于太阳的物质组成

编者按

太阳的热量是从哪里来的？这篇评论的作者从赫尔曼·冯·亥姆霍兹和威廉·汤姆孙最近对太阳物质组成的研究中发现他们提出的观点中有一个与太阳的能量来源有关。他们提出的“收缩论”为太阳光和热的输出提供了一个可信的解释。他们猜测：如果太阳真的在不断收缩，那么在温度不发生变化时，每年也会损失一定的能量。亥姆霍兹曾经证明：哪怕太阳直径每年只收缩 0.1%，也会释放出比观测到的太阳辐射能高 20,000 倍以上的能量。他们的假说对组成太阳的各种可能的物质释放的能量给出了看似可靠的估计。

谷德博士已经在《弗兰克兰研究院院刊》上发表了一篇有关上述论题的重要报道。在文章的第一部分，他参考了由洛克耶先生的观测结果引发的关于太阳物质组成的最新观点，并且对洛克耶和弗兰克兰博士关于太阳色球层和光球层本身的吸收作用以及日冕源于地球的观点表示赞同。

而后他进一步分析了太阳的大致年龄：

“亥姆霍兹和汤姆孙关于太阳作为宇宙热源的时限的结果给我们提供了一个研究范围，在缺少更明确证据的前提下，我们必须把我们关于太阳照射地球的时间长度的理论限制在这一时间范围内。这些杰出的物理学家已经对收缩论进行了非常深入的讨论，并提出了一个看上去能够解释太阳光和热的唯一令人满意的模式，因为我们知道流星通常沿着彗星轨道运转，也知道地球的可居住性，以及行星之间总是相互吸引，这些都证明流星理论是错误的。根据普耶的观测数据（来自于一些实验，我们应该在若干年后当太阳黑子数达到极小值时重复这些实验），亥姆霍兹指出：即便假设太阳密度是均匀的，那么它的直径收缩 1/1,000 也将导致太阳的年辐射量达到目前的 20,000 倍。但如果太阳比现在更热的话，同样比例的收缩将使太阳放出更多的热；因此我们必须把上面的估计看作是一个最小值。

“氢气在 100℃时的膨胀系数是 0.3661，任何气体的膨胀系数都不可能小于 0.360，这相当于线性膨胀系数为 0.108。玻璃的膨胀系数约为氢气的 1/150，据我所知这是最小的，即使对于固体也是一样的，即体积膨胀系数为 0.00244，线性膨胀系数为 0.00081。因此，即使是玻璃，直径收缩 1% 也意味着温度下降 1,230℃且比热达到 218。看起来这肯定是最小值。

"But if we suppose the expansion coefficient to be as large as that of hydrogen, a contraction of 1 percent would correspond to a change of temperature by 8.2 ℃ or a mean specific heat of 32,700, if equivalent to 20,000 years' supply. This is out of the question.

"Now Thomson has computed that bodies smaller than the sun, falling from a state of relative rest at mutual distances which are large in comparison with their diameters, and forming a globe equal to the sun, would generate 20,000 times the present annual supply. This would be greater did we consider the unquestionable increase of the sun's density towards his centre. And since it seems out of the question that resistance and previous minor impacts could have consumed more than one-half the heat, he inferred ten million times a year's supply to be the lowest, and one hundred million times to be the highest, estimate of the sun's initial heat.

"Now we have every reason for the belief that radiation is proportional to temperature. Assuming this and taking the temperature of the sun's photosphere as 14,000 ℃,
10,000,000 times the present annual supply would be radiated
in 3,650,000 years if the specific heat were 218,
in 7,280,000 ,, ,, ,, ,, 1,000.
100,000,000 times the present annual supply would be radiated
in 8,250,000 years if the specific heat were 218,
in 25,500,000 ,, ,, ,, ,, 1,000
500,000,000 times the present annual supply would be radiated
in 11,700,000 years if the specific heat were 218,
in 38,900,000 ,, ,, ,, ,, 1,000

"For vapours, other than hydrogen, the greatest known specific heat, so far as I am aware, is 0.508 (ammonia); and hydrogen, which has less than 3.5, cannot form any considerable portion of the sun's mass.* A specific heat so high as 1,000 seems altogether out of the question; yet it will be seen that, even on this supposition, an amount of initial heat equal to 500,000,000 the present annual supply, would have been radiated in less than forty million years, were the sun's radiative capacity proportional to his temperature. Taking the more probable age, 10,000,000 years, we should find 226 million times the present annual supply to have been radiated within this period if the specific heat were not greater than 218; and even were the specific heat 1,000, the total radiation would have been eighteen million times a year's radiation at present.

"Thus the limit given by Thomson, although so vastly below that afforded by the speculations of some geologists, would appear itself to demand a considerable additional reduction. And I cannot see how we can well suppose the sun in its present form to have

---

* It seems to form certainly not more than the 18, 000th part of the mass of the earth.

“但是，如果我们假设膨胀系数和氢一样大，那么1%的收缩相当于温度变化8.2℃，或者平均比热为32,700，等价于20,000年的能量供给。这是不可能的。

“现在，汤姆孙通过计算得出，一些比太阳小的天体，当它们之间的距离与它们的直径相比很大时，会从相对静止的状态开始聚集并形成一个类似于太阳的球体，那么这些天体产生的能量将是目前年辐射量的20,000倍。如果我们认为太阳的密度越靠近中心一定越大的话，这个数还会更大一些。因为阻力和之前的小碰撞不可能消耗一半以上的热量，他估计太阳最初的热量至少是年热辐射量的1,000万倍，最多是1亿倍。

“现在，我们有充分的理由相信辐射正比于温度。按照这个假设并假定太阳光球层的温度为14,000℃，则：

如果比热为218，	太阳在3,650,000年内	辐射的热量将是目前年辐射量的10,000,000倍。
如果比热为1,000，	太阳在7,280,000年内	
如果比热为218，	太阳在8,250,000年内	辐射的热量将是目前年辐射量的100,000,000倍。
如果比热为1,000，	太阳在25,500,000年内	
如果比热为218，	太阳在11,700,000年内	辐射的热量将是目前年辐射量的500,000,000倍。
如果比热为1,000，	太阳在38,900,000年内	

“对于气体而言，除了氢气之外，就我所知，已知的最小比热是0.508（氨气）；氢气的比热低于3.5，不可能是组成太阳大气的主要成分。* 看起来比热高达1,000是完全不可能的；即便就按这样的假设，那么，如果太阳的辐射量与它的温度成正比的话，500,000,000倍于目前年辐射量的初始热量也将在不到4,000万年的时间里被辐射掉。在1,000万年这个更可能的时间长度内，我们会发现，如果比热不高于218，那么将有2.26亿倍于目前年辐射量的热量在这段时间内被消耗掉；即使比热能够达到1,000，总辐射量也将是目前年热辐射量的1,800万倍。

“因此，汤姆孙给出的极限虽然已经大大低于一些地质学家的推测，但该值似乎仍然需要一定程度的下调。我认为按照目前的情况来看，太阳向外辐射的时间不可能超过2,000万年，300万年或400万年可能是一个更合理的估计，除非我们假设在

* 氢在地球中的含量肯定不超过地球质量的1/18,000。

radiated heat for more than twenty millions of years, while three or four millions would seem to be a far more probable estimate, unless the thermic laws be totally different in those exalted temperatures which we must suppose to have existed at some past epoch.

"The very great diversity of the limiting values for the specific heat seems to afford ample scope for every needful allowance on account of the natural action of the particles within the body of the sun, even conceding to this the immense effect (analogous to the increase of specific heat) which has been assigned to it by some investigators. Even did we conceive a primitive heat equal to 200,000,000 times the amount now yearly radiated, and a specific heat 10,000 times as great as is possessed by any known gaseous body excepting hydrogen, we could not deduce so long a period as 80,000,000 of years for the past duration of the sun's heat."

(**2**, 34; 1870)

过去曾经存在过一段温度很高的时期，在那样的高温下热学定律完全不同。

“即使承认这个曾被一些研究者考虑过的巨大效应（类似于比热的增加），比热极限值之间的巨大差异也能为太阳内部粒子在各种必要情形下的自然作用提供极大的可能性。就算我们假设太阳的初始热量等于其目前年辐射量的 200,000,000 倍，比热是除氢以外现在已知的任意一种气体的 10,000 倍，那也不可能推出太阳在过去的 80,000,000 年中一直在辐射自己的热量。”

（魏韧 翻译；邓祖淦 审稿）

Spectroscopic Observations of the Sun

J. N. Lockyer

Editor's Note

Here *Nature*'s editor J. Norman Lockyer describes his recent presentation at the Royal Society on spectroscopic observations of the Sun. Lockyer used spectroscopy to deduce the Sun's composition, and the preceding year he had discovered a new element—helium—by this means. Doppler shifts in atomic emission lines enabled him to deduce motions of material at the Sun's surface, and in this way Lockyer observed the dynamical behaviour of sunspots and solar eruptions. Here he records his latest observations on that topic.

ROYAL Society, May 19.—"Spectroscopic Observations of the Sun". No. VI. By J. Norman Lockyer, F.R.S.

The weather has lately been fine enough and the sun high enough, during my available observation-time, to enable me to resume work. The crop of new facts is not very large, not so large as it would have been had I been working with a strip of the sun, say fifty miles or a hundred miles wide, instead of one considerably over 1,000—indeed, nearer 2,000 in width; but in addition to the new facts obtained, I have very largely strengthened my former observations, so that the many hours I have spent in watching phenomena, now perfectly familiar to me, have not been absolutely lost.

The negative results which Dr. Frankland and myself have obtained in our laboratory-work in the matter of the yellow bright line, near D, in the spectrum of the chromosphere being a hydrogen line, led me to make a special series of observations on that line, with a view of differentiating it, if possible, from the line C.

It had been remarked, some time ago, by Prof. Zollner, that the yellow line was often less high in a prominence than the C line; this, however, is no evidence (bearing in mind our results with regard to magnesium). The proofs I have now to lay before the Royal Society are of a different order, and are, I take it, conclusive:—

1. With a tangential slit I have seen the yellow line bright below the chromosphere, while the C line has been dark; the two lines being in the same field of view.

2. In the case of a bright prominence over a spot on the disc, the C and F lines have been seen bright, while the yellow line has been invisible.

3. In a high-pressure injection of hydrogen, the motion indicated by change of wavelength has been less in the case of the yellow line than in the case of C and F.

4. In a similar quiescent injection the pressure indicated has been less.

观测太阳光谱

洛克耶

编者按

在这篇文章中，《自然》的主编诺曼·洛克耶阐述了他最近在英国皇家学会所作的关于观测太阳光谱的报告。洛克耶根据光谱推测出了太阳的化学组成，此前一年他还用这种方法发现了一种新元素——氦。根据原子发射谱线的多普勒频移他推断出物质在太阳表面的移动，并用这种方式观测到了太阳黑子和太阳爆发的动力学特征。在这篇文章中，他给出了相关的最新观测结果。

皇家学会，5 月 19 日——《观测太阳光谱》，第 6 篇，作者为诺曼·洛克耶，英国皇家学会会员。

最近天气一直很好，而且太阳也达到了足够的高度，这使我在可用于观测的时间内可以重新开始我的工作。观察到的新现象不是很多，如果我进行观测研究的是太阳的某一狭长范围，比方说是 50~100 英里的高度范围，而不是远大于 1,000 英里——实际上接近 2,000 英里的高度，则得到的新的结果将比现在多得多。不过，除了观测到一些新现象外，我还极大地拓展了我以前的观测工作，因此，我虽然花了很多时间去观测那些现在已经很熟知的现象，但也并非是白白浪费了时间。

弗兰克兰博士和我本人在实验室工作中得到靠近 D 线的黄色明亮谱线，这在色球层光谱中属于氢线的部分中没有找到。这促使我对该谱线进行了一系列的观测，尝试性地从另一个角度，把它同 C 线区分开来。

不久前，策尔纳教授已经注意到，在日珥中这条黄线通常不及 C 线高；但这没有被证实（请注意我们的结果只是对镁而言）。我现在提交给皇家天文学会的一些观测证据，思路有所不同但我认为至关重要：

1. 当利用一个切向狭缝观察时，我看到色球层底部有一条明亮的黄线，但是 C 线却是暗的，虽然这两条线都处在同一个视场中。

2. 当日面上有明亮的日珥遮住黑子时，看到的 C 线和 F 线都很亮，却看不到黄线。

3. 在氢的高压喷流中，由波长变化表示的该运动，在黄线的情况下不及在 C 线和 F 线的情况下大。

4. 在近似静止的喷流中，显示的压力变小了。

5. In one case the C line was seen long and unbroken, while the yellow line was equally long, but broken.

The circumstance that this line is so rarely seen dark upon the sun makes me suspect a connection between it and the line at 5,015 Ångström, which is also a bright line, and often is seen bright in the chromosphere, and then higher than the sodium and magnesium lines, when they are visible at the same time; and the question arises, must we not attribute these lines to a substance which exists at a higher temperature than those mixed with it, and to one of very great levity? for its absorption line remains invisible, as a rule, in spot-spectra.

I have been able to make a series of observations on the fine spot which was visible when I commenced them on April 10th, not far from the centre of its path over the disc. At this time, the spot, as I judged by the almost entire absence of indications of general absorption in the penumbral regions, was shallow, and this has happened to many of the spots seen lately. A few hours' observation showed that it was getting deeper apparently, and that the umbrae were enlarging and increasing in number, as if a general down-sinking were taking place; but clouds came over, and the observations were interrupted.

By the next day (April 11) the spot had certainly developed, and now there was a magnificently bright prominence, completely over the darkest mass of umbra, the prominence being fed from the penumbra or very close to it, a fact indicated by greater brilliancy than in the bright C and F lines.

April 12. The prominence was persistent.

April 15. Spot nearing the limb, prominence still persistent over spot. At eleven I saw no prominence of importance on the limb, but about an hour afterwards I was absolutely startled by a prominence not, I think, depending upon the spot I have referred to, but certainly near it, more than 2′ high, showing a tremendous motion towards the eye. There were light clouds, which reflected to me the solar spectrum, and I therefore saw the black C line at the same time. The prominence C line (on which changes of wavelength are not so well visible as in the F line) was only coincident with the absorption-line for a few seconds of arc!

Ten minutes afterwards the thickness of the line towards the right was all the indication of motion I got. In another ten minutes the bright and dark lines were coincident.

And shortly afterwards what motion there was was towards the red!

I pointed out to the Royal Society, now more than a year age*, that the largest prominences, as seen at any one time, are not necessarily those in which either the intensest action or the most rapid change is going on. From the observations made on this and the following day, I think that we may divide prominences into two classes:—

* *Proc. Roy. Soc.* 1869, p. 354, Mar. 17.

5. 只有一次观测中，C 线看起来长且没有间断，而黄线虽然与 C 线具有相同的长度，但中间有间断。

太阳光谱中这条线很暗的时候是非常少的，这使我怀疑这条谱线是否与 5,015 Å 的谱线有联系。这条 5,015 Å 的谱线也是一条亮线，经常可以在色球层的光谱中看到，而且当它与钠线和镁线同时出现时，它的亮度高于钠线和镁线。于是问题就产生了，难道我们就不能认为，产生这些谱线的物质与其周围混合在一起的物质相比温度更高而且非常不稳定吗？因为它的吸收谱线在黑子光谱中通常是看不见的。

4 月 10 日，我开始对一个小黑子进行一系列的观测，这个黑子距离其通过日面路径的中点不远。在这次观测中，黑子半影区内几乎完全没有出现吸收现象，我据此判断这是一个很浅的黑子。最近观测到的很多黑子都是这样。几个小时的观测过程中，这个黑子显然在逐渐变深，本影部分变大，且数目增加，就好像有某种下沉运动正在进行；但是这时有云飘过来，我的观测被迫停止。

等到了第二天（4 月 11 日），黑子显然已经变大了，而且这时出现了一个非常明亮的日珥，完全遮住了黑子本影最暗的区域，这个日珥从黑子的半影或者是很靠近半影的地方出现，这一点根据其亮度比明亮 C 线和 F 线更高可知。

4 月 12 日，日珥仍然停留在那里。

4 月 15 日，黑子接近日面的边缘，日珥仍然在黑子上方。到 11 时，在日面边缘处我没有看到明显的日珥，但是大约一个小时以后，我就完全被一个日珥惊呆了，这个日珥与我观测的那个黑子没有关系，但是确实和它靠得很近，其高度超过 2 角秒，并且朝视线方向有巨大的运动。这时天上有薄云，它影响了太阳的光谱，这时我完全看不到C线。日珥光谱中的C线(在这个波段上的波长变化不像F线那么明显)与吸收谱线的重合只有几个角秒。

10 分钟后，谱线向右侧展宽，这就是我提到的运动的全部征兆。再过 10 分钟，亮线与暗线又重合在一起了。

过了很短的时间，那里的运动移向了红端！

一年多以前我曾向皇家学会报告过 *，在任何一次观测中，最大的日珥未必运动得最强烈或变化得最快。根据这次和第二天的观测，我认为可以将日珥分为两类：

---

*《皇家学会学报》，1869 年 3 月 17 日，第 354 页。

1. Those in which great action is going on, lower vapours being injected; in the majority of cases these are not high, they last only a short time—are throbs, and are often renewed, and are not seen so frequently near the sun's poles as near the equator. They often accompany spots, but are not limited to them. These are the intensely bright prominences of the American photographs.

2. Those which are perfectly tranquil, so far as wavelength evidence goes. They are often high, are persistent, and not very bright. These do not, as a rule, accompany spots. These are the "radiance" and dull prominences shown in the American photographs.

I now return to my observations of the spot. On the 16th, the last of the many umbrae was close to the limb, and the most violent action was indicated occasionally. I was working with the C line, and certainly never saw such rapid changes of wavelength before. The motion was chiefly horizontal, or nearly so, and this was probably the reason why, in spite of the great action, the prominences, three or four of which were shut out, never rose very high.

I append some drawings, made, at my request, by an artist, Mr. Holiday, who happened to be with me, and who had never seen my instrument or the solar spectrum widely dispersed before. I attach great importance to them, as they are the untrained observations of a keen judge of form.

The appearances were at times extraordinary and new to me. The hydrogen shot out rapidly, scintillating as it went, and suddenly here and there the bright line, broad and badly defined, would be pierced, as it were, by a line of intensely brilliant light parallel to the length of the spectrum, and at times the whole prominence spectrum was built up of bright lines so arranged, indicating that the prominence itself was built up of single discharges, shot out from the region near the limb with a velocity sometimes amounting to 100 miles a second. After this had gone on for a time, the prominence mounted, and the cyclonic motion became evident; for away from the sun, as shown in my sketch, the separate masses were travelling away from the eye; then gradually a background of less luminous hydrogen was formed, moving with various velocities, and on this background the separate "bombs" appeared (I was working with a vertical spectrum) like exquisitely jewelled ear-rings.

It soon became evident that the region of the chromosphere just behind that in which the prominence arose, was being driven back with a velocity something like twenty miles a second, the back-rush being so local that, with the small image I am unfortunately compelled to use, both the moving and rigid portions were included in the thickness of the slit. I saw the two absorption-lines overlap.

These observations were of great importance to me; for the rapid action enabled me to put together several phenomena I was perfectly familiar with separately, and see their connected meaning.

1. 一类是活动性强，气体喷射较低的日珥。在大多数情况下，这种日珥不高，仅仅持续比较短的时间，并且会有跳动，形状经常翻新，这种日珥在太阳两极区域出现的频率没有在赤道附近出现的频率高。它们经常与黑子一起出现，但也不全是这样。美国人拍摄的照片中那些极其明亮的日珥就属于这类。

2. 另一类日珥非常稳定，这从其谱线的波长总是不变可以看出。这类日珥通常很高、持续时间很长，但不是特别亮。原则上，它们并不伴随黑子出现。这些就是美国人拍摄的照片里那些“深粉红色”并且暗淡的日珥。

我现在要回到我对黑子的观测上来。到 16 日，许多持续的本影区已经接近于日面边缘了，非常剧烈的活动只是偶尔才出现。我主要观测的是 C 线，以前从未出现过如此快速的波长变化。其运动方向基本上是水平的，或者是接近水平的。这也许就是为什么有三四个日珥，虽然运动剧烈，向外抛射，但总也升不太高的原因。

我附上了一些插图，这些插图是我邀请艺术家霍利迪先生画的，他当时恰好与我在一起，但他以前从来没有见过我的观测仪器和这么庞杂的太阳光谱。我非常重视这些插图，因为它们出自一位未受过专业训练但具有敏锐判断力的观察者之手。

有几次我观察到了一些反常的新现象。氢物质快速向外喷射，同时发出闪光，突然，到处都有谱线很宽的亮线，而且难以辨认，就像是被一束与波长方向平行的特别强的光穿过一样。有时候整个日珥光谱都由这样的亮线组成，这表明日珥本身是由单一的放电形成的，从靠近日面边缘的区域向外抛射，速度有时可以达到每秒 100 英里。当这些现象过去一段时间以后，日珥又出现了，并显示出气旋式的运动；从太阳上向外喷出，看上去是一些分散运动的物质，如我的附图所示。然后逐渐地，形成了一个不太明亮的氢的背景，并以各种不同的速度运动。在这个背景上，呈现出一系列分离的“爆炸”（我的光谱是垂直的），就像镶着宝石的精美耳环一样。

很快我又发现，日珥出现之处背后的色球区域又以每秒约 20 英里的速度被拉回来，这种反冲仅限于局部区域，我手头只有一幅小图，但又不得不用它，在这张图中，运动部分和固定部分都处在狭缝宽度内。我看到了两条吸收谱线的重叠。

这些观测结果对我来说是很重要的；因为这些快速运动可以让我将一些我非常熟悉的分立现象联系在一起，看一看把它们联系起来有什么意义。

They may be summarised as follows, and it will be seen that they teach us much concerning the nature of prominences. When the air is perfectly tranquil in the neighbourhood of a large spot, or, indeed, generally in any part of the disc, we see absorption-lines running along the whole length of the spectrum, crossing the Frauenhofer lines, and they vary in depth of shade and breadth according as we have pore, corrugation, or spot under the corresponding part of the slit—a pore, in fact, is a spot. Here and there, where the spectrum is brightest (where a bright point of facula is under the slit) we suddenly see this appearance—an interesting bright lozenge of light. This I take to be due to bright hydrogen at a greater pressure than ordinary, and this then is the reason of the intensely bright points seen in ranges of faculae observed near the limb.

The appearance of this lozenge in the spectroscope, which indicates a diminution of pressure round its central portion, is the signal for some, and often all, of the following phenomena:—

1. A thinning and strange variation in the visibility and thickness of the hydrogen absorption-line under observation.

2. The appearance of other lozenges in the same locality.

3. The more or less decided formation of a bright prominence on the disk.

4. If near the limb, this prominence may extend beyond it, and its motion-form will then become more easy of observation. In such cases the motion is cyclonic in the majority of cases, and generally very rapid, and—another feature of a solar storm—the photospheric vapours are torn up with the intensely bright hydrogen, the number of bright lines visible determining the depth from which the vapours are torn, and varying almost directly with the amount of motion indicated.

Here, then, we have, I think, the chain that connects the prominences with the brighter points of the faculae.

These lozenge-shaped appearances, which were observed close to the spot on the 16th, were accompanied by the "throbs" of the eruption, to which I have before referred; while Mr. Holiday was with me—a space of two hours—there were two outbursts, separated by a space of almost rest, and each outburst consisting of a series of discharges, as I have shown. I subsequently witnessed a third outburst. The phenomena observed on all three were the same in kind.

On this day I was so anxious to watch the various motion-forms of the hydrogen-lines, that I did not use the tangential slit. This I did the next day (the 17th of April) in the same region, when similar eruptions were visible, though the spot was no longer visible.

下面将这些现象综合起来，从中我们会看到这些现象将告诉我们更多有关日珥的特性。在大的黑子周围，或者一般来说，实际上在日面的任何一部分，只要气体处于完全的静止状态，我们会看到吸收谱线沿着整个光谱的波长方向移动，横穿夫琅和费线，吸收谱线的明暗程度和宽度都会发生变化，这取决于狭缝下面对应的是小黑点、褶皱还是黑子，小黑点其实也是黑子。在光谱最亮的地方（在狭缝下面，该位置有一个明亮光斑），我们突然发现了一个有趣的菱形亮光。我认为这是由于明亮的氢处在高于正常压力的状态下所致，这也就是我们在日面边缘附近的光斑中看见亮点的原因。

分光镜中出现的这种菱形表明它中心位置周围压力的减小，它的出现标志着下述某些现象的发生，而且往往是下述各种现象的同时发生。

1. 在观测期间，氢吸收谱线的能见度和宽度出现了微弱而奇异的变化。

2. 在同一位置出现了其他菱形。

3. 可以比较肯定地确认在日面上形成了一个明亮日珥。

4. 如果靠近日面边缘，这个日珥会延伸到日面以外，这样它的运动形态就更容易被观测。在这种情况下，这类日珥运动大多会呈气旋状，运动速度一般都很快，另外光球大气被撕裂，同时伴有非常明亮的氢线（这正是太阳风暴的另一特征），可见的明亮谱线的数目反映了气体被撕裂位置的深度，并且这个数目的变化直接反映了那里的运动状况。

我认为我们现在已经建立了日珥与光斑中较明亮点之间的联系。

我们在 16 日观测到的在太阳黑子附近出现的菱形亮点是伴随着“脉动”式的喷发出现的，我在前面已经提到过这种脉动；霍利迪先生当时正和我在一起，两个小时内我们看到了两次喷发，在两次喷发之间有一段几乎完全宁静的间歇期。每次喷发都伴有一系列放电现象，就像我前面所说的那样。随后我又见证了第三次喷发。这三次喷发都属于同一类现象。

那天我因为急于观测对应于各种不同运动形式的氢线而没来得及使用切向的狭缝。第二天（4 月 17 日），我使用切向狭缝对同一区域进行了观测，这时看到了同样的喷发现象，但那个黑子已经看不见了。

Judge of my surprise and delight, when upon sweeping along the spectrum, I found hundreds of the Frauenhofer lines beautifully bright at the base of the prominence!!!

The complication of the chromosphere spectrum was greatest in the regions more refrangible than C, from E to long past *b*, and near F, and high pressure iron vapour was one of the chief causes of the phenomenon.

I have before stated to the Royal Society that I have seen the chromosphere full of lines; but the fulness then was as emptiness compared with the observation to which I now refer.

A more convincing proof of the theory of the solar constitution, put forward by Dr. Frankland and myself, could scarcely have been furnished. This observation not only endorses all my former work in this direction, but it tends to show the shallowness of the region on which many of the more important solar phenomena take place, as well as its exact locality.

The appearance of the F line, with a tangential slit at the base of the prominence, included two of the lozenge-shaped brilliant spots to which I have before referred; they were more elongated than usual—an effect of pressure, I hold, greater pressure and therefore greater complication of the chromosphere spectrum; this complication is almost impossible of observation on the disc.

It is noteworthy that in another prominence, on the same side of the sun, although the action was great, the erupted materials were simple, *i.e.*, only sodium and magnesium, and that a moderate alteration of wavelength in these vapours was obvious. Besides these observations on the 17th, I also availed myself of the pureness of the air to examine telescopically the two spots on the disc, which the spectroscope reported tranquil as to up and down rushes. I saw every cloud-dome in their neighbourhood perfectly, and I saw these domes drawn out, by horizontal currents, doubtless, in the penumbrae, while on the floors of the spots, here and there, were similar cloud-masses, the distribution of which varied from time to time, the spectrum of these masses resembling that of their fellows on the general surface of the sun.

I have before stated that the region of a spot comprised by the penumbrae appears to be shallower in the spots I have observed lately (we are now nearing the maximum period of sun spots); I have further to remark that I have evidence that the chromosphere is also shallower than it was in 1868.

I am now making special observations on these two points, as I consider that many important conclusions may be drawn from them.

(**2**, 131-132; 1870)

当我沿着光谱扫视时，发现在日珥底部出现了几百条非常明亮的夫琅和费线！！！可想而知我是多么地惊喜。

色球层的光谱在折射率大于 C 线的区域最为复杂——从 E 线经过 *b* 一直延伸到 F 线附近。高压的铁蒸气是产生这种谱线的主要因素之一。

我以前曾向皇家学会陈述过我看到了色球层的所有谱线，但是那些谱线与我现在观测到的谱线数目相比是完全不值一提的。

由弗兰克兰博士和我提出的关于太阳结构理论的更有说服力的证据可以算是刚刚完成。这项观测不仅支持了以前我在这个方向上的所有工作成果，同时还表明太阳上很多更加重要的现象都是发生在活动区的浅层，并且完全是局部性的。

当我使用切向狭缝观察日珥底部时，发现 F 线处有两个前面提到过的菱形亮点；它们比通常情况下出现的亮点长一些，我认为这是由于压力的影响，压力越高，色球层的光谱越复杂；在日面上几乎不可能观察到这么复杂的光谱。

值得一提的是，在太阳的同一侧有另一个日珥，虽然它的活动性很强，但喷出的物质成分却比较简单，只有钠和镁，在这些气体中谱线波长的不断变化也很明显。除了这些在 17 日得到的观测结果以外，我还在大气比较纯净的时候用望远镜仔细观察了日面上的两个黑子，分光仪显示向上的和向下的活动都是平稳的。我仔细观察了这两个黑子附近的每一个云状隆起结构，发现在黑子半影部分的隆起结构无疑是被水平方向的气流拉长了，而在黑子的底层，到处是类似的云状物，其分布情况不时地发生变化，这些云状物的光谱与位于太阳表面的云状物的光谱是类似的。

我以前曾陈述过，在我最近观察过的黑子中，黑子的半影区域看上去比原来更浅（我们现在临近太阳黑子活动的极大期）；我还要进一步说明，有证据表明现在的色球层厚度比 1868 年时薄一些。

我现在正在对上述的两点情况进行专门的观测，因为我认为根据这些观测可以得出许多重要的结论。

（史春晖 翻译；何香涛 审稿）

The Unit Of Length

Editor's Note

This essay announces the outcome of a battle in Great Britain over the standard of length. A Royal Commission on the issue had concluded that the metric system should be introduced. The advantages of the metre—then defined as one ten-millionth of the distance between the Earth's equator and North pole on a line through Paris—lie in its wide international familiarity and in the simplicity of the metric system. However, the essay notes, much work remains to define the metre and eliminate small but potentially significant variations in standards used in different nations. Only in 1983 would physicists redefine the metre in more precise terms by using optical means.

THE battle of the Standards is over, and we may say the Metre has gained the victory. The need of a new system of weights and measures to amend the strange diversities which disfigure our practice being admitted, the question has once more been started— Should we once for all found our system on a natural basis? The pendulum vibrating seconds in a certain latitude, was long ago proposed as a universal basis of linear measure, and the House of Commons somewhat countenanced it years ago, by prescribing that the length of the yard shall be determined by the length of the second's pendulum. But the action of gravitation on which the terms of the vibration depends, is subject to so many variations and disturbances, that the quantity sought cannot, even on the same spot, be absolutely the same at all times. The real length of a normal pendulum is almost unattainable, so limited is our knowledge of the force of gravity on land and at sea. A more certain basis for a natural unit has been found in the polar axis, the length of which, according to Sir John Herschel, bears a close relation to our imperial inch, and has the advantage of avoiding the many causes of error resulting from the physical peculiarities of the countries through which any measured arc may happen to pass. But are our physicists agreed as to the real length of the polar axis, and would it be worth while to make any alteration in our weights and measures for the sole purpose of attaining some scientific correspondence between the unit in use and a unit founded on nature?

The advocates of the metre rest their arguments on a much broader basis. They do not assert that the metre is absolutely and mathematically the ten millionth part of the quadrant of the earth; they know that the meridians of places differing in longitude are not all precisely of the same length; and they admit that were we now to make a new measurement with our better instruments and more extended information, we might attain much greater accuracy than was arrived at by the French philosophers at the end of the eighteenth century. What commends the metre above any other unit, is the fact, that it is already a cosmopolitan unit, widely recognised, and in general use among many nations; and that whilst other units remain as philosophical abstractions, the metre is the basis of a system, not only perfectly complete,

长度单位

编者按

这篇文章公布了发生在英国的一场关于长度标准的争论的结果。负责这一领域的皇家委员会最后决定应该采用米制。当时定义的一米是指沿着通过巴黎的经线从地球赤道到北极点之间距离的千万分之一。米制的优点在于其简单而且全球范围内的人都熟悉。不过，这篇文章指出，还需要做很多工作来确切地定义米并消除不同国家使用的标准中存在的可能会造成显著差异的微小区别。后来直到 1983 年，物理学家才通过光学方法以更精确的方式重新定义了米。

关于标准的战争终于结束，我们可以说，米最终获得了胜利。为了消除多种单位制给实践带来的不便而产生的对新度量衡系统的需求，使这一问题再一次摆在了我们面前——我们是否需要将度量衡系统完全建立在自然的基础之上？很久以前，曾有人提出用某个纬度上秒摆的长度作为线性测量的通用基础，数年前这一提议还得到了国会下议院的支持，规定一码的长度由秒摆的长度决定。但振动周期受重力影响，而重力依赖于很多变量和扰动，即便是在同一个地点，这个值也无法一直保持绝对不变。标准秒摆的真实长度几乎无法获得，因为我们对海上和陆上的重力情况了解有限。还有一个更容易确定的自然单位就是地轴的长度，按照赫歇尔爵士的说法，地轴的长度和我们英制的英寸长度密切相关，并且它不会像弧度的测量那样因不同国家的地理特征不同而产生误差。但物理学家们对地轴长度的真实值达成一致了吗？仅仅为了在实用单位和自然单位之间建立一种科学上的联系，就去改变我们现有的度量衡，这值得吗？

米的支持者将他们的论据建立在更加广泛的基础之上。他们并不坚持米在数学上绝对是四分之一个地球周长的千万分之一；他们知道，子午线的长度随着经度而改变，长度并不完全一致；他们承认，如果现在用更好的仪器和更广博的知识去建立一种新的测量方法，那么我们测量得到的结果可以比 18 世纪末法国科学家得到的结果更加精确。为什么米比其他单位更有优越性呢？因为它已经成了一个国际性的单位，被广泛地认可，许多国家都在使用；虽然其他单位还有科学上的意义，但是米却是一个体系的基础，不仅完备、均一、科学，而且简单实用。在定义米时

homogeneous, and scientific, but simple and practical in all its parts. Any slight error in the determination of the metre, is more than counterbalanced by the extreme simplicity, symmetry, and convenience of the metric system; and not the least of its recommendations are, that the unit of linear measure applied to matter in its three forms of extension, viz., length, breadth, and thickness, is the standard of all measures of length, surface, and solidity; and that the cubic contents of the linear measure in distilled water at a temperature of great contraction, furnish at once the standard weight and measure of capacity.

When we said that the battle of the Standards is over and that the Metre has gained the victory, it was meant that, for practical purposes, all opposition to the introduction of the metric system has been abandoned, and that Parliament and the Government are now left to introduce it in such a way and at such a time as may be found at once practicable and satisfactory. The use of the metric system has been legalised for the last half-dozen years, but it was not till quite lately that the whole question was submitted to the calm deliberation of a Royal Commission. The Standard Commissioners, who included among their members the Astronomer Royal, the President of the Royal Society, and the late Master of the Mint, considered the question of the introduction of metric weights and measures, in any form, *ab initio.* And after careful examination they gave their verdict in its favour in the following terms: —

> "Considering the information which has been laid before the Commission, —
>
> "Of the great increase during late years of international communication, especially in relation to trade and commerce;
>
> "Of the general adoption of the metric system of weights and measures in many countries, both in Europe and other parts of the world, and more recently in the North German Confederation and in the United States of America;
>
> "Of the progress of public opinion in this country in favour of the metric system as a uniform international system of weights and measures;
>
> "And of the increasing use of the metric system in scientific researches and in the practice of accurate chemistry and engineering construction;
>
> "We are of opinion that the time has now arrived when the law should provide, and facilities be afforded by the Government, for the introduction and use of metric weights and measures in the United Kingdom."

The Commissioners further recommended that metric standards, accurately verified in relation to the primary metric standards at Paris, should be legalised; that verified copies of the official metric standards should be provided by the local authorities for inspectors of such districts as may require them; and that the French nomenclature, as well as the decimal scale of the metric system, should be introduced in this country. The Commissioners, whatever might have been their predilections, could not resist the fact that the civilised world pronounced itself for the metre, and they sanctioned its legalisation. What is to be regretted is that they stopped there. Since the complete substitution of the metric for the present practice is now practically certain, would it not be much better to prepare for the change and carry it into effect as speedily as possible? No advantage can come from a policy of indecision, and we trust that the Legislature may adopt a more definite course than the

出现的微小误差，都远不足以抵消米制单位的简洁、对称和便利性；还有很重要的一点是，这种用来测量物体长、宽、高的线性单位是所有长度、面积、体积测量的标准；而且用线性测量方法测得蒸馏水在使其密度极大的温度下的容积，就马上可以得到标准的重量和容积。

我们说，关于标准的战争终于结束，米获得了最终胜利，这意味着，从实用角度出发，所有反对引入米制系统的意见都被抛弃，国会和政府接下来将在适当的时间，以适当的方式将其推广，并将很快得到可行且满意的结果。虽然米制系统在6年前已经合法化，但是直到最近某皇家委员会才开始慎重地考虑这个问题。由皇家天文学家、皇家学会主席、前任铸币大臣等人组成的标准委员会开始考虑重新（无论以何种方式）引入米制度量衡的问题。经过仔细的考量，他们通过了以下决议：

“根据委员会掌握的信息：

近年来国际交往迅速增加，尤其是商贸上的往来；

在欧洲和世界其他地区的许多国家，米制度量衡体系已经得到广泛的应用，最近北德联盟和美国也开始使用这种体系；

在本国，公众的意见也更加倾向于支持米制系统作为标准的国际度量衡系统；

在科学研究、精细化工和工程建设等领域米制系统得到越来越多地运用；

我们认为，在英国，现在是由政府立法引入和实施米制系统并推广的时候了。”

标准委员会进一步建议，在巴黎确认的基本米制标准应该得到法律上的认可；地方政府在当地检查员有需求的时候，应能提供通过验证的官方米制标准件；米制相关的法语专有名词以及米制系统的十进制都应引入本国。不管委员们自己的偏好如何，他们都不能阻止文明世界选择米制单位并将其合法化。他们要为自己的止步不前感到后悔。既然用米制单位代替现有单位已经在实践上得到了肯定，那么为什么不做好改变的准备尽快让它得到有效执行呢？优柔寡断是不会有任何进步的，我们相信，立法机构可以提出比皇家委员们的草案更加明确的方案。不要指望人们会不怕麻烦地学习新单位，无论它如何完美和简单，除非新单位的使用是绝对必要的。

one sketched out by the Royal Commissioners. Let it not be imagined that the people will give themselves the trouble of learning the new system, however beautiful and easy, so long as its use is not absolutely necessary. With all the desire of the teachers to introduce it in the schools, they find that they cannot teach the old and the new tables. They cannot afford the time. A compulsory measure is the only method of dealing with the question.

The Warden of the Standards being now employed in procuring Metric Standards, it may be well to add that the mode of constructing them, either from the original Metre at the Archives, or from the copy at the Conservatoire des Arts et Métiers, has been much debated. The International Statistical Congress, held at Berlin, decided "That the care of preparing and putting into execution the regulations to be followed in the construction of the standards, and of the system itself, should be entrusted to an International Commission, which will also see to the correction of the small scientific defects of the system." The International Geodesical Conference held at Berlin in 1867 decided: "In order to define the common unit of measures for all the countries of Europe, and for all times, with as much exactness as possible, the Conference recommends the construction of a new prototype European Metre. The length of this European Metre should differ as little as possible from that of the Metre of the Archives in Paris, and should in all cases be compared with the greatest exactness. In the construction of the new prototype standard, care should be taken to secure the facility and exactness of the necessary comparisons." And "the construction of the new prototype metre, as well as the preparation of the copies destined for different countries, should be confided to an International Commission, in which the States interested should be represented." Since then, the Imperial Academy of Science of St. Petersburg has taken the matter in hand, and a committee of the Physico-Mathematical class, consisting of MM. Struve, Wild, and Jacobi, has made a report on the subject, observing that the standard metric weights and measures of the various countries of Europe and of the United States differ by sensible though small quantities from one another. They expressed their opinion that the continuation of these errors would be highly prejudicial to science. They believed that the injurious effects could not be guarded against by private labour, however meritorious, and they recommended that an International Commission should be appointed by the countries interested to deal with the matter. This suggestion was approved by the French Government, and consequently the Conference will take place in Paris in August next, when the Astronomer Royal, Professor William H. Miller, and the Warden of the Standards, will represent this country. Everything seems thus tending towards the early realisation of the great scheme of uniformity of weights and measures throughout the world.

(**2**, 137-138; 1870)

教师们希望把新单位制引入学校，但是他们不可能同时教授新旧两种体系，他们没那么多时间。强制推行是解决该问题的唯一办法。

标准委员会的主席目前正致力于建立米制标准的工作，这里最好说明一下，关于米制标准如何建立，是使用巴黎档案馆中的米原器，还是使用巴黎工艺技术学院的拷贝，这个问题还在争论之中。在柏林召开的国际统计大会决定："应委托一个国际委员会负责制定和执行米制度量衡标准以及在建立这个标准过程中的一些规则，而这个委员会还要负责修正这个系统在科学上的微小误差。"1867年在柏林召开的国际测量学大会决定："为了给欧洲各国确定一个在任何时间都尽可能准确的通用度量单位，大会建议建立一个新的欧洲米原器。这个欧洲米原器的长度要尽可能地与巴黎档案馆中的米原器相符，且在任何情况下都能保持最高的精度。在建立这个新的原器的过程中，要兼顾便利和精确两个要素。"而且，"新的米原器的建立以及交送到各国的拷贝的制作，都应委托给一个国际委员会，而这个委员会要由相关国家的代表组成。"而后，圣彼得堡皇家科学院接受了这个工作，由斯特鲁韦、维尔德和雅可比组成的数理委员会就此问题作了一个报告，指出欧美各国使用的米制度量衡标准之间存在微小却可以观察到的差异。他们认为这些误差会导致科学上的严重偏差。他们确信：个人的研究工作，不管多么有价值，也不能避免出现不利的影响，因此他们建议相关国家委派一个国际委员会来处理这个问题。法国政府批准了这个建议，今年8月将在巴黎举行一个会议，届时皇家天文学家威廉·米勒教授和标准委员会主席都将出席这个会议。一切似乎都在朝着早日实现世界度量衡统一的伟大计划前进。

（王静 翻译；李淼 审稿）

Pasteur's Researches on the Diseases of Silkworms

J. Tyndall

Editor's Note

This is an account of French chemist Louis Pasteur's first proof that an infectious disease is caused by the physical transmission of minute entities or "germs" (bacteria). It is written by John Tyndall of the Royal Institution in London, one of the leading British scientists of his time. Pasteur had been asked by the French chemist Jean-Baptiste Dumas to investigate the cause of a disease ravaging France's silkworms. Pasteur focused on the microscopic "corpuscles" seen by others in silkworm blood but previously thought to be native to the creatures. Pasteur's initial presentation in France of his theory on infectious germs met with strong criticism, but his famously systematic experiments eventually made the case irrefutable.

I have recently received from M. Pasteur a copy of his new work, "Sur la Maladie des vers à soie", a notice of which, however brief and incomplete, will, I am persuaded, interest a large class of the readers of *Nature*. The book is the record of a very remarkable piece of scientific work, which has been attended with very remarkable practical results. For fifteen years a plague had raged among the silkworms of France. They had sickened and died in multitudes, while those that succeeded in spinning their cocoons furnished only a fraction of the normal quantity of silk. In 1853 the silk culture of France produced a revenue of one hundred and thirty millions of francs. During the twenty previous years the revenue had doubled itself, and no doubt was entertained as to its future augmentation. "Unhappily, at the moment when the plantations were most flourishing, the prosperity was annihilated by a terrible scourge." The weight of the cocoons produced in France in 1853 was twenty-six millions of kilogrammes; in 1865 it had fallen to four millions, the fall entailing in the single year last mentioned a loss of one hundred millions of francs.

The country chiefly smitten by this calamity happened to be that of the celebrated chemist Dumas, now perpetual secretary of the French Academy of Sciences. He turned to his friend, colleague, and pupil, Pasteur, and besought him with an earnestness which the circumstances rendered almost personal, to undertake the investigation of the malady. Pasteur at this time had never seen a silkworm, and he urged his inexperience in reply to his friend. But Dumas knew too well the qualities needed for such an inquiry to accept Pasteur's reason for declining it. "Je met," said he, "un prix extrême à voir votre attention fixée sur la question qui interesse mon pauvre pays; la misère surpasse tout ce que vous pouvez imaginer." Pamphlets about the plague had been showered upon the public, the monotony of waste paper being broken at rare intervals by a more or less useful publication. "The Pharmacopoeia of the Silkworm," wrote M. Cornalia in 1860,

巴斯德对家蚕疾病的研究

廷德尔

编者按

法国化学家路易·巴斯德提出，感染性疾病是由很小的实体或者说“病菌”（细菌）的物理传播造成的，这篇文章报道的正是巴斯德关于此项研究的首例证据。文章作者是伦敦皇家研究院的约翰·廷德尔，他是当时英国第一流的科学家之一。早先，法国化学家让–巴蒂斯特·杜马邀请巴斯德研究当时给法国家蚕带来极大破坏的一种疾病的起因。巴斯德把重点放在了一种微观“粒子”上。早先就有人在家蚕的体液中发现了这种粒子，但先前人们认为这种粒子是家蚕本身就有的。巴斯德的感染性病菌理论一经在法国公布便遭到了强烈的批判，不过，通过出色的系统实验，他最终无可辩驳地证明了其理论的正确。

最近我收到了一本巴斯德先生寄来的他的最新著作，书名是《关于家蚕疫病》，这里将对这本书作个介绍，我想尽管这个介绍非常简要甚至不太完整，但一定有很多《自然》的读者们对此很感兴趣。这本书中记录了一段不平凡的科学研究，并附有一些极具价值的应用成果。在法国，一种家蚕疫病肆虐了 15 年之久，导致大量家蚕生病死亡，而那些成功结茧的家蚕提供的丝只有正常产量的一小部分。1853 年，法国丝绸业创收 1.3 亿法郎，较 20 年前翻了一番，毫无疑问丝绸业将继续保持上升势头。“可很不幸的是，就在种桑养蚕最为兴旺的时候，一场严重的灾难突然降临，完全破坏了整个行业繁荣的局面。”1853 年，法国的蚕茧产量是 2,600 万千克，到了 1865 年则下降到了 400 万千克，仅 1865 年这一年，蚕茧数量减少导致的丝绸业损失就高达 1 亿法郎。

这一灾难使法国受到重创，时任法国科学院常务秘书长的著名化学家杜马向他的朋友、同事兼学生巴斯德求助，他几乎是以个人的名义真诚地恳求巴斯德来研究这种家蚕疾病的。当时的巴斯德根本就没见过家蚕，他极力强调自己的经验不足，并以此答复了杜马。不过杜马很清楚巴斯德具备承担这项研究所需的能力，因而没有接受巴斯德经验不足的拒绝理由。他说：“我们国家的状况可谓惨不忍睹，我愿出巨资请您来研究并解决这个问题。”当时，关于这种家蚕疾病的小册子大量地出现，除了偶尔有那么一两本还算有些用处外，其他大部分都是单调空洞的废纸。1860 年，科尔纳利阿先生曾写道：“眼下，治疗家蚕疾病的药就像人吃的药一样复杂，有气体的，有液体的，也有固体的。从盐酸到硫酸，从硝酸到朗姆酒，从糖类到硫酸奎宁，

"is now as complicated as that of man. Gases, liquids, and solids have been laid under contribution. From chlorine to sulphurous acid, from nitric acid to rum, from sugar to sulphate of quinine, —all has been invoked in behalf of this unhappy insect." The helpless cultivators, moreover, welcomed with ready trustfulness every new remedy, if only pressed upon them with sufficient hardihood. It seemed impossible to diminish their blind confidence in their blind guides. In 1863 the French Minister of Agriculture himself signed an agreement to pay 500,000 francs for the use of a remedy which its promoter declared to be infallible. It was tried in twelve different departments of France and found perfectly useless. In no single instance was it successful. It was under these circumstances that M. Pasteur, yielding to the entreaties of his friend, betook himself to Alais in the beginning of June 1865. As regards silk husbandry, this was the most important department in France, and it was also that which had been most sorely smitten by the epidemic.

The silkworm had been previously attacked by *muscardine*; a disease proved by Bassi to be caused by a vegetable parasite. Muscardine, though not hereditary, was propagated annually by the parasitic spores, which, wafted by winds, often sowed the disease in places far removed from the centre of infection. According to Pasteur, muscardine is now very rare; but for the last fifteen or twenty years a deadlier malady has taken its place. A frequent outward sign of this disease are the black spots which cover the silkworms, hence the name pébrine, first applied to the plague by M. de Quatrefages, and adopted by Pasteur. Pébrine also declares itself in the stunted and unequal growth of the worms, in the languor of their movements, in their fastidiousness as regards food, and in their premature death. The discovery of the inner workings of the epidemic may be thus traced. In 1849 Guerin Méneville noticed in the blood of certain silkworms vibratory corpuscles which he supposed to be endowed with independent life, and to which he gave a distinctive name. As regards the motion of the particles, Filippi proved him wrong; their motion was the well-known Brownian motion. But Filippi himself committed the error of supposing the corpuscles to be normal to the life of the insect. They are really the cause of its mortality—the form and substance of its disease. This was studied and well described by Cornalia; while Lebert and Frey subsequently found the corpuscles not only in the blood, but in all the tissues of the silkworm. Osimo, in 1857, discovered the corpuscles in the eggs, and on this observation Vittadiani founded, in 1859, a practical method of distinguishing healthy from diseased eggs. The test often proved fallacious, and it was never extensively applied.

The number of these corpuscles is sometimes enormous. They take possession of the intestinal canal, and spread thence throughout the body of the worm. They fill the silk cavities, the stricken insect often going through the motions of spinning without any material to answer to the act. Its organs, instead of being filled with the clear viscous liquid of the silk, are packed to distension by these corpuscles. On this feature of the plague Pasteur fixed his attention. He pursued it with the skill which appertains to his genius, and with the thoroughness that belongs to his character. The cycle of the silkworm's life is briefly this:—From the fertile egg comes the little worm, which grows, and after some time casts its skin. This process of moulting is repeated two or three times at subsequent

所有这些药品都被用来处理患病的家蚕。”无助的家蚕养殖者们大胆地尝试着各种新的治疗方法，尽管盲目却信心十足。1863 年，法国农业部部长亲自签署了一项协议，花费 50 万法郎推行一个据称是绝对有效的新治疗方案，该方案在法国 12 个不同的地区试行，可惜的是，试行结果表明该方案完全无效。当时所有试图治疗家蚕疾病的尝试没有一例成功。在这种情况下，巴斯德只好答应朋友的请求，并于 1865 年 6 月初在艾雷斯开展研究。艾雷斯是法国最重要的生产丝绸的地区，同时也是疫情最严重的地区。

之前，家蚕曾受到**白僵病**的攻击，巴锡研究表明，这是一种由蔬菜寄生虫引起的疫病。白僵病虽然不具有遗传性，但是通过随风飘浮的寄生虫孢子，它也会传播到远离疫情中心的地区。巴斯德说，白僵病现在已经非常罕见了，但最近的 15~20 年间出现了一种更严重的疾病。这种疫病的一个鲜明标志是患病的蚕身上通常有许多黑点，根据这一特征，德卡特勒法热先生最先将这种病命名为微粒子病，巴斯德沿用了这一名称。患微粒子病的家蚕生长受到阻碍且体型发育不均，蠕动缓慢无力，挑食且易早死。关于传染性疾病内在机制的研究从此揭开了序幕。1849 年，介朗·莫奈维勒发现一些蚕的体液中存在以振动形式运动的微粒，他认为这些微粒拥有独立的生命并给它们取了一个特别的名字。关于微粒的运动方式，后来菲利皮证明介朗·莫奈维勒是错误的，这些微粒的运动方式正是著名的布朗运动。不过，菲利皮本人错误地认为这些微粒的出现对于蚕来说没什么不正常。其实，正是这些微粒导致了疫情中蚕的不正常死亡——这种疾病的表现和本质。科尔纳利阿仔细地研究并描述了这种微粒，而莱贝特和弗赖最终发现，不仅蚕的体液中存在这些微粒，而且其他所有组织中都有。1857 年，奥希姆在蚕卵中也发现了这种运动微粒，基于这一发现维塔迪尼在 1859 年建立了区分健康蚕卵与患病蚕卵的操作方法。不过，这个区分方法时常会出错，所以从来没有得到广泛应用。

有时候，蚕体中这种微粒的数量大得惊人。它们占据肠道并且向蚕体其他部位蔓延，有时甚至填满蚕的腹腔，使得蚕在整个吐丝过程中动弹不得并且吐不出丝。蚕体内不是充满了洁净蚕丝黏液，而是胀满了这些微粒。巴斯德把注意力集中在疾病的这一特征上。他靠那种似乎只有他才具有的天赋抓住了这一重点，而他深入钻研的科学作风又使他能对此进行彻底的研究。家蚕的生命周期大概如下：从受精卵长成幼虫，幼虫逐渐长大，随后开始蜕皮。在家蚕的整个生命中，蜕皮过程会重复 2~3 次。最后一次蜕皮结束后，家蚕会爬到提前放好的荆棘丛中吐丝织茧。这样家

intervals during the life of the insect. After the last moulting the worm climbs the brambles placed to receive it, and spins among them its cocoon. It passes thus into a chrysalis; the chrysalis becomes a moth, and the moth when liberated lays the eggs which form the starting-point of a new cycle. Now Pasteur proved that the plague-corpuscles might be incipient in the egg, and escape detection; they might also be germinal in the worm, and still baffle the microscope. But as the worm grows, the corpuscles grow also, becoming larger and more defined. In the aged chrysalis they are more pronounced than in the worm; while in the moth, if either the egg or the worm from which it comes should have been at all stricken, the corpuscles infallibly appear, offering no difficulty of detection. This was the first great point made out in 1865 by Pasteur. The Italian naturalists, as aforesaid, recommended the examination of the eggs before risking their incubation. Pasteur showed that both eggs and worms might be smitten and still pass muster, the culture of such eggs or such worms being sure to entail disaster. He made the moth his starting-point in seeking to regenerate the race.

And here is to be noted a point of immense practical importance. The worms issuing from the eggs of perfectly healthy moths may afterwards become themselves infected through contact with diseased worms, or through germs mixed with the dust of the rooms in which the worms are fed. But though the moths derived from the worms thus infected may be so charged with corpuscles as to be totally unable to produce eggs fit for incubation, still Pasteur shows that the worms themselves, in which the disease is not hereditary, never perish before spinning their cocoons. This, as I have said, is a point of capital importance; because it shows that the moth-test, if acted upon, even though the worms during their "education" should contract infection, secures, at all events, the next subsequent crop.

Pasteur made his first communication on this subject to the Academy of Sciences in September 1865. It raised a cloud of criticism. Here forsooth was a chemist rashly quitting his proper *métier* and presuming to lay down the law for the physician and biologist on a subject which was eminently theirs. "On trouva étrange que je fusse si peu au courant de la question; on m' opposa des travaux qui avaient paru depuis longtemps en Italie, dont les resultats montraient I'inutilité de mes efforts, et I'impossibilité d'arriver à un resultat pratique dans la direction que je m'étais engagé. Que mon ignorance fut grande au sujet des recherches sans nombre qui avaient paru depuis quinze années." Pasteur heard the buzz, but he continued his work. In choosing the eggs intended for incubation, the cultivators selected those produced in the successful "educations" of the year. But they could not understand the frequent and often disastrous failures of their selected eggs; for they did not know, and nobody prior to Pasteur was competent to tell them, that the finest cocoons may envelop doomed corpusculous moths. It was not, however, easy to make the cultivators accept new guidance. To strike their imagination and if possible determine their practice, Pasteur hit upon the expedient of prophecy. In 1866 he inspected at St. Hippolyte-du-Fort fourteen different parcels of eggs intended for incubation. Having examined a sufficient number of the moths which produced these eggs, he wrote out the prediction of what would occur in 1867, and placed the prophecy as a sealed letter in the hands of the Mayor of St. Hippolyte.

蚕就成了被蚕丝包裹的蚕蛹，随后蚕蛹变成蛾子，最后蛾子破茧而出并产下蚕卵，这样就又开始新一轮的生命周期。现在巴斯德证明蚕卵中的微粒还处在繁殖的最初阶段，所以无法检测到，在幼虫体内也只是萌芽期，所以通过显微观察还很难确认。而随着幼虫逐渐长大，这些微粒也开始变大从而更容易被确认。在蚕蛹期，这些颗粒比在幼虫期时更加清晰可辨。如果一开始的蚕卵或幼虫就受到感染，那么在相应的蛾子体内就可以很容易地准确检测到颗粒的存在。这是巴斯德在 1865 年提出的关于这一疫病的第一个重要观点。前面已经提过，一些意大利的博物学家建议在孵化前先对蚕卵进行检测。而巴斯德则认为，这些蚕卵和幼虫都可能是染病的但仍能通过检测，这些通过患病的蚕卵和幼虫发育而来的家蚕显然还是染病的。为了筛选出不染病的家蚕品种，他把蛾子作为检测控制的起点。

还有一点在实际应用中非常重要。由完全健康的蛾子产的卵孵化而来的健康家蚕随后也会受到感染，比如通过与病蚕的接触，或者通过饲养蚕的房间内混在灰尘中的病菌而感染。巴斯德通过研究表明，尽管家蚕在养殖过程中可能会染病，从而使蛾子体内充满微粒而完全不能产生适于孵化的健康卵，但这些家蚕本身在吐丝之前从来不会死亡，同时这种病在家蚕中并不会发生遗传。正如前文已经说过的，这一点是相当重要的，它表明即便家蚕在“成长”过程中被感染，但只要对蛾子进行切实的检测控制，至少接下来的蚕丝收成是有保障的。

1865 年 9 月，巴斯德就此课题与法国科学院的成员进行了首次交流，结果招来一片指责。在这里巴斯德被认为是一位**不务正业的**化学家，对从事本专业研究的医生和生物学家指手画脚。“认为我不了解眼下的问题，这是个奇怪的想法；他们认为：我很早之前在意大利发表的研究结果毫无意义，我在自己的研究领域内没有得到过实际有效的成果，而过去 15 年来我表现出来的愚昧无知更是数不胜数。”巴斯德当然听到了那些非议，不过他还是继续他的研究。在选择适于养殖的蚕卵方面，养殖者们一般选择那些当年正常“成长”的蚕产的卵。不过他们不明白为什么精心挑选出来的蚕卵时常会导致惨痛的损失，因为在巴斯德之前还没有人能告诉他们为什么看起来最为完好的蚕茧却可能包裹着带病的蛾子。不过，要想说服家蚕养殖者们接受新的方法并非易事，为了给养殖者们直观的印象并尽可能改变他们的做法，巴斯德想到了通过对孵育结果进行预测来证实自己理论的正确性。1866 年，巴斯德在圣希波利特堡检查了 14 份装有适于孵育的蚕卵的包裹。在检测完产生这些蚕卵的蛾子后，他写下了对 1867 年孵育结果的预测，并将预测结果密封后交给了圣希波利特市的市长。

In 1867 the cultivators communicated to the mayor their results. The letter of Pasteur was then opened and read, and it was found that in twelve out of fourteen cases, there was absolute conformity between his prediction and the observed facts. Many of the educations had perished totally; the others had perished almost totally; and this was the prediction of Pasteur. In two out of the fourteen cases, instead of the prophesied destruction, half an average crop was obtained. Now, the parcels of eggs here referred to were considered healthy by their owners. They had been hatched and tended in the firm hope that the labour expended on them would prove remunerative. The application of the moth-test for a few minutes in 1866 would have saved the labour and averted the disappointment. Two additional parcels of eggs were at the same time submitted to Pasteur. He pronounced them healthy; and his words were verified by the production of an excellent crop. Other cases of prophecy still more remarkable, because more circumstantial, are recorded in the work before us.

These deadly corpuscles were found by Leydig in other insects than the silkworm moth. He considers them to belong to the class of psososperms founded by J. Müller. "This," says Pasteur, "is to regard the corpuscular organism as a kind of parasite, which propagates itself after the manner of parasites of its class." Pasteur subjected the development of the corpuscles to a searching examination. With admirable skill and completeness he also examined the various modes by which the plague is propagated. He obtained perfectly healthy worms from moths perfectly free from corpuscles, and selecting from them 10, 20, 30, 50, as the case might be, he introduced into the worms the corpusculous matter. It was first permitted to accompany the food. Let us take a single example out of many. Rubbing up a small corpusculous worm in water, he smeared the mixture over the mulberry leaves. Assuring himself that the leaves had been eaten, he watched the consequences from day to day. Side by side with the infected worms he reared their fellows, keeping them as much as possible out of the way of infection. These constituted his "lot temoign", his standard of comparison. On the 16th of April, 1868, he thus infected thirty worms. Up to the 23rd they remained quite well. On the 25th they seemed well, but on that day corpuscles were found in the intestines of two of the worms subjected to microscopic examination. The corpuscles begin to be formed in the tunic of the intestine. On the 27th, or eleven days after the infected repast, two fresh worms were examined, and not only was the intestinal canal found in each case invaded, but the silk organ itself was found charged with the corpuscles. On the 28th the twenty-six remaining worms were covered by the black spots of pébrine. On the 30th the difference of size between the infected and non-infected worms was very striking, the sick worms being not more than two-thirds of the size of the healthy ones. On the 2nd of May a worm which had just finished its fourth moulting was examined. Its whole body was so filled with corpuscles as to excite astonishment that it could live. The disease advanced, the worms died and were examined, and on the 11th of May only six out of the thirty remained. They were the strongest of the lot, but on being searched they also were found charged with corpuscles. Not one of the thirty worms had escaped; a single corpusculous meal had poisoned them all. The standard lot, on the contrary, spun their fine cocoons, and two only of their moths were found to contain any trace of corpuscles. These had doubtless been introduced during the rearing of the worms.

到了1867年，家蚕养殖者们先将实际的养殖结果报告给市长。然后当巴斯德的预测结果被打开并宣读之后，人们发现14份样本中有12份的养殖结果和巴斯德的预测完全相同。其中大部分在养殖时彻底被疾病摧毁了，其余的也几乎完全被疾病摧毁。这恰恰是巴斯德的预测结果。14份样本中的另外2份没有被疾病毁掉，最终大概有将近一半的收成。而当初这些蚕卵的主人都认为他们的蚕卵是健康的。他们悉心地孵化，仔细地照料幼蚕，殷切地期望着付出的劳动能得到回报。如果当初他们在1866年进行几分钟的蛾子检测，就能够节省劳动并避免失望一场了。当时，另外还有2份样本也寄来让巴斯德检测。他检测后宣布这2份是健康的，来年丰硕的收成证明了他的预言完全正确。当然，除此之外还有一些间接的但同样具有重要意义的预测此前已经报道过了。

在此之前，莱迪希在其他昆虫体内也发现过类似的致命性微粒。他认为这些微粒属于米勒发现的寄生病菌家族。巴斯德指出，“这等于是将微粒看作是寄生虫，它们在完成寄生行为的同时还繁殖后代。”他通过实验研究了微粒的发育过程，熟练而全面地检查了疾病的各种传播途径。他还筛选出了由完全不携带病原微粒的蛾子产的卵孵化的完全健康的幼蚕，然后根据实验需要，他又从这些完全健康的幼蚕中选出10条、20条、30条或50条，向其体内引入那些微粒物质。一开始这些微粒是混在食物中喂给家蚕的。我们就说说其中一个简单的实例吧。巴斯德先把携带微粒的蚕体在水中磨碎，然后将所得的混合物涂在桑叶上，再让家蚕吃下这些混有微粒的桑叶，然后每天追踪观察结果。另外，他还在这些受感染的蚕的附近同时饲养健康蚕，并尽量让它们保持距离不相互感染。这就组成了他“大量检验”的对比标准。就是通过这样的方法，他在1868年4月16日得到了30条受感染的家蚕。到23日它们状态仍然良好。到25日它们看起来也还可以，不过通过显微镜在2条家蚕的小肠内观察到了微粒，可以发现在小肠膜上微粒已经形成。到27日，也就是健康蚕吃桑叶而发生感染的11天后，检查发现有2条家蚕不仅在它们的肠道内而且在吐丝器官上出现了微粒。到28日，剩下的26条家蚕全身布满了微粒子病特有的黑点。到30日时，受感染的蚕的蚕体大小已经和没受感染的蚕有了明显的差异，病蚕蚕体不超过健康蚕蚕体的2/3。到5月2日，经检查发现，一条刚完成第四次蜕皮的成蚕的体内充满了大量病原微粒，这样的蚕居然还能存活真是令人惊奇！随着病情的发展，病蚕开始死亡，巴斯德对它们都进行了检查。一开始的30条家蚕到5月11日只有6条还活着，它们是这30条蚕中最强壮的，不过观察发现它们体内同样充满了病原微粒。最终，30条家蚕都死掉了，无一幸免。只因为吃了一次带病原微粒的桑叶，它们就统统丢了性命。相比之下，没受感染的对照组则顺利地吐丝结茧，随后只在其中2个蛾子中发现有少量的病原微粒。毫无疑问，这是在饲养幼虫时不小心引入的。

As his acquaintance with the subject increased, Pasteur's desire for precision augmented, and he finally gives the growing number of corpuscles seen in the field of his microscope from day to day. After a contagious repast the number of worms containing the parasite gradually augmented until finally it became cent per cent. The number of corpuscles would at the same time rise from 0 to 1, to 10, to 100, and sometimes even to 1,000 or 1,500 for a single field of his microscope. He then varied the mode of infection. He inoculated healthy worms with the corpusculous matter, and watched the consequent growth of the disease. He showed how the worms inoculate each other by the infliction of visible wounds with their "crochets". In various cases he washed the "crochets", and found corpuscles in the water. He demonstrated the spread of infection by the simple association of healthy and diseased worms. In fact, the diseased worms sullied the leaves by their dejections, they also used their crochets, and spread infection in both ways. It was no hypothetical infected medium that killed the worms, but a definitely-organised and isolated thing. He examined the question of contagion at a distance, and demonstrated its existence. In fact, as might be expected from Pasteur's antecedents, the investigation was exhaustive, the skill and beauty of his manipulation finding fitting correlatives in the strength and clearness of his thought.

Pébrine was an enigma prior to the experiments of Pasteur. "Place," he says, "the most skilful educator, even the most expert microscopist, in presence of large educations which present the symptoms described in our experiments; his judgment will necessarily be erroneous if he confines himself to the knowledge which preceded my researches. The worms will not present to him the slightest spot of pébrine; the microscope will not reveal the existence of corpuscles; the mortality of the worms will be null or insignificant; and the cocoons leave nothing to be desired. Our observer would, therefore, conclude without hesitation that the eggs produced will be good for incubation. The truth is, on the contrary, that all the worms of these fine crops have been poisoned; that from the beginning they carried in them the germ of the malady; ready to multiply itself beyond measure in the chrysalides and the moths, thence to pass into the eggs and smite with sterility the next generation. And what is the first cause of the evil concealed under so deceitful an exterior? In our experiments we can, so to speak, touch it with our fingers. It is entirely the effect of a single corpusculous repast; an effect more or less prompt according to the epoch of life of the worm that has eaten the poisoned food."

It was work like this that I had in view when, in a lecture which has brought me much well-meant chastisement from a certain class of medical men, and much gratifying encouragement from a different class, I dwelt on the necessity of experiments of physical exactitude in testing medical theories. It is work like this which might be offered as a model to the physicians of England, many indeed of whom are pursuing with characteristic skill and energy the course marked out for them by this distinguished master. Prior to Pasteur, the most diverse and contradictory opinions were entertained as to the contagious character of pébrine; some stoutly affirmed it, others as stoutly denied it. But on one point all were agreed. "They believed in the existence of a deleterious medium, rendered epidemic by some occult and mysterious influence, to which was attributed the cause of

随着巴斯德对这一领域的不断熟悉，他逐渐期望能提高实验的精确性，后来他开始记录每天用显微镜观察时视野中不断增加的微粒数目。给家蚕喂食一顿带病原微粒的食物之后，体内出现病原微粒的家蚕数量就开始逐渐增加，直到最终所有家蚕体内都出现病原微粒。与此同时，在显微镜单个视野中观察到的病原微粒数目也从 0 增加到 1、10 以至 100，有时甚至高达 1,000 或者 1,500。接着，巴斯德又改用另一种方式感染家蚕。他直接把微粒物质注入健康家蚕体内，然后观察病情的变化。他发现，家蚕会通过自己的“腹足趾钩”对在一起生活的其他家蚕造成外伤，从而使病原颗粒在家蚕之间互相传染。在多次实验中，他用水清洗这些“腹足趾钩”后在水中都发现了微粒物质。因此，他提出只是把健康蚕和病蚕放在一起饲养就会造成疾病在家蚕之间传染。事实上，疫病在家蚕之间传染有两条途径，一个是病蚕的粪便污染桑叶，另一个是通过腹足趾钩。杀死家蚕的并不是假想的什么传染介质，而是确实存在并能分离出来的物质。另外巴斯德还研究了远距离传染的问题，并证实远距离传染确实可能发生。实际上，由巴斯德之前的研究过程可以想到，这项研究是非常繁杂的，巴斯德高超的实验技能和艺术性的实验操作反映了他思考的深度和清晰的思路。

在巴斯德做这些实验前，人们对微粒子病的了解是一团迷雾。巴斯德说道：“即使是让那些训练有素的专业人员甚至是非常杰出的显微镜专家来分析我们的实验，如果他们局限在我们实验之前的那些知识而并不了解我们在实验中发现的关于这种疾病的许多信息，那他们的判断也将必错无疑。他们不会看到得微粒子病的蚕的躯体上很小的斑点，通过显微镜观察也根本看不到微粒，而病蚕的死亡率接近零或者非常低，至于蚕茧就更看不出有什么问题了。那么观察者可能就会毫不犹豫地认为，这样的蚕卵是适合孵育的。但是事实正好相反，所有这些正常吐丝的蚕都被喂食了带病原微粒的食物，从一开始它们产的卵就携带着死亡的种子，这些病原微粒在蚕蛹和蛾体内准备进行疯狂的自身繁殖，然后传入蚕卵并彻底杀死下一代，导致下一代无法产卵。那么在这种欺骗性的表象之下导致这种灾难的首要原因是什么呢？在我们的实验中，我们是亲手触摸般地发现这一原因的。其实这完全就是吃入带病原微粒的食物引起的，结果的严重程度一定程度上和家蚕吃这些带病食物时所处的生命周期有关。”

像这样的研究我曾经也想到过，那是在一次报告上，一部分医生对我提出了并无恶意的责备，而另一部分医生则充分鼓励我，当时我就想到，实验的物理精度对于医学理论的检验是很必要的。我们应该把巴斯德这样的研究工作作为范例提供给英国的医生们，实际上许多既有专业技能又有热情的医生正期望着能接受大师为他们定制的课程指导。在巴斯德之前，关于微粒子病传播特征的观点形形色色、相互矛盾，有些人轻率固执地肯定它，也有些人同样轻率固执地否定它。不过有一点当时是达成共识的，“他们相信必定存在某种有毒的媒介物质以未知的神秘的方式传播疫情，从而导致了疾病灾难。”我想，任何一个头脑清晰的人，都会在这种观念和巴斯德的研究结果之间毫不犹豫地作出正确的选择。

the malady." Between such notions and the work of Pasteur, no physically-minded man will, I apprehend, hesitate in his choice.

Pasteur describes in detail his method of securing healthy eggs, which is nothing less than a mode of restoring to France her ancient prosperity in silk husbandry. And the justification of his work is to be found in the reports which reached him of the application, and the unparalleled success of his method, at the time he was putting his researches together for final publication. In France and Italy his method has been pursued with the most surprising results. It was an up-hill fight which led to this triumph, but it is consoling to think that even the stupidities of men may be converted into elements of growth and progress. Opposition stimulated Pasteur, and thus, without meaning it, did good service. "Ever," he says, "since the commencement of these researches, I have been exposed to the most obstinate and unjust contradictions; but I have made it a duty to leave no trace of these contests in this book." I have met with only a single allusion to the question of spontaneous generation in M. Pasteur's work. In reference to the advantage of rearing worms in an isolated island like Corsica, he says:—"Rien ne serait plus facile que d'éloigner, pour ainsi dire, d'une manière absolue la maladie des corpuscles. Il est au pouvoir de l'homme de faire disparaître de la surface du globe les maladies parasitaires, si, comme c'est ma conviction, la doctrine des générations spontanées est une chimère." It is much to be desired that some really competent person in England should rescue the public mind from the confusion now prevalent regarding this question.

M. Pasteur has investigated a second disease, called in France *flacherie*, which has co-existed with pébrine, but which is quite distinct from it. Enough, I trust, has been said to send the reader interested in these questions to the original volumes for further information. I report with deep regret the serious illness of M. Pasteur; an illness brought on by the labours of which I have tried to give some account. The letter which accompanied his volumes ends thus:—"Permettez-moi de terminer ces quelques lignes que je dois dicter, vaincu que je suis par la maladie, en vous faisant observer que vous rendiez service aux Colonies de la Grande Bretagne en repandant la connaissance de ce livre, et des principes que j'établis touchant la maladie des vers à soie. Beaucoup de ces colonies pourraient cultiver le mûrier avec succès, et en jetant les yeux sur mon ouvrage vous vous convaincrez aisement qu'il est facile aujourdhui, non seulement d'éloigner la maladie régnante, mais en outre de donner aux récoltes de la soie une prospérité qu'elles n'ont jamais eue."

(**2**, 181-183; 1870)

John Tyndall: Royal Institution, 30th June.

巴斯德详细地叙述了他保证蚕卵处于健康状态的方法，用这种方法完全能够恢复法国丝绸业以往的繁荣。描述巴斯德工作的应用并介绍其方法所获得的空前成功的报告合理地解释了他的研究工作，当时巴斯德还正在整理他的研究工作以供最终出版。巴斯德的方法在法国和意大利取得了惊人的成果。这是持续不断的努力换来的最终胜利，令人欣慰的是，即便是人类的那些愚蠢意见后来也变成了发展进步大道上的铺路石。那些批驳的意见激励着巴斯德，无意中起到了推动作用。巴斯德说："从我开始这项研究起，我就一直受到极为顽固而不公的批判，不过我在书中不会提起这些事情。"的确，在巴斯德的著作中我只发现有一处暗示性地提了一下。当谈到在类似科西嘉岛这样的孤立岛屿上养殖家蚕的好处时，他说："可以说，没有什么比完全清除致病微粒更容易解决这个问题了。如果如我所坚信的那样，自然发生学说只是不切实际的空想，那么就只有通过人类的力量才能消灭地球上的各种寄生性疾病。"而现在在英国，还真需要一批有识之士向民众澄清关于这些问题的种种流言。

巴斯德先生已经开始研究另一种疾病了。这种病在法国叫做**软腐病**，它与微粒子病共存，不过与微粒子病又完全不同。我相信，我所写的这些内容已经足以吸引那些对这一系列问题感兴趣的读者去阅读巴斯德先生的原著以获得更多的信息。这里我很遗憾地告诉读者，巴斯德先生现在得了重病，他是在进行前面详细介绍了的实验中染病的。巴斯德先生在随著作一起给我寄来的信的末尾写道："由于我身患重病，这封短信是我口述完成的。在信的末尾，我想说，你那些来自英国的桑树种子不仅有利于这本书中知识的传播，而且有利于我在对影响家蚕的疾病的研究中得到相关的原理。那些种子大部分都顺利地长成了桑树，从我的工作中你可以很容易地看到，现在可以很容易地消灭这种流行的疾病，并恢复以往丝绸业的繁荣了。"

（高如丽 翻译；刘力 审稿）

Spontaneous Generation

J. A. Wanklyn

Editor's Note

The interest of this comment from English chemist James Alfred Wanklyn on the topic of spontaneous combustion, discussed recently by H. Christian Bastian, is that it shows how tangled up the issue was becoming with the germ theory of disease. Wanklyn implies that the "vitalists" now invoked invisible airborne "germs" to explain spontaneous generation in water. A specialist in the analysis of water quality, Wanklyn complains that there is simply not enough nitrogen-rich organic matter in air to make that plausible. But of course microscopic airborne germs were precisely what Louis Pasteur was then correctly advancing as the cause of many diseases. Bastian himself was later to have an acerbic exchange in *Nature* with Thomas Huxley over spontaneous generation.

DR. H. C. Bastian, who has recently called attention to the nature of the evidence before scientific men in favour of the theory of so-called spontaneous generation, has supplemented it by fresh experiments of his own. The dilemma in which the opponents of this doctrine are now placed is that they must either admit it, or else allow that a temperature of 150℃ maintained for four hours, and applied by means of liquid, is incapable of killing the germs of infusoria. Many, doubtless, of these opponents will courageously mount this horn of the dilemma, and make the requisite enlargement of their ideas on the subject of vital resistance to change. There are, however, other difficulties in the way. For instance, great difficulties are involved in the assumption that the atmosphere constitutes a storehouse of germs of all kinds ready to burst out into life on the occurrence of suitable conditions.

However small these germs may be, still they must weigh something. And there must be very many of them, seeing that there must be an immense number of kinds of germs, if a volume of air is to supply to any given infusion precisely the right kinds of germs suitable to the conditions provided by the infusion.

Now chemists are in possession of data showing that the possible amount of organic nitrogenous matter in common clear water and common good air is remarkably small—so small, indeed, that the question may fairly be asked—Is it large enough to admit of the requisite number of germs, the existence of which the vitalists assume in water and air?

By the employment of our ammonia method, Chapman, Smith, and myself have shown that the organic ammonia from a kilogramme of good filtered water often falls as low as 0.05 milligramme, and Dr. Angus Smith has shown that a kilogramme of good air sometimes contains as little as 0.085 milligramme of organic ammonia.

A gramme of air—that is about 700 cubic centimetres—contains only 0.000085

自然发生学说

万克林

编者按

英国化学家詹姆斯·阿尔弗雷德·万克林对最近克里斯蒂安·巴斯蒂安讨论的自然发生学说作了评论，这篇评论的意义在于，它表明了该学说是怎样与疾病的细菌理论纠缠起来的。万克林暗示，“活力论者”现在要通过引入一个看不见的在空气中传播的“细菌”来解释水中的自然发生现象。作为一名水质分析专家，万克林抗议说，空气中根本没有足量的含氮有机物，因此该观点是不合理的。但恰恰是这些只有在显微镜下才可以看到的在空气中传播的细菌，被路易·巴斯德正确地判定为造成很多疾病的根源。后来巴斯蒂安本人在《自然》上围绕自然发生学说与托马斯·赫胥黎展开了一场激烈的争辩。

巴斯蒂安博士最近呼吁科学家们关注那些呈现在他们眼前的支持所谓自然发生学说的证据的本质，现在他通过自己刚做的实验对这个问题进行了补充。这一学说的反对者们目前所处的困境是：他们必须承认这一学说，或者承认将某种液体的温度维持在 150℃并持续 4 小时也不能杀死其中的纤毛虫类细菌。无疑，许多反对者将勇敢地面对这一困境，并且对他们的有关对外界变化的重要抗性的观点进行必要的扩展。然而，他们还面临着许多其他的困难。例如，关于大气是由各种各样随时准备好在适宜条件下萌发生命的细菌组成的大仓库这一假说就面临着许多巨大的困难。

不管这些细菌有多么小，它们肯定是有重量的。细菌的种类也一定非常多，因为，如果大量空气能让其中任意的内含物中都含有细菌，而这些细菌又适于在该内含物所提供的条件下生长的话，那么空气中就一定有大量不同种类的细菌存在。

现在化学家们所拥有的确切数据显示，有机含氮物质在普通清洁水和普通新鲜空气中的含量非常少，少得以至于人们很可能要问：其含量是否足够大到能容许活力论者所假定的在水和空气中存在的细菌的必需数目？

通过使用我们的测氨法，查普曼、史密斯和我本人已经发现，从 1 千克优质过滤水中得到的有机氨经常少到只有 0.05 毫克。安格斯·史密斯博士也发现 1 千克新鲜空气中有时仅含有少至 0.085 毫克的有机氨。

1 克空气大约是 700 立方厘米，其中仅含有 0.000085 毫克有机氨。将有机氨的

milligramme of organic ammonia. Expressing the organic ammonia in its equivalent of dry albumen we have in 700 cubic centimetres of air 0.00085 milligramme of dry albumen. Translated into volume this 0.00085 milligramme of dry albumen will fall short of a cube, the face of which is 1/10 millimetre in diameter.

Expressed in English measures, the result is, that rather more than one pint of average atmospheric air does not contain so much organic nitrogenous matter as corresponds to a cube of dry albumen of the 1/250 part of an inch in diameter.

Now is this quantity adequate to admit of the existence of the immense multitudes of germs, the existence of which in atmospheric air is assumed by the vitalists?

(**2**, 234-235; 1870)

含量用与其等量的干燥蛋白来表示的话，就是 700 立方厘米空气中有 0.00085 毫克干燥蛋白。换算成体积的话，这 0.00085 毫克干燥蛋白还装不满一个表面直径只有 1/10 毫米的立方体。

以英制单位表示的话，结果是，多于 1 品脱的普通大气中含有的有机含氮物质对应的干燥蛋白还不到直径为 1/250 英寸的一个立方体那么多。

现在这一数量结果能否足以让我们承认，大气中确实存在如活力论者所假定的那样大量的细菌吗？

（刘皓芳 翻译；刘力 审稿）

Spontaneous Generation

C. Ekin

Editor's Note

Earlier in *Nature*, pugnacious chemist James Alfred Wanklyn had questioned whether the low amounts of nitrogenous matter in the air were really enough to account for the multitude of invisible microorganisms being proposed as the cause of many diseases. Instead, he favoured spontaneous generation, the idea that living forms could emerge from purely inorganic building blocks. Wanklyn's letter had needled one reader to suggest his calculations supported rather than undermined the germ theory of disease. Here Charles Ekin expresses surprise that there is anything to argue about, citing results published by Louis Pasteur almost a decade earlier. In spite of the growing evidence in support of germ theory, some, like Wanklyn, were to dispute it for several decades to come.

IF there is one thing more curious than another in the "Spontaneous Generation" theory, it is the way in which so-called matters of fact, as proved by careful experiment, are brought forward by the one side to be disproved by the other, one need only instance Pasteur's famous flask experiments, which were thought to be so overwhelming at the time, but which were afterwards refuted, I think by Frémy and others.

I notice with surprise the letters of Prof. Wanklyn and Dr. Lionel Beale in *Nature*, with regard to the presence of germs in the air; there is an experiment of Pasteur's, given in his "Memoirs upon the organised Corpuscles which exist in the Atmosphere, 1862", and which I have never seen disproved, and if not disproved it must surely settle at least this part of the question. He passed a quantity of air, taking various precautions to eliminate error, which I need not here detail, by means of an aspirator through a plug of gun-cotton; he then dissolved the gun-cotton in ether, and on examining the sediment which subsided in the course of an hour or two, he found abundant evidence of the presence of organised corpuscles.

(**2**, 296; 1870)

Charles Ekin: Bath.

自然发生学说

伊金

编者按

好辩的化学家詹姆斯·阿尔弗雷德·万克林曾在《自然》上发表文章提出，鉴于空气中含氮物质的含量很低，是否真的可能存在那么多被认为足以导致疾病的看不见的细菌？而他更倾向于自然发生说，即生命体可以从纯无机物中生成。一位读者读了万克林的快报文章后提出，万克林的计算实际上是对疾病的细菌学说的支持而不是反对。在这篇文章中查尔斯·伊金表示对还有人在争论以上两种理论感到吃惊，因为路易·巴斯德早在近十年前就已经提出了这些观点。尽管有越来越多的证据支持了疾病的细菌学说，但有些人，如万克林在后来的几十年中还在不断地对这个理论提出质疑。

如果在“自然发生”学说中存在最令人好奇的事情的话，那就是由某一方提出的所谓经过精细实验证实的事实证据却被另一方证伪的过程了，这里只需要举出巴斯德著名的烧瓶实验一例即可，这一实验当时被认为是非常具有说服力的，但是，我记得后来却被弗雷米等人驳倒了。

我很惊奇地注意到，在万克林教授和莱昂内尔·比尔博士发表于《自然》上的关于空气中存在细菌这一主题的快报文章中提到了巴斯德的一个实验，该实验在巴斯德于 1862 年发表的《对存在于大气中的有序微粒子的研究》中介绍过。我到现在还没看到有报道对该实验表示异议，如果没有人表达过不同意的意见，那么可以肯定这一实验至少解决了相应的问题。巴斯德借助抽气机使得一定量的空气通过一块火棉，同时他采取了各种预防措施以排除误差，具体的我想就不需要在这里详细介绍了。然后，他将火棉溶解在乙醚中，检测在一两个小时这么长的时间里沉积到火棉上的沉淀物，通过这一实验他得到了大量的能够证明有序微粒子存在的证据。

（刘皓芳 翻译；刘力 审稿）

Colour of the Sky

E. R. Lankester

Editor's Note

The colour of smoke from a cigar is strikingly blue in comparison to the smoke a person exhales. This is because newly formed smoke consists of very fine particles, which scatter light to give a bluish colour, while exhaled smoke, its particles made larger by condensed vapour, appears whiter. Might there be a similar explanation for the colour of seas and lakes, asks E. Ray Lankester here? He acknowledges that water's colour ranges from the vibrant blues of the Mediterranean to the pale blue of sea near chalk cliffs and the intense green of water in abandoned copper mines, and he suggests that the differences must be due to different kinds of suspended particles.

WITH reference to Mr. Brett's observations on the colour of sea and sky, I have one or two remarks to offer which I think may be of interest. Smokers have all noticed that the smoke from the end of a pipe or cigar is bluer than that which they puff from the mouth, and many may have wondered, as I did for a long time, what the reason of this could be. The contrast may be well seen on a bright sunny day. This is, in fact, the simplest form of the experiment of the condensation of vapours causing them to pass through a fine blue to a white condition, which Professor Tyndall exhibited about two years ago, and which he employed to explain the blue colour of the sky, and the remarkable polarisation of its light. The finer state of division in the freshly-formed smoke gives it its bright blue colour, as does the finely divided aqueous vapour give to the blue sky; the smoke which has passed through the pipe-stem and mouth has become more condensed, and consequently gives a whiter cloud.

The colour of water is, it appears, to a great degree dependent on the same cause as that of the sky. The investigations which Mr. Brett asks for have been already commenced. M. Soret, of Geneva, soon after Professor Tyndall's researches on the cause of the blueness of the sky were published, made similar researches on the waters of the lake of Geneva, and found that the light from the water, when blue, was polarised as the light from the sky, and, so far, there was the probability of the cause of the colour being similar in the two cases (See *Comptes Rendus* (Paris), April, 1869). That particles in a fine state of division are the cause of the blueness of water as well as of sky is also made evident from a comparison of the waters of different lakes, seas, and rivers. There are two popular theories as to the cause of the colour of masses of water, which have very deep root, and yet must, it seems, be abandoned. One is that seas or lakes are blue by reflecting the blue sky. On this ground I have heard Mr. Brett's picture in the Academy this year of a deep blue sea, severely criticised, because the sky, which he has painted with it, is not correspondingly blue, and could not furnish the sea's tint by reflection. Mr. Brett is, however, quite right in his fact, as many people know well enough; and the criticism was misplaced, if the blue colour

天空的颜色

兰克斯特

编者按

显然从雪茄冒出来的烟比一个人吐出来的烟更蓝。这是因为在新形成的烟雾中有很多微小的颗粒，这些微粒对光的散射使之显现蓝色；而对于吐出来的烟圈，其中的颗粒则会因水蒸气在其上凝结而变得更大，所以显得更白一些。在本文中，雷·兰克斯特提出了这样一个问题：对于海水和湖水的颜色是否也可以这样解释？他注意到水的颜色并不完全相同：地中海的海水是明亮的蓝色，白垩崖附近的海水呈现出淡蓝色，而废弃铜矿旁边的水洼则是深绿色。他认为这些差异一定是由悬浮颗粒的种类不同造成的。

关于布雷特先生对海洋和天空颜色的观察，我想作一两点评论，借此谈谈我自认为还算有意义的观点。吸烟的人都会注意到，从烟斗或雪茄的末端冒出的烟比直接从嘴里吐出的烟更蓝。可能很多人想知道为什么会这样，对这个问题我也想了很长时间。这种对比在阳光明媚的晴天可能更容易被观察到。事实上，这是水蒸气凝结实验最简单的形式，水蒸气凝结使它们从深蓝色变成白色，廷德尔教授两年前就展示过这个实验，并且据此解释了天空的蓝色以及光的奇异偏振现象。刚刚形成的烟雾的细分状态使它呈现为明亮的蓝色，这就类似于细分的水蒸气使天空显现为蓝色；只不过从烟斗管和嘴中喷出的烟雾凝结得更厉害，因此看起来就更泛白了。

看起来，水的颜色成因很大程度上和天空的颜色成因是一样的。布雷特先生倡议的调查研究已经在进行了。就在廷德尔教授发表了关于天空蓝色成因的研究之后不久，日内瓦的索雷特就对日内瓦湖的湖水进行了相似的研究，结果发现当湖面呈现蓝色时，来自湖水的光线会表现出和来自天空的光线一样的偏振性。至此，索雷特认为，这两种情况中蓝色的成因可能是相似的（参见《法国科学院院刊》（巴黎），1869 年 4 月）。通过对不同湖泊、海洋以及河流的水进行比较，也可以非常清楚地看到，造成水面和天空都呈现蓝色的原因正是细分状态的颗粒。目前有两种流行的理论用于解释水体的颜色，尽管这两种理论都有很深的根基，但是看起来它们都必须被抛弃。其中一种理论认为，海洋和湖泊的蓝色是由于反射天空的蓝色而产生的。我听说，就是因为这种理论，布雷特先生今年在科学院展示的深蓝海洋的图画受到了强烈的批评，因为他并没有将图画中的天空绘成相应的蓝色，因而不可能通过反射让海洋呈现出深蓝色。但是正如许多人知晓的那样，布雷特先生提供的事实是

of a mass of water is dependent on the reflection of light from within water containing finely-divided particles—not from the surface only—as explained above. The second popular theory which seems to be ill-founded is that the green colour of lakes, rivers, and seas is due to plants growing on the bottoms and giving their colour by reflection. The green colour is produced in the same way as the blue in all probability, and may be due to a yellowness of the water in some cases, but it is less easily accounted for than the blue colour. M. Sainte-Claire Deville is quoted by M. Soret as stating that waters which give a white residue on evaporation are *blue*, whilst those which give a yellow residue are *green*. Reflection of the colour of the sky, and of the plant colour from the bottom, does no doubt produce colour of water in some cases, but it is only in shallow pools that the latter can have any effect, or through perfectly smooth surfaces that the former can be effective. Some cases of water-coloration which I have noted will be not out of place here:— 1. Intensely blue on a bright day, with pale sky and large cumulous clouds, was the colour of water in reservoirs twenty feet deep at Plumstead, depositing chalk (by means of which the water is softened according to a patent process). 2. Intensely blue (the bluest here noted)—Mediterranean at Marseilles. 3. Bright blue—Lake of Geneva. 4. Darker blue, tending to Indigo—sea near Guernsey; also the Laacher See, in the Eifel. 5. Pale blue—sea near chalk cliffs, being at a little distance from the coast green or greyish. 6. Pale blue or greyish blue—the Rhone, the Mosel, glacier streams, &c. 7. Green—the Rhine, the Scheldt (very markedly so at Antwerp, as testified in Belgian pictures), the Seine, Thames Estuary, &c. 8. Intense green—in patches on the Lake of Geneva; in the evening, when the sun was just below the mountains, more frequently on the Lakes of Thun and Lucerne. 9. Bright green—the sea, on a windy day, with bright sun, off the Isle of Man. 10. The sea round the coral reefs of Florida is said to be intensely green; when away from the coast it is deep blue. 11. On a heavy, clouded day, with rain, gleams of sunshine out at sea give patches of green colour and reddish brown. 12. Water standing in an old copper mine at Killarney was intensely green, whilst the water in the lake at the side was black in the mass. 13. Red colour is produced in some seas by algae, in others and in some rivers by the breaking up of soil coloured red by iron. 14. Opaque green colour is produced in ponds (Serpentine and ornamental waters) by unicellular organisms, which sometimes swarm in these waters. They may similarly become red. Perhaps the most remarkable instance of blue colour, due to the optical properties of water, is the blue grotto of Caprera, where, at any rate, the reflection of the sky is eliminated. A similar phenomenon is the glorious blue and green of the glacier fissures.

Leaving the question of surface reflection aside, which can only come into play in the case of road-side pools and such mirror-like waters, and also leaving aside the appearance of vegetation in clear shallow streams and ponds, it seems that at the present time we may ascribe the blue colour of masses of water to a peculiar reflection of the light from within the water, accompanied with polarisation, and depending on suspended particles. Blackish, brownish, and yellow colour is due to vegetable matter in solution; reddish brown to iron, sometimes; green, sometimes, to copper or algae, but the green commonly seen on seas, lakes, and rivers, like that of glacier-fissures, probably admits of a similar explanation to that of the blue. I trust some physicist may be induced to enter into the subject in these

完全正确的。如果大量水呈现出蓝色的原因正如前面解释的那样，即是由其内部包含的细小颗粒对光线的反射造成，而非仅仅由其表面对光线的反射造成，那么人们对布雷特先生的批评就是有误的。第二种比较流行的理论看起来没有什么根据。该理论认为，湖泊、河流以及海洋的绿色是由于生长在水底的植物反射了它们自己的颜色。其实，绿色的成因很可能和蓝色是一样的，或者某些情况下也可能是因为水发黄而引起的。解释绿色的成因的确比解释蓝色的成因要更困难些。索雷特引述了圣克莱尔·德维尔的话，认为蒸发时呈白色的水在液态时是**蓝色的**，而蒸发时呈黄色的水在液态时是**绿色的**。在某些情况下，无疑确实是对天空颜色以及水底植物颜色的反射使水呈现出一定的颜色，但是，对水底植物颜色的反射通常只在浅水池里才有作用，而对天空颜色的反射也只是对于非常平静的水面有作用。我自己也记录过一些关于水的颜色的例子，放在这里应该是合适的：(1) 晴朗的日子里，在灰白的天空和大块积云下，普勒斯台德的 20 英尺深的水库里用沉积石灰石处理（一种根据专利技术软化水的方法）过的水是湛蓝色的。(2) 马赛的地中海是深蓝色的（这些记录中最深的蓝色）。(3) 日内瓦湖是蔚蓝色的。(4) 位于格恩西附近的海和位于艾费尔的拉赫湖，颜色都更蓝一些，接近于靛蓝色。(5) 白垩崖附近的海水是淡蓝色的，而离海岸不远处的海水是绿色或者浅灰色的。(6) 罗纳河、摩泽尔河和冰河等都是淡蓝色或者灰蓝色的。(7) 莱茵河、斯海尔德河（正如比利时绘画描绘的那样，在安特卫普非常明显）、塞纳河和泰晤士河河口等都是绿色的。(8) 日内瓦湖呈现出一片片的深绿色。在傍晚太阳刚刚落山时，这种颜色在图恩湖和卢塞恩湖上更加常见。(9) 在阳光明媚又有风的日子，马恩岛附近的海域是鲜绿色的。(10) 佛罗里达珊瑚礁附近的海域据说是深绿色的，而远离海岸的地方则是深蓝色的。(11) 在阴云密布的雨天，微弱的阳光使海呈现出一块块绿色和红棕色。(12) 基拉尼一座古老的铜矿矿井里的积水是深绿色的，而它旁边湖泊中的水基本上是黑色的。(13) 一些海域因为藻类而呈现出红色，还有一些海域和河流则因为含铁的红色土壤被分解而呈现出红色。(14) 一些池塘（蛇形湖和许多观赏性水池）由于单细胞有机体而呈现出不透明的绿色，这些有机体有时甚至会充满整个池塘。类似的，这些池塘也可能会变成红色。也许，由水的光学属性而呈现出蓝色的最显著的例子是卡普雷拉岛的蓝洞，那里至少可以排除天空的反射这一因素的影响。另一个类似的现象是冰川裂缝中壮丽的蓝色和绿色。

下面我们就不再考虑表面反射，因为它只对路边的池塘和那些像镜面一样的水面起作用，另外也不再考虑清澈见底的浅溪和水池中出现的植物。这样看起来，目前我们可能要将大量水的蓝色归因于一种来自水内部的奇特反射。这种反射伴随着偏振，依赖于悬浮的颗粒。黑色、棕色和黄色的出现都是由于水里溶解了植物中的物质，某些情况下，红棕色是因为铁；绿色有时是因为铜或海藻，但是在海洋、湖泊以及河流中经常见到的，类似冰川裂缝中颜色的那种绿，则很可能具有和蓝色相似的成因。我相信，一些物理学家可能会被吸引而开始研究本文所述的问题。日落时产生的一系列色彩中会不会就包含着能够解释蓝色和绿色的水会呈现出各种色彩

pages. Has not the production of a series of tints at sunset an origin which may tend to explain the various tints of blue and green waters? I find that Mr. Sorby in the *Philosophical Magazine*, November, 1867, ascribed the blue colour of the sky and the successive yellow orange and red tints of the setting sun to the absorption of the red rays more than the blue, by the fine aqueous vapour of the higher regions of the atmosphere, and of the blue rays more than the red by the coarser vapours near the earth's surface—as *e.g.* a fog.

The foregoing notes may suggest to others similar observations of greater importance, which it would be interesting to collect. It would be very satisfactory, and of interest to many readers, if some one who could speak with authority on the physics of light, would discuss these phenomena, however suggestively, in your pages.

(**2**, 235; 1870)

的原理呢？我发现，在 1867 年 11 月的《哲学杂志》上索比先生就将天空的蓝色以及日落时相继出现的橙黄和红色解释为：高层大气中细小的水蒸气颗粒对红光的吸收多于对蓝光的吸收，而地面附近像雾这样比较粗大的水蒸气颗粒对蓝光的吸收多于对红光的吸收。

上述说明可能会使人想起一些类似的但更为重要的观察结果，收集这些结果应该是件非常有趣的事情。如果有哪位光物理学方面的权威人士能够发表文章来讨论这些现象，哪怕在文中给出的只是提示性的讨论，那也会是一件令人欣慰并且深受读者欢迎的事情。

（金世超 翻译；李军刚 审稿）

The Source of Solar Energy

R. P. Greg

Editor's Note

With no knowledge of nuclear reactions, scientists in the late nineteenth century speculated freely and imaginatively on the source of energy driving the Sun. One idea was that continually infalling meteors might supply the energy through impacts. Yet here Robert Greg points out that this hypothesis conflicts with most of what was then known about meteors. The vast majority of these are objects in regular orbit around the Sun, and barely come any closer to the Sun than does the Earth, so there is no reason to expect them to fall into it. Indeed, do any meteors do so? Some meteors almost certainly do, says Greg, but surely not enough to account for the enormous energy of the Sun.

IT is, I think, rather unfortunate that Mr. Proctor, in his recent work entitled "More Worlds than One", should have re-advocated the earlier and now discarded views of Sir W. Thomson concerning the source of solar heat or energy by *meteoric percussion*. That theory, however ingenious as advanced by the physicist, is surely hardly one to be admitted by the astronomer. Nothing less than an intense desire or necessity for finding some solution to the problem, whence or how the solar heat is maintained, could have encouraged scientific men seriously to advance or support so plausible and unsatisfactory a doctrine, or one, when examined, so little supported by what we really know either of meteors or of nature's laws. Having given much attention to *meteoric* astronomy, may I be permitted briefly to state what I hold are serious and practical objections to the validity of the meteoric or dynamical theory as applied to the conservation of solar heat and energy.

1. Because meteors and aerolites are known to impinge and strike the earth in her orbit, *ergo*, as I understand Mr. Proctor, numbers infinitely greater must no doubt be constantly rushing into the sun, as a body at once far larger, and much nearer to myriads of such bodies than the earth herself; but which, at a much smaller distance, are more likely to be drawn into the sun. Now, all that we really do know about meteors amounts to this, that by far the greater number of shooting stars visible in our atmosphere, in size no larger than a bean, and really separated from each other by thousands of miles, belong to fixed and definite systems or rings, having fixed radiant points for certain epochs or periods, showing clearly that these bodies are revolving round the sun, in courses as true and regular as the planets themselves, and are no more eddying or rushing into the sun, merely because they are so insignificant, than is the earth herself. Having projected upon celestial charts the apparent courses or tracks of nearly 5,000 meteors, observed during every part of the year, I feel I am justified in stating that not more than seven or eight percent of the shooting stars observed on any clear night throughout the year, are *sporadic*, or do not belong to meteor systems at present known to us. More than one hundred meteor systems

太阳能量的来源

格雷格

编者按

19 世纪后期的科学家还不知道核反应为何物，他们可以自由地发挥自己的想象力去推测太阳能量的来源。其中的一个想法是不断陨落的流星通过撞击太阳而提供能量。不过罗伯特·格雷格在这篇文章里指出，这一假说不符合大部分当时所知的有关流星的理论。绝大多数流星都是以固定轨道围绕太阳运转的天体，它们几乎不可能比地球更靠近太阳，所以没有理由相信它们会落到太阳上。是否真的有流星落到太阳上呢？格雷格说，也许有些流星确实落到太阳上了，但肯定不足以用来解释太阳的巨大能量。

普洛克特先生在他最近的题为《不止一个世界》的著作中，再次推崇了现在已经被抛弃了的汤姆孙爵士早期的理论，我认为这是相当令人遗憾的。汤姆孙爵士认为，太阳的热或者能量是由**流星撞击**形成的。然而由物理学家提出的这一创新理论，对于天文学家而言显然是难以接受的。太阳的热量究竟源自哪里，又是如何维持的呢？正是寻觅这一问题的答案的迫切需要和强烈愿望激励了科学家，使他们认真地提出或者支持某种理论，这里就有上述那种看似合理实际上很难令人满意的学说，另外还有一种经检验发现很难被我们知晓的关于流星和自然定律的知识支持的学说。我对**流星**天文学下过不少工夫，请允许我再简要地对我坚持的观点加以阐述：无论从重要性上还是实用性上，用流星及其动力学理论去解释太阳上热量和能量的守恒都是不够的。

1. 众所周知，流星和陨石在地球的轨道内可能会撞击地球，因此，在我看来普洛克特先生的说法也就意味着：数量极其庞大的流星和陨石确定无疑会持续不断地撞击在太阳上，因为和地球相比，太阳更为巨大，也更接近这些小天体。不过，当距离很小的时候，这些流星和陨石很可能会坠入太阳内部。然而，迄今为止我们对流星的全部了解都表明：从大气中划过的大量流星，用肉眼看起来仅仅宛若豆子大小，彼此相隔数千英里，各自具有自己固定的运动系统，在某些时段还会呈现出固定的辐射点。这些都清楚地表明：它们和其他行星一样在规则的轨道上绕着太阳旋转，而仅仅因为它们实在是有些微不足道，使得它们和地球一样并不会漩涡式地或者直线式地冲向太阳。我曾经把近 5,000 颗常年都能观察到的流星的视轨迹投影在天球上，结果表明，对于所有那些在全年任何一个晴朗的夜晚都能观察到的流星，其中最多只有 7% 或 8% 的流星是**偶然出现的**或者不属于任何已知的流星群。目前

are now recognised, several of which appear most certainly to be connected with known comets; and from a paper I have just received from Professor Schiaparelli, of Milan, it would appear that the approximate average *perihelion* distance for 44 of these meteor systems is not less than 0.7, the earth's distance from the sun being 1.0; whilst of these 44 systems, only 4, or about 10 percent have their *perihelion* distance under 0.1, that is, approach the sun nearer than nine millions of miles! Now, it is pretty well admitted that meteors are intimately connected with comet systems, yet out of some 200 comets, the elements of whose orbits have been calculated with tolerable precision, only 5 percent have their *perihelion* distance under 0.1. The same argument holds good also for planets, whose numbers also diminish after a certain considerable mean distance from the sun. Are these facts, then, in accordance with the notion that meteoric bodies either increase in number as we approach the sun, or that meteors are so constantly losing their senses, or sense of gravity, as to be ever rushing into or against the sun? I might almost ask, do *any* meteors rush or fall into the sun? Is it probable that the *mass* of all "the countless myriads of meteors" in the solar system exceeds that of a single planet? whether that of Mercury or Jupiter does not much signify. When we take into consideration the gigantic amount of meteoric deposits required to maintain the solar heat for hundreds of millions of years, in the meteoric theory, surely the supply of meteors would long since have been exhausted, were the supply at least confined merely to the meteors under a mean distance of 0.1 belonging to our own solar system! The argument, to begin with, is in a great degree fallacious, *e.g.*, because meteors frequently strike the earth, they must, it is argued, strike the sun in vastly greater numbers, and with far greater velocities. But it is forgotten that the meteors themselves, like the earth, are revolving round the sun as a common centre, in regular orbits, and only by accident, as it were, come into mutual collision, just as the tail of a comet might pass through the system of Jupiter and his satellites; while to the end of time neither the earth nor the meteors need necessarily come into contact with the sun.

2. But it is not merely meteors belonging to the solar system which are taxed to provide fuel for our sun; *space* itself may be filled with meteors ready to impinge upon the sun. The arguments against this are: (1) judging from analogy as well as from facts, comparatively few meteors are *sporadic*, consequently the majority cannot belong to stellar space, but to our own system; (2) granting that space itself is really more or less filled with meteors, these would not necessarily rush straight into the sun, unless, as would very unlikely be the case, they had no proper motion of their own. They might be drawn into or enter our system, it is true, but, according to Schiaparelli, only to circulate like comets in definite orbits.

3. The *zodiacal light* is another victim to the emergencies of the *meteoric* theory of solar energy. Whether composed of myriads of small meteors, or merely a nebulous appendage, or atmospheric emanation belonging to the sun, is it credible that for hundreds of millions of years there could, physically speaking, be sufficient material in the zodiacal light to maintain the sun's heat and supply all the fuel required? Has it ever yet been proved that the entire mass of matter constituting the zodiacal light, is either composed of matter in a solid state, or, if it were, that its mass would be equal to that of our own earth? If

人们已经确认了 100 多个流星群，基本上可以确信其中一些是和已知彗星有关联的。从我刚刚收到的来自米兰的斯基亚帕雷利教授的文章可知，如果将地球与太阳之间的距离定义为 1 个单位，那么这些流星群中有 44 个的平均**近日点**距离不小于 0.7 个单位。另外，在这 44 个群中只有 4 个群（即大约 10%）的**近日点**距离小于 0.1 个单位，也就是说与太阳的距离小于 900 万英里！目前，在很大程度上人们已经认可了流星与彗星系统密切相关的观点，尽管对于 200 多颗已经被颇为精确地计算出轨道参数的彗星来说，其中只有 5% 的彗星的**近日点**距离小于 0.1 个单位。对于行星来说情况也是如此，在与太阳的间距超出某一个平均距离后，行星的数目也会减少。有观点认为，或者是越靠近太阳流星体的数量越多，或者是流星体不断失去万有引力的影响而冲向太阳，或者与太阳背道而驰，但以上的事实与这些观点中的哪一个一致呢？我还想要问，真的有**任何**流星群冲向或者坠入了太阳吗？太阳系中所有的“数不尽的流星”的**质量总和**能超过一颗行星的质量吗？这颗行星是水星还是木星无关紧要。当我们按照流星理论来考虑维持太阳数亿年的发热所必需的巨大数量的流星沉积时，如果我们将流星的供给最小程度地局限在太阳系中距离太阳的平均距离小于 0.1 个单位的范围内，那么很显然流星的供给在很早很早以前就已经被消耗殆尽了！这一理论从一开始就是相当荒谬的。比如，这一理论声称，因为流星频繁地撞击地球，所以也一定有数量更为庞大的流星以极快的速度撞击太阳。但是该理论却忽略了流星本身也是像地球一样在规则的轨道上围绕太阳这一公共中心旋转，只是偶尔才会与其他星体发生碰撞，这就像彗尾偶尔可能会扫过木星及其卫星组成的系统一样，但是无论什么时候，地球和流星都绝不会与太阳发生接触。

2. 但是也许，为太阳提供燃料的不只是太阳系内的流星，**太空**中可能到处都有会撞击太阳的流星。不过，以下的证据是反对这一观点的：(1) 从事实和推理可以判断，只有极少数流星是**偶现**流星，绝大多数流星不可能属于其他恒星空间，只能是属于我们的太阳系；(2) 即使太空中确实散布着一些流星，那么它们也不太可能直接冲向太阳，除非在极特殊的情况下，这些流星没有自己固有的运动方向。这样的流星或许有可能被吸引或进入到我们的太阳系内，但是，根据斯基亚帕雷利的观点，它们只能是像彗星一样在一定的轨道上绕太阳运行。

3. 关于太阳能量的**流星**理论出现后，**黄道光**就成了另一个受害者。不管它是由无数小流星组成的，还是由某种星云物质或者来自太阳的大气组成的，对于黄道光中存在的物质能够维持数亿年间太阳的发热，并且为太阳提供全部所需燃料，我们能够相信吗？另外，是否已经有人证明过组成黄道光的所有物质都是固态的？即便如此，它们的总质量是否能抵得上我们地球的质量？如果黄道光是由许多彼此独立

composed of separate meteors, are they not each individually revolving round the sun, rather than occupied in being gradually drawn into it as a vortex? *

Of course I do not say that meteors are *never* drawn into the sun, or that they may not occasionally and by accident enter the solar atmosphere; I have merely endeavoured to show that, from what we really do know about meteors and the laws of nature, it is highly improbable that our sun could derive, in sufficient quantity, a needful supply of fuel from meteoric sources. The comet of 1843, which approached the sun within 550,000 miles, was not sensibly deflected from its course; it is just possible that so small a thing as an aërolite might at that distance have been drawn into the sun; but is it not also possible, from what we know of comet and meteor systems, it may be wisely ordained that the smaller bodies of our solar system, such as meteors, do not as a rule approach the sun too closely; and they probably do not, if their *perihelia* distances are rarely under 10,000,000 of miles?

Aërolites are doubtless of larger size and weight than shooting stars, and as far as is yet known, not so regular in their appearance as shooting stars; but even with that class of phenomena, we notice a certain degree of periodicity in *maxima* and *minima* for certain times of the year, tending to show that they also may be subject to regular laws, and not fall so frequently or promiscuously upon the sun's surface as has been sometimes supposed. If they do not fall in vastly greater numbers, area for area, upon the sun than they do upon our earth, certainly the dynamical effect would be very minute! I may here also observe that even these bodies generally fall to the earth without being consumed, and with a very moderate velocity; their original cosmical velocity having been lost before reaching the surface of the earth. In the case of an aërolite falling upon the sun's surface, its original velocity may similarly have been gradually checked in its passage through the solar atmosphere, and a considerable amount therefore of the calculated mechanical effect lost. Small meteors would probably be consumed thousands of miles from the real body of the sun, seeing that the sun's inflamed atmosphere is now known to extend at times some 50,000 miles. It might almost be a question whether the sun's proper heat may not even be greater than that caused by the simple friction of a meteor through the solar atmosphere!

I merely allude to these minor matters, however, in order to point out some of the numerous uncertainties and difficulties connected with this meteoric or mechanical theory of the origin and conservation of solar heat, in addition to those already alluded to, bearing more especially upon the astronomical bearings of the question. For the present it must still remain a mystery, whence or how the solar heat is maintained, or to what extent really wasted.

(**2**, 255; 1870)

Robert P. Greg: Prestwich, Manchester, July 11.

* We beg to refer our readers to Jones' and Liais' observations of the Zodiacal Light. They certainly have not received the attention in this country that they deserve.—ED.

的流星组成的话，那么这些流星为什么不会各自独立地围绕太阳运行，而非要旋涡式地逐步坠落到太阳里面去呢？*

当然，我并不是说流星**从来都不会**坠落到太阳里，也不是说他们不可能偶然地进入太阳大气。我只是想尽力说明，根据我们确确实实掌握的关于流星和自然定律的知识，太阳所需的数量巨大的燃料真的不太可能来自流星。1843 年，彗星靠近太阳时与太阳的间距小于 550,000 英里，但它的轨道并没有发生明显的偏转，在这样的距离下可能只有像陨石那么小的物体才会被拽入太阳吧。但是根据我们对彗星和流星群的了解，太阳系中像流星这样的小天体并不会十分靠近太阳。如果这些小天体的**近日点**距离很少小于 10,000,000 英里的话，那它们怎么可能会是太阳燃料的主要来源呢？

就目前所知，**陨石**的尺寸和重量无疑比流星大，但不像流星那样有规律地出现。我们注意到，陨石这类现象在一年内出现的频率有**大**有**小**，存在一定程度的周期性，这说明陨石现象可能也是遵循一定的规律的，而不是像人们曾经猜测的那样频繁而又杂乱地坠落到太阳表面。如果坠落在太阳上的陨石数量并不比坠落在地球上的陨石数量大很多的话，那么产生的动力学影响必将是非常微小！这里我还想指出，陨石坠落到地球上时一般并没有燃烧耗尽，并且落地的速度是比较慢的，在到达地球表面之前，它们原初的宇宙速度已经被消耗了。对于坠落到太阳表面的陨石来说，情况可能是类似的，在陨石通过太阳大气层时它们的速度就会逐渐降低，因此原先估算的力学效应也随之大幅降低了。我们知道太阳燃烧的大气有时会向外延伸 50,000 英里以上，因此小流星很可能在距离太阳实体数千英里之外就已经燃烧耗尽了。难道太阳本身的热量还不及流星穿越太阳大气时摩擦产生的热量吗？这恐怕是个问题吧！

这里，我仅仅提到了一些细节以指出用流星及其动力学理论解释太阳能量的来源和守恒时存在各种不确定因素和困难，除此之外，先前已经有人对该理论提出过一些疑问，我在这里则更多的是从天文学的角度对该理论提出质疑。至于太阳的热量究竟源自哪里以及是怎样维持和被消耗的，现在仍然是个谜。

（金世超 翻译；何香涛 审稿）

* 请读者参考琼斯和莱斯对于黄道光的观测，可惜他们并没有在这个国家受到应有的重视。——编辑注。

The Source of Solar Energy

R. A. Proctor

Editor's Note

Richard Proctor writes here to correct an earlier essay of a Mr. Greg, which he felt misrepresented his views regarding the energy source of the Sun. Greg implied incorrectly that Proctor believed infalling meteors to be the Sun's sole source of energy, and to act by impact at the solar surface. Proctor here claims that he has no firm opinion on the matter, although many facts do point to infalling meteors as an energy source. They would still deliver their kinetic energy, he says, regardless of whether they reach the Sun in a solid, fluid or vaporous state. Proctor also argues that many facts support the conjecture that the zodiacal light is best explained by a mass of bodies in the neighbourhood of the Sun.

MR. Greg ascribes to me views I do not hold, and then employs my own reasoning to overthrow them. He must have formed his conceptions of my theories from Prof. Pritchard's critique of my "Other Worlds"—a most unreliable source.

To begin with,—I do *not* believe that the solar heat supply is solely derived from the downfall of meteors. I have impressed this very clearly at p. 54 of my "Other Worlds".

I do not believe that *any part whatever* of the solar heat supply is derived from meteoric *percussion*, nor that any meteor ever comes within tens of thousands of miles of the sun's surface in the solid state.

Mr. Greg is very careful to show me that the meteor-systems encountered by the earth cannot fall into the sun. I dwell on this very fact at p. 203 of "Other Worlds"—I say, *totidem verbis*, that no known meteoric system can form a hail of meteors upon the sun. "It is forgotten," says Mr. Greg, "that the meteors themselves revolve round the sun," &c. If *he* has at any time forgotten this, I certainly have not.

"Has it ever been proved," he asks me, "that the entire mass of meteors constituting the zodiacal light, is either composed of matter in a solid state, or, if it were, that its mass would be equal to that of our own earth?" I answer, as Mr. Greg would—"No, it has not been proved, nor is it by any means probable."

There is nothing new to me in Mr. Greg's letter, and little which I have not described myself long ago in the *Intellectual Observer and Student* of 1867, 1868, and 1869. To suppose that I should venture to treat at all of meteoric astronomy, in ignorance of such elementary facts—the very ABC of the science—is not complimentary. Mr. Greg might,

太阳能量的来源

普洛克特

编者按

理查德·普洛克特写这篇文章的目的是为了纠正格雷格先生在以前发表的一篇论文中的错误观点，他认为格雷格先生误解了他对太阳能量来源问题的认识。格雷格先生错误地指出普洛克特认定陨落的流星是太阳能量的唯一来源，而且流星还会与太阳表面发生碰撞。普洛克特在这里声明，他在该问题上还没有十分明确的看法，尽管有许多事实确实证明了陨落的流星可以作为一种能量来源。他说，不管流星是以固态、液态或气态方式降落到太阳上，它们都会释放自己的动能。普洛克特还争辩说，许多事实都支持黄道光形成的最佳解释是它来自于太阳附近的一些小天体。

格雷格先生将一些本来不属于我的观点强加在我身上，然后又运用我的推理去推翻这些观点。他一定是从普里查德教授对我的《其他世界》一书的评论中形成了对我的理论的看法，然而这一评论是站不住脚的。

首先，我**不**认为太阳热量的供给完全来自坠落的流星。对于这一点，我在《其他世界》一书的第 54 页已经非常清楚地阐述过了。

其次，我不认为会有**任何一部分**太阳热量源自流星**撞击**，也不认为任何流星会以固态的形式到达距离太阳表面数万英里的范围内。

格雷格先生非常仔细地向我说明了与地球遭遇的流星群不可能坠入太阳。其实，我在《其他世界》一书的第 203 页已经对这一事实进行了详述。在那段叙述中，我正是这样指出的，没有任何已知的流星群可以在太阳表面形成流星雨。格雷格先生在他的文章中提到“流星本身围绕太阳旋转的事实被遗忘了”，即使**他**在某些时候忘了这一点，我也不会忘的。

他还问我“是否已经有人证明过组成黄道光的所有流星都是固态的？即便如此，它们的总质量是否能抵得上我们地球的质量？”我的回答是“没有，这一点还没有被证明过，而且现在还没有办法去证明这一点”，也许格雷格先生的答案也是如此吧。

对我来说，格雷格先生的文章中并没有什么新内容，几乎全部都是很久以前我在 1867、1868 和 1869 年的《聪明的观察家与学生》上阐述过的。假如有人认为我是在完全不了解如此基本的科学常识的情况下无知无畏地研究流星天文学的话，那

at least, have examined what I have written before assigning to me the absurdities he attacks so successfully.

The fact is, this matter of the solar energy only comes in *par parenthèse* in my "Other Worlds". I express no confident opinion whatever about it. I point to some deductions from known facts, and respecting *them* express a certain feeling of confidence. It is not my fault (nor, indeed, can I blame Mr. Greg) if Prof. Pritchard has tacked my words "I am certain" (used with reference to reliable inferences) to a theory respecting which I have distinctly written, that "I should not care positively to assert" its truth. Even that theory is not the absurd one attacked (very properly) by Mr. Greg.

For the rest, most of Mr. Greg's letter is sufficiently accurate, but there are two mistakes in it.

1. We have abundant evidence that the density of the aggregation of cometic perihelia increases rapidly near the sun. For example, whereas between limits of distance 40,000,000 and 60,000,000 miles from the sun this density is represented by the number 1.06, it is represented by the number 1.67 for limits 20,000,000 and 40,000,000 miles, and by the number 8.65 within the distance 20,000,000 miles. The evidence derived from this observed increase of aggregation is not affected by what we know of those cometic or meteoric systems whose orbits nearly intersect the earth's (for they must form but the minutest fraction of the total number) nor by the observed minimum perihelion distance of cometic orbits (for observed comets are but the minutest fraction of the total number).

2. It makes no difference whatever as regards the force-supply of the solar system, whether the substance of a meteor reaches the sun in the solid, fluid, or vaporous state. Given that the substance of a meteor, moving at one time with a certain velocity at a certain distance from the sun, is at another time (after whatever processes) brought to rest upon or within the sun's substance, then either the "force-equivalent" of its motion has been already distributed or the substance of the meteor is in a condition to distribute that "force-equivalent" mediately or directly. In other words, either heat and light have been already distributed, or the central energy has been recruited to the full extent corresponding to the mass, motion, and original distance of the meteor.

I may express here my agreement with the opinion of the Editor of *Nature* that the observations made on the zodiacal light by Lieut. Jones and M. Liais ought to be taken into account in any theory of that mysterious object. Taken in conjunction with the other known phenomena of the zodiacal light, they admit of but one interpretation as to the position, dimensions, and general characteristics of the object. Taken alone, we might infer from them that the zodiacal light is a ring of bodies or vapours travelling around the earth (at a considerable distance); other phenomena suggest that the zodiacal light is a disc of bodies or vapours travelling around the sun; yet others suggest that the zodiacal light is a phenomenon of our own atmosphere. But the only theory which accounts at once

我要说我实在很不欢迎这样的臆测。格雷格先生将那些谬论强加于我并进行如此成功的攻击之前，至少应该先检查一下我到底写过什么吧。

事实上，在我的《其他世界》一书中只是**附带性**地谈了一些关于太阳能量的事情。我只是根据已知事实提出了一些推论，对于**它们**我并不确信，也没有十足的把握。如果普里查德教授将我说过的“我确信”（这个词是用来表述一些可靠的推论的，而且使用时都附注了参考文献）强加到某个我确实写出来过但“我并没有担保”其正确性的理论上的话，那这并不是我的过错（当然，我也不会责备格雷格先生），不过即便这个被普里查德教授强行认为是我确信的理论也并不是格雷格先生（非常合理地）攻击过的那条荒谬的理论。

除此之外，格雷格先生的文章中大部分内容都是非常准确的，不过其中也有两处错误：

1. 我们有充分的证据表明，随着彗星近日点靠近太阳，其聚集密度迅速增大。例如，在距离太阳 40,000,000~60,000,000 英里的范围内，这一密度为 1.06；在 20,000,000~40,000,000 英里的范围内，这一密度为 1.67；而在 20,000,000 英里以内，这一密度增加到 8.65。这些观测得到的聚集度的增加，并不受我们已知的某些彗星或流星群的轨道几乎与地球轨道相交的影响（这样的群肯定会出现，但只占总数中极小的一部分），也与观测到的彗星轨道的最小近地点的距离无关（人们观测到的彗星仅仅是全部彗星中极小的一部分）。

2. 从考虑太阳系的支撑力的角度来看，组成流星的物质到底是以固态、液态还是气态形式到达太阳并没有什么区别。给定某种流星物质状态，在距太阳某一距离以一定的速度运动，而在另一时刻（不管经历了什么过程）被吸引到太阳上的，甚至成为太阳的组成物质，那么，流星运动的“等效力”要么已经转化，要么流星的组成物质正处在间接或直接地转化为这种“等效力”的过程中。换句话说，要么已经转化成光和热，要么中心能量已经有所增加，但总是与由流星的质量、运动以及初始距离确定的能量总量相一致。

《自然》的编辑认为，关于黄道光这种神秘物体的任何理论都应该考虑到琼斯中尉和里阿斯对黄道光的观测结果，对此我表示同意。结合其他关于黄道光的已知现象，他们认为对于黄道光这种物体的位置、尺度以及一般特性只有一种解释。单独来看，根据他们的观测结果我们可能会推测出黄道光是由围绕地球运行的天体或者蒸气组成的圆环（在距离地球相当远的位置上），但另一些现象则暗示黄道光是由围绕太阳运行的天体或蒸气组成的圆盘，同时还有其他现象表明黄道光是由地球大气本身造成的一种现象。不过只有一种理论可以同时解释人们观测到的**所有**现象，那就是：黄道光不过是由于太阳周围连续存在许多小天体碎块或各种蒸气（诸

for *all* observed phenomena, is that which regards the zodiacal light as simply due to the continual presence in the sun's neighbourhood of bodies or vapours (meteoric or cometic, or both) which come there from very far beyond the earth's orbit, and pass away again on their eccentric orbits. A disc thus formed of continually varying constituents would shift in position, and would wax and wane in extent as well as splendour, precisely as the zodiacal light is observed to do.

(**2**, 275-276; 1870)

如流星或彗星，或者两者同时存在）而形成的，这些小天体碎块和蒸气来自非常遥远的地方，远远超出了地球轨道的范围，之后，他们会沿着自己的偏心轨道远离而去。这样一个由持续变化的成分形成的圆盘，其位置自然会发生漂移，而其大小和绽放出的光辉也会时增时减，这和人们观测到的黄道光的表现一模一样。

（金世超 翻译；何香涛 审稿）

The Source of Solar Energy

R. A. Proctor

Editor's Note

Still feeling misinterpreted, Richard Proctor tries once more to clarify his views. He asserts that he believed infalling meteors to supply a portion, but not all, of the Sun's energy. Moreover, he says these meteors are not from the same source as those which strike the Earth, as these cannot reach the Sun. Mr. Greg also made a mistake, Proctor pointed out, in his considerations of how an infalling body might affect the energy of the Sun. Were a large mass of solid iron to enter into the Sun, it would not decrease the Sun's overall energy, as Greg had asserted, though it might decrease the solar temperature. The total solar energy would still increase, just as it would if the earth itself fell into the Sun.

MR. Greg still misses my meaning. I do believe that meteors supply a portion of the solar energy, and I also believe that they fall in enormous quantities into the sun; what I do not believe is that the whole solar energy is derived from meteors, or that any meteors fall in a solid state upon the sun (whose surface is also certainly not solid, even if any part of his mass be).

Mr. Greg's reasoning only proves what I have already pointed out, that none of the meteor systems our earth encounters can supply a meteoric downfall on the sun. This is, however, so obvious as to need no enforcing.

The reasoning by which I show that enormous quantities of meteors must fall upon the sun is wholly untouched by Mr. Greg's arguments, and is, so far as I can see, simply incontrovertible.

Surely Mr. Greg is not in earnest in saying that there would be a loss of solar energy if a large mass of iron fell on the sun before it was quite melted (any conceivable mass would, by the way, be vaporised), *because the sun would have to melt the portion which remained solid.* That solar energy would be consumed in the process is true enough; but if Mr. Greg supposes that the total solar energy would be diminished, he altogether misapprehends the whole subject he is dealing with. If the action of the solar energy in changing the condition of matter forming (as the imagined meteorite would) part of the sun's substance had to be counted as loss of energy, the sun would be extinguished in a very short time indeed. Such processes involve exchange, not loss.

If the earth could be placed on the sun's surface, the action of the sun in melting and vaporising the earth, and producing the dissociation of all compound bodies in the earth's substance, would involve an enormous expenditure of energy, yet the solar energy,

太阳能量的来源

普洛克特

编者按

理查德·普洛克特还是认为他的观点被别人误解了，他又一次发表文章澄清自己的观点。他宣称自己的观点是陨落的流星是太阳能量的一部分来源，而不是全部。他还认为这些流星与撞击地球的流星的来源不同，因为撞击地球的流星无法到达太阳。普洛克特指出，格雷格先生在考虑下落物如何影响太阳的能量时犯了一个错误。如果大量的固态铁落到太阳上，尽管它们可能会降低太阳的温度，但不会像格雷格所说的那样降低太阳的整体能量。太阳的总能量是会升高的，就像如果地球坠入太阳也会提升太阳的能量一样。

格雷格先生仍然误解了我的意思。我确实相信太阳能量的一部分来自流星，并且我也相信有数量庞大的流星坠入太阳。但是，我不相信太阳的全部能量都是来自流星，也不相信流星会以固态形式坠落在太阳上（即使太阳的一部分物质是固态的，但其表面肯定不是固态的）。

格雷格先生的推理只是证明了我先前已经指出的，即地球遭遇的流星群不可能形成太阳上的流星雨。然而这一道理太明显了，没必要再次强调。

我还通过推理指出必然有数量庞大的流星坠落在太阳上，但是格雷格先生在申辩中完全没有谈及这一条。据我看来，这一推理是无可置疑的。

显然，格雷格先生并没有坦诚地指出：如果有大量的铁在熔化之前坠落到太阳上（顺便提一下，任何可以想象到的物质都会被汽化），太阳的能量必然会有损失，**因为太阳必然会熔化掉那些还是固态的物质**。在这一过程中太阳的能量会被消耗，这是完全正确的。但是，如果格雷格先生由此而认为太阳的总能量也会减少，那他就完全误解了他所讨论的整个内容。如果那些即将成为太阳的一部分的物质在发生状态改变（那些假想中的流星就会经历这一过程）时会造成太阳能量的损失，那么太阳肯定会在很短的时间内熄灭。因此，这样的过程中会有能量的交换，但太阳的能量不会损失。

假设把地球放到太阳表面，那么，在太阳使得地球熔化、汽化，并使组成地球的所有物质发生分解的过程中，肯定会有巨大的能量消耗。但是，从整体上来看太阳的能量应该会有所增加的，即便我们不考虑地球可以为太阳提供燃料这一事实。

considered as a whole, would be recruited, even apart from the fact that the earth would serve as fuel. The absolute temperature of the sun would, I grant, be diminished in this imaginary case, though quite inappreciably, but his total heat would be increased by whatever heat exists in the earth's substance.

Apart from this, however, if the minimum velocity with which a meteor or other body can reach the sun, is such as would—if wholly applied to heating the body—completely melt it, then the size of the body makes no difference whatever in the result. The meteor might not be melted if enormously large, but in that case the balance of heat would be communicated to the sun. In reality, of course, the heat corresponding to meteoric motion near the sun is very far greater than is here implied.

But I really must apologise for bringing before your readers considerations depending on the most elementary laws of the conservation of energy.

(**2**, 315; 1870)

我承认，在这个虚构的例子中太阳的绝对温度可能会下降，不过下降的幅度应该是非常微小的，而太阳的总热量应该会由于地球物质中存在的热量而增加。

然而除此之外，如果一颗流星或者其他天体到达太阳表面时的最低速度用来加热该天体就足以使其完全熔化的话，那么天体本身的尺寸差异对于结果就没有什么影响了。如果流星过于巨大的话，也许有可能不会被完全熔化，但这种情况下太阳仍会达到热平衡。当然，实际情况是流星靠近太阳时，其运动对应的热量远比这里假设的大得多。

不过在这里，我也确实要向贵刊的读者们表示歉意，以上只是我基于能量守恒的一些最基本定律所作的一些个人思考。

（金世超 翻译；何香涛 审稿）

The Coming Transits of Venus*

Editor's Note

This comment points out that the next transit of the planet Venus across the Sun's disk is expected in four years. The author calls for scientists to begin developing plans for observations and cooperation. One key challenge is to improve on the error of as much as twenty seconds still present in estimates of the moment when the planet "touches" the edge of the Sun, an error that came from the fuzzy boundaries of the Sun's atmosphere.

TRANSITS of Venus over the disc of the sun have more than any other celestial phenomena occupied the attention and called forth the energies of the astronomical world. In the last century they furnished the only means known of learning the distance of the sun with an approach to accuracy, and were therefore looked for with an interest corresponding to the importance of this element. Although other methods of arriving at this knowledge with equal accuracy are now known, the rarity of the phenomenon in question insures for it an amount of attention which no other system of observation can command. As the rival method, that of observations of Mars at favourable times, requires, equally with this, the general co-operation of astronomers, the power of securing this co-operation does in itself give the Transits of Venus an advantage they would not otherwise possess.

Although the next transit does not occur for four years, the preliminary arrangements for its observation are already being made by the governmental and scientific organisations of Europe. It is not likely that our Government will be backward in furnishing the means to enable its astronomers to take part in this work. The principal dangers are, I apprehend, those of setting out with insufficient preparation, with unmatured plans of observation, and without a good system of co-operation among the several parties. For this reason I beg leave to call the attention of the Academy to a discussion of the measures by which we may hope for an accurate result.

In planning determinations of the solar parallax from the Transits of Venus, it has hitherto been the custom to depend entirely upon the observations of the internal contact of the limbs of the sun and planet proposed by Halley. It is a little remarkable, that while astronomical observations in general have attained a degree of accuracy wholly unthought of in the time of Halley, this particular observation has never been made with a precision at all approaching that which Halley believed that he himself had actually attained. In his paper he states that he was sure of the time of the internal contact of Mercury and the

* Substance of a paper read before the Thirteenth Annual Session of the American Academy of Sciences, held at Washington, by Prof. Simon Newcomb. (The original paper was illustrated by diagrams.)

即将到来的金星凌日*

编者按

这篇文章指出下一次金星凌日（金星划过日面）将发生在4年后。作者呼吁科学家们应该开始着手制定观测和协作的计划。一直以来，由于太阳大气的边界比较模糊，所以在判断行星“接触”日面边界的时刻上一直存在长达20秒的误差。对此作出改进是一件极具挑战的事情。

金星从日面划过而发生的凌日现象比其他任何天文现象都更能引起人们的关注，天文界投入了巨大的精力去研究这一现象。在上个世纪，研究凌日现象是精确获得日地距离的唯一方法，凌日现象的这种重要性使得人们对它的出现相当期待。尽管现在也有其他的方法能以同样的精度确定日地距离，但金星凌日这种天象的罕见性使它受到了比其他观测更多的关注。和观测金星凌日一样，对火星的某些特殊观测同样需要天文学家们的集体协作，而确保这次协作的能力为观测金星凌日提供了在火星的观测中不曾拥有的优势。

尽管下一次凌日是在4年后，但是欧洲的政府和科学机构已经开始着手准备观测的前期工作了。在给天文学家们提供所需的各种支持以便他们能够参与这项观测方面，本国政府不见得会落后。我最担心的问题是出发前的准备工作不够充分，观测计划不够完善，以及各观测组之间不能密切协作。因此我恳请科学院能对如何观测才能期望得到精确的结果进行讨论给予关注。

利用金星凌日确定太阳视差的方法最早是由哈雷提出的，这种方法依赖于对太阳和行星内切的观测。这里尚有一点值得商榷，就是他在论文中说他确定的水星和太阳内切的时间可以精确到一秒以内。尽管现在的天文观测已经达到了哈雷时代无法想象的精度，但实际测量依然没能达到哈雷认为他自己曾经实际达到过的精度。最近一次（1868年11月）对水星凌日的观测，内切时间仍有几秒的误差。而且大家也都知道在对1769年6月的上一次金星凌日观测时，并没有以预期的精度确定太

* 这是在华盛顿举行的美国科学院第13次年会上西蒙·纽科姆教授所宣读文章的要点（原始文章有图示）。

sun within a second. The latest observations of a transit of Mercury, made in November 1868, are, as we shall presently see, uncertain by several seconds. It is also well known that the observations of the last transit of Venus, that of June 1769, failed to fix the solar parallax with the certainty which was looked for, the result of the standard discussion being now known to be erroneous by one-thirtieth of its entire amount. One of the first steps to carry out the object of the present paper will be an inquiry into the causes of this failure, and into the different views which have been held respecting it.

The discrepancies which have always been found in the class of observations referred to, when the results of different observers have been compared, has been generally attributed to the effect of irradiation. The phenomenon of irradiation presents itself in this form: When we view a bright body, projected upon a dark ground, the apparent contour of the bright body projects beyond its actual contour. The highest phenomenal generalisation of irradiation which I am aware of having been reached is this: A lucid point, however viewed, presents itself to the sense, not as a mathematical point, but as a surface of appreciable extent. A bright body being composed of an infinity of lucid points, its apparent enlargement is an evident result of the law just cited.

[The speaker here drew a number of diagrams for the purpose of illustrating his theory.]

The following diagrams show the effect of this law upon the time of interval contact of a planet with the disc of the sun. The planet being supposed to approach the solar disc, Fig. 1 shows the geometrical form of a portion of the apparent surface of the sun, or the phenomenon as it would be if there were no irradiation immediately before the moment of internal contact. Fig. 2 shows the corresponding appearance immediately after the contact. To indicate the effect of irradiation, or to show the phenomenon as it will actually appear on the theory of irradiation, we have only to draw an infinity of minute circles for each point of the sun's disc visible around the planet to indicate the apparent phenomenon. The effect of this is shown in Figs. 1*a* and 2*a*. The exceedingly thin thread shown in Fig. 1 is thus thickened as in 1*a*, and the sharp cusps of Fig. 2 are rounded off as shown in Fig. 2*a*. The apparent radius of the planet is diminished by an amount equal to the radius of the circle of irradiation, and the radius of the sun is increased by the same amount. Comparing Figs. 1*a* and 2*a*, it will be seen that the moment of internal contact is marked by the formation of a ligament, or "black drop," between the limbs of the sun and the planet. This formation is of so marked a character that it has been generally supposed there could be little doubt of the moment of its occurrence. The remarks of the observers have given colour to this supposition, the black drop being generally described as appearing suddenly at a definite moment.

Examining Fig. 2*a*, it will be seen that the planet still appears entirely within the disc of the sun. The geometrical circle which bounds the latter, and that which bounds the planet, instead of touching, are separated by an amount equal to double the irradiation. And, when they finally do touch, neither of them will be visible at the point of contact. The estimate of the moment of contact must therefore be very rough, the means of estimating

阳的视差，而现在知道了当时的结果有 1/30 的误差。因此实现本文目标的第一步就是研究这次失败的原因，以及关于这次失败的各种不同的观点。

在比较不同观测者的结果时，经常会发现存在差异，这通常会被认为是由一种辐射效应引起的。这种辐射效应通常表述为：在观察投射到昏暗背景上的明亮物体时，它的轮廓看起来比实际的大，我能想到的最明显的辐射效应就是一个亮点，无论怎么看，它呈现出的都不是一个数学上的点，而是有扩展的圆面。按照这个规律，一个由无限多个亮点组成的物体看起来也应该有更大的轮廓。

[报告人这时拿出几张图来解释他的理论。]

这几张图给出了按照这个规律当行星和日面内切时应呈现的效果。假如行星接近日面，图 1 是不存在辐射效应时内切前太阳表面一部分的几何图形，图 2 是刚内切后相应的图像。要表示出辐射效应，展示由辐射理论得到的实际景象，只需在行星周围的日面上的每一点画出小圆，效果如图 1*a*、2*a* 所示，图 1 中的细线在图 1*a* 中变粗了，图 2 中相对明显的切点在图 2*a* 中变得圆滑，行星的可视半径也减小到辐射圆半径的大小，而太阳的半径增加了同样多的量。比较图 1*a* 和 2*a*，可以看出内切时刻的标示是太阳和行星边缘处形成的一条切线，也就是常说的“黑滴”现象。通常认为这种“黑滴”现象具有十分明显的特征，以至于一般认为，它的发生时刻是无可置疑的，当然观测者的描述也给这个解释增添了许多色彩，他们通常认为“黑滴”是在某个确切的时刻突然出现的。

检查图 2*a* 可以发现，行星看起来仍然全在日面中，但太阳的实际轮廓与行星的并没有相切，而是在两倍辐射效应半径处分开了。当它们真正相接时，在切点处都看不见。因此对内切时间的估计成了一件相当困难的事情，估计的精度比用通常的准线测微计得到的精度低得多。在实际观测中，眼睛必须持续观测两个圆的切点，

being far less accurate than those afforded by a common filar micrometer. In the actual case the eye has to continue the two circles to the point of contact by estimation, through a distance depending on the amount of irradiation, while measures with a micrometer are made by actual contact of a wire with a disc. Such estimates have, therefore, been generally rejected by investigators, not only from their necessary inaccuracy, but because the time of "apparent contact" depends upon the amount of irradiation, which varies with the observer and the telescope. If there is no irradiation at all, the time of apparent contact and that of true contact will be the same, as shown in Fig. 2, while, when the cusps are enlarged by irradiation, apparent contact will not occur until the planet has moved through a space equal to double the irradiation.

Let us return to the phenomenon at actual contact. According to the theory as it has been presented, the formation or rupture of the black ligament connecting the dark body of Venus with the dark ground of the sky is a well-marked phenomenon, occurring at the moment of true internal contact. This was, I believe, the received theory until Wolf and André made their experiments on artificial transits in the autumn and winter of 1868 and 1869. They announced, as a result of these experiments, that the formation of the ligament was not contemporaneous with the occurrence of internal contact, but followed it at the ingress of the planet, and preceded it at egress. In other words, it appeared while the thread of light was still complete. They furthermore announced that with a good telescope the ligament did not appear at all, but the thread of light between Venus and the sun broke off by becoming indefinitely thin.

The result is not difficult to account for. Irradiation has already been described as a spreading of the light emitted from each point of the surface viewed, so that every such point appears as a small circle. The obvious effect of this spreading is a dilution of the light emitted by a luminous thread, whenever the diameter of the thread is less than that of the circle of irradiation. In consequence of this dilution, the thread may be invisible while it is really of sensible thickness, a given amount of light producing a greater effect on the eye the more it is concentrated. Since the thread of light must seem to break when it becomes invisible at its thinnest point, the formation or rupture of the thread marks, not the moment of actual contact, but the moment at which the thread of light becomes so thick as to be visible, or so thin as to be invisible. The greater the irradiation, and the worse the definition, the thicker will be the thread at this moment.

An interesting observation, illustrative of this point, was made by Liais at Rio Janciro, during the transit of Mercury of November 1, 1868. He had two telescopes, one much smaller than the other. He watched the planet in the small one till it seemed to touch the disc of the sun, then looking into the large one, he saw a thread of light distinctly between the planet and the sun, and they did not really touch until several seconds later.

而切点是通过由辐射量大小决定的一段距离来判断的，同时用测微计对日面相切进行测量。这样的目测显然不够精确，依赖于辐射度的“视觉接触时间”，也会因不同的观测者和望远镜而不同，因此这种方法不被研究者接受。如果完全没有辐射效应，视觉接触时间就会如图 2 所示和实际相切时间一致，而有了被辐射效应放大的交切点，行星运行到相当于两倍辐射效应的距离处才会出现可分辨的视觉接触。

现在来看实际的相切情况。根据上文的理论，在真正的内切发生时，金星黑色轮廓与昏暗的天空背景间黑色接点的形成或破裂是很显著的。在沃尔夫和安德烈于 1868 年和 1869 年的秋天和冬天进行一系列人造凌日实验之前，这一直是标准理论，至少我个人是这么认为的。他们的实验结果表明切线并不是在内切发生时形成的，而是在行星入凌之后或者出凌之前。换句话说，它在光线仍然完整时就出现了。他们进一步指出，这种现象在好的望远镜中根本不会出现，取而代之的只是在金星和太阳之间的光线会变得无限细而消失。

这个结果不难解释。辐射是由被观测表面上每点发出的光扩展而成的，因此每个点看起来都是个小圆。每当光线直径小于辐射圆时，这种明显的扩散效应是一种由明亮的光线发生散射而变弱的效应，正是因为这种扩散，切线在仍然很宽时就看不见了，一定量的光汇聚得越集中，眼睛观察到的这种效应就越明显。光线一定是在它最细的地方断裂，因此它的形成和断裂都不是发生在实际相切的时候，而是发生在光线粗得可以看见了，或者细得看不见了的时候。辐射效应越强，清晰度就越差，光线出现时也就越粗。

在 1868 年 11 月 1 日水星凌日时，里阿斯在里约热内卢进行了一项有趣的观测，其中就展示了这一点。他有两个望远镜，其中一个比另一个口径小很多（口径相差很多）。开始用小望远镜观测，在看到行星接触日面后立刻改用大口径的观测，他发现这时水星和太阳之间仍有光线存在，直到几秒后才相切。

Reference to the figures will make it clear that there is no generic difference between the phenomenon commonly called the rupture of the black drop and that of the formation of the thread of light. If the bright cusps are much rounded, as in Fig. 1*a*, the appearance between them is necessarily that of a drop, while if they are seen in their true sharpness, as in Fig. 1, the form of the drop will not appear. It has been shown that with different instruments the phenomenon of contact may exhibit every gradation between these extremes. The only well-defined phenomenon which all can see is the meeting of the bright cusps and the consequent formation of the thread of light at ingress, and the rupture of the thread at egress.

To recapitulate our conclusions—

1. The movement of observed internal contact at ingress is that at which the thread of light between Venus and the sun becomes thick enough to be visible.
2. The least visible thickness varies with the observer and the instrument, and, perhaps, with the state of the atmosphere.
3. The apparent initial thickness of the thread varies with the irradiation of the telescope.

Two questions are now to be discussed. The observed times varying with the observer and the instrument, we must know how wide the variation may be. If it be wide enough to render uncertain the results of observation, we shall inquire how its injurious effects may be obviated.

The first question can be decided only by comparison of the observations of different observers upon one and the same phenomenon. For such comparison I shall select the observations of the egress of Mercury on the occasion of its last transit over the disc of the sun. This selection is made for the reason that this egress was observed by a great number of experienced observers with the best instruments, while former transits, whether of Venus or Mercury, have been observed less extensively or at a time when practical astronomy was far from its present state of perfection, and that the transit in question would therefore furnish much better data of judging what we might expect in future observations. The comparison was made in the following way:—I selected from the "Astronomische Nachrichten", the "Monthly Notices of the Royal Astronomical Society", and the "Comptes Rendus", all the observations of internal contact at egress which there was reason to believe related to the breaking of the thread of light, and which were made at stations of known longitude. Each observation was then reduced to Greenwich time, and to the centre of the earth.

From these comparisons it appears that the contact was first seen by Le Verrier, at Marseilles, at two seconds before nine o'clock, Greenwich time. In one second more it was seen by Rayet at Paris, Oppolzer at Vienna, Lynn at Greenwich, and Kaiser at Leyden. The times, noted by twenty other observers, are scattered very evenly over the following fifteen seconds. Kam and Kaiser, at Leyden, did not see the contact until nineteen and twenty-four seconds past nine.

借助这些图示，我们可以清楚地知道黑滴断裂和光线形成之间没有什么不同，如果明亮的边缘像图 1*a* 中显示的那样再圆一点，它们之间就会出现黑滴一样的图像，而当它们像图 1 中的那么尖锐时，则不会有黑滴出现。就像上文提到的那样，用不同的设备可以看见这两种极端情况之间的任何阶段。唯一能够很好确定的现象就是亮切线相遇及入凌时光线的形成，以及出凌时光线的断裂。

下面总结我们的结论；

1、观测到的入凌时的内切运动是处于金星和太阳间的光线强到能够被看出的位置。

2、最低可见度随观测者和仪器的不同而有所不同，可能还受大气状况的影响。

3、光线的初始视强度与望远镜的辐射度有关。

下面讨论两个问题。既然观测到的入凌时间与观测者和观测仪器都有关，我们就必须知道；（1）这种误差的幅度到底有多大？（2）如果幅度太大，影响到观测结果的确定，那我们如何才能避免这种误差？

第一个问题只能通过比较不同观测者对同一天象的观测结果来确定。为进行这种比较，我选择了上次水星凌日时出凌的观测数据。之所以这样选择是因为；首先，之前的无论是水星还是金星凌日，要么观测范围太小，要么当时的实测天文学还没有发展到现在的高度，而相对于这些情况，这次对水星凌日的观测都有所改善，并且它是由许多经验丰富的观测人员用最先进的仪器记录下的观测结果；其次，我们所讨论的这次凌日观测也会为以后的观测结果提供参考数据。比较是这样进行的：我从《天文通报》、《皇家天文学会月刊》以及《法国科学院院刊》中找出并统计与光线断裂有关的出凌时的内切数据，这些数据的观测经度已知，所有观测时间均归算到格林尼治时间，地点归算到地球中心，之后加以比较。

比较后我们发现，位于马赛的勒威耶在格林尼治时间差 2 秒 9 点时最早看到内切，在随后的一秒内，巴黎的拉耶、维也纳的奥波尔策、格林尼治的林恩、莱顿的凯泽也都看到了，另外 20 个观测者记录的时间平均分布在接下来的 15 秒内，而莱顿的卡姆和凯泽直到 9 点 19 秒和 24 秒时才分别看到。

It thus appears that among the best observers, using the best instruments, there is a difference exceeding twenty seconds between the times of noting contact. This difference corresponds to more than a second of arc, so that really these observations were scarcely made with more accuracy than measures under favourable circumstances with a micrometer, and are not therefore to be relied on. But a great addition to the accuracy of the determination could be made by measures of the distance of the cusps, while the planet was entering upon the disc of the sun. It would tend greatly to the accuracy of the results, if the observers should meet beforehand with the telescopes they were actually to use in observing the transit and make observations in common on artificial transits. It would be a comparatively simple operation to erect an artificial representation of the sun's disc at the distance of a few hundred yards, and to have an artificial planet moved over it by clock-work. The actual time of contact could be determined by electricity, and the relative positions of the planet and the disc by actual measurement. With this apparatus it would be easy to determine the personal errors to which each observer was liable, and these errors would approximately represent those of the observations of actual transit.

Still it would be very unsafe to trust mainly to any determination of internal contact. Understanding the uncertainty of such determinations, the German astronomers have proposed to trust to measures with a heliometer, made while the planet is crossing the disc. The use of a sufficient number of heliometers would be both difficult and expensive, and I think we have an entirely satisfactory substitute in photography. Indeed, Mr. De la Rue has proposed to determine the moment of internal contact by photography. But the result would be subject to the same uncertainty which affects optical observations—the photograph which first shows contact will not be that taken when the thread of light between Venus and the sun's disc was first completed, but the first taken after it became thick enough to affect the plate, and this thickness is more variable and uncertain than the thickness necessary to affect the eye. We know very well that a haziness of the sky which very slightly diminishes the apparent brilliancy of the sun, will very materially cut off the actinic rays, and the photographic plate has not the power of adjustment which the eye has.

But, although we cannot determine contacts by photography, I conceive that we may thereby be able to measure the distance of the centres of Venus and the sun with great accuracy. Having a photograph of the sun with Venus on its disc, we can, with a suitable micrometer, fix the position of the centre of each body with great precision. We can then measure the distance of the centers in inches with corresponding precision. All we then want is the value in arc of an inch on the photograph plate. This determination is not without difficulty. It will not do to trust the measured diameters of the images of the sun, because they are affected by irradiation, just as the optical image is. If the plates were nearly of the same size, and the ratio of the diameters of Venus and the sun the same in both plates, it would be safe to assume that they were equally affected by irradiation. But should any show itself, it would not be safe to assume that the light of the sun encroached equally upon the dark ground of Venus and upon the sky, because it is so much fainter near the border.

可以看到即便是最好的观测者使用最好的观测设备，记录到的内切时间也会存在超过 20 秒的误差，这相当于一弧秒多的视差，因此这样的观测不会比在好的条件下用千分尺测得的结果更精确，也并不可靠。但是在行星进入太阳圆面时测量切线的距离会极大地提高精度，让观测者们预先用凌日时将要使用的望远镜观测人造凌日也会提高精度。比较简单的方法是在几百码远的地方竖一个人造太阳面，然后用发条装置带动人造行星从表面经过，由电子装置确定真实的相切时刻，而行星和圆面的相对位置可以通过实际测量得到。用这样一套装置能够很容易地确定每个观测人员的个人误差，这些误差大致能够代表实际观测时的误差。

但主要依赖于内切时间的测定仍是不保险的。德国天文学家们意识到这种测量的不确定性，并提出在行星划过日面时用太阳仪进行测量。但是用足够多的太阳仪进行测量会很困难，成本也会很高，我们可以用照相法代替。其实德拉鲁先生提出过用照相法确定内切时刻的方案，但由于所有的光学观测都有同样的不确定性，结果仍会受到影响。第一张显示相切图像的照片记录的其实不是金星和日面之间的光线完全消失的瞬间，而是光线暗到刚好能影响底片成像的时刻。这种暗度比眼睛感受到的更加不确定，而且大气会降低太阳的可视亮度，同样也会削弱底片上的光化学反应效应，而照相底片也不像人眼那样能够自动适应环境。

虽然无法用照相法确定内切时间，但我们依然可以以很高的精度来测量两个天体中心的距离。拍摄一张金星影像位于日面内的太阳照片，就能精确地用千分尺确定二者中心的位置，然后以合适的精度测量出两个中心之间以英寸为单位的距离。接下来就需要知道照片上的一英寸对应多少弧度。做这件事情并不容易，不能简单地相信测量出的太阳影像的直径，因为这是光学图像，会受到辐射效应的影响。如果照相底片大小相近，且它们上面的太阳和金星图像的直径比相同，则可以假定两张图片所受的辐射效应影响相同。但是不论图像看起来如何，都不应当假定日光对金星的暗背景和对天空的影响是一样的，因为太阳光在接近边缘处暗得多。

If the photographic telescope were furnished with clock-work, it would be advisable to take several photographs of the Pleiades belt, before and after the transit, to furnish an accurate standard of comparison free from the danger of systematic error. There is little doubt that if the telescopes and operators practise together, either before or after the transit, data may be obtained for a satisfactory solution of the problem in question.

To attain the object of the present paper, it is not necessary to enter into details respecting choice of stations and plans of observation. I have endeavoured to show that no valuable result is to be expected from hastily-organised and hurriedly-equipped expeditions; that every step in planning the observations requires careful consideration, and that in all the preparatory arrangements we should make haste very slowly. I make this presentation with hope that the Academy will take such action in the matter as may seem proper and desirable.

(**2**, 343-345; 1870)

如果照相用的望远镜装了钟表的机械装置，可以考虑在凌日之前和之后对昴星团天区先拍几张照片作为比对标准，从而消除系统误差。如果观测者能够在凌日前后多和望远镜进行磨合，则能够获得更令人满意的数据，从而解决以上这些问题。

我报告的目标并不是要深入到选址和制定观测计划之类的细节中，而是要尽力向大家表明，一个仓促组织、临时装备的队伍是不会取得有价值的结果的，计划中的每一步都需要认真考虑，初步准备工作也都需要慎重考虑。我这次的介绍就是希望科学院能够在此事上准备充分，不负众望。

（余恒 翻译；蒋世仰 审稿）

Address of Thomas Henry Huxley

T. H. Huxley

Editor's Note

In 1870, Thomas Henry Huxley was president of the British Association for the Advancement of Science, and used the occasion for an eloquent (and long) paper arguing that life is never generated spontaneously but only from pre-existing living things. This appears to have been a direct reply to a paper published in instalments in *Nature* on 30 June, 7 July and 14 July 1870 by H. Charlton Bastian describing experiments to observe the minute organisms that appeared in apparently sterile liquids apparently left to stand for long periods of time and from which air was not excluded. Bastian's paper had been submitted to the Royal Society but withdrawn by its author after a long delay and, apparently, accepted for publication in *Nature*. The paper occupied 26 pages of the journal.

MY Lords, Ladies, and Gentlemen,—It has long been the custom for the newly installed President of the British Association for the Advancement of Science to take advantage of the elevation of the position in which the suffrages of his colleagues had, for the time, placed him, and, casting his eyes around the horizon of the scientific world, to report to them what could be seen from his watch-tower; in what directions the multitudinous divisions of the noble army of the improvers of natural knowledge were marching; what important strongholds of the great enemy of us all, ignorance, had been recently captured; and, also, with due impartiality, to mark where the advanced posts of science had been driven in, or a long-continued siege had made no progress.

I propose to endeavour to follow this ancient precedent, in a manner suited to the limitations of my knowledge and of my capacity. I shall not presume to attempt a panoramic survey of the world of science, nor even to give a sketch of what is doing in the one great province of biology, with some portions of which my ordinary occupations render me familiar. But I shall endeavour to put before you the history of the rise and progress of a single biological doctrine; and I shall try to give some notion of the fruits, both intellectual and practical, which we owe, directly or indirectly, to the working out, by seven generations of patient and laborious investigators, of the thought which arose, more than two centuries ago, in the mind of a sagacious and observant Italian naturalist.

It is a matter of every-day experience that it is difficult to prevent many articles of food from becoming covered with mould; that fruit, sound enough to all appearance, often contains grubs at the core; that meat, left to itself in the air, is apt to putrefy and swarm with maggots. Even ordinary water, if allowed to stand in an open vessel, sooner or later becomes turbid and full of living matter.

托马斯·亨利·赫胥黎的致词

赫胥黎

编者按

1870 年，时任英国科学促进会主席的托马斯·亨利·赫胥黎借致词的机会宣读了一篇非常雄辩有力（而且很长）的论文。在文章中他指出，生命体绝不可能自发产生，而只能由先前已经存在的生物体得到。此前，查尔顿·巴斯蒂安在 1870 年 6 月 30 日、7 月 7 日以及 7 月 14 日的《自然》上分期发表了一篇描述实验结果的论文，声称在长时间放置的与空气隔绝的无菌液体中发现了很小的生物体。对于巴斯蒂安的系列论文来说，赫胥黎的这篇致词可以说是针锋相对的回应。早先，巴斯蒂安曾将其论文提交给皇家学会，不过在被搁置了很长一段时间后作者自己就将提交的论文撤销了。后来《自然》接受了巴斯蒂安的论文，该论文占据了《自然》的 26 个页面。

尊敬的各位来宾、女士们、先生们——长久以来，对于新上任的英国科学促进会主席，有一个传统是：他要利用职位的升迁这次机会，向那些为使其此次能够处于主席之位而投票的同事们作报告，报告人应放眼于科学世界，告诉大家从他的高度可以看到些什么：例如，为了丰富自然知识的贵族军队，其大部队是朝什么方向前进的；我们的大敌（愚昧无知）中有哪些重要据点最近被攻克；此外，本着公平公正的原则，他要标出何处是科学的前沿，或者在长久持续的围攻下都没有取得进展的领域。

我打算以一种适合于我有限的知识和能力的方式来尽力遵循这一古老的惯例。我不会冒昧地试图去统览整个科学世界的全貌，甚至不会去概括生物学某一重大领域里大家都在研究些什么，尽管我平时的工作使我熟悉其中的部分内容。但是我会尽力将一个生物学法则的产生与发展历史呈现在你们面前，并给出一些关于智力成果和实践成果的看法，这些都直接或间接地归功于七代孜孜不倦的研究者对两个多世纪前一位聪明睿智且具有敏锐观察力的意大利博物学家的思想的研究与解读。

通过日常经验我们知道，很难阻止许多食物表面上长出霉菌；外观十分完整的水果却经常在果核里生了虫；放在空气中的肉很容易腐烂、生满蛆虫。即使是普通的水，如果将其放在敞口的容器里，迟早会变得浑浊，并充满各种生物。

The philosophers of antiquity, interrogated as to the cause of these phenomena, were provided with a ready and a plausible answer. It did not enter their minds even to doubt that these low forms of life were generated in the matters in which they made their appearance. Lucretius, who had drunk deeper of the scientific spirit than any poet of ancient or modern times except Goethe, intends to speak as a philosopher, rather than as a poet, when he writes that "with good reason the earth has gotten the name of mother, since all things are produced out of the earth. And many living creatures, even now, spring out of the earth, taking form by the rains and the heat of the sun." The axiom of ancient science, "that the corruption of one thing is the birth of another," had its popular embodiment in the notion that a seed dies before the young plant springs from it; a belief so wide spread and so fixed, that Saint Paul appeals to it in one of the most splendid outbursts of his fervid eloquence:—

"Thou fool, that which thou sowest is not quickened, except it die."

The proposition that life may, and does, proceed from that which has no life, then, was held alike by the philosophers, the poets, and the people, of the most enlightened nations, eighteen hundred years ago; and it remained the accepted doctrine of learned and unlearned Europe, through the middle ages, down even to the seventeenth century.

It is commonly counted among the many merits of our great countryman, Harvey, that he was the first to declare the opposition of fact to venerable authority in this, as in other matters; but I can discover no justification for this wide spread notion. After careful search through the "Exercitationes de Generatione", the most that appears clear to me is, that Harvey believed all animals and plants to spring from what he terms a "*primordium vegetale*", a phrase which may nowadays be rendered "a vegetative germ;" and this, he says, is "oviforme", or "egg-like"; not, he is careful to add, that it necessarily has the shape of an egg, but because it has the constitution and nature of one. That this "*primordium oviforme*" must needs, in all cases, proceed from a living parent is nowhere expressly maintained by Harvey, though such an opinion may be thought to be implied in one or two passages; while, on the other hand, he does, more than once, use language which is consistent only with a full belief in spontaneous or equivocal generation. In fact, the main concern of Harvey's wonderful little treatise is not with generation, in the physiological sense, at all, but with development; and his great object is the establishment of the doctrine of epigenesis.

The first distinct enunciation of the hypothesis that all living matter has sprung from pre-existing living matter, came from a contemporary, though a junior, of Harvey, a native of that country, fertile in men great in all departments of human activity, which was to intellectual Europe, in the sixteenth and seventeenth centuries, what Germany is in the nineteenth. It was in Italy, and from Italian teachers, that Harvey received the most important part of his scientific education. And it was a student trained in the same schools, Francesco Redi—a man of the widest knowledge and most versatile abilities, distinguished alike as scholar, poet, physician, and naturalist—who, just two hundred and

古代的哲学家们询问起这些现象的成因时，得到的是经过精心准备的、貌似合理的答案。他们甚至不想去质疑这些低级的生物是否是从那些它们所出现的物质中产生的。卢克莱修是一位比其他任何古代以及现代诗人（除了歌德）都更充分地领会了科学精神的人。当他写下“万物都是由地球而生，所以地球具有充分的理由获得母亲这一称呼。即便现在，也有许多生物从地球涌出，然后在雨水的滋润和阳光的照耀下成长”的诗句时，他是试图以一位哲学家而非诗人的口吻说话。“一种事物的灭亡意味着另一种事物的诞生”，这一古代科学的公理自有其能够为大家广泛接受的具体化身，即种子会在幼苗萌发之前死掉。这一观念传播得如此广泛并且根深蒂固，以至于圣保罗在一篇极具爆发力的狂热雄辩中呐喊道：

“无知的人啊！除非你播种的植物死掉，否则它不会长出新的生命。”

生命可能并且确实由无生命之物产生，这一命题与 1,800 年前那些最开明国家的哲学家、诗人以及平民所持的观点相同；并且，整个中世纪，甚至直到 17 世纪，这一主张依然是整个欧洲不管是博学的人还是平民都接受的信条。

在我们的伟大同胞哈维的众多功绩之中，经常被提到的是，正如在其他一些观点上一样，他是第一个就此观点向德高望重的权威们提出相反事实的人；但是我发现这一广泛传播的观念并没有任何合理性可言。在仔细查阅了《论生物的发生》之后，我最清楚的一点就是：哈维相信所有动植物都起源于他称为“**植物原基**”的东西，这一短语现在可以被表述成“植物性胚芽”；他说这种植物性胚芽是“卵形的”；并谨慎地补充说，它不一定具有卵的形状，这样称呼只是因为它具有卵的组成和性质。无论在什么情况下，这种“**卵形原基**”肯定需要由活着的亲本来产生，这一观点尽管可能被认为曾在一两个段落里有所暗示，但是哈维并没有明确地这样宣称过；另一方面，他不止一次地表达出完全相信自然发生或者不明确发生的想法。事实上，哈维精彩小论文的主要关注点根本不是生理意义上的发生，而是发育；他的伟大目标是建立渐成论。

所有生命物质都是从已存在的生命物质中产生而来的，第一次清楚阐明这一假说的是与哈维同时代的一个年轻人。这个年轻人所在的国家在人类活动的各个方面都涌现出了非常伟大的人物，16 世纪和 17 世纪时这个国家对于开明的欧洲而言，就像是 19 世纪的德国。这个国家就是意大利，哈维就是在意大利师从意大利的教师，接受了他科学教育培养中最重要的部分。弗朗切斯科·雷迪也是同一所学校培育出来的学生。他是一个具有非常渊博的知识和极其多样的才能的人，以学者、诗人、内科

two years ago, published his "Esperienze intorno alla Generazione degl' Insetti", and gave to the world the idea, the growth of which it is my purpose to trace. Redi's book went through five editions in twenty years; and the extreme simplicity of his experiments, and the clearness of his arguments, gained for his views, and for their consequences, almost universal acceptance.

Redi did not trouble himself much with speculative considerations, but attacked particular cases of what was supposed to be "spontaneous generation" experimentally. Here are dead animals, or pieces of meat, says he; I expose them to the air in hot weather, and in a few days they swarm with maggots. You tell me that these are generated in the dead flesh; but if I put similar bodies, while quite fresh, into a jar, and tie some fine gauze over the top of the jar, not a maggot makes its appearance, while the dead substances, nevertheless, putrefy just in the same way as before. It is obvious, therefore, that the maggots are not generated by the corruption of the meat; and that the cause of their formation must be a something which is kept away by gauze. But gauze will not keep away aëriform bodies, or fluids. This something must, therefore, exist in the form of solid particles too big to get through the gauze. Nor is one long left in doubt what these solid particles are; for the blowflies, attracted by the odour of the meat, swarm round the vessel, and, urged by a powerful but in this case misleading instinct, lay eggs out of which maggots are immediately hatched upon the gauze. The conclusion, therefore, is unavoidable; the maggots are not generated by the meat, but the eggs which give rise to them are brought through the air by the flies.

These experiments seem almost childishly simple, and one wonders how it was that no one ever thought of them before. Simple as they are, however, they are worthy of the most careful study, for every piece of experimental work since done, in regard to this subject, has been shaped upon the model furnished by the Italian philosopher. As the results of his experiments were the same, however varied the nature of the materials he used, it is not wonderful that there arose in Redi's mind a presumption, that in all such cases of the seeming production of life from dead matter, the real explanation was the introduction of living germs from without into that dead matter. And thus the hypothesis that living matter always arises by the agency of pre-existing living matter, took definite shape; and had, henceforward, a right to be considered and a claim to be refuted, in each particular case, before the production of living matter in any other way could be admitted by careful reasoners. It will be necessary for me to refer to this hypothesis so frequently, that, to save circumlocution, I shall call it the hypothesis of *Biogenesis*; and I shall term the contrary doctrine—that living matter may be produced by not living matter—the hypothesis of *Abiogenesis*.

In the seventeenth century, as I have said, the latter was the dominant view, sanctioned alike by antiquity and by authority; and it is interesting to observe that Redi did not escape the customary tax upon a discoverer of having to defend himself against the charge of impugning the authority of the Scriptures; for his adversaries declared that the generation of bees from the carcase of a dead lion is affirmed, in the Book of Judges, to have been the origin of the famous riddle with which Samson perplexed the Philistines:—

医师和博物学家著称。他恰于202年前出版了《关于昆虫世代的实验》一书，向世人展示了他的观点，该观点的发展历程也是我决定要探寻的。雷迪的书在20年间出了5版；他的实验极其简单易懂，并且论证清晰，这为他的观点以及实验结果几乎赢得了普遍的认同。

雷迪并没有使自己过于被推理性因素羁绊，而是通过实验攻克了那些被认为是“自然发生”的几个特殊的案例。他说，这些案例中有死亡的动物或者碎肉块；在炎热的天气里，我把它们暴露在空气中，几天后它们就会长满蛆虫。你们告诉我这些蛆虫是在死肉中产生的；但是如果我把类似的只是非常新鲜的死肉放进一个坛子里，然后将坛子的顶部用干净的纱布扎起来，那么就不会有任何蛆虫出现，尽管死肉仍会以一种与暴露在空气中时同样的方式腐烂掉。因此很明显，蛆虫不是由肉的腐烂产生的；它们产生的原因肯定是某种被纱布隔离掉的东西。但是，纱布不能隔离掉气体和液体。因此，这里的某种东西肯定是以固体颗粒形式存在的，并且由于太大而不能通过纱布。那么，这些固体颗粒是什么？这个长久以来一直困扰着我们的问题，已经得到了解决。答案是：绿头苍蝇被肉的臭味吸引，云集在容器周围，受一种强大的但在本例中却是产生误导的本能驱使，将卵产在纱布之外，于是这些卵立刻在纱布之上被孵化成蛆虫。因此，必然的结论就是，蛆虫并非由肉产生，而是那些苍蝇通过空气带来的卵产生的。

这些实验好像简单得近乎幼稚，有人会好奇为什么以前没有人想到过这些。然而，尽管这些实验很简单，但是却值得对它们进行最认真的研究，因为从那时开始针对这一问题所做的每项实验工作，都被这位意大利哲学家设计的模型定型了。不管他使用的材料的性质如何变化，他的实验结果都一样，因此，雷迪的脑海中出现如下设想并不奇怪，他认为在所有这些好像是从死物质产生了生命的例子中，其真正的原因是将有生命的微生物引入了原本没有这些微生物的死物质当中。因此生命物质总是借助于已存在的生命物质而产生这一假说就明确成形了；于是，从此以后，在每一个特殊具体的例子中，在通过谨慎的推理能够确认有任何其他方式产生生命物质之前，大家都有权进行考虑并拒绝接受某种假说。我必须频繁提到以上的假说，为了节省迂回累赘的陈述，我称之为**生源论**假说；并且将与此相反的学说——即认为生命物质可以由非生命物质产生，称为**无生源论**假说。

正如我已说过的，无生源论假说是17世纪的主流观点，被旧势力和权威们认可；有趣的是我们发现雷迪并没有逃脱作为一个因抨击经典权威、受到控诉而进行中辩的发现者通常需要承受的压力；因为他的对手声称，死亡狮子的尸体能够产生蜜蜂，这已经在《士师记》一书中被证实是萨姆森用来难住腓力斯人的著名谜语的起源：

"Out of the eater came forth meat,
And out of the strong came forth sweetness."

Against all odds, however, Redi, strong with the strength of demonstrable fact, did splendid battle for Biogenesis; but it is remarkable that he held the doctrine in a sense which, if he had lived in these times, would have infallibly caused him to be classed among the defenders of "spontaneous generation." "Omne vivum ex vivo", "no life without antecedent life", aphoristically sums up Redi's doctrine; but he went no further. It is most remarkable evidence of the philosophic caution and impartiality of his mind, that although he had speculatively anticipated the manner in which grubs really are deposited in fruits and in the galls of plants, he deliberately admits that the evidence is insufficient to bear him out; and he therefore prefers the supposition that they are generated by a modification of the living substance of the plants themselves. Indeed, he regards these vegetable growths as organs, by means of which the plant gives rise to an animal, and looks upon this production of specific animals as the final cause of the galls and of at any rate some fruits. And he proposes to explain the occurrence of parasites within the animal body in the same way.

It is of great importance to apprehend Redi's position rightly; for the lines of thought he laid down for us are those upon which naturalists have been working ever since. Clearly, he held *Biogenesis* as against *Abiogenesis*; and I shall immediately proceed, in the first place, to inquire how far subsequent investigation has borne him out in so doing.

But Redi also thought that there were two modes of Biogenesis. By the one method, which is that of common and ordinary occurrence, the living parent gives rise to offspring which passes through the same cycle of changes as itself—like gives rise to like; and this has been termed *Homogenesis*. By the other mode the living parent was supposed to give rise to offspring which passed through a totally different series of states from those exhibited by the parent, and did not return into the cycle of the parent; this is what ought to be called *Heterogenesis*, the offspring being altogether, and permanently unlike the parent. The term Heterogenesis, however, has unfortunately been used in a different sense, and M. Milne-Edwards has therefore substituted for it *Xenogenesis*, which means the generation of something foreign. After discussing Redi's hypothesis of universal Biogenesis, then, I shall go on to ask how far the growth of science justifies his other hypothesis of Xenogenesis.

The progress of the hypothesis of Biogenesis was triumphant and unchecked for nearly a century. The application of the microscope to anatomy in the hands of Grew, Lecuwenhoek, Swammerdam, Lyonet, Vallisnieri, Réaumur, and other illustrious investigators of nature of that day, displayed such a complexity of organisation in the lowest and minutest forms, and everywhere revealed such a prodigality of provision for their multiplication by germs of one sort or another, that the hypothesis of Abiogenesis began to appear not only untrue, but absurd; and, in the middle of the eighteenth century, when Needham and Buffon took up the question, it was almost universally discredited.

“吃的从吃者出来，
甜的从强者出来。”

尽管如此，雷迪还是凭借可论证的事实的力量，为生源论进行了精彩的斗争；但是值得注意的是，如果他生活在这种时代，那么从某种意义上说，他坚持的学说将肯定会使他被归为“自然发生学说”的拥护者。“生命源于生命”，“没有先前的生命就不会有新的生命产生”，这些以警句的形式概括了雷迪的学说；但是他没有更进一步。以下证据最显著地证明了他头脑中对待自然科学的谨慎和公正：尽管他曾经推测性地预见了存在于水果以及植物虫瘿中的幼虫产生的方式，但是他谨慎地承认那些证据不足以证实自己的观点，因此他更倾向于推测这些幼虫是通过对植物自身的生命物质加以修饰而产生的。实际上，他将这些植物的生长视为器官的生长，植物借助于器官产生动物，并且将这样产生的特殊动物看成是瘤而且至少是某些果实产生的最终根源。他提议以同样方式解释动物体内寄生虫的出现。

正确理解雷迪的观点具有重要的意义，因为从那以后博物学家一直依据他为我们呈现的思想脉络进行研究工作。很明显，他坚持**生源论**，反对**无生源论**。首先，我会立即着手考查后续的研究能在多大程度上证实他的想法。

但是雷迪也认为有两种生源论模式。一种模式是普通平常的发生方法，活着的亲本产生后代，后代经历同亲本一样的变化周期，即同类型产生同类型，这被称为**同型生殖**。另一种模式认为活着的亲本产生的后代经历与亲本表现出的完全不同的一系列状态，并不会回到亲本经历的周期中去，其后代全然、永远地与亲本不同，这应该被称为**异型生殖**。然而，不幸的是，异型生殖一词已经在一个不同的意义下被使用了，因此米尔恩·爱德华兹用**异源发生**来代替异型生殖，其意思是外源事物的产生。在讨论了雷迪的普遍生源论假说之后，我想再问一个问题，那就是科学的发展能在多大程度上证明其另一个假说——异源发生假说的合理性？

生源论假说在近一个世纪里的发展是顺利的，没有受到任何遏制。格鲁、列文虎克、斯瓦默丹、莱尔尼特、瓦利斯涅里、雷奥米尔及其他同时代的著名自然科学研究者将手中的显微镜应用到解剖学，展示了最低级、最微小形式的生物体组织结构竟是如此复杂，而且在所有例子中都揭示了通过某种微生物进行繁殖的规律，以致无生源论假说开始变得不仅仅不正确，而且很荒谬。18世纪中期，当尼达姆和布丰着手研究这一问题时，无生源论假说几乎已经被普遍放弃了。

But the skill of the microscope-makers of the eighteenth century soon reached its limit. A microscope magnifying 400 diameters was a *chef d'oeuvre* of the opticians of that day; and at the same time, by no means trustworthy. But a magnifying power of 400 diameters, even when definition reaches the exquisite perfection of our modern achromatic lenses, hardly suffices for the mere discernment of the smallest forms of life. A speck, only 1/25 of an inch in diameter, has, at 10 inches from the eye, the same apparent size as an object 1/10,000 of an inch in diameter, when magnified 400 times; but forms of living matter abound, the diameter of which is not more than 1/40,000 of an inch. A filtered infusion of hay, allowed to stand for two days, will swarm with living things, among which, any which reaches the diameter of a human red blood-corpuscle, or about 1/3,200 of an inch, is a giant. It is only by bearing these facts in mind, that we can deal fairly with the remarkable statements and speculations put forward by Buffon and Needham in the middle of the eighteenth century.

When a portion of any animal or vegetable body is infused in water, it gradually softens and disintegrates; and, as it does so, the water is found to swarm with minute active creatures, the so-called Infusorial Animalcules, none of which can be seen, except by the aid of the microscope; while a large proportion belong to the category of smallest things of which I have spoken, and which must have all looked like mere dots and lines under the ordinary microscopes of the eighteenth century.

Led by various theoretical considerations which I cannot now discuss, but which looked promising enough in the lights of that day, Buffon and Needham doubted the applicability of Redi's hypothesis to the infusorial animalcules, and Needham very properly endeavoured to put the question to an experimental test. He said to himself, if these infusorial animalcules come from germs, their germs must exist either in the substance infused, or in the water with which the infusion is made, or in the superjacent air. Now the vitality of all germs is destroyed by heat. Therefore, if I boil the infusion, cork it up carefully, cementing the cork over with mastic, and then heat the whole vessel by heaping hot ashes over it, I must needs kill whatever germs are present. Consequently, if Redi's hypothesis hold good, when the infusion is taken away and allowed to cool, no animalcules ought to be developed in it; whereas, if the animalcules are not dependent on pre-existing germs, but are generated from the infused substance, they ought, by-and-by, to make their appearance. Needham found that, under the circumstances in which he made his experiments, animalcules always did arise in the infusions, when a sufficient time had elapsed to allow for their development.

In much of his work Needham was associated with Buffon, and the results of their experiments fitted in admirably with the great French naturalist's hypothesis of "organic molecules", according to which, life is the indefeasible property of certain indestructible molecules of matter, which exist in all living things, and have inherent activities by which they are distinguished from not living matter. Each individual living organism is formed by their temporary combination. They stand to it in the relation of the particles of water to a cascade, or a whirlpool; or to a mould, into which the water is poured. The form of the

但是18世纪时，显微镜制造技术达到了极限。一台能够放大400倍的显微镜就算是当时光学仪器商的杰作了；与此同时，绝对不可以完全信赖这种仪器。但是即使显微镜的分辨率达到了我们现代的消色差透镜的敏锐精细程度，放大400倍的能力几乎连仅仅想辨别最微小的生命形式的要求都满足不了。一个直径仅仅1/25英寸的灰尘，在距离观察者的眼睛10英寸时，与将一个直径是1/10,000英寸的物体放大400倍，具有相同的表观尺寸；但是生物的形式丰富多样，很多生物的直径都不足1/40,000英寸。放置了两天的干草过滤浸液会长满有生命的物质，其中任何达到人类血红细胞直径大小或者大约1/3,200英寸大小的物质都算得上是庞然大物了。只有记住这些事实，我们才能公平地对待18世纪中期由布丰和尼达姆提出的引人注目的言论和推测。

将任何动物体或植物体的一部分浸泡在水中时，这部分就会逐渐软化、分解；此时，会发现水中充满微小的活动着的生物，即所谓的藻类微生物，除非借助于显微镜，否则仅仅依靠肉眼是看不到它们的；它们大部分属于我已经讲过的最小生物一类，用18世纪的普通显微镜来观察的话，它们看起来肯定都不过是些小点和线状物。

在一些这里我不会讨论但是当时看起来非常有前景的各种理论思考的引领下，布丰和尼达姆对雷迪的假说是否适用于藻类微生物表示怀疑，并且尼达姆尽力用一个非常恰当的实验来检验这一问题。他对自己说，如果这些藻类微生物来源于细菌，那么这些细菌肯定存在于浸泡的物质当中，或者存在于制取浸液用的水中，或者存在于上方的空气当中。而所有细菌的生命力都可以用热来摧毁。因此，如果我把浸液煮沸，将它仔细地用软木塞塞住，并用树脂封好软木塞，然后通过在容器上堆积热灰来加热整个容器，势必会杀死所有存在的细菌。于是，如果雷迪的假说是正确的话，那么当取走浸液、冷却下来之后，浸液中应该不会再出现新的微生物；反过来，如果微生物并不依赖于已存在的细菌，而是从浸泡的物质中产生的话，那么不久之后它们就应该会出现。尼达姆发现，在他的实验环境下，当经过足够满足微生物的发育要求的时间后，浸液中确实总会出现微生物。

尼达姆的许多工作都与布丰有关，他们的实验结果与法国著名博物学家的“有机分子”假说非常吻合。根据该假说，生命是某些不可毁灭的分子物质的固有属性，存在于所有生命体当中，具有内在的区别于非生命物质的活动。每个有生命的生物个体都由临时的组合构成，如同水滴与小瀑布、或者水滴与漩涡，或者水与注入了水的模子的关系，生物体坚守着自己的临时组合。因此，生物的形式是由外界条件和构成它的有机分子的内在活动之间的相互作用共同决定的。如同漩涡的中断

organism is thus determined by the reaction between external conditions and the inherent activities of the organic molecules of which it is composed; and, as the stoppage of a whirlpool destroys nothing but a form, and leaves the molecules of the water, with all their inherent activities intact, so what we call the death and putrefaction of an animal, or of a plant, is merely the breaking up of the form, or manner of association, of its constituent organic molecules, which are then set free as infusorial animalcules.

It will be perceived that this doctrine is by no means identical with *Abiogenesis*, with which it is often confounded. On this hypothesis, a piece of beef, or a handful of hay, is dead only in a limited sense. The beef is dead ox, and the hay is dead grass; but the "organic molecules" of the beef or the hay are not dead, but are ready to manifest their vitality as soon as the bovine or herbaceous shrouds in which they are imprisoned are rent by the macerating action of water. The hypothesis therefore must be classified under Xenogenesis, rather than under Abiogenesis. Such as it was, I think it will appear, to those who will be just enough to remember that it was propounded before the birth of modern chemistry, and of the modern optical arts, to be a most ingenious and suggestive speculation.

But the great tragedy of Science—the slaying of a beautiful hypothesis by an ugly fact—which is so constantly being enacted under the eyes of philosophers, was played, almost immediately, for the benefit of Buffon and Needham.

Once more, an Italian, the Abbé Spallanzani, a worthy successor and representative of Redi in his acuteness, his ingenuity, and his learning, subjected the experiments and the conclusions of Needham to a searching criticism. It might be true that Needham's experiments yielded results such as he had described, but did they bear out his arguments? Was it not possible, in the first place, that he had not completely excluded the air by his corks and mastic? And was it not possible, in the second place, that he had not sufficiently heated his infusions and the superjacent air? Spallanzani joined issue with the English naturalist on both these pleas, and he showed that if, in the first place, the glass vessels in which the infusions were contained were hermetically sealed by fusing their necks, and if, in the second place, they were exposed to the temperature of boiling water for three-quarters of an hour, no animalcules ever made their appearance within them. It must be admitted that the experiments and arguments of Spallanzani furnish a complete and a crushing reply to those of Needham. But we all too often forget that it is one thing to refute a proposition, and another to prove the truth of a doctrine which, implicitly or explicitly, contradicts that proposition, and the advance of science soon showed that though Needham might be quite wrong, it did not follow that Spallanzani was quite right.

Modern chemistry, the birth of the latter half of the eighteenth century, grew apace, and soon found herself face to face with the great problems which biology had vainly tried to attack without her help. The discovery of oxygen led to the laying of the foundations of a scientific theory of respiration, and to an examination of the marvellous interactions of organic substances with oxygen. The presence of free oxygen appeared to be one of

只是一种形式的中止，并不会损坏任何东西，水分子中所有的内在活动都是完好的，我们所说的动物或植物的死亡和腐烂只是构成它们的有机分子形式的分解或者结合方式的破裂，它们随后都会以藻类微生物的形式释放出来。

经常会有人将这一学说与**无生源论**混淆，实际上两者是截然不同的。根据这一假说，一块牛肉或者一把干草只是在有限的意义上是死亡的。牛肉是死亡的牛，而干草是死亡的草；但是牛肉或干草中的“有机分子”并没有死，而是随时准备好，一旦其受到的束缚被水的浸离作用破坏掉，它们就会展现自己的生命力。因此该假说肯定应该被归入异源发生，而非无生源论。尽管这一假说可能不太准确，但对于那些刚好能记住它是在现代化学及现代光学技术诞生之前就被提出来了的人而言，我认为这会是一个非常具有独创性和启示性的推测。

但是为了布丰和尼达姆的利益，科学的巨大灾难——使用一个丑陋的事实来残杀一个美丽的假说——几乎立即上演，这种灾难非常频繁地在哲学家的眼皮底下上演。

再一次地，一个意大利人，阿贝·斯帕兰扎尼，一个值得尊敬的雷迪的继承者和代表者，以其独有的敏锐性、独创性和学识，将尼达姆的实验和结论置于一个有待于评论的境地。也许尼达姆的实验结果真如他自己描述的那样，但是它们证实他的论点了吗？首先，有没有可能通过软木塞和树脂他并没有完全排除空气呢？其次，有没有可能他没有对浸液和上面的空气进行足够的加热呢？斯帕兰扎尼在这两个问题上与英国的博物学家存在争议。他指出，如果首先将盛有浸液的玻璃器皿通过热熔其颈部来达到密封的目的，然后再将它们暴露于沸水中达 45 分钟的话，那么浸液中就不会再有微生物出现。必须承认斯帕兰扎尼的实验和论证对尼达姆的实验和论证给予了彻底的、粉碎性的反击。但是我们也经常忘记驳斥一个命题是一件事，含蓄或明确地证明一个与此命题相矛盾的另一学说的正确性又是另一回事。科学的进步不久就表明，尽管尼达姆可能是非常错误的，但并不表示斯帕兰扎尼就是非常正确的。

现代化学于 18 世纪下半叶诞生，并且发展迅速，它很快就发现自己面对着巨大的难题，这些难题都是生物学家曾经试图在没有化学的帮助下进行攻克却徒劳无获的。氧气的发现奠定了呼吸这一科学理论的基础，引发了对有机物质与氧气之间奇妙的相互作用的研究。自由氧的存在似乎是生命存在的条件之一，也是有机物发生

the conditions of the existence of life, and of those singular changes in organic matters which are known as fermentation and putrefaction. The question of the generation of the infusory animalcules thus passed into a new phase. For what might not have happened to the organic matter of the infusions, or to the oxygen of the air, in Spallanzani's experiments? What security was there that the development of life which ought to have taken place had not been checked or prevented by these changes?

The battle had to be fought again. It was needful to repeat the experiments under conditions which would make sure that neither the oxygen of the air, nor the composition of the organic matter, was altered in such a matter as to interfere with the existence of life.

Schulze and Schwann took up the question from this point of view in 1836 and 1837. The passage of air through red-hot glass tubes, or through strong sulphuric acid, does not alter the proportion of its oxygen, while it must needs arrest or destroy any organic matter which may be contained in the air. These experimenters, therefore, contrived arrangements by which the only air which should come into contact with a boiled infusion should be such as had either passed through red-hot tubes or through strong sulphuric acid. The result which they obtained was that an infusion so treated developed no living things, while if the same infusion was afterwards exposed to the air such things appeared rapidly and abundantly. The accuracy of these experiments has been alternately denied and affirmed. Supposing them to be accepted, however, all that they really proved was that the treatment to which the air was subjected destroyed *something* that was essential to the development of life in the infusion. This "something" might be gaseous, fluid, or solid; that it consisted of germs remained only an hypothesis of greater or less probability.

Contemporaneously with these investigations a remarkable discovery was made by Cagniard de la Tour. He found that common yeast is composed of a vast accumulation of minute plants. The fermentation of must or of wort in the fabrication of wine and of beer is always accompanied by the rapid growth and multiplication of these *Torulae*. Thus fermentation, in so far as it was accompanied by the development of microscopical organisms in enormous numbers, became assimilated to the decomposition of an infusion of ordinary animal or vegetable matter; and it was an obvious suggestion that the organisms were, in some way or other, the causes both of fermentation and of putrefaction. The chemists, with Berzelius and Liebig at their head, at first laughed this idea to scorn; but in 1843, a man then very young, who has since performed the unexampled feat of attaining to high eminence alike in Mathematics, Physics, and Physiology—I speak of the illustrious Helmholtz—reduced the matter to the test of experiment by a method alike elegant and conclusive. Helmholtz separated a putrefying or a fermenting liquid from one which was simply putrescible or fermentable by a membrane which allowed the fluids to pass through and become intermixed, but stopped the passage of solids. The result was, that while the putrescible or the fermentable liquids became impregnated with the results of the putrescence or fermentation which was going on on the other side of the membrane, they neither putrefied (in the ordinary way) nor fermented; nor were any of the organisms which abounded in the fermenting or putrefying liquid generated in them.

发酵和腐烂等异常变化的条件之一。因此藻类微生物如何发生的问题进入了一个新的阶段。在斯帕兰扎尼的实验中，为什么浸液中的有机物质或者空气中的氧气似乎并没有发生什么变化呢？斯帕兰扎尼实验中的变化提供了什么安全措施使得本应发生的生命发育过程没有被检测到或者被阻止了呢？

这场战斗不得不再打一次。有必要再在确保空气中的氧气和有机物质的构成成分不改变，并且没有干扰生命存在的物质的情况下，再次重复这些实验。

舒尔策和施旺在1836年和1837年开始从这一角度出发进行研究。通过灼热的玻璃管或通过浓硫酸之后，空气中所含氧气的比例都不会改变，但是这样肯定会抑制或者破坏空气中可能含有的任何有机物质。因此，这些实验被设计成，只有那些通过了炽热的玻璃管或者浓硫酸的空气才会与沸腾的浸液相接触。他们得到的结果是，这样处理的浸液并没有产生任何有生命的物质，而如果过后将相同的浸液暴露在空气中，那么很快就会产生大量的有生命的物质。这些实验的准确性时而被否决，时而被证实。然而，假设被接受了，那么这些实验真正证明的就是，对空气的处理毁掉了对于浸液中生命发育必不可少的**某种物质**。这里所说的“某种物质”可能是气态的、液态的或者固态的；认为其中含有微生物还只是一个有一定可能性的假说。

在得到这些研究结果的同时，卡尼亚尔·德拉图尔取得了一项引人注目的发现。他发现普通酵母菌由大量微小的植物积累构成。在酿造葡萄酒和啤酒的过程中，葡萄汁或麦芽汁的发酵总是伴随着这些**圆酵母**的快速生长和增殖。因此，伴随着大量微生物的发育过程的发酵作用，类似于普通动物性物质或植物性物质的浸液的腐烂过程；而且在某种意义上，这明确提示了是微生物引起了发酵和腐烂。以伯齐利厄斯和李比希为首的化学家们首先嘲笑了这个想法并表示不屑；但是在1843年，一个当时还非常年轻的人通过精巧而又令人信服的方法来减少检测实验中的物质。我说的这个年轻人就是著名的亥姆霍兹，他曾在数学、物理学和生理学领域都作出了极其卓越的无可比拟的业绩。亥姆霍兹使用了一种薄膜将正在腐烂的或者正在发酵的液体与仅仅可能会腐烂或可能会发酵的液体分开，该薄膜允许液体通过并发生混合，但是可以阻止固体通过。结果是，当膜的另一侧的正在腐烂或者发酵的产物渗入到可能会腐烂或者可能会发酵的液体中时，它们既不以通常的方式腐烂也不发酵；并且它们中也不会产生任何充满于发酵液或者腐烂液中的微生物。因此这些微生物发育的原因肯定在于某种不能通过膜的物质；由于亥姆霍兹的研究比格雷姆对胶体进行的研究早很多，所以他自然而然地得出结论，认为被拦截的媒介肯定是固体物质。

Therefore the cause of the development of these organisms must lie in something which cannot pass through membranes; and as Helmholtz's investigations were long antecedent to Graham's researches upon colloids, his natural conclusion was that the agent thus intercepted must be a solid material. In point of fact, Helmholtz's experiments narrowed the issue to this: that which excites fermentation and putrefaction, and at the same time gives rise to living forms in a fermentable or putrescible fluid, is not a gas and is not a diffusible fluid; therefore it is either a colloid, or it is matter divided into very minute solid particles.

The researches of Schroeder and Dusch in 1854, and of Schroeder alone, in 1859, cleared up this point by experiments which are simply refinements upon those of Redi. A lump of cotton-wool is, physically speaking, a pile of many thicknesses of a very fine gauze, the fineness of the meshes of which depends upon the closeness of the compression of the wool. Now, Schroeder and Dusch found, that, in the case of all the putrefiable materials which they used (except milk and yolk of egg), an infusion boiled, and then allowed to come into contact with no air but such as had been filtered through cotton-wool, neither putrefied nor fermented, nor developed living forms. It is hard to imagine what the fine sieve formed by the cotton-wool could have stopped except minute solid particles. Still the evidence was incomplete until it had been positively shown, first, that ordinary air does contain such particles; and, secondly, that filtration through cotton-wool arrests these particles and allows only physically pure air to pass. This demonstration has been furnished within the last year by the remarkable experiments of Professor Tyndall. It has been a common objection of Abiogenists that, if the doctrine of Biogeny is true, the air must be thick with germs; and they regard this as the height of absurdity. But Nature occasionally is exceedingly unreasonable, and Professor Tyndall has proved that this particular absurdity may nevertheless be a reality. He has demonstrated that ordinary air is no better than a sort of stirabout of excessively minute solid particles; that these particles are almost wholly destructible by heat; and that they are strained off, and the air rendered optically pure by being passed through cotton-wool.

But it remains yet in the order of logic, though not of history, to show that among these solid destructible particles there really do exist germs capable of giving rise to the development of living forms in suitable menstrua. This piece of work was done by M. Pasteur in those beautiful researches which will ever render his name famous; and which, in spite of all attacks upon them, appear to me now, as they did seven years ago, to be models of accurate experimentation and logical reasoning. He strained air through cotton-wool, and found, as Schroeder and Dusch had done, that it contained nothing competent to give rise to the development of life in fluids highly fitted for that purpose. But the important further links in the chain of evidence added by Pasteur are three. In the first place he subjected to microscopic examination the cotton-wool which had served as strainer, and found that sundry bodies clearly recognizable as germs, were among the solid particles strained off. Secondly, he proved that these germs were competent to give rise to living forms by simply sowing them in a solution fitted for their development. And, thirdly, he showed that the incapacity of air strained through cotton-wool to give rise

事实上，亥姆霍兹的实验将问题缩小到了下述范围：引起发酵和腐烂作用的、同时在可发酵或可腐烂液体中产生生命的既不是气体也不是扩散性的液体；因此这种物质要么是胶体，要么是一种被分成非常微小的固体颗粒的物质。

施罗德和杜施在1854年进行的研究以及施罗德自己在1859年进行的研究都通过实验澄清了这个问题，这些实验都只是在雷迪实验的基础上进行了简单的改良。从物理学角度而言，一块脱脂棉就是一堆不同厚度的纱布，纱布的网眼细度取决于毛料压缩的密实度。现在，施罗德和杜施发现，在他们使用的所有会腐烂的材料中（除了牛奶和蛋黄），如果煮沸后的浸液不与空气接触，而且用脱脂棉进行过滤，那么也是既不会腐烂，也不会发酵，且没有任何生命形式发育出来。很难想象脱脂棉形成的精细筛子除了阻止微小的固体颗粒外还阻止了别的什么东西。证据要得以完整，需要确定以下两点：第一，普通空气中确实含有这种颗粒；第二，脱脂棉的过滤作用阻止了这些颗粒的通过，而只允许物理意义上纯净的空气通过。廷德尔教授已于去年通过自己的著名实验为此提供了实证。这招来了无生源论者的普遍异议，即如果生源论法则是正确的，那么空气中肯定充满了细菌；他们认为这荒唐透顶。但是自然界有时就是非常不合常理，廷德尔教授已经证明了这个特别的谬论确实是真实的。他证明了普通空气几乎跟一种含有大量微小固体颗粒的稀饭一样；通过加热，可以使这些颗粒几乎完全被破坏掉；也可以通过过滤除掉它们，所以使空气通过脱脂棉后，从视觉上看它就变得纯净了。

尽管根据历史记录细菌已经被除掉了，但是依逻辑学次序考虑，仍然表明在这些固态可破坏颗粒中，确实存在能够在适当溶剂下引发生命发育的细菌。巴斯德通过一些完美的研究完成了这一工作，并使他的名字为众人所知。尽管存在各种各样的攻击，但是正如七年前这些研究给我的感觉那样，我现在仍觉得它们是准确的实验和逻辑推理的典范。巴斯德使空气通过脱脂棉从而被过滤，结果跟施罗德和杜施的发现一样，即这样处理过的空气不含有那些特别的液体中所含有的任何能够产生生命发育的物质。但是巴斯德进一步补充了这一证据链中的3个重要环节。首先他借助显微镜检测了作为过滤器的脱脂棉，发现被过滤掉的固体颗粒中，有各种可以被明确识别为细菌的物体。其次，他证明了把这些微生物散播到适合它们发育的溶液中能够产生生命。再次，他证明了通过脱脂棉过滤的空气不能产生生命，并不是由于脱脂棉本身影响空气组分发生了任何超自然变化。如果完全不用脱脂棉，那么

to life, was not due to any occult change effected in constituents of the air by the wool, by proving that the cotton-wool might be dispensed with altogether, and perfectly free access left between the exterior air and that in the experimental flask. If the neck of the flask is drawn out into a tube and bent downwards; and if, after the contained fluid has been carefully boiled, the tube is heated sufficiently to destroy any germs which may be present in the air which enters as the fluid cools, the apparatus may be left to itself for any time and no life will appear in the fluid. The reason is plain. Although there is free communication between the atmosphere laden with germs and the germless air in the flask, contact between the two takes place only in the tube; and as the germs cannot fall upwards, and there are no currents, they never reach the interior of the flask. But if the tube be broken short off where it proceeds from the flask, and free access be thus given to germs falling vertically out of the air, the fluid which has remained clear and desert for months, becomes, in a few days turbid and full of life.

These experiments have been repeated over and over again by independent observers with entire success; and there is one very simple mode of seeing the facts for oneself, which I may as well describe.

Prepare a solution (much used by M. Pasteur, and often called "Pasteur's solution") composed of water with tartrate of ammonia, sugar, and yeast-ash dissolved therein. Divide it into three portions in as many flasks; boil all three for a quarter of an hour; and, while the steam is passing out, stop the neck of one with a large plug of cotton-wool, so that this also may be thoroughly steamed. Now set the flasks aside to cool, and when their contents are cold, add to one of the open ones a drop of filtered infusion of hay which has stood for twenty-four hours, and is consequently full of the active and excessively minute organisms known as *Bacteria*. In a couple of days of ordinary warm weather the contents of this flask will be milky from the enormous multiplication of *Bacteria*. The other flask, open and exposed to the air, will, sooner or later, become milky with *Bacteria*, and patches of mould may appear in it; while the liquid in the flask, the neck of which is plugged with cotton-wool, will remain clear for an indefinite time. I have sought in vain for any explanation of these facts, except the obvious one, that the air contains germs competent to give rise to *Bacteria*, such as those with which the first solution has been knowingly and purposely inoculated, and to the mould-*Fungi*. And I have not yet been able to meet with any advocate of Abiogenesis who seriously maintains that the atoms of sugar, tartrate of ammonia, yeast-ash, and water, under no influence but that of free access of air and the ordinary temperature, rearrange themselves and give rise to the protoplasm of *Bacterium*. But the alternative is to admit that these *Bacteria* arise from germs in the air; and if they are thus propagated, the burden of proof that other like forms are generated in a different manner, must rest with the assertor of that proposition.

To sum up the effect of this long chain of evidence: —

It is demonstrable that a fluid eminently fit for the development of the lowest forms of life, but which contains neither germs, nor any protein compound, gives rise to living things

就在外界空气与实验中所用的长颈瓶中的空气之间形成了完全自由的通道。如果将长颈瓶的颈部拉长成一个管状并向下弯曲，然后将其中盛入的液体小心煮沸，之后如果将管子也充分加热以破坏当液体冷却时进入管子的空气中的所有细菌的话，那么无论将装置放置多久，其中的液体都不会产生任何生命。原因很简单。尽管充满着微生物的大气和长颈瓶中的无菌空气间存在自由的流通，但是这两个地方的空气只能在管子里发生接触。因为细菌不能朝上落，也没有气流，所以它们永远都到达不了长颈瓶的内部。但是如果从长颈瓶延伸处将管子截短，这样外界空气中的细菌就可以自由地垂直下降，那么原本放置数月仍然保持清澈且无生命的液体在几天内就会变浑并且充满生命。

这些实验已经由不同的观察者独立重复了一遍又一遍，并且都取得了成功；有一套非常简单的，可以亲自观察这些事实的模式，这里我不妨也描述一下。

准备一份溶有酒石酸铵、糖和酵母粉的水溶液。（由于这是巴斯德先生经常采用的溶液，所以通常被称为“巴斯德溶液”。）将该溶液分成三等份分别盛入 3 个长颈瓶中；将 3 瓶溶液都煮沸一刻钟；当看见有蒸汽冒出时，将其中一个长颈瓶的颈部用一大块脱脂棉堵住，这样这一瓶就可以被彻底蒸透了。现在将旁边的长颈瓶放置至冷却，当其中的溶液冷却后，向开口的长颈瓶中加入一滴已放置了 24 小时的过滤了的干草浸液，因而这瓶溶液中会充满大量活动的微小生物，即通常所说的细菌。在经历几天普通的温暖天气后，这只长颈瓶中的内含物将会因细菌的大量繁殖而变成乳状。另一个开口并暴露于空气中的长颈瓶迟早也会长满细菌而变成乳状，也可能出现多片霉菌；而颈部被脱脂棉塞住的长颈瓶中的液体无论放置多长时间都是澄清的。我曾经试图解释这些现象，但是除了空气中含有能够产生细菌（比如故意向第一种溶液中接种的那些）和霉菌的微生物这个明显的原因外，别无所获。无生源论的支持者坚持认为，在没有受到任何影响、只是在空气的自由流通和普通温度下，糖、酒石酸铵、酵母粉和水的原子发生了重排，从而产生了细菌的原生质。对于他们的观点，我到目前都不能苟同。但是另一种观点承认这些细菌是由空气中的微生物产生的；如果它们是这样繁殖的，那么举证责任，即举出其他类似的形式是以不同方式产生的，就必须得靠那一命题的主张者了。

这一长串证据的影响可以概括如下：

可以证实：一种不含有微生物和任何蛋白质复合物的，特别适合于最低级形式的生命发育的液体，如果暴露于普通空气中就可以产生大量的生物；而如果通过机

in great abundance if it is exposed to ordinary air, while no such development takes place if the air with which it is in contact is mechanically freed from the solid particles which ordinarily float in it and which may be made visible by appropriate means.

It is demonstrable that the great majority of these particles are destructible by heat, and that some of them are germs or living particles capable of giving rise to the same forms of life as those which appear when the fluid is exposed to unpurified air.

It is demonstrable that inoculation of the experimental fluid with a drop of liquid known to contain living particles gives rise to the same phenomena as exposure to unpurified air.

And it is further certain that these living particles are so minute that the assumption of their suspension in ordinary air presents not the slightest difficulty. On the contrary, considering their lightness and the wide diffusion of the organisms which produce them, it is impossible to conceive that they should not be suspended in the atmosphere in myriads.

Thus the evidence, direct and indirect, in favour of *Biogenesis* for all known forms of life must, I think, be admitted to be of great weight.

On the other side the sole assertions worthy of attention are that hermetically sealed fluids, which have been exposed to great and long-continued heat, have sometimes exhibited living forms of low organization when they have been opened.

The first reply that suggests itself is the probability that there must be some error about these experiments, because they are performed on an enormous scale every day with quite contrary results. Meat, fruits, vegetables, the very materials of the most fermentable and putrescible infusions are preserved to the extent, I suppose I may say, of thousands of tons every year, by a method which is a mere application of Spallanzani's experiment. The matters to be preserved are well boiled in a tin case provided with a small hole, and this hole is soldered up when all the air in the case has been replaced by steam. By this method they may be kept for years without putrefying, fermenting, or getting mouldy. Now this is not because oxygen is excluded, inasmuch as it is now proved that free oxygen is not necessary for either fermentation or putrefaction. It is not because the tins are exhausted of air, for *Vibriones* and *Bacteria* live, as Pasteur has shown, without air or free oxygen. It is not because the boiled meats or vegetables are not putrescible or fermentable, as those who have had the misfortune to be in a ship supplied with unskilfully closed tins well know. What is it, therefore, but the exclusion of germs? I think that Abiogenists are bound to answer this question before they ask us to consider new experiments of precisely the same order.

And in the next place, if the results of the experiments I refer to are really trustworthy, it by no means follows that Abiogenesis has taken place. The resistance of living matter to heat is known to vary within considerable limits, and to depend, to some extent, upon

械方法除去与液体接触的空气中的固体颗粒（这些固体颗粒通常悬浮于空气中，并且可以通过适当的方式使其可见）的话，那么生物便不会产生。

这些颗粒中的绝大部分可以通过加热破坏掉，其中有些颗粒是微生物或者有生命的颗粒，它们能够产生的生命形式与液体暴露于未净化的空气中时产生的生命形式相同，这一点也是可以证实的。

将已知含有有生命颗粒的一滴液体接种到实验液体中，能够产生与暴露于未净化空气中同样的现象，这一点同样是可以证实的。

此外，可以进一步肯定的是，这些有生命的颗粒太小了，因此假设它们悬浮于普通空气中时，不存在任何困难。相反，考虑到它们的轻微以及产生它们的有机体的广泛扩散，必然能想到它们会以极大的数量悬浮于大气之中。

因此我认为必须承认，那些支持对于所有已知生命形式都成立的**生源论**的直接和间接证据，是非常重要的。

另一方面，值得注意的独特论断是，对密封液体进行长时间、充分的加热后，有时也会产生低级的生命形式。

对于此论断，第一个回复是这些实验本身可能有某种误差，因为这些实验每天都在很大的范围内进行着，并且得到了相当矛盾的结果。我估计，每年光是运用斯帕兰扎尼的实验方法腌制的肉、水果和蔬菜，以及最易发酵和腐烂的浸液材料就有成千上万吨。将要腌制的物质放在一个有小孔的铁罐里充分煮沸，当罐里的所有空气都被蒸汽替换的时候，将这个孔焊住。通过这种办法，这些材料可以保存多年而不会腐烂、发酵或发霉。这不是因为排除了氧气，而是因为氧气并不是发酵或腐烂的必要条件，这一点现在已经得到了证明。正如巴斯德表明的，这并不是由于罐子里的空气被排除了，因为弧菌和细菌的生存并不需要空气或氧气。这也不是因为煮熟的肉或者蔬菜不会腐烂或者发酵，那些在船上不幸食用了没有密闭好的罐子中的肉和蔬菜的人们都清楚地知道这一点。那么到底是什么排除了微生物呢？我认为无生源论者一定会在要求我们想出同样精准、有条理的新实验之前就对这一问题作出回答。

第二，如果我引用的实验结果真的值得信任，那么得出的观点绝对不会与无生源论的观点相同。我们知道生命物质对热的耐受在一定范围内变化，并在某种程度

the chemical and physical qualities of the surrounding medium. But if, in the present state of science, the alternative is offered us, either germs can stand a greater heat than has been supposed, or the molecules of dead matter, for no valid or intelligible reason that is assigned, are able to rearrange themselves into living bodies, exactly such as can be demonstrated to be frequently produced in another way, I cannot understand how choice can be, even for a moment, doubtful.

But though I cannot express this conviction of mine too strongly, I must carefully guard myself against the supposition that I intend to suggest that no such thing as Abiogenesis ever has taken place in the past or ever will take place in the future. With organic chemistry, molecular physics, and physiology yet in their infancy, and every day making prodigious strides, I think it would be the height of presumption for any man to say that the conditions under which matter assumes the properties we call "vital" may not, some day, be artificially brought together. All I feel justified in affirming is that I see no reason for believing that the feat has been performed yet.

And looking back through the prodigious vista of the past, I find no record of the commencement of life, and therefore I am devoid of any means of forming a definite conclusion as to the conditions of its appearance. Belief, in the scientific sense of the word, is a serious matter, and needs strong foundations. To say, therefore, in the admitted absence of evidence, that I have any belief as to the mode in which the existing forms of life have originated, would be using words in a wrong sense. But expectation is permissible where belief is not; and if it were given me to look beyond the abyss of geologically recorded time to the still more remote period when the earth was passing through physical and chemical conditions, which it can no more see again than a man can recall his infancy, I should expect to be a witness of the evolution of living protoplasm from not living matter. I should expect to see it appear under forms of great simplicity, endowed, like existing fungi, with the power of determining the formation of new protoplasm from such matters as ammonium carbonates, oxalates and tartrates, alkaline and earthy phosphates, and water, without the aid of light. That is the expectation to which analogical reasoning leads me; but I beg you once more to recollect that I have no right to call my opinion anything but an act of philosophical faith.

So much for the history of the progress of Redi's great doctrine of Biogenesis, which appears to me, with the limitations I have expressed, to be victorious along the whole line at the present day.

As regards the second problem offered to us by Redi, whether Xenogenesis obtains, side by side with Homogenesis; whether, that is, there exist not only the ordinary living things, giving rise to offspring which run through the same cycle as themselves, but also others, producing offspring which are of a totally different character from themselves, the researches of two centuries have led to a different result. That the grubs found in galls are no product of the plants on which the galls grow, but are the result of the introduction

上依赖于周围介质的化学和物理性质。但是在目前的科学水平下，我们面对的是二选一，不是微生物可以承受超过我们预期的更充分的加热，就是死亡物质的分子能够将自己重排成新的有生命的物质（这一点不能用任何有效的或可理解的已知原因来解释），但是事实证明有生命的物质经常是以其他方式产生的，于是我不明白在二者中进行选择怎么会令人生疑，哪怕只是一时的怀疑。

尽管我不能将自己的这种信念表达得过于强烈，但我必须谨慎地使自己不去猜想——我觉得无生源论这种事情根本就没有发生过，将来也不会发生。有机化学、分子物理学和生理学尚处于初期阶段，每天都在取得巨大的进步，我认为任何人都可能会极为自以为是地认为物质表现出来的我们称为“有生命的”属性的条件终有一天不再被人为地整合到一起。我有依据能确定迄今尚无证据表明这一假设已经奏效。

回顾过去的奇异景象，我发现没有关于生命开端的记录，因此对于生命出现的条件，我找不到任何依据来形成确定的结论。从词语的科学意义上说，信念是一个严肃的事情，它需要坚实的基础。因此，如果要在公认缺乏证据的情况下让我说出我相信生命形式通过哪种方式产生而来，那么我只能说我无从相信任何一种对现存生命形式的产生模式的假设，否则这种信念就是对词语意义的误用。但是在信念不存在的地方却是允许期望存在的。如果让我向地质记录年代更加久远的深渊望去，甚至是地球正在经历物理化学变化的那些更遥远的时期——当然这些情景如同一个人要回想起他的婴儿时期一样都是不可能再看到的事情，那么我希望自己能够见证生命原生质如何从没有生命的物质进化而来。我希望看到生命物质以非常简单的形式出现，就像现存的真菌，它们被赋予了决定如何在不依靠光的情况下，从碳酸铵、草酸盐和酒石酸盐、碱土金属磷酸盐和水等物质中形成新的原生质的力量。这就是类推推理引导我想到的期望。但是我再次请求你们记住我只能将我的观点称为为哲学信仰的结果。

对雷迪的伟大生源论法则的进展历史，我就介绍到此。尽管它存在我前面已经提到的局限性，但是即使在今天，在我看来它也取得了彻底的胜利。

雷迪为我们提出的第二个问题是异源发生是否伴随同型生殖一同存在，或者说是否不仅存在普通的生物，其产生的后代与自身经历同样的生命周期，而且存在其他生物，能够产生具有完全不同于自己特征的后代。至于这一问题，两个世纪以来的研究已经产生了不同的结果。虫瘿中发现的幼虫不是长了虫瘿的植物产生的，而是昆虫的卵进入植物体内的结果。这一现象是由瓦利斯涅里、雷奥米尔等人于

of the eggs of insects into the substance of these plants, was made out by Vallisnieri, Reaumur, and others, before the end of the first half of the eighteenth century. The tapeworms, bladderworms, and flukes continued to be a stronghold of the advocates of Xenogenesis for a much longer period. Indeed, it is only within the last thirty years that the splendid patience of Von Siebold, Van Beneden, Leuckart, Küchenmeister, and other helminthologists, has succeeded in tracing every such parasite, often through the strangest wanderings and metamorphoses, to an egg derived from a parent, actually or potentially like itself; and the tendency of inquiries elsewhere has all been in the same direction. A plant may throw off bulbs, but these, sooner or later, give rise to seeds or spores, which develop into the original form. A polype may give rise to Medusae, or a pluteus to an Echinoderm, but the Medusa and the Echinoderm give rise to eggs which produce polypes or plutei, and they are therefore only stages in the cycle of life of the species.

But if we turn to pathology it offers us some remarkable approximations to true Xenogenesis.

As I have already mentioned, it has been known since the time of Vallisnieri and of Reaumur, that galls in plants, and tumours in cattle, are caused by insects, which lay their eggs in those parts of the animal or vegetable frame of which these morbid structures are outgrowths. Again, it is a matter of familiar experience to everybody that mere pressure on the skin will give rise to a corn. Now the gall, the tumour, and the corn are parts of the living body, which have become, to a certain degree, independent and distinct organisms. Under the influence of certain external conditions, elements of the body, which should have developed in due subordination to its general plan, set up for themselves and apply the nourishment which they receive to their own purposes.

From such innocent productions as corns and warts, there are all gradations to the serious tumours which, by their mere size and the mechanical obstruction they cause, destroy the organism out of which they are developed; while, finally, in those terrible structures known as cancers, the abnormal growth has acquired powers of reproduction and multiplication, and is only morphologically distinguishable from the parasite worm, the life of which is neither more nor less closely bound up with that of the infested organism.

If there were a kind of diseased structure, the histological elements of which were capable of maintaining a separate and independent existence out of the body, it seems to me that the shadowy boundary between morbid growth and Xenogenesis would be effaced. And I am inclined to think that the progress of discovery has almost brought us to this point already. I have been favoured by Mr. Simon with an early copy of the last published of the valuable "Reports on the Public Health", which, in his capacity of their medical officer, he annually presents to the Lords of the Privy Council. The appendix to this report contains an introductory essay "On the Intimate Pathology of Contagion", by Dr. Burdon Sanderson, which is one of the clearest, most comprehensive, and well-reasoned discussions of a great question which has come under my notice for a long time. I refer you to it for details and for the authorities for the statements I am about to make.

18世纪上半叶末之前搞清楚的。绦虫、囊尾幼虫和吸虫类在很长一段时期内都一直是异源发生论的拥护者捍卫的要塞。实际上，就是在过去的短短30年间，冯西博尔德、范贝内登、洛伊卡特、库彻梅斯特和其他蠕虫学家通过不懈的努力才成功地将上述每一种寄生虫追溯到与其自身存在实际的或者是潜在的相似性的亲本的卵，这些卵通常经过了最奇特的游走和变态。其他方面的研究趋势与此一致。一株植物可能会脱掉球茎，但是这些球茎迟早会产生种子或孢子，它们可以发育成原来的形式。水螅体可以产生水母，长腕幼虫可以长成棘皮动物，而水母和棘皮动物产生的卵又可以长成水螅体或长腕幼虫，因此它们（水螅体和长腕幼虫）只是这些物种生命周期中经历的阶段而已。

但是当我们转向病理学时就会发现它给我们提供了一些与真正的异源发生非常接近的情形。

正如我已提到的，从瓦利斯涅里和雷奥米尔时代起人们就已经知道植物的虫瘿和牛的瘤都是由昆虫引起的。昆虫将自己的卵产在动物或植物的某些部位，于是这些部位就会长成病态结构。另外，对每个人来说都很熟悉的一种经历是，皮肤受到压力后会角质化。植物的虫瘿、动物的瘤及角质化结构都是生命体的一部分，但在某种程度上，它们已经成为了独立的、截然不同的生命体。在一定外界条件的影响下，本应按照总体规划按部就班发育的身体的某些部分，会自行调整，并且按照自己的目的来利用吸收到的营养。

诸如角质化结构和疣等这些无害的产物都有可能渐变成严重的肿瘤，这些肿瘤会由于其大小和引起的机械性阻塞而破坏培育出它们的机体；最终，在那些可怕的结构（即众所周知的肿瘤）中，不正常的生长已经获得了复制和增殖的能力，因此肿瘤只是在形态学上不同于寄生虫，而其生命与被感染的生物体的生命恰好是密切相关的。

如果说存在一种病态结构，即能够脱离机体而独立存在的组织结构，那么对我而言似乎病态生长和异源发生之间的模糊界限就不存在了。我倾向于认为发现的进展差不多已经将我们带到了这一步。西蒙先生在他最后出版的极具价值的《公共卫生报告》的早期版本中对我的观点进行了有力的支持。作为一名卫生官员，他有责任每年向枢密院的高级官员呈交一份这样的报告。这份报告的附录中包含一篇入门短文，由伯登·桑德森撰写，题目是《探秘传染病的病理学》，这篇小文章是长久以来最清晰、最易理解、最合理地讨论了我关注的这一重大问题的作品之一。为了便于我进行接下来的陈述，在此特向你们提一下其中的细节及权威性的引文。

You are familiar with what happens in vaccination. A minute cut is made in the skin, and an infinitesimal quantity of vaccine matter is inserted into the wound. Within a certain time a vesicle appears in the place of the wound, and the fluid which distends this vesicle is vaccine matter, in quantity a hundred or a thousandfold that which was originally inserted. Now what has taken place in the course of this operation? Has the vaccine matter, by its irritative property, produced a mere blister, the fluid of which has the same irritative property? Or does the vaccine matter contain living particles, which have grown and multiplied where they have been planted? The observations of M. Chauveau, extended and confirmed by Dr. Sanderson himself, appear to leave no doubt upon this head. Experiments, similar in principle to those of Helmholtz on fermentation and putrefaction, have proved that the active element in the vaccine lymph is non-diffusible, and consists of minute particles not exceeding 1/20,000 of an inch in diameter, which are made visible in the lymph by the microscope. Similar experiments have proved that two of the most destructive of epizootic diseases, sheep-pox and glanders, are also dependent for their existence and their propagation upon extremely small living solid particles, to which the title of *microzymes* is applied. An animal suffering under either of these terrible diseases is a source of infection and contagion to others, for precisely the same reason as a tub of fermenting beer is capable of propagating its fermentation by "infection", or "contagion", to fresh wort. In both cases it is the solid living particles which are efficient; the liquid in which they float, and at the expense of which they live, being altogether passive.

Now arises the question, are these microzymes the results of *Homogenesis*, or of *Xenogenesis*; are they capable, like the *Torulae* of yeast, of arising only by the development of pre-existing germs; or may they be, like the constituents of a nutgall, the results of a modification and individualisation of the tissues of the body in which they are found, resulting from the operation of certain conditions? Are they parasites in the zoological sense, or are they merely what Virchow has called "heterologous growths"? It is obvious that this question has the most profound importance, whether we look at it from a practical or from a theoretical point of view. A parasite may be stamped out by destroying its germs, but a pathological product can only be annihilated by removing the conditions which give rise to it.

It appears to me that this great problem will have to be solved for each zymotic disease separately, for analogy cuts two ways. I have dwelt upon the analogy of pathological modification, which is in favour of the xenogenetic origin of microzymes; but I must now speak of the equally strong analogies in favour of the origin of such pestiferous particles by the ordinary process of the generation of like from like.

It is, at present, a well-established fact that certain diseases, both of plants and of animals, which have all the characters of contagious and infectious epidemics, are caused by minute organisms. The smut of wheat is a well-known instance of such a disease, and it cannot be doubted that the grape-disease and the potato-disease fall under the same category. Among animals, insects are wonderfully liable to the ravages of contagious and infectious diseases caused by microscopic *Fungi*.

接种疫苗过程中发生的情况你们肯定都很熟悉。在皮肤上开一个微小的口，然后将极微量的疫苗物质注入伤口。一定时间内，在伤口处会出现一个小水泡，其中充满的液体就是疫苗物质，数量上相当于最初注入的一百倍或一千倍。那么在这一过程中发生了什么？是具有刺激性的疫苗物质导致了水泡的产生吗？水泡中的液体是否具有同疫苗物质相同的刺激性？或者疫苗物质中含有有生命的颗粒吗？这些颗粒会在它们被接种的地方生长并繁殖吗？肖沃通过观察确定无疑地回答了这些问题，桑德森博士自己又对此观察结果进行了扩充和证实。与亥姆霍兹进行的发酵和腐烂实验依据的原理相似，这些实验证明了痘浆中的活性成分是不可扩散的，由直径不超过 1/20,000 英寸的微小颗粒构成，可以在显微镜下观察到。相似的实验证明，最具破坏性的两种动物流行病，即绵羊痘和马鼻疽，其存在和传播也依赖于极小的有生命的固体颗粒，这些小的固体颗粒被称为**酵母菌**。动物如果染上了这两种可怕疾病中的任何一种，对其他动物来说都会成为间接传染源和接触传染源，原因与一桶正在发酵的啤酒能够通过“间接传染”或“接触传染”使其他新鲜的麦芽汁也发酵一样。这两个例子中，发生作用的都是固体生命颗粒；这些颗粒漂浮于其中的液体只是提供颗粒生存的环境，除了颗粒以外的物质都是不起作用的。

于是就有了如下的问题：这些酵母菌是**同型生殖**的结果还是**异源发生**的结果？这些酵母菌能够像**圆酵母**一样，只有通过已存在的微生物才可以发育出来吗？或者它们就像五倍子的组分一样，是通过对特定条件的操作，由它们所在的有机体的组织经过修饰和个性化作用而产生的结果？这些寄生菌是动物学意义上的，还是仅仅是菲尔绍所谓的“异源发生”的产物？很明显，无论我们从实践的角度还是理论的角度来思考这一问题，它都具有至关重要的意义。通过破坏宿主微生物可以消灭某种寄生菌，但是病理学的产物则只能通过破坏其产生条件来消除。

我觉得这个大问题需要通过分别解决每种发酵病来各个击破，根据相似性可以分为两种方式。我已经详细叙述了病理学修饰方式，它支持酵母菌的异源发生起源；但是我现在必须要提及与此同等重要的方式，即支持这样的传染病颗粒是通过同类型产生同类型的一般过程产生的。

现在非常确定的事实是植物和动物的某些具有间接传染性和接触传染性的流行病学特征的疾病都是由微小的有机体引起的。小麦的黑粉病就是这种病的著名例子，毋庸置疑的是葡萄短节病和马铃薯腐烂病也属于这一类。在动物界，昆虫很容易受到微观真菌引起的间接传染性疾病和接触传染性疾病的攻击。

In autumn, it is not uncommon to see flies, motionless upon a window-pane, with a sort of magic circle, in white, drawn round them. On microscopic examination, the magic circle is found to consist of innumerable spores, which have been thrown off in all directions by a minute fungus called *Empusa muscae*, the spore-forming filaments of which stand out like a pile of velvet from the body of the fly. These spore-forming filaments are connected with others which fill the interior of the fly's body like so much fine wool, having eaten away and destroyed the creature's viscera. This is the full-grown condition of the *Empusa*. If traced back to its earlier stages, in flies which are still active, and to all appearance healthy, it is found to exist in the form of minute corpuscles which float in the blood of the fly. These multiply and lengthen into filaments, at the expense of the fly's substance; and when they have at last killed the patient, they grow out of its body and give off spores. Healthy flies shut up with diseased ones catch this mortal disease and perish like the others. A most competent observer, M. Cohn, who studied the development of the *Empusa* in the fly very carefully, was utterly unable to discover in what manner the smallest germs of the *Empusa* got into the fly. The spores could not be made to give rise to such germs by cultivation; nor were such germs discoverable in the air, or in the food of the fly. It looked exceedingly like a case of Abiogenesis, or, at any rate, of Xenogenesis; and it is only quite recently that the real course of events has been made out. It has been ascertained, that when one of the spores falls upon the body of a fly, it begins to germinate and sends out a process which bores its way through the fly's skin; this, having reached the interior cavities of its body, gives off the minute floating corpuscles which are the earliest stage of the *Empusa*. The disease is "contagious," because a healthy fly coming in contact with a diseased one, from which the spore-bearing filaments protrude, is pretty sure to carry off a spore or two. It is "infections" because the spores become scattered about all sorts of matter in the neighbourhood of the slain flies.

The silkworm has long been known to be subject to a very fatal and infectious disease called the *Muscardine*. Audouin transmitted it by inoculation. This disease is entirely due to the development of a fungus, *Botrytis Bassiana*, in the body of the caterpillar; and its contagiousness and infectiousness are accounted for in the same way as those of the fly-disease. But of late years a still more serious epizootic has appeared among the silkworms; and I may mention a few facts which will give you some conception of the gravity of the injury which it has inflicted on France alone.

The production of silk has been for centuries an important branch of industry in Southern France, and in the year 1853 it had attained such a magnitude that the annual produce of the French sericulture was estimated to amount to a tenth of that of the whole world, and represented a money-value of 117,000,000 of francs, or nearly five millions sterling. What may be the sum which would represent the money-value of all the industries connected with the working up of the raw silk thus produced is more than I can pretend to estimate. Suffice it to say that the city of Lyons is built upon French silk as much as Manchester was upon American cotton before the civil war.

秋天时，苍蝇落在玻璃窗上一动不动，同时还有一种白色的幻圈围绕着它们，这种现象很常见。镜检时，发现该幻圈是由无数的孢子构成的，它们是由被称为蝇单枝虫霉的一种微小的真菌向四面八方散发出来的，这种霉菌的孢子形成的长纤丝就像从苍蝇身体上长出的一堆天鹅绒一样竖立着。这些孢子形成的长纤丝互相连接在一起，就像很多细羊绒一样充满了苍蝇的身体内部，吞食并破坏着苍蝇的内脏。这是蝇疫霉菌发育完成时的状态。如果追溯到其早期生长阶段，在尚且活跃、外观看起来完全健康的苍蝇体内，可以看到它们以微粒子形式漂浮在苍蝇的血液中。这些微粒子靠消耗苍蝇的体质成分来繁殖并延伸成长纤丝；当最终杀死宿主后，它们从宿主的身体中长出来并释放出孢子。健康的苍蝇如果患上这一致命疾病就会像其他有病的苍蝇一样死亡。一位非常杰出的观察者——科恩非常认真地研究了苍蝇中的蝇疫霉菌，却全然没有发现蝇疫霉菌的最小菌体是以何种方式进入到苍蝇体内的。通过培养孢子并不能产生这种细菌，在空气以及苍蝇的食物中也都没有发现这种细菌。这看起来太像是无生源论的例子了，即使不是无生源论，至少也算是异源发生的例子；直到最近才弄清楚这些事件的真正过程。已经确认的是当一个孢子落到苍蝇的身体上时，它就开始萌芽，并在苍蝇皮肤上钻孔而进入到苍蝇体内；到达苍蝇内腔后，排出微小的悬浮粒子，这些粒子就是蝇疫霉菌的最早期阶段。这种疾病是“接触传染性的”，这是因为如果健康的苍蝇与一只伸出含有孢子的长纤丝的患病苍蝇接触的话，那么这只健康的苍蝇肯定也会带走一两个孢子。这种疾病也是“间接传染性的”，这是因为孢子会在被杀死的苍蝇附近散布得到处都是。

我们早就知道，蚕一直遭受着一种非常致命且具有传染性的疾病的困扰，这种疾病被称为**白僵病**。奥杜安通过接种完成了这种疾病的传染。该病完全是由于毛虫体内感染了一种被称为白僵菌的真菌；这种疾病也具有接触传染性和间接传染性，原因与上述苍蝇疾病的相同。但是近年来，在蚕中出现了一种更严重的家畜流行病。我会提到一些事例以使大家了解这种伤害对法国带来了多大的痛苦。

在法国南部，丝绸的生产作为工业体系的一个重要分支已经有几个世纪了。1853 年，法国该产业的年产量已经达到了占据整个世界该产业年产量的 1/10 的程度，这代表着 117,000,000 法郎的货币价值，或者说将近 500 万英镑的货币价值。所有产业中与生丝的生产相关的行业创造的货币价值的总额，我想比我能够估计出来的还要多得多。可以这么说，里昂是靠法国的蚕丝建立起来的，就像内战前的曼彻斯特是靠美国的棉花建立起来的一样。

Silkworms are liable to many diseases; and even before 1853 a peculiar epizootic, frequently accompanied by the appearance of dark spots upon the skin (whence the name of "Pébrine" which it has received), has been noted for its mortality. But in the years following 1853 this malady broke out with such extreme violence, that, in 1858, the silk-crop was reduced to a third of the amount which it had reached in 1853; and, up till within the last year or two, it has never attained half the yield of 1853. This means not only that the great number of people engaged in silk growing are some thirty millions sterling poorer than they might have been; it means not only that high prices have had to be paid for imported silkworm eggs, and that, after investing his money in them, in paying for mulberry-leaves and for attendance, the cultivator has constantly seen his silkworms perish and himself plunged in ruin; but it means that the looms of Lyons have lacked employment, and that for years enforced idleness and misery have been the portion of a vast population which, in former days, was industrious and well to do.

In 1858 the gravity of the situation caused the French Academy of Sciences to appoint Commissioners, of whom a distinguished naturalist, M. de Quatrefages, was one, to inquire into the nature of this disease, and, if possible, to devise some means of staying the plague. In reading the Report made by M. de Quatrefages in 1859, it is exceedingly interesting to observe that his elaborate study of the Pébrine forced the conviction upon his mind that, in its mode of occurrence and propagation, the disease of the silkworm is, in every respect, comparable to the cholera among mankind. But it differs from the cholera, and so far is a more formidable disease, in being hereditary, and in being, under some circumstances, contagious as well as infectious.

The Italian naturalist, Filippi, discovered in the blood of the silkworms affected by this strange disease a multitude of cylindrical corpuscles, each about 1/6,000 of an inch long. These have been carefully studied by Lebert, and named by him *Panhistophyton*; for the reason that in subjects in which the disease is strongly developed, the corpuscles swarm in every tissue and organ of the body, and even pass into the undeveloped eggs of the female moth. But are these corpuscles causes, or mere concomitants, of the disease? Some naturalists took one view and some another; and it was not until the French Government, alarmed by the continued ravages of the malady, and the inefficiency of the remedies which had been suggested, dispatched M. Pasteur to study it, that the question received its final settlement; at a great sacrifice, not only of the time and peace of mind of that eminent philosopher, but, I regret to have to add, of his health.

But the sacrifice has not been in vain. It is now certain that this devastating, cholera-like Pébrine is the effect of the growth and multiplication of the *Panhistophyton* in the silkworm. It is contagious and infectious because the corpuscles of the *Panhistophyton* pass away from the bodies of the diseased caterpillars, directly or indirectly, to the alimentary canal of healthy silkworms in their neighbourhood; it is hereditary, because the corpuscles enter into the eggs while they are being formed, and consequently are carried within them when they are laid; and for this reason, also, it presents the very singular peculiarity of being inherited only on the mother's side. There is not a single one of all the apparently

蚕容易得很多种疾病。有一种经常伴随着皮肤上出现暗色斑点（因此该病被称为“微粒子病”）的特殊家畜流行病，甚至在 1853 年以前，这种病就以其高致死率而著称。在 1853 年之后的几年中，这种疾病的爆发带来了极大的破坏力。1858 年，蚕作物的产量减少到了 1853 年年产量的 1/3；直到过去的一两年，其产量都没有再达到过 1853 年年产量的一半。这意味着不只是为数众多的从事养蚕的人的总收入比过去减少了约 3,000 万英镑；也不仅仅意味着人们不得不花高价进口蚕种，而且在投入这些钱之后，还要付钱买桑叶以及付费进行照料，但是养殖者却总是看到他的蚕死掉了，自己也破产了；而且还意味着里昂的织造业缺少就业机会，数年来，无奈的失业和痛苦曾一度成为那些昔日勤劳而又富裕的人们生活的一部分。

1858 年，形势的严峻性迫使法国科学院指派专员去调查这种疾病的原因，并且如果有可能的话，就设计一些方法来控制疫情。著名的博物学家德卡特勒法热就是专员之一。读到德卡特勒法热先生于 1859 年递交的报告时，我发现了非常有趣的一点，即他对微粒子病的精细研究使他坚信：蚕的这种疾病的发生和传播模式，从任何一方面看，都与人类中出现的霍乱有类似之处。但它又不同于霍乱，迄今为止，它是一种更难对付的疾病，因为它是可遗传的，并且在某些情况下，它是可以接触传染以及间接传染的。

意大利自然学家菲利皮发现，受此奇怪疾病影响的蚕的血液中出现了大量柱状小体，每个小体长约 1/6,000 英寸。莱贝特曾经仔细研究了这些小体，并且将其命名为微粒子。这种疾病在机体内猛烈地发作，这些小体会在体内的每个组织和器官中大量出现，甚至会进入雌蛾尚未发育完全的卵子中。但是，究竟是这些小体还是其附随物引起了疾病？自然学家们各持己见。直到法国政府因持续的疾病攻击而感到惊慌并且所有建议的补救措施都无济于事，于是委托巴斯德来研究这种疾病时，这一问题才最终得以解决；为此也付出了巨大的代价，牺牲了卓越的哲学家的时间和内心的平静，我很遗憾地补充一句，而且也牺牲了他的健康。

但是这些牺牲并没有白费。现在已经确定，这种破坏性类似霍乱的微粒子病就是微粒子在蚕体内生长和繁殖造成的。这种病是可以接触传染及间接传染的，因为造成微粒子病的粒子通过得病蚕的身体直接或间接传递到邻近的健康蚕的消化管中；这种病也是可遗传的，因为这种粒子可以进入到正在形成的卵中，于是在产卵时就被卵子携带；也因为这个原因，这种疾病表现出母系遗传的独特性质。这并不是微粒子病表现出的所有多变、难以解释的现象中的唯一一个，但是这一问题已经通过

capricious and unaccountable phenomena presented by the Pébrine, but has received its explanation from the fact that the disease is the result of the presence of the microscopic organism, *Panhistophyton*.

Such being the facts with respect to the Pébrine, what are the indications as to the method of preventing it? It is obvious that this depends upon the way in which the *Panhistophyton* is generated. If it may be generated by Abiogenesis, or by Xenogenesis, within the silkworm or its moth, the extirpation of the disease must depend upon the prevention of the occurrence of the conditions under which this generation takes place. But if, on the other hand, the *Panhistophyton* is an independent organism, which is no more generated by the silkworm than the mistletoe is generated by the oak or the apple-tree on which it grows, though it may need the silkworm for its development in the same way as the mistletoe needs the tree, then the indications are totally different. The sole thing to be done is to get rid of and keep away the germs of the *Panhistophyton*. As might be imagined, from the course of his previous investigations, M. Pasteur was led to believe that the latter was the right theory; and, guided by that theory, he has devised a method of extirpating the disease, which has proved to be completely successful wherever it has been properly carried out.

There can be no reason, then, for doubting that, among insects, contagious and infectious diseases, of great malignity, are caused by minute organisms which are produced from preexisting germs, or by homogenesis; and there is no reason, that I know of, for believing that what happens in insects may not take place in the highest animals. Indeed, there is already strong evidence that some diseases of an extremely malignant and fatal character to which man is subject, are as much the work of minute organisms as is the Pébrine. I refer for this evidence to the very striking facts adduced by Professor Lister in his various well-known publications on the antiseptic method of treatment. It seems to me impossible to rise from the perusal of those publications without a strong conviction that the lamentable mortality which so frequently dogs the footsteps of the most skilful operator, and those deadly consequences of wounds and injuries which seem to haunt the very walls of great hospitals, and are, even now, destroying more men than die of bullet or bayonet, are due to the importation of minute organisms into wounds, and their increase and multiplication; and that the surgeon who saves most lives will be he who best works out the practical consequences of the hypothesis of Redi.

I commenced this Address by asking you to follow me in an attempt to trace the path which has been followed by a scientific idea, in its long and slow progress from the position of a probable hypothesis to that of an established law of nature. Our survey has not taken us into very attractive regions; it has lain, chiefly, in a land flowing with the abominable, and peopled with mere grubs and mouldiness. And it may be imagined with what smiles and shrugs, practical and serious contemporaries of Redi and of Spallanzani may have commented on the waste of their high abilities in toiling at the solution of problems which, though curious enough in themselves, could be of no conceivable utility to mankind.

如下事实得到了解释，即这种疾病是微生物——微粒子的存在造成的结果。

这就是微粒子病的情况，那么这些情况对于预防该病的方法有什么启示呢？很明显这依赖于微粒子产生的方式。如果它是通过无生源方式产生的或者是通过异源发生方式产生的，那么要想消灭蚕或蛾的这种疾病就一定要阻止能够产生这种疾病的条件的出现。但是另一方面，如果微粒子是一种独立的生物，它不是由蚕产生的，而是像槲寄生是由其所寄生的橡树或苹果树产生的一样，那么尽管它的发育可能与槲寄生需要树的方式相同，它们也会需要蚕，但是所给出的启示就完全不同了。这时，唯一要做的事情就是除去微粒子这种微生物并远离它们。正如由巴斯德以前的研究过程可以想象到的，巴斯德相信后者才是正确的理论；于是，在此理论的指导下，他发明了一种消灭这种疾病的方法，事实已经证明，无论在什么地方，只要准确操作，这种方法都能取得圆满的成功。

没有理由怀疑，昆虫界极具毒害性的接触传染性疾病和间接传染性疾病都是由先前存在的微生物产生的微小生命体引起的，或者由同源生殖引起的；我也知道没有理由去相信昆虫中发生的这些疾病并不会发生在高等动物中。事实上，已经有有力的证据表明人类所遭受的极具危害性和致命性的一些疾病正如微粒子病一样，也是由微生物的作用引起的。我这里提到这个证据是想引出利斯特教授在其论述杀菌处理方法的几篇知名的著作中援引的一些非常令人震惊的事例。我研读这些著作后深信，经常与技术最好的外科医生形影相随的令人惋惜的死亡，以及那些萦绕在各大医院的围墙之内的创伤和损伤造成的致命后果，即使现在，都在摧毁着很多人，其杀伤力超过子弹和刺刀，这些都是由于微小的生物进入了创口，而后不断增加、繁殖引起的；拯救了众生的外科医生将是最能够搞清楚雷迪假说的实践结果的人。

在从一个可能的假说发展成一条确定的自然法则的漫长而缓慢的过程中，科学的思想有其发展所遵循的路径，我在这篇致词一开始就邀请你们随我一起尝试追踪这条路径。我们的研究并没有将我们带入非常引人入胜的领域；简单地说，这是一片充满着令人憎恶的蛆虫和霉菌的土地。可以想象一下，与雷迪以及斯帕兰扎尼同时代的那些实际而严肃的人可能如何微笑着耸耸肩，去评论他们将自己的优秀能力浪费在苦苦寻求这些问题的解决办法上，尽管这些问题本身非常令人好奇，但是对人类而言，它们当时没有任何可以想象到的实用性。

Nevertheless you will have observed that before we had travelled very far upon our road there appeared, on the right hand and on the left, fields laden with a harvest of golden grain, immediately convertible into those things which the most sordidly practical of men will admit to have value—viz., money and life.

The direct loss to France caused by the Pébrine in seventeen years cannot be estimated at less than fifty millions sterling; and if we add to this what Redi's idea, in Pasteur's hands, has done for the wine-grower and for the vinegar-maker, and try to capitalise its value, we shall find that it will go a long way towards repairing the money losses caused by the frightful and calamitous war of this autumn. And as to the equivalent of Redi's thought in life, how can we over-estimate the value of that knowledge of the nature of epidemic and epizootic diseases, and consequently of the means of checking, or eradicating, them, the dawn of which has assuredly commenced?

Looking back no further than ten years, it is possible to select three (1863, 1864, and 1869) in which the total number of deaths from scarlet-fever alone amounted to ninety thousand. That is the return of killed, the maimed and disabled being left out of sight. Why, it is to be hoped that the list of killed in the present bloodiest of all wars will not amount to more than this! But the facts which I have placed before you must leave the least sanguine without a doubt that the nature and the causes of this scourge will, one day, be as well understood as those of the Pébrine are now; and that the long-suffered massacre of our innocents will come to an end.

And thus mankind will have one more admonition that "the people perish for lack of knowledge;" and that the alleviation of the miseries, and the promotion of the welfare, of men must be sought, by those who will not lose their pains, in that diligent, patient, loving study of all the multitudinous aspects of Nature, the results of which constitute exact knowledge, or Science. It is the justification and the glory of this great meeting that it is gathered together for no other object than the advancement of the moiety of science which deals with those phenomena of nature which we call physical. May its endeavours be crowned with a full measure of success!

(**2**, 400-406; 1870)

然而，你会发现我们在自己的路上长途跋涉以前，道路的左右两边都出现了长满金黄稻谷的田地，但是它们会立刻变成最肮脏的只讲实际的人类才承认有价值的那些东西——就是金钱和生命。

17 年来微粒子病对法国造成的直接损失估计至少有 5,000 万英镑；如果我们把巴斯德掌握的雷迪的想法对葡萄酒栽培者和醋酿造者已造成的影响加上去，并且试着计算它的价值时，我们会发现要想弥补发生在这个秋天的由可怕的灾难性战争引起的金钱损失，恐怕将有很长的路要走。至于雷迪对生命的想法的等价物，我们怎么会高估对传染病和家畜流行病的本质的了解，以及检测或者消灭它们的方法的价值呢？可以肯定在这些方面我们已经迎来了曙光。

回顾最近 10 年之内，我们可以选出 3 年（1863 年、1864 年和 1869 年），这 3 年中仅仅由猩红热导致死亡的总共就有 9 万人。曾经远离我们视线的那些被杀死的、受重伤的和变残疾的人又重新出现在我们的面前。我们怎么可以期望，在人类与疾病的历次斗争中，伤亡最惨重的这次斗争中死亡的人数将不会超过以上的数据！你们可能幻想着，就像现在对微粒子病的了解一样，总有一天我们会充分了解这一灾害的本质和原因，我们无辜的患者长期遭受的大屠杀也将结束，然而毫无疑问，我这里向你们陈述的事实一定会令人们失去所有乐观的希望。

因此人类还有一个训诫，即“人类因知识的贫乏而灭亡”；人类痛苦的减轻和福利的提高都要由那些愿意不辞辛劳的人来努力达成，因为他们勤奋、耐心且忠诚地研究大自然各个方面而得到的结果构成了真正的知识或者叫科学。我们聚集在这里，没有其他目的，只是希望能推动研究自然界现象的科学的发展，这是本次大会召开的理由，也是本次大会的光荣。祝愿本次大会圆满成功！

（刘皓芳 翻译；刘力 审稿）

Mathematical and Physical Science

J. C. Maxwell

Editor's Note

James Clerk Maxwell, probably the most perceptive physical scientist in the second half of the nineteenth century, was president of the physics section of the British Association for the Advancement of Science in 1870. His presidential address, he explained, was stimulated by the previous year's address by the mathematician J. J. Cayley. It dealt with the role of molecules in the explanation of physical phenomena.

The president delivered the following address: —

At several of the recent meetings of the British Association the varied and important business of the Mathematical and Physical Section has been introduced by an Address, the subject of which has been left to the selection of the president for the time being. The perplexing duty of choosing a subject has not, however, fallen to me. Professor Sylvester, the president of Section A at the Exeter meeting, gave us a noble vindication of pure mathematics by laying bare, as it were, the very working of the mathematical mind, and setting before us, not the array of symbols and brackets which form the armoury of the mathematician, or the dry results which are only the monuments of his conquests, but the mathematician himself, with all his human faculties directed by his professional sagacity to the pursuit, apprehension, and exhibition of that ideal harmony which he feels to be the root of all knowledge, the fountain of all pleasure, and the condition of all action. The mathematician has, above all things, an eye for symmetry; and Professor Sylvester has not only recognised the symmetry formed by the combination of his own subject with those of the former presidents, but has pointed out the duties of his successor in the following characteristic note:—

"Mr. Spottiswoode favoured the Section, in his opening address, with a combined history of the progress of mathematics and physics; Dr. Tyndall's address was virtually on the limits of physical philosophy; the one here in print," says Professor Sylvester, "is an attempted faint adumbration of the nature of mathematical science in the abstract. What is wanting (like a fourth sphere resting on three others in contact) to build up the ideal pyramid is a discourse on the relation of the two branches (mathematics and physics) to, and their action and reaction upon, one another—a magnificent theme, with which it is to be hoped that some future president of Section A will crown the edifice, and make the tetralogy (symbolisable by A+A′, A, A′, AA′) complete."

The theme thus distinctly laid down for his successor by our late President is indeed a magnificent one, far too magnificent for any efforts of mine to realise. I have endeavoured

数学和物理科学

麦克斯韦

编者按

詹姆斯·克拉克·麦克斯韦可以算是19世纪后半叶最敏锐的物理学家了，他是英国科学促进会1870年会议的物理分会主席。他认为自己的主席演说受到了数学家凯莱在前一年发表的演说的启发。这篇演讲稿论述了在解释物理现象时分子所起到的作用。

主席发表了如下演讲：

在最近几次英国科学促进会的会议上，有一次演讲指出了数学和物理分会中各种重要的事务，因而这一演讲主题被纳入了现任主席的选择。不过，选择主题这一艰巨的任务并没有落在我身上。埃克塞特会议中担任第一分会主席的西尔维斯特教授为我们带来了对纯数学的庄严拥护，通过直截了当地提出数学头脑所从事的真正工作，他呈现给我们的不仅仅是从数学家的武器库中取出的符号和括号的堆砌品，也不仅仅是只有数学家自得其乐的干巴巴的结果，而是一位数学家本身，在职业睿智的指引下，将全部的个人才华都用于追求、理解和展示在他看来是一切知识之本、快乐之源与行为之因的理想的和谐一致。最重要的是，数学家有着调和的眼光；西尔维斯特教授不仅考虑到将他本人的主题与此前各位主席所演讲的主题相结合，还在下面这段典型的文字中指出了后继者的职责：

"斯波蒂斯伍德先生在他的公开演讲中称这一分会为数学与物理发展历史的结合；廷德尔博士的演讲将其本质界定为物理哲学；在这份出版物中，"西尔维斯特教授讲到，"它是一种尝试性的、对于数学科学抽象本质的模糊概括。建立这座理想的金字塔还需要的（如同置于三个彼此接触的球之上的第四个球）是一场涉及两大分支学科（数学和物理）的关系及其相互作用与反作用的讨论——这是一个宏大的主题，期望第一分会未来的某位主席能够为这座大厦剪彩，并完成这出四部曲（可以符号化地表示为A+A′，A，A′，AA′）。"

我们的前任主席为其继任者清晰规划出的这一主题的确是宏伟的，以至于尽我所有的努力都无法将其实现。我曾努力追随斯波蒂斯伍德先生，他以深远的见地将

to follow Mr. Spottiswoode, as with far-reaching vision he distinguishes the systems of science into which phenomena, our knowledge of which is still in the nebulous stage, are growing. I have been carried by the penetrating insight and forcible expression of Dr. Tyndall into that sanctuary of minuteness and of power where molecules obey the laws of their existence, clash together in fierce collision, or grapple in yet more fierce embrace, building up in secret the forms of visible things. I have been guided by Professor Sylvester towards those serene heights

> "Where never creeps a cloud, or moves a wind,
> Nor ever falls the least white star of snow,
> Nor ever lowest roll of thunder moans,
> Nor sound of human sorrow mounts, to mar
> Their sacred everlasting calm."

But who will lead me into that still more hidden and dimmer region where Thought weds Fact; where the mental operation of the mathematician and the physical action of the molecules are seen in their true relation? Does not the way to it pass through the very den of the metaphysician, strewed with the remains of former explorers, and abhorred by every man of science? It would indeed be a foolhardy adventure for me to take up the valuable time of the section by leading you into those speculations which require, as we know, thousands of years even to shape themselves intelligibly.

But we are met as cultivators of mathematics and physics. In our daily work we are led up to questions the same in kind with those of metaphysics; and we approach them, not trusting to the native penetrating power of our own minds, but trained by a long-continued adjustment of our modes of thought to the facts of external nature. As mathematicians, we perform certain mental operations on the symbols of number or of quantity, and, by proceeding step by step from more simple to more complex operations, we are enabled to express the same thing in many different forms. The equivalence of these different forms, though a necessary consequence of self-evident axioms, is not always, to our minds, self-evident; but the mathematician, who, by long practice, has acquired a familiarity with many of these forms, and has become expert in the processes which lead from one to another, can often transform a perplexing expression into another which explains its meaning in more intelligible language.

As students of physics, we observe phenomena under varied circumstances, and endeavour to deduce the laws of their relations. Every natural phenomenon is, to our minds, the result of an infinitely complex system of conditions. What we set ourselves to do is to unravel these conditions, and by viewing the phenomenon in a way which is in itself partial and imperfect, to piece out its features one by one, beginning with that which strikes us first, and thus gradually learning how to look at the whole phenomenon so as to obtain a continually greater degree of clearness and distinctness. In this process, the feature which presents itself most forcibly to the untrained inquirer may not be that which is considered most fundamental by the experienced man of science; for the success of any physical investigation depends on the judicious selection of what is to be observed as of primary importance, combined with a voluntary abstraction of the mind from those

科学系统归类成各种不断增长的现象，而我们对于这些现象的认识还处于蒙昧状态。我也曾被廷德尔博士那敏锐的洞察力和有力的陈述带入那个力量与精微的圣殿，在那里，分子遵循着自身存在的定律，猛烈地撞在一起，或者在更为热烈的拥抱中纠缠，就这样悄然形成了事物的可见形式。我曾被西尔维斯特教授引领向那些宁静的峰巅

“那里从没有云的踪迹，或风的迹象，
从不曾有些微雪花的斑痕，
从不曾有丝毫雷电的呼啸，
或是人类的悲怨之声，能够破坏，
他们那庄严持久的宁静。”

然而，谁能引领我进入那更为隐蔽与晦暗的思想与事实交汇的地带，数学家的头脑运算与分子间的物理作用呈现出其真实关联的所在在哪里呢？难道这条路不会经过那遍布着早期探索者的遗迹并被每一位科研工作者所痛恨的形而上学家的巢穴吗？对于我来说，占用整个分会的宝贵时间将各位引入那些就我们所知需要用几千年时间才能建立成型的思考中，实在是一种莽撞的冒险。

然而，我们是别人眼中数学和物理的耕耘者。在日常工作中，我们向形而上学家遇到的同类型的问题进军；我们着手处理那些问题，但并不完全寄希望于我们与生俱来的洞察力，而是通过长期持续地调整自身思维模式使之符合客观自然现象来训练自己。作为数学家，我们对数字与数量符号进行特定的大脑运算，并且通过从易到难的运算一步一步地推导，我们可以用多种不同形式表达同一事物。尽管这些形式之间的等价性是不证自明的公理的必然推论，但是对我们的头脑来说却并不总是不言而喻的；而数学家们，经过长期的实践已经对其中诸多形式颇为熟悉，并且已经成为将一种形式转变为另一种形式这一过程的专家，他们经常能够将一种复杂的表达形式转变为另外一种能以更容易被理解的语言解释其含义的形式。

作为物理研究者，我们在各种条件下观察现象，并致力于归纳出表达这些现象之间关联的定律。对我们的头脑来说，每一种自然现象都是一个无限复杂的条件体系的结果。我们努力做的就是将这些条件分解，用本身就是局部的、不完全的方式来观察现象，然后从我们最初遇到的那些开始，将现象的特征一个接一个地拼凑起来，并由此逐渐了解到该如何看待整个现象才能使明确度与清晰度不断增加。在这一过程中，那些在未经训练的研究者看来表现得最为强烈的特征，可能在有经验的科学工作者看来并不是最根本的性质；因为任何物理研究的成功都取决于在所观察到的一切中对于首要因素的明智选择，还要结合对那些尽管看来很诱人却还未能充分有

features which, however attractive they appear, we are not yet sufficiently advanced in science to investigate with profit.

Intellectual processes of this kind have been going on since the first formation of language, and are going on still. No doubt the feature which strikes us first and most forcibly in any phenomenon, is the pleasure or the pain which accompanies it, and the agreeable or disagreeable results which follow after it. A theory of nature from this point of view is embodied in many of our words and phrases, and is by no means extinct even in our deliberate opinions. It was a great step in science when men became convinced that, in order to understand the nature of things, they must begin by asking, not whether a thing is good or bad, noxious or beneficial, but of what kind is it? and how much is there of it? Quality and quantity were then first recognised as the primary features to be observed in scientific inquiry. As science has been developed, the domain of quantity has everywhere encroached on that of quality, till the process of scientific inquiry seems to have become simply the measurement and registration of quantities, combined with a mathematical discussion of the numbers thus obtained. It is this scientific method of directing our attention to those features of phenomena which may be regarded as quantities which brings physical research under the influence of mathematical reasoning. In the work of the section we shall have abundant examples of the successful application of this method to the most recent conquests of science; but I wish at present to direct your attention to some of the reciprocal effects of the progress of science on those elementary conceptions which are sometimes thought to be beyond the reach of change.

If the skill of the mathematician has enabled the experimentalist to see that the quantities which he has measured are connected by necessary relations, the discoveries of physics have revealed to the mathematician new forms of quantities which he could never have imagined for himself. Of the methods by which the mathematician may make his labours most useful to the student of nature, that which I think is at present most important is the systematic classification of quantities. The quantities which we study in mathematics and physics may be classified in two different ways. The student who wishes to master any particular science must make himself familiar with the various kinds of quantities which belong to that science. When he understands all the relations between these quantities, he regards them as forming a connected system, and he classes the whole system of quantities together as belonging to that particular science. This classification is the most natural from a physical point of view, and it is generally the first in order of time. But when the student has become acquainted with several different sciences, he finds that the mathematical processes and trains of reasoning in one science resemble those in another so much that his knowledge of the one science may be made a most useful help in the study of the other. When he examines into the reason of this, he finds that in the two sciences he has been dealing with systems of quantities, in which the mathematical forms of the relations of the quantities are the same in both systems, though the physical nature of the quantities may be utterly different. He is thus led to recognise a classification of quantities on a new principle, according to which the physical nature of the quantity is subordinated to its mathematical form. This is the point of view which is characteristic

益于科学研究的想法的自觉提炼。

这种理性过程自从语言形成以来就一直在进行着，并且还要进行下去。无疑，在任何现象中，最先刺激我们并表现得最为强烈的特征，就是伴随该现象而来的喜悦或烦恼，以及接踵而至的一致或不一致的结果。一种基于此看法的自然理论体现在很多词汇和短语中，而且即使在经过深思熟虑的观点中也不会消逝。当人们最终深信，为了理解事物的本质，他们一开始必须询问的不是该事物是好还是坏、是有害的还是有益的，而是它属于哪一类别、具体有多少时，科学便迈出了伟大的一步。自此,定性和定量第一次被认为是科学研究中要观察的首要特征。科学一旦建立起来，定量的疆域就会从各个角落侵占定性的疆域，直到科学研究的过程逐渐变为只是数量的测量与记录再加上对由此获得的数字的数学讨论。正是这种将我们的注意力引向自然现象中那些可以被视为量的特征的科学方法，把我们带入了数学推理影响下的物理研究中。在本分会所涉及的研究工作中，我们有丰富的实例可以说明这种方法在最近的科学成就中的成功应用；但是现在，我希望可以将各位的注意力引向科学发展对那些有时被认为是亘古不变的基本概念的相反的影响。

如果说数学家的技艺使实验家看到了他已测量的量之间有着必然的联系，物理的发现则已向数学家揭示出他们自己绝对无法想象出来的量的新形式。在数学家作为自然界的研究者付出辛劳时所用到的最有益的方法中，我认为当前最重要的方法就是量的系统分类。我们在数学和物理中所研究的量可以用两种不同的方式分类。想要掌握任何一门特定科学的学生必须使自己熟悉该科学的各种量。一旦他理解了这些量之间的全部关联，就会认为它们形成了一个联结起来的体系，并把整个量的体系归在一起作为属于该特定科学的类。从物理的观点来看这种分类法是最自然的，并且在时间上一般也是最先出现的。但是当这名学生逐渐通晓了若干个不同学科时，他会发现在一门科学中的数学过程和推理训练与另外一门中的十分相似，以至于一门科学中的知识对于学习另外一门科学极有帮助。在分析其中的道理时，他会发现，在需要处理包含各自量的系统的两门科学中，尽管各自的量的物理本质可能是完全不同的，但两个系统中量之间关系的数学形式是一样的。由此，他认识到基于一种新原则的量的分类法，根据这一原则，量的物理本质服从于其数学形式。这是带有数学家特征的观点；不过在时间顺序上它位于物理观点之后，因为，为了使人类的头脑能够想象出不同种类的量，首先它们本身就必须是确实存在的。这里我并没有提及这样的事实，即所有的量本身都必须服从代数和几何规则，并因此服从于很多

of the mathematician; but it stands second to the physical aspect in order of time, because the human mind, in order to conceive of different kinds of quantities, must have them presented to it by nature. I do not here refer to the fact that all quantities, as such, are subject to the rules of arithmetic and algebra, and are therefore capable of being submitted to those dry calculations which represent, to so many minds, their only idea of mathematics. The human mind is seldom satisfied, and is certainly never exercising its highest functions, when it is doing the work of a calculating machine. What the man of science, whether he be a mathematician or a physical inquirer, aims at is, to acquire and develop clear ideas of the things he deals with. For this purpose he is willing to enter on long calculations, and to be for a season a calculating machine, if he can only at last make his ideas clearer. But if he finds that clear ideas are not to be obtained by means of processes, the steps of which he is sure to forget before he has reached the conclusion, it is much better that he should turn to another method, and try to understand the subject by means of well-chosen illustrations derived from subjects with which he is more familiar. We all know how much more popular the illustrative method of exposition is found, than that in which bare processes of reasoning and calculation form the principal subject of discourse. Now a truly scientific illustration is a method to enable the mind to grasp some conception or law in one branch of science, by placing before it a conception or a law in a different branch of science, and directing the mind to lay hold of that mathematical form which is common to the corresponding ideas in the two sciences, leaving out of account for the present the difference between the physical nature of the real phenomena. The correctness of such an illustration depends on whether the two systems of ideas which are compared together are really analogous in form, or whether, in other words, the corresponding physical quantities really belong to the same mathematical class. When this condition is fulfilled, the illustration is not only convenient for teaching science in a pleasant and easy manner, but the recognition of the mathematical analogy between the two systems of ideas leads to a knowledge of both, more profound than could be obtained by studying each system separately.

There are men who, when any relation or law, however complex, is put before them in a symbolical form, can grasp its full meaning as a relation among abstract quantities. Such men sometimes treat with indifference the further statement that quantities actually exist in nature which fulfil this relation. The mental image of the concrete reality seems rather to disturb than to assist their contemplations. But the great majority of mankind are utterly unable, without long training, to retain in their minds the unembodied symbols of the pure mathematician; so that if science is ever to become popular and yet remain scientific, it must be by a profound study and a copious application of those principles of truly scientific illustration which, as we have seen, depend on the mathematical classification of quantities. There are, as I have said, some minds which can go on contemplating with satisfaction pure quantities presented to the eye by symbols, and to the mind in a form which none but mathematicians can conceive. There are others who feel more enjoyment in following geometrical forms, which they draw on paper, or build up in the empty space before them. Others, again, are not content unless they can project their whole physical energies into the scene which they conjure up. They learn at what a rate the planets rush through space, and they experience a delightful feeling of exhilaration. They calculate the

人认为的是代表其唯一数学观念的那些干巴巴的计算。人类的头脑在从事计算机器的工作时，很少会得到满足，当然也不会用到其最高级的功能。无论是数学家还是物理学者，科学工作者所希求的，是获得和发展有关他所研究的事物的清晰观念。为了这个目的，他甘愿投身于漫长的计算，并暂时充当一台计算机器，只要最终能使自己的观念变得清晰一些。不过，要是他发现通过某种过程（在得到计算结果之前就必然已把前面的步骤忘光的过程）不可能获得清晰的观念，那么他最好还是换一种方法，并通过在他比较熟悉的学科衍生出的种种解释中仔细挑选出的解释，来理解这个学科。我们都知道，直观的解释方法比从讨论主题出发进行干巴巴的推导和计算的过程受欢迎得多。那么，一个真正的科学解释是使头脑能够掌握某一科学分支中的一些概念和定律的方法，这是通过以下方式实现的：在头脑中呈现出另一科学分支中的一个概念或一条定律，并指引头脑去把握两个科学分支的对应观念中所共有的数学形式，而不去考虑那些真实现象的物理本质之间的不同。这样一种解释的正确性取决于两个放在一起比较的观念体系是否真的在形式上相似，或者换句话说，相应的物理量是否真的属于同一数学类。一旦满足了这一条件，不仅这种解释便于以轻松愉快的方式进行科学教学，而且对两个观念体系之间数学上的相似性的认识还会促进这两个系统的知识的发展，这就比我们单独研究任何一门科学所能获得的知识更加深刻。

任何一种无论多么复杂的关系或定律以符号形式呈现在人们面前时，总有人能够把握到它作为抽象的量之间关系的全部含义。有时，这些人对那些满足这一关系的自然界真实存在的量的进一步阐述毫不关心。具体现实在头脑中的印象似乎对他们思考的干扰多于帮助。但是，若非经过长期训练，绝大多数人绝对没法在其头脑中记住纯数学家的抽象符号。因此，如果科学想要广受欢迎同时保持科学性，它就必须经历深入的研究，以及能用真正的科学进行解释的原理的大量应用。而这，如同我们已经看到的，有赖于量的数学分类。正如我在前面提到的，有些头脑满足于利用那些纯粹的量进行思考，这些量以符号的形式呈现在他们眼前，并以某种除了数学家之外无人可以理解的形式展示于他们的头脑中。而另外一些人则以研究几何形式为乐，他们将几何形式描绘在纸上，或搭建在面前的空间中。还有一些人，只满足于将其全部精力投入到他们在头脑中幻想出的景象。他们研究行星掠过太空时的速度，从中体会喜悦的满足感。他们计算天体彼此之间的牵引力，并感觉自己的

forces with which the heavenly bodies pull at one another, and they feel their own muscles straining with the effort.

To such men impetus, energy, mass, are not mere abstract expressions of the results of scientific inquiry. They are words of power which stir their souls like the memories of childhood. For the sake of persons of these different types, scientific truths should be presented in different forms, and should be regarded as equally scientific, whether it appears in the robust form and the vivid colouring of a physical illustration, or in the tenuity and paleness of a symbolical expression. Time would fail me if I were to attempt to illustrate by examples the scientific value of the classification of quantities. I shall only mention the name of that important class of magnitudes having direction in space which Hamilton has called Vectors, and which form the subject-matter of the Calculus of Quaternions—a branch of mathematics which, when it shall have been thoroughly understood by men of the illustrative type, and clothed by them with physical imagery, will become, perhaps under some new name, a most powerful method of communicating truly scientific knowledge to persons apparently devoid of the calculating spirit. The mutual action and reaction between the different departments of human thought is so interesting to the student of scientific progress, that, at the risk of still further encroaching on the valuable time of the Section, I shall say a few words on a branch of science which not very long ago would have been considered rather a branch of metaphysics: I mean the atomic theory, or, as it is now called, the molecular theory of the constitution of bodies. Not many years ago, if we had been asked in what regions of physical science the advance of discovery was least apparent, we should have pointed to the hopelessly distant fixed stars on the one hand, and to the inscrutable delicacy of the texture of material bodies on the other. Indeed, if we are to regard Comte as in any degree representing the scientific opinion of his time, the research into what takes place beyond our own solar system seemed then to be exceedingly unpromising, if not altogether illusory. The opinion that the bodies which we see and handle, which we can set in motion or leave at rest, which we can break in pieces and destroy, are composed of smaller bodies which we cannot see or handle, which are always in motion, and which can neither be stopped nor broken in pieces, nor in any way destroyed or deprived of the least of their properties, was known by the name of the Atomic Theory. It was associated with the names of Democritus and Lucretius, and was commonly supposed to admit the existence only of atoms and void, to the exclusion of any other basis of things from the universe.

In many physical reasonings and mathematical calculations we are accustomed to argue as if such substances as air, water, or metal, which appear to our senses uniform and continuous, were strictly and mathematically uniform and continuous. We know that we can divide a pint of water into many millions of portions, each of which is as fully endowed with all the properties of water as the whole pint was, and it seems only natural to conclude that we might go on subdividing the water for ever, just as we can never come to a limit in subdividing the space in which it is contained. We have heard how Faraday divided a grain of gold into an inconceivable number of separate particles, and we may see Dr. Tyndall produce from a mere suspicion of nitrite of butyle an immense cloud, the

肌肉在这种作用下的变形。

对这样的人来说，冲量、能量和质量不仅仅是科学研究结果的抽象表达，它们还是如同童年记忆一样可以触动这些人灵魂的有力文字。因为有上述不同类型的人，所以科学真理应该以不同的形式表达，而且无论它出现在充实的表格和生动多彩的物理图解中，还是单调苍白的方程中，都应该被视为具有同等的科学性。时间不允许我在这里通过举例来说明量的分类的科学意义。我只能稍微说说被哈密顿称为矢量的一类具有空间方向的重要量，这正是四元数微积分的主题。四元数微积分是这样一个数学分支：当它完全被善于作举例说明的人所理解并被他们赋予丰富的物理想象的时候，它（也可能起了新名字）就会成为向那些看似缺乏计算头脑的人们传达真正科学知识的最有力的方法。人类思维的各个不同部分之间的相互作用和反作用对科学过程的研究者来说是如此有趣，以至于即使冒着继续侵占整个分会宝贵时间的风险，我也要说一说一个在不久之前还被看作是形而上学的科学分支，即，关于物体构成的原子理论或者说分子理论（现在是这样叫的）。几年前，如果有人问物理学中哪个领域进展最缓慢，我们会一手指向令人绝望的远地星体，另一手指向神秘的物质结构。的确，如果我们认为在任何情况下孔德都可以代表他所在时代的科学观点，那么对于发生在太阳系之外的事物的研究即使不是幻想，也只是毫无希望的事情。我们可以看到和触碰的、可以发动和制动的、可以打碎和毁灭的物体都是由我们看不到也触碰不到、一直处于运动状态并且不能被我们阻止、打碎、毁灭或剥夺其任何性质的更小的物体组成的，这种观点就是所谓的原子理论。这一理论与德谟克利特和卢克莱修这两个名字相联系，并且一般假定只承认原子和空隙的存在，而排除宇宙中其他一切物质基础的存在。

在很多物理推导和数学计算中，我们习惯地认为诸如空气、水或金属这些给我们均匀、连续的感觉的物质在数学上是严格均匀和连续的。我们知道，我们可以将一品脱水分成数百万份，每一份水都具有这一品脱水作为一个整体时所具有的所有性质，很自然地我们会得出这样的结论：我们可以无限细分这些水，就如同我们可以无限细分盛有这些水的空间一样。我们已经听说法拉第是如何将一粒金子分成不计其数的分离颗粒，而且我们可能会看到廷德尔博士用少量的丁基亚硝酸盐生成了一个巨大的云团，这个大云团的可见的微小部分仍是云团，因此一定包含很多丁基

minute visible portion of which is still cloud, and therefore must contain many molecules of nitrite of butyle. But evidence from different and independent sources is now crowding in upon us which compels us to admit that if we could push the process of subdivision still further we should come to a limit, because each portion would then contain only one molecule, an individual body, one and indivisible, unalterable by any power in nature. Even in our ordinary experiments on very finely divided matter we find that the substance is beginning to lose the properties which it exhibits when in a large mass, and that effects depending on the individual action of molecules are beginning to become prominent. The study of these phenomena is at present the path which leads to the development of molecular science. That superficial tension of liquids which is called capillary attraction is one of these phenomena. Another important class of phenomena are those which are due to that motion of agitation by which the molecules of a liquid or gas are continually working their way from one place to another, and continually changing their course, like people hustled in a crowd. On this depends the rate of diffusion of gases and liquids through each other, to the study of which, as one of the keys of molecular science, that unwearied inquirer into nature's secrets, the late Prof. Graham, devoted such arduous labour.

The rate of electrolytic conduction is, according to Wiedemann's theory, influenced by the same cause; and the conduction of heat in fluids depends probably on the same kind of action. In the case of gases, a molecular theory has been developed by Clausius and others, capable of mathematical treatment, and subjected to experimental investigation; and by this theory nearly every known mechanical property of gases has been explained on dynamical principles, so that the properties of individual gaseous molecules are in the fair way to become objects of scientific research. Now Sir William Thomson has shown by several independent lines of argument, drawn from phenomena so different in themselves as the electrification of metals by contact, the tension of soap-bubbles, and the friction of air, that in ordinary solids and liquids the average distance between contiguous molecules is less than the hundred-millionth, and greater than the two-thousand-millionth of a centimetre. This of course is an exceedingly rough estimate, for it is derived from measurements, some of which are still confessedly very rough; but if, at the present time, we can form even a rough plan for arriving at a result of this kind, we may hope that as our means of experimental inquiry become more accurate and more varied, our conception of a molecule will become more definite, so that we may be able at no distant period to estimate its weight. A theory which Sir W. Thomson has founded on Helmholtz's splendid hydrodynamical theorems, seeks for the properties of molecules in the ring-vortices of a uniform, frictionless, incompressible fluid. Such whirling rings may be seen when an experienced smoker sends out a dexterous puff of smoke into the still air, but a more evanescent phenomenon it is difficult to conceive. This evanescence is owing to the viscosity of the air; but Helmholtz has shown that in a perfect fluid such a whirling ring, if once generated, would go on whirling for ever, would always consist of the very same portion of the fluid which was first set whirling, and could never be cut in two by any natural cause. The generation of a ring-vortex is of course equally beyond the power of natural causes, but once generated, it has the properties of individuality, permanence in quantity, and indestructibility. It is also the recipient of impulse and of energy, which is all

亚硝酸盐分子。然而，从各个独立的渠道所获得的证据正朝我们涌来，这迫使我们承认如果将细分过程继续推进，我们将会遇到一个极限，因为这时每一部分只包含一个分子，这是一种不能被任何自然界的力量分割或改变的独立个体。即使是在平时关于分得很细的物质的实验中，我们也会发现这些物质已经开始失去作为大块材料时所具有的一些性质,产生这一结果的原因是分子的个别作用开始变得显著。目前，研究这种现象是分子科学发展的一种途径。被称为毛细引力的液体表面张力就是这种现象中的一个。另一类重要的现象是那些由激发运动产生的现象，由于激发运动，液态或气态分子不断从一个地方运动到另一个地方，并且不断变化它们的航向，就如同人群中奔忙的人那样。这种运动决定了气体和液体相互扩散的速度，对此类现象的研究正是永不停歇地探寻自然之谜的分子科学的关键点之一，已故的格雷姆教授为此付出了非常艰辛的努力。

根据维德曼的理论，电解导电速率也受到相同原因的影响；流体中的热传导可能也取决于同类作用。气体分子理论已经由克劳修斯和其他研究者发展成熟，可以用数学方法处理，并且与实验研究结果相符。应用这个理论，气体的几乎所有已知的力学性质都可以用动力学原理解释，因此，单个气体分子的性质就理所当然地成了科学研究的目标。现在，威廉·汤姆孙爵士通过一些各自独立的现象，例如金属的接触带电、肥皂泡的张力和空气的摩擦力，从不同角度证明了在普通的固体和流体中近邻分子之间的平均距离在 $5\times10^{-10}\sim1\times10^{-8}$ 厘米之间。当然这是一个非常粗略的估算，因为导出这个估算值的测量中包含着一些仍旧被公认为非常粗略的结果，但是如果到目前为止，可以建立一套能够获得这类结果的哪怕只是粗略的计划，我们可能就会希望随着实验研究手段的越来越精确和多样化，我们对单个分子的概念会越来越清晰，这样，在不远的将来，我们就能够估算它的质量。基于亥姆霍兹精妙的流体动力学定理，汤姆孙爵士建立了一个研究分子在均匀、无摩擦且不可压缩的液体环形漩涡中的性质的理论。当一个有经验的吸烟者向静止的空气中吐出一个烟圈时就能够看到这样的涡流环，但是这是一个非常短暂的现象，很难去想象。这种短暂性是空气的黏性造成的；但是，亥姆霍兹认为在完全流体中，这样的涡流环一旦产生就会永远涡旋下去，而且一直是由开始成为涡旋的那部分流体组成，决不会被任何自然力分成两半。当然，环形漩涡的产生也同样不是出于自然原因，但是一旦产生,它就具有个体特征、量的恒定和不可摧毁性。我们能够断言的仅仅是漩涡，也就是冲量和能量的接受者，可以认为是一种物质；这些环形漩涡具有如此多样的关系和自卷结，以至于不同的涡流结一定像不同的分子一样属于不同的类型。

we can affirm of matter; and these ring-vortices are capable of such varied connections, and knotted self-involutions, that the properties of differently knotted vortices must be as different as those of different kinds of molecules can be.

If a theory of this kind should be found, after conquering the enormous mathematical difficulties of the subject, to represent in any degree the actual properties of molecules, it will stand in a very different scientific position from those theories of molecular action which are formed by investing the molecule with an arbitrary system of central forces invented expressly to account for the observed phenomena. In the vortex theory we have nothing arbitrary, no central forces or occult properties of any other kind. We have nothing but matter and motion, and when the vortex is once started its properties are all determined from the original impetus, and no further assumptions are possible. Even in the present undeveloped state of the theory, the contemplation of the individuality and indestructibility of a ring vortex in a perfect fluid cannot fail to disturb the commonly received opinion that a molecule, in order to be permanent, must be a very hard body. In fact one of the first conditions which a molecule must fulfil is, apparently, inconsistent with its being a single hard body. We know from those spectroscopic researches which have thrown so much light on different branches of science, that a molecule can be set into a state of internal vibration, in which it gives off to the surrounding medium light of definite refrangibility—light, that is, of definite wave-length and definite period of vibration. The fact that all the molecules, say of hydrogen, which we can procure for our experiments, when agitated by heat or by the passage of an electric spark, vibrate precisely in the same periodic time, or, to speak more accurately, that their vibrations are composed of a system of simple vibrations having always the same periods, is a very remarkable fact. I must leave it to others to describe the progress of that splendid series of spectroscopic discoveries by which the chemistry of the heavenly bodies has been brought within the range of human inquiry. I wish rather to direct your attention to the fact that not only has every molecule of terrestrial hydrogen the same system of periods of free vibration, but that the spectroscope examination of the light of the sun and stars shows that in regions the distance of which we can only feebly imagine there are molecules vibrating in as exact unison with the molecules of terrestrial hydrogen as two tuning forks tuned to correct pitch, or two watches regulated to solar time. Now this absolute equality in the magnitude of quantities, occurring in all parts of the universe, is worth our consideration. The dimensions of individual natural bodies are either quite indeterminate, as in the case of planets, stones, trees, &c., or they vary within moderate limit, as in the case of seeds, eggs, &c.; but, even in these cases, small quantitive differences are met with which do not interfere with the essential properties of the body. Even crystals, which are so definite in geometrical form, are variable with respect to their absolute dimensions. Among the works of man we sometimes find a certain degree of uniformity. There is a uniformity among the different bullets which are cast in the same mould, and the different copies of a book printed from the same type. If we examine the coins, or the weights and measures, of a civilised country, we find a uniformity, which is produced by careful adjustment to standards made and provided by the State. The degree of uniformity of these national standards is a measure of that spirit of justice in the nation which has enacted laws to

如果在解决了这个问题的大量数学疑难之后，可以找到一个这种类型的理论来表示分子的所有真实性质，那么这个理论将处于与那些在研究具有主观中心力系统（这个系统是为了解释宏观现象而臆造的）的分子时形成的分子相互作用理论非常不同的科学地位。在涡旋理论中，没有什么是主观的，没有中心力或其他任何类型的超自然性质。只存在物质和运动，而且一旦涡旋产生，它的性质就完全取决于初始的动力，可以不用做任何进一步的假设。即使是在目前这种理论尚未发展成熟的情况下，关于完全流体中环形漩涡的个别性和不可毁灭性的思考也会扰乱人们的普遍观点：为了具有永久性，分子必须是非常坚硬的个体。事实上，分子必须满足的一个首要条件似乎与它是一个单一坚硬个体这一观点是不一致的。我们从已经惠及多门科学分支的光谱研究可以知道，分子能够调节到内部振动状态，在这种状态下，分子向周围介质辐射具有确定折射率的光（即，该光具有确定的波长和确定的振动周期）。所有可以为我们的实验服务的分子，比如氢分子，当它们受热激发时或在电火花经过而引起激发时，都会精确地以相同的周期振动，或者更准确地说，它们的振动是由一个具有相同振动周期的简单振动系统组成的，事实的确如此。我不得不把一系列精彩的光谱学发现留给别人去描述，尽管正是这些精彩的发现将天体化学这一领域纳入了人类探索的范围。我宁愿将你们的注意力引到这样的事实中：不仅地球上每一个氢分子都有相同的自由振动周期系统，而且对太阳和行星发出的光的光谱检测结果表明，在那些与地球之间的距离大到无法想象的行星上，也有与地球上氢分子的振动完全一致的分子振动，就好像两个被调节到相同音调的音叉，或者两块校准到太阳时的表。那么这种发生在宇宙各个部分的量在量值上的绝对等价性就值得我们思考。自然个体的尺度不是非常不确定（比如，行星、石块、树木等），就是能够在适度的范围内变化（比如，种子、卵等）；但是即使在这些例子中所遇到的量值上微小的不同也不会影响物体的本质特征。即使是具有非常确定的几何形状的晶体，其绝对尺寸也是可变的。在人类的工作中我们时常会发现一定程度的一致性。比如，同一模子中铸造出的不同子弹具有一致性，同一版次印刷出的书籍的不同副本具有一致性。如果我们观察一个文明国家的货币和度量衡，也会发现一致性，一种产生于按国家规定的标准仔细调节的一致性。这些国家标准的一致程度是对这个制定了法律来规范各种标准并且委派了官员来检测各种标准的国家的公平意识程度的衡量。作为科学个体来说，这个主题是我们很感兴趣的问题，并且大家都意识到大量科学工作已经投入并且有利地投入到为商业或科学目的提供度量衡之中。地球作为长度的永久基准已经得到了测量，同时金属的每一个性质都已经得到了研究，

regulate them and appointed officers to test them. This subject is one in which we, as a scientific body, take a warm interest, and you are all aware of the vast amount of scientific work which has been expended, and profitably expended, in providing weights and measures for commercial and scientific purposes. The earth has been measured as a basis for a permanent standard of length, and every property of metals has been investigated to guard against any alteration of the material standards when made. To weigh or measure anything with modern accuracy, requires a course of experiment and calculation in which almost every branch of physics and mathematics is brought into requisition.

Yet, after all, the dimensions of our earth and its time of rotation, though, relatively to our present means of comparison, very permanent, are not so by any physical necessity. The earth might contract by cooling, or it might be enlarged by a layer of meteorites falling on it, or its rate of revolution might slowly slacken, and yet it would continue to be as much a planet as before. But a molecule, say of hydrogen, if either its mass or its time of vibration were to be altered in the least, would no longer be a molecule of hydrogen. If, then, we wish to obtain standards of length, time, and mass which shall be absolutely permanent, we must seek them not in the dimensions, or the motion, or the mass of our planet, but in the wave-length, the period of vibration, and the absolute mass of these imperishable and unalterable and perfectly similar molecules. When we find that here, and in the starry heavens, there are innumerable multitudes of little bodies of exactly the same mass, so many, and no more, to the grain, and vibrating in exactly the same time, so many times, and no more, in a second, and when we reflect that no power in nature can now alter in the least either the mass or the period of any one of them, we seem to have advanced along the path of natural knowledge to one of those points at which we must accept the guidance of that faith by which we understand that "that which is seen was not made of things which do appear." One of the most remarkable results of the progress of molecular science is the light it has thrown on the nature of irreversible processes,—processes, that is, which always tend towards, and never away from, a certain limiting state. Thus if two gases be put into the same vessel they become mixed, and the mixture tends continually to become more uniform. If two unequally heated portions of the same gas are put into the vessel, something of the kind takes place, and the whole tends to become of the same temperature. If two unequally heated solid bodies be placed in contact, a continual approximation of both to an intermediate temperature takes place. In the case of the two gases, a separation may be effected by chemical means; but in the other two cases the former state of things cannot be restored by any natural process. In the case of the conduction or diffusion of heat the process is not only irreversible, but it involves the irreversible diminution of that part of the whole stock of thermal energy which is capable of being converted into mechanical work. This is Thomson's theory of the irreversible dissipation of energy, and it is equivalent to the doctrine of Clausius concerning the growth of what he calls Entropy. The irreversible character of this process is strikingly embodied in Fourier's theory of the conduction of heat, where the formulae themselves indicate a possible solution of all positive values of the time which continually tends to a uniform diffusion of heat. But if we attempt to ascend the stream of time by giving to its symbol continually diminishing values, we are led up to a state of things in which the

目的是为了在制造过程中避免材料标准的任何改变。以现在的精度去称量或测量任何东西，都需要经过一个实验和计算的过程，这个过程几乎会用到物理和数学的所有分支。

虽然地球的尺寸及其自转的时间相对于我们现有的比较方法来说是永恒不变的，但毕竟不是任何情况下都是这样。地球可能会因为冷却而收缩，可能会因为落在其表面的陨石层而增大，它的公转速度可能会逐渐变缓，尽管这样地球仍然和以前一样是一个星球。但是，如果一个分子（如氢分子）的质量或者振动周期发生轻微的变化，那么它将不再是氢分子。因此，如果我们想要得到绝对永恒的长度、时间和质量基准，就不能在行星的尺寸、运动和质量中寻找答案，而只能在这些不灭、不变，并且完全相同的分子的波长、振动周期和绝对质量中寻求。当我们发现在地球上和星空中存在无数质量完全相同的小物体以及无数周期完全相同的振动，并且到目前为止，没有任何自然力可以使它们之中任何一个的质量或者周期发生丝毫的改变，我们似乎是沿着自然知识的道路向那些观点中的一点前进的。从这种观点来看，我们必须接受这种信仰的指引，通过这个信仰我们明白了“我们看见的事物并不是由可见的事物组成的。”分子科学进程中最显著的成就之一就是对不可逆过程（一种一直向特定最终状态发展，从不偏离的过程）本质的解释。因此，如果将两种气体装入同一个容器中，它们会混合在一起，并且混合得越来越均匀。如果将两部分温度不同的同种气体装入这个容器中，类似的过程将会发生，容器中的气体将趋于相同的温度。如果两块温度不同的固体相互接触，它们的温度将会向着一个中间的温度转变。在有两种气体的例子中，可以通过化学方法将混合在一起的两种气体再次分开，但是在另外两种情况中，不可能通过任何自然方法将系统再恢复到初始状态。热传导或热扩散过程不仅是不可逆的，而且还伴随着可以转变为机械能的储热的不可逆减少。这就是汤姆孙的能量不可逆损耗理论，它等价于克劳修斯的熵增加理论。这一过程的不可逆性显著地体现在傅立叶的热传导理论中，其方程本身就表明了热扩散达到均匀状态所需时间的所有可能的正值解。然而，如果我们试图通过给符号赋予不断减小的值来增大时间的流动，便会得到事物的一个状态，在这个状态中，公式具有所谓的临界值。而如果我们研究这个状态之前的一个状态，就会发现这个公式变得很荒谬。因此，我们得出这样的观点：事物的状态并不能看作是前一状态的物理结果。我们还发现临界状态实际上并不存在于过去的一个时期，而是被一个有限的间隔与现在的时间分开。这个新起点的观念是最近物理研究带给我们的，是任

formula has what is called a critical value; and if we inquire into the state of things the instant before, we find that the formula becomes absurd. We thus arrive at the conception of a state of things which cannot be conceived as the physical result of a previous state of things, and we find that this critical condition actually existed at an epoch not in the utmost depths of a past eternity, but separated from the present time by a finite interval. This idea of a beginning is one which the physical researches of recent times have brought home to us, more than any observer of the course of scientific thought in former times would have had reason to expect. But the mind of man is not like Fourier's heated body, continually settling down into an ultimate state of quiet uniformity, the character of which we can already predict; it is rather like a tree shooting out branches which adapt themselves to the new aspects of the sky towards which they climb, and roots which contort themselves among the strange strata of the earth into which they delve. To us who breathe only the spirit of our own age, and know only the characteristics of contemporary thought, it is as impossible to predict the general tone of the science of the future as it is to anticipate the particular discoveries which it will make. Physical research is continually revealing to us new features of natural processes, and we are thus compelled to search for new forms of thought appropriate to these features. Hence the importance of a careful study of those relations between mathematics and physics which determine the conditions under which the ideas derived from one department of physics may be safely used in forming ideas to be employed in a new department. The figure of speech or of thought by which we transfer the language and ideas of a familiar science to one with which we are less acquainted may be called scientific metaphor. Thus the words velocity, momentum, force, &c., have acquired certain precise meanings in elementary dynamics. They are also employed in the dynamics of a connected system in a sense which, though perfectly analogous to the elementary sense, is wider and more general. These generalised forms of elementary ideas may be called metaphorical terms in the sense in which every abstract term is metaphorical. The characteristic of a truly scientific system of metaphors is that each term in its metaphorical use retains all the formal relations to the other terms of the system which it had in its original use. The method is then truly scientific, that is, not only a legitimate product of science, but capable of generating science in its turn. There are certain electrical phenomena, again, which are connected together by relations of the same form as those which connect dynamical phenomena. To apply to these the phrases of dynamics with proper distinctions and provisional reservations is an example of a metaphor of a bolder kind; but it is a legitimate metaphor if it conveys a true idea of the electrical relations to those who have been already trained in dynamics. Suppose, then, that we have successfully introduced certain ideas belonging to an elementary science by applying them metaphorically to some new class of phenomena. It becomes an important philosophical question to determine in what degree the applicability of the old ideas to the new subject may be taken as evidence that the new phenomena are physically similar to the old. The best instances for the determination of this question are those in which two different explanations have been given of the same thing. The most celebrated case of this kind is that of the corpuscular and the undulatory theories of light. Up to a certain point the phenomena of light are equally well explained by both; beyond this point one of them fails. To understand the true relation of these theories in that part of the field where they

何之前的具有科学思维的观察者所不能想象的。但是人类的思维并不像傅立叶的热体那样，一直向着完全均匀的最终状态发展，而我们已经可以预测这个状态的性质。这就好像一棵树伸展枝杈，去适应一片新的生长天空，而根则在陌生的地层中委屈自身，因为那才是它们钻研的目标。对于我们这些只呼吸同龄人的精神、只知道当代思维特征的人，是不可能预言未来科学的普遍状况和特别发现的。物理研究不断向我们揭示自然过程的新性质，这也迫使我们寻求可以适应这些性质的新的思维方式。因此，对数学和物理学之间关系的细致研究的重要性在于可以决定从物理的一个领域推导出的观点被安全地用于将在新领域中使用而形成的观点。通过演讲或者思考中的符号，我们可以将一门熟悉的科学中的语言和观点迁移到我们认识较少的科学中，这些符号可以称为科学比喻。因此，速度、动量、力这些词在基础动力学中获得了特定的精确含义。它们也被用在连通系统的动力学中，虽然与其基本含义完全类似，但却有了更加广泛和普遍的意义。这些基本概念的推广形式可以称之为比喻性的词汇，因为其中每一个抽象词汇都是比喻的。比喻这个真正科学系统的特征就是比喻时所用到的每一个词都保留着自己在最初被使用的时候与系统中其他词汇之间的形式关系。这样，这种方法是真正科学的，即，不仅是科学合法的产物，而且能够继续产生科学。一些电现象之间的相互联系和动力学现象之间的相互联系具有相同的形式。在作出恰当区分和提出临时条件后，将这些动力学词汇应用到电现象中就是一个大胆的比喻的例子，但是如果能够向受过动力学方面培训的人传达真正的电学关系，这就是一个合理的比喻。假设我们通过比喻的方式，成功地将某些属于一门基础科学的观点引入到某类新现象，那么旧观点在新课题上的应用能在多大程度上被当作新现象在物理上与旧观点相似的证据，这就成了一个重要的哲学问题。解决这一问题的最好例子就是，对于同种事物存在两种不同的解释。最著名的当属光的粒子理论和波动理论了。在某些情况下，光现象可以用两种理论很好地解释，而在这些情况之外，其中一种理论就会失效。为了理解两种理论在它们同样适用的领域中的真正关系，我们必须看到哈密顿对它们作出的解释，他发现任何瞬时问题都相应于一个自由运动问题，虽然这个自由运动问题会涉及不同的速度和时间，但是会导致相同的几何路径。泰特教授曾经写了一篇关于这个主题的有趣的论文。根据一个在德国取得重大进步的电学理论，两个带电粒子会相隔一定距离直接相互作用，但是据韦伯所说，相互作用力取决于它们的相对速度，而根据高斯提出并由黎曼、洛伦茨和纽曼发展的理论，相互作用不是即时的，而是发生在由距离决定的一段时间之后。为了得到认同，这些杰出的研究者所支持的这一理论解释每

seem equally applicable we must look at them in the light which Hamilton has thrown upon them by his discovery that to every brachystochrone problem there corresponds a problem of free motion, involving different velocities and times, but resulting in the same geometrical path. Professor Tait has written a very interesting paper on this subject. According to a theory of electricity which is making great progress in Germany two electrical particles act on one another directly at a distance, but with a force which, according to Weber, depends on their relative velocity, and according to a theory hinted at by Gauss, and developed by Riemann, Lorenz, and Neumann, acts not instantaneously, but after a time depending on the distance. The power with which this theory, in the hands of these eminent men, explains every kind of electrical phenomena must be studied in order to be appreciated. Another theory of electricity which I prefer denies action at a distance and attributes electric action to tensions and pressures in an all-pervading medium, these stresses being the same in kind with those familiar to engineers, and the medium being identical with that in which light is supposed to be propagated. Both these theories are found to explain not only the phenomena by the aid of which they were originally constructed, but other phenomena which were not thought of, or perhaps not known at the time, and both have independently arrived at the same numerical result which gives the absolute velocity of light in terms of electrical quantities. That theories, apparently so fundamentally opposed, should have so large a field of truth common to both is a fact the philosophical importance of which we cannot fully appreciate till we have reached a scientific altitude from which the true relation between hypotheses so different can be seen.

I shall only make one more remark on the relation between mathematics and physics. In themselves, one is an operation of the mind, the other is a dance of molecules. The molecules have laws of their own, some of which we select as most intelligible to us and most amenable to our calculation. We form a theory from these partial data, and we ascribe any deviation of the actual phenomena from this theory to disturbing causes. At the same time, we confess that what we call disturbing causes are simply those parts of the true circumstances which we do not know or have neglected, and we endeavour in future to take account of them. We thus acknowledge that the so-called disturbance is a mere figment of the mind, not a fact of nature, and that in natural action there is no disturbance. But this is not the only way in which the harmony of the material with the mental operation may be disturbed. The mind of the mathematician is subject to many disturbing causes, such as fatigue, loss of memory, and hasty conclusions; and it is found that from these and other causes mathematicians make mistakes. I am not prepared to deny that, to some mind of a higher order than ours, each of these errors might be traced to the regular operation of the laws of actual thinking; in fact we ourselves often do detect, not only errors of calculation, but the causes of these errors. This, however, by no means alters our conviction that they are errors, and that one process of thought is right and another process wrong. One of the most profound mathematicians and thinkers of our time, the late George Boole, when reflecting on the precise and almost mathematical character of the laws of right thinking as compared with the exceedingly perplexing, though perhaps equally determinate, laws of actual and fallible thinking, was led to

一种电学现象的能力必须得到验证。另一种我更喜欢的电学理论，否认相隔一定距离的相互作用，它将电作用归因于扩散介质中的张力和压力，这些应力与工程师所熟悉的力属于相同的类型，而其中的介质与光传播的介质一样。这两种理论不仅能够解释那些它们最初建立时要解释的现象，还能够解释一些没有想到或者现在还不知道的现象，并且这些理论都各自独立地根据电学量得到了相同的光的绝对速度的数值结果。其实这些看上去完全相反的理论，在很大程度上都是真实的，我们只有在达到可以看清不同假说之间的真实关系的科学高度时，才能理解其中的哲学价值。

我应该再次强调数学和物理学之间的关系。就它们本身而言，一个是大脑的运算，另一个是分子的舞蹈。分子有其自身的法则，我们只是选择其中一些最容易的去理解和计算。我们根据这些片面的数据建立理论，还将任何偏离这个理论的实际现象归因于干扰因素。同时，我们也承认所谓的干扰因素只是真实环境中我们不知道或者已经忽略的那部分，并且在将来我们会努力对这些因素进行全面考虑。因此，我们认识到，所谓的干扰仅仅是大脑的虚构，而不是自然事实，在自然作用中不存在干扰。然而，这并不是大脑运算中物质受干扰的唯一途径。数学家的大脑受到很多干扰因素的支配，如疲劳、失忆和犹豫，并且人们发现数学家会因为这些或其他因素犯错。我并不准备否认，对于那些比我们高级的大脑，这些错误都会被归咎于实际思维法则的规则运算。实际上，我们自身也会经常发现，不只是计算的错误，还有导致这些错误发生的原因。然而，这决不会改变我们的判断：它们是错误，一种思维过程是正确的，而另一种则是错误的。我们这个时代最深奥的数学家和思想家——已故的乔治·布尔，当他考虑到正确思维（相对于极其复杂的，虽然可能是同等正确的，真实的且易错的思维来说），其法则的精确性和近乎数学性时，也被引向了另一种观点，这种观点中科学似乎寻求一个不属于她的区域。“我们必须承认，”他说，“确实存在（思维的）法则，即使是它们严格的数学形式也没能免于被破坏。

another of those points of view from which science seems to look out into a region beyond her own domain. "We must admit," he says, "that there exist laws" (of thought) "which even the rigour of their mathematical forms does not preserve from violation. We must ascribe to them an authority, the essence of which does not consist in power, a supremacy which the analogy of the inviolable order of the natural world in no way assists us to comprehend."

(**2**, 419-422; 1870)

我们必须把它们归因于权威，这个原因的本质并不在于力量，神圣不可破坏的自然界秩序的类比不会帮助我们理解至高法则。”

（王耀杨 翻译；江丕栋 审稿）

Fuel of the Sun

J. J. Murphy

Editor's Note

Here Joseph John Murphy argues that perpetually infalling meteors keep the Sun hot on geological timescales. Such meteor-like phenomena have been observed, he says, citing two meteor-like bodies bright enough to stand out against the Sun's disc, which were seen to move across the Sun from west to east before disappearing. If there is a constant supply of meteors, says Murphy, wouldn't they, like all bodies in the solar system, move around the Sun from west to east, and be found in greater numbers near the Sun's equatorial regions, making these hotter than its poles? Murphy is wrong about the source of solar energy, which comes from nuclear fusion, but his speculations testify to the magnitude of the puzzle.

I am not mathematician enough to form any opinion on the merits of the controversy as to the "fuel of the sun"; that is to say, I am not able to decide whether it is consistent with the conditions of the equilibrium of the solar system that the sun's heat should have been kept up through the ages of geological time by the falling in of meteors. But I wish to state some evidence which proves that meteors are constantly falling in, though it does not touch the question whether this source is sufficient to account for the whole or any large part of the total supply of heat radiated away by the sun.

In the first place, the meteors have been seen. On Sept. 1, 1859, Mr. Carrington and another observer simultaneously observed two meteor-like bodies, of such brightness as to be bright against the sun's disc, suddenly appear, move rapidly across the sun from west to east, and disappear.

The fact that their motion was from west to east is important. If the supply of meteors to the sun is constant and tolerably regular, it is scarcely possible to doubt that the meteors, like the entire solar system, move round the sun from west to east, and occupy a space of the form of a very oblate spheroid, having its equator nearly coincident with the sun's equator.

If this is the case, the meteors ought to fall in greater numbers near the sun's equator than near his poles, making the equator hotter than the poles. Such is the fact. Secchi, without having any theory to support, has ascertained that the sun's equator is sensibly hotter than his poles. The instrument used was an electric thermo-multiplier, and the indications show, not the ratio, but the difference of the heat from the two sources compared.

太阳的燃料

墨菲

编者按

约瑟夫·约翰·墨菲在这篇文章中列举理由以证明不断陨落的流星在地质时间尺度中一直在维持着太阳的热度。他说，这种类似于流星的现象曾经被人观察到过，据说有两个像流星一样的天体亮得可以在日面上显现出来，在消失前它们从太阳的东边移动到西边。墨菲说，如果流星以固定的数量不断地落到太阳上，它们就会像太阳系中的所有其他天体一样，围绕太阳自西向东运动，在太阳的赤道附近数量比较集中，以至于使这些区域的温度高于两极。墨菲的观点当然是错误的，现在大家都知道太阳的能量来自核聚变反应，但是从他的推理中我们可以看出人们曾被这个难题深深困扰过。

我并不是个数学家，因此不足以评论关于“太阳的燃料”的争论的价值。换句话说，我不能确定“由于流星的坠落，太阳的热量在漫长的地质年代中应该一直维持在某个水平”这一观点与太阳系的平衡条件是否相一致。不过，我还是想阐述一些能够证明流星确实在持续不断地坠落到太阳上的证据，至于太阳热辐射的全部或大部分能量是否来源于此，那就是和这些证据无关的另一个问题了。

首先，人们已经观测到了流星。1859 年 9 月 1 日，卡林顿先生和另一位观测者同时观测到两个类似流星的天体，在日面的背景上仍然显得很明亮，它们自西向东迅速地横跨过太阳，突然出现然后又突然消失。

流星自西向东运动的事实是非常重要的。如果坠落到太阳上的流星是持续不断而且坠落密度差不多是均匀的话，那么几乎不用怀疑：流星肯定会像整个太阳系一样围绕着太阳自西向东运动，从空间上来说它们应该是非常扁圆的球体，而且其赤道面应该几乎和太阳的赤道面相重合。

如果事实的确如此，那么大量的流星就会坠落到太阳的赤道附近而不是两极，从而使太阳的赤道比两极更热。实际上，事实正是如此。在没有任何理论支撑的情况下，塞奇已经通过实验确认了太阳的赤道明显比两极热。实验中使用了电热倍增器，该仪器显示的结果是来自太阳赤道和两极这两个热源的热量差值，而不是其比率。

It can scarcely he doubted that the meteors must enter the sun's atmosphere with a velocity not much less than that of a planet, revolving at the distance at which they enter. We know that the sun's rotatory motion is incomparably less than this, and consequently the meteors, revolving from west to east, ought to make the sun's atmosphere move round his body in the same direction, and with greater velocity in the equatorial regions, where most meteors fall in. This is what is observed. Mr. Carrington, also without any theory to support, has shown that the motion of the solar spots from west to east is most rapid in the latitudes nearest the equator. We cannot compare the motion of the spots with that of the sun's body, as we do not see his body. But the fact that the motion from west to east is most rapid in the equatorial latitudes proves that these motions are not due to any cause like that which produces trade-winds and "counter-trades" of our planet; for, supposing the sun or any planet to rotate from west to east, in any circulation that could be produced in its atmosphere by unequal heating at different latitudes, the relative motions of the atmospheric currents in high and low latitudes would be similar to that of the trade-winds and "counter-trades", and opposite to that which the motions of the spots indicate in the atmosphere of the sun. This will be true at all depths in the atmosphere.

(**2**, 451; 1870)

毫无疑问，流星必然是以一定的速度进入太阳大气，这一速度与行星的速度相比并不会小很多，流星进入太阳大气后会在相应的距离上围绕太阳旋转。我们也知道，太阳的自转与这种旋转相比慢得多，因此，绕着太阳自西向东旋转的流星应该会带动太阳大气也绕着太阳实体沿相同方向旋转，而且在大多数流星坠入其中的太阳赤道区域，太阳大气绕转的速度会更大。观测到的结果正是如此。同样是在没有任何理论支持的情况下，卡林顿先生已经指出，太阳上纬度最靠近赤道的区域中太阳黑子自西向东运动的速度最快。由于看不到太阳实体，因此我们无法比较太阳实体的运动和黑子的运动，但是，太阳黑子自西向东的运动在赤道附近最为迅速这一事实就能证明，这些运动并不是由造成我们星球上的信风或反信风现象的那类原因引起的。这是因为，假设太阳或行星都是自西向东旋转的话，那么由于不同纬度上热量的不平衡而引起其大气发生的任何循环中，气流在高纬度和低纬度的相对运动都将与信风和“反信风”的运动情况类似，而与黑子运动所暗示的太阳大气的运动情况相反。对于太阳大气来说，在任何高度上都存在这一矛盾。

（金世超 翻译；蒋世仰 审稿）

Dr. Bastian and Spontaneous Generation

T. H. Huxley

Editor's Note

Thomas Huxley's address to the British Association in 1870, published earlier in *Nature*, was a critique of the assertion by H. Charlton Bastian (itself made in a long paper in *Nature*) that micro-organisms could be spontaneously generated in sterile water. Bastian had responded to Huxley at length, and here Huxley engages in the next round of battle. His letter, magisterially dismissive of Bastian's claims, is a fine example of Victorian polemic. Huxley was quite correct to be sceptical, but the debate shows that spontaneous generation was still entertained at that time in serious scientific circles. It was eventually ruled out, thanks in part to careful experiments by Louis Pasteur.

I find that the "Address" which it was my duty to deliver at Liverpool, fills thirteen columns of *Nature*. The "Reply" with which Dr. Bastian has favoured you occupies fifteen columns, and yet professes to deal with only the first portion of the "Address". Between us, therefore, I should imagine that both you and your readers must have had enough of the subject; and, so far as my own feeling is concerned, I should be disposed to leave both Dr. Bastian and his reply to the benign and Lethean influences of Time.

But I am credibly informed that there are persons upon whom Dr. Bastian's really wonderful effluence of words weighs as much as if it were charged with solid statements and accurate reasonings; and I am further told that it is my duty to the public to state why such distinguished special pleading makes not the least impression on my mind. With your permission, therefore, I will do so in the briefest possible manner.

The first half of Dr. Bastian's "Reply" occupies seven columns of your number for the 22nd of September. In all this wilderness of words there is but one paragraph which appears to me to be worth serious notice. It is this: —

"In the first place, he does not attempt to deny—he does not even allude to the fact—that *living things may and do arise as minutest visible specks, in solutions in which, but a few hours before, no such specks were to be seen*. And this is in itself a very remarkable omission. The statement must be true or false—and if true, as I and others affirm, the question which Professor Huxley has set himself to discuss is no longer one of such a simple nature as he represents it to be. It is henceforth settled that as far as *visible* germs are concerned, living beings can come into being without them."

If I did not allude to the assertion which Dr. Bastian has put in italics—it is because it bears absurdity written upon its face to any one who has seriously considered the

巴斯蒂安博士与自然发生学说

赫胥黎

编者按

早些时候，《自然》发表过托马斯·赫胥黎于1870年向英国科学促进会作的报告。该报告批评了查尔顿·巴斯蒂安主张的无菌水中可能会自发产生微生物的观点（这是在一篇发表于《自然》上的长篇论文中提出来的）。巴斯蒂安详尽地回复了赫胥黎的意见，于是赫胥黎又发起了新一轮辩论。赫胥黎的文章权威地驳斥了巴斯蒂安的观点，可以说是维多利亚时代学术争辩的典范。赫胥黎的观点是非常正确无可怀疑的，不过，通过他们之间的辩论可以看到，当时在严肃的学术圈内也还有人坚持自然发生学说的观点。一定程度上来说，是因为路易·巴斯德的一系列精细实验，自然发生学说才最终被人们抛弃。

我发现，我因个人职责而在利物浦发表的“致词”居然占据了《自然》13栏的版面。巴斯蒂安博士支持贵杂志而作出的“回复”又占据了15栏的版面，但却声称只针对我的“致词”中的第一部分进行了讨论。因此，我应该想象得出《自然》及其读者对于我们俩之间讨论的话题已经有了充分的了解；另外，就我个人情感而言，我愿意让巴斯蒂安博士和他的回复给时代留下一个友好的、能够遗忘过去的影响。

但是我据可靠消息得知，有一些人认为巴斯蒂安博士的精彩论断就像充满了可靠的论述和精准的推理一样有分量；我还被进一步告知，我有责任对公众说明为什么这么卓越非凡的抗辩却没有给我留下一丝一毫的印象。因此，如果您许可，我将尽可能以最简洁的方式在此做一说明。

在贵杂志9月22日那一期上，巴斯蒂安博士的“回复”的前半部分就占了7栏。在我看来，这些长篇大论中只有一段值得认真注意一下。该段原文如下：

“首先，他并没有试图否认，甚至没有提及如下事实——**溶液中可能并且确实出现了生物，那是一些极小的可见斑点，而在几个小时前这些溶液中还看不到这种斑点**。这本身是一个非常值得注意的遗漏，因为该陈述肯定非对即错。如果是对的，那么就像我和其他人断言的那样，赫胥黎教授自己着力讨论的问题并不像他描绘出来的那样简单。从今以后我们可以确定，生物可以在不存在**可见**细菌的情况下产生。”

假如我没有提及巴斯蒂安博士那些突出表示的论断——那是因为对于任何一个认真考虑了显微观察条件的人来说，这种论断都相当于在自己脸上写下荒谬的印记。

conditions of microscopic observation. I have tried over and over again to obtain a drop of a solution which should be optically pure, or absolutely free from distinguishable solid particles, when viewed under a power of 1,200 diameters in the ordinary way. I have never succeeded; and, considering the conditions of observation, I never expect to succeed. And though I hesitate to speak with the air of confident authority which sits so well on Dr. Bastian, I venture to doubt whether he ever has prepared, or ever will prepare, a solution, in a drop of which no "minutest visible specks" are to be seen by a careful searcher. Suppose that the drop, reduced to a thin film by the cover-glass, occupies an area 1/3 of an inch in diameter; to search this area with a microscope in such a way as to make sure that it does not contain a germ 1/40,000 of an inch in diameter, is comparable to the endeavour to ascertain with the unassisted eye whether the water of a pond, a hundred feet in diameter is or is not absolutely free from a particle of duckweed. But if it is impossible to be sure that there is no germ 1/40,000 of an inch in diameter in a given fluid, what becomes of the proposition so valuable to Dr. Bastian that he has made your printer waste special type upon it ?

I now pass to the second part of the "Reply", which, though longer than the first, is really more condensed, inasmuch as it contains two important statements instead of only one.

The first is, that Dr. Bastian has found *Bacterium* and *Leptothrix* in some specimens of preserved meats. I should have been very much surprised if he had not. If Dr. Bastian will boil some hay for an hour or so, and then examine the decoction, he will find it to be full of *Bacteria* in active motion. But the motion is a modification of the well-known Brownian movement, and has not the slightest resemblance to the very rapid motion of translation of active living *Bacteria*. The *Bacteria* are just as dead as those which Dr. Bastian has seen in the preserved meats and vegetables; and which were, I doubt not, as much put in with the meat, as they are with the hay, in the experiment to which I invite his attention.

The second important statement in the second part of the "Reply" is: —

"Professor Huxley is inclined to believe that there has been some error about the experiments recorded by myself and others."

In this I cordially concur. But I do not know why Dr. Bastian should have expressed this my conviction so tenderly and gently as regards his own experiments; inasmuch as I thought it my duty to let him know both orally and by letter, in the plainest terms, six months ago, not only that I conceived him to be altogether in the wrong, but why I thought so.

Any time these six months Dr. Bastian has known perfectly well that I believe that the organisms which he has got out of his tubes are exactly those which he has put into them; that I believe that he has used impure materials, and that what he imagines to have been

我曾经用普通的可以放大 1,200 倍的显微镜进行观察，反复尝试以期得到一滴视觉上纯净的溶液，或完全不含可辨别固体颗粒的溶液。但都失败了；考虑到这种观察条件，我也没指望过会成功。尽管我不愿像巴斯蒂安博士那样以自信的权威口吻讲话，但我还是冒昧地对其表示怀疑——他是否已经制备好了，或者将会制备出一种溶液，即使是很仔细的观察者也看不到这样的一滴溶液中有“极小的可见斑点”。假如通过盖玻片将这滴溶液压成薄膜后，只占据直径为 1/3 英寸的区域，那么以这种方法用显微镜搜索这一区域，以确定该区域是否含有直径为 1/40,000 英寸的细菌，就如同仅靠肉眼来确定直径为 100 英尺的一方池塘的水中是否绝对不含一片浮萍一样。但如果不可能确定给定的液体中不含直径为 1/40,000 英寸的细菌的话，那么巴斯蒂安博士的哪些主张如此有价值以至你们的印刷机肯为其浪费呢？

现在我要开始讨论“回复”的第二部分，尽管这部分比第一部分长，但是其实这部分内容已经更加精简了，因为它包含的是两个重要的陈述而不是一个。

第一个重要陈述是，巴斯蒂安博士已经在某些腌肉样本中发现了杆菌和纤毛菌。如果他没有发现的话，那就太令我惊讶了。如果巴斯蒂安博士将一些干草煮沸一小时左右，然后再观察煮出来的草汁，那么他将发现草汁中充满了活跃运动着的杆菌。但是这种运动只是著名的布朗运动的一种变体，与活跃的、有生命的杆菌的快速运动方式没有一点相像之处。这些杆菌与巴斯蒂安博士看到的腌肉和咸菜中的杆菌一样都是死的。并且，我相信这些细菌是在实验中同肉一起放进去的，就像它们随干草一起被放进去一样，对于这一点，我希望能够引起他的注意。

“回复”第二部分的第二个重要陈述是：

“赫胥黎教授倾向于认为由我自己和其他人记录的实验中存在错误。”

我由衷地赞成这一点。但是我不知道巴斯蒂安博士为什么将我对他的实验的这种指责表达得如此温和婉转；因为我觉得通过口头和书面形式让他知道这一点是我的职责，所以 6 个月前，我就以最简明的措辞让他不仅知道他是完全错误的，而且告诉了他我为什么这么认为。

这 6 个月里，无论何时，巴斯蒂安博士都已经清楚地知道，我认为他试管中的生物其实正是被他放进试管中去的那些；我还认为他使用了不纯的物质，而他所想

the gradual development of life and organisation in his solutions, is the very simple result of the settling together of the solid impurities, which he was not sufficiently careful to see, in their scattered condition when the solutions were made.

Any time these six months Dr. Bastian has known why I hold this opinion. He will recollect that he wrote to me asking permission to bring for my examination certain preparations of organic structures, which he declared he had clear and positive evidence to prove to have been developed in his closed and digested tubes. Dr. Bastian will remember that when the first of these wonderful specimens was put under my microscope, I told him at once that it was nothing but a fragment of the leaf of the common Bog Moss (*Sphagnum*); he will recollect that I had to fetch Schacht's book "Die Pflanzenzelle", and show him a figure which fitted very well what we had under the microscope, before I could get him to listen to my suggestion; and that only actual comparison with *Sphagnum*, after he had left my house, forced him to admit the astounding blunder which he had made.

To any person of critical mind, versed in the preliminary studies necessary for dealing with the difficult problem which Dr. Bastian has rashly approached—the appearance of a scarlet geranium, or of a snuff-box, would have appeared to be hardly more startling than this fragment of a leaf, which no one even moderately instructed in vegetable histology could possibly have mistaken for anything but what it was; but to Dr. Bastian, agape with speculative expectation, this miracle was no wonder whatever. Nor does Dr. Bastian's chemical criticality seem to be of a more susceptible kind. He sees no difficulty in the appearance of living things in potash-alum, until Dr. Sharpey puts the not unimportant question, whence did they get their nitrogen? And then it occurs to him to have the alum analysed and he finds ammonia in it.*

And as to the elementary principles of physics—in his last communication to you, Dr. Bastian shows, that he is of opinion that water in a vessel with a hole in it, from which the steam freely issues, may be kept at a temperature of "230° to 235° F for more than an hour and a half."† I hope that Professor Tyndall, whom Dr. Bastian scolds as authoritatively and as unsparingly as he does me, will take note of this revolutionary thermotic discovery, in the next edition of his work on Heat.

It is no fault of mine if I am compelled to write thus of Dr. Bastian's labours. I have been blamed by some of my friends for remaining silent as long as I have done concerning them. But when, because I have preserved a silence, which was the best kindness I could show to Dr. Bastian, he presumes to accuse me publicly of unfairness, and to tell your readers that my Address "is calculated to mislead" them, I have no alternative left but to give them the means of judging of the competency of my assailant.

(**2**, 473; 1870)

T. H. Huxley: Jermyn Street, Oct. 10.

---

* See *Nature*, No. 36, p. 198.

† Ibid, No. 48, p. 433.

象的那些在他的溶液中经历了生命和组织结构的逐步发育的东西，其实就是配制溶液时以分散状态一起放入试管中的固体杂质，他只是没有足够认真地去观察而已。

这6个月里，无论何时，巴斯蒂安博士都应该已经清楚我为什么会持有这种观点。他应该记得，他曾写信请我对他的某些具有有机结构的制品进行检查，并声称有清楚确凿的证据证明这些有机物是在他的密闭消化管中得到的。他也应该记得，当这些极好的样品中的第一个被放在我的显微镜下时，我立刻告诉他这只不过是普通泥炭藓的叶子碎片而已；他还应该记得，当我不能让他信服我的意见时，我不得不找出沙赫特的《植物细胞》一书，给他指出其中的一幅图画，那与我们在显微镜下所看到的完全吻合；离开我家后，他又将自己的实验结果与真实的泥炭藓进行了比较，才被迫承认自己所犯的惊人错误。

对于任何具有批判思维的人，在对待巴斯蒂安博士鲁莽提出的难题时，都会谨慎地进行必要的初步研究——出现一株绯红色的天竺葵或者一个鼻烟盒恐怕都不会比这片叶子碎片更令人吃惊，并且，即使是仅仅受过中等植物组织学教育的人，也不可能将它误认为其他东西；但是对于这位对根据推理而得出的预期都会瞠目结舌的巴斯蒂安博士来说，这样的奇事也就不足为奇了。巴斯蒂安博士对化学制品的知识好像也不是那么敏感。他不费吹灰之力就在明矾中发现了生物，直到沙比博士提出了如下重要问题，即，它们是从哪里得到所需的氮气的？这时候他才开始分析明矾，并且又从中发现了氨。*

至于物理学的基本原理——在与贵刊的最后一次通信中，巴斯蒂安博士表明，他的观点是对于有孔容器中的水，由于水蒸气可以从小孔处自由排出，所以水温可以保持在“华氏230~235度之间达一个半小时以上。”† 巴斯蒂安博士曾像斥责我一样以一种权威的口吻毫不宽容地斥责过廷德尔教授，我希望廷德尔教授能够在他下一项关于热量的研究工作中留意这一革命性的热学发现。

我不得不将巴斯蒂安博士的工作以这种方式写出来，这并非我的过错。我的一些朋友已经责备过我，埋怨我自从关注到这些工作以来就一直在保持沉默。我已经保持了沉默，这是我对巴斯蒂安博士表现出的最友好的态度，但是当他不正当地妄自公开控诉我，并且告诉《自然》的读者我的致词“旨在欺骗误导”他们的时候，除了给读者们判断攻击我的人的能力素质的方法之外，我别无选择。

（刘皓芳 翻译；刘力 审稿）

*《自然》，第36期，第198页。

† 同上，第48期，第433页。

On the Colour of the Lake of Geneva and the Mediterranean Sea

J. Tyndall

Editor's Note

Here physicist John Tyndall adds to the topical debate over the cause of water's blueness. Supplied with bottles of water from Lake Geneva and the Mediterranean, he reports experiments to test the hypothesis that water's blue colour arises from light scattering by suspended particles. Illuminating the samples, Tyndall found that the transmitted beam was blue, suggesting that something in the water is scattering the shorter (bluer) wavelengths more strongly. Using crystals to probe the light's polarisation, he found it to be most polarised perpendicular to the beam, like light scattered in the atmosphere. Yet Tyndall also notes the prescient suggestion of a French physicist, M. Lallemand, that molecules of water themselves, not foreign particles, may be scattering the shorter wavelengths of light.

THROUGH kindness for which I have reason to feel both proud and grateful, I have had placed in my hands two bottles of water taken from the Mediterranean Sea, off the coast of Nice. To my friend M. Soret I am also indebted for two other bottles taken from the Lake of Geneva. The friendly object in each case was to enable me to examine whether the colour of the water could in any way be connected with the scattering of light by minute foreign particles, which is found so entirely competent to produce and explain the colour of the sky. In the open Lake of Geneva, Soret himself had studied this question with considerable success*, and my desire was to apply to it other methods of examination.

The bottles, as they reached me, and with their stoppers unmoved, were placed in succession in the convergent beam of an electric lamp. Water optically homogeneous would have transmitted the beam without revealing its track. In such water the course of the light would be no more seen than in optically pure air. The cone of light, however, which traversed the liquid, was in both cases distinctly blue, the colour from the Lake of Geneva water being especially rich and pure. Something, therefore, existed in the liquid which intercepted and scattered, in excess, the shorter waves of the beam. The longer waves were also scattered, but in proportions too scanty to render the track of the beam white. The action, in fact, was identical with that of the sky. Viewed through a Nicol's prism the light was found polarised, and the polarisation along the perpendicular to the illuminating beam was a maximum. In this direction, indeed, the polarisation was sensibly perfect. A crystal of tourmaline placed with its axis perpendicular to the beam was transparent; with its axis parallel to the beam it was opaque. By shaking the liquid larger

* See his letter to me, *Philosophical Magazine*, May, 1869.

日内瓦湖和地中海的色彩

廷德尔

编者按

在这篇文章中，物理学家约翰·廷德尔针对海水为什么显现蓝色这一论题陈述了自己的观点。他拿来了日内瓦湖和地中海的水样，并指出实验结果已经证明悬浮颗粒所引起的光散射是海水呈现蓝色的原因。在照亮这些样品的时候，廷德尔发现透过的光束是蓝色的，这说明水中必定存在某种更容易散射较短波长的光（蓝光）的物质。通过用晶体检测光的偏振，他还发现垂直于光束方向的光的偏振效应最显著，这和空气中的光散射是一样的。廷德尔还特别提到了法国物理学家拉勒芒的预测，即认为散射较短波长的光的正是水分子本身而不是外来颗粒。

我手头上保存着两瓶取自尼斯海岸附近地中海的水样，这两瓶水样蕴含着朋友们的热心帮助，对此我感到非常自豪和感激；我也非常感谢我的朋友索雷特，是他帮助我采集了另外两瓶日内瓦湖的水样。这些透着浓浓友情的样品使我能够检验水的色彩是否与外来微粒的光散射有关，而这又可以很好地解释天空的颜色。在广阔的日内瓦湖面上，索雷特本人已经相当成功地研究了这个问题*，所以我希望可以用其他方法来研究这个问题。

这些瓶子到我手上的时候连瓶盖都还没有动过，我把它们按照次序放在电灯的会聚光束之下。光线将会在这些看起来均质的水中传播而不显示轨迹。在水中看不到光路，这一点和光在看上去很纯的空气中的传播一样。不过穿透液体的光锥在这两种水样中都是很明显的蓝色，而从日内瓦湖的水样中穿过的光的色彩特别饱满和纯正。因此，水中应该存在某种物质拦截并散射了很大部分的短波光线；长波光线也会被散射，但是由于比例太小不足以产生白色光迹。事实上，微粒对光线的这种作用和在天空中的效果是相同的。透过尼科尔棱镜观察，可以发现光发生了偏振，而且在与发光光束垂直的方向上偏振达到了最大值，在此方向上，确实是明显的全偏振。如果将电气石晶体的晶轴垂直于光线放置，光线就可以通过晶体，如果平行放置，光线则不能通过晶体。如果摇动水样，较大的颗粒就会漂浮并在光束中闪耀。

* 参见他给我的来信，《哲学杂志》，1869 年 5 月。

particles could be caused to float and sparkle in the beam. The delicate blue light between these particles could be quenched by the Nicol while they were left shining in the darkened field. A concave plate of selenite, placed between the Nicol and the water, showed a system of vividly coloured rings. They were most brilliant when the vision was at right angles to the beam, just as they are most brilliant when the blue sky is regarded at right angles to the rays of the sun. In no respect could I discover that the blue of the water was different from that of the firmament. The colour presented by the Mediterranean water was a good sky-blue, while that presented by the Geneva water matched a sky of exceptional purity.

My interest was long ago excited by the attempts made to account for the colour of the Lake of Geneva, and continued observation in 1857 impressed me more and more with the notion that the blue was mainly that of a turbid medium. Soon afterwards I wrote thus regarding this colour:—

"Is it not probable that this action of finely divided matter may have some influence on the colour of some of the Swiss lakes—on that of Geneva for example? This lake is simply an expansion of the river Rhone, which rushes from the end of the Rhone glacier. Numerous other streams join the Rhone right and left during its downward course, and these feeders being almost wholly derived from glaciers, carry with them the fine matter ground by the ice from the rocks over which it has passed. Particles of all sizes must be thus ground off, and I cannot help thinking that the finest of them must remain suspended in the lake throughout its entire length. Faraday has shown that a precipitate of gold may be so fine as to require a month so sink to the bottom of a bottle five inches high; and in all probability it would require *ages* of calm subsidence to bring all the particles in the Lake of Geneva to its bottom. It seems certainly worthy of examination whether such particles, suspended in the water, do not contribute to that magnificent blue which has excited the admiration of all who have seen it under favourable circumstances."*

Through the observations of Soret, and through those here recorded, the surmise of thirteen years ago has become the verity of today.

But though in the action of small particles we have a cause demonstrably sufficient to produce the blueness referred to, it is not the only cause operative. In the Lake of Geneva we have not only the blue of scattering by small particles, but also the blue arising from true molecular absorption. Indeed, were it not for this, the light *transmitted* by a column of the water would be yellow, orange, or red, like the light of sunrise or sunset.† Not only then is the light mainly blue from the first moment of its reflection from the minute particles, but the less refrangible elements which always accompany the blue are still further abstracted during the transmission of the scattered light. Through the action of both these causes, scattering and absorption, the intense and exceptional blueness both of the Lake of Geneva and the Mediterranean Sea I hold completely accounted for.

---

* Glaciers of the Alps (1860), p. 261.

† In fact, we have a dichroitic action of this kind exerted by glacier water when the subsidence is less complete than in the Lake of Geneva.

透过尼科尔棱镜观察，粒子在暗场处发光，但粒子之间的微弱蓝色光线进一步弱化。若在尼科尔棱镜和水样之间放置一个透明石膏制成的凹面盘，则会产生色彩鲜艳的光环。在与光束垂直的方向观察时光环最明亮，就像在与太阳光线垂直的方向观察时蓝天最明亮一样。从任何方面来看，水的蓝色与苍穹的蓝色都没有区别。地中海海水的色彩是晴天时的那种天蓝色，而日内瓦湖湖水的色彩则是天空在格外纯净时的那种颜色。

我多年来一直试图解释日内瓦湖的色彩，这个想法鞭策着我不断努力。1857 年，持续的观察越来越强烈地促使我形成这样一种推测：这种蓝色主要是由一种浑浊的介质形成的。关于这种色彩，后来我是这样写的：

“这种细分物质的作用会不会影响瑞士湖泊的色彩？比如日内瓦湖，它是罗纳河扩展而形成的，罗纳河则发源自罗纳河冰川。大量支流在罗纳河下游两岸汇入其中，这些支流几乎全部发源于冰川，这些支流带走了冰川流过岩石时侵蚀产生的细碎物质。在搬运和迁移中，各种大小的颗粒被磨圆，所以我认为，最细小的颗粒在湖泊中一直保持悬浮状态。法拉第已经证明，很小的金沉淀物需要一个月时间才能沉淀到 5 英寸高的瓶子的底部。所以要想使日内瓦湖中所有颗粒以静水沉积的方式都沉淀到湖底，很可能需要**相当长的时间**。所以，我觉得研究一下是否是这些悬浮在水中的颗粒引起了这种神奇的、令人赞叹的蓝色是非常有价值的。”*

通过索雷特的观察和我在这里的讲述可以看到，13 年前的推测到今天已经变成了事实。

尽管我们已经有理有据地阐明微小颗粒的作用足以形成蓝色，但这并不是唯一起作用的原因。日内瓦湖的蓝色不仅源于微小颗粒引起的蓝色散射，也有分子吸收而产生的蓝色。如若不然，在水中**传播的**光将是黄、橙、或红色，有如旭日东升或者夕阳西下时的色彩。†不仅微小颗粒最初反射的结果是蓝光，而且在散射光的传播过程中，那些总是伴随着蓝光而又不易被折射的成分被进一步吸收。我认为，通过色散和吸收这两种作用，日内瓦湖和地中海独一无二的浓厚的蓝色可以得到完整的解释。

---

*《阿尔卑斯山的冰川》(1860 年)，第 261 页。

† 实际上，我们发现，沉淀并不像日内瓦湖中那么严重时，冰川水会表现出这种二色性作用。

During the year 1869, M. Lallemand communicated to the Paris Academy of Sciences some interesting papers on the optical phenomena exhibited by certain liquids and solids when illuminated like the actinic clouds in my experiments. I also, in 1868, had examined a great number of liquids in the same manner, and a brief reference to these experiments will be found towards the end of a paper on the blue colour of the sky and the polarisation of its light, published in the *Proceedings of the Royal Society* for the 16th of December, 1868. M. Lallemand supposed the scattering of the light to be effected not by foreign particles but by the molecules of the liquids with which he experimented. M. Soret, on the other hand, contends against this novel view, maintaining that the scattering of the light is an affair of particles and not of molecules. While admiring the skill and learning displayed by the young French physicist, I am forced to take the side of Soret in this discussion M. Lallemand assumes a purely hypothetical cause while a true cause is at hand. He bases his case mainly on clear glass and distilled water. But the clearness is that observed in ordinary daylight, which is a very deceptive test. Glass exhibits the phenomena of scattering in every degree of intensity. Exceedingly fine examples of dichroitic action on the part of this substance are to be seen in Salviati's window in St. James's Street.* By reflected light the dishes and vases there exposed exhibit a beautiful blue—by transmitted light, a ruddy brownish yellow. The change of colour is very striking when, having seen the blue, a white cloud is regarded through the glass. Where the opalescence is strongest, the transmitted light, as might be expected, is most deeply tinged. From these examples, where the foreign ingredient is intentionally introduced, we may pass by insensible gradations to M. Lallemand's glass. The difference between them is but one of degree. Many of the bottles of our laboratory show substantially the same effect as the glass of Salviati. We can hardly ascribe to molecular action, which is constant, an effect so variable as this. It is also a significant fact that, in the case of pellucid bodies—rock salt, for example—where the powerfully cleansing force of crystallisation has come into play, M. Lallemand himself found the scattering to be *nil*. Under severe examination, rock salt itself would probably be found not altogether devoid of scattering power. I have examined many fine specimens of this substance, and have not succeeded in finding a piece of any size absolutely free from defect. A common form of turbidity exhibited by clear rock salt, when severely tested, resembles on a small scale "a mackerel sky". Nor have the specimens of Iceland spar that I have hitherto examined proved absolutely wanting in this internal scattering power.

In relation to this question, which is one of the first importance, the deportment of ice is exceedingly instructive. As a rule the concentrated beam may be readily tracked through ice, at least at this season of the year, when the substance shows signs of breaking up internally. In some cases the sparkle of motes, which are evidently spots of optical rupture, reveals the track of the beam. In other cases the track appears bluish, though rarely of a uniform blue. By causing a previously sifted beam to traverse lake ice in various directions, we are soon made aware of remarkable variations in the intensity of the scattering, and

* Mentioned to me by a correspondent.

1869 年，拉勒芒向巴黎科学院提交了几篇很有趣的关于光学现象的论文，这些现象是某些液体和固体在受到光照时产生的，就像在我的实验中出现的光化云雾。1868 年，我也用同一种方法检验了大量的液态物质，并在一篇论文的结束部分简要地提到了这些实验的情况，主要是关于天空的蓝色及其光的偏振现象，这篇论文发表在 1868 年 12 月 16 日的《皇家学会学报》上。拉勒芒认为，光的色散并不是由外部粒子引起的，而是由他在实验中所用液态的分子引起的。索雷特反对这种新奇的观点，认为光的色散是粒子作用的结果，而非分子作用。尽管我也尊重这位年轻的法国物理学家所表现出的技能和学识，但我不得不在讨论中支持索雷特的观点。拉勒芒提出了一种纯粹假设性的理由，而真正的理由却唾手可得。他的研究只是建立在洁净的玻璃杯和蒸馏水上，但这个洁净程度是在普通光线下观察到的，这种实验不一定靠得住。玻璃在任何光强下都能够发生色散，在圣詹姆斯大街上的萨尔维亚蒂的橱窗里可以看到非常好的二色性作用。* 通过反射光线的照射，那里展示的盘子和花瓶呈现出亮丽的蓝色；通过透射光线的照射，则呈现出黄褐色。当看到蓝色后，我们就会注意到玻璃的白色云雾，这时这种颜色的变化是非常明显的。在乳白色的光最强的地方，透射光线是最弱的，这和我们的预期是一致的。通过这些特意引入了外部因素的实验，我们就可以忽略拉勒芒的玻璃瓶实验中不易被察觉的颜色层次的渐变，各实验结果之间的区别就只是一维的了。我们实验室中的许多瓶子实际上表现出了和萨尔维亚蒂的玻璃窗一样的效果。我们不可能将这种不同的效果归因于通常的分子作用。另一个重要的事实是，在透明体（如岩盐）的实例中，其强烈的结晶净化起了作用，拉勒芒发现色散现象**完全消失**。经过严格的检验，我们发现岩盐本身可能并不是完全没有色散能力。我检验了很多这种物质的样本，并发现它们都或多或少存在晶体缺陷。一块具有普通浑浊形态的岩盐，经严格检测后可以发现，它很像小范围的“鱼鳞天”。迄今为止我检测过的冰洲石晶体，没有一块是完全没有内在色散力的。

对于这个重要问题，冰的形态很有启示作用。通常情况下，会聚光束很容易透过冰，至少在每年冰将要从内部开始破碎的这个季节是这样。在某些情况下，微尘颗粒能够揭示光的轨迹，因为它们很明显是光路的断点。在其他情形下，轨迹略显浅蓝色，尽管很少一成不变。让先前射入的光束以不同方向通过湖冰，我们立刻可以发现色散强度的变化，并可以发现有些地方光迹完全消失。会聚光束有时被横断面分开，一半光锥可见，另一半不可见。其他情形下，光锥被从顶端到底端的平面

* 一名通讯员告诉我的。

we find some places where the track of the beam wholly disappears. The convergent beam is sometimes divided by a transverse plane, one half of the cone being visible and the other invisible. In other cases the cone is divided by a plane passing from apex to base, one half shining with scattered light, and the other showing the darkness of true transparency. Now, if the scattering were molecular, it ought to occur everywhere, but it does not so occur, therefore it is not molecular. The scattering is, perhaps, in most cases due to the entanglement in the ice, when the freezing is rapid, of the ultra-microscopic particles abounding in the water. It is only by excessively slow freezing that such particles could be excluded from the ice. Purely optical ruptures of the substance itself, if minute and numerous enough, would also produce the observed effect.

The liquids which I examined in 1868 all showed in a greater or less degree the scattering of light, to which was added in many cases strong fluorescence. In no respect did the deportment of the non-fluorescent liquids which showed a blue track differ from that of the blue actinic clouds with which I was then occupied. I examined water from various sources and found it uniformly charged, not only with particles small enough to scatter blue light, but with far grosser particles. Tested by the concentrated beam, our ordinary drinking water presents a by no means agreeable appearance; some of the water with which London is supplied is exceedingly thick and muddy. Nor does distillation entirely remove the suspended matter. Soret vainly tried to get rid of it; he diminished its effect, but he did not abolish it. I was favoured a few days ago with specimens of distilled water from four of the principal London laboratories. Looked at in ordinary daylight the liquid in each case would, in ordinary parlance, be pronounced "as clear as crystal", but when placed in the concentrated beam of the electric lamp, the notion of purity became simply ludicrous. No one who had not seen it would be prepared for the change produced by the concentrated illumination. There were differences of purity among the specimens, arising, doubtless, from the different modes of distillation, but to an eye capable of seeing in ordinary light what was revealed by the concentrated beam each of the specimens would appear as muddy water. I also examined a specimen of extra purity distilled from the permanganate of potash and liquefied in a glass condenser. It contained a large amount of foreign particles; not of those which scatter blue light, but grosser ones. Such must ever be the case with water distilled in the laden air of cities and collected in vessels contaminated by such air. These facts amply justify the language applied by Mr. Huxley to the statement that solutions without particles can be obtained by the processes hitherto pursued. Such a statement could only be based upon defective observation. In the number of this journal for the 17th of March, an experiment is described in which water was obtained from the combustion of hydrogen in air, the aqueous vapour arising from the combustion being condensed by a silver surface of unimpeachable purity. In this case, though the floating particles of the air were, in the first instance, totally consumed, the water was still well laden with foreign matter. The method of obtaining water here referred to had been resorted to by M. Pouchet with a view of utterly destroying all germs, and my especial object in repeating the experiment was to reveal the dangers incident to the inquiries on which M. Pouchet and others were then engaged. But the warning was unheeded. It is not for the purpose of adding to the weight of calamities, already sufficiently heavy, that

分开，一半闪烁色散光，另一半完全黑暗。所以，如果色散是分子运动引起的，各部分就应该都有这种现象发生，但事实并非如此，因此，色散不是分子运动的结果。色散或许在多数情形下是由被冰包围着的粒子引起的，当结冰很迅速时，水中存在的大量超微粒子也被夹入其中。只有极其缓慢的结冰过程才可以将这样的粒子排除在冰块之外。这种物质本身的纯粹光学裂缝，如果足够小而且数量足够多的话，也会产生观测到的这种结果。

1868 年我所检测的液态物质都或多或少有光的色散现象，很多情形下都出现了很强的荧光现象。从任何方面看，那些可以显现出蓝色轨迹的非荧光液体的行为与我当时研究的蓝色光化云雾都没有什么不同。我所检测的水样的来源各不相同，但这些水样都带有电荷，不仅含有可以发散蓝色光的微小粒子，也含有相对较大的颗粒。用会聚光束进行检测，普通饮用水给出的结果也很不相同，伦敦地区的一些水样非常混浊不清，即便蒸馏也不能完全除去悬浮物，索雷特试图彻底除去这些悬浮物，但没能成功，他的确降低了悬浮物的影响，但是并没有完全消除。几天前，我获得了伦敦 4 个最重要的实验室的蒸馏水水样，在日光下，这些液体看上去“就像水晶一样晶莹透亮”。但放置于普通电灯的会聚光束下，纯净这一概念就变得有些滑稽可笑了。没有看过这种现象的人绝不会对这种会聚光照射下的变化做好心理准备。尽管由不同的蒸馏方式产生的水样的纯净程度各不相同，但是对于能在普通光线下看清由会聚光束照射而揭示出的情景的眼睛来说，各组样品简直都是一团混浊。我也检测了一种非常纯的样品，它是通过蒸馏高锰酸钾溶液而后在玻璃冷凝管中冷凝得到的，其中含有大量的外来粒子，但这些并不是散射蓝光的粒子，而是比较粗大的粒子。如果城市空气中富含杂质，而蒸馏水又被放置在被空气污染的容器里，那这样的样品给出的结果也一定跟高锰酸钾冷凝液给出的结果差不多。这些事实充分证明了赫胥黎对于可以用已有方法获得不含粒子的溶液这一论述的看法。显然这样的说法只是基于不真实的观察。在 3 月 17 日的《自然》上，有一个实验描述了水可以通过氢气燃烧生成，燃烧所产生的水蒸气在纯银表面冷凝。在这个实验中，尽管空气中的飘浮粒子完全被消耗掉了，但产生的水中仍然混有外部物质。这里所提到的获取水的方法曾经被普歇采用，并被认为可以完全消灭各种细菌，而我重复这个实验的特殊目的是要揭示那些研究的危险，恐怕普歇和其他一些人正深陷其中。但这种警示没有引起人们的注意。我在这里提到这个问题，并不是要有意扩大危险的

I allude to this, but rather to advertise the adventurous neophyte, who may be disposed to rush into inquiries which have taxed the skill of the greatest experimenters, of some of the snares and pitfalls that lie in his way.

(**2**, 489-490; 1870)

John Tyndall: Royal Institution, Oct. 18.

影响程度，这已经够严重了，而是要提醒那些爱冒险的年轻学者们他们的研究道路上存在一些陷阱，他们被安排去做的可能是已经耗费了许多伟大的研究人员的能力的课题。

（孙惠南 翻译；张泽渤 审稿）

The Evolution of Life: Professor Huxley's Address at Liverpool

H. C. Bastian

Editor's Note

This is the fifth part of an exchange between H. Charlton Bastian and Thomas Huxley, president of the British Association, on the topic of the spontaneous generation of life in water. Bastian supported the idea; Huxley did not. Here Bastian continues to assert his case, with perhaps more rhetoric than real evidence. Huxley was eventually proved correct; but a point of interest in Bastian's reply here is his argument over whether tiny particles seen in the water are self-propelled and thus living, or just moving by Brownian motion caused by random impacts of water molecules. Brownian motion was soon to emerge as central to the molecular theory of matter.

BELIEVING that readers of *Nature* can feel no interest in the extended personalities with which Prof. Huxley almost fills his letter this week, and believing also that such matters are little worthy of occupying your columns, I shall only allude to that part of his letter which contains statements having a scientific bearing.

The distinct issue raised in my experiments was, were *living* things to be found in the fluids of my flasks? If so, such living things must either have braved a higher degree of heat than had been hitherto thought possible, or else they had been evolved *de novo*.

The effect of the very high temperature upon pre-existing living things, which were purposely exposed thereto, was shown by their complete disorganisation in an experiment which is recorded in *Nature*, No. 37, p. 219, and to this I would especially direct Prof. Huxley's attention.

Prof. Huxley advances an explanation of the mode of origin of the distinct fungi, bearing masses of fructification (*Nature*, No. 36, figs. 12, 14, and 17) and of the inextricably tangled coils of spiral fibres (figs. 13 and 15) found in my flasks after exposure to temperatures at and beyond Pasteur's standard of destructive heat; his theory is entirely novel, apparently extemporised for the occasion, and is very startling. He says, and in justice to Prof. Huxley I quote the passage in full, "Any time these six months Dr. Bastian has known perfectly well that I believe that the organisms which he got out of his tubes are exactly those which he has put into them; that I believe that he has used impure materials, and that what he imagines to have been the gradual development of life and organisation in his solution is *the very simple result of the settling together of the solid impurities; which he was not sufficiently careful to see when in their scattered condition when the solutions were made.*"

生命的进化：赫胥黎教授在利物浦的演说

巴斯蒂安

编者按

这是查尔顿·巴斯蒂安和英国科学促进会的主席托马斯·赫胥黎就生命是否能在水中自然生成这一论题进行的一次激烈争论的第五部分。巴斯蒂安赞同这一论点，而赫胥黎则持反对意见。在这篇文章中，巴斯蒂安仍然坚持自己的观点，但更多的是卖弄言辞技巧，而不是提出真凭实据。人们最终认识到赫胥黎的看法是正确的；但令人感兴趣的是巴斯蒂安在回复中提到的问题：在水中看到的小颗粒的运动究竟是有生命物体的自发运动，还是由水分子的无规则碰撞导致的布朗运动？此后不久布朗运动就成了物质分子理论的中心话题。

赫胥黎教授本周的信件中几乎都是些无关的人身攻击，我相信《自然》的读者们对此毫无兴趣，我也相信，这种事不值得占用贵刊的版面，因此我将只提及他信中具有科学依据的部分言论。

我的实验引发的一个很清楚的问题是：在我的长颈瓶中的液体里会不会发现**有生命的**物质？如果会的话，那么这种生物要么一定能够承受高于我们目前所认为生物可能承受的高温，要么它们就是在实验过程中新进化出来的。

将溶液中已存在的生物故意暴露在高温下以研究其影响的实验已经被报道过了（《自然》，第37期，第219页），结果表明生物组织被完全破坏了。我特别想请赫胥黎教授注意这一点。

赫胥黎教授提出了一种对不同真菌的起源模式的解释，当我按照巴斯德的破坏性热量标准以及高于这个标准的温度处理真菌时，我发现它们在长颈瓶中产生了大量实体（《自然》，第36期，图12、14和17）及纠缠在一起的螺旋纤维团（图13和15）；而赫胥黎教授的理论完全令人耳目一新，很明显那是当场的即兴演说，相当令人吃惊。为了对赫胥黎教授公正起见，我将完整引用该段，他说“这6个月里，无论何时，巴斯蒂安博士都已经清楚地知道，我认为他试管中的生物其实正是被他放进试管中去的那些；我还认为他使用了不纯的物质，而他所想象的那些在他的溶液中经历了生命和组织结构的逐步发育的东西，**其实就是配制溶液时以分散状态一起放入试管中的固体杂质，他只是没有足够认真地去观察而已。**”

Now, although it was quite true that minute portions of *Sphagnum* leaf were found in two unpublished experiments, it seems very marvellous that on this slender foundation Prof. Huxley should hazard such a purely imaginative and unprecedented hypothesis as to the mode of production of fungi.

I have, moreover, not been able to see why the occurrence of the incident to which he refers should make him repudiate a number of experiments in which *unmistakeably living things* were found in fluids from hermetically sealed flasks after these and their contents had been exposed to temperatures higher than those which living things are known to be capable of resisting.

Following a precept more honoured in dialectics than in science, Prof. Huxley has attacked his opponent rather than the arguments which he affects to destroy. He objects to only one passage in my "Reply", and this he thinks was not worthy of the special type in which it was printed; and yet, notwithstanding its special type, I can only conclude from his reply that Prof. Huxley has failed to appreciate its meaning. My words were: "Living things may and do arise as minutest visible specks, in solutions in which, but a few hours before, no *such* specks were to be seen." The word which now alone stands in italics was ignored by Prof. Huxley. I had no wish to tell him that certain refractive particles, or foreign bodies, might not be visible in the thin film of fluid to which I referred. I alluded to the gradual and equable development of living specks throughout a fluid containing no apparent germs. His retort that some unobserved visible germs might have become centres of development is a *contre-sens*. It does not apply to the gradual appearance of myriads of *equally diffused* motionless particles in a motionless film of fluid.

The very authoritative tone which Prof. Huxley has lately assumed in his remarks concerning Brownian movements and those of living organisms, fails to impress me very much. His knowledge about these movements, as I have good reason to know, is of quite recent growth. Movements which, in the month of March of the present year, Prof. Huxley did not regard as Brownian, he now does believe to have been of this nature. If he is now right, what value is to be set upon his knowledge of Brownian movements six months ago; and what guarantee have we that in another six months Prof. Huxley may not again take a different view?

Let me assure Prof. Huxley, however, that the duty which he is "credibly informed" he owes to the public remains still undischarged. I protested against his "Address" on scientific grounds which are fully stated, and those who have read my protest will see that Prof. Huxley cannot dispose of the question really at issue by recounting any mistakes of mine, whether real or imaginary. If, as I believe, he has failed to give any worthy or serious view of the question, this could have been in no way necessitated by a disbelief, however strong, in my experiments. The labours of Profs. Wyman, Mantegazza, and Cantoni had already taken the question into regions never attained by M. Pasteur, and therefore they demanded a fair consideration. Is Prof. Huxley, in his capacity of President of the British Association, warranted in ignoring their labours, and therefore in misrepresenting the

尽管现在我已经在两个尚未发表的实验中发现了泥炭藓叶子的一小部分，这确实是事实，但是不可思议的是，赫胥黎教授居然凭借不实依据而对真菌产生模式得出一个纯粹想象的和史无前例的假说。

此外，我并不明白他提到的那些事件的发生为什么会令他否认许多如下情况的实验，在那些实验中，当密封的长颈瓶及其内部容纳的物质被加热到高于已知的生物能够承受的温度时，仍然可以在其中的液体里发现**生物**，**这是确定无疑的**。

赫胥黎教授更推崇的是辩证法而非科学原则，即他攻击的是他的反对者而非他佯装想要驳倒的论点。他只反对我的“回复”中的一个段落，而且他认为这部分不值得出现在用来发表它的特殊版面上；然而，且不论是不是特殊版面，我从他的回复中得到的唯一结论是：赫胥黎教授根本不理解它的意义。我的话如下：“溶液中可能并且确实出现了生物，那是一些极小的可见斑点，而在几个小时前这些溶液中还看不到**这种**斑点。”此句中突出强调的词被赫胥黎教授忽视了。我原本不想告诉他在我所用的液体薄膜上可能看不到某些折光颗粒或者外来物质。我说的是不含有明显的微生物的液体中有生命的颗粒可以逐渐、温和地进行发育。他在反驳中提到的某些未观察到的可见微生物可能变成了发育的主要物质，是与本来的意义相反的。这并不适用于一张静止的液体薄膜上逐渐出现大量**同等分散的**静止颗粒的情况。

赫胥黎教授在他最近发表的关于布朗运动和生物的布朗运动的言论中所使用的权威腔调并没有给我留下太深的印象。他只是最近才对这些运动有所了解，对于这一点我很清楚。今年 3 月赫胥黎教授还不认为是布朗运动的那些运动，他现在却相信它们具有这样的属性。如果他现在是正确的，那么对于 6 个月之前他对布朗运动的了解，又该赋予什么样的价值呢？另外，怎么能保证再过 6 个月赫胥黎教授不会又有不同的观点呢？

不过，让我来向赫胥黎教授保证，他依然承担着“据可靠消息得知”的对公众所负有的职责。我对他的“致词”提出的抗议是站在科学立场上的，这已经表述得很充分了，那些读过我的抗议的人会看到，就算赫胥黎教授重提我的那些真实的或者想象中的错误，也不能逃避那些存在争议的问题。因为我相信，如果他无法给出关于此问题的任何有价值的或严肃的观点，那么就无法强迫别人不相信我的实验。怀曼教授、曼泰加扎教授和坎托尼教授的工作已经将该问题归入到巴斯德先生从未涉足过的领域之内，他们也需要公平的对待。难道在英国科学促进会主席的能力范围内，赫胥黎教授有权忽视他们的劳动，乃至可以对这一问题在科学领域的当前状

present state of science on the subject, because, owing to two errors among my many experiments, he declares himself to have altogether lost faith in my skill or capacity as an investigator? The answer cannot be doubtful. Is it, again, consistent with his high responsibilities that he should pervert the real issues, and should do a grave injustice to others, in order that he might preserve a "silence" which should be his "best kindness" to me? Let me tell Prof. Huxley that I repudiate such "kindness", as any honest man would who is simply seeking after truth, and relegate it to the same regions as I would that indescribable air of restrained omniscience whereby he endeavours to crush arguments and facts, to which he altogether fails to reply.

(**2**, 492; 1870)

H. Charlton Bastian: University College, Oct. 17.

态加以曲解吗？他有权可以因为我众多实验中存在的两个错误就声称自己对我作为一位研究者的技术和能力完全失去信任了吗？答案是不容置疑的。那么他曲解真正的争议、对其他人严重不公，保持“沉默”来作为他“最友好的态度”，这些难道与他的高度责任心是相符的吗？让我来告诉赫胥黎教授，我与任何只追求真理的诚实的人一样，拒绝这样的“友好”，我认为他是试图凭借他有限的全知全能来造成莫可名状的氛围，借以不遗余力地打压争论和事实，因为他根本无法对这些作出回应。

（刘皓芳 翻译；刘力 审稿）

Progress of Science in 1870

J. P. E.

Editor's Note

This editorial suggests that 1870 was not a vintage year for science, being characterised more by consolidation than discovery. Astronomers' alleged focus on the Sun may partly reflect the preoccupations of *Nature*'s editor Norman Lockyer. The synthesis of alizarin by Liebermann and Graebe was an important step in organic chemistry, but was actually accomplished in 1868, while the indigo synthesis mentioned here must have been incomplete—that didn't happen in full until 1877. And while deep-sea dredging is highlighted, the most fruitful example of that was the Challenger expedition of 1872–1876. But "Bathybius" protoplasm turned out to be an experimental artefact. The germ theory of disease, due mostly to Louis Pasteur and Robert Koch, did however mature in 1870.

THE year which has just come to a close has neither been characterised by many new and striking scientific discoveries, nor have any novel applications of Science to ordinary industry and manufacture attracted special attention. The work done has been more a strengthening of that of past years, and a confirming or a disproving of theories and experiments, than the inventing of new ones. In one branch of Science only has any great advance been made, and that, as we shall presently show, we believe to have taken place in Geology. But this advance is one somewhat overlooked at present; but still of so important a character that, when once fully recognised in all its bearings, it may tend to disprove much of the geological teaching of the present day.

Taking the various Sciences as much as possible separately, we will begin with Astronomy. Here attention has been chiefly directed, as has been the case for so many years past, to the sun. Since it is now generally understood that when once the nature of this vast self-luminous body is accurately made out, much light will be thrown on many now perplexing and strange phenomena, the Eclipse of the 22nd of December last was anxiously watched for, and all possible observations were taken here by those who were unable to take part in the Government Expedition to Spain and Sicily. It is to be hoped that the labours of this Expedition, in spite of accident both on land and sea, and the unsatisfactory state of the weather at the time of observation, will yet yield results of great importance. At any rate we may fairly congratulate ourselves that at last we have a Government which has shown itself in other instances besides this special one, not unmindful of the claims of Science and of the value of accurate scientific investigation.

Mr. Lockyer and Mr. Huggins have continued their spectroscopic observations of the sun, and Prof. Zöllner has published a very valuable paper on the solar prominences, theorising

1870 年的科学发展状况

J. P. E.

编者按

这篇评论文章认为，1870 年对于科学界来说并不是一个丰收的年份，因为这一年主要是对已有成果的巩固而没有多少新的发现。天文学家们声称要专注于对太阳的研究，这一定程度上反映了《自然》的主编诺曼·洛克耶最为关注的问题。利伯曼和格雷贝合成了茜素，这是有机化学领域很重要的成果，不过实际上在 1868 年就已经完成了这个工作。文章中提到的靛青的合成当时尚未完成，直到 1877 年才完全取得成功。深海挖掘是个亮点，该领域最富成效的工作是 1872~1876 年挑战者号的考察。不过，后来证明“深海”原生质实际上是实验假象。关于细菌致病理论则确实是 1870 年发展成熟的，这主要有赖于路易·巴斯德和罗伯特·科赫的工作。

在刚刚过去的一年里，不仅没有出现大量新的引人注目的科学发现，而且在将科学技术应用于普通工业和制造业方面也没有出现特别值得关注的新进展。去年的工作主要是巩固以前的成果，进一步证实或反驳之前的理论或实验结果，而非创造新的理论。科学界有一个分支取得了比较大的进展，就是下面我们将要谈到的地质学上的发现。不过目前这项新进展多少有点被忽视了，但它仍不失为一个重大发现，一旦人们认识到了它的价值，就将证明当今地质教学中的许多理念都是错误的。

下面要把尽可能多的不同门类的学科分别列出，我们将从天文学入手。目前天文学的研究方向和若干年前一样，重点是对太阳的研究。因为目前人们普遍认为，一旦了解了太阳这个巨大发光体的特征，许多困扰我们的令人难以理解的现象就会变得更加明朗。在去年 12 月 22 日，大家观测到了盼望已久的日食，那些留在这里没能参加政府组织的赴西班牙和西西里岛考察活动的人尽其所能对日食进行了全面的观察。尽管在考察时可能会遇到陆地和海面上的偶发事件，观测的天气情况也未必尽如人意，我们仍希望考察活动能取得重大突破。无论如何我们应该为自己感到庆幸，因为我们拥有一个关注科学界呼声和重视实地考察活动的政府，它善于处理各类事务，同时也包括这项特殊活动。

洛克耶先生和哈金斯先生仍在继续观测太阳光谱，策尔纳教授发表了一篇对日

very boldly as to the temperature and pressure at the sun's surface;* while in America Prof. Young has worked with good results at the same subject. Before leaving this branch of our subject, we would mention that Mr. Proctor has published some novel views as to the constitution of the stellar systems, which, under the somewhat fanciful titles of "star-drift" and "star-mist" must be familiar to most of our readers.

Whilst the vast domain of Organic Chemistry has been still further widened by the innumerable workers who plunge into this branch of the subject and neglect the many untrodden paths in Inorganic Chemistry, nevertheless no special or important discoveries are to be chronicled, unless we may mention the beautiful process by which Indigo has been synthetically constructed by M. M. Emmerling and Engler, following closely on the artificial manufacture of Alizarine by M. M. Liebermann and Graebe.

Molecular Physics has occupied a large share of attention, and the discussion before the Chemical Society on the existence, or non-existence, of Atoms and Molecules, has only too clearly shown how doctors differ amongst themselves, and that the very foundations of a Science, considered so essential by some, are utterly repudiated by others. A very remarkable paper on the Size of Atoms, originally published in these columns (*Nature*, vol. I, p. 551) by Sir William Thomson, in which he gives four distinct trains of reasoning by which he arrives at a proof of their absolute magnitude, has attracted much attention, and has been translated and copied into most of the continental and American scientific journals. Dr. Thomas Andrews has also pursued his remarkable investigations on the Continuity of the liquid and gaseous states of matter. The death of Prof. Wm. Allen Miller, F.R.S., and Dr. Matthiessen, F.R.S., have left sad voids in the ranks of our English experimental chemists.

In Biology, the investigations of Prof. Tyndall, "On Atmospheric Germs, and the Germ Theory of Disease" †, have contributed to a clearer knowledge of the nature of some of the most virulent of our infectious diseases, and have caused those diseases to be studied in a much more scientific manner than before.

The theory of Spontaneous Generation, which has been very prominently before the scientific world for the last ten years, has, during the past year, been very strongly attacked on the one hand by Prof. Huxley, and defended on the other by Dr. Bastian and Dr. Child. In his Inaugural Address to the British Association meeting at Liverpool, Prof. Huxley gave a long review of all the researches on the subject, from the time of Spallanzani and Needham to the present day, and declared his belief, after carefully weighing the evidence on both sides, that all life has its origin in some pre-existing life, and that Spontaneous Generation, or, as he termed it, Abiogenesis, is not now proved to take place. The investigations of Dr. Bastian, originally intended to have been read before the Royal Society, were published instead in these columns, in a series of three long

* Translation in full, *Nature*, vol. II, pp. 522-526.

† See *Nature*, vol. I, pp. 327, 351, 499, &c.

珥研究有重大价值的论文，大胆地推测出日表温度和压力的理论值。* 在美国，扬教授也在同一课题中取得了良好的进展。在结束对这一学科的介绍之前，我们还要提及普洛克特先生就星系构成发表的一些新奇见解，诸如“星流”和“星团”之类富有想象力的名词想必已为我们广大的读者所熟悉。

同时，大量从事有机化学研究的工作者们仍在进一步拓展这一广阔的领域，但他们忽视了在无机化学领域还有许多小径无人涉足，尽管如此，除了利伯曼和格雷贝人工合成了茜素，以及随后爱默林和恩格勒合成靛蓝染料的完美工艺值得一提外，再没有什么特别的或重要的发现值得载入史册了。

分子物理学已经引起了大部分人的关注，化学学会正在就原子或分子是否存在的问题进行讨论，讨论充分显示出学者们在观点上的重大分歧，而且一部分人认为非常基本的学科基础却被另一部分人彻底否定。威廉·汤姆孙爵士通过一系列专栏发表了一篇关于原子尺寸的著名论文（《自然》，第 1 卷，第 551 页），在论文中他提出了 4 个不同系列的理由作为原子有绝对尺寸的证据。该论文引起了极大的轰动，被大多数欧美科技期刊争相翻译和转载。托马斯·安德鲁斯博士正在完善他关于物质液态和气态的连续性的非凡研究。艾伦·米勒教授（皇家学会会员）和马西森博士（皇家学会会员）的去世是英国实验化学界的重大损失。

在生物学方面，廷德尔教授《论空气中存在的细菌，以及细菌致病论》† 的研究工作使人们对一些致命传染病的致病机理有了更加明确的认识，也使人们对这些疾病的研究方式比以往更加科学。

自然发生学说是最近 10 年内科学界面临的突出问题。去年，该学说一方面遭到了赫胥黎教授的极力反对，另一方面又为巴斯蒂安博士和蔡尔德博士所支持。在利物浦举行的英国科学促进会会议上，赫胥黎教授在致开幕辞时对长期以来有关这一主题的所有研究作了一个回顾，从斯帕兰扎尼和尼达姆的时代一直到现在，在仔细衡量了双方的论据之后，他宣布了自己的观点，认为所有生物都源于先前就已出现的某种生物，因此，自然发生学说，或他所称的“无生源论”，现在已被证明不能成立。巴斯蒂安博士的研究论文，原打算在皇家学会宣读，后改为以一系列专栏的形式发表，在三篇成系列的长文章（《自然》，第 2 卷，第 170、193 和 219 页）中，他

* 全文翻译，见《自然》，第 2 卷，第 522~526 页。

† 《自然》，第 1 卷，第 327、351 和 499 等页。

articles (*Nature*, vol. II, pp. 170, 193, 219), in which he gave the reasons for his belief that Spontaneous Generation certainly does occur. Feeling himself attacked and his experiments somewhat underrated by Prof. Huxley in his Address, he criticised it at considerable length, and detailed the results of some new experiments (*Nature*, vol. II, pp. 410, 431, and 492) which confirmed his previous deductions.

The Darwinian theory of Natural Selection has been attacked by Mr. A. W. Bennett and Mr. Murray*, and defended by Mr. A. R. Wallace and others; Mr. Wallace having also vindicated his claims to priority in this question, since he published many of the now-recognised theories and speculations on the subject of Natural Selection, at a time when he was resident in the East Indies, and entirely unacquainted with what Mr. Darwin had written on the same subject.

As respects Geology, during the past year the Government has continued its grants of money for the purpose of Deep Sea Dredgings, and at present the report of the most recent Expedition is anxiously looked forward to. The results of the Expedition in the autumn of 1869, as given to the public by Dr. Carpenter, Prof. Wyville Thomson, and Mr. Gwyn Jeffreys during the past year, have been of the greatest possible interest and importance. They found that on the same level, at the bottom of the deep sea, two different deposits are in process of formation side by side, each characterised by a distinct Fauna, and yet apparently produced under perfectly similar conditions of land and sea, area, depth of water, &c. On investigating this curious result, however, it was found that the temperature of the water circulating over these two areas is very different, and that this mere difference of temperature is capable of entirely changing the character of the fauna of the simultaneously formed deposits. Thus an entirely new element is brought into geological speculations, since it is shown that at one and the same time strata may be accumulated containing widely different organic remains. In addition to this, they have shown that the calcareous deposit known to us as chalk is now being deposited all over the bed of the Atlantic Ocean, and there are many weighty reasons for believing that this deposit has gone on steadily ever since the time during which we imagined the cretaceous rocks of the world to have begun and ended. Many organisms formerly supposed entirely extinct have been re-discovered in these deep-sea dredgings; and, in short, much has been done to show that our past geological reasoning requires thorough and careful revision. Prof. Gümbel's discovery of the existence of Bathybius and similar organisms at all depths, and stretching over an indefinite period of geological time, is of the greatest importance in relation to this subject. Prof. Agassiz, on the other side of the Atlantic, has published reports of the deep-sea dredging off the Florida Coast, and has stated that the results of his researches, and those of others, both English and Scandinavian, have convinced him that there is life all over the sea bottom, and that where evidence of marine life cannot be found, we are justified in calling in the agency of the sea to explain certain obscure facts. These conclusions cannot be without their important bearing on many commonly received geological theories.†

---

* *Nature*, vol. III, pp. 30, 49, 65, and 154.

† During the past year all the most important papers on Deep-Sea Dredging have appeared in these columns, and we would refer our readers to vol. I, pp. 135, 166, 267, 612, 657; vol II, pp. 257, 513, &c.

给出了自己之所以相信自然发生必然存在的理由。他感到在赫胥黎教授的演讲中自己受到了攻击，他的实验也在一定程度上被低估，因而他用比较长的篇幅评论了这一事件，并详细地描述了一些新的实验结果（《自然》，第 2 卷，第 410、431 和 492 页）来进一步证实他之前的论点。

达尔文的自然选择理论遭到了贝内特先生和默里先生的反对*，但得到了华莱士先生以及其他学者的支持。华莱士先生还争辩到，是自己首先提出了这一理论，因为他在东印度群岛居住的时候，曾在发表的文章中提出过许多现在已被认可的关于自然选择学说的理论和推测，那时他完全不知道达尔文先生已经就同一主题撰写了著作。

至于地质学，政府在过去的一年中继续提供资金援助深海挖掘工作，现在我们正期待最近一次探险的报告。1869 年秋季考察的结果是由卡彭特博士、怀韦尔 · 汤姆森教授和格温 · 杰弗里斯先生于去年向公众展示的，这篇报告也许是最为重要且最具价值的。他们发现，在深海底部的同一高度，两个紧挨着的不同的沉积层正在形成，每个沉积层都有自己独特的动物群，尽管它们显然是在非常相似的陆海环境、地域和水深等条件下形成的。不过，人们在研究这一奇特现象时发现，这两个区域的水温差异很大，说明仅由温度的差别就能够完全改变同时期形成的沉积层中的动物区系特征。既然在同一时间，地层中可能积累完全不同的有机体残遗物，那么就可以在地质学的研究体系中引入一个全新的要素。除此之外，他们还指出我们所熟知的石灰质沉积物，如白垩，现在分散沉积在大西洋底的各个地方，我们有足够的理由确信，在地球上白垩纪岩石从形成到终结的这段时间内，这种沉积物在稳步地增加。许多之前认为已完全绝迹的生物体在这次的深海挖掘中又被重新发现。简而言之，很多已完成的工作都表明我们之前确立的地质学理论还需要全面细致的修订。冈贝尔教授发现，在不同深度、不同地质年代的海洋中都有深水鱼类以及类似的生物存在，这一发现对该学科具有重大意义。阿加西斯教授在大西洋的另一端发表了有关佛罗里达沿岸深海挖掘物的报告，并表示，他的研究结果以及其他英国人和斯堪的纳维亚人的研究结果使他确信海底各处都是有生命存在的，在那些没有发现海洋生物的地方，我们有必要提请海事机构来解释某些模糊的事实。以上这些论点对于人们普遍接受的地质学理论必然具有重大的意义。†

---

* 《自然》第 3 卷，第 30、49、65 和 154 页。

† 去年有关深海挖掘的所有重要文章都出现在这些专栏中，建议读者参考第 1 卷的第 135、166、267、612 和 657 页，以及第 2 卷的第 257、513 等页。

In Botany many very careful series of observations have been made in the physiological department. Among the most important we may mention those of Prillieux and Duchartre in France, confirmed by Dr. M'Nab in this country, that, contrary to the previously accepted hypothesis, plants do not absorb any appreciable amount of aqueous vapour through their leaves; and those previously announced by M. Dehérain, that the evaporation of water from the leaves of plants is due to sunlight rather than to heat, and proceeds independently of the degree of saturation of the atmosphere. Much attention has also been paid in Germany, Italy, and England, to the fertile field of the phenomena of fertilisation, opened out by Mr. Darwin's observations.

In Meteorology there is no great advance to chronicle. It still remains a Science without a head, a chaotic mass of facts with no definite order or arrangement; for though many are working at this subject, and some valuable papers on the Origin of Winds and Storms have been published, still no definite progress can be ascertained.

The splendid appearances of the Aurora Borealis, visible all over the British Isles in September and October, have directed public attention to those unmistakeably magnetic phenomena, and to the connection which exists between their appearance, great magnetical perturbations, and large solar spots. They have been examined very frequently during the past year by means of the spectroscope, and there is distinct evidence of lines in the green and red portion of the spectrum, the latter presumably due to hydrogen. We would direct attention to our desire to publish a complete tabular list of the more remarkable meteorological phenomena of the past year, so as to be serviceable to observers in all parts of the world. To render this as perfect as possible, we would invite the kind co-operation of all those interested in the subject who can forward us any data.

We cannot conclude without noticing how much Science has lost during the latter half of the year just ended by the fearful struggle that has taken place between France and Germany, where each nation has brought into requisition all the resources of Science only to inflict as much injury as possible on the other. For nearly six months we have witnessed the sad sight of workshops shut up, laboratories closed, universities and public schools wanting both professors and students, and the friendly emulation of similar tastes and pursuits turned to the fierce rivalry of the sword. Science will have to deplore the untimely loss of many of her most attached workers, and their country will have lost those who would in happier times have done her as much honour at home as they have shown bravery in the field. Whilst the French Academy, shut up in besieged Paris, has brought the art of ballooning to its present state of perfection, so that now it is used as a means of communication with the outside world, the result of the subtle strategy of the Germans, and the scientific education they so generally possess, has been to give them advantages which have, to the present time, baffled their adversaries.

(**3**, 181-182; 1871)

在植物学方面，人们进行了许多植物生理学领域的细致观察，其中最重要的应属法国普利略和迪沙尔特的研究成果，该成果已被我国的蒙纳布博士证实，即认为植物通过叶片吸收的水蒸气微不足道，这与之前公认的假说正好相反，也与德埃兰的理论相反。德埃兰先前提出，水从植物叶片上蒸发的过程取决于阳光而不是受热，并与大气的饱和度无关。德国、意大利和英国的科学家还对生殖学领域的受精现象给予了许多关注，该问题是由达尔文先生的观察引发的。

在气象学领域，没有值得记载的重大进展。气象学仍然是一个没有领头人的学科，从一大堆杂乱无章的现象中找不到明确的顺序和分类；尽管有许多人在这一领域从事研究工作，也有一些关于风和暴风雨的起源的有价值的论文发表，但仍然没有显著的进展。

9 月和 10 月，在大不列颠全岛都可见到壮观的北极光，北极光将公众的注意力转移到了明确的磁力现象以及北极光外观和剧烈的磁扰动、太阳黑子活动增强之间的关系上。在去年一整年的时间里，人们通过光谱学的方法频繁地观测了太阳黑子，在光谱的绿光区和红光区有明确的谱线证据，而后者很可能是由氢引起的。我们希望能发布一个完整的表单，列出去年发生的值得关注的气象学现象，以便更好地服务于世界各地的观测者。为了让资料尽可能完善，我们将邀请所有爱好这一学科并能为我们提供数据的人士与我们一起合作。

在文章的结尾，我们不能不提及去年后半年发生在德法之间的可怕争斗给科学界带来的重大损失，两个国家都在征用所有的科技资源，只是为了给对手造成尽可能大的伤害。近 6 个月以来，我们目睹了工厂倒闭、实验室关门、大学和公立中小学缺少教授和学生这样的悲惨景象，而品位和追求相似的人们也由友好竞争转为持戈相向。科学界将不得不为过早地失去众多忠实于她的工作者而感到悲痛，他们的祖国也将失去那些在和平年代里同样能为国争光的人们，而这些人现在却正在战场上勇敢拼杀。同时，在被围困的巴黎，已使气球飞行技术达到目前的完美状态的法国科学院也被关闭，现在气球被用作与外界联系的工具，这是德国人的狡诈战略造成的结果，而他们所拥有的十分普及的科学教育，现在却成了挫败对手的有利条件。

（何铭 翻译；赵见高 审稿）

The Descent of Man

P. H. Pye-Smith

Editor's Note

"If Mr. Darwin had closed his rich series of contributions to Science by the publication of the 'Origin of Species', he would have made an epoch in Natural History like that which Socrates made in philosophy, or Harvey in Medicine." Such extravagant plaudits opened physiologist Philip Henry Pye-Smith's two-part essay on Darwin's two-volume work in which he presented evidence for the biological origin of humanity, attempted to trace our evolution from lower forms of life, and then advanced his pioneering ideas on sexual selection as a potent driver of evolutionary change: "But though in the lists of Love the battle is often to the strong," Pye-Smith writes, "even more frequently it is to the beautiful."

The Descent of Man, and Selection in relation to Sex.
By Charles Darwin, M. A., F.R.S., &c.
In two volumes, pp. 428, 475. (Murray, 1871)

I

IF Mr. Darwin had closed his rich series of contributions to Science by the publication of the "Origin of Species", he would have made an epoch in Natural History like that which Socrates made in philosophy, or Harvey in medicine. The theory identified with his name has stimulated ethnological and anatomical inquiries in every direction; it has been largely adopted and followed out by naturalists in this country and America, but most of all in the great work-room of modern science, whence a complete literature on "Darwinismus" has sprung up, and there disciples have appeared who stand in the same relation to their master as Muntzer and the Anabaptists did to Luther. Like most great advances in knowledge, the theory of Evolution found everything ripe for it. This is shown by the well-known fact that Mr. Wallace arrived at the same conclusion as to the origin of species while working in the Eastern Archipelago, and scarcely less so by the manner in which the theory has been worked out by men so distinguished as Mr. Herbert Spencer and Prof. Haeckel. But it was known when the "Origin of Species" was published, that instead of being the mere brilliant hypothesis of a man of genius, of which the proofs were to be furnished and the fruits gathered in by his successors, it was really only a summary of opinions based upon the most extensive and long-continued researches. Its author did not simply open a new province for future travellers to explore, he had already surveyed it himself, and the present volumes show him still at the head of his followers. They are written in a more popular style than those on "Animals and Plants under Domestication", as they deal with subjects of more general interest; but all the great

人类的由来

派伊–史密斯

编者按

生理学家菲利普·亨利·派伊–史密斯发表了一篇由两部分组成的文章，对达尔文的两卷本著作《人类的由来及性选择》给予了极高的称赞。史密斯在开篇写道，“如果《物种起源》为达尔文先生丰富的科学著述画上句号的话，那么他在自然史上的里程碑作用就如同苏格拉底在哲学领域、哈维在医学领域一样。”达尔文在《人类的由来及性选择》中展示了关于人类的生物学起源的证据，试图寻找人类从低等生命形式进化而来的足迹，并开创性地提出了性选择是人类进化的驱动力的观点。派伊–史密斯认为，“尽管在爱的竞技场上，胜利经常属于强者，但更多时候是美丽者获胜。”

《人类的由来及性选择》

作者：查尔斯·达尔文，文学硕士，皇家学会会员。

共两卷，428 页，475 页。（默里，1871 年）

I

如果《物种起源》为达尔文先生丰富的科学著述画上句号的话，那么他在自然史上的里程碑作用就如同苏格拉底在哲学领域、哈维在医学领域一样。这项用他的名字命名的理论激发了学者对人种学和解剖学各方面问题的讨论，得到了我国和美国的自然学家的广泛接受和推崇。但最主要的是在现代科学的大工作室里诞生了关于“达尔文主义”的一套完整学说，达尔文主义的信奉者们与达尔文的关系，就像闵采尔和再浸礼论者与路德之间的那种复杂关系。正如知识方面的许多巨大进展一样，进化论在各方面都具备了成熟的条件。可证明这一点的著名事例有，华莱士先生在东部群岛工作的时候对物种起源问题得到了同样的结论；著名的赫伯特·斯宾塞先生和海克尔教授也提出了十分相似的理论。但《物种起源》出版时，它还只是各种基于广泛且长期的研究而得出的观点的综述，并非某个天才的杰出理论假说；该假说的证据将由后来的追随者提供，最终成果也由他们获得。实际上，其作者不只是简单地为未来的探索者开辟了一个新的研究领域，他本人已经亲自进行了调查，而且这两卷著作表明他至今仍然先于他的追随者。该书的书写风格比《动物和植物在家养下的变异》更通俗，因为讨论了更多人们普遍感兴趣的话题。但是，在研究的勤勉与准确、构建假说的能力以及判断的公正性这些重要方面，则一如达尔文先

qualities of industry and accuracy in research, of fertility in framing hypotheses, and of impartiality in judgment, are as apparent in this as in Mr. Darwin's previous works. To one who bears in mind the too frequent tone of the controversies these works have excited, the turgid rhetoric and ignorant presumption of those "who are not of his school—or any school", and the still more lamentable bad taste which mars the writings of Vogt and even occasionally of Haeckel, it is very admirable to see the calmness and moderation (for which philosophical would be too low an epithet) with which the author handles his subject. If prejudice can be conciliated, it will surely be by a book like this.

It consists of two parts. The first treats of the origin of man, his affinities to other animals, and the formation of the races (or sub-species) of the human family. Besides the obvious interest to all Mr. Darwin's readers of a discussion on the subject of their "proper knowledge", naturalists will find the detailed application of the laws of natural selection to a single common and well-known species an excellent test of their truth and illustration of their difficulties. It is in dealing with the latter, which are never extenuated or passed by, that the author introduces the subject of sexual selection. This is dealt with in the second part, which forms more than two-thirds of the work, and that not only as it affects man, but in its entire range. Reserving this division of the book for a future article, we will endeavour here to give a summary of the course of argument in the earlier portion.

The author, justly assuming that the general principles of natural selection are admitted by all who have examined the evidence on the subject, with the exception of many of "the older and honoured chiefs in natural science", proceeds at once to discuss the proofs of the origin of man considered apart from those affecting all animals in common. The first group of facts adduced to show his kinship with other forms of animal life, relate to the strict correspondence of his bodily parts with those of other mammalia. To say that these structures are the same because they have the same uses, is untrue, for many of them have no use in the sense of active function, and we constantly find the same structures in animals turned to different uses, and the same uses subserved by different structures. To say that the bodies of men and animals are alike because they are formed on the same plan, or because they are the realisation of the same idea in the Creator, is true enough, but is beside the mark; for natural science inquires how or by what steps these things have become so, not why and from what first cause. If one sees two men very much alike, one naturally supposes that they are brothers; if they are rather less so, they may be cousins; if only agreeing in general characters, we recognise them as at least belonging to the same race or nation; and so, when the facts to be accounted for are once ascertained, nothing but prejudice or repugnance to acknowledge our true relations, can explain why it was so long before naturalists admitted the hypothesis of community of origin between men and other animals. What is called the Darwinian theory accounts for the way in which diversities have arisen, and thus has converted an apparently obvious hypothesis into a well-grounded theory. But in expounding the likeness between men and animals, the author does not confine himself to anatomical structure, but shows how the same resemblance extends to the laws of disease, the distribution of parasites, and other minute particulars.

生先前的著作。如果一个人牢记这些著作所引发的太过频繁的争议，“非达尔文学派或其他任何学派的人”的浮夸修辞和愚昧假设，以及更令人遗憾的损害沃格特作品，甚至有时是海克尔作品的低品位论调，那么作者在处理书中问题时所表现出的镇定与谦和（对此使用冷静这个词远远不够）就值得人们敬仰了。如果偏见可以被纠正，那肯定是像这样的书才能够做到。

这本书包括两部分。第一部分讲述人类的起源、人类与其他动物的亲缘关系以及人科的种（亚种）的形成。除了讨论他们自己的“专业知识”以引起所有达尔文先生的读者的兴趣之外，自然学家也将通过对真实性的精确检验和对难点问题的图解说明来找到自然选择法则在一种简单而又为人所熟知的物种上的具体应用。在处理从未被藐视或忽视的自然选择的具体应用时，作者引入了性选择的观点。这在文章的第二部分中谈到，内容超过整个著作的2/3，它不仅谈论人类，还涉及其生活环境。该书的这部分内容我们会在以后的文章中谈及，这里我们将尽可能对前一部分提到的争论进行总结。

作者理所当然地认为，自然选择的总法则得到了除一些“长者和受人尊敬的领导们”之外的所有分析过自然选择论据的人的承认，因此，他只讨论人类起源的直接证据，而没有谈论对所有动物都有影响的共同因素。第一组用来说明人类与其他动物关系的亲缘证据与人类和其他哺乳动物在身体结构上的严格对应有关，认为功能相同从而结构也相同，这种看法是不正确的，因为这些结构中有很多并没有实际的功能，而且我们也经常能在动物中发现，相同的结构实现不同的功能，而相同的功能又可以由不同的结构来完成。人类的身体结构与动物相似是因为他们是由同一方案设计形成的，或者说是按照造物主的同一个想法实现出来的，这种说法有足够的理由，但无法得到检验，因为自然科学要弄清楚的是生物如何或者说通过什么步骤变成现在的样子，而不是为何这样以及最初的起因。如果看到两个人长得非常相像，很自然就会认为他们是兄弟；如果不是特别相像，则可能是堂（表）兄弟；如果只是一般特征相似，则认为他们至少属于同一种族或民族。因此，一旦要说明的事物得到证实，只有在我们承认真正起源时出现的偏见或嫌恶才能解释为何过了那么久自然学家才承认人类与其他动物具有共同起源的假说。达尔文理论解释了多样性是如何产生的，并将显而易见的假说转化为证据充分的理论。在详细说明人类与动物的相似性时，作者并没有将自己局限于解剖结构方面，而是充分阐述了在疾病、寄生虫的分布及其他细节方面同样具有的相似之处。

The next argument brought forward is the equally familiar one drawn from the likeness of the human embryo to that of other vertebrata. Then follows an account of the rudimentary organs in man, which in all other species are justly held among the most important indications of affinities. One such rudiment is mentioned which is, we believe, hitherto unrecorded. It is a slight projection of the rim of the helix of the auricle, which would correspond when unfolded to the point of an erect ear. (See illustration.) This occasional abnormity may, perhaps, be recognised by future anatomists as the *Angulus Woolnerii* after its first observer.

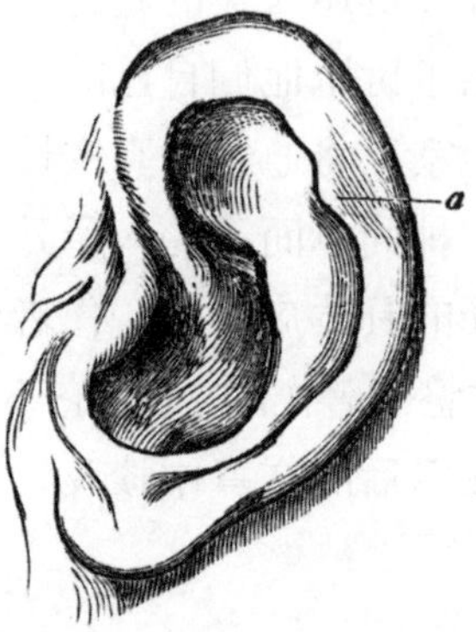

Fig. 1. Human Ear, Modelled and Drawn by Mr. Woolner. *a.* The projecting point.

In the second chapter Mr. Darwin shows that a consideration of the mental faculties of man, including the use of language, which has been held the greatest difficulty to admitting his kinship to other animals, may rather strengthen than weaken the arguments derived from his bodily structure. Memory and curiosity, jealousy and friendship, and even the power of correct reasoning, and of communication by sounds, are shown to belong to many of the lower animals, while the faculty of reflection and self-consciousness, and "the ennobling belief in the existence of an Omnipotent God", cannot be ascribed to the lowest tribes of the human family. At the same time it is argued that the use of articulate language, the power of forming abstract ideas, and even the sense of right and wrong, may have been gradually acquired by steps which here and there it is not impossible to trace. The question of the origin of the moral sense leads to the proposition of the following theory. Some natural emotions are of great intensity but short endurance, and their force is not easily recalled by memory; others, though less powerful at certain times, exert a constant influence, or one which is only interrupted by being overpowered for a time by the former. Accordingly, during the greater part of life, and always when there is leisure for reflection, the gratification connected with the more violent passions, such as hunger, sexual desire, and revenge, appears small, whereas the social instincts of sympathy and the pleasures of benevolence exert their full power. Hence we find social virtues, as courage, fidelity, obedience, among savages and even animals, long before the "self-regarding" virtues begin to appear. This theory is analogous to that by which Mr. Bain explains the higher character of the pleasures of sight compared with those of smell; they can be more easily recalled; and corresponds to the distinction drawn by the same writer between the acute and the more "massive" and permanent pleasures.

随后提出的论据同样是大家都熟悉的，即人类胚胎与其他脊椎动物胚胎的相似性。接着又对人类的退化器官进行了说明，这些也存在于其他物种中的器官是判断亲缘关系的最重要的指示之一。作者提到了一个这样的痕迹器官，我们相信，至今尚无记载。这个痕迹器官是外耳耳轮边缘的一个微小突起（结节），它显示时表现为一竖直耳朵的耳尖（见图示）。这一偶然的不规则形状有可能被今后的解剖学家根据其第一个发现者的名字而命名为**伍尔纳角**。

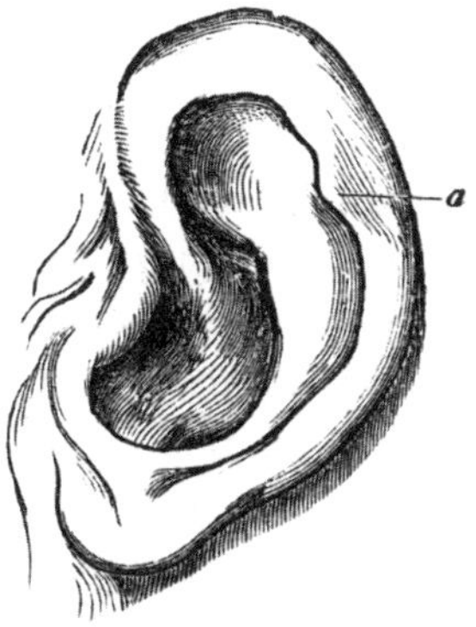

图 1. 该图是由伍尔纳先生构思并绘制的人耳。*a*. 结节

在第二章中达尔文先生指出，一项关于人类智力水平的研究可能会加深而不是削弱由身体结构引发的争论。该研究包括了语言的使用，而这被认为是承认达尔文提出的人类与其他动物的亲缘关系的最大困难。现已发现，记忆和好奇，妒嫉和友谊，甚至正确推理的能力和用声音沟通的能力在许多低等动物中都存在，但是思考的能力和主动意识，甚至“坚信存在万能上帝的高尚信念”却并不为人类家族中最低级的类群所拥有。与此同时，使用有音节的语言、形成抽象的概念，乃至形成是非观都是循序渐进地获得的，而且是可以追溯其起源的。道德观念的起源问题引出了下述理论。自然情绪中有一些来势凶猛但持续时间短，它们的影响力不容易通过记忆回想起来；而其他情绪，尽管在某些时间不如前者那么强烈，但能产生持久的影响力，或者产生一种只会因无法忍受前一种情绪的影响而被打断的影响力。相应地，生命的大部分时间通常是有闲暇进行思考的，这时，与非常强烈的感情如饥饿、性欲和报复等有关的个人满足感就显得微不足道，而同情这种社会本能和行善的快乐一直都在发挥它们全部的力量。因此我们发现，野人甚至动物中的社会道德，如勇气、忠诚和顺从，在“利己主义”品德开始出现以前已存在很久。这一看法与贝恩先生的视觉愉悦比嗅觉愉悦更高级的说法相似，视觉愉悦更易于被记起，这与该作者所说的短暂快乐与更“大众化的”、更永久性的快乐之间的区别是相对应的。

In the fourth chapter Mr. Darwin discusses the manner in which man was developed. It is shown that the broad facts on which the theory of Natural Selection rests apply to him. He is prolific enough to share in the struggle for existence. In him, as in all organic forms, there is a constant tendency to growth, which being checked and modified by external influences, proceeds in the direction of least resistance, and so produces the variations which are often ascribed to an assumed inherent tendency. Among the various forms produced, those will survive which are best fitted for the surrounding conditions, and they will transmit their character to their descendants, still subject to the same liability to vary. Next the author argues that the mental endowments of man, including language, his social habits, his upright position, and perfect hands, are of direct advantage to him in the struggle with other animals and with his fellows. It has always appeared that the difficult point in the development of man by Natural Selection is at the period when he was more defenceless than an anthropoid ape and less intelligent than the lowest savage; but Mr. Darwin thinks that the transition may have been safely made in some large tropical island where there was abundance of forest and of fruit. That man, once developed, can maintain himself, is obvious from his present existence. The arguments in favour of civilised man being the descendant of savages, which have been so admirably developed by Sir John Lubbock and Mr. Tylor, are of course brought forward in support of the author's view, and the important question is discussed how far we may hope for future improvement in the race by means of continued Natural Selection. Thus, while admitting that the process undergoes many checks and complications among human beings, the author does not assent to the arguments urged by Mr. Wallace that it would cease to operate as soon as the moral faculties came into play.* One human peculiarity which is apparently inexplicable by Natural Selection, the nakedness of the body and presence of a beard, is referred by Mr. Darwin to the operation of Sexual Selection. To this same agency is attributed the origin of the so-called Races of Man, which is discussed with admirable clearness and impartiality in the last chapter, and this leads to the complete exposition of the theory of Sexual Selection which occupies the second part of this work, and must be considered in a future article.

It only remains here to add a word on the account of the affinities and genealogy of man contained in the sixth chapter. As a kind of retribution for the attempt to raise Cuvier's order *Bimana* into a sub-class, not only have most naturalists now reverted to a modified definition of the *Primates* of Linnaeus, but Mr. Darwin shows reasons for refusing to the genus *Homo* even the rank of a family in this order, which Prof. Huxley admits, and regards it simply as an aberrant member of the Catarrhine division of the *Simiadae*. This conclusion, which seems to us to be a just one, will only be distasteful to those who so little appreciate the true characters of man as a spiritual being, that they could feel self-complacency in the brevet-rank of a sub class.

---

* In reviewing in these columns the contributions of the latter eminent writer, we took occasion to quote the estimate he expresses of Mr. Darwin's claims. Should anyone be disposed to overlook the original value of Mr. Wallace's work, he will be corrected by a somewhat similar passage in the present volume. See pp. 137, note, and 416.

在第四章中，达尔文先生探讨了人类的发育模式。文中表明，自然选择理论所依赖的大量事实也是达尔文先生本人的生活写照。他在为生存而斗争方面积累了大量的生活经验。他就像所有的生物体一样，始终有生长的趋势，这一生长受到外界影响的检验和修饰，朝着阻力最小的方向发展，并最终产生了变异，通常假设这种变异是由内在的趋势引起的。在产生的各种变异类型中，只有最能适应周围环境的类型才能存活下来，并将自己的性状传递给后代，再由后代进行新的变异。随后作者又探讨了人类的心智天赋，包括语言、社会习惯、直立行走和完美的双手，这些都是人类与其他动物或同类争斗的有利条件。在人类的发展过程中，自然选择最难解释的阶段就是当他们的防御能力不及类人猿、聪明程度不及最低级野人的阶段。不过，达尔文先生认为这一过渡时期可以在有丰富森林和水果的大型热带岛屿上安全度过。人类一旦发展成熟，便能够立足于世，这从人类的现存状态就能明显地看出来。书中引用了约翰·卢伯克爵士和泰勒先生提出的文明人是野人的后代的观点，以支持作者的看法，同时还讨论了一个重要问题，即通过不断的自然选择，我们可以预期人类种族能发展到什么程度。因此，尽管作者承认人类发展过程经受了许多考验和复杂情况，但并不赞成华莱士先生提出的如下论断：一旦道德职能开始起作用，自然选择就会失灵。* 人类的某些特性，比如躯体裸露和长有胡子，很难用自然选择来解释，达尔文先生将其解释为性选择的作用。最后一章清楚而公正地讨论了所谓的人类种族的起源问题。这就引出了对该工作的第二部分内容——性选择理论进行完整说明的必要性，这将在下一篇中阐述。

第六章对人类的亲缘关系和系谱关系作了进一步的阐述。为了反对居维叶试图将他命名的二手目提升为一个亚纲，不仅大多数自然学家重新采用林耐对灵长目的修正定义，而且达尔文先生也给出了拒绝使用该目中人属甚至是科级阶元名称的理由，并得到了赫胥黎教授的认可，赫胥黎认为它只是旧大陆猿中狭鼻猿的一个异常成员而已。我们可能觉得这一结论是公正的，但是对于那些没有认识到人类作为智慧生物的真实特性的人来说，却是令他们不悦的，这些人只会自我满足于亚纲地位提升的虚衔。

* 在查阅后者这位著名作家的专栏文章时，我们趁机引用了他对达尔文先生的观点的评价。没有人存心忽略华莱士先生的工作的原创价值，本卷书中一段有点类似的内容纠正了他的观点。参见第 137 页和第 416 页。

Mr. Darwin mentions Africa as the possible seat of the Catarrhine progenitors of man, but shows the futility of speculations on this point, until we know more of the recent changes of the earth, the records of palaeontology, and the laws affecting the rapidity of animal modifications. He does not advert to Prof. Haeckel's hypothesis of a "Lemuria" in the Indian Ocean, but agrees with him in next tracing the phylum of man to the *Prosimiae*. These again were developed from "forms standing very low in the deciduate mammalian series" (possibly, as Prof. Huxley suggests, most nearly allied to the existing *Insectivora*), and thus, through the Marsupials and Monotremes from the Reptilian stock, and thence through the *Dipnoi* and Ganoids from the *Urtyhus* of the vertebrate series, represented by the Lancelet alone. Nor does Mr. Darwin stop here, but adds the weight of his judgment to the theory based on the observations of Kowalewsky and Kuppfer, which deduces the primeval *Vertebrata* from a form resembling a Tunicate larva. Perhaps the most brilliant of the many new suggestions in these volumes is one thrown out incidentally in a note to p. 212, and based upon this supposed relation of man to the Ascidians. Beyond the organic world Mr. Darwin does not attempt to trace the genealogy of man. Considering how essential this extension of the theory of evolution is held by men so distinguished as Haeckel, and how keenly the question of Abiogenesis has recently been discussed, the reticence shown in avoiding allusion to the subject is perhaps the most remarkable among the many remarkable characters of this great work.

(**3**, 442-444; 1871)

II

That selection in relation to sex has been an important factor in the formation of the present breeds of animals was more than indicated in the "Origin of Species", and the theory has since been especially worked out by Professor Haeckel. It includes two distinct hypotheses. One is that in contests between males, the weakest would go to the wall, and thus either be killed outright, or at least debarred more or less completely from transmitting their characters to another generation. This may be regarded as a particular case of Natural Selection, and may be compared with the theory of protection by mimicry, suggested by Mr. Bates, and carried out by him and by Mr. Wallace. But though in the lists of Love the battle is often to the strong, even more frequently it is to the beautiful. This introduces a new process, of which the effects are not nearly so obvious as those of Natural Selection, either in its simplest form or in the more complicated cases of mimicry, and of sexual selection by battle. Many circumstances must combine in order that the most successful wooers shall have a larger and more vigorous progeny than the rest. In the first place, all hermaphrodite and all sessile animals may be excluded, and also those cases in which sexual differences depend on different habits of life. Mr. Darwin then shows that secondary sexual characters are eminently variable, and that males vary more than females from the standard of the species, a standard determined by the young, by allied forms, and sometimes by the character of the male himself when his peculiar functions are only periodical, or when they have been artificially prevented. Moreover it is the males who

达尔文先生提到非洲可能是狭鼻猿祖先的起源地，但他也表示目前还无法证实这一推测，这需要对地球近期的变化、古生物学记录以及诱发动物变异的速度的规律有更多的了解。他没有提及海克尔教授的“利莫里亚”大陆沉入印度洋的假说，而是对其另一个观点表示认同，即将人类追溯到原猴亚目。原猴亚目由“低等蜕膜哺乳动物”（据赫胥黎教授提示，最可能与现存的食虫目同属一类）发育而来，即从爬行类经过有袋类和单孔类发育而来，而爬行类则从以文昌鱼为代表的脊椎动物头索动物亚门经过肺鱼和硬鳞鱼发育而来。达尔文先生并没有停止于此，他还评价了柯瓦莱夫斯基和库普费尔通过观察建立的理论，这一理论推断原始的脊椎动物是从一种类似于被囊动物幼虫的生物形式进化而来的。可能在众多新建议中最杰出的就是基于人类与海鞘类的这种猜测性联系而在第 212 页所作的一条附带说明。达尔文先生追溯人类谱系自始至终都没有超出有机世界。考虑到像海克尔这么著名的人物对人类进化理论所作的扩充的重要性以及最近围绕无生源论的讨论的激烈程度，那么，作者谨慎避免提及这个话题可能是这一伟大著作众多特点中最重要的一点。

II

性选择是现代动物养殖的重要依据，这一理论早在《物种起源》中就提到过，之后海克尔教授对其进行了详细研究并作为理论提出。该理论包括两个不同的假说。假说一是雄性间的竞争。失败的弱者将被直接杀死，或者至少得不到交配机会，从而不能将性状传递给下一代。性选择可以看作自然选择的一个特例，可以与拟态保护理论相比，后者是由贝茨提出并与华莱士共同完成的。尽管在爱的竞技场上，胜利经常属于强者，但更多时候是美丽者获胜。这里涉及一个新过程，拟态无论是简单形式还是复杂形式，其影响都不如自然选择和通过争斗进行的性选择明显。许多情况必须结合起来以便最成功的求爱者拥有更多更强壮的后代。首先，所有的雌雄同体动物和固着动物可以被排除，性别差异受不同生活习性影响的动物也被排除在外。达尔文先生随后指出第二性征的变化是非常明显的，从物种的标准来看，雄性比雌性发生的变化更多，这种标准由幼体和联姻形式决定，当雄性功能是周期性的或者被人为阻止时，则由雄性的自身特点决定。此外，在交配中主动的一方通常是雄性，它们不仅要进行战斗来争夺配偶，还要展示它们的颜色、声音或者其他有独特魅力的方面来吸引异性。这一规律得到了食火鸡和一些其他物种的实例的证实：

take the active part in pairing, and who not only fight for the possession of their mates, but display their colours, their voice, of whatever be their peculiar attractions, in order to gain the same end. This rule is confirmed by the exceptional case of the cassowary and a few other species in which the hens court the male birds, fight together in rivalry, and accordingly assume the brighter colours and more attractive shape usually worn by the male. Not only the parental and incubating instincts, but the usual moral qualities of the two sexes are in these cases reversed: "the females being savage, quarrelsome, and noisy, the males gentle and good." But it is further necessary to show that the females exert a choice among the males, and that the latter are polygamous, or arrive earlier at the place of pairing, as is the case with some birds, or else exceed in numbers, at least when both sexes are mature. On this point a series of observations is recorded relating chiefly to man, to domesticated mammals, and to insects. The rule as to transmission of male characters to both sexes appears to be that when variations appear late in life they are usually developed in the same sex only of the next generation, although they are, of course, transmitted in a latent condition through both; while, on the other hand, the differences which appear before maturity in the parent are equally developed in both sexes when transmitted to the offspring. The numerous apparent exceptions to these laws of inheritance and of sexual selection are examined with wonderful fairness and fertility in resource. I may particularly refer to the discussion of the ways in which the young and adults of both sexes differ among birds. The extreme intricacy of some of the questions considered is best shown by a postscript in which, with characteristic candour, the author corrects "a serious and unfortunate error" in the eighth chapter.

The remainder of the first and the greater portion of second volume are occupied by a survey of sexual variations throughout the animal kingdom. Passing rapidly over the other invertebrate classes, the author devotes two chapters to the secondary sexual characters of insects. The weapons, the ornaments, and the sounds peculiar to the males of this vast group of animals are briefly described, and the remarkable analogy between insects and birds which is seen in so many other particulars is traced here also. The brilliant colours of many caterpillars, which, of course, cannot be due to sexual selection, offer one of the many difficulties which are faced, and this is explained by the aid of what the author terms Mr. Wallace's "innate genius for solving difficulties", as being due to natural selection. The bright colours warn the enemies of the caterpillars that they are unfit for food, and so benefit the latter, "on nearly the same principle that certain poisons are coloured by druggists for the good of man." Many cases are probably further complicated by mimicry, savoury caterpillars assuming the colours of distasteful ones so as to share in their immunity, in the same way that a druggist might label his bottles of sweetmeats "poison," to keep them from the shop boy.

In the frigid classes of the lower Vertebrata one would think that sexual selection would have little play; yet Mr. Darwin gives several instances among fishes, amphibians, and reptiles in which weapons or ornaments, peculiar to the males, appear to have been acquired by this means. (See Fig. 2.) But it is in the great class of birds that the most complete series of examples is found, and our advanced knowledge of the habits of this

雌性食火鸡互相打斗以追求雄鸟，因此雌鸟也拥有更鲜艳的色彩和更迷人的体型，而这些通常是雄性的特征。在这些例子中，不仅双亲本能和孵育本能在两性中是相反的，就连一般的品行也是相反的，“雌性野蛮、爱争吵、聒噪，而雄性则绅士而友好。”但是有必要进一步说明的是，雌性对雄性有选择权，而它们实行一妻多夫制；或者像一些鸟类那样，雄性先到达交配地点，或者至少在两性成熟时数量更多。在这一点上，一系列与人类、家养哺乳动物及昆虫有关的观察已经被记录下来。雄性特征向两种性别的后代传递的法则应该是，出现在生命后期的变异通常只会传递给下一代中的同性，尽管这些变异肯定是在某种潜在的条件下传递给两种性别的子代；另一方面，父母亲成熟前产生的变异可以平等地传递给两种性别的子代。在资源丰富而又公平分配的条件下，也观察到了大量并不符合遗传和性选择法则的明显例外。我要特别提到关于鸟类中两种性别的幼鸟和成鸟之间不同点的讨论。其中有些非常复杂的问题，作者在附言中对其进行了很好的阐述，同时特别坦诚地对第八章中“一系列严重的和令人遗憾的错误”进行了纠正。

第一卷其余部分及第二卷的大部分内容是关于整个动物王国中普遍存在的性别变异的调查。作者用两章内容来阐述昆虫的第二性征，而对其他无脊椎动物则快速带过。该部分对雄性昆虫特有的武器、装饰结构和声音作了简要描述，同时还描绘了昆虫与鸟类在很多细节方面的相似之处。许多蝴蝶幼虫具有明丽的颜色，当然用性选择理论是无法解释的，这成为该研究面临的困难之一，而作者凭借华莱士先生所称的“先天就具有的解决困难的天赋”对该现象作出了解释，认为这是自然选择的结果。蝴蝶幼虫明丽的颜色意在警告天敌它们不适合作为食物，从而对后者也是有益的，“与此同理，药剂师将某些毒药染上颜色以利于人们识别。”伪装使许多情况变得复杂，具有伪装本领的、美味可口的毛毛虫把自己伪装成看起来味道很差的样子从而避免被吃掉，同样，药剂师可能在盛有甜食的瓶子上贴上“毒药”的标签，以免被店里的男孩吃掉。

人们可能会认为，性选择在脊椎动物的几个低等冷血的纲中几乎没有作用，然而达尔文先生给出了鱼、两栖动物和爬行动物的例子，它们的雄性特有的武器或装饰结构都是通过性选择作用获得的（见图 2）。最完整的例子是在鸟类中发现的，由于我们对鸟的习性了解得比较多，使得鸟类最可能成为揭示整个理论的领域。当在

class renders it the best possible field for the exposition of the whole theory. Again and again our author forestalls the evidence adduced in the chapters on sexual selection among birds, when tracing its first obscure operation among lower classes, and falls back on the same stronghold when explaining its less obvious working in the mammalia.

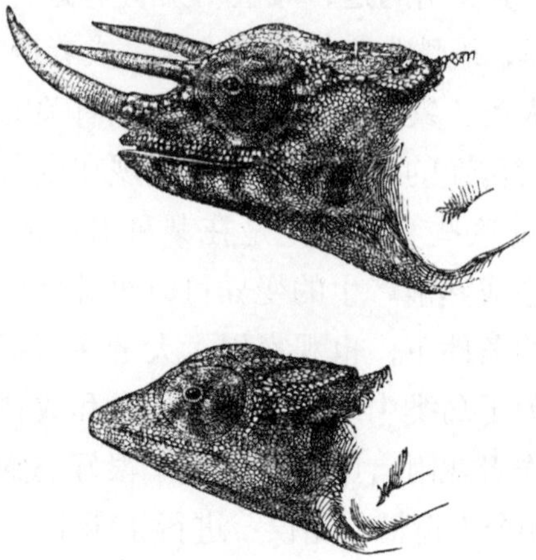

Fig. 2. *Chamaeleon Owenii.* Upper figure, male; lower figure, female.

Among birds the rivalry of beauty has led to far more striking results than has the rivalry of strength. Foremost of these is the power of song, which, in accordance with the law of the least waste, is usually confined to birds of inconspicuous colours, while the combination of the harsh note with the magnificent plumage of the peacock is a familiar converse example. The object of the adornment of birds is conclusively proved by its being, as a rule, confined to males, and often to them only during the breeding season, as well as by the pains they take to exhibit their beauties to the hens. The difficulty is to show the precise way in which the results have been attained by gradual selection. In two remarkable instances, the wings of the Argus pheasant and the train of the peacock, Mr. Darwin succeeds in tracing the gradations in the same bird or the same family by which these wonderful and elaborate ornaments have been brought to their present perfection. The woodcut which illustrate these gradations are unfortunately too numerous to be reproduced here; they are admirably drawn, and convey the impression of the feathers as nearly as is possible by the means employed. Indeed, we may here remark that throughout these volumes the original cuts generally of details of structure, contrast very favourably with the figures of species taken from Brehm's "Thierleben", which are feebly drawn and ill-engraved.

Sexual selection has, of course, been continually checked and modified by the never-ceasing influence of natural selection, sometimes, as in the case of the horns of stags, being only somewhat diverted, but often directly opposed, as when it produces dangerously conspicuous colours, and dangerously cumbersome ornaments. In the case of birds, Mr. Darwin holds that the usual tendency of sexual selection being to produce variation in males, its transmission to hen birds has been checked by natural selection. Mr. Wallace, on the other hand, believes that both tendencies have generally operated together, in opposite directions, so as to make successive generations of males more and more conspicuous

低级动物中探索这一理论的运作机制时，作者屡屡避开在讨论鸟类性选择时所采用的举证环节，在解释哺乳动物更不明显的性选择作用时也采用了同样的方式。

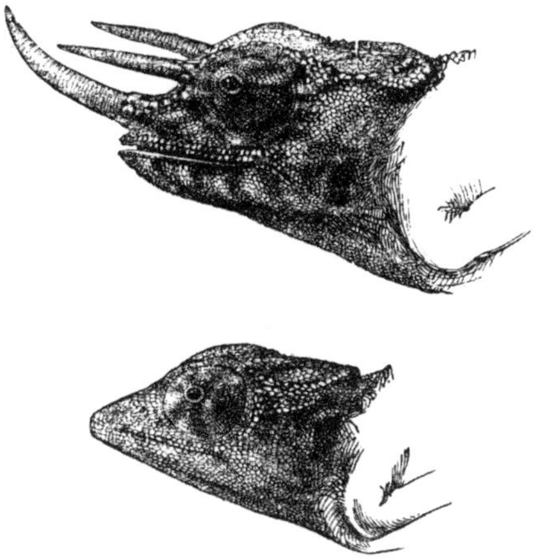

图 2. 奥云变色龙。上图：雄性；下图：雌性。

鸟类的美丽之争比力量之争产生的结果更具影响力。其中最重要的是歌声的力量，那些没有亮丽颜色的鸟儿往往具备美妙的歌声，这也与最少消耗原则相吻合；而与之相反的如孔雀拥有美丽的翅膀但声音却很刺耳也是我们非常熟悉的例子。鸟类装饰结构的作用通常完全可由其表现证实为，只出现在雄性身体上，而且只出现于繁殖期间雄鸟不辞辛苦地向雌鸟展示美丽的时候。研究的难点在于弄清楚通过逐步选择从而获得最后结果的具体方式。两个具有代表性的例子是阿耳戈斯雉的翅膀和孔雀的长尾，达尔文先生成功地追踪到同种鸟或同一家族鸟的演变过程，现在的装饰结构已变得更加华丽而精致。很遗憾，描绘这一渐变过程的木刻画太复杂而无法在此复制。这些图画尽可能真实地表达出了各个阶段的羽毛结构。的确，这里我们可以注意到，达尔文的卷册里的原创细节结构图与布雷姆《动物的生命》中苍白有误的物种形象形成了鲜明的对比。

当然，性选择一直以来就受到自然选择的影响而不断被检验和被修正；有时只是略微修改，如雄鹿的角的例子，而通常则是直接向相反方向转变，例如出现危险性的醒目颜色和笨重装饰的例子。达尔文先生坚持认为，鸟类性选择的一般趋势是雄性发生变异，这些变异向雌鸟的传递受到自然选择的控制。另一方面，华莱士先生则相信，两种趋势通常是从不同方向共同起作用的，以使雄性子代具有比亲代更加显著的性状，而雌性子代则更不明显。但是作为一般规则，幼鸟的羽毛都与雌鸟

than the primitive type, and those of females less so. The fact that, as a rule, young birds resemble hens in their plumage, is a strong argument for the former opinion since most naturalists admit that early characters are the most trustworthy guide to natural alliances, *i.e.*, to true genealogy. To explain the transmission in some cases of brilliant colours (acquired probably by sexual selection, and therefore properly a male character) to both sexes indiscriminately, Mr. Wallace has framed the ingenious hypothesis, that the females have been protected from the dull uniformity threatened by natural selection, by their very general habit of building covered nests. Our author looks at the facts in a reversed way, and supposes that in most cases these hen birds, having inherited bright colours from the males, were led to the habit of building covered nests for the sake of protection.

Among mammals sexual selection has chiefly operated by increasing the size and strength of the males, and furnishing them with weapons of offence;* but besides allurements to the senses of smell and hearing, this class offers not a few instances, especially among the Quadrumana, of brilliant colouring being developed as a secondary sexual character. Here also we have the most striking instances of the production of defensive organs by the same process, as in the manes of lions, the cheekpads of some of the *Suidae*, and possibly the upper tusks of that ancient enigma, the barbirusa. Lastly, it is in the class of mammals that we meet with cases of what may be called primary sexual ornament, as in *Cercopithecus cynosurus*, which make one wonder, with a thankful wonder, why such apparently obvious results are not more common. We must, however, admit that such adornment is not more disgusting, nor that of which we copy a figure more ludicrous, than the personal decorations of savages. Sir Joshua Reynolds says that if a European in full dress and pigtail were to meet a Red Indian in his warpaint, the one who showed surprise or a disposition to laugh would be the barbarian.† But who could stand this test when meeting *Semnopithecus rubicundus* or *Pithecia satanas*?

Fig. 3. Head of *Semnopithecus rubicundus*. This figure (from Prof. Gervais) is given to show the odd arrangement and development of the hair on the head.

* The very general transmission of such weapons to both sexes may, perhaps, be explained by the need females have of means to defend their young.

† Discourse delivered at the Royal Academy, December 10, 1776.

相像，这和前面的观点了形成了强烈的冲突，因为大多数自然学家都认为早期性状是研究自然联姻也就是真正的谱系的最可靠向导。为了解释在一些例子中，亮丽的颜色（可能通过性选择获得，因此是雄性性状）可以不加区别地传递给两种性别的后代，华莱士先生提出了一个独特的假说，即雌性可通过建造有盖巢穴的普遍习性来保护自己不受自然选择的威胁。该书作者则从相反的角度认为，大多数继承了雄性的明亮颜色的雌鸟都会为了保护自己而获得建造有盖巢穴的习性。

哺乳动物的性选择主要通过增加雄性个体的大小和力量以及装备抵御的武器来起作用。* 除了在嗅觉和听觉的诱惑特征，此纲中有许多的例子表明，发育形成的亮丽色彩成了第二性征，尤其是在灵长类动物中。这里介绍了一些引人注目的通过性选择产生防御器官的例子，如狮子的鬃毛、猪科动物的颊垫以及远古之谜巨獠猪的上牙。最后，在哺乳动物纲（如长尾猴）中，我们发现了被称之为原始性装饰结构的例子，这一发现令我们好奇而欣慰，好奇的是为什么这种明显的装饰作用不具有普遍性。不过，我们必须承认，这样的装饰以及我们复制的图示与野人的装饰一样，不但滑稽可笑，而且令人难以接受。乔舒亚·雷诺兹爵士说，如果一个身着正式礼服、扎着辫子的欧洲人要会见一个涂了涂料的印第安人，那么那个表现出惊讶或者大笑的人就是没有教养的人。† 但是在看到婆罗州长尾猴或者黑色狐尾猴时，谁又能忍住不笑呢？

图 3. 婆罗州长尾猴。此图（引自热尔韦教授）用于显示头上毛发的奇怪布局和生长情况。

* 这些抵御武器可传递给两种性别的后代，可以解释为，雌性需要用它们来保护幼体。

† 参见 1776 年 12 月 10 日皇家学会会议上所作的报告。

We must admit, notwithstanding such anomalies, that, on the whole, birds and other animals admire the same forms and colours which we admire, and this, perhaps, may be admitted as an additional argument in favour of their kinship with us. Some of the ugliest creatures (like the hippopotamus) appear to have been quite uninfluenced by sexual selection, while the magnificent plumes of pheasants and birds of paradise are undoubtedly due to its operation. That it has occasionally led to unpleasing results in birds and monkeys of aberrant taste, is no more strange than that all savages do not carve and colour as well as the New Zealanders, or that most Englishmen admire ugly buildings and vulgar pictures. The prevailing aspect of nature is beauty, and the prevailing taste of man is for beauty also. The *means* by which natural beauty has been attained are various. Natural selection is one, by which the healthiest, and therefore the most symmetrical forms survive the rest. Protective mimicry is another, by which fishes have assumed the bright colours of a coral garden and butterflies the delicate venation of leaves. Flowers again have in many cases obtained their gay petals and fantastic shapes from the advantage thus gained for fertilisation by insects. The successive steps which have led to the graceful forms and brilliant tints of shells, to the intricate symmetry of an echinus-spine or a nummulite, these are as yet untraced even in imagination.

But that many of the most striking ornaments of the higher animals, and almost all those which are peculiar to one sex, have been developed by means of sexual selection, is a conclusion which can no longer be distrusted. There remain doubtless many exceptions to be accounted for, many modifying influences to be discovered; but the existence of a new principle has been established which has helped to guide the organic world to its present condition. Side by side with the struggle for existence has gone on a rivalry for reproduction, and the survival of the fittest has been tempered by the success of the most attractive.

(**3**, 463-465; 1871)

尽管有些反常情况，但我们必须承认，总体上鸟类和其他动物喜好的形式和颜色也是我们人类所欣赏的，这或许可以作为支持它们与我们具有亲缘关系的佐证。有些丑陋的生物（如河马）好像并未受到性选择的影响，但是雉和极乐鸟具有的华丽羽毛，肯定是性选择的作用。有时品味怪异的鸟类和猴子也会令人不快，但这并不比并非所有的野人都像新西兰土著居民那样擅长雕刻粉饰、许多英国人喜欢丑陋的建筑和粗俗的图画更奇怪。自然的主流是美丽，人类的主流品位也是追求美丽。达到自然美的**途径**是多种多样的。自然选择就是其中之一，它使得最健康也是最对称的类型更易于存活。保护性的拟态是另外一种途径，它使得鱼儿呈现出珊瑚丛般明亮的色彩，使得蝴蝶拥有树叶般精美的翅脉。在许多情况下，花朵都因其鲜艳的花瓣和漂亮的外形而优先获得昆虫的帮助从而完成受精。贝壳具有优雅的外形和鲜艳的色彩，货币虫具有复杂而对称的海胆棘，它们的发生过程让人难以琢磨，甚至无法想象。

但是，高等动物中许多醒目的装饰结构，以及仅为一种性别所特有的结构，都是通过性选择的方式发育而来，这已经成为毋庸置疑的结论。诚然，还有许多例外的情况尚待解释，许多修饰作用的影响尚待发现，但一个新法则已经建立起来，在它的帮助指导下，有机世界才成为现在的样子。伴随生存竞争的是繁衍竞争，最具吸引者的成功繁衍则调和了适者生存的法则。

（刘皓芳 翻译；冯兴无 审稿）

Pangenesis

C. Darwin

Editor's Note

Despite receiving much praise, Charles Darwin's evolutionary theory was at the same time under attack by scientific colleagues. Francis Galton, from University College London, had carried out experiments to test Darwin's theory of heredity, called pangenesis. In this scheme, Darwin had supposed that the germ cells which unite in the fertilisation of the ovum of an animal or plant are physically made from entities called "gemmules" derived from their organs or structures. *Nature* published Darwin's comments on Galton's experiments. He almost admits in his closing sentence that his theory is indefensible.

IN a paper, read March 30, 1871, before the Royal Society, and just published in the Proceedings, Mr. Galton gives the results of his interesting experiments on the inter-transfusion of the blood of distinct varieties of rabbits. These experiments were undertaken to test whether there was any truth in my provisional hypothesis of Pangenesis. Mr. Galton, in recapitulating "the cardinal points", says that the gemmules are supposed "to swarm in the blood". He enlarges on this head, and remarks, "Under Mr. Darwin's theory, the gemmules in each individual must, therefore, be looked upon as entozoa of his blood", &c. Now, in the chapter on Pangenesis in my "Variation of Animals and Plants under Domestication", I have not said one word about the blood, or about any fluid proper to any circulating system. It is, indeed, obvious that the presence of gemmules in the blood can form no necessary part of my hypothesis; for I refer in illustration of it to the lowest animals, such as the Protozoa, which do not possess blood or any vessels; and I refer to plants in which the fluid, when present in the vessels, cannot be considered as true blood. The fundamental laws of growth, reproduction, inheritance, &c., are so closely similar throughout the whole organic kingdom, that the means by which the gemmules (assuming for the moment their existence) are diffused through the body, would probably be the same in all beings; therefore the means can hardly be diffusion through the blood. Nevertheless, when I first heard of Mr. Galton's experiments, I did not sufficiently reflect on the subject, and saw not the difficulty of believing in the presence of gemmules in the blood. I have said (Variation, &c., vol. II, p. 379) that "the gemmules in each organism must be thoroughly diffused; nor does this seem improbable, considering their minuteness, and the steady circulation of fluids throughout the body." But when I used these latter words and other similar ones, I presume that I was thinking of the diffusion of the gemmules through the tissues, or from cell to cell, independently of the presence of vessels,—as in the remarkable experiments by Dr. Bence Jones, in which chemical elements absorbed by the stomach were detected in the course of some minutes in the crystalline lens of the eye; or again as in the repeated loss of colour and its recovery after a few days by the hair,

泛生论

达尔文

编者按

当时，尽管获得了许多称赞，查尔斯·达尔文的进化论还是受到了科学界同行的攻击。伦敦大学学院的弗朗西斯·高尔顿进行了相关实验来检验被达尔文称为泛生论的遗传学理论。达尔文在这一理论中假设生殖细胞（动物或植物的卵子受精后的结合）是由来源于生物器官或组织的被称为“微芽”的实体组成的。《自然》发表了达尔文对高尔顿的实验的评论。达尔文在其评论的结尾几乎承认了自己的理论还很脆弱。

在1871年3月30日的皇家学会会议上，高尔顿先生公布了一篇论文，该论文刚被发表在《皇家学会学报》上。在这篇论文中，高尔顿先生给出了他的实验结果，是关于不同种兔子间输血的有趣实验。这些实验是用来检验我提出的关于泛生论的暂定假说是否可信。高尔顿先生在概括“最重要的几点”时说，微芽被认为是“充满在血液中的”。他对此进行了详细阐述，并评论说，“因此，根据达尔文先生的理论，每个个体的微芽必须被看作是其血液内的寄生物。”现在我要说，在我的《动物和植物在家养下的变异》一书介绍泛生论的章节中，我没有提到关于血液或任何循环系统中的任何液体的内容。很明显，血液中微芽的存在实际上并不是我的假说的必要组成部分。因为我在说明该假说时提到了最低级的动物，例如没有血液或任何脉管的原生动物，另外我也提到了植物，在它们的脉管中的液体并不能被当作是真正的血液。整个有机王国普遍存在的生长、繁殖、遗传等基本法则是非常相似的，因此微芽（暂时假定是存在的）在体内扩散的方式在所有生物中可能都是一样的，那么就不大可能是通过血液扩散的。不过，当我第一次听说高尔顿先生的实验时，我并没有充分思考这个问题，也没有想到要相信血液中存在微芽有什么困难。我已经说过（见《动物和植物在家养下的变异》，第2卷，第379页），“每种生物的微芽一定是完全分散的，考虑到微芽的微小及液体在全身的稳定循环，这好像不无可能。”但当我说这些话和其他类似的话时，我认为我考虑的是微芽在细胞间的扩散或穿过组织的扩散，这与有无脉管无关——正如本斯·琼斯博士所做的了不起的实验，即被胃吸收的化学物质经过一段时间后可以在眼睛的晶状体中观察到，或者又如佩吉特先生记录的一个患有神经痛的妇女的特例，其中头发会发生重复性掉色并且几天后又可以自行恢复。另外，也不能否认微芽不能通过组织或细胞壁，因为每种花粉

in the singular case of a neuralgic lady recorded by Mr. Paget. Nor can it be objected that the gemmules could not pass through tissues or cell-walls, for the contents of each pollen-grain have to pass through the coats, both of the pollen-tube and embryonic sack. I may add, with respect to the passage of fluids through membrane, that they pass from cell to cell in the absorbing hairs of the roots of living plants at a rate, as I have myself observed under the microscope, which is truly surprising.

When, therefore, Mr. Galton concludes from the fact that rabbits of one variety, with a large proportion of the blood of another variety in their veins, do not produce mongrelised offspring, that the hypothesis of Pangenesis is false, it seems to me that his conclusion is a little hasty. His words are, "I have now made experiments of transfusion and cross circulation on a large scale in rabbits, and have arrived at definite results, negativing, in my opinion, beyond all doubt the truth of the doctrine of Pangenesis." If Mr. Galton could have proved that the reproductive elements were contained in the blood of the higher animals, and were merely separated or collected by the reproductive glands, he would have made a most important physiological discovery. As it is, I think every one will admit that his experiments are extremely curious, and that he deserves the highest credit for his ingenuity and perseverance. But it does not appear to me that Pangenesis has, as yet, received its death blow; though, from presenting so many vulnerable points, its life is always in jeopardy; and this is my excuse for having said a few words in its defence.

(**3**, 502-503; 1871)

粒的内含物都必须穿过花粉管和胚囊共同构成的外被。对于液体穿越细胞膜，我想补充的是，正如我自己在显微镜下观察到的那样，它们以一定的速率在活体植物的吸收根毛的细胞间穿越，这是非常令人惊讶的。

因此，当高尔顿先生根据“静脉血管中含有大量的另一种兔子血液的兔子并没有产生混血后代”这一事实而断定泛生论假说是错误的时候，我觉得这个结论下得有点轻率。他的原话如下，“我目前已经用兔子进行了大规模的输血和交叉循环实验，并且已经得到了确定的阴性结果。在我看来，这些结果毫无疑问地否定了泛生论学说的真实性。”如果高尔顿先生可以证实高等动物的生殖物质包含在血液里，仅由生殖腺隔开或聚集，那么他可就是取得了一个非常重大的生理学发现。如果真是这样，我觉得每个人都会承认他的实验是非常奇妙的，他也将值得凭借其独到的实验设计和锲而不舍的毅力获得最高荣誉。但到目前为止，尽管泛生论暴露出了许多弱点，就此而言，其生存一直处于危险之中，但我认为泛生论并没有受到致命的打击，而且我也有理由为它辩护。

（刘皓芳 翻译；刘力 审稿）

Pangenesis

F. Galton

Editor's Note

In response to comments from Charles Darwin on his recent experiments on heredity, his cousin Francis Galton offered for publication the letter below, in which he acknowledges that his work with rabbits did not amount to a disproof of Darwin's notion of pangenesis, which, he suggested, had now been redefined by Darwin.

IT appears from Mr. Darwin's letter to you in last week's *Nature**, that the views contradicted by my experiments, published in the recent number of the "Proceedings of the Royal Society", differ from those he entertained. Nevertheless, I think they are what his published account of Pangenesis (Animals, &c., under Domestication, II, 374, 379) are most likely to convey to the mind of a reader. The ambiguity is due to an inappropriate use of three separate words in the only two sentences which imply (for there are none which tell us anything definite about) the *habitat* of the Pangenetic gemmules; the words are "circulate", "freely", and "diffused". The proper meaning of circulation is evident enough—it is a re-entering movement. Nothing can justly be said to circulate which does not return, after a while, to a former position. In a circulating library, books return and are re-issued. Coin is said to circulate, because it comes back into the same hands in the interchange of business. A story circulates, when a person hears it repeated over and over again in society. Blood has an undoubted claim to be called a circulating fluid, and when that phrase is used, blood is always meant. I understood Mr. Darwin to speak of blood when he used the phrases "circulating freely", and "the steady circulation of fluids", especially as the other words "freely" and "diffusion" encouraged the idea. But it now seems that by circulation he meant "dispersion", which is a totally different conception. Probably he used the word with some allusion to the fact of the dispersion having been carried on by eddying, not necessarily circulating, currents. Next, as to the word "freely". Mr. Darwin says in his letter that he supposes the gemmules to pass through the solid walls of the tissues and cells; this is incompatible with the phrase "circulate freely". Freely means "without retardation"; as we might say that small fish can swim freely through the larger meshes of a net; now, it is impossible to suppose gemmules to pass through solid tissue without *any* retardation. "Freely" would be strictly applicable to gemmules drifting along with the stream of the blood, and it was in that sense I interpreted it. Lastly, I find fault with the use of the word "diffused", which applies to movement in or with fluids, and is inappropriate to the action I have just described of solid boring its way through solid. If Mr. Darwin had given in his work an additional paragraph or two to a description of the whereabouts of the gemmules which, I must remark, is a cardinal point of his theory, my

* *Nature* vol. III, p. 502.

泛生论

高尔顿

编者按

查尔斯·达尔文解释了他最近关于遗传学方面的实验，作为回应，他的表弟弗朗西斯·高尔顿寄来如下的回信要求发表。在这封回信中他提出，他用兔子进行研究的结果并不是对达尔文已经重新定义过的泛生论观点的反驳。

从达尔文先生上周写给《自然》的信来看*，他持有的观点与我发表在最近一期《皇家学会学报》上的实验结果所反驳的观点似乎并不相同。我觉得，我的结果反驳的不过是那些他发表的关于泛生论的说明（见《动物和植物在家养下的变异》,第2卷，第374、379页）中很可能会传递给读者的观点。这一歧义是由两个句子中三个使用不当的词引起的,这两个句子暗示了(因为并没有明确地告诉我们)泛生微芽的**聚集地**。这三个词是“循环”，“自由地”和“分散的”。循环的恰当含义是非常明显的——它是一种再次进入的运动。任何东西如果没有在后来返回到原先的位置就不能称为循环。在一个循环的图书馆里，书籍能够返回并被再利用。钱币被说成是循环流通的，是因为在商业交换中它能返回到同一个人手中。故事是循环传播的，是因为当某个人听到它时，该故事在社会上已经被重复叙述过许多遍了。毫无疑问，血液可以被称作是一种循环的液体，而且当人们说到循环液体的时候，通常就是指血液。因此当达尔文先生使用“自由地循环”、“液体的稳定循环”这些词时，我认为他就是在说血液，尤其是“自由地”和“分散的”这两个词更坚定了我的理解。但现在看来，好像他是在用循环来表示“散布”的意思，但这是个完全不同的概念。可能他使用这个词时想的是散布是由涡流造成的，而未必是环流。接下来是第二个词“自由地”。达尔文先生在他的信中说，他认为微芽可以穿过坚固的组织壁和细胞壁。这与“自由地循环”这一表达是不相符的。“自由地”就意味着“没有阻碍”，例如我们可以说小鱼能“自由地”游过渔网上的大网眼，但我们现在不可能认为微芽可以没有**任何**阻碍地穿越坚固的组织。严格来说，“自由地”适用于微芽随着血流漂移的情况，我就是从这个意义上来理解这个词的。我要挑的最后一个毛病是“分散的”这个词的使用。这个词适用于液体中的或与液体一起进行的运动，而不适用于我刚刚描述过的固体穿越固体的运动。如果达尔文先生在其著作中再用一两个段落来描述微芽的所在之处，那么我就不会像

---

* 参见《自然》，第3卷，第502页。

misapprehension of his meaning could hardly have occurred without more hesitancy than I experienced, but I certainly felt and endeavoured to express in my memoir some shade of doubt; as in the phrase, p. 404, "that the doctrine of Pangenesis, pure and simple, as I have interpreted it, is incorrect."

As I now understand Mr. Darwin's meaning, the first passage (II, 374), which misled me, and which stands: "...minute granules...which circulate freely throughout the system" should be understood as "minute granules...which are dispersed thoroughly and are in continual movement throughout the system"; and the second passage (II, 379), which now stands: "The gemmules in each organism must be thoroughly diffused; nor does this seem improbable, considering...the steady circulation of fluids throughout the body", should be understood as follows: "The gemmules in each organism must be dispersed all over it, in thorough intermixture; nor does this seem improbable, considering...the steady circulation of the blood, the continuous movement, and the ready diffusion of other fluids, and the fact that the contents of each pollen grain have to pass through the coats, both of the pollen tube and of the embryonic sack." (I extract these latter *addenda* from Mr. Darwin's letter.)

I do not much complain of having been sent on a false quest by ambiguous language, for I know how conscientious Mr. Darwin is in all he writes, how difficult it is to put thoughts into accurate speech, and, again, how words have conveyed false impressions on the simplest matters from the earliest times. Nay, even in that idyllic scene which Mr. Darwin has sketched of the first invention of language, awkward blunders must of necessity have often occurred. I refer to the passage in which he supposes some unusually wise, ape-like animal to have first thought of imitating the growl of a beast of prey so as to indicate to his fellow monkeys the nature of expected danger. For my part, I feel as if I had just been assisting at such a scene. As if, having heard my trusted leader utter a cry, not particularly well articulated, but to my ears more like that of a hyena than any other animal, and seeing none of my companions stir a step, I had, like a loyal member of the flock, dashed down a path of which I had happily caught sight, into the plain below, followed by the approving nods and kindly grunts of my wise and most-respected chief. And I now feel, after returning from my hard expedition, full of information that the suspected danger was a mistake, for there was no sign of a hyena anywhere in the neighbourhood. I am given to understand for the first time that my leader's cry had no reference to a hyena down in the plain, but to a leopard somewhere up in the trees; his throat had been a little out of order—that was all. Well, my labour has not been in vain; it is something to have established the fact that there are no hyenas in the plain, and I think I see my way to a good position for a look out for leopards among the branches of the trees. In the meantime, *Vive* Pangenesis.

(**4**, 5-6; 1871)

之前那样没有太过犹豫就误解了他的意思，另外我必须指出，微芽的所在之处是其理论的一个关键点。我确实有一些怀疑并且尽力在我的文章里表达出来，例如第 404 页写道的那句“如我所解释的单纯而简单的泛生论是不正确的。”

因为我现在理解了达尔文先生的意思，所以曾误导过我的表述为“……小颗粒……在整个系统中自由地循环”的第一段（第 2 卷，第 374 页）应当理解成“……小颗粒……在整个系统中完全散布并不停地移动”。误导我的第二段（第 2 卷，第 379 页），原描述为“每种生物的微芽一定是完全分散的；考虑到……液体在全身的稳定循环，这好像不无可能，”应理解为“每种生物的微芽一定是以充分混合的状态散布在全身各处的；考虑到……血液稳定的循环，其他液体连续的移动和充分的分散，以及每种花粉粒的内含物都必须穿过花粉管和胚囊共同构成的外被，这好像不无可能。”（后半部分**附加的**内容摘自达尔文先生的信件。）

我并不是要过多地抱怨模棱两可的语言使我产生了误解，因为很早之前我就知道达尔文先生在写出所有这些话时是多么认真尽责，我也知道要将自己的思想准确地付诸语言是多么困难，以及文字曾经怎样在最简单的事情上传递了错误的观念。不但如此，就算是在达尔文先生曾经描绘过的，语言刚被发明出来的田园时代，一些笨拙的错误也一定不可避免地经常发生。在某段文章中，达尔文先生指出有些非常聪明的猿类动物能够一下子想到通过模仿野兽捕食猎物时的咆哮声来提示他的猴子同伴们他所预料到的危险的性质，在这里我要提一下这段内容。对我来说，我觉得我好像也一直处在这样的情境中。就好像是听到我信任的领袖发出一声呼喊，尽管这种声音的意义不是特别明确，但对我的耳朵来说这种声音最像土狼的叫声，当我发现我的同伴都没有动身的意向时，我已经像一个群体中的一名忠实的成员一样，沿着我先前很幸运地发现的一条通往下方平原的小路飞奔出了一段距离，我非常敬重的英明领袖给了我赞许的点头和友好的啧啧声。当我从自己的艰难探险中回来之后，我才发现我错误地判断了危险，因为周围任何地方都没有土狼出现的迹象。我第一次明白，我的领袖的喊声并不是提示平原上有土狼，而是树上某处有豹。他的嗓音有点失控了——这就是全部。总之，我的劳动并非徒然，因为我的劳动确定了平原上没有土狼的事实，另外我觉得我也知道了用以留神藏于树枝间的豹的好方法。同时，祝泛生论万岁。

（刘皓芳 翻译；刘力 审稿）

On Colour Vision*

J. C. Maxwell

Editor's Note

Among the lesser known scientific contributions of James Clerk Maxwell is his work on colour theory. Maxwell showed how three primary colours of light—red, green and blue—can be mixed to generate almost any colour, paving the way for colour photography and projection. In this contribution based on a talk at London's Royal Institution, he reviews this work and expands on its implications for colour vision. Thomas Young had proposed in 1801 that this involved three kinds of light receptor in the eye. Maxwell shows that these receptors have now been identified as rod- and cone-shaped cells in the retina. The existence of three distinct types of colour-sensitive cone cells had not yet been proven, but Maxwell clearly suspects it.

ALL vision is colour vision, for it is only by observing differences of colour that we distinguish the forms of objects. I include differences of brightness or shade among differences of colour.

It was in the Royal Institution, about the beginning of this century, that Thomas Young made the first distinct announcement of that doctrine of the vision of colours which I propose to illustrate. We may state it thus:—We are capable of feeling three different colour-sensations. Light of different kinds excites these sensations in different proportions, and it is by the different combinations of these three primary sensations that all the varieties of visible colour are produced. In this statement there is one word on which we must fix our attention. That word is, Sensation. It seems almost a truism to say that colour is a sensation; and yet Young, by honestly recognising this elementary truth, established the first consistent theory of colour. So far as I know, Thomas Young was the first who, starting from the well-known fact that there are three primary colours, sought for the explanation of this fact, not in the nature of light, but in the constitution of man. Even of those who have written on colour since the time of Young, some have supposed that they ought to study the properties of pigments, and others that they ought to analyse the rays of light. They have sought for a knowledge of colour by examining something in external nature—something out of ourselves.

Now, if the sensation which we call colour has any laws, it must be something in our own nature which determines the form of these laws; and I need not tell you that the only evidence we can obtain respecting ourselves is derived from consciousness.

* Lecture delivered before the Royal Institution, March. 24th.

论色觉*

麦克斯韦

编者按

詹姆斯·克拉克·麦克斯韦在颜色理论方面的研究是其不太著名的科学贡献之一。早先，麦克斯韦就向人们展示了如何通过混合光的三原色（红、绿、蓝）来得到几乎所有的颜色，这为彩色照相技术和投影技术铺平了道路。在这篇以一次在伦敦皇家研究院所作的报告为基础而形成的文稿中，麦克斯韦回顾了这项研究工作并进一步阐述了其对色觉的意义。1801 年，托马斯·杨就提出眼睛中有三种光感受体。麦克斯韦在这里指出，人们已经确认这些光感受体就是视网膜上的杆状细胞和锥形细胞。尽管当时还没有证实存在三种不同类型的对颜色敏感的锥形细胞，但麦克斯韦仍然明确地支持这种猜想。

所有的视觉都是色觉，因为我们只有通过观察颜色的差别才能区分物体的形态。我把明暗的区别也包含在了颜色的区别当中。

约在本世纪初，托马斯·杨在皇家研究院第一次明确地宣布了这个关于色觉的学说，这里，我将要对它进行阐述。我们可以这样来表述：我们能够体验到三种不同的颜色感觉。不同类型的光会以不同比例激发三种颜色感觉，所有可见的颜色就是由这三种基本感觉经过不同的组合而形成的。在这里，有一个词是值得我们注意的，那就是“感觉”。说颜色是一种感觉简直就是一个起码的常识；但杨真正确认了这个基本事实，首先建立了与之一致的关于颜色的理论。据我所知，托马斯·杨是第一个从人类的知觉而不是从光的本质来解释众所周知的三原色的人。即便是那些在杨以后撰写了有关颜色的著作的人们，不是认为应该去研究颜料的特性就是认为应该去分析光线。他们试图用人类自身之外的那些外在本质来揭示颜色的奥秘。

现在，如果说我们称之为颜色的这种感觉遵循某种规律的话，那么一定是我们自身的本质决定了这种规律的形式。无需由我来告诉你，我们所能获得的关于自身的唯一证据就来自于我们的意识。

* 这是 3 月 24 日向皇家研究院所作的演讲。

The science of colour must therefore be regarded as essentially a mental science. It differs from the greater part of what is called mental science in the large use which it makes of the physical sciences, and in particular of optics and anatomy. But it gives evidence that it is a mental science by the numerous illustrations which it furnishes of various operations of the mind.

In this place we always feel on firmer ground when we are dealing with physical science. I shall therefore begin by showing how we apply the discoveries of Newton to the manipulation of light, so as to give you an opportunity of feeling for yourselves the different sensations of colour.

Before the time of Newton, white light was supposed to be of all known things the purest. When light appears coloured, it was supposed to have become contaminated by coming into contact with gross bodies. We may still think white light the emblem of purity, though Newton has taught us that its purity does not consist in simplicity.

We now form the prismatic spectrum on the screen. These are the simple colours of which white light is always made up. We can distinguish a great many hues in passing from the one end to the other; but it is when we employ powerful spectroscopes, or avail ourselves of the labours of those who have mapped out the spectrum, that we become aware of the immense multitude of different kinds of light, every one of which has been the object of special study. Every increase of the power of our instruments increases in the same proportion the number of lines visible in the spectrum.

All light, as Newton proved, is composed of these rays taken in different proportions. Objects which we call coloured when illuminated by white light, make a selection of these rays, and our eyes receive from them only a part of the light which falls on them. But if they receive only the pure rays of a single colour of the spectrum, they can appear only of that colour. If I place a disc containing alternate quadrants of red and green paper in the red rays, it appears all red, but the red quadrants brightest. If I place it in the green rays both papers appear green, but the red paper is now the darkest. This, then, is the optical explanation of the colours of bodies when illuminated with white light. They separate the white light into its component parts, absorbing some and scattering others.

Here are two transparent solutions. One appears yellow, it contains bichromate of potash; the other appears blue, it contains sulphate of copper. If I transmit the light of the electric lamp through the two solutions at once, the spot on the screen appears green. By means of the spectrum we shall be able to explain this. The yellow solution cuts off the blue end of the spectrum, leaving only the red, orange, yellow, and green. The blue solution cuts off the red end, leaving only the green, blue, and violet. The only light which can get through both is the green light, as you see. In the same way most blue and yellow paints, when mixed, appear green. The light which falls on the mixture is so beaten about between the yellow particles and the blue, that the only light which survives is the green. But yellow and blue light when mixed do not make green, as you will see if we allow them to fall on the same part of the screen together.

因此，色彩学在本质上应该被当作是一种精神科学。它与大多数所谓的精神科学有很大的区别，它要用到物理学，特别是要用到光学和解剖学。但是，种种精神活动提供了大量的例证，可以证明色彩学是一种精神科学。

当我们利用物理学来处理这一问题的时候，总是感到有更坚实的理论基础。因此，我将从说明我们如何把牛顿的发现运用于光线的操控入手，以此给你一个机会，让你知道你对颜色的不同感觉。

在牛顿之前，白光被认为是所有已知物质中最纯粹的物质。有色光被认为是白光接触到物体而受到了污染。我们也许仍然可以认为白光象征着纯粹，但是牛顿已经告诉我们，白光的纯粹并不意味着简单。

现在，我们在屏幕上呈现棱镜光谱，得到的就是构成白光的基本颜色。当我们从一端向另一端观察的时候，可以分辨出很多不同的色彩；但是当我们使用功能更为强大的分光镜，或者利用别人已经制好的光谱时，我们就会发现大量不同种类的光线，每一种都值得专门研究。光谱中可分辨谱线数量增加的比例与仪器分辨率提高的比例是一致的。

牛顿已经证实，所有的光都是由上面所提到的光线以不同的比例组合而成的。当我们所谓的有色物体被白光照亮的时候，它会选择光线，而我们的眼睛能接受到的只是照射在其上的一部分光线。如果物体只被光谱中纯粹的单色光线所照射，那么它就只能呈现出那种颜色。如果我把红纸和绿纸交替放在一个盘子的不同象限里，用红光照射，整个盘子都会呈现出红色，但是红纸所在部分最明亮。如果把盘子放在绿光中，那么红纸和绿纸都会呈现出绿色，但这一次红纸部分是最暗的。这就是物体被白光照射时所呈现的颜色的光学解释。它们把白光拆分成不同的组成部分，然后吸收一部分，反射另外的部分。

这里有两种透明的溶液。一种是黄色的重铬酸钾溶液，另一种是蓝色的硫酸铜溶液。如果我让电灯发射的光通过这两种溶液，那么投射到屏幕上的是绿色光斑，这可以用光谱来解释。黄色溶液将光谱中的蓝色一端切断，只剩下了红色、橙色、黄色和绿色；蓝色溶液则将光谱中的红色一端切断，只剩下绿色、蓝色和紫色。正如你看到的那样，只有绿色光才能通过两种溶液。同样的道理，蓝色和黄色的颜料混合在一起通常会呈现出绿色。光照射在混合颜料上，黄色和蓝色的颜料颗粒吸收各自范围的光线，只有绿色光线可以反射出来。但是黄光和蓝光却不能混合成绿光，如果我们把它们投射到屏幕上的同一个区域，就可以看出来。

It is a striking illustration of our mental processes that many persons have not only gone on believing, on the evidence of the mixture of pigments, that blue and yellow make green, but that they have even persuaded themselves that they could detect the separate sensations of blueness and of yellowness in the sensation of green.

We have availed ourselves hitherto of the analysis of light by coloured substances. We must now return, still under the guidance of Newton, to the prismatic spectrum. Newton not only

Untwisted all the shining robe of day,

but showed how to put it together again. We have here a pure spectrum, but instead of catching it on a screen, we allow it to pass through a lens large enough to receive all the coloured rays. These rays proceed, according to well-known principles in optics, to form an image of the prism on a screen placed at the proper distance. This image is formed by rays of all colours, and you see the result is white. But if I stop any of the coloured rays, the image is no longer white, but coloured; and if I only let through rays of one colour, the image of the prism appears of that colour.

I have here an arrangement of slits by which I can select one, two, or three portions of the light of the spectrum, and allow them to form an image of the prism while all the rest are stopped. This gives me a perfect command of the colours of the spectrum, and I can produce on the screen every possible shade of colour by adjusting the breadth and the position of the slits through which the light passes. I can also, by interposing a lens in the passage of the light, show you a magnified image of the slits, by which you will see the different kinds of light which compose the mixture.

The colours are at present red, green, and blue, and the mixture of the three colours is, as you see, nearly white. Let us try the effect of mixing two of these colours. Red and blue form a fine purple or crimson, green and blue form a sea-green or sky-blue, red and green form a yellow.

Here again we have a fact not universally known. No painter, wishing to produce a fine yellow, mixes his red with his green. The result would be a very dirty drab colour. He is furnished by nature with brilliant yellow pigments, and he takes advantage of these. When he mixes red and green paint, the red light scattered by the red paint is robbed of nearly all its brightness by getting among particles of green, and the green light fares no better, for it is sure to fall in with particles of red paint. But when the pencil with which we paint is composed of the rays of light, the effect of two coats of colour is very different. The red and the green form a yellow of great splendour, which may be shown to be as intense as the purest yellow of the spectrum.

I have now arranged the slits to transmit the yellow of the spectrum. You see it is similar in colour to the yellow formed by mixing red and green. It differs from the mixture, however, in being strictly homogeneous in a physical point of view. The prism, as you see, does not

这是一个与我们的精神活动过程有关的惊人例子：根据颜料混合物的实验结果，许多人不仅相信蓝色加上黄色会呈现绿色，而且还认为自己可以从绿色的视觉感受中分离出黄色和蓝色的部分。

到目前为止，我们都在用有色的物质来分析光学问题。现在，我们仍然需要按照牛顿的理论回到棱镜光谱上来。牛顿不仅

解开了日光那耀眼的罩袍，

而且还展示了如何把它重新整合起来。我们有一束纯的分光光谱，但没有将它投射到屏幕上，而是让它通过一个足够大的棱镜以便接收各种颜色的光。依照我们熟知的光学原理，这些光线在其前方一定距离处的屏幕上会形成分光光谱的图像。这个图像由各种颜色的光线组成，而你看到的结果是白色的。但如果我挡住任何一种颜色的光，图像将不再是白色的，而是有色的；如果我只让一种颜色的光通过，那么分光光谱的图像上所呈现的就是那种颜色。

这里，我可以利用设置狭缝的方法选择光谱中的一部分、两部分或三部分光谱线，使其成像，而其他部分则被挡住。这样我就可以极好地控制光谱中的颜色了，通过调整光路中各狭缝的宽度和位置，可以使屏幕上呈现出每一种颜色的图像。我还可以在光路中插入透镜，使你看到狭缝的像，这样你就可以观察到混合在一起的不同种类的光。

现在，选取红、绿、蓝三种颜色，正如你所见，它们混合在一起后几乎是白色的。我们也可以尝试一下混合三种当中的任意两种颜色。红色和蓝色形成纯紫色或者深红色，绿色和蓝色形成海绿色或者天蓝色，红色和绿色形成黄色。

这里，我们又得到了一个并非被广泛了解的事实。没有哪一个画家会用他的红颜料和绿颜料混合在一起调出纯黄色。这样做只能得到一种很脏的灰黄色。他自己本来就有鲜亮的黄颜料，用这个就行了。当他混合红绿两种颜料的时候，红颜料颗粒所反射出来的红光因为被绿颜料颗粒吸收而几乎失去了全部亮度，绿光的情况也好不了多少，因为绿颜料颗粒反射出的绿光也会被红颜料颗粒吸收。但是如果我们作画时所用的笔是由光线组成的，那么涂覆两种颜色得到的效果就会完全不同。红光和绿光会形成非常漂亮的黄色，和光谱中最纯的黄光一样鲜艳。

我现在调整狭缝，选取光谱中的黄光。你会发现它与红光和绿光混合在一起的颜色非常相似。然而，用物理学的观点来看，它与混合物不同，因为它是严格均质的。正如你所见，棱镜并没有像对待混合光线那样把它分成两部分。让我们把这束黄光

divide it into two portions as it did the mixture. Let us now combine this yellow with the blue of the spectrum. The result is certainly not green; we may make it pink if our yellow is of a warm hue, but if we choose a greenish yellow we can produce a good white.

You have now seen the most remarkable of the combinations of colours—the others differ from them in degree, not in kind. I must now ask you to think no more of the physical arrangements by which you were enabled to see these colours, and to concentrate your attention upon the colours you saw, that is to say, on certain sensations of which you were conscious. We are here surrounded by difficulties of a kind which we do not meet with in purely physical inquiries. We can all feel these sensations, but none of us can describe them. They are not only private property, but they are incommunicable. We have names for the external objects which excite our sensations, but not for the sensations themselves.

When we look at a broad field of uniform colour, whether it is really simple or compound, we find that the sensation of colour appears to our consciousness as one and indivisible. We cannot directly recognise the elementary sensations of which it is composed, as we can distinguish the component notes of a musical chord. A colour, therefore, must be regarded as a single thing, the quality of which is capable of variation.

To bring a quality within the grasp of exact science, we must conceive it as depending on the values of one or more variable quantities, and the first step in our scientific progress is to determine the number of these variables which are necessary and sufficient to determine the quality of a colour. We do not require any elaborate experiments to prove that the quality of colour can vary in three and only in three independent ways.

One way of expressing this is by saying, with the painters, that colour may vary in hue, tint, and shade.

The finest example of a series of colours varying in hue, is the spectrum itself. A difference in hue may be illustrated by the difference between adjoining colours in the spectrum. The series of hues in the spectrum is not complete; for, in order to get purple hues, we must blend the red and the blue.

Tint may be defined as the degree of purity of a colour. Thus, bright yellow, buff, and cream-colour, form a series of colours of nearly the same hue, but varying in tint. The tints corresponding to any given hue form a series, beginning with the most pronounced colour, and ending with a perfectly neutral tint.

Shade may be defined as the greater or less defect of illumination. If we begin with any tint of any hue, we can form a gradation from that colour to black, and this gradation is a series of shades of that colour. Thus we may say that brown is a dark shade of orange.

和光谱中的蓝光混合起来。结果当然不是绿光；如果采用暖色调的黄光，我们得到的将是粉色，但如果我们选的是偏绿的黄光，就会得到很好的白色。

你已经看到了一些最显著的颜色组合，其他颜色组合与这些相比只有程度上的差别，而没有本质上的差别。现在，我请你别去考虑那些让你看到颜色的实验装置，而把你的注意力集中在你所看到的颜色上，也就是说，把注意力集中在你的感受上。我们遇到的困难是，我们无法进行纯粹的物理意义上的研究。我们都能感觉得到，但是谁也无法描述这一切。感觉不仅是个人的感受，而且难以表达出来。我们能说出那些刺激我们感受的外部物体的名字，但是无法描述那种感受本身。

当我们注视一大块均匀的颜色时，不论这种颜色是简单的还是复合的，我们发现在我们的意识中对颜色的感觉是一个不可分割的整体。我们无法像分辨和弦中的音符那样，直接把构成这种感受的元素分离出来。所以颜色应该被看作是一种单一的东西，而它的性质可以改变。

为了能用精确的科学术语描述一个量，我们必须建立这个量与一个或几个变量之间的依赖关系，我们研究工作的第一步是确定能充分必要地决定一种颜色性质的变量的数目。我们不需要任何复杂的实验就能证明颜色的性质依赖且只依赖于三个独立的变量。

画家们对此有一种说法，即颜色之间的区别是：色相、纯度和明度。

一系列色彩依色相变化的最好例子就是光谱本身。光谱中相邻颜色的差异可以用来说明色相的区别。光谱中的色相系列并不完全；因为要想得到紫色色相，我们必须混合红光和蓝光。

纯度可以被定义为一种颜色纯净的程度。这样，明黄色、浅黄色和奶黄色形成了一个具有几乎相同色相的系列，但纯度不同。对应于某个给定的色相，纯度不同的一组颜色可以形成一个系列，从最浓的颜色开始，到最淡的颜色结束。

明度可以被定义为光照程度的多少。如果我们从任意色相的任意纯度出发，就可以在这种颜色和黑色之间形成一个渐变，这个渐变就是这种颜色的一个明度序列。这样，我们就可以说，棕色是橙色明度变暗得到的结果。

The quality of a colour may vary in three different and independent ways. We cannot conceive of any others. In fact, if we adjust one colour to another, so as to agree in hue, in tint, and in shade, the two colours are absolutely indistinguishable. There are therefore three, and only three, ways in which a colour can vary.

I have purposely avoided introducing at this stage of our inquiry anything which may be called a scientific experiment, in order to show that we may determine the number of quantities upon which the variation of colour depends by means of our ordinary experience alone.

Here is a point in this room: if I wish to specify its position. I may do so by giving the measurements of three distances—namely, the height above the floor, the distance from the wall behind me, and the distance from the wall at my left hand.

This is only one of many ways of stating the position of a point, but it is one of the most convenient. Now, colour also depends on three things. If we call these the intensities of the three primary colour sensations, and if we are able in any way to measure these three intensities, we may consider the colour as specified by these three measurements. Hence the specification of a colour agrees with the specification of a point in the room in depending on three measurements.

Let us go a step farther, and suppose the colour sensations measured on some scale of intensity, and a point found for which the three distances, or co-ordinates, contain the same number of feet as the sensations contain degrees of intensity. Then we may say, by a useful geometrical convention, that the colour is represented to our mathematical imagination by the point so found in the room; and if there are several colours, represented by several points, the chromatic relations of the colours will be represented by the geometrical relations of the points. This method of expressing the relations of colours is a great help to the imagination. You will find these relations of colours stated in an exceedingly clear manner in Mr. Benson's "Manual of Colour", one of the very few books on colour in which the statements are founded on legitimate experiments.

There is a still more convenient method of representing the relations of colours, by means of Young's triangle of colours. It is impossible to represent on a plane piece of paper every conceivable colour, to do this requires space of three dimensions. If, however, we consider only colours of the same shade, that is, colours in which the sum of the intensities of the three sensations is the same, then the variations in tint and in hue of all such colours may be represented by points on a plane. For this purpose we must draw a plane cutting off equal lengths from the three lines representing the primary sensations. The part of this plane within the space in which we have been distributing our colours will be an equilateral triangle. The three primary colours will be at the three angles, white or gray will be in the middle, the tint or degree of purity of any colour will be expressed by its

一种颜色的性质可以按三种不同且独立的方式变化。我们想不出任何其他的变化方式。事实上，如果我们将一种颜色调整为另一种颜色，使得前者和后者在色相、纯度、明度上都一致，那么这两种颜色绝对没有任何差异。因此一种颜色的变化方式有且只有三种。

我有意避免在科学实验的层面上去介绍我们的研究，这样做是为了表明，我们仅凭日常经验就可以确定描述颜色变化的变量数目。

房间里有一个点，如果我想确定这个点的位置，我可以给出三个距离的测量值：即相对于地面的高度、到我身后的墙的距离以及到我左手边的墙的距离。

这只是描述一个点位置的多种方法中的一种，但它是最方便的一种。现在，颜色也同样依赖于三个变量。如果我们把这些称作三种原色的感觉强度，并且如果我们能用某种方法测量出这三种强度，那么我们就可以认为颜色能够通过对这三者的测量而被确定下来。因此，描述一种颜色和描述空间某点的位置一样，都要依赖于对三个变量的测量。

让我们再深入一步。假设用强度等级来衡量的色觉所包含的强度数值，与空间中某点所包含的用英尺数表示的三个距离，或三个坐标的数值相同，那么我们就可以说，通过实用几何学的常规做法，我们可以在数学上将色觉想象成空间中的某一点；如果有多种颜色，就用多个点来表示，那么这些颜色之间的关系也可以用点之间的几何学关系来描述。这样的描述对于我们想象不同颜色之间的关系大有帮助。在本森先生所著的《颜色手册》一书中，你会发现颜色之间的这些关系被叙述得非常清晰。基于正规实验的关于颜色方面的著作少之又少，而这本书就是其中之一。

还有一种更方便的描述颜色之间关系的方法，即杨的颜色三角形法。我们无法在一张纸的平面上描述任何一种可见的颜色，要做到这一点需要三维的空间。但是，如果我们只考虑具有相同明度的颜色，也就是说，在这些颜色中三种色觉的强度之和是一样的，那么纯度和色相的变化就可以用平面上的点来描述了。为此，我们用三条等长的、代表原色感觉的线来切割一个平面。它们所围的区域是一个等边三角形，我们将在这个区域中分配我们的颜色。三种原色将位于三个顶角，中间是白色或者灰色，颜色的纯度用这种颜色到中点的距离来表示，颜色的色相则取决于它与中点连线的角度。

distance from the middle point, and its hue will depend on the angular position of the line which joins it with the middle point.

Thus the ideas of tint and hue can be expressed geometrically on Young's triangle. To understand what is meant by shade, we have only to suppose the illumination of the whole triangle increased or diminished, so that by means of this adjustment of illumination Young's triangle may be made to exhibit every variety of colour. If we now take any two colours in the triangle and mix them in any proportions, we shall find the resultant colour in the line joining the component colours at the point corresponding to their centre of gravity.

I have said nothing about the nature of the three primary sensations, or what particular colours they most resemble. In order to lay down on paper the relations between actual colours, it is not necessary to know what the primary colours are. We may take any three colours, provisionally, as the angles of a triangle, and determine the position of any other observed colour with respect to these, so as to form a kind of chart of colours.

Of all colours which we see, those excited by the different rays of the prismatic spectrum have the greatest scientific importance. All light consists either of some one kind of these rays, or of some combination of them. The colours of all natural bodies are compounded of the colours of the spectrum. If, therefore, we can form a chromatic chart of the spectrum, expressing the relations between the colours of its different portions, then the colours of all natural bodies will be found within a certain boundary on the chart defined by the positions of the colours of the spectrum.

But the chart of the spectrum will also help us to the knowledge of the nature of the three primary sensations. Since every sensation is essentially a positive thing, every compound colour-sensation must be within the triangle of which the primary colours are the angles. In particular, the chart of the spectrum must be entirely within Young's triangle of colours, so that if any colour in the spectrum is identical with one of the colour-sensations, the chart of the spectrum must be in the form of a line having a sharp angle at the point corresponding to this colour.

I have already shown you how we can make a mixture of any three of the colours of the spectrum, and vary the colour of the mixture by altering the intensity of any of the three components. If we place a compound colour side by side with any other colour, we can alter the compound colour till it appears exactly similar to the other. This can be done with the greatest exactness when the resultant colour is nearly white. I have therefore constructed an instrument which I may call a colour-box, for the purpose of making matches between two colours. It can only be used by one observer at a time, and it requires daylight, so I have not brought it with me to-night. It is nothing but the realisation of the construction of one of Newton's propositions in his "Lectiones Opticae", where he shows how to take a beam of light, to separate it into its components, to deal with these

这样，纯度和色相就可以利用杨氏三角形得到一个几何上的描述。为了理解明度的含义，我们只需增加或者减弱整个三角的照明度就可以了。因此用调整照明度的方法，杨氏三角形可以表示所有的颜色。如果我们从杨氏三角形中选取任意两种颜色，然后把二者以任意比例混合在一起，混合后的颜色对应于这两种颜色连线上重心的位置。

我没有做任何有关三种原色感觉本质的说明，也没有说它们与哪些颜色更接近。要在本文中解释清楚实际颜色之间的关系，不需要知道三原色到底是什么。我们可以选取任意的三种颜色，暂且把它们放在杨氏三角形的三个顶点上，然后就可以确定其他可见颜色与它们的相对位置，这样就得到了一种色卡。

我们见到的所有被分光光谱中的不同光线所激发的颜色在科学上都具有极高的重要性。所有的光都是其中的一种光线，或者是其中几种光线的组合。自然界中实物的颜色都是由光谱中的颜色构成的。因此，如果我们能构造一个光谱的色卡，用颜色的不同位置来表示它们之间的关系，那么，自然界中所有物体的颜色都可以在用光谱中颜色的位置界定的色卡上找到它们的位置。

色卡还有助于我们了解三原色的本质。由于每一种感觉都是一种实实在在的东西，每一种复合的色觉都必然包含在以三原色为顶角的三角形中。特别是光谱的色卡一定完全包含在杨氏三角形内部，这样，如果光谱中任何一种颜色和某种色觉相一致，那么光谱在杨氏三角形中的形式一定是一条和这种颜色所在的点成很小角度的直线。

我已经告诉大家怎样将光谱中的任意三种颜色混合，并且用改变颜色三分量中任意一个的强度来改变这种混合的颜色。如果我们把一种混合颜色和另外一种颜色并列在一起，我们可以调整这种混合颜色，直到它和另外一种颜色完全相同为止。当最终要得到的颜色接近白色时，这个过程可以最为精确地完成。于是我构造了一种我称作色箱的装置，用来匹配两种颜色。这个装置每次实验只能允许一个人进行观察，而且需要在日光下进行，所以今晚我没有把它带来。这个装置没什么大不了，只不过实现了牛顿在《光学讲义》中谈到的一个构想而已，牛顿告诉我们如何获得一束光，并将其分离成不同组分，以及如何用狭缝来获取这些组分，然后再把它们

components as we please by means of slits, and afterwards to unite them into a beam again. The observer looks into the box through a small slit. He sees a round field of light, consisting of two semicircles divided by a vertical diameter. The semicircle on the left consists of light which has been enfeebled by two reflexions at the surface of glass. That on the right is a mixture of colours of the spectrum, the positions and intensities of which are regulated by a system of slits.

The observer forms a judgment respecting the colours of the two semicircles. Suppose he finds the one on the right hand redder than the other, he says so, and the operator, by means of screws outside the box, alters the breadth of one of the slits, so as to make the mixture less red; and so on, till the right semicircle is made exactly of the same appearance as the left, and the line of separation becomes almost invisible.

When the operator and the observer have worked together for some time they get to understand each other, and the colours are adjusted much more rapidly than at first.

When the match is pronounced perfect, the positions of the slits, as indicated by a scale, are registered, and the breadth of each slit is carefully measured by means of a gauge. The registered result of an observation is called a "colour equation". It asserts that a mixture of three colours is, in the opinion of the observer (whose name is given), identical with a neutral tint, which we shall call Standard White. Each colour is specified by the position of the slit on the scale, which indicates its position in the spectrum, and by the breadth of the slit, which is a measure of its intensity.

In order to make a survey of the spectrum we select three points for purposes of comparison, and we call these the three Standard Colours. The standard colours are selected on the same principles as those which guide the engineer in selecting stations for a survey. They must be conspicuous and invariable, and not in the same straight line.

In the chart of the spectrum you may see the relations of the various colours of the spectrum to the three standard colours, and to each other. It is manifest that the standard green which I have chosen cannot be one of the true primary colours, for the other colours do not all lie within the triangle formed by joining them. But the chart of the spectrum may be described as consisting of two straight lines meeting in a point. This point corresponds to a green about a fifth of the distance from *b* towards F. This green has a wavelength of about 510 millionths of a millimetre by Ditscheiner's measure. This green is either the true primary green, or at least it is the nearest approach to it which we can ever see. Proceeding from this green towards the red end of the spectrum, we find the different colours lying almost exactly in a straight line. This indicates that any colour is chromatically equivalent to a mixture of any two colours on opposite sides of it and in the same straight line. The extreme red is considerably beyond the standard red, but it is in the same straight line, and therefore we might, if we had no other evidence, assume the extreme red as the true primary red. We shall see, however, that the true primary red is

重新整合成一束光。观察者通过一个狭缝来观察箱子的内部。他将看到一个圆形的发光区域，一条垂直方向的直径把它分割成左右两个半圆。左边半圆是经过两次镜面反射而减弱的光；右边半圆是由光谱中的颜色混合而成的，其位置和强度都可以通过狭缝来调节。

观测者将对两边半圆的颜色进行判断。假如他认为右边的光比左边的光更红，那么他就可以让色箱的操作者通过拧紧箱外的螺丝来调节某个狭缝的宽度，使得混合光线的红色变浅，如此这般，直到左右两个半圆看起来完全相同，中间的分界线几乎看不出来为止。

操作者和观察者在一起工作过一段时间以后，他们的合作会更加默契，调整颜色的速度也会比初次合作时更快。

当颜色匹配完成之后，每一个狭缝位置的刻度都被记录下来，狭缝的宽度用刻度尺仔细测量。一次观察的记录结果被称为一个“颜色方程”。它说明观察者（他的名字将被记录下来）认为，三种颜色混合而成的颜色是一种中性色，我们称之为标准白色。每一种颜色在光谱中的位置都由狭缝的位置确定，而狭缝的宽度则表示了它的强度。

为了考察光谱的特性，我们选择三个点用以比较，我们称它们为三个标准色。标准色的选择原则与工程师选取观测点的原则相同，这些点必须既突出又稳定，且不在同一条直线上。

在光谱的色卡上，你可以看到光谱中的不同颜色和三个标准色之间的关系，以及不同颜色之间的关系。事实表明，我选择的标准绿色不可能是三原色之一，因为其他颜色并不全在三点之间的区域中。但是光谱色卡可以描述成两条相交的直线。相应的交点对应一种绿色，它到标准绿色的距离为 *b* 到 F 距离的 1/5。根据迪特沙纳的测量，这种绿色的波长是 510 纳米。这种绿色即便不是真正的原色，至少也是我们曾经见到过的最接近原色的颜色。从这种绿色向光谱中的红色一端连线，我们发现不同的颜色几乎都落在了这条直线上。这表明，从色度上来说，任一种颜色都等价于位于该直线两端的两种颜色的混合。极端红色应该在比标准红色更远的位置，但是它与标准红色在同一条直线上，因此，如果没有相反的证据，我们可以将极端红色视为原红。然而，我们可以看到，真正的原红并没有出现在光谱的任何部分。它在比极端红色更远的位置，但仍然在同一条直线上。

not exactly represented in colour by any part of the spectrum. It lies somewhat beyond the extreme red but in the same straight line.

On the blue side of primary green the colour equations are seldom so accurate. The colours, however, lie in a line which is nearly straight. I have not been able to detect any measurable chromatic difference between the extreme indigo and the violet. The colours of this end of the spectrum are represented by a number of points very close to each other. We may suppose that the primary blue is a sensation differing little from that excited by the parts of the spectrum near G.

Now, the first thing which occurs to most people about this result is that the division of the spectrum is by no means a fair one. Between the red and the green we have a series of colours apparently very different from either, and having such marked characteristics that two of them, orange and yellow, have received separate names. The colours between the green and the blue, on the other hand, have an obvious resemblance to one or both of the extreme colours, and no distinct names for these colours have ever become popularly recognised.

I do not profess to reconcile this discrepancy between ordinary and scientific experience. It only shows that it is impossible, by a mere act of introspection, to make a true analysis of our sensations. Consciousness is our only authority; but consciousness must be methodically examined in order to obtain any trustworthy results.

I have here, through the kindness of Professor Huxley, a picture of the structure upon which the light falls at the back of the eye. There is a minute structure of bodies like rods and cones or pegs, and it is conceivable that the mode in which we become aware of the shapes of things is by a consciousness which differs according to the particular rods on the ends of which the light falls, just as the pattern on the web formed by a Jacquard loom depends on the mode in which the perforated cards act on the system of movable rods in that machine. In the eye we have on the one hand light falling on this wonderful structure, and on the other hand we have the sensation of sight. We cannot compare these two things; they belong to opposite categories. The whole of Metaphysics lies like a great gulf between them. It is possible that discoveries in physiology may be made by tracing the course of the nervous disturbance

> Up the fine fibres to the sentient brain;

but this would make us no wiser than we are about those colour-sensations which we can only know by feeling them ourselves. Still, though it is impossible to become acquainted with a sensation by the anatomical study of the organ with which it is connected, we may make use of the sensation as a means of investigating the anatomical structure.

A remarkable instance of this is the deduction of Helmholtz's theory of the structure of the retina from that of Young with respect to the sensation of colour. Yong asserts that there are three elementary sensations of colour; Helmholtz asserts that there are three systems of nerves in the retina, each of which has for its function, when acted on by light or any other disturbing agent, to excite in us one of these three sensations.

在原绿的蓝端，颜色方程就不是那么精确了。但色点近似分布于一条直线上。我现在还无法测量出极端靛青和紫色的区别。在光谱中这一端的颜色是用许多非常接近的点来表示的。我们可以假设原蓝这种色觉略微区别于由光谱中靠近 G 的部分所激发的感觉。

现在，面对这样的结果，摆在大家面前的首要问题是，光谱的划分并不公平。红绿之间的一系列颜色都有明显的区别，它们的区分非常明显，以至于黄色和橙色需要有各自不同的名字。反之，绿蓝之间的颜色却与这两种极端颜色或其中之一很相似，这些被广泛认可的颜色也没有自己的名字。

我并非是要调和这种一般经验和科学实验之间的差异和矛盾。只是事实表明，仅仅用自省的方式不可能对我们的感觉作出正确的分析。感觉是我们唯一的凭据，但是感觉必须经过系统的检验才能得到可靠的结果。

我从赫胥黎教授那里得到了一张描述光线落在眼睛后部成像的结构图。这里有很多棒状、锥状、钉状的微结构。我们很可能就是通过确定光线到底落在哪些棒状体的末端而感觉到物体的形状的，就像提花织布机织出什么样的花纹取决于打孔卡作用于机器中可移动棒的方式。在眼睛里，一方面光线照射在这种精密的结构之上，另一方面我们有视觉感受。我们无法比较这两个方面，因为它们属于不同的范畴。形而上学就是二者之间的鸿沟。跟踪
从神经纤维到大脑之间
的神经扰动可能会在生理学上得到一些发现；但是，这些对我们关于色觉的认识并没什么帮助，因为我们只能靠自己去感受颜色。虽然我们不可能通过解剖相关的器官来增加对色觉的了解，但是我们可以利用我们的感觉，把它作为研究组织结构的一种手段。

这里有一个著名的例子，就是从杨的色觉理论推出亥姆霍兹关于视网膜结构的理论。杨声称有三种基本的色觉；而亥姆霍兹则声称在视网膜里有三种神经系统，每一种系统都有自己的功能，当有光照或者其他扰动作用时，每一种系统都会激发出我们这三种感觉中的一种。

No anatomist has hitherto been able to distinguish these three systems of nerves by microscopic observation. But it is admitted in physiology that the only way in which the sensation excited by a particular nerve can vary is by degrees of intensity. The intensity of the sensation may vary from the faintest impression up to an insupportable pain; but whatever be the exciting cause, the sensation will be the same when it reaches the same intensity. If this doctrine of the function of a nerve be admitted, it is legitimate to reason from the fact that colour may vary in three different ways, to the inference that these three modes of variation arise from the independent action of three different nerves or sets of nerves.

Some very remarkable observations on the sensation of colour have been made by M. Sigmund Exner in Prof. Helmholtz's physiological laboratory at Heidelberg. While looking at an intense light of a brilliant colour, he exposed his eye to rapid alternations of light and darkness by waving his fingers before his eyes. Under these circumstances a peculiar minute structure made its appearance in the field of view, which many of us may have casually observed. M. Exner states that the character of this structure is different according to the colour of the light employed. When red light is used a veined structure is seen; when the light is green, the field appears covered with minute black dots, and when the light is blue, spots are seen, of a larger size than the dots in the green, and of a lighter colour.

Whether these appearances present themselves to all eyes, and whether they have for their physical cause any difference in the arrangement of the nerves of the three systems in Helmholtz's theory I cannot say, but I am sure that if these systems of nerves have a real existence, no method is more likely to demonstrate their existence than that which M. Exner has followed.

Colour Blindness

The most valuable evidence which we possess with respect to colour vision is furnished to us by the colour-blind. A considerable number of persons in every large community are unable to distinguish between certain pairs of colours which to ordinary people appear in glaring contrast. Dr. Dalton, the founder of the atomic theory of chemistry, has given us an account of his own case.

The true nature of this peculiarity of vision was first pointed out by Sir John Herschel in a letter written to Dalton in 1832, but not known to the world till the publication of "Dalton's Life" by Dr. Henry. The defect consists in the absence of one of the three primary sensations of colour. Colour-blind vision depends on the variable intensities of two sensations instead of three. The best description of colour-blind vision is that given by Prof. Pole in his account of his own case in the "Phil. Trans.", 1859.

In all cases which have been examined with sufficient care, the absent sensation appears to resemble that which we call red. The point P on the chart of the spectrum represents the

到目前为止，还没有任何一个解剖学家能够在微观尺度的观察中分辨出这三种神经系统。但是生理学上却认为，由某种神经激发的感觉只能在强度上有所变化。感觉强度的变化可以从最微弱的触感到难以忍受的疼痛；但不论激发感觉的原因是什么，只要激发的强度相同，那么感受也相同。如果这种神经功能的学说得到认可，那么从颜色能够以三种不同的方式变化这个事实中，就可以推断出这三种不同的色觉模式起源于三种不同的神经或者神经集合。

在亥姆霍兹教授位于海德堡的生理学实验室里，西格蒙德·埃克斯纳作出了一些非常引人注目的关于色觉的观察结果。当注视一种色彩耀眼的强光时，他不断地在眼前挥动手指，使眼睛迅速地在明亮和黑暗之间切换。在这种情况下，一种奇异的微结构出现在了视野当中，可能很多人都曾偶然发现过这个现象。埃克斯纳声称，该结构的特征随着光源颜色的不同而变化。红光照射时，见到的是叶脉结构；绿光照射时，视野中好像布满了小黑点；蓝光照射时，看到的斑点比绿光中的更大，颜色也更淡。

我不知道是否每个人都能感受到这些现象，也不知道亥姆霍兹理论中所说的三种神经系统的排布是否会由于个体之间的差异而在不同人中有所不同，但是我确信，如果这样的神经系统真的存在的话，那么埃克斯纳所用的方法就是最好的证明办法。

色 盲

色盲现象为我们提供了关于色觉的最有价值的证据。在每一个大型社区中都有相当多的人无法分辨出在正常人看来区别很明显的一些颜色。化学原子理论的奠基人道尔顿博士本人就为我们提供了例子。

1832 年，约翰·赫歇尔爵士在他写给道尔顿博士的一封信中第一次指出了这种异常色觉现象的本质，但是直到亨利博士出版《道尔顿生平》一书时，这封信的内容才公之于世。这种缺陷是由于缺少三种原色感觉中的一种而造成的。色盲者的视觉只依赖于两种色觉的强度变化，而不是三种。波尔教授在 1859 年的《自然科学会报》上给出了他自己的亲身体验，这是迄今为止对色盲现象的最佳描述。

在所有经过精心检验的例子中，我们发现，那种缺失的色觉好像类似于我们所称的红色。光谱色卡上的 P 点代表了缺失的色觉和光谱中颜色的关系，这是根据波

relation of the absent sensation to the colours of the spectrum, deduced from observations with the colour box furnished by Prof. Pole.

If it were possible to exhibit the colour corresponding to this point on the chart, it would be invisible, absolutely black, to Prof. Pole. As it does not lie within the range of the colours of the spectrum we cannot exhibit it; and, in fact, colour-blind people can perceive the extreme end of the spectrum which we call red, though it appears to them much darker than to us, and does not excite in them the sensation which we call red. In the diagram of the intensities of the three sensations excited by different parts of the spectrum, the upper figure, marked P, is deduced from the observations of Prof. Pole; while the lower one, marked K, is founded on observations by a very accurate observer of the normal type.

The only difference between the two diagrams is that in the upper one the red curve is absent. The forms of the other two curves are nearly the same for both observers. We have great reason therefore to conclude that the colour sensations which Prof. Pole sees are what we call green and blue. This is the result of my calculations; but Prof. Pole agrees with every other colour-blind person whom I know in denying that green is one of his sensations. The colour-blind are always making mistakes about green things and confounding them with red. The colours they have no doubts about are certainly blue and yellow, and they persist in saying that yellow, and not green, is the colour which they are able to see.

To explain this discrepancy we must remember that colour-blind persons learn the names of colours by the same method as ourselves. They are told that the sky is blue, that grass is green, that gold is yellow, and that soldiers' coats are red. They observe difference in the colours of these objects, and they often suppose that they see the same colours as we do, only not so well. But if we look at the diagram we shall see that the brightest example of their second sensation in the spectrum is not in the green, but in the part which we call yellow, and which we teach them to call yellow. The figure of the spectrum below Prof. Pole's curves is intended to represent to ordinary eyes what a colour-blind person would see in the spectrum. I hardly dare to draw your attention to it, for if you were to think that any painted picture would enable you to see with other people's vision I should certainly have lectured in vain.

On the Yellow Spot

Experiments on colour indicate very considerable differences between the vision of different persons, all of whom are of the ordinary type. A colour, for instance, which one person on comparing it with white will pronounce pinkish, another person will pronounce greenish. This difference, however, does not arise from any diversity in the nature of the colour sensations in different persons. It is exactly of the same kind as would be observed if one of the persons wore yellow spectacles. In fact, most of us have near the middle of the retina a yellow spot through which the rays must pass before they reach the sensitive

尔教授提供的色箱进行观察而推出的结果。

如果可以将色卡上这一点所代表的颜色呈现在波尔教授眼前，那么他将什么也看不见，或者说眼前一片黑暗。由于这种颜色不在光谱的范围之内，所以我们无法将其呈现出来。事实上，色盲者能够看到光谱的红端，尽管他们看到的红色比我们看到的暗很多，因而无法激发出我们正常人所感受到的红色。在由光谱不同部分激发的三种基本色觉的强度图中，上图中标为 P 的点是从波尔教授的观察结果推导出来的，而下图中标为 K 的点是一个色觉正常的观察者经过精确实验得到的结果。

两张图之间的唯一区别是，上图中缺少红色的曲线。对这两个观察者来说，另外两条曲线的形状几乎是完全相同的。因此，我们可以非常肯定地说，波尔教授能感受到的颜色是我们所称的绿色和蓝色。这就是我的计算结果，但是波尔教授，还有我认识的其他色盲者都不承认他们能感受到绿色。色盲者经常会把绿色的东西看错或者把红色和绿色搞混。色盲者肯定能看到的颜色是蓝色和黄色，他们坚持认为，他们能看到的颜色是黄色而不是绿色。

要想解释这个矛盾，我们必须知道，色盲者了解颜色名字的方式和我们正常人相同。别人告诉他们，天空是蓝色的，草地是绿色的，金子是黄色的，军装是红色的。他们观察这些物体在颜色上的差别，他们以为他们看到的颜色与我们看到的一样，只是不那么清楚而已。但如果我们看一看这张图，就会发现，在他们的第二种色觉中，最明亮的部分并不是光谱中的绿色，而是我们称为黄色的部分，我们告诉他们这就是黄色。波尔教授所绘曲线下面的光谱图向色觉正常的人展示了一个色盲者的眼睛所看到的光谱。其实我不敢让大家注意它，因为要是你认为自己可以用别人的视觉来看一幅画的话，那我前面所说的就都白费了。

关于黄色斑点

关于颜色的种种实验表明，人与人之间在视觉上的差别很明显，虽然他们都是视觉正常的人。比如，把一种颜色和白色作比较时，有人认为它偏粉色，有人认为它偏绿色。然而，这种差别并不能说明每个人的色觉有本质上的不同。这就好像有人戴上了黄色眼镜观察事物。事实上，我们中的大多数人在接近视网膜中部的地方都有一个黄色的斑点，光线必须穿过这个黄斑才能到达感觉器官。这个斑点之所以

organ: this spot appears yellow because it absorbs the rays near the line F, which are of a greenish-blue colour. Some of us have this spot strongly developed. My own observations of the spectrum near the line F are of very little value on this account. I am indebted to Professor Stokes for the knowledge of a method by which any one may see whether he has this yellow spot. It consists in looking at a white object through a solution of chloride of chromium, or at a screen on which light which has passed through this solution is thrown. This light is a mixture of red light with the light which is so strongly absorbed by the yellow spot. When it falls on the ordinary surface of the retina it is of a neutral tint, but when it falls on the yellow spot only the red light reaches the optic nerve, and we see a red spot floating like a rosy cloud over the illuminated field.

Very few persons are unable to detect the yellow spot in this way. The observer K, whose colour equations have been used in preparing the chart of the spectrum, is one of the very few who do not see everything as it through yellow spectacles. As for myself, the position of white light in the chart of the spectrum is on the yellow side of true white even when I use the outer parts of the retina; but as soon as I look direct at it, it becomes much yellower, as is shown by the point W C. It is a curious fact that we do not see this yellow spot on every occasion, and that we do not think white objects yellow. But if we wear spectacles of any colour for some time, or if we live in a room lighted by windows all of one colour, we soon come to recognise white paper as white. This shows that it is only when some alteration takes place in our sensations that we are conscious of their quality.

There are several interesting facts about the colour sensation which I can only mention briefly. One is that the extreme parts of the retina are nearly insensible to red. If you hold a red flower and a blue flower in your hand as far back as you can see your hand, you will lose sight of the red flower, while you still see the blue one. Another is, that when the light is diminished red objects become darkened more in proportion than blue ones. The third is, that a kind of colour blindness in which blue is the absent sensation can be produced artificially by taking doses of santonine. This kind of colour blindness is described by Dr. Edmund Rose, of Berlin. It is only temporary, and does not appear to be followed by any more serious consequences than headaches. I must ask your pardon for not having undergone a course of this medicine, even for the sake of becoming able to give you information at first hand about colour-blindness.

(**4**, 13-16; 1871)

呈现黄色，是因为它吸收了F线附近蓝绿色的光线。有些人的这个斑点非常粗大。由于这个原因，我自己看到的光谱在F线附近的区域就很微弱。我要感谢斯托克斯教授教给我判断一个人是否长有这种黄斑的方法。方法如下：让观察者通过氯化铬溶液观察一个白色物体，或者将一束光透过氯化铬溶液投射到屏幕上，让观察者去看这个屏幕。这束光是由红光和会被黄斑强烈吸收的光混合而成。当这束光线投射到正常人的视网膜上时，看到的会是一种中性色；但当它投射到黄斑上时，只有红光能够到达视神经，于是我们在被照亮的区域将看到一团像红云一般浮动着的红色斑点。

用这种方法检测发现只有极少数人没有这个黄斑。观察者K，就是其颜色方程曾被用于制作光谱色卡的那位，是极少数不用透过黄色眼镜看世界的人之一。至于我，在我的色卡上，白光的位置在真正的白色偏黄的一侧，即便我用视网膜的外侧观察也是如此；不过，当我直视的时候，我看到的白色位置就更加偏黄了，正如W C点所示。奇怪的是，我们并不是在所有情况下都能看到这个黄斑，而且我们也不会认为白色物体是黄色的。但如果我们戴上任意颜色的眼镜后过一段时间，或者我们所住房间的窗户都是同一种颜色时，我们很快就会认出白纸是白色的。这表明，只有当我们的感觉发生了一些改变的时候，我们才能意识到它们的性质。

最后，我只能简单介绍几个关于色觉的有趣事例。一个是，视网膜最外面的部分几乎感受不到红色。如果你手里拿着一枝红花和一枝蓝花，然后把手放到你身后几乎看不见的位置，这时你可能就看不见红花了，但仍然能看见蓝花。另一个是，当光减弱的时候，红色物体将比蓝色物体更快地变暗。第三，服用大剂量的山道年能人为地造成一种不能识别蓝色的假性色盲状态。柏林的埃德蒙·罗泽博士谈到过这种色盲。这只是一种暂时的状态，除了头痛以外不会有其他更严重的后果。我必须请求大家的原谅，因为我没有服用过这种药物，尽管我知道这样做可以给你们提供关于色盲的第一手资料。

（王静 翻译；江丕栋 审稿）

A New View of Darwinism

Editor's Note

While bowing to Darwin's authority, correspondent Henry Howorth here advanced what he saw as a problem with Darwin's idea of evolution by natural (and sexual) selection. If survival goes to the strongest such that they reproduce, how is it that plantsmen and animal-breeders conspire to induce reproduction in their charges by pruning them and even actively reducing their condition? The next issue of *Nature* contained a brief and courteous response from Darwin referring to his refutation of the idea in his published work, followed by a longer and altogether far less courteous letter from Alfred Russel Wallace.

I have noticed that *Nature* is very catholic in its sympathies, and allows all views which are not palpably absurd to be discussed in its pages, and I therefore venture to ask for some space in which to present a few of the difficulties which have been suggested by Mr. Darwin's theory of Natural Selection, and which have not, so far as I know, been as yet discussed. I have not the taste for the language nor the arguments which were used by a *Times* reviewer, and I have much too great a reverence for one of the most fearless, original, and accurate investigators of modern times, to speak of Mr. Darwin and his theory in the terms used by that very ignorant person. Approaching the subject in this spirit, and knowing how very small a section of biologists are now opposed to Mr. Darwin, I may be very rash, but hardly impertinent, in stating my difficulties.

I cannot dispute the validity and completeness of many of Mr. Darwin's proofs to account for individual cases of variation and isolated changes of form. Within the limits of these proofs it is impossible to deny his position. But when he heaves these individual and often highly artificial cases, and deduces a general law from them, it is quite competent for me to quote examples of a much wider and more general occurrence that tell the other way. In this communication I shall confine myself to Mr. Darwin's theory, and shall not trepass upon the doctrine of evolution, with which it is not to be confounded.

The theory of Natural Selection has been expressively epitomised as "the Persistence of the Stronger", "the Survival of the Stronger". Sexual selection, which Mr. Darwin adduces in his last work as the cause of many ornamental and other appendages whose use in the struggle for existence is not very obvious, is only a by-path of the main conclusion. Unless by the theory of the struggle for existence is meant the purely identical expression that those forms of life survive which are best adapted to survive, I take it that it means in five words the Persistence of the Stronger.

对达尔文学说的新看法

编者按

通讯员亨利·霍沃思在认可达尔文的权威性的同时，也在此对达尔文关于自然选择（性选择）的进化论观点提出了一点质疑。如果物种延续都是靠强者的繁殖，那为什么栽培植物的园丁和养殖动物的饲养员要通过修剪枝干甚至主动限制被养生物的生活空间的方法来促使它们繁殖后代呢？在后一期的《自然》杂志上，达尔文简短而客气地回应了霍沃思对其著作的驳斥，紧随其后的是阿尔弗雷德·拉塞尔·华莱士写的一封毫不客气的长信。

我发现《自然》对所有不是特别荒唐的观点都有广博的宽容心，兼收并蓄，在自己的版面上允许出现对这些观点的探讨，因此我在此冒昧地借贵刊对达尔文先生的自然选择学说提出几点疑问。据我所知，到目前为止，这些疑问还未曾被讨论过。我不太喜欢《泰晤士报》的评论员在谈及达尔文先生及其学说时所使用的语言和持有的观点，觉得他所使用的那些用语很无知，而我对这位现代最无畏、最具独创性、最准确无误的研究者充满着无限敬意。我知道现在只有很少的生物学家反对达尔文先生，但我是抱着这样一种精神来讨论这一话题的，因此虽然我在陈述自己的疑问时可能有些急躁，但并非鲁莽。

我不想争论达尔文先生用来说明变异和隔离变化形式的个案时所用的许多证据的有效性和完整性。仅仅用这些证据是不足以否认达尔文学说的地位的。但是当他举出这些个别的而且往往在很大程度上是人为造成的事件，并从中推导出一条普遍规律时，我就有权引用更普遍、更广泛发生的事件来得出其他的观点。在这封信里，我会将问题局限在达尔文先生的学说范围内，不会涉及关于进化的学说，以防混淆。

自然选择学说已经被概括表述为“强者的延续”、“强者生存”。达尔文先生在其最近的一部著作中提出，性选择是许多装饰性的以及其他的附属部分的成因，这些附属部分在生存斗争中的用处并不是很明显，性选择在这里只是主要结论的一个分支而已。生存斗争理论除了意味着幸存的生命形式是最适于生存的这条完全一致的表述之外，我对它的理解就是那五个字：强者的延续。

Among the questions which stand at the very threshold of the whole inquiry, and which I have overlooked in Mr. Darwin's books if it is to be found there, is a discussion of the causes which produce sterility and those which favour fertility in races. He no doubt discusses with ingenuity the problem of the sterility of mules and of crosses between different races, but I have nowhere met with the deeper and more important discussion of the general causes that induce or check the increase of races. The facts upon which I rely are very common-place, and are furnished by the smallest plot of garden or the narrowest experience in breeding domestic animals. The gardener who wants his plants to blossom and fruit takes care that they shall avoid a vigorous growth. He knows that this will inevitably make them sterile; that either his trees will only bear distorted flowers, that they will have no seed, or bear no blossoms at all. In order to induce flowers and fruit, the gardener checks the growth and vigour of the plant by pruning its roots or its branches, depriving it of food, &c., and if he have a stubborn pear or peach tree which has long refused to bear fruit, he adopts the hazardous, but often most successful, plan of ringing its bark. The large fleshy melons or oranges have few seeds in them. The shrivelled starvelings that grow on decaying branches are full of seed. And the rule is universally recognised among gardeners as applying to all kinds of cultivated plants, that to make them fruitful it is necessary to check their growth and to weaken them. The law is no less general among plants in a state of nature, where the individuals growing in rich soil, and which are well-conditioned and growing vigorously, have no flowers, while the starved and dying on the sandy sterile soil are scattering seed everywhere.

On turning to the animal kingdom, we find the law no less true. "Fat hens won't lay", is an old fragment of philosophy. The breeder of sheep and pigs and cattle knows very well that if his ewes and sows and cows are not kept lean they will not breed; and as a startling example I am told that to induce Alderney cows, which are bad breeders, to be fertile they are actually bled, and so reduced in condition. Mr. Doubleday, who wrote an admirable work in answer to Malthus, to which I am very much indebted, has adduced overwhelming evidence to show that what is commonly known to be true of plants and animals is especially true of man. He has shown how individuals are affected by generous diet and good living, and also how classes are so affected. For the first time, so far as I know, he showed why population is thin and the increase small in countries where flesh and strong food is the ordinary diet, and large and increasing rapidly where fish or vegetable or other weak food is in use; that everywhere the rich, luxurious, and well-fed classes are rather diminishing in numbers or stationary; while the poor, under-fed, and hard-worked are very fertile. The facts are excceedingly numerous in support of this view, and shall be quoted in your pages if the result is disputed. This was the cause of the decay of the luxurious power of Rome, and of the cities of Mesopotamia. These powers succumbed not to the exceptional vigour of the barbarians, but to the fact that their populations had diminished, and were rapidly being extinguished from internal causes, of which the chief was the growing sterility of their inhabitants.

在提出所有这些正处于全部调查出发点的问题，以及我所忽略但可以从达尔文先生的书中发现的问题之前，我们要先讨论是什么原因导致了不育以及有哪些因素可以促进物种的繁殖。毫无疑问，达尔文先生对骡子的不育和不同物种间杂交问题的探讨很有独创性，但是我没有在该书其他部分读到更深刻、更重要的有关导致或阻碍种群数目增加的普遍原因的讨论。我所信赖的事实其实是老生常谈的问题，这些事实可以通过花园里最细微的情节或者仅通过饲养家畜的极为有限的经历来提供。一名园丁如果想要让自己种植的植物开花结果，就会注意避免他的植物旺盛生长。他知道旺盛生长肯定会导致植物不育；一旦生长过于旺盛，那么他的树要么只开出畸形的不会结种子的花朵，要么根本就不开花。为了诱导出花和果实，园丁会通过修剪树木的根部和枝干、减少施肥等措施来阻止植物的生长和活力。假如他有一棵很久都没有结过果实的顽固的梨树或桃树，他会采用冒险但通常会很成功的环状剥皮措施。园丁界流传着一条公认的规律，即认为大个的甜瓜或桔子几乎没有种子，而缺少养分只能依靠腐坏枝叶生长的干枯植株上面的果实通常结满了种子，园丁们认为这一法则适用于所有的栽培植物，并且认为要使各种栽培植物产生多汁的果实，就需要阻碍或者削弱他们的生长。这一法则对处于自然状态下的植物也一样适用，那些生长在肥沃土壤里、状态良好、生长旺盛的植物都不开花，而那些生长在贫瘠的沙地中营养不良的垂死植物却会将它们的种子散播到各处。

再来看看动物界，我们发现这一规律同样存在。“肥鸡不下蛋”，是一句古老的哲语。羊、猪、牛的饲养者都清楚地知道，如果不让母羊、母猪和母牛保持瘦弱的话，它们就不会繁殖；别人告诉过我一个惊人的例子，即为了使难以繁殖的奥尼德尼母牛能够多产，它们竟然被放血来使其身体状况变差。作为对马尔萨斯的回答，道布尔迪先生写了一部令人敬佩的著作，这部著作也使我受益匪浅，其中援引了大量证据来表明对动植物普遍适用的法则对人类也同样适用。他向我们说明了，个体是如何受丰富的饮食和舒适的生活条件影响的，以及社会阶层是如何受此影响的。据我所知，他首次解释了为什么在以肉和高能量食物为日常饮食的国家里人口数量少并且增长幅度小，而在以鱼或蔬菜或其他低能量食物为日常饮食的国家里，人口数量却很大并且增长迅速；无论在何处，富有的、奢侈的、营养好的种群的数量都是处于减少的状态或保持稳定不变；而贫穷的、营养不良的和辛苦工作的种群则往往大量繁殖。支持这种观点的事实不计其数，如果有人对这一观点还存在异议的话，我可以在贵刊上引述这些事实。这就是使罗马的强大力量衰退的原因，也是令美索不达米亚城没落的原因。这些力量不是被野蛮人的异常体力打垮的，而是被自己群体不断缩小的现实击败的，他们是由于内因而迅速灭亡，而其中主要的内因就是其居民的不育情况在不断增加。

The same cause operated to extinguish the Tasmanians and other savage tribes which have decayed and died out, when brought into contact with the luxuries of civilisation, notwithstanding every effort having been made to preserve them. In a few cases only have the weak tribes been supplanted by the strong, or weaker individuals by stronger; the decay has been internal, and of remoter origin. It has been luxury and not want; too much vigour and not too little, that has eviscerated and destroyed the race. If this law then be universal both in the vegetable and animal kingdoms, a law too, which does not operate on individuals and in isolated cases only, but universally, it is surely incumbent upon the supporters of the doctrine of Natural Selection, as propounded by Mr. Darwin, to meet and to explain it, for it seems to me to cut very deeply into the foundations of their system. If it be true that, far from the strong surviving the weak, the tendency among the strong, the well fed, and highly favoured, is to decay, become sterile, and die out, while the weak, the under-fed, and the sickly are increasing at a proportionate rate, and that the fight is going on everywhere among the individuals of every race, it seems to me that the theory of Natural Selection, that is, of the persistence of the stronger, is false, as a general law, and true only of very limited and exceptional cases. This paper deals with one difficulty only, others may follow if this is acceptable.

Henry H. Howorth

(**4**, 161-162; 1871)

* * *

I am much obliged to Mr. Howorth for his courteous expressions towards me in the letter in your last number. If he will be so good as to look at p. 111 and p. 148, vol. II of my "Variation of Animals and Plants under Domestication", he will find a good many facts and a discussion on the fertility and sterility of organisms from increased food and other causes. He will see my reasons for disagreeing with Mr. Doubleday, whose work I carefully read many years ago.

Charles Darwin

(**4**, 180-181; 1871)

* * *

The very ingenious manner in which Mr. Howorth first misrepresents Darwinism, and then uses an argument which is not even founded on his own misrepresentation, but on a quite distinct fallacy, may puzzle some of your readers. I therefore ask space for a few lines of criticism.

Mr. Howorth first "takes it" that the struggle for existence "means, in five words, the persistence of the stronger". This is a pure misrepresentation. Darwin says nothing of the kind. "Strength" is only one out of the many and varied powers and faculties that lead

同样的原因使塔斯马尼亚和其他的野蛮部落灭绝了，进入奢侈的文化状态时，尽管他们做了一切努力来保护自己，但还是衰退并最终消亡了。只有在少数的几个例子中，弱小的部落才被强壮的部落取代，或者较弱的个体被较强的取代；衰退起源于远古时代传承下来的内因。种族元气大伤或毁灭是由于奢侈而非欲望，是由于他们的精力太充沛而非太少。如果这一规律在植物界和动物界都是普遍的，它们不是只对个体起作用或者只存在于极个别的案例中，而是普遍存在的，那么这就是支持达尔文先生提出来的自然选择学说的人应该解决的问题了，因为这似乎已经触及到了自然选择学说体系的基础。如果这是真的，那么远非强者生存弱者淘汰，而是强壮的、营养充足的、非常有利的群体趋向于衰退、不育，直至灭绝，而瘦弱的、营养不良的、多病的群体却以适当的比率在逐渐增加，并且各个地方的每个种族内个体间的战斗一直在进行着，我觉得作为一般规律而言，自然选择理论所认为的强者的延续这一观点是错误的，这种现象只在几个有限的、个别的例子中才存在。这篇文章旨在讨论这一疑问，如果您接受该观点，可以继续讨论下去。

亨利 · 霍沃思

* * *

对于霍沃思先生在贵刊上一期刊登的信中对我的恭维，我很感激。我想恳请他看一下我的《动物和植物在家养下的变异》第 2 卷，第 111、148 页，他就会发现许多事实，以及对食物增加和其他原因导致的生物繁殖和不育性进行的讨论。关于道布尔迪先生，我在许多年前就仔细读过了他的著作，在上述两页中，霍沃思先生也会看到我不同意道布尔迪先生的观点的原因所在。

查尔斯 · 达尔文

* * *

霍沃思先生首次以非常独创的方式对达尔文学说进行了歪曲，他使用了一个论据，该论据甚至不是建立在他自己曲解的基础上，而是建立在一个非常明显的谬论的基础上，这些有可能会误导贵刊的读者。因此我请求贵刊允许我在此占用一点空间作几句评论。

霍沃思先生首先将生存斗争“理解”为“五个字：强者的延续”。这是一个纯粹的曲解。达尔文没有说过任何这种意思的话。“强”只是许多不同的有助于在生存斗

to success in the battle for life. Minute size, obscure colours, swiftness, armour, cunning, prolificness, nauseousness, or bad odour, have any one of them as much right to be put forward as the cause of "persistence". The error is so gross that it seems wonderful that any reader of Darwin could have made it, or, having made it, could put it forward deliberately as a fair foundation for a criticism. He says, moreover, that the theory of Natural Selection "has been expressively epitomised" as "the persistence of the stronger", "the survival of the stronger". By whom? I should like to know. I never saw the terms so applied in print by any Darwinian. The most curious and even ludicrous thing, however, is that, having thus laid down his premisses. Mr. Howorth makes no more use of them, but runs off to something quite different, namely, that *fatness* is prejudicial to fertility. "Fat hens won't lay", "overgrown melons have few seeds", "overfed men have small families",—these are the *facts* by which he seeks to prove that the *strongest* will not survive and leave offspring! But what does nature tell us? That the strongest and most vigorous plants *do* produce the most flowers and seed, not the weak and sickly. That the strongest and most healthy and best fed wild animals *do* propagate more rapidly than the starved and sickly. That the strong and thoroughly well-fed backwoodsmen of America increase more rapidly than any half-starved race of Indians upon earth. No *fact*, therefore, has been adduced to show that even "the persistence of the stronger" is not true; although, if this had been done, it would not touch Natural Selection, which is the "survival of the fittest".

Alfred R. Wallace

(**4**, 181; 1871)

Henry H. Howorth: Derby House, Eccles.
Charles Darwin: Down, Beckenham, Kent, July 1.

争中取得成功的力量和能力中的一种。微小的体型、晦暗的颜色、敏捷、防护具、狡猾、大量繁殖、腐臭或者不好的气味，拥有这些特点中的任何一点都可以被当作是“延续”下来的原因。达尔文学说的读者中竟然有人会犯这种显而易见的错误，或者已经犯了，并且还故意地将这种错误作为一种公正的批评根据提出来，这实在太让人吃惊了。此外，他还说自然选择理论“已经被概括表述”为“强者的延续”和“强者生存”。是谁这样概括的？我倒是很想知道。我从没有看到任何达尔文主义者在任何出版物中使用过这样的词语。然而，最令人好奇和最为荒谬的事情是，霍沃思先生在已经作出这样的谬论铺垫之后却再没有使用这些前提，而是转到了别的毫不相关的问题上，也就是，**肥胖**有损于生育能力。“肥鸡不下蛋”，“生长过度的甜瓜几乎没有种子”，“吃得太多的人往往只有很小的家庭”——这些就是他想用来证明**最强者**将不能幸存并且不能留下后代的**事实**！但是自然界告诉我们的又是什么？产出了最多的花朵和种子的**正是**那些最强壮、最有活力的植物，而不是那些虚弱多病的植物。同样最强壮、最健康、吃得最好的野生动物**确实**比那些挨饿、多病的动物繁殖得更快。那些住在美洲大陆森林地区中的强壮且营养非常充足的人也比那片土地上总是处于半饥饿状态的印第安人增加得更快。因此没有任何**事实**可以用来说明“强者的延续”是不真实的，退一步说，即使存在这样的例子，那也撼动不了自然选择学说所认为的“适者生存”的法则。

阿尔弗雷德·华莱士

（刘皓芳 翻译；刘力 审稿）

The Copley Medalist of 1870

J. Tyndall

Editor's Note

James Prescott Joule was born in the city of Salford in Lancashire and worked either there or in the adjacent city of Manchester. His scientific work centred on electricity and the rules by which energy of one kind—a current of electricity, for example—may be converted into energy of some other kind (such as mechanical work). Joule is now commemorated by the use of his name as the unit of energy in the MKS system. The Copley medal, which was awarded to Joule in 1870, is the Royal Society's most venerable award. The author of this article, John Tyndall, was a professor at the institution in Manchester that eventually became its first university, but had by 1871 become a professor at the Royal Institution in London.

THIRTY years ago Electro-magnetism was looked to as a motive power which might possibly compete with steam. In centres of industry, such as Manchester, attempts to investigate and apply this power were numerous, as shown by the scientific literature of the time. Among others Mr. James prescott Joule, a resident of Manchester, took up the subject, and in a series of papers published in Sturgeon's "Annals of Electricity" between 1839 and 1841, described various attempts at the construction and perfection of electro-magnetic engines. The spirit in which Mr. Joule pursued these inquiries is revealed in the following extract: "I am particularly anxious," he says, "to communicate any new arrangement in order, if possible, to forestall the monopolising designs of those who seem to regard this most interesting subject merely in the light of pecuniary speculation." He was naturally led to investigate the laws of electro-magnetic attractions, and in 1840 he announced the important principle that the attractive force exerted by two electro-magnets, or by an electro-magnet and a mass of annealed iron, is directly proportional to the square of the strength of the magnetising current; while the attraction exerted between an electro-magnet and the pole of a permanent steel magnet varies simply as the strength of the current. These investigations were conducted independently of, though a little subsequently to, the celebrated inquiries of Henry, Jacobi, and Lenz and Jacobi on the same subject.

On the 17th of December, 1840, Mr. Joule communicated to the Royal Society a paper on the production of heat by Voltaic electricity; in which he announced the law that the calorific effects of equal quantities of transmitted electricity are proportional to the resistance overcome by the current, whatever may be the length, thickness, shape, or character of the metal which closes the circuit; and also proportional to the square of the quantity of transmitted electricity. This is a law of primary importance. In another paper, presented to but declined by the Royal Society, he confirmed this law by new experiments, and materially extended it. He also executed experiments on the heat consequent on the passage of Voltaic electricity through electrolytes, and found in all cases that the heat

1870 年的科普利奖章获得者

廷德尔

编者按

詹姆斯·普雷斯科特·焦耳出生在兰开夏郡的索尔福德，他在他的家乡（或者说曼彻斯特临近的城市）工作。他的研究工作集中在电学和一种能量形式（比如电流）转化成另外的能量形式（比如动能）的规律上。为了纪念焦耳，米–千克–秒单位制系统采用他的名字作为能量单位。1870 年，焦耳被授予科普利奖章，这是皇家学会的最高荣誉。这篇文章的作者约翰·廷德尔是曼彻斯特研究所的教授，该研究所最终成为了当地的第一所大学。但在 1871 年之前作者就已经成为伦敦皇家研究院的一名教授。

30 年前，人们希望电磁能能够成为一种可以和蒸汽相媲美的动力形式。在像曼彻斯特那样的工业中心，人们对这种动力的研究和应用十分广泛，这可以从当时的科学文献中看出来。和其他人一样，曼彻斯特人詹姆斯·普雷斯科特·焦耳也研究了这个课题，并于 1839～1841 年间在斯特金的《电学年鉴》上发表了一系列文章，描述了设计和完善电磁发动机的各种尝试。焦耳先生的探索精神可以从下面的话中体现出来，他说，“我非常急切地要把所有适宜的新成果公之于众，为的是尽量抢在那些仅仅为了金钱而想把这个极为有趣的课题垄断掉的人前面。”他很自然地把研究转向了电磁吸引作用的规律，并在 1840 年宣布了一个重要的定律：两个电磁体或者电磁体和退火铁之间的吸引力正比于磁化电流强度的平方；电磁体和永磁体磁极之间的相互作用只与电流强度成正比。这些研究是焦耳独立进行的，尽管比亨利、雅各比以及楞次和雅各比通过著名实验得出的相同结果略晚一些。

1840 年 12 月 17 日，焦耳先生在皇家学会宣读了他关于伏打电流发热的论文；在论文中，他宣称，等量的传输电流产生的热效应跟电流所克服的电阻成正比，与闭合电路中金属的长度、厚度、形状和特性无关，并且与电流的平方成正比。这是一个非常重要的定律。在另一篇提交给皇家学会但遭退稿的文章中，焦耳用新的实验证实了这个定律，并将其大大扩展。他还针对伏打电流通过电解液时的热效应进行了实验研究，最终发现，在所有情况下，伏打电流产生的热量都正比于电流强度的平方与电阻的乘积。通过这个定律，他推出了一些在电化学领域极为重要的结论。

evolved by the proper action of any Voltaic current is proportional to the square of the intensity of that current multiplied by the resistance to conduction which it experiences. From this law he deduced a number of conclusions of the highest importance to electro-chemistry.

It was during these inquiries, which are marked throughout by rare sagacity and originality, that the great idea of establishing quantitative relations between Mechanical Energy and Heat arose and assumed definite form in his mind. In 1843 Mr. Joule read before the meeting of the British Association at Cork a paper "On the Calorific Effects of Magneto-Electricity and on the Mechanical Value of Heat". Even at the present day this memoir is tough reading, and at the time it was written it must have appeared hopelessly entangled. This I should think was the reason why Prof. Faraday advised Mr. Joule not to submit the paper to the Royal Society. But its drift and results are summed up in these memorable words by its author, written some time subsequently: "In that paper it was demonstrated experimentally that the mechanical power exerted in turning a magneto-electric machine is converted into the heat evolved by the passage of the currents of induction through its coils, and on the other hand, that the motive power of the electro-magnetic engine is obtained at the expense of the heat due to the chemical reaction of the battery by which it is worked."* It is needless to dwell upon the weight and importance of this statement.

Considering the imperfections incidental to a first determination, it is not surprising that the "mechanical values of heat," deduced from the different series of experiments published in 1843, varied somewhat widely from each other. The lowest limit was 587, and the highest 1,026 foot-pounds for 1° F of temperature.

One noteworthy result of his inquiries, which was pointed out at the time by Mr. Joule, had reference to the exceedingly small fraction of the heat which is actually converted into useful effect in the steam-engine. The thoughts of the celebrated Julius Robert Mayer, who was then engaged in Germany upon the same question, had moved independently in the same groove; but to his labours due reference will doubtless be made on a future occasion. In the memoir now referred to Mr. Joule also announced that he had proved heat to be evolved during the passage of water through narrow tubes; and he deduced from these experiments an equivalent of 770 foot-pounds, a figure remarkably near to the one now accepted. A detached statement regarding the origin and convertibility of animal heat strikingly illustrates the penetration of Mr. Joule and his mastery of principles at the period now referred to. A friend had mentioned to him Haller's hypothesis, that animal heat might arise from the friction of the blood in the veins and arteries. "It is unquestionable," writes Mr. Joule, "that heat is produced by such friction, but it must be understood that the mechanical force expended in the friction is a part of the force of affinity which causes the venous blood to unite with oxygen, so that the whole heat of the system must still be referred to the chemical changes. But if the animal were engaged in turning a piece of machinery, or in ascending a mountain, I apprehend that in proportion

---

* *Phil. Mag.* May 1845.

在从事这些充满超人智慧和独创性的研究过程中，焦耳萌生并逐渐形成了在机械能和热量之间建立定量关系的想法。1843 年焦耳先生在英国科学促进会于科克召开的会议上宣读了一篇《论磁电的热效应和热的机械值》的论文。即便在今天，这个研究报告都是很难读懂的，在当时要读懂它简直是毫无希望的。我想这就是法拉第教授建议焦耳先生不要将它提交给皇家学会的原因。但隔了一段时间以后，焦耳对文章的主旨和结论作出了如下的概括，给人留下极深的印象："那篇文章用实验证明了施加于磁电式电机上的机械能被转化成线圈中感应电流所释放的热量，另一方面，电磁发动机的动力则来源于电池化学反应所提供的热量。"* 这个结论的重要性和价值也就不必在此赘述了。

在 1843 年发表的论文中，一系列实验所得到的"热的机械值"差别较大，考虑到首次得出结论的实验所要面对的各种偶然因素，这也就不足为怪了。对应于温度上升华氏 1° 所做的功，测量得到的最低值是 587 英尺磅，最高值是 1,026 英尺磅。

当时焦耳先生指出，在他的研究中有一个著名的结论，即蒸汽机的热能转化为有效能量的效率非常低。著名的尤利乌斯·罗伯特·迈尔后来在德国也研究了这个问题，并独立地得到了同样的结论，但是他的观点还需要在未来的实验中去确证。焦耳先生在他的研究报告中还指出，他发现水流过细管子时会放出热量，通过实验他得到的数值是 770 英尺磅，这个值非常接近于今天我们承认的数值。另一份关于动物热量起源和转化的报告突出地表明了焦耳先生敏锐的洞察力和他在那个年代对规律的掌控能力。一位朋友曾经向他提到过哈勒尔的假说，该假说认为动物的热量可能起源于血液在静脉和动脉中流动时的摩擦。"这是毫无疑问的，"焦耳在回信中写道，"热量就是由这样的摩擦产生的，但是要知道，在摩擦中消耗的机械力也是使静脉血与氧结合的亲和力引发的，所以动物系统的全部热量必然来自于化学反应。但动物在搬运东西或者上山的时候，我认为，动物将消耗其通过化学反应释放出的热量，而系统热量的**减少**与动物肌肉为完成这项工作所做的功成正比。"句子中突出强调的部分是焦耳在 1843 年的原稿中就标注了的。

*《哲学杂志》，1845 年 5 月。

to the muscular effort put forth for the purpose, a *diminution* of the heat evolved in the system by a given chemical action would be experienced." The italics in this memorable passage, written it is to be remembered in 1843, are Mr. Joule's own.

The concluding paragraph of this British Association paper equally illustrates his insight and precision regarding the nature of chemical and latent heat. "I had," he writes, "endeavoured to prove that when two atoms combine together, the heat evolved is exactly that which would have been evolved by the electrical current due to the chemical action taking place, and is therefore proportional to the intensity of the chemical force causing the atoms to combine. I now venture to state more explicitly, that it is not precisely the attraction of affinity, but rather the mechanical force expended by the atoms in falling towards one another, which determines the intensity of the current, and, consequently, the quantity of heat evolved; so that we have a simple hypothesis by which we may explain why heat is evolved so freely in the combination of gases, and by which indeed we may account 'latent heat' as a mechanical power prepared for action as a watch-spring is when wound up. Suppose, for the sake of illustration, that 8 lbs. of oxygen and 1 lb. of hydrogen were presented to one another in the gaseous state, and then exploded; the heat evolved would be about 1° F in 60,000 lbs. of water, indicating a mechanical force expended in the combination equal to a weight of about 50,000,000 lbs. raised to the height of one foot. Now if the oxygen and hydrogen could be presented to each other in a liquid state, the heat of combination would be less than before, because the atoms in combining would fall through less space." No words of mine are needed to point out the commanding grasp of molecular physics, in their relation to the mechanical theory of heat, implied by this statement.

Perfectly assured of the importance of the principle which his experiments aimed at establishing, Mr. Joule did not rest content with results presenting such discrepancies as those above referred to. He resorted in 1844 to entirely new methods, and made elaborate experiments on the thermal changes produced in air during its expansion: firstly, against a pressure, and therefore performing work; secondly, against no pressure, and therefore performing no work. He thus established anew the relation between the heat consumed and the work done. From five different series of experiments he deduced five different mechanical equivalents; the agreement between them being far greater than that attained in his first experiments. The mean of them was 802 foot-pounds. From experiments with water agitated by a paddle-wheel, he deduced, in 1845, an equivalent of 890 foot-pounds. In 1847 he again operated upon water and sperm-oil, agitated them by a paddle-wheel, determined their elevation of temperature, and the mechanical power which produced it. From the one he derived an equivalent of 781.5 foot-pounds; from the other an equivalent of 782.1 foot-pounds. The mean of these two very close determinations is 781.8 foot-pounds.

At this time the labours of the previous ten years had made Mr. Joule completely master of the conditions essential to accuracy and success. Bringing his ripened experience to bear upon the subject, he executed in 1849 a series of 40 experiments on the friction of

焦耳对于化学能与潜热的精确洞察力在这篇英国科学促进会论文的结束语中也有所体现。他写道，“我曾经竭力去证明两个原子结合时所释放的热量与化学反应产生的电流所释放的热量相等，也就是与使原子结合的化学力的强度成正比。现在，我要大胆地说得更明确一点，吸引力这个说法并不确切，应该说是两个原子在相互靠近时所克服的机械力，这个力决定了电流的强度，也决定了释放的热量。这样，我们就得到了一个简单的假说，可以用它来解释为什么在气体发生化合反应的过程中热量可以自由地释放，根据这个假说我们真的可以把‘潜热’看作是一种储备的机械力，就像上好的手表发条一样。例如，假设 8 磅氧和 1 磅氢在气态下混合在一起，然后发生了爆炸。爆炸所释放的热量差不多可以使 60,000 磅水的温度上升华氏 1°，说明化合反应释放的机械能相当于把 50,000,000 磅的重物从地面提升 1 英尺所做的功。如果氧和氢可以在液态下发生反应，那么化合反应放出的热将小于气态的情况，因为原子之间的结合只需克服更小的距离。”这些陈述高屋建瓴地把握了分子物理学和热力学之间的关系，对此我无需再赘述了。

焦耳先生极好地证实了他的实验所要建立的原理的重要性，但是他并没有满足于此，因为正如上面所提到的那样，实验结果还存在较大的差异。他于 1844 年采用全新的方法，精确地测量了空气在膨胀过程中的热量变化：首先，在抵抗外部压力的膨胀中，空气对外做功；其次，空气自由膨胀时，不对外做功。他用这种方式重新建立了热量消耗和做功之间的关系。他从 5 组不同的实验中推出了 5 个不同的热功当量；这些数据的吻合程度远远好于他以前的实验结果。测量得到的平均值是 802 英尺磅。1845 年，焦耳用叶轮搅动水的实验测得的热功当量值是 890 英尺磅。1847 年，他又用叶轮搅动水和鲸油，重复了上面的实验，测定了温度升高的度数，也就得到了致使温度升高的机械能的大小。一次实验得到的结果是 781.5 英尺磅，另一次实验的结果是 782.1 英尺磅。两者非常接近，它们的平均值是 781.8 英尺磅。

到目前为止，焦耳前 10 年的工作经验已经使他能够相当熟练地掌控与实验精度和成败密切相关的各种条件。1849 年焦耳将他的成熟经验应用于热功当量的测定之中，做了 40 组水的摩擦实验，50 组水银的摩擦实验和 20 组铸铁盘的摩擦实验。他

water, 50 experiments on the friction of mercury, and 20 experiments on the friction of plates of cast-iron. He deduced from these experiments our present mechanical equivalent of heat, justly recognised all over the world as "Joule's equivalent".

There are labours so great and so pregnant in consequences, that they are most highly praised when they are most simply stated. Such are the labours of Mr. Joule. They constitute the experimental foundation of a principle of incalculable moment, not only to the practice, but still more to the philosophy of Science. Since the days of Newton, nothing more important than the theory of which Mr. Joule is the experimental demonstrator has been enunciated.

I have omitted all reference to the numerous minor papers with which Mr. Joule has enriched scientific literature. Nor have I alluded to the important investigations which he has conducted jointly with Sir William Thomson. But sufficient, I think, has been here said to show that, in conferring upon Mr. Joule the highest honour of the Royal Society, the Council paid to genius not only a well-won tribute, but one which had been fairly earned twenty years previously.*

Comparing this brief history with that of the Copley Medalist of 1871, the differentiating influence of "environment" on two minds of similar natural cast and endowment comes up in an instructive manner. Withdrawn from mechanical appliances, Mayer fell back upon reflection, selecting with marvellous sagacity from existing physical data the single result on which could be founded a calculation of the mechanical equivalent of heat. In the midst of mechanical appliances, Joule resorted to experiment, and laid the broad and firm foundation which has secured for the mechanical theory the acceptance it now enjoys. A great portion of Joule's time was occupied in actual manipulation; freed from this, Mayer had time to follow the theory into its most abstruse and impressive applications. With their places reversed, however, Joule might have become Mayer, and Mayer might have become Joule.

(**5**, 137-138; 1871)

* Had I found it in time, this notice should have preceded that of the Copley Medalist of 1871.

从这些实验中推出了我们现在所使用的热功当量，即全世界都认可的“焦耳当量”。

这些工作取得的成果非常重要也非常出名，每逢被人提及时总能得到高度评价。这就是焦耳先生的工作，它们构成了一个意义重大的定理的实验基础，不仅仅在实践层面上意义重大，对于科学哲学更是如此。自牛顿时代以来，还没有什么科学成果能比焦耳先生用实验所阐明的理论更重大。

我没有提及焦耳先生科学著作中很多的小文章，也没有介绍他和威廉·汤姆孙爵士一同完成的重要研究。但是，我觉得我在这里介绍的内容已经足以说明：授予焦耳先生皇家学会的最高奖项，不仅仅是委员会给天才颁发的恰如其分的奖励，也是他在 20 年以前就应当得到的嘉奖。*

将焦耳工作的简短历史和 1871 年科普利奖章获得者的相比，就可以从“环境”对两个具有相同资质的人的不同影响中得到启发。迈尔在放弃了机械设备的研究之后，经过深思熟虑颇有远见地选择了利用现有数据探索出一个计算热功当量的简便算法。而焦耳身处机械设备当中，致力于实验工作，为我们今天所接受的力学理论打下了广泛而坚实的基础。焦耳的大部分时间都花在了实验操作上；而迈尔则从繁琐的工作中解放出来，他有充足的时间将理论带入最深奥和最有价值的应用之中。如果将他们的位置交换一下，也许焦耳会成为迈尔，而迈尔则会成为焦耳。

（王静 翻译；李军刚 审稿）

* 如果我及时地发现这一点，那么这段说明应该早于 1871 年科普利奖。

Periodicity of Sun-spots*

Editor's Note

Rudolph Wolf's observations on the variations of sunspot activity had both estimated their average period—roughly 11.1 years—and shown an interesting contrast between the increasing and decreasing parts of the cycle. Wolf had found that, on average, activity increased over 3.7 years but then decreased more slowly over 7.4 years. Here other scientists present new observations that largely agree with Wolf's, but which also add important details, such as a fixed ratio of about 1 : 2.1 between the duration of any increasing period and the period of the immediately preceding decrease. Modern mathematical methods show that sunspots in fact cycle irregularly, with an average periodicity of about 10.5 years.

IN the short account of some recent investigations by Prof. Wolf and M. Fritz on sun-spot phenomena, which has been published lately in the "Proceedings of the Royal Society" (No. 127, 1871), it was pointed out that some of Wolf's conclusions were not quite borne out by the results which we have given in our last paper on Solar Physics in the *Philosophical Transactions* for 1870, pp. 389–496. A closer inquiry into the cause of this discrepancy has led us to what appears a definite law, connecting numerically the two branches of the periodic sun-spot curve, viz., the time during which there is a regular diminution of spot-production, and the time during which there is a constant increase.

It will be well, for the sake of clearness, to allude here again, as briefly as possible, to Prof. Wolf's results before stating those at which we have arrived.

Prof. Wolf has previously devoted the greater part of his laborious researches to a precise determination of the mean *length* of the whole sun-spot period, but latterly he has justly recognised the importance of obtaining some knowledge of the average character of the periodic increase and decrease. Hence he has, as far as he has been able to do so by existing series of observations, and his peculiar and ingenious method of rendering observations made at different times and by different observers comparable with each other, endeavoured to investigate more closely the nature of the periodic sun-spot curve, by tabulating and graphically representing the monthly means taken during two and a half years before and after the minimum. and applying this method to five distinct minimum epochs, which he has fixed by the following years:—

---

* Abstract of paper read before the Royal Society December 21, 1871. "On some recent Researches in Solar Physics, and a Law regulating the time of duration of the Sun-spot Period". By Warren De La Rue, F. R. S., Balfour Stewart, F. R. S., and Benjamin Loewy, F. R. A. S.

太阳黑子的周期*

编者按

鲁道夫·沃尔夫在对太阳黑子活动性的观察过程中不仅估算出了它们的平均周期——约 11.1 年，还发现太阳黑子活动性的上升周期和下降周期之间的比值存在一定的规律。沃尔夫发现，平均地说，太阳黑子活动性的上升周期为 3.7 年，而下降过程比上升过程慢，平均周期是 7.4 年。本文中指出：其他科学家的最新观测结果也在很大程度上证实了沃尔夫的基本论点，并得到了一些很重要的数据，比如任一上升周期和随后的下降周期之间的比值固定为 1∶2.1。现代数学方法证明太阳黑子活动周期实际上并不是恒定的，平均周期大约是 10.5 年。

在沃尔夫教授和弗里茨先生最近发表在《皇家学会学报》（第 127 期，1871 年）上的关于太阳黑子现象的近期研究简报中，我们发现沃尔夫教授的有些结论与我们在《自然科学会报》（1870 年，第 389～496 页）上最新发表的有关太阳物理方面的研究结果并不十分相符。通过对这种差异的原因做进一步研究，我们似乎找到了更加明确的规律，从而在数值上建立起太阳黑子周期曲线中的两条分支之间，也就是黑子数目逐渐减少阶段与黑子数目持续增加阶段之间的联系。

在阐述我们自己的结论之前，为了清楚起见，在这里再次简要地介绍一下沃尔夫教授的研究结果。

沃尔夫教授早先曾主要致力于精确测定太阳黑子周期平均**长度**的工作，但之后他意识到获得一些与太阳黑子数目周期性增减的一般特性有关的知识也是非常重要的。因此他利用现有的观测数据，并用自己独创的特殊方法使在不同时间段由不同观测者观测得到的数据可以相互比较，为的是更准确地研究太阳黑子曲线的周期特性。他用图和表来表示最小值前后两年半时间内每月观测到的黑子数量的平均值，并将这种方法应用于 5 个最小值周期中，这 5 个最小值所在的年份如下：

* 本文摘自由沃伦·德拉鲁（皇家学会会员）、鲍尔弗·斯图尔特（皇家学会会员）和本杰明·洛伊（皇家天文学会会员）在 1871 年 12 月 21 日召开的皇家学会会议上宣读的题为《近期太阳物理与太阳黑子周期规律的研究》的文章。

1823.2
1833.8
1844.0
1856.2
1867.2

In a table he gives their mean numbers, expressing the solar activity, arranged in various columns; and arrives at the following results:—

(1) It is shown now with greater precision than was previously possible, that the curve of sun-spots ascends with greater rapidity than it descends. The fact is shown in the subjoined diagram, which it may be of interest to compare with the curves given previously by ourselves in the above-mentioned place. The zero-point in this diagram corresponds to the minimum of each period; the abscissae give the time before and after it, viz., two and a half years, or thirty months; the ordinates express the amount of spot-production in numbers of an arbitrary scale. The two finely dotted curves are intended to show the actual character of a portion of two periods only, viz., those which had their minima in 1823.2 and 1867.2; the strongly dotted curve, however, gives the mean of all periods (five) over which the investigation extends.

(2) Denoting by x the number of years during which the curve ascends, and presuming that the behaviour is approximately the same throughout the whole period of 11.1 years as during the five years investigated, we have the proportion

$$\frac{x}{11.1 - x} = \frac{1}{2},$$

whence $$x = 3.7,$$

or the average duration of an ascent is 3.7 years, that of a descent 7.4 years.

(3) The character of a single period may essentially differ from the mean, but on the whole it appears that a $\left\{\begin{matrix}\text{retarded}\\ \text{accelerated}\end{matrix}\right\}$ descent corresponds to a $\left\{\begin{matrix}\text{retarded}\\ \text{accelerated}\end{matrix}\right\}$ ascent. Thus the minimum of 1844.0 behaved very normally; but that of 1856.2, and still more that of 1823.2, shown in the following diagram, presents a retarded ascent and descent; on the other hand, the minimum of 1833.8, and still more in that of 1867.2, also shown in the diagram, both ascent and descent are accelerated.

Finally Prof. Wolf arranged in the manner shown in the following table the successive minima and maxima, in order to arrive at some generalisation which might enable him to foretell the general character and length of a future period. Taking the absolute differences in time of every two successive maxima, and the mean differences of every two alternating minima, he shows that the greatest acceleration of both maximum and minimum happens

1823.2
1833.8
1844.0
1856.2
1867.2

在一张表的不同列里他列出了各月太阳黑子数目的平均值，以说明太阳的活动情况，并且得出了以下一些结论：

(1) 现在这张曲线图比以往更精确地显示出，太阳黑子数量上升的速度高于下降的速度。这个事实也可以从附表中看出，而这也可以与我们之前得出的结论（在前面提到的论文中）进行比较。图的零点对应于每个周期的最小值，横坐标给出了最小值之前与之后的时间，即，两年半或者 30 个月。纵坐标则用任意选定的尺度标示黑子产生的数量。两条由细点组成的曲线只显示出了最小值分别在 1823.2 和 1867.2 的两个周期中部分时间的实际特征。而那条由粗点组成的曲线则表示研究工作延伸的所有（5 个）周期内的平均值。

(2) 用 x 来表示曲线上升的年份数，并且假设在 5 年观察期内太阳黑子的活动规律与整个周期 11.1 年内的规律大致相同，则我们可以列出下面的比例式：

$$\frac{x}{11.1-x}=\frac{1}{2},$$

从而得到
$$x=3.7,$$
或者说曲线上升期的平均年份为 3.7 年，下降期的平均年份为 7.4 年。

(3) 对于某一个具体的周期而言，可能和平均值有一定的偏差；但对于整体而言，一个$\begin{Bmatrix}\text{减速}\\\text{加速}\end{Bmatrix}$的下降对应着一个$\begin{Bmatrix}\text{减速}\\\text{加速}\end{Bmatrix}$的上升。所以最小值为 1844.0 的曲线非常正常；最小值为 1856.2 和 1823.2 的曲线上升和下降的速度都比较慢，尤其是 1823.2，见下图；另一方面，最小值为 1833.8 和 1867.2 的曲线则又显示出了上升和下降速度都很快的趋势，尤其是 1867.2，见下图。

最后，沃尔夫教授用下表中的方式列出了几个连续的最小值和最大值，为的是得到一些规律性的结论，使他能够预言太阳黑子的一般特征和未来周期的长度。通过研究与每两个连续最大值对应的时间绝对差和与每两个次近邻最小值对应的时间平均差，他得出了这样的结论：黑子最大值和最小值的最大加速过程是同时发生的。

together. This result strengthens our own conclusions, to be immediately stated, by new evidence, as it is derived from observations antecedent to the time over which our researches extend.

Minima	Differences of alternating Minima	Means	Maxima	Differences of successive Maxima
1810.5				
			1816.8	
1823.2	23.3	11.65		12.7
			1829.5	
1833.8	20.8	10.4		7.7
			1837.2	
1844.0	22.4	11.2		11.4
			1848.6	
1856.2	23.2	11.6		11.6
			1860.2	
1867.2				

From this Prof. Wolf predicts for the present period a very accelerated maximum—a prediction which seems likely to be fulfilled.

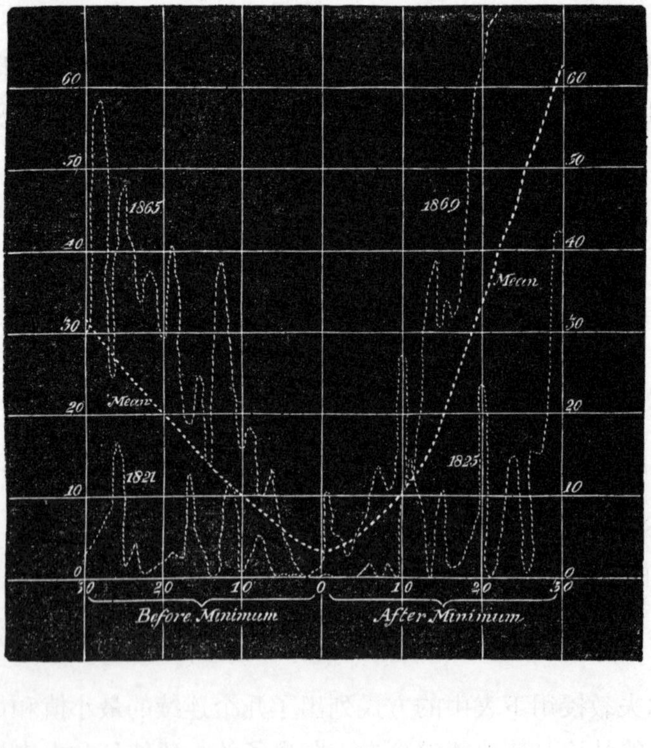

Comparing now M. Wolf's results with our own, it must not be overlooked, in judging of the agreement or discrepancy of these two independently obtained sets, that our facts have been derived from the actual measurement and subsequent calculation of the spotted area from day to day since 1833, recorded by Schwabe, Carrington, and the Kew

通过对以前观察结果的新一轮论证，这一结果也证实了下面马上要谈到的我们自己的论点。

最小值	次近邻最小值的差	平均值	最大值	连续最大值的差
1810.5				
			1816.8	
1823.2	23.3	11.65		12.7
			1829.5	
1833.8	20.8	10.4		7.7
			1837.2	
1844.0	22.4	11.2		11.4
			1848.6	
1856.2	23.2	11.6		11.6
			1860.2	
1867.2				

沃尔夫教授由此预言，我们现在的这个周期是一个加速上升到极大的周期，这个预言似乎很可能会实现。

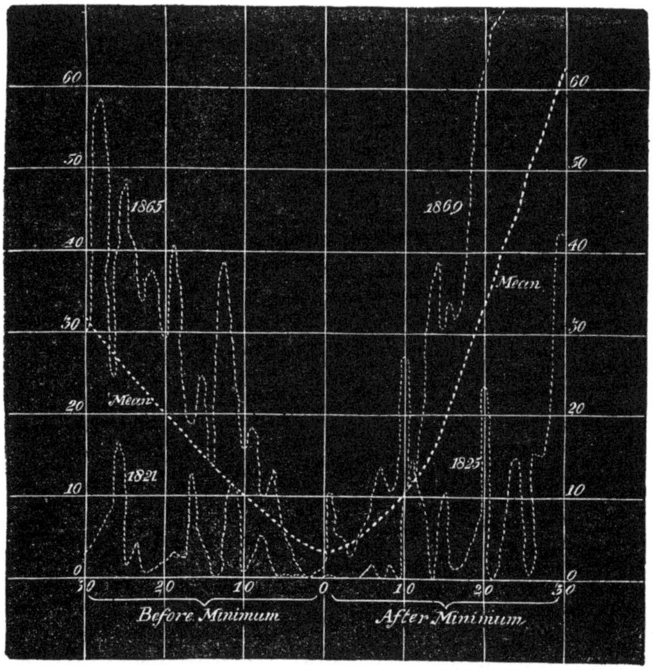

现在比较一下沃尔夫先生的结果和我们自己的结果，不要过分看重对这两个独立研究得出的结论的相同点和不同点的判断，因为我们不能忽视以下事实：我们的结论来自 1833 年以后每天对黑子区域的实际测量和计算，而这些测量数据是由施

solar photograms, which measurements are expressed as millionths of the sun's visible hemisphere, while the conclusions of M. Wolf are founded on certain "relative numbers", which give the amount of observed spots on an arbitrary scale, chiefly designed to make observations made at different times and by various observers comparable with each other. This will obviously, in addition to the sources of error to which our own method is liable, introduce an amount of uncertainty arising from errors of estimation, and the possibility of using for a whole series an erroneous factor of reduction. Nevertheless we shall find a very close agreement in various important results, and this seems a sufficient proof of the great value and reliability of M. Wolf's "relative numbers", especially for times previous to the commencement of regular sun observations.

The following is a comparison of the date of periodic epochs, as fixed by ourselves and M. Wolf: —

Minima epochs	I	II	III	IV
De La Rue, Stewart, and Loewy	1833.92	1843.75	1856.31	1867.12
Rudolf Wolf	1833.8	1844.0	1856.2	1867.2
Maxima epochs	I	II	III	
De La Rue, Stewart, and Loewy	1836.98	1847.87	1859.69	
Rudolf Wolf	1837.2	1846.6	1860.2	

It will be seen from this comparison that only one appreciable difference occurs, viz., in the maximum of 1847, which M. Wolf fixes nearly one and a quarter years before our date.

The mean length of a period is found by us to be 11.07 years, which agrees very well with M. Wolf's value, viz., 11.1 years.

We found the following times for the duration of increase of spots during the three periods, and for the corresponding decrease, or for ascent and descent of the graphic curve, beginning with the minimum of 1833: —

	Time of ascent	Time of descent
I	3.06 years	6.77 years
II	4.12 ,,	8.44 ,,
III	3.37 ,,	7.43 ,,
Mean	3.52 ,,	7.55 ,,

Prof. Wolf gives 3.7 years and 7.4 years for the ascent and descent respectively; and considering that he derived these numbers only from an investigation of a portion of each period, the agreement is indeed surprising, and would by itself suggest that the times of ascent and descent are connected by a definite law.

M. Wolf has expressed in general terms the following law with reference to this relation of increase and decrease of spots: —

瓦贝和卡林顿两人以及邱天文台的太阳照相法所记录的，并以可见太阳半球的百万分之一为单位进行测量；而沃尔夫先生的论断是建立在某些“相对数”的基础上的，他用任选的尺度给出了观测到的黑子总数，主要为了使不同时间，不同观测者的观测结果可以相互比较。很显然，沃尔夫的方法除了存在那些和我们的方法一样的错误源以外，还引入了由估算方法不正确产生的大量不确定性，而且他可能为整个推算过程选了一个错误的因子。然而我们发现两者的重要结论是非常一致的，而这似乎为沃尔夫教授的“相对数”的巨大价值和可靠性提供了强有力的佐证，尤其是在对太阳的常规观测开始之前。

下面是一些对黑子周期历元时期的比较，综合了我们和沃尔夫教授的数据：

最小值时期	I	II	III	IV
德拉鲁，斯图尔特和洛伊	1833.92	1843.75	1856.31	1867.12
鲁道夫·沃尔夫	1833.8	1844.0	1856.2	1867.2
最大值时期	I	II	III	
德拉鲁，斯图尔特和洛伊	1836.98	1847.87	1859.69	
鲁道夫·沃尔夫	1837.2	1846.6	1860.2	

从这些数据的比较中我们可以看出只有一处有明显的差异，即 1847 年的最大值，沃尔夫确定的时间大概比我们的提前了 1.25 年。

我们发现一个周期的平均长度为 11.07 年，与沃尔夫先生的周期 11.1 年非常接近。

我们找出黑子数量增长的 3 个时间段和黑子数量下降的 3 个时间段，对应于图中曲线中的上升阶段和下降阶段，最小值出现在 1833 年。

	上升时间	下降时间
I	3.06 年	6.77 年
II	4.12 年	8.44 年
III	3.37 年	7.43 年
平均	3.52 年	7.55 年

沃尔夫教授分别给出 3.7 年的上升期和 7.4 年的下降期，考虑到他仅仅是在考查每个周期的部分时间后得出的这些数字，这种一致性确实令人惊讶，这本身也说明了黑子增多期和减少期确实会与某些特定的规律相联系。

沃尔夫先生用大众化的语言描述了关于黑子数量增长期和下降期之间的规律：

"The character of a single period may essentially differ from the mean behaviour, but on the whole it appears that a $\left\{\begin{matrix}\text{retarded}\\ \text{accelerated}\end{matrix}\right\}$ descent corresponds to a $\left\{\begin{matrix}\text{retarded}\\ \text{accelerated}\end{matrix}\right\}$ ascent."

We, on the other hand, have, by an inspection of our curves (*vide Phil. Trans.* 1870, p. 393), been induced to make the following remark on the same question: —

"We see that the second curve, which was no longer in period as a whole than either of the other two, manifests this excess in each of its branches, that is to say, its left or ascending branch is larger as a whole than the same branch of the two other curves, and the same takes place for the second or descending branch. On the other hand, the maximum of this curve is not so high as that of either of the other two—in fact, the curve has the appearance as if it were pressed down from above and pressed out laterally so as to lose in elevation what it gains in time."

Although both statements appear to lead up to the same conclusion—viz., that ascent and descent are connected by law—still they differ essentially in this respect, that if A, B, C represent the three following consecutive events, descent, ascent, descent, Prof. Wolf's law refers to the connection between A and B, while our remark refers to B and C. We consider two successive minima as the beginning and end of a single period, while M. Wolf, at least in this particular research, places the minimum within the period, and compares the descent from the preceding maximum with the ascent to the next one.

We have considered the connection thus indicated of sufficient importance to apply to it the following test. If, using the previous notation, a definite relation exists between A and B, the *ratio* of the times which the events occupy in every epoch ought to be approximately constant; similarly with respect to B and C; and this ratio should not be influenced by the *absolute* duration of the two successive events. It is clear that the greater uniformity of these ratios will be a test of their interdependence. The following is the result of the comparison: —

a. Prof. Wolf's law: comparison of A and B

	Periods	Duration of descent (A)	Periods	Duration of ascent (B)
I	1829.5 to 1833.8	4.3 years	1833.8 to 1837.2	3.4 years
II	1837.2 to 1844.0	6.8 ,,	1844.0 to 1846.6	2.0 ,,
III	1846.6 to 1856.2	9.6 ,,	1856.2 to 1860.2	4.0 ,,

“对于某一个具体的周期而言可能和平均值有一定的偏差，但对于整体而言一个$\left\{\begin{array}{c}\text{减速}\\\text{加速}\end{array}\right\}$的下降对应着一个$\left\{\begin{array}{c}\text{减速}\\\text{加速}\end{array}\right\}$的上升。”

另一方面，通过研究我们自己的曲线（参阅《自然科学会报》，1870 年，第 393 页），针对上述问题，我们可以得出以下一些结论：

“我们看到第二条曲线不像另外两条曲线那样具有一个整体的周期，它的每一条分支都要长一些，也就是说它的左分支或者上升分支从总体上看要比另外两条曲线的对应分支大，而另一个分支或者说下降分支也遵循同样的规律。另一方面，这条曲线的最大值也没有另两条曲线高——实际上，这条曲线就好像被从顶端压下来然后从两侧被拉伸了，以至于失去了它应有的高度。”

虽然，似乎这两方面的陈述都能导出同样的结论，即上升和下降都有一定的规律性，但是我们之间又有本质上的不同，如果用 A，B，C 来分别代表三个连续的发展阶段，下降，上升，下降。沃尔夫教授的规则适用于 A 与 B 之间的关系，而我们的结论则适用于 B 与 C 之间的联系。我们用两个连续的最小值作为一个周期的起点和终点，而沃尔夫先生，至少在这项特殊的研究中，把这个最小值放在一个周期的内部，并且比较了之前从最大值下降到这个最小值的过程及从这个最小值上升至下一个最大值的过程。

我们认为这种联系用下面的测试来检验是非常必要的。假如，按照以前的想法，A 和 B 之间存在某种确定的联系，则每个周期中两个发展阶段（指 A 和 B）所经历的时间之**比**应该是近似恒定的数。类似的关系也应适用于 B 和 C 之间。这个比值不会受到两个相继的发展阶段持续时间的**绝对值**的影响。很显然比值越恒定越说明这种关系是存在的，下面列出的是对比的结果：

a. 沃尔夫教授的定律：A 与 B 的比值

	周期	下降期（A）	周期	上升期（B）
I	1829.5~1833.8	4.3 年	1833.8~1837.2	3.4 年
II	1837.2~1844.0	6.8 年	1844.0~1846.6	2.0 年
III	1846.6~1856.2	9.6 年	1856.2~1860.2	4.0 年

	Ratio $\frac{A}{B}$		Difference from mean
I	1.265	Mean 2.093	− 0.728
II	2.615		+ 0.522
III	2.400		+ 0.307

These differences from the mean are so considerable that in the present state of the inquiry a connection between any descent and the immediately *succeeding* ascent appears highly improbable. A very new and apparently important relation seems, however, to result from a similar comparison of any ascent and the immediately succeeding descent, or between B and C.

b. Comparison of B and C

	Periods	Duration of ascent (B)	Periods	Duration of descent (C)
I	1833.92 to 1836.98	3.06 years	1836.98 to 1843.75	6.77 years
II	1843.75 to 1847.87	4.12 ,,	1847.87 to 1856.31	8.44 ,,
III	1856.31 to 1859.69	3.38 ,,	1859.69 to 1867.12	7.43 ,,

	Ratio $\frac{C}{B}$		Difference from mean
I	2.212	Mean 2.151	+ 0.061
II	2.044		− 1.107
III	2.198		+ 0.047

(**5**, 192-194; 1872)

	比率 $\frac{A}{B}$		与均值的差
I	1.265	均值 2.093	− 0.728
II	2.615		+ 0.522
III	2.400		+ 0.307

这些比值与平均值的差异如此之大，以至于在这样的情况下研究下降期与**紧随其后的**上升期之间的关系几乎不可能。但是一种新发现的、重要的联系，却可以用类似的方法（即一个上升期与随之而来的下降期之间的比率，或者说 B 与 C 之间的比率）推算出来。

b. B 与 C 的比值

	周期	上升期（B）	周期	下降期（C）
I	1833.92~1836.98	3.06 年	1836.98~1843.75	6.77 年
II	1843.75~1847.87	4.12 年	1847.87~1856.31	8.44 年
III	1856.31~1859.69	3.38 年	1859.69~1867.12	7.43 年

	比率 $\frac{C}{B}$		与均值的差
I	2.212	均值 2.151	+ 0.016
II	2.044		− 0.107
III	2.198		+ 0.047

（刘东亮 翻译；蒋世仰 审稿）

Clerk Maxwell's Kinetic Theory of Gases

J. C. Maxwell

Editor's Note

James Clerk Maxwell's recent kinetic theory of gases gave theoretical foundation to the macroscopic properties of gases. Here Maxwell defends his theory against a recent criticism. The theory predicts that a vertical column of gas should have the same temperature at all heights. However, the action of gravity implies that any molecule moving downward on some path should pick up energy, while those going upward should lose it. So shouldn't there be a net flow of energy downward? This argument, says Maxwell, assumes that particles projected upwards tend to have the same mean energy as those projected downward. Yet this cannot be so: at equilibrium, more must be ejected downwards, recovering the agreement with observations.

YOUR correspondent, Mr. Guthrie, has pointed out an, at first sight, very obvious and very serious objection to my kinetic theory of a vertical column of gas. According to that theory, a vertical column of gas acted on by gravity would be in thermal equilibrium if it were at a uniform temperature throughout, that is to say, if the mean energy of the molecules were the same at all heights. But if this were the case the molecules in their free paths would be gaining energy if descending, and losing energy if ascending. Hence, Mr. Guthrie argues, at any horizontal section of the column a descending molecule would carry more energy down with it that an ascending molecule would bring up, and since as many molecules descend as ascend through the section, there would on the whole be a transfer of energy, that is, of heat, downwards; and this would be the case unless the energy were so distributed that a molecule in any part of its course finds itself, on an average, among molecules of the same energy as its own. An argument of the same kind, which occurred to me in 1866, nearly upset my belief in calculation, and it was some time before I discovered the weak point in it.

The argument assumes that, of the molecules which have encounters in a given stratum, those projected upwards have the same mean energy as those projected downwards. This, however, is not the case, for since the density is greater below than above, a greater *number* of molecules come from below than from above to strike those in the stratum, and therefore a greater number are projected from the stratum downwards than upwards. Hence since the total momentum of the molecules temporarily occupying the stratum remains zero (because, as a whole, it is at rest), the smaller number of molecules projected upwards must have a greater initial velocity than the larger number projected downwards. This much we may gather from general reasoning. It is not quite so easy, without calculation, to show that this difference between the molecules projected upwards and downwards from the same stratum exactly counteracts the tendency to a downward transmission of energy pointed out by Mr. Guthrie. The difficulty lies chiefly in forming

克拉克·麦克斯韦的气体动力学理论

麦克斯韦

编者按

詹姆斯·克拉克·麦克斯韦最近提出的气体动力学理论为研究气体的宏观性质奠定了理论基础。在这篇文章中麦克斯韦坚持自己的理论，反驳了最近对其理论提出的批评。麦克斯韦的理论认为一个垂直气体柱在各个高度的温度应该保持一致。然而重力作用会使沿某一路径向下运动的分子获得能量，使向上运动的粒子失去能量。所以似乎应该出现一个向下的净能流。麦克斯韦说，上述论点假设向上运动的粒子和向下运动的粒子平均能量相同。但这是不可能的：在准静态，向下运动的粒子更多，这样和观测结果就不会矛盾了。

贵刊的通讯员格思里先生指出了我的垂直气体柱理论中一个初看上去非常明显而又严重的缺陷。根据那个理论，重力场中的一个垂直气体柱，如果各处温度相同，或者说如果处于不同高度的气体分子的平均动能相同，那么它将处于热平衡状态。但是如果事实确实如此，则自由运动的分子在向下运动的过程中动能将增加，向上运动的时候动能将减少。格思里先生论证说，如此一来，在气体柱的任何一个水平截面上，由于向上和向下运动的分子数目相同，向下运动的分子带走的动能将超过向上运动的分子带来的动能，所以整体上就会出现一个动能的转移，也就是热量的下移过程。要避免格思里所说的这个矛盾，只能假定动能沿高度的分布能够保证一个分子不管运动到哪里都与周边分子的动能相同。我在1866年也遇到了同样的问题，当时几乎动摇了我对计算结果的信心。一段时间之后我才发现这个观点中站不住脚的地方。

上述观点假定：对于任一个给定水平层上发生碰撞的分子，向上运动的分子和向下运动的分子平均动能相同。这一点并不符合实际情况。由于水平层下的分子密度大于水平层以上的，所以水平层下与水平层中分子发生碰撞的分子**数目**要大于水平层上与水平层中分子发生碰撞的分子数目，相应地，在水平层中发生碰撞而向下运动的分子数目也要大于发生碰撞而向上运动的分子数目。因此，既然该时刻水平层中的分子总动量保持为零（因为，从总体上来看水平层是静止的），那么相对于数目较大的向下运动的分子，数目较少的向上运动的分子就必须拥有更大的初始速度。从普通的推理中我们只能了解到这么多。不经过计算，想要说明从同一个水平层中向上和向下运动的分子在数目和速度上的差别正好能够抵消格思里先生指出的

exact expressions for the state of the molecules which instantaneously occupy a given stratum in terms of their state when projected from the various strata in which they had their last encounters. In my paper in the *Philosophical Transaction*, for 1867, on the "Dynamical Theory of Gases", I have entirely avoided these difficulties by expressing everything in terms of what passes through the boundary of an element, and what exists or takes place inside it. By this method, which I have lately carefully verified and considerably simplified, Mr. Guthrie's argument is passed by without ever becoming visible. It is well, however, that he has directed attention to it, and challenged the defenders of the kinetic theory to clear up their ideas of the result of those encounters which take place in a given stratum.

(**8**, 85; 1873)

那个向下的动能转移趋势，是非常困难的。主要困难在于需要找到一个确切的表达式，将某一时刻各水平层中刚发生碰撞而运动到某一特定水平层上所有分子的状态描述出来。在我在 1867 年的《自然科学会报》上发表的一篇题为《论气体动力学》的文章中，我把一切都用穿过一个单元边界的通量和单元内部本来就存在或者生成的量来表达，这样就回避了上面提到的种种麻烦。不久前我曾仔细检验并大大地简化了这个方法，利用这个方法，格思里先生提出的问题被绕过去了，没有表现出来。但是，现在他将其提出来引起大家的注意是很有益处的，因为这对分子动力论的支持者是一个挑战，促使他们整理自己关于一个特定水平层上分子碰撞结果的构想。

（何钧 翻译；鲍重光 审稿）

Molecules*

J. C. Maxwell

Editor's Note

James Clerk Maxwell was sometimes accused of being a poor lecturer, but this talk delivered to the British Association offers a lucid, engaging picture of the current understanding of the molecular nature of matter. Maxwell made a decisive contribution himself with his kinetic theory of gases, which explained how the macroscopic properties of gases, such as the laws relating pressure, volume and temperature, could be explained from the microscopic motions of the constituent particles. Maxwell's estimate of the size of a hydrogen molecule is only slightly bigger than the modern view. And his discussion of molecular diffusion anticipates the work of Albert Einstein and Jean Perrin on Brownian motion that provided the first real evidence for molecules as physical entities.

AN atom is a body which cannot be cut in two. A molecule is the smallest possible portion of a particular substance. No one has ever seen or handled a single molecule. Molecular science, therefore, is one of those branches of study which deal with things invisible and imperceptible by our senses, and which cannot be subjected to direct experiment.

The mind of man has perplexed itself with many hard questions. Is space infinite, and if so in what sense? Is the material world infinite in extent, and are all places within that extent equally full of matter? Do atoms exist, or is matter infinitely divisible?

The discussion of questions of this kind has been going on ever since men began to reason, and to each of us, as soon as we obtain the use of our faculties, the same old questions arise as fresh as ever. They form as essential a part of the science of the nineteenth century of our era, as of that of the fifth century before it.

We do not know much about the science organisation of Thrace twenty-two centuries ago, or of the machinery then employed for diffusing an interest in physical research. There were men, however, in those days, who devoted their lives to the pursuit of knowledge with an ardour worthy of the most distinguished members of the British Association; and the lectures in which Democritus explained the atomic theory to his fellow-citizens of Abdera realised, not in golden opinions only, but in golden talents, a sum hardly equalled even in America.

* Lecture delivered before the British Association at Bradford, by Prof. Clerk Maxwell, F. R. S.

分 子*

麦克斯韦

编者按

有人认为詹姆斯·克拉克·麦克斯韦缺乏演讲天赋，但这篇在英国科学促进会所作的报告中，麦克斯韦对物质分子的那些得到普遍接受的性质描述得非常透彻，给人留下了深刻的印象。麦克斯韦在分子学方面作出了有决定意义的贡献，他提出的气体动力学理论能够说明如何用组成粒子的微观运动来解释气体的宏观性质，如与压力、体积和温度相关的定律。麦克斯韦对氢分子大小的估计只比现在的公认值略大一些。他对分子扩散现象的讨论促使阿尔伯特·爱因斯坦和让·佩兰开始了关于布朗运动的研究，这使人们第一次认识到分子是一种物理实体。

原子是不能被一分为二的实体。分子是组成物质的最小单位。没有人看见或者摆弄过单个分子。因此，分子科学是研究不可见也不可感觉的事物的一门学问，我们无法对它进行直接实验。

人类经常思索很多难以回答的问题。空间是无限的吗？如果是，是从什么意义上讲的？物质世界的范围是无限的吗？在这个范围内是不是每个地方都同等地充满了物质？原子存在吗？或物质是否无限可分？

自从人类开始理性思考以来，关于这类问题的讨论就一直没有停止过。对于我们每个人来说，一旦开始用心智思考，那些古老的问题就会像从前一样令人觉得新奇。不论是在我们所处的19世纪，还是在公元前5世纪，这些问题都构成了科学的基本部分。

我们对2,200年前位于色雷斯的科学组织所知甚少，也不知道他们用何种方式来传播对自然研究的兴趣。不过那时候确实有人毕生追求知识，热情不亚于英国科学促进会中最杰出的成员。当德谟克利特向他的阿布德拉市民开设讲座讲解自己的原子理论时，他获得的高度评价和丰厚报酬即使在今天的美国也很少有人能比得上。

* 皇家学会会员克拉克·麦克斯韦教授在布拉德福德对英国科学促进会作的报告。

To another very eminent philosopher, Anaxagoras, best known in the world as the teacher of Socrates, we are indebted for the most important service to the atomic theory, which, after its statement by Democritus, remained to be done. Anaxagoras, in fact, stated a theory which so exactly contradicts the atomic theory of Democritus that the truth or falsehood of the one theory implies the falsehood or truth of the other. The question of the existence or non-existence of atoms cannot be presented to us this evening with greater clearness than in the alternative theories of these two philosophers.

Take any portion of matter, say a drop of water, and observe its properties. Like every other portion of matter we have ever seen, it is divisible. Divide it in two, each portion appears to retain all the properties of the original drop, and among others that of being divisible. The parts are similar to the whole in every aspect except in absolute size.

Now go on repeating the process of division till the separate portions of water are so small that we can no longer perceive or handle them. Still we have no doubt that the sub-division night be carried further, if our senses were more acute and our instruments more delicate. Thus far all are agreed, but now question arises. Can this sub-division be repeated for ever?

According to Democritus and the atomic school, we must answer in the negative. After a certain number of sub-divisions, the drop would be divided into a number of parts each of which is incapable of further sub-division. We should thus, in imagination, arrive at the atom, which, as its name literally signifies, cannot be cut in two. This is the atomic doctrine of Democritus, Epicurus, and Lucretius, and, I may add, of your lecturer.

According to Anaxagoras, on the other hand, the parts into which the drop is divided, are in all respects similar to the whole drop, the mere size of a body counting for nothing as regards the nature of its substance. Hence if the whole drop is divisible, so are its parts down to the minutest sub-divisions, and that without end.

The essence of the doctrine of Anaxagoras is that the parts of a body are in all respects similar to the whole. It was therefore called the doctrine of Homoiomereia. Anaxagoras did not of course assert this of the parts of organised bodies such as men and animals, but he maintained that those inorganic substances which appear to us homogeneous are really so, and that the universal experience of mankind testifies that every material body, without exception, is divisible.

The doctrine of atoms and that of homogeneity are thus in direct contradiction.

But we must now go on to molecules. Molecule is a modern word. It does not occur in *Johnson's Dictionary*. The ideas it embodies are those belonging to modern chemistry.

另一位杰出的哲学家，即以身为苏格拉底的老师而闻名于世的阿那克萨哥拉在德谟克利特之后对原子学说作出了最重要的贡献。实际上，阿那克萨哥拉和德谟克利特两人的原子学说是如此地针锋相对，一方正确则另一方必错。今晚我们对原子到底是否存在这一问题的讨论，用这两位哲学家的对立理论来表达，是最清楚不过了。

随便取一份物质，比如一滴水，来观察它的性质。就像我们看到的其他物质一样，它是可以分割的。把它分成两份，每一份都保持原来那滴水的所有性质，其他可以分割的物质也是一样。每一个部分除了尺寸比整体小些，其他各方面都和整体相似。

就这么一直分下去，直到分出来的水滴小到我们再也看不见，也无法对它们进行操作。但是大家都明白，如果我们的感官更敏锐，我们的设备更精密，细分过程还是可以接着进行下去的。到此为止不会有什么异议，但是现在问题就来了：这样的细分过程可以永远继续下去吗？

德谟克利特和原子学派的回答是否定的。经过一定次数的分割后，水滴就被分成很多很小的部分，每一部分都不能进一步细分了。也就是我们已经细分到了想象中的原子，原子这个名字的字面含义就是不可分的意思。这就是德谟克利特、伊壁鸠鲁、卢克莱修还有我，你们的演讲者，所赞成的原子论。

另一方面，阿那克萨哥拉认为，水滴被分割成的各个部分，除了和物质性质无关的物体尺寸发生了变化以外，其他一切方面都和整个的水滴类似。因此，如果原来的水滴是可分的，那么分割得到的各个部分也应该是可分的，哪怕分到极小，永无止境。

阿那克萨哥拉学说的根本点是：一个物体的部分和它的整体在所有方面都是相似的，所以被称为同质性学说。阿那克萨哥拉当然没有把它应用到人和动物这样的有机体的身上，但是他认为那些看上去是均质的无机物的确是同质的，并且人类的普遍经验也能证实每一种物质实体毫无例外都是可分的。

原子学说和同质性学说就是这样地针锋相对。

但是现在我们必须转入对分子这个现代名词的讨论，它在《约翰逊词典》里是没有的。分子所包含的概念属于现代化学的范畴。

A drop of water, to return to our former example, may be divided into a certain number, and no more, of portions similar to each other. Each of these the modern chemist calls a molecule of water. But it is by no means an atom, for it contains two different substances, oxygen and hydrogen, and by a certain process the molecule may be actually divided into two parts, one consisting of oxygen and the other of hydrogen. According to the received doctrine, in each molecule of water there are two molecules of hydrogen and one of oxygen. Whether these are or are not ultimate atoms I shall not attempt to decide.

We now see what a molecule is, as distinguished from an atom.

A molecule of a substance is a small body such that if, on the one hand, a number of similar molecules were assembled together they would form a mass of that substance, while on the other hand, if any portion of this molecule were removed, it would no longer be able, along with an assemblage of other molecules similarly treated, to make up a mass of the original substance.

Every substance, simple or compound, has its own molecule. If this molecule be divided, its parts are molecules of a different substance or substances from that of which the whole is a molecule. An atom, if there is such a thing, must be a molecule of an elementary substance. Since, therefore, every molecule is not an atom, but every atom is a molecule, I shall use the word molecule as the more general term.

I have no intention of taking up your time by expounding the doctrines of modern chemistry with respect to the molecules of different substances. It is not the special but the universal interest of molecular science which encourages me to address you. It is not because we happen to be chemists or physicists or specialists of any kind that we are attracted towards this centre of all material existence, but because we all belong to a race endowed with faculties which urge us on to search deep and ever deeper into the nature of things.

We find that now, as in the days of the earliest physical speculations, all physical researches appear to converge towards the same point, and every inquirer, as he looks forward into the dim region towards which the path of discovery is leading him, sees, each according to his sight, the vision of the same quest.

One may see the atom as a material point, invested and surrounded by potential forces. Another sees no garment of force, but only the bare and utter hardness of mere impenetrability.

But though many a speculator, as he has seen the vision recede before him into the innermost sanctuary of the inconceivably little, has had to confess that the quest was not for him, and though philosophers in every age have been exhorting each other to direct their minds to some more useful and attainable aim, each generation, from the earliest dawn of science to the present time, has contributed a due proportion of its ablest intellects to the quest of the ultimate atom.

回到我们原来的例子，一滴水能够最大限度地被分割成一定数量的彼此相似的部分。每一个这样的部分都是现代化学家称作的水分子。水分子包含两种不同的物质，氧和氢，所以它绝不是一个原子。通过一定的处理方式，确实可以把水分子分解成两部分，一部分含氢，另一部分含氧。根据公认的理论，每一个水分子都包含两个氢分子和一个氧分子。至于这些氢分子和氧分子是不是不能再分解的原子，我这里先不去确定。

现在我们明白了分子是什么，它和原子有什么不同。

一种物质的分子是很小的，一方面，如果许多同样的分子聚合在一起，就会形成大量这种物质；另一方面，如果这些分子的某个部分缺失，它们与其他经过相同处理的分子聚合在一起也不能形成原来的物质。

任何物质，无论是简单的还是复合的，都有自己的分子。如果这个分子再被分割，形成的部分就是其他物质的分子。一个原子，如果确实存在的话，应该是一种基本物质的分子。不是每个分子都是原子，但是每个原子都是分子，所以我将使用含义更广的分子这个术语。

我不想浪费时间去详细解释现代化学关于各种物质分子的理论。我作这个报告的目的是讲述分子科学中普遍的而不是具体的问题。不是因为我们恰好是化学家、物理学家或者其他某个领域的专家才对这个与所有物质息息相关的中心问题感兴趣，而是因为我们都属于人类这个物种，其所具备的资质促使我们不断地深入研究事物的本质。

现在我们发现，就像早期物理猜想时代一样，所有的物理研究似乎都汇集到了同一点上；每一个探求者，在眺望发现之途指向的茫茫区域时，虽然目力各有不同，看到的却都是同一件宝物的幻象。

有些人眼中的原子是一个物质点，被有势力场包围着。另一些人则看不到力的存在，只看到裸露而坚硬的不可穿透的实体。

很多人看到幻象在眼前消退而去，躲进那不可思议的渺小之物最隐秘的庇护所之后，不得不承认宝物非他所属；各个时代的哲人们互相劝诫对方去追求更实际、更容易达到的目标。虽然如此，自从科学的启蒙时期直到今天，每一代都不乏最富才智的人投身于对最终原子的探求。

Our business this evening is to describe some researches in molecular science, and in particular to place before you any definite information which has been obtained respecting the molecules themselves. The old atomic theory, as described by Lucretius and revived in modern times, asserts that the molecules of all bodies are in motion, even when the body itself appears to be at rest. These motions of molecules are in the case of solid bodies confined within so narrow a range that even with our best microscopes we cannot detect that they alter their places at all. In liquids and gases, however, the molecules are not confined within any definite limits, but work their way through the whole mass, even when that mass is not disturbed by any visible motion.

This process of diffusion, as it is called, which goes on in gases and liquids and even in some solids, can be subjected to experiment, and forms one of the most convincing proofs of the motion of molecules.

Now the recent progress of molecular science began with the study of the mechanical effect of the impact of these moving molecules when they strike against any solid body. Of course these flying molecules must beat against whatever is placed among them, and the constant succession of these strokes is, according to our theory, the sole cause of what is called the pressure of air and other gases.

This appears to have been first suspected by Daniel Bernoulli, but he had not the means which we now have of verifying the theory. The same theory was afterwards brought forward independently by Lesage, of Geneva, who, however, devoted most of his labour to the explanation of gravitation by the impact of atoms. Then Herapath, in his "Mathematical Physics", published in 1847, made a much more extensive application of the theory to gases, and Dr. Joule, whose absence from our meeting we must all regret, calculated the actual velocity of the molecules of hydrogen.

The further development of the theory is generally supposed to have been begun with a paper by Krönig, which does not, however, so far as I can see, contain any improvement on what had gone before. It seems, however, to have drawn the attention of Prof. Clausius to the subject, and to him we owe a very large part of what has been since accomplished.

We all know that air or any other gas placed in a vessel presses against the sides of the vessel, and against the surface of any body placed within it. On the kinetic theory this pressure is entirely due to the molecules striking against these surfaces, and thereby communicating to them a series of impulses which follow each other in such rapid succession that they produce an effect which cannot be distinguished from that of a continuous pressure.

If the velocity of the molecules is given, and the number varied, then since each molecule, on an average, strikes the side of the vessel the same number of times, and with an impulse of the same magnitude, each will contribute an equal share to the whole pressure. The pressure in a vessel of given size is therefore proportional to the number of molecules in it, that is to the quantity of gas in it.

我们今天晚上的任务是介绍分子科学中的一些研究成果，特别是向你们展示那些关于分子本身的现在已经比较确定的认识。卢克莱修所描述的旧原子论到了现代重获新生。他认为所有物体中的分子都在不停地运动，即便当物体本身处于静止状态时也不例外。在固体中，分子的运动只局限于一个很小的范围，哪怕是利用当前最好的显微镜我们也察觉不到它们的移动。但是对于液体和气体的情况，分子的运动没有受到确切的范围限制，可以在整个物质中移动，哪怕这个整体没有受到任何可见的运动的干扰。

这个过程被称为扩散。它在气体、液体甚至一些固体中持续进行着，可以由实验验证，同时它也是分子运动论最有力的证明之一。

现在分子科学的最新进展，是从研究这些运动着的分子碰撞固体表面的机械效应开始的。飞行中的分子一定会撞击所有置于其中的物质。根据我们的理论，这种持续不断的撞击，就是产生所谓的空气和其他气体压力的唯一原因。

丹尼尔·伯努利可能是第一个想到这一点的人，但是他当时没有我们今天的实验手段来验证他的理论。后来日内瓦的勒萨热也独立提出过这个理论，但是他的工作主要是用原子碰撞来解释重力现象。赫拉帕斯在他 1847 年出版的《数学物理学》一书中，将这个理论更广泛地应用于各种气体。焦耳博士计算了氢分子的实际速度，今天他没有在场实在让人感到遗憾。

大家普遍认为这一理论的进一步发展是从克勒尼希的一篇论文开始的。但在我个人看来，这篇文章本身并没有在前人工作的基础上作出什么改进。不过它引起了克劳修斯教授对这个问题的关注，而后者对以后的理论发展起了很大作用。

我们都知道放置在一个容器中的空气或其他气体会对容器壁以及放置在其中的其他物体表面产生压力。根据气体动力学理论，这个压力完全是由分子碰撞这些表面产生的。这种碰撞带给表面的一系列冲击之间的间隔非常小，产生的效应和连续的压力没有什么两样。

假定分子的速率一定，但数量不同。那么平均来说，每个分子碰撞器壁的次数相同，产生的冲击强度也相同，因此对总压力的贡献也相同。这样一来，一个固定容积的容器承受的压强和其中分子的总数量，也就是其中的气体总量，成正比。

This is the complete dynamical explanation of the fact discovered by Robert Boyle, that the pressure of air is proportional to its density. It shows also that of different portions of gas forced into a vessel, each produces its own part of the pressure independently of the rest, and this whether these portions be of the same gas or not.

Let us next suppose that the velocity of the molecules is increased. Each molecule will now strike the sides of the vessel a greater number of times in a second, but besides this, the impulse of each blow will be increased in the same proportion, so that the part of the pressure due to each molecule will vary as the *square* of the velocity. Now the increase of the square of velocity corresponds, in our theory, to a rise of temperature, and in this way we can explain the effect of warming the gas, and also the law discovered by Charles that the proportional expansion of all gases between given temperatures is the same.

The dynamical theory also tells us what will happen if molecules of different masses are allowed to knock about together. The greater masses will go slower than the smaller ones, so that, on an average, every molecule, great or small, will have the same energy of motion.

The proof of this dynamical theorem, in which I claim the priority, has recently been greatly developed and improved by Dr. Ludwig Boltzmann. The most important consequence which flows from it is that a cubic centimetre of every gas at standard temperature and pressure contains the same number of molecules. This is the dynamical explanation of Gay Lussac's law of the equivalent volumes of gases. But we must now descend to particulars, and calculate the actual velocity of a molecule of hydrogen.

A cubic centimetre of hydrogen, at the temperature of melting ice and at a pressure of one atmosphere, weighs 0.00008954 grammes. We have to find at what rate this small mass must move (whether altogether or in separate molecules makes no difference) so as to produce the observed pressure on the sides of the cubic centimetre. This is the calculation which was first made by Dr. Joule, and the result is 1,859 metres per second. This is what we are accustomed to call a great velocity. It is greater than any velocity obtained in artillery practice. The velocity of other gases is less, as you will see by the table, but in all cases it is very great as compared with that of bullets.

We have now to conceive the molecules of the air in this hall flying about in all directions, at a rate of about seventeen miles in a minute.

If all these molecules were flying in the same direction, they would constitute a wind blowing at the rate of seventeen miles a minute, and the only wind which approaches this velocity is that which proceeds from the mouth of a cannon. How, then, are you and I able to stand here? Only because the molecules happen to be flying in different directions, so that those which strike against our backs enable us to support the storm which is beating against our faces. Indeed, if this molecular bombardment were to cease, even for an instant, our veins would swell, our breath would leave us, and we should, literally,

这就是罗伯特·玻意耳发现的空气压强正比于其密度这一事实的完整动力学解释。它还表明，如果我们将不同批次的气体加入容器，则无论它们的种类是否相同，每个部分都将独立地产生自己的分压强。

下一步让我们假定分子速率增加的情形。因为每秒钟内，每个分子碰撞器壁的次数相应增加，同时每次碰撞的冲击强度也成比例增加，所以每个分子对压强的贡献和它的速度的**平方**成正比。在我们现在的理论中，分子速率平方的增加和温度的增加相对应。这样我们就能解释加热气体导致压强增加的效应，以及随温度增加各种气体体积同比增大的查理定律。

动力论还告诉我们不同质量的气体分子互相碰撞会发生什么结果。质量大的分子比质量小的分子运动速率要小一些，所以平均来说，每个分子，不论质量大小，其动能都相同。

对这一动力学定律的证明是我最先提出的，近来被路德维希·玻尔兹曼博士加以改进和发展。它的一个重要推论就是在标准温度和压强下，1 立方厘米的任何气体都含有同等数量的分子，这就是盖·吕萨克定律关于相同体积气体的动力学解释。现在我们举一个具体的例子来计算一个氢分子的实际速率。

在温度为冰的熔点，压力为一个大气压时，1 立方厘米氢气的质量是 0.00008954 克。这么小的质量的运动（是合在一起还是分散到各个分子，对结果没有影响）到底需要多大速率，才能在 1 立方厘米的容器壁上产生测量到的压强呢？焦耳博士首先对此进行了计算，他的结果是每秒 1,859 米。通常对我们来说，这是一个很大的速率，比任何炮弹的速率都要大。从附表中大家可以看到，其他气体的速率要小一些，但是无论如何都远大于子弹的速率。

现在我们要设想一下这个大厅里的空气分子以每分钟 17 英里的速率向各个方向飞行。

如果所有分子都向同一个方向飞行，它们就会形成速率为每分钟 17 英里的强风，只有从加农炮炮口出膛的风速能够接近这个速率。那么，你我怎么能够在这里保持站立？这只是因为这些分子飞行的方向各不相同，在前面和后面撞击我们的分子冲击作用互相抵消。事实上，如果这种分子碰撞哪怕停止一刻，我们也会静脉肿胀，不能呼吸，一命呜呼。这些分子不只是撞击我们和房间四周的墙壁。考虑到它

expire. But it is not only against us or against the walls of the room that the molecules are striking. Consider the immense number of them, and the fact that they are flying in every possible direction, and you will see that they cannot avoid striking each other. Every time that two molecules come into collision, the paths of both are changed, and they go off in new directions. Thus each molecule is continually getting its course altered, so that in spite of its great velocity it may be a long time before it reaches any great distance from the point at which it set out.

I have here a bottle containing ammonia. Ammonia is a gas which you can recognise by its smell. Its molecules have a velocity of six hundred metres per second, so that if their course had not been interrupted by striking against the molecules of air in the hall, everyone in the most distant gallery would have smelt ammonia before I was able to pronounce the name of the gas. But instead of this, each molecule of ammonia is so jostled about by the molecules of air, that it is sometimes going one way and sometimes another. It is like a hare which is always doubling, and though it goes a great pace, it makes very little progress. Nevertheless, the smell of ammonia is now beginning to be perceptible at some distance from the bottle. The gas does diffuse itself through the air, though the process is a slow one, and if we could close up every opening of this hall so as to make it air-tight, and leave everything to itself for some weeks, the ammonia would become uniformly mixed through every part of the air in the hall.

This property of gases, that they diffuse through each other, was first remarked by Priestley. Dalton showed that it takes place quite independently of any chemical action between the inter-diffusing gases. Graham, whose researches were especially directed towards those phenomena which seem to throw light on molecular motions, made a careful study of diffusion, and obtained the first results from which the rate of diffusion can be calculated.

Still more recently the rates of diffusion of gases into each other have been measured with great precision by Prof. Loschmidt of Vienna.

He placed the two gases in two similar vertical tubes, the lighter gas being placed above the heavier, so as to avoid the formation of currents. He then opened a sliding valve, so as to make the two tubes into one, and after leaving the gases to themselves for an hour or so, he shut the valve, and determined how much of each gas had diffused into the other.

As most gases are invisible, I shall exhibit gaseous diffusion to you by means of two gases, ammonia and hydrochloric acid, which, when they meet, form a solid product. The ammonia, being the lighter gas, is placed above the hydrochloric acid, with a stratum of air between, but you will soon see that the gases can diffuse through this stratum of air, and produce a cloud of white smoke when they meet. During the whole of this process no currents or any other visible motion can be detected. Every part of the vessel appears as calm as a jar of undisturbed air.

们数量巨大，正在四处乱飞，你会发现它们不可能不互相碰撞。每当两个分子撞到一起，其轨道就都会改变，而它们会飞向新的方向。这样每个分子都在不断地改变轨道，因而虽然其速率很快，但要从出发点移开一定距离也需要不少时间。

我这里有一个里面装着氨气的瓶子。氨气的味道大家可以闻得出来。它的分子运动速率是每秒 600 米，因而如果它们的运动轨迹没有因为和大厅里的空气分子碰撞而改变，就算坐得最远的听众都会在我说出氨气这个名称之前闻到它。但是实际上并非如此，每个氨气分子都被空气分子撞来撞去，一会儿向东一会儿向西，就像一只野兔，老是改变方向，虽然步子很快，但是跑不了多远。虽然如此，在离瓶子一定距离以内的听众还是开始闻到氨气的味道了。氨气确实在空气中不断地扩散着，只是速率较慢而已，如果我们把大厅的所有出口都封起来，不让空气流走，并保持几个星期不动它，那么氨气就会均匀地散布在大厅的每个角落。

气体的这个相互扩散的性质是由普里斯特利最先指出的。道尔顿指出这种扩散的发生与相互扩散的气体间发生的具体化学反应无关。格雷姆的研究工作集中在那些能为分子运动提供线索的现象上。他仔细研究了扩散现象，最早得到了可以用来计算扩散速率的结果。

再晚些时候，维也纳的洛施密特教授精确地测量了气体分子相互扩散的速率。

他把两种不同气体分别装入两个相似的竖直试管中，轻的气体放在重的气体上面以防止对流产生。随后打开滑动阀门让两个试管相通。将它们静置大约一个小时以后，他关上阀门，然后测量每种气体中有多少已经扩散到了另一种气体中。

因为大多数气体都是不可见的，所以我得用氨气和盐酸这两种相遇后能生成固体反应物的气体来展示气体的扩散过程。氨气比较轻，所以放在盐酸上面，中间隔着一层空气。但是你们马上就可以看到，这两种气体能通过扩散穿过空气层而相遇，并产生一股白烟。整个过程看不到气体流动或者其他任何可见的运动。容器的每个部分都像一罐未受扰动的空气一样平静。

But, according to our theory, the same kind of motion is going on in calm air as in the inter-diffusing gases, the only difference being that we can trace the molecules from one place to another more easily when they are of a different nature from those through which they are diffusing.

If we wish to form a mental representation of what is going on among the molecules in calm air, we cannot do better than observe a swarm of bees, when every individual bee is flying furiously, first in one direction, and then in another, while the swarm, as a whole, either remains at rest, or sails slowly through the air.

In certain seasons, swarms of bees are apt to fly off to a great distance, and the owners, in order to identify their property when they find them on other people's ground, sometimes throw handfuls of flour at the swarm. Now let us suppose that the flour thrown at the flying swarm has whitened those bees only which happened to be in the lower half of the swarm, leaving those in the upper half free from flour.

If the bees still go on flying hither and thither in an irregular manner, the floury bees will be found in continually increasing proportions in the upper part of the swarm, till they have become equally diffused through every part of it. But the reason of this diffusion is not because the bees were marked with flour, but because they are flying about. The only effect of the marking is to enable us to identify certain bees.

We have no means of marking a select number of molecules of air, so as to trace them after they have become diffused among others, but we may communicate to them some property by which we may obtain evidence of their diffusion.

For instance, if a horizontal stratum of air is moving horizontally, molecules diffusing out of this stratum into those above and below will carry their horizontal motion with them, and so tend to communicate motion to the neighbouring strata, while molecules diffusing out of the neighbouring strata into the moving one will tend to bring it to rest. The action between the strata is somewhat like that of two rough surfaces, one of which slides over the other, rubbing on it. Friction is the name given to this action between solid bodies; in the case of fluids it is called internal friction or viscosity.

It is in fact only another kind of diffusion—a lateral diffusion of momentum, and its amount can be calculated from data derived from observations of the first kind of diffusion, that of matter. The comparative values of the viscosity of different gases were determined by Graham in his researches on the transpiration of gases through long narrow tubes, and their absolute values have been deduced from experiments on the oscillation of discs by Oscar Meyer and myself.

Another way of tracing the diffusion of molecules through calm air is to heat the upper stratum of the air in a vessel, and so observe the rate at which this heat is communicated

但是根据我们的理论，这种气体间相互扩散的运动进程，同样在平静的空气中发生着。区别只是当扩散发生在不同气体间时，追踪分子从一处到另一处的移动要容易一些。

如果我们想要在头脑中构思出一幅表现分子在平静空气中运动的图像，最好去观察一群蜜蜂，每只蜜蜂都拼命地飞来飞去，先朝一个方向飞，然后再朝另一个方向飞，但是整个蜂群不是停着不动，就是在空中缓慢地移动。

有的季节蜂群可以飞得很远，养蜂人为了能在别人的地盘上也能认出自己的蜂群，有时会向蜂群洒一把面粉。现在我们假定面粉正好把蜂群下面一半的蜜蜂染白，而上面一半没有沾上面粉。

如果这群蜜蜂继续散乱地飞来飞去，蜂群上半部就会有越来越多的沾上面粉的蜜蜂，直到它们均匀地分布于上下两部分。但是这种扩散的原因不是因为蜜蜂沾上了面粉,而是因为它们到处乱飞。沾面粉标记的目的只是为了帮助我们识别特定的蜜蜂。

我们没有办法标记一定数目的空气分子，使得它们在扩散到其他分子中之后还能被追踪到。但是我们可以用某些参数的传递来证明它们的扩散。

比如说，如果一个水平空气层向水平方向移动，那么从该层扩散到上下两个相邻层的分子将携带水平动量，并把这个动量传递给相邻的上下空气层。而从相邻空气层中扩散进入这个水平移动的水平层的分子，会减慢水平层的移动。相邻水平层之间的作用，就像两个粗糙固体表面之间的滑动和摩擦。固体之间的这种作用叫做摩擦，在流体的情况下则称为内摩擦或者黏滞力。

实际上这只不过是另一种类型的扩散——动量的横向扩散。其大小可以通过观察第一种扩散，也就是物质扩散的数据计算出来。研究气体通过长细管的蒸腾过程，格雷姆确定了不同气体的相对黏滞系数。黏滞系数的绝对值由奥斯卡·迈耶和我通过圆盘振动的实验结果推导得到。

另一种跟踪分子在平静空气中扩散的方法是加热容器顶层的空气，然后观察热量向下层传递的速率。这实际上是第三种类型的扩散——能量扩散。在直接进行热

to the lower strata. This, in fact, is a third kind of diffusion—that of energy, and the rate at which it must take place was calculated from data derived from experiments on viscosity before any direct experiments on the conduction of heat had been made. Prof. Stefan, of Vienna, has recently, by a very delicate method, succeeded in determining the conductivity of air, and he finds it, as he tells us, in striking agreement with the value predicted by the theory.

All these three kinds of diffusion—the diffusion of matter, of momentum, and of energy—are carried on by the motion of the molecules. The greater the velocity of the molecules and the farther they travel before their paths are altered by collision with other molecules, the more rapid will be the diffusion. Now we know already the velocity of the molecules, and therefore by experiments on diffusion we can determine how far, on an average, a molecule travels without striking another. Prof. Clausius, of Bonn, who first gave us precise ideas about the motion of agitation of molecules, calls this distance the mean path of a molecule. I have calculated, from Prof. Loschmidt's diffusion experiments, the mean path of the molecules of four well-know gases. The average distance travelled by a molecule between one collision and another is given in the table. It is a very small distance, quite imperceptible to us even with our best microscopes. Roughly speaking, it is about the tenth part of the length of a wave of light, which you know is a very small quantity. Of course the time spent on so short a path by such swift molecules must be very small. I have calculated the number of collisions which each must undergo in a second. They are given in the table and are reckoned by thousands of millions. No wonder that the travelling power of the swiftest molecule is but small, when its course is completely changed thousands of millions of times in a second.

The three kinds of diffusion also take place in liquids, but the relation between the rates at which they take place is not so simple as in the case of gases. The dynamical theory of liquids is not so well understood as that of gases, but the principal difference between a gas and a liquid seems to be that in a gas each molecule spends the greater part of its time in describing its free path, and is for a very small portion of its time engaged in encounters with other molecules, whereas in a liquid the molecule has hardly any free path, and is always in a state of close encounter with other molecules.

Hence in a liquid the diffusion of motion from one molecule to another takes place much more rapidly than the diffusion of the molecules themselves, for the same reason that it is more expeditious in a dense crowd to pass on a letter from hand to hand than to give it to a special messenger to work his way through the crowd. I have here a jar, the lower part of which contains a solution of copper sulphate, while the upper part contains pure water. It has been standing here since Friday, and you see how little progress the blue liquid has made in diffusing itself through the water above. The rate of diffusion of a solution of sugar has been carefully observed by Voit. Comparing his results with those of Loschmidt on gases, we find that about as much diffusion takes place in a second in gases as requires a day in liquids.

传导的实验之前，这种传递的速率是由黏滞系数的实验结果计算出来的。维也纳的斯特藩教授最近通过一个极其精巧的方法，成功地确定了空气的传导系数，他发现测量值和理论预测值惊人地符合。

所有这三种扩散——物质、动量和能量的扩散，都是由分子的运动完成的。分子的运动速率越大，在它与其他分子碰撞而使其运动方向发生改变之前所走的行程就越长，扩散的速率就越快。既然分子速率已知，通过扩散速率的实验，我们就可以确定一个分子在两次碰撞之间所走的平均距离。波恩的克劳修斯教授是第一个给出分子受激运动精确构想的人，他把这个距离叫做分子的平均自由程。根据洛施密特教授的扩散实验，我计算了 4 种常见气体的分子平均自由程。附表中列出了一个分子在两次碰撞之间的平均距离，这个距离非常之小，即使用最好的显微镜也不能分辨。它大概是光波长的 1/10，这样说你们就知道它有多小了。以分子的运动速率之大，行走这么短的距离，所用的时间当然是非常之短。我曾计算过一个分子每秒钟内碰撞的次数，结果也列在表中，都是几十亿次的水平。一秒钟之内方向要被改变几十亿次，难怪分子的运动速率虽大，却走不了多远。

这三种扩散过程在液体中也会发生，但是它们之间的速率关系就不像在气体中那么简单。液体的动力学理论不像气体的理论那么完善。看上去它们之间的根本区别是：相对而言，气体分子大部分时间是自由飞行，和其他分子发生碰撞的时间不多；液体分子则正好相反，平均自由程很短，总是在和附近的分子发生密切接触。

这样一来，液体中的动量扩散就比液体分子本身的扩散要快很多，这就好比在密集的人群中递送一封信，通过众人之手相传要比找一个专门的送信人穿越人群快一样。我这里有一个罐子，下半部装的是硫酸铜溶液，上半部装的是纯水。这个罐子从星期五就一直放在这里，而现在你们几乎看不出蓝色的硫酸铜扩散到上面的纯水中。沃伊特仔细计算了一种蔗糖溶液的扩散速率。把他的结果和洛施密特得到的气体扩散速率进行对比，我们发现在气体中一秒钟就能完成的扩散过程，在液体中则需要一整天。

The rate of diffusion of momentum is also slower in liquids than in gases, but by no means in the same proportion. The same amount of motion takes about ten times as long to subside in water as in air, as you will see by what takes place when I stir these two jars, one containing water and the other air. There is still less difference between the rates at which a rise of temperature is propagated through a liquid and through a gas.

In solids the molecules are still in motion, but their motions are confined within very narrow limits. Hence the diffusion of matter does not take place in solid bodies, though that of motion and heat takes place very freely. Nevertheless, certain liquids can diffuse through colloid solids, such as jelly and gum, and hydrogen can make its way through iron and palladium.

We have no time to do more than mention that most wonderful molecular motion which is called electrolysis. Here is an electric current passing through acidulated water, and causing oxygen to appear at one electrode and hydrogen at the other. In the space between, the water is perfectly calm, and yet two opposite currents of oxygen and of hydrogen must be passing through it. The physical theory of this process has been studied by Clausius, who has given reasons for asserting that in ordinary water the molecules are not only moving, but every now and then striking each other with such violence that the oxygen and hydrogen of the molecules part company, and dance about through the crowd, seeking partners which have become dissociated in the same way. In ordinary water these exchanges produce, on the whole, no observable effect, but no sooner does the electromotive force begin to act than it exerts its guiding influence on the unattached molecules, and bends the course of each toward its proper electrode, till the moment when, meeting with an unappropriated molecule of the opposite kind, it enters again into a more or less permanent union with it till it is again dissociated by another shock. Electrolysis, therefore, is a kind of diffusion assisted by electromotive force.

Another branch of molecular science is that which relates to the exchange of molecules between a liquid and a gas. It includes the theory of evaporation and condensation, in which the gas in question is the vapour of the liquid, and also the theory of the absorption of a gas by a liquid of a different substance. The researches of Dr. Andrews on the relations between the liquid and the gaseous state have shown us that though the statements in our own elementary text-books may be so neatly expressed that they appear almost self-evident, their true interpretation may involve some principle so profound that, till the right man has laid hold of it, no one ever suspects that anything is left to be discovered.

These, then, are, some of the fields from which the data of molecular science are gathered. We may divide the ultimate results into three ranks, according to the completeness of our knowledge of them.

To the first rank belong the relative masses of the molecules of different gases, and their velocities in metres per second. These data are obtained from experiments on the pressure and density of gases, and are known to a high degree of precision.

液体中动量的扩散速率也比气体中动量的扩散速率慢，但是没有差那么多。等量的运动在水中的衰减时间大概是空气中的 10 倍。让我来搅动一下这两个罐子，一个装的是水，另一个只有空气，大家看看结果是什么？液体和气体在温度传递上的速率差别还要更小一些。

固体中的分子也在不停地运动，但是它们的运动被限制在很小的范围内。所以物质扩散不会发生在固体中，但动量和能量扩散可以非常自由地进行。不过某些液体可以渗过像果冻和树胶一类的胶质固体，而氢气能够透过铁和钯。

时间有限，我们只能简单提一下电解这个最奇妙的分子运动现象。这里有电流通过酸性的水时，两个电极分别产生氧气和氢气。电极之间的水是完全静止的，但是其中必然有氧和氢的相对流动。克劳修斯研究了这一过程的物理学原理，给出理由认为普通水中的分子不但是运动的，而且相互碰撞的力量还很大，造成水分子中氧和氢的分离，分离的氧和氢在水中四处游动，寻找因同样原因而游离的其他对象来结合。在普通水中，这个交换过程在整体上没有造成可观测的效应。但是一旦电动势开始起作用，就对独立分子的运动产生定向影响，驱使游离分子向相应的电极运动，直到碰上异性游离分子而形成相对稳定的结合体。这个结合体还有可能再次被冲击离解。这样来看，电解是一种电动势协助下的分子扩散过程。

分子科学的另一个分支是关于液体和气体之间的分子交换。它既包括同一种物质气液两态之间的蒸发和凝结理论，也包括液体吸收不同种物质气体分子的理论。安德鲁斯博士关于气液两态之间关系的研究表明，虽然我们基本教科书中的结论看上去如此简洁，似乎是天经地义，但其实际的意义可能包含着非常深奥的原理。在有合适的人将其搞清楚之前，人们认为一切已经尽善尽美了。

以上就是分子科学取得数据成果的一些领域。我们根据完整的相关知识，把这些最终的成果分成三个等级。

第一个等级中有不同气体分子的相对质量以及它们以米每秒为单位的速度。这些数据是根据气体压强和密度的实验结果得出的，精度很高。

In the second rank we must place the relative size of the molecules of different gases, the length of their mean paths, and the number of collisions in a second. These quantities are deduced from experiments on the three kinds of diffusion. Their received values must be regarded as rough approximations till the methods of experimenting are greatly improved.

There is another set of quantities which we must place in the third rank, because our knowledge of them is neither precise, as in the first rank, nor approximate, as in the second, but is only as yet of the nature of a probable conjecture. These are the absolute mass of a molecule, its absolute diameter, and the number of molecules in a cubic centimetre. We know the relative masses of different molecules with great accuracy, and we know their relative diameters approximately. From these we can deduce the relative densities of the molecules themselves. So far we are on firm ground.

The great resistance of liquids to compression makes it probable that their molecules must be at about the same distance from each other as that at which two molecules of the same substance in the gaseous form act on each other during an encounter. This conjecture has been put to the test by Lorenz Meyer, who has compared the densities of different liquids with the calculated relative densities of the molecules of their vapours, and has found a remarkable correspondence between them.

Now Loschmidt has deduced from the dynamical theory the following remarkable proportion:—As the volume of a gas is to the combined volume of all the molecules contained in it, so is the mean path of a molecule to one-eighth of the diameter of a molecule.

Assuming that the volume of the substance, when reduced to the liquid form, is not much greater than the combined volume of the molecules, we obtain from this proportion the diameter of a molecule. In this way Loschmidt, in 1865, made the first estimate of the diameter of a molecule. Independently of him and of each other, Mr. Stoney in 1868, and Sir W. Thomson in 1870, published results of a similar kind, those of Thomson being deduced not only in this way, but from considerations derived from the thickness of soap bubbles, and from the electric properties of metals.

According to the table, which I have calculated from Loschmidt's data, the size of the molecules of hydrogen is such that about two million of them in a row would occupy a millimetre, and a million million million million of them would weigh between four and five grammes.

In a cubic centimetre of any gas at standard pressure and temperature there are about nineteen million million million molecules. All these numbers of the third rank are, I need not tell you, to be regarded as at present conjectural. In order to warrant us in putting any confidence in numbers obtained in this way, we should have to compare together a greater number of independent data than we have as yet obtained, and to show that they lead to consistent results.

在第二个等级中，我们应该归入的是不同种气体分子的相对尺寸、平均自由程和每秒钟的碰撞次数。这些参数的量值是从三种扩散的实验结果推导出来的。在实验方法得到极大改进之前，这些公认的结果只能被看作是大概的近似。

还有一批参数只能放进第三个等级，这是因为我们在这个等级中的相关知识只是基于可能的推测，既不像第一等级中的那么精确，也不像第二等级中的那样近似精确。这里面包括分子的绝对质量、绝对直径和 1 立方厘米中的分子数目。我们有不同分子的相对质量的准确结果，也大概知道它们的相对直径。由此我们可以得到分子的相对密度。这些都是确实有据的结果。

液体强大的抗压缩性似乎表明：其分子之间的距离已经很近，和同一物质的两个气态分子在碰撞时发生相互作用的距离差不多。洛伦茨 · 迈耶想了一个办法验证这个猜想，他对各种液体的密度和相应液体蒸气的相对密度的计算值进行比较，发现它们明显相关。

现在洛施密特根据动力学原理推导出一个不寻常的比例关系：气体的体积和该气体中所有分子体积之和的比值，等于分子平均自由程与其直径之比的 1/8。

假定气体液化后的体积，比相应分子的体积之和大不了多少，我们就可以通过这个比例关系得到分子的直径。1865 年洛施密特由此第一次估计出了分子的直径。斯托尼先生在 1868 年，汤姆孙爵士于 1870 年也都各自独立发表了类似的结果。后者除了以上的考虑，还考虑了与肥皂泡厚度以及金属电特性有关的结果。

表中的结果是我根据洛施密特的数据计算得到的。按照这个表，200 万个氢分子排成一列，只有 1 毫米长。100 万的四次方那么多的氢分子合起来的质量只有 4~5 克。

在标准压强和温度下，1 立方厘米的任何气体都含有 1.9×10^{19} 那么多的分子。这些量值都属于前面所说的第三个等级，不用我说你也知道，它们只是当前的推论。除非由独立方法得到的比现在多得多的数据经过比较后都趋向一致的结果，否则我们不能轻易接受以上的结果。

Thus far we have been considering molecular science as an inquiry into natural phenomena. But though the professed aim of all scientific work is to unravel the secrets of nature, it has another effect, not less valuable, on the mind of the worker. It leaves him in possession of methods which nothing but scientific work could have led him to invent, and it places him in a position from which many regions of nature, besides that which he has been studying, appear under a new aspect.

The study of molecules has developed a method of its own, and it has also opened up new views of nature.

When Lucretius wishes us to form a mental representation of the motion of atoms, he tells us to look at a sunbeam shining through a darkened room (the same instrument of research by which Dr. Tyndall makes visible to us the dust we breathe,) and to observe the motes which chase each other in all directions through it. This motion of the visible motes, he tells us, is but a result of the far more complicated motion of the invisible atoms which knock the motes about. In his dream of nature, as Tennyson tells us, he

> "saw the flaring atom-streams
> And torrents of her myriad universe,
> Ruining along the illimitable inane,
> Fly on to clash together again, and make
> Another and another frame of things
> For ever."

And it is no wonder that he should have attempted to burst the bonds of Fate by making his atoms deviate from their courses at quite uncertain times and places, thus attributing to them a kind of irrational free will, which on his materialistic theory is the only explanation of that power of voluntary action of which we ourselves are conscious.

As long as we have to deal with only two molecules, and have all the data given us, we can calculate the result of their encounter, but when we have to deal with millions of molecules, each of which has millions of encounters in a second, the complexity of he problem seems to shut out all hope of a legitimate solution.

The modern atomists have therefore adopted a method which is I believe new in the department of mathematical physics, though it has long been in use in the Section of Statistics. When the working members of Section F get hold of a Report of the Census, or any other document containing the numerical data of Economic and Social Science, they begin by distributing the whole population into groups, according to age, income-tax, education, religious belief, or criminal convictions. The number of individuals is far too great to allow of their tracing the history of each separately, so that, in order to reduce their labour within human limits, they concentrate their attention on a small number of artificial groups. The varying number of individuals in each group, and not the varying state of each individual, is the primary datum from which they work.

到现在为止，我们一直认为分子科学是对自然现象的一种探索。虽然所有科学工作的目的都是为了揭示自然的秘密，但是它还有另外一个至少同样重要的作用，就是对科学工作者心灵的影响。只有科学工作才能让他们发明并掌握新的方法，这使他们站在了一个新的高度，看到除了自己的研究领域之外，其他许多自然领域也呈现出新的面貌。

分子领域的研究发展了一套自己的方法，也打开了一扇观察自然的窗户。

为了让我们在头脑中建立一幅原子运动的图像，卢克莱修让我们观察一束射入暗室的日光（廷德尔博士用同样的设备展示了我们呼吸的灰尘的运动情况），在这束日光中我们可以看到向各个方向追逐乱窜的尘埃。他告诉我们，这些可见尘埃的运动，是更加复杂的不可见的原子运动不断将它们撞来撞去的结果。就像丁尼生告诉我们的一样，在卢克莱修梦想的自然中，他

“看见闪耀的原子流
和她在无穷宇宙中的激流，
在无尽的虚空中衰耗，
又飞来撞在一起，
创造一个又一个事物的构架，
永不止息。”

难怪他试图打破必然的枷锁，让他的原子在很不确定的时刻和地点改变轨迹，因而赋予它们一种非理性的自由意志，这就是他在物质理论中对我们所知的随机行为的产生所做的唯一解释。

如果我们只需要研究两个分子，而且知道所有的数据，我们就可以计算它们相互碰撞的结果。但实际上有极多的分子，每个分子每秒钟都要经历极多的碰撞，问题的复杂性似乎超越了任何合理解答的可能性。

现代原子论者采取了一种我认为在数学物理学科中是全新的方法，尽管它在统计部门中已经应用很久了。当F部门的工作人员得到一份人口普查报告，或者是其他包含经济和社会科学统计数据的文件时，他们先把整个人口按照年龄、所得税、教育水平、宗教信仰或犯罪记录分组。人口的数量太大，很难一一跟踪每个人的实际情况。为了在人力资源有限的情况下把工作量控制在合理范围，他们把注意力集中在少数几个人为划分出来的组群上，每个组群个体人数的变化，而非每个个体的状态变化是他们进行研究工作的最初数据。

This, of course, is not the only method of studying human nature. We may observe the conduct of individual men and compare it with that conduct which their previous character and their present circumstances, according to the best existing theory, would lead us to expect. Those who practise this method endeavour to improve their knowledge of the elements of human nature, in much the same way as an astronomer corrects the elements of a planet by comparing its actual position with that deduced from the received elements. The study of human nature by parents and schoolmasters, by historians and statesmen, is therefore to be distinguished from that carried on by registrars and tabulators, and by those statesmen who put their faith in figures. The one may be called the historical, and the other the statistical method.

The equations of dynamics completely express the laws of the historical method as applied to matter, but the application of these equations implies a perfect knowledge of all the data. But the smallest portion of matter which we can subject to experiment consists of millions of molecules, not one of which ever becomes individually sensible to us. We cannot, therefore, ascertain the actual motion of any one of these molecules, so that we are obliged to abandon the strict historical method, and to adopt the statistical method of dealing with large groups of molecules.

The data of the statistical method as applied to molecular science are the sums of large numbers of molecular quantities. In studying the relations between quantities of this kind, we meet with a new kind of regularity, the regularity of averages, which we can depend upon quite sufficiently for all practical purposes, but which can make no claim to that character of absolute precision which belongs to the laws of abstract dynamics.

Thus molecular science teaches us that our experiments can never give us anything more than statistical information, and that no law deduced from them can pretend to absolute precision. But when we pass from the contemplation of our experiments to that of the molecules themselves, we leave the world of chance and change, and enter a region where everything is certain and immutable.

The molecules are conformed to a constant type with a precision which is not to be found in the sensible properties of the bodies which they constitute. In the first place the mass of each individual molecule, and all its other properties, are absolutely unalterable. In the second place the properties of all molecules of the same kind are absolutely identical.

Let us consider the properties of two kinds of molecules, those of oxygen and those of hydrogen.

We can procure specimens of oxygen from very different sources—from the air, from water, from rocks or every geological epoch. The history of these specimens has been very different, and if, during thousands of years, difference of circumstances could produce difference of properties, these specimens of oxygen would show it.

这显然不是研究人类性质的唯一方法。我们可以观察个体的行为，也可以按照当前最好的理论，根据以往的特征预测目前情况下的行为，并把观察结果和预测的行为相比较。人们采用这种方法以增进他们对人类性质各个方面的认识，就像天文学家通过对比行星的实际位置和依据已知参数预测的位置之间的差别来修正轨道参数一样。父母、校长、历史学家、政治家对人性的研究，和登记员、制表人还有注重数据的政治家的研究是不同的。一个可以叫历史方法，另一个则是统计方法。

动力学方程完全是历史方法定律在物质研究上的应用，但应用这些方程需要知道所有的数据，然而哪怕是实验中物质最小的部分也包含着极多我们看不见的分子。我们不可能确定任何一个分子的实际运动状况，所以必须放弃严格的历史方法，采用统计方法来处理大量分子的情况。

应用在分子科学中的统计方法牵涉的数据是大量分子的参量的总和。在研究这类参量之间的关系时，我们碰上了一种新的规律，也就是平均值的规律。对于实际应用，依赖这些规律就足够了，但是它们不能给出理论动力学定律所具有的绝对精确性。

分子科学告诉我们：实验永远只能提供统计信息，从实验中总结的定律都没有绝对的精确性。但是当我们的关注点从实验转向分子本身时，我们就离开了充满偶然性和变化的世界，进入到一切都是确定不变的领域。

分子都精确地具有恒定不变的属性，这种属性是分子所组成的物体的可测宏观参数所不具备的。首先，所有分子的质量和所有其他性质都是绝对不变的。其次，同种物质所有分子的所有性质都是绝对相同的。

让我们来考虑一下氧和氢这两种分子的性质。

我们可以从空气、水和各个地质时代的岩石等不同来源制备氧气样品。这些样品形成的历史完全不同。如果在几千年的时间里，不同的环境能造就不同的性质，这些氧气样品应该就能表现出这些不同的性质。

In like manner we may procure hydrogen from water, from coal, or, as Graham did, from meteoric iron. Take two litres of any specimen of hydrogen, it will combine with exactly one litre of any specimen of oxygen, and will form exactly two litres of the vapour of water.

Now if, during the whole previous history of either specimen, whether imprisoned in the rocks, flowing in the sea, or careering through unknown regions with the meteorites, any modification of the molecules had taken place, these relations would no longer be preserved.

But we have another and an entirely different method of comparing the properties of molecules. The molecule, though indestructible, is not a hard rigid body, but is capable of internal movements, and when these are excited it emits rays, the wavelength of which is a measure of the time of vibration of the molecule.

By means of the spectroscope the wavelengths of different kinds of light may be compared to within one ten-thousandth part. In this way it has been ascertained, not only that molecules taken from every specimen of hydrogen in our laboratories have the same set of periods of vibration, but that light, having the same set of periods of vibration, is emitted from the sun and from the fixed stars.

We are thus assured that molecules of the same nature as those of our hydrogen exist in those distant regions, or at least did exist when the light by which we see them was emitted.

From a comparison of the dimensions of the buildings of the Egyptians with those of the Greeks, it appears that they have a common measure. Hence, even if no ancient author had recorded the fact that the two nations employed the same cubit as a standard of length, we might prove it from the buildings themselves. We should also be justified in asserting that at some time or other a material standard of length must have been carried from one country to the other, or that both countries had obtained their standards from a common source.

But in the heavens we discover by their light, and by their light alone, stars so distant from each other that no material thing can ever have passed from one to another, and yet this light, which is to us the sole evidence of the existence of these distant worlds, tells us also that each of them is built up of molecules of the same kinds as those which we find on earth. A molecule of hydrogen, for example, whether in Sirius or in Arcturus, executes its vibrations in precisely the same time.

Each molecule, therefore, throughout the universe, bears impressed on it the stamp of a metric system as distinctly as does the metre of the Archives at Paris, or the double royal cubit of the Temple of Karnac.

同样地，我们可以从水、煤炭或者像格雷姆那样从陨铁中制备出氢气样品。任取两升氢气样品，它都可以和任取的一升氧气样品反应，正好生成两升水蒸气。

如果在任何一个样品的整个历史中，不论它是固锁在岩石中，还是漂流在海洋里，或者是跟着陨石穿越着未知的区域，只要分子发生了任何变化，上面的关系都不能再保持。

我们有另一种完全不同的办法来比较分子的性质。分子虽然不能被毁灭，但也不是刚性实体，它有内部运动，当内部运动被激发时，分子发出射线，这个射线的波长表征了分子振动的周期。

光谱仪可以比较不同光的波长差别，精度可达万分之一。用这个办法，我们可以确认，不但我们实验室中每一个氢气样品中的分子具有同样的振动周期组合，就连太阳和其他恒星发射的光也存在同样的振动周期组合。

这样我们就确信了在茫茫的宇宙中也存在和我们的氢分子一样的分子，或者说至少在那些我们用来观察分子的光线放射出时，它们是存在的。

比较古埃及和古希腊的建筑规模，可以感觉到它们似乎使用了同样的度量标准。因此，尽管历史上没有留下两个国家都使用同样的肘长作为长度标准的记载，我们也许可以通过建筑本身证明这一点。我们也有理由认为，肯定是某个实物长度标准在某个时候被人从一个国家带到了另一个国家，或者两个国家从同一个来源得到了这个标准。

对天空中的星体，我们只是通过它们发射的光线证实它们的存在，它们之间的距离如此遥远，物质从一个恒星到另一个恒星的传递是完全不可能的，但是它们发射的光线，除了作为证明这些遥远天体存在的唯一证据以外，还告诉我们，组成每一个天体的分子都和我们在地球上发现的分子相同。比如，不论是天狼星还是大角星，上面的氢分子都以同样的周期振动。

宇宙中的每一个分子，都铭刻着一个度量系统的痕迹，它的独特清晰，就如同巴黎档案局的米原尺，或者卡尔纳克神庙的皇家双肘尺一样。

No theory of evolution can be formed to account for the similarity of molecules, for evolution necessarily implies continuous change, and the molecule is incapable of growth or decay, of generation or destruction.

None of the processes of Nature, since the time when Nature began, have produced the slightest difference in the properties of any molecule. We are therefore unable to ascribe either the existence of the molecules or the identity of their properties to the operation of any of the causes which we call natural.

On the other hand, the exact quality of each molecule to all others of the same kind gives it, as Sir John Herschel has well said, the essential character of a manufactured article, and precludes the idea of its being eternal and self existent.

Thus we have been led, along a strictly scientific path, very near to the point at which Science must stop. Not that Science is debarred from studying the internal mechanism of a molecule which she cannot take to pieces, any more than from investigating an organism which she cannot put together. But in tracing back the history of matter Science is arrested when she assures herself, on the one hand, that the molecule has been made, and on the other that it has not been made by any of the processes we call natural.

Science is incompetent to reason upon the creation of matter itself out of nothing. We have reached the utmost limit of our thinking faculties when we have admitted that because matter cannot be eternal and self-existent it must have been created.

It is only when we contemplate, not matter in itself, but the form in which it actually exists, that our mind finds something on which it can lay hold.

That matter, as such, should have certain fundamental properties—that it should exist in space and be capable of motion, that its motion should be persistent, and so on, are truths which may, for anything we know, be of the kind which metaphysicians call necessary. We may use our knowledge of such truths for purposes of deduction but we have no data for speculating as to their origin.

But that there should be exactly so much matter and no more in every molecule of hydrogen is a fact of a very different order. We have here a particular distribution of matter—a *collocation*—to use the expression of Dr. Chalmers, of things which we have no difficulty in imagining to have been arranged otherwise.

The form and dimensions of the orbits of the planets, for instance, are not determined by any law of nature, but depend upon a particular collocation of matter. The same is the case with respect to the size of the earth, from which the standard of what is called the metrical system has been derived. But these astronomical and terrestrial magnitudes are far inferior in scientific importance to that most fundamental of all standards which forms

任何进化论都不能解释分子的相似性，因为进化隐含着不断地变化，而分子既不生长也不腐朽，不能被制造也不能被摧毁。

自从自然界形成以来，自然界的所有过程都没能使任何分子的性质有丝毫的改变。所以我们不能把分子的存在或其性质的同一性归结为任何自然原因的作用。

另一方面，正像约翰·赫歇尔爵士说的那样，同类分子精确相同的性质是其所构成的物质的根本特征，这就排除了分子的永恒自存在性。

这样一来，我们就被引领着，沿着一条严格的科学道路，走到了科学的尽头。这不是说科学手段不能分解分子就不能用来研究分子的内部机制，就像不能因为科学不能组合成生物就不用她来研究生物一样。但在追溯物质的历史时，科学被困住了，她一方面确认分子已被创造，另一方面又确认任何自然界的过程都不能创造分子。

科学不能理解物质如何被无中生有地创造出来。当我们承认由于物质不能永恒自存在，因此它一定已经被创造出来的时候，我们已经到了自己思维能力的极限。

只有当我们思考物质的存在形式，而不是物质本身的时候，我们的思维才算找到了一些可以解决问题的线索。

我们知道，形而上学者认为必要的真理是：物质必须具有一些基本的性质——它应该存在于空间中，能够运动，运动应该是持续的等等。我们可以利用已知的与这些真理有关的知识进行推导，但是并无数据可以用来猜测这些真理的本源。

但是每一个氢分子的质量都正好是这么多，这个事实就完全是另一个性质了。这种质量的分布，或者用查默斯博士的词语——这种**配置**，实在是太特别了，让我们都不习惯。

比如，行星轨道的形式和尺度，也不是由任何自然定律决定的，而是取决于特殊的质量的布置。作为米制计量体系标准的地球的尺寸也是这样。但是这些天文和地理的量值在科学上的重要性都远不能和构成分子系统基础的那些最基本的标准相比。我们知道自然界过程一直在起作用，就算最后不摧毁，它们至少也要改变地球

the base of the molecular system. Natural causes, as we know, are at work, which tend to modify, if they do not at length destroy, all the arrangements and dimensions of the earth and the whole solar system. But though in the course of ages catastrophes have occurred and may yet occur in the heavens, though ancient systems may be dissolved and new systems evolved out of their ruins, the molecules out of which these systems are built—the foundation stones of the material universe—remain unbroken and unworn.

They continue this day as they were created, perfect in number and measure and weight, and from the ineffaceable characters impressed on them we may learn that those aspirations after accuracy in measurement, truth in statement, and justice in action, which we reckon among our noblest attributes as men, are ours because they are essential constituents of the image of Him who in the beginning created, not only the heaven and the earth, but the materials of which heaven and earth consist.

Table of Moncular Date

		Hydrogen	Oxygen	Carbonic oxide	Carbonic acid
Rank I	Mass of molecule (hydrogen = 1)	1	16	14	22
	Velocity (of mean square), metres per second at 0℃	1859	465	497	396
Rank II	Mean path, tenth-metres.	965	560	482	379
	Collisions in a second, (millions)	17750	7646	9489	9720
Rank III	Diameter, tenth-metre	5.8	7.6	8.3	9.3
	Mass, twenty-fifth-grammes.	46	736	644	1012

Table of Diffusion: $\frac{(\text{centimetre})^2}{\text{second}}$ measure

	Calculated	Observed	
H& O	0.7086	0.7214	Diffusion of matter observed by Loschmidt.
H& CO	0.6519	0.6422	
H& CO_2	0.5575	0.5558	
O& CO	0.1807	0.1802	
O& CO_2	0.1427	0.1409	
CO & CO_2	0.1386	0.1406	
H	1.2990	1.49	Diffusion of momentum Graham and Meyer.
O	0.1884	0.213	
CO	0.1748	0.212	
CO_2	0.1087	0.117	
Air		0.256	Diffusion of temperature observed by Stefan.
Copper		1.077	
Iron		0.183	
Cane sugar in water		0.00000365	Voit
Diffusion in a day		0.3144	
Salt in water		0.00000116	Fick

(**8**, 437-441; 1873)

和整个太阳系的秩序和尺寸。然而，尽管在历史上宇宙曾经发生过灾变，以后也可能再发生，虽然旧的体系可能被消灭，新的体系在旧体系的废墟上演化发展，但作为这些系统组成基础的分子——物质宇宙的基石，却不会被破坏和磨损。

它们被创造的时候是什么样，现在还是什么样，数量、大小和质量都没有丝毫改变，从铭记在它们身上的不可磨灭的特征，我们也许能够明白，我们对测量的精准、言语的真切、行为的正义这些人类最高尚品质的追求，是因为它们也是造物主形象的根本组成部分。他当初不仅创造了天和地，也创造了组成天和地的所有物质。

分子数据表

		氢气	氧气	一氧化碳	二氧化碳
第一等级	分子质量（氢气=1）	1	16	14	22
	0℃下的速度（均方根），米 / 秒	1859	465	497	396
第二等级	平均自由程，10^{-10}米	965	560	482	379
	每秒碰撞次数（百万次）	17750	7646	9489	9720
第三等级	直径，10^{-10}米	5.8	7.6	8.3	9.3
	质量，10^{-25}克	46	736	644	1012

扩散数据表：单位是 $\frac{(\text{厘米})^2}{\text{秒}}$

	计算值	测量值	
氢 & 氧	0.7086	0.7214	洛施密特观察的物质扩散
氢 & 一氧化碳	0.6519	0.6422	
氢 & 二氧化碳	0.5575	0.5558	
氧 & 一氧化碳	0.1807	0.1802	
氧 & 二氧化碳	0.1427	0.1409	
一氧化碳 & 二氧化碳	0.1386	0.1406	
氢	1.2990	1.49	格雷姆和迈耶的动量扩散
氧	0.1884	0.213	
一氧化碳	0.1748	0.212	
二氧化碳	0.1087	0.117	
空气		0.256	斯特藩观察的温度扩散
铜		1.077	
铁		0.183	
蔗糖在水溶液中		0.00000365	沃伊特
一天后扩散的距离		0.3144	
食盐在水溶液中		0.00000116	菲克

（何钧 翻译；鲍重光 审稿）

On the Dynamical Evidence of the Molecular Constitution of Bodies*

J. C. Maxwell

Editor's Note

Here James Clerk Maxwell offers an update on efforts to understand the properties of gases and liquids from first principles—what today is called statistical mechanics. Maxwell had made a decisive contribution with his kinetic theory of gases. Here he reports on other work, such as that of Rudolph Clausius, who introduced a theoretical quantity known as the "virial", representing an average over all particle pairs of the product of the distance between the particles and the force of attraction or repulsion. Maxwell also mentions the work of two other figures, an unknown graduate student named Johannes Diderik van der Waals and an American physicist, Josiah Willard Gibbs. Both later became leaders in understanding the fundamental behavior of material systems.

I

WHEN any phenomenon can be described as an example of some general principle which is applicable to other phenomena, that phenomenon is said to be explained. Explanations, however, are of very various orders, according to the degree of generality of the principle which is made use of. Thus the person who first observed the effect of throwing water into a fire would feel a certain amount of mental satisfaction when he found that the results were always similar, and that they did not depend on any temporary and capricious antipathy between the water and the fire. This is an explanation of the lowest order, in which the class to which the phenomenon is referred consists of other phenomena which can only be distinguished from it by the place and time of their occurrence, and the principle involved is the very general one that place and time are not among the conditions which determine natural processes. On the other hand, when a physical phenomenon can be completely described as a change in the configuration and motion of a material system, the dynamical explanation of that phenomenon is said to be complete. We cannot conceive any further explanation to be either necessary, desirable, or possible, for as soon as we know what is meant by the words configuration, motion, mass, and force, we see that the ideas which they represent are so elementary that they cannot be explained by means of anything else.

The phenomena studied by chemists are, for the most part, such as have not received a complete dynamical explanation.

---

* A lecture delivered at the Chemical Society, Feb. 18, by Prof. Clerk Maxwell, F. R. S.

关于物体分子构成的动力学证据*

麦克斯韦

编者按

在这篇文章中，克拉克·麦克斯韦介绍了从基本原理（现在的名字是统计力学）出发研究气体和液体性质的最新进展。此前，麦克斯韦提出了他的气体动力学理论，这是他作出的决定性的贡献。在本文中他报道了鲁道夫·克劳修斯等其他研究人员的工作。克劳修斯引入了“位力”这个理论物理量，用来表示所有粒子对的粒子间距离与粒子间引力或斥力乘积的总平均。麦克斯韦在这篇文章中还提到了另外两个人的工作，其中一个是当时还没什么名气的研究生约翰内斯·迪德里克·范德瓦尔斯，另一个是美国物理学家约西亚·威拉德·吉布斯，这两人后来都成了研究物质体系基本行为特征的领袖人物。

I

任何一种现象，一旦可以被视为某种适用于其他现象的普遍原理的实例，我们就可以说这种现象得到了解释。然而，解释包括很多不同层面，要视所采用原理的普适程度而定。因此，第一个观察到将水泼入火中这种现象的人，在发现结果总是相似的而且这些结果并不取决于水和火之间那瞬息万变、反复无常的不相容性时，便会获得一定程度的精神满足。这是一种最低层次的解释，在这种解释中，以上现象所涉及的分类还包含其他现象，而这些现象只是发生的时间和地点不同，所涉及的原理却是非常普适的，因此时间和地点并不属于能够决定自然过程的条件。另一方面，当某一物理现象能够被完全描述为物质系统结构或运动的变化时，我们就可以说这种现象得到了完整的动力学解释。我们想不出还有什么更进一步的解释是必要的、有用的或者可能的，因为一旦我们了解了结构、运动、质量和力这些词的意义，便会发现它们所表示的意义如此基本，以至于无法再用其他概念来解释它们。

化学家们研究的现象，大部分还没有得到完整的动力学解释。

* 2月18日，克拉克·麦克斯韦教授（皇家学会会员）在化学学会所作的讲座。

Many diagrams and models of compound molecules have been constructed. These are the records of the efforts of chemists to imagine configurations of material systems by the geometrical relations of which chemical phenomena may be illustrated or explained. No chemist, however, professes to see in these diagrams anything more than symbolic representations of the various degrees of closeness with which the different components of the molecule are bound together.

In astronomy, on the other hand, the configurations and motions of the heavenly bodies are on such a scale that we can ascertain them by direct observation. Newton proved that the observed motions indicate a continual tendency of all bodies to approach each other, and the doctrine of universal gravitation which he established not only explains the observed motions of our system, but enables us to calculate the motions of a system in which the astronomical elements may have any values whatever.

When we pass from astronomical to electrical science, we can still observe the configuration and motion of electrified bodies, and thence, following the strict Newtonian path, deduce the forces with which they act on each other; but these forces are found to depend on the distribution of what we call electricity. To form what Gauss called a "construirbar Vorstellung" of the invisible process of electric action is the great desideratum in this part of science.

In attempting the extension of dynamical methods to the explanation of chemical phenomena, we have to form an idea of the configuration and motion of a number of material systems, each of which is so small that it cannot be directly observed. We have, in fact, to determine, from the observed external actions of an unseen piece of machinery, its internal construction.

The method which has been for the most part employed in conducting such inquiries is that of forming an hypothesis, and calculating what would happen if the hypothesis were true. If these results agree with the actual phenomena, the hypothesis is said to be verified, so long, at least, as some one else does not invent another hypothesis which agrees still better with the phenomena.

The reason why so many of our physical theories have been built up by the method of hypothesis is that the speculators have not been provided with methods and terms sufficiently general to express the results of their induction in its early stages. They were thus compelled either to leave their ideas vague and therefore useless, or to present them in a form the details of which could be supplied only by the illegitimate use of the imagination.

In the meantime the mathematicians, guided by that instinct which teaches them to store up for others the irrepressible secretions of their own minds, had developed with the utmost generality the dynamical theory of a material system.

很多化合物分子的图示或模型已经建立。它们都是化学家根据可以说明或解释的化学现象的几何关系努力猜想物质系统的结构而得到的结果。不过，这些图示只不过是对构成分子的不同成分束缚在一起的各种不同的紧密程度的符号表示而已，没有一位化学家敢说他能从中看出更多的东西。

另一方面，在天文学中，天体的结构和运动所具有的尺度是我们可以通过直接观测来确定的。牛顿证明了观测到的运动显示出所有物体之间都有不断彼此靠拢的趋势，他建立的万有引力定律不仅解释了观测到的我们自身所在系统的运动，而且使我们能够计算出具有任意量值的天文学对象所组成的系统的运动情况。

从天文学转向电学，我们仍然可以观测带电体的结构和运动，并遵循严格的牛顿式方法来推导出它们的相互作用力。但是我们发现这种力依赖于所谓的电的分布。因此，建立一种关于电作用不可见过程的、被高斯称为“统一表象”的理论，就成为这门科学的迫切需要。

在试图将动力学方法推广以用于对化学现象的解释时，我们必须形成关于大量物质系统的结构和运动的观念，而其中每一个物质系统都小到无法直接观测的程度。事实上，我们只能通过观测那些看不见的结构部分的外部作用来确定其内部结构。

主要用来进行此类研究的方法是先建立一个假说，然后计算如果该假说成立会产生什么结果。若这些结果与实际现象相吻合，我们就说这个假说得到了确证，至少在没有其他人提出与客观现象更一致的另一种假说之前，是可以这样说的。

之所以有如此多的物理理论依靠假说的方式建立起来，是因为理论家早期没有获得具有足够普适性的方法和术语来表达他们归纳出的结果。受此限制，他们要么将观点表达得含糊不清以至于毫无用处，要么用另一种形式表达观点，而这种形式的细节只能靠非常规的想象力来弥补。

与此同时，数学家在其本能（这种本能使他们将自身才智孕育出的丰硕果实供给他人）的驱使下已经发展出了具有充分普适性的关于物质系统的动力学理论。

Of all hypotheses as to the constitution of bodies, that is surely the most warrantable which assumes no more than that they are material systems, and proposes to deduce from the observed phenomena just as much information about the conditions and connections of the material system as these phenomena can legitimately furnish.

When examples of this method of physical speculation have been properly set forth and explained, we shall hear fewer complaints of the looseness of the reasoning of men of science, and the method of inductive philosophy will no longer be derided as mere guess-work.

It is only a small part of the theory of the constitution of bodies which has as yet been reduced to the form of accurate deductions from known facts. To conduct the operations of science in a perfectly legitimate manner, by means of methodised experiment and strict demonstration, requires a strategic skill which we must not look for, even among those to whom science is most indebted for original observations and fertile suggestions. It does not detract from the merit of the pioneers of science that their advances, being made on unknown ground, are often cut off, for a time, from that system of communications with an established base of operations, which is the only security for any permanent extension of science.

In studying the constitution of bodies we are forced from the very beginning to deal with particles which we cannot observe. For whatever may be our ultimate conclusions as to molecules and atoms, we have experimental proof that bodies may be divided into parts so small that we cannot perceive them.

Hence, if we are careful to remember that the word particle means a small part of a body, and that it does not involve any hypothesis as to the ultimate divisibility of matter, we may consider a body as made up of particles, and we may also assert that in bodies or parts of bodies of measurable dimensions, the number of particles is very great indeed.

The next thing required is a dynamical method of studying a material system consisting of an immense number of particles, by forming an idea of their configuration and motion, and of the forces acting on the particles, and deducing from the dynamical theory those phenomena which, though depending on the configuration and motion of the invisible particles, are capable of being observed in visible portions of the system.

The dynamical principles necessary for this study were developed by the fathers of dynamics, from Galileo and Newton to Lagrange and Laplace; but the special adaptation of these principles to molecular studies has been to a great extent the work of Prof. Clausius of Bonn, who has recently laid us under still deeper obligations by giving us, in addition to the results of his elaborate calculations, a new dynamical idea, by the aid of which I hope we shall be able to establish several important conclusions without much symbolical calculation.

在所有关于物体构成的假说中，最可信的无疑是这样一种假说：它除了假定物体构成物质系统之外绝无更多假定，并仅就观测到的现象在合理范围内所提供的有关物质系统的条件和联系等信息进行推导。

在合理地阐明和解释这种物理假说方法的实例之后，我们受到的对科学界人士推理不严谨的抱怨将会变得很少，并且相应的归纳哲学方法也不会再被人嘲笑为凭空臆测了。

迄今为止，物体构成理论中只有一小部分已经通过已知的事实被归纳为精确的结论。要通过合理的实验和严格的论证引导科学活动沿着十分合理的道路前进，就需要一种战略技巧，这种技巧可遇不可求，即使对于那些凭借原始观测与丰富联想进行科学研究的人来说也是一样。下面的事实并不会贬低科学先驱的成就，即他们在未知领域的开拓工作，这些开拓工作曾经一度被隔离于有着已确定的作用基础的交流体系之外，而这个交流体系正是任何科学持续发展的唯一保障。

要研究物体构成，我们从一开始就不得不处理看不见的微粒。因为不管我们对于分子和原子得到的最终结论将是什么，现在我们已经得到了实验上的证据，可以证明物体可以被分割成小到无法察觉的部分。

由此，如果我们细心地注意到微粒这个词只表示物体的一小部分，而且毫不涉及关于物质终极可分性的假设，我们就可以认为物体是由微粒组成的，我们还可以断言，在物体或者具有可观测尺度的物体的某一部分中，微粒的数量是非常大的。

下面我们还需要一种可以研究由大量微粒组成的物质系统的动力学方法，这就要形成一种关于微粒的结构、运动以及作用在粒子上的力的概念，并从动力学理论中推导出在系统的可见部分中可观测到的现象，尽管这些现象是由不可见微粒的运动和结构决定的。

进行这种研究必不可少的动力学原理是由多位动力学奠基人发展起来的，从伽利略、牛顿到拉格朗日和拉普拉斯；而将这些原理应用于分子研究，则很大程度上是波恩的克劳修斯教授的贡献，除了复杂的计算结果之外，他最近提出的一种新的动力学观点，给我们的研究工作带来了很大的帮助。借助这种观点，我希望我们可以不用大量的符号演算也能得到一些重要的结论。

The equation of Clausius, to which I must now call your attention, is of the following form: —

$$pV = \frac{2}{3}T - \frac{2}{3}\Sigma\Sigma\left(\frac{1}{2}Rr\right)$$

Here p denotes the pressure of a fluid, and V the volume of the vessel which contains it. The product pV, in the case of gases at constant temperature, remains, as Boyle's Law tells us, nearly constant for different volumes and pressures. This member of the equation, therefore, is the product of two quantities, each of which can be directly measured.

The other member of the equation consists of two terms, the first depending on the motion of the particles, and the second on the forces with which they act on each other.

The quantity T is the kinetic energy of the system, or, in other words, that part of the energy which is due to the motion of the parts of the system.

The kinetic energy of a particle is half the product of its mass into the square of its velocity, and the kinetic energy of the system is the sum of the kinetic energy of its parts.

In the second term, r is the distance between any two particles, and R is the attraction between them. (If the force is a repulsion or a pressure, R is to be reckoned negative.)

The quantity $\frac{1}{2}Rr$, or half the product of the attraction into the distance across which the attraction is exerted, is defined by Clausius as the virial of the attraction. (In the case of pressure or repulsion, the virial is negative.)

The importance of this quantity was first pointed out by Clausius, who, by giving it a name, has greatly facilitated the application of his method to physical exposition.

The virial of the system is the sum of the virials belonging to every pair of particles which exist in the system. This is expressed by the double sum $\Sigma\Sigma(\frac{1}{2}Rr)$, which indicates that the value of $\frac{1}{2}Rr$ is to be found for every pair of particles, and the results added together.

Clausius has established this equation by a very simple mathematical process, with which I need not trouble you, as we are not studying mathematics tonight. We may see, however, that it indicates two causes which may affect the pressure of the fluid on the vessel which contains it: the motion of its particles, which tends to increase the pressure, and the attraction of its particles, which tends to diminish the pressure.

We may therefore attribute the pressure of a fluid either to the motion of its particles or to a repulsion between them.

现在我必须提醒大家注意，克劳修斯方程具有如下形式：

$$pV = \frac{2}{3}T - \frac{2}{3}\Sigma\Sigma\left(\frac{1}{2}Rr\right)$$

其中，p 表示某种流体的压力，V 表示盛放流体的容器的体积。由玻意耳定律可知，对于恒温条件下的气体，乘积 pV 在不同的体积和压力下基本保持不变。可见，方程的左边，是两个直接观测量的乘积。

方程的另一边由两项组成，第一项由微粒的运动决定，第二项由微粒间的相互作用力决定。

T 这个量指系统的动能，或者换一种说法，是由于系统各部分的运动而具有的那部分能量。

微粒的动能是其质量与速率平方的乘积的一半，而系统的动能则是其各部分动能的总和。

在第二项中，r 表示任意两个微粒之间的距离，R 表示它们之间的吸引力。（如果作用力为排斥力或者压力，则 R 取负值。）

$\frac{1}{2}Rr$ 这个量，即引力与彼此吸引的微粒之间距离的乘积的一半，克劳修斯将其定义为引力的位力。（对于压力和排斥力，位力取负值。）

克劳修斯首次指出了这一物理量的重要性。通过给这一物理量指定名字，克劳修斯大大推动了他的方法在物理解释中的应用。

系统的位力是系统中每一对微粒之间位力的总和，可以表示为双重求和 $\Sigma\Sigma(\frac{1}{2}Rr)$，意思是要得到每一对微粒之间的 $\frac{1}{2}Rr$ 值，并将结果累加起来。

克劳修斯通过简单的数学过程建立了这一方程，这里我不需要以此来烦扰各位，因为今晚我们不是在研究数学问题。不过我们还是可以看到，方程指出有两个因素可能会影响流体对容纳它的容器的压力：流体中微粒的运动倾向于增大压力，而微粒间的吸引力倾向于减小压力。

由此，我们可以将流体的压力归结为流体中微粒的运动或者微粒间的排斥力。

Let us test by means of this result of Clausius the theory that the pressure of a gas arises entirely from the repulsion which one particle exerts on another, these particles, in the case of gas in a fixed vessel, being really at rest.

In this case the virial must be negative, and since by Boyle's Law the product of pressure and volume is constant, the virial also must be constant, whatever the volume, in the same quantity of gas at constant temperature. It follows from this that Rr, the product of the repulsion of two particles into the distance between them, must be constant, or in other words that the repulsion must be inversely as the distance, a law which Newton has shown to be inadmissible in the case of molecular forces, as it would make the action of the distant parts of bodies greater than that of contiguous parts. In fact, we have only to observe that if Rr is constant, the virial of every pair of particles must be the same, so that the virial of the system must be proportional to the number of pairs of particles in the system—that is, to the square of the number of particles, or in other words to the square of the quantity of gas in the vessel. The pressure, according to this law, would not be the same in different vessels of gas at the same density, but would be greater in a large vessel than in a small one, and greater in the open air than in any ordinary vessel.

The pressure of a gas cannot therefore be explained by assuming repulsive forces between the particles.

It must therefore depend, in whole or in part, on the motion of the particles.

If we suppose the particles not to act on each other at all, there will be no virial, and the equation will be reduced to the form

$$Vp = \frac{2}{3}T$$

If M is the mass of the whole quantity of gas, and c is the mean square of the velocity of a particle, we may write the equation—

$$Vp = \frac{1}{3}Mc^2,$$

or in words, the product of the volume and the pressure is one-third of the mass multiplied by the mean square of the velocity. If we now assume, what we shall afterwards prove by an independent process, that the mean square of the velocity depends only on the temperature, this equation exactly represents Boyle's Law.

But we know that most ordinary gases deviate from Boyle's Law, especially at low temperatures and great densities. Let us see whether the hypothesis of forces between the particles, which we rejected when brought forward as the sole cause of gaseous pressure, may not be consistent with experiment when considered as the cause of this deviation from Boyle's Law.

When a gas is in an extremely rarefied condition, the number of particles within a given distance of any one particle will be proportional to the density of the gas. Hence the virial

让我们用克劳修斯的结果来验证一下下面的理论：气体压力完全源于一个微粒施加在另一个微粒上的排斥力，而就恒容容器中的气体而言，这些微粒是真正静止的。

在上述情况中，位力必定为负值，又由玻意耳定律可知压力与体积的乘积为常数，则等量恒温的气体，不论其体积是多少，位力也一定是常数。由此可知，两个微粒之间的排斥力与距离的乘积 Rr 必为常数，换言之，排斥力必定与距离成反比。然而，牛顿已经指出这一定律不适用于分子间作用力的情况，因为它将使物体中相隔较远的部分间的相互作用比相邻的部分间的相互作用更强。实际上，我们只需注意到，如果 Rr 是常数，那么每一对微粒的位力都一定是相同的，于是系统的位力一定会正比于系统中微粒的对数——即正比于微粒数的平方，或者说正比于容器中气体数量的平方。根据这一定律，相同密度的气体在不同容器中的压力也是不同的，在较大容器中的压力会比在较小容器中的压力更大，而在开放的大气中的压力比在任何普通容器中的压力都大。

因此，气体的压力无法通过假定微粒间存在排斥力进行解释。

这样一来，气体的压力就一定是完全地或部分地依赖于微粒的运动。

如果我们假定微粒间没有任何相互作用，那么就没有位力存在，方程可简化为如下形式：

$$Vp = \frac{2}{3}T$$

如果用 M 表示气体的总质量，c^2 表示微粒速率的均方，我们就可以将方程写作——

$$Vp = \frac{1}{3}Mc^2,$$

或者用文字表述为，体积与压力的乘积是质量与速率均方乘积的 1/3。现在我们先假设，速率的均方仅取决于温度，那么这个方程恰好就是玻意耳定律，后面我们将通过独立的过程来证明。

但是我们知道大部分常见气体都会偏离玻意耳定律，尤其是在低温和高密度的状态下。现在我们再来看看刚才在作为产生气体压力的唯一原因被提出时，已经被我们否定的微粒间存在作用力的假说，在作为偏离玻意耳定律的原因时是否会与实验不一致。

当气体处于极稀薄的状态时，在任一微粒周围给定距离之内的微粒数量将正比于气体密度。因此，一个微粒作用于其他微粒而产生的位力将随密度变化，单位体

arising from the action of one particle on the rest will vary as the density, and the whole virial in unit of volume will vary as the square of the density.

Calling the density ρ, and dividing the equation by V, we get—

$$p = \frac{1}{3}\rho c^2 - \frac{2}{3}A\rho^2,$$

where A is a quantity which is nearly constant for small densities.

Now, the experiments of Regnault show that in most gases, as the density increases the pressure falls below the value calculated by Boyle's Law. Hence the viral must be positive; that is to say, the mutual action of the particles must be in the main attractive, and the effect of this action in diminishing the pressure must be at first very nearly as the square of the density.

On the other hand, when the pressure is made still greater the substance at length reaches a state in which an enormous increase of pressure produces but a very small increase of density. This indicates that the virial is now negative, or, in other words, the action between the particles is now, in the main, repulsive. We may therefore conclude that the action between two particles at any sensible distance is quite insensible. As the particles approach each other the action first shows itself as an attraction, which reaches a maximum, then diminishes, and at length becomes a repulsion so great that no attainable force can reduce the distance of the particles to zero.

The relation between pressure and density arising from such an action between the particles is of this kind.

As the density increases from zero, the pressure at first depends almost entirely on the motion of the particles, and therefore varies almost exactly as the pressure, according to Boyle's Law. As the density continues to increase, the effect of the mutual attraction of the particles becomes sensible, and this causes the rise of pressure to be less than that given by Boyle's Law. If the temperature is low, the effect of attraction may become so large in proportion to the effect of motion that the pressure, instead of always rising as the density increases, may reach a maximum, and then begin to diminish.

At length, however, as the average distance of the particles is still further diminished, the effect of repulsion will prevail over that of attraction, and the pressure will increase so as not only to be greater than that given by Boyle's Law, but so that an exceedingly small increase of density will produce an enormous increase of pressure.

Hence the relation between pressure and volume may be represented by the curve *A B C D E F G*, where the horizontal ordinate represents the volume, and the vertical ordinate represents the pressure.

积内的总位力将随密度的平方变化。

设密度为ρ，将方程两边同除以 V，我们得到——

$$p = \frac{1}{3}\rho c^2 - \frac{2}{3}A\rho^2,$$

其中 A 这个量在密度很小时近似为常数。

勒尼奥的实验表明，大多数气体在密度增大时，其压力会下降到低于玻意耳定律的计算值。因此位力必定是正的；也就是说，微粒间的相互作用必定以引力为主，而且这种作用减小压力的效果在开始的时候一定是近似与密度的平方有关。

另一方面，当压力继续增大，物质最终将达到另一种状态，即压力急剧增大但密度只有很小的增加。这表明此时位力为负值，或者换句话说，此时微粒间的相互作用主要为斥力。我们可以由此确定，在任何可感知的尺度中，两个微粒之间的相互作用都是十分微弱的。随着微粒彼此靠近，它们之间的相互作用首先表现为引力，引力达到最大值后开始减小，最终转变为极大的斥力，以至于没有任何可获得的力能够将微粒间的距离减小到零。

源于微粒间这种相互作用的压力与密度之间的关系就是这样的。

随着密度从零开始逐渐增大，一开始压力几乎完全取决于微粒的运动，因此几乎与玻意耳定律计算得到的压力精确地吻合。随着密度继续增加，微粒间相互吸引力的影响逐渐体现出来，这使得压力的增大小于根据玻意耳定律预计的值。如果温度很低，吸引力的影响会大到可以与运动的影响相抗衡的程度，那么压力就不再总是随着密度的增加而增大，而是在达到一个最大值后开始减小。

最终，随着微粒的平均距离继续减小，斥力的影响将胜过引力的影响，而压力将会增大到超出玻意耳定律所预计的值，而且此时很小的密度增加也会导致压力的急剧增大。

由此，压力与体积之间的关系可以用曲线 $A\ B\ C\ D\ E\ F\ G$ 来表示，其中横坐标代表体积，纵坐标代表压力。

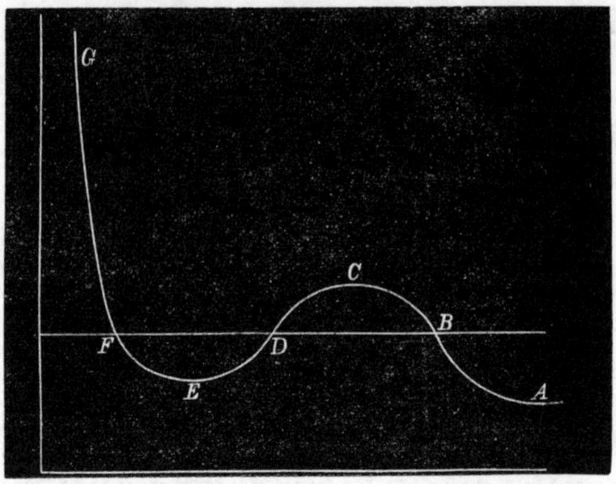

Fig. 1

As the volume diminishes, the pressure increases up to the point *C*, then diminishes to the point *E*, and finally increases without limit as the volume diminishes.

We have hitherto supposed the experiment to be conducted in such a way that the density is the same in every part of the medium. This, however, is impossible in practice, as the only condition we can impose on the medium from without is that the whole of the medium shall be contained within a certain vessel. Hence, if it is possible for the medium to arrange itself so that part has one density and part another, we cannot prevent it from doing so.

Now the points *B* and *F* represent two states of the medium in which the pressure is the same but the density very different. The whole of the medium may pass from the state *B* to the state *F*, not through the intermediate states *C D E*, but by small successive portions passing directly from the state *B* to the state *F*. In this way the successive states of the medium as a whole will be represented by points on the straight line *B F*, the point *B* representing it when entirely in the rarefied state, and *F* representing it when entirely condensed. This is what takes place when a gas or vapour is liquefied.

Under ordinary circumstances, therefore, the relation between pressure and volume at constant temperature is represented by the broken line *A B F G*. If, however, the medium when liquefied is carefully kept from contact with vapour, it may be preserved in the liquid condition and brought into states represented by the portion of the curve between *F* and *E*. It is also possible that methods may be devised whereby the vapour may be prevented from condensing, and brought into states represented by points in *B C*.

The portion of the hypothetical curve from *C* to *E* represents states which are essentially unstable, and which cannot therefore be realised.

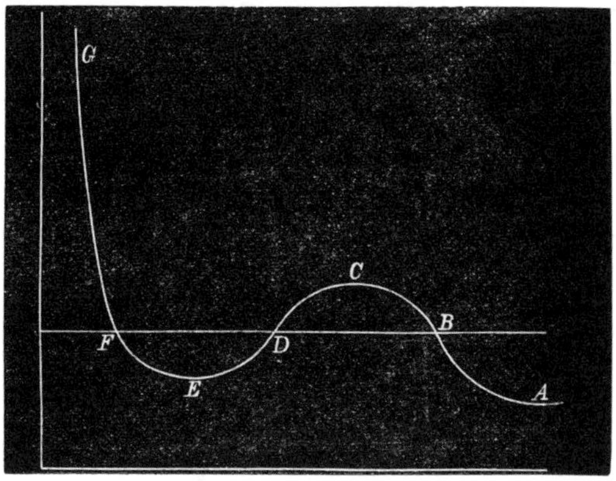

图 1

随着体积缩小，压力先增大到 C 点，然后减小到 E 点，最终将随体积缩小而无限增大。

到目前为止，我们始终假设在实验过程中介质的每一部分都具有相同的密度。不过，这在实际中是不可能的，我们能够从外部施加于介质的唯一限制，仅仅是确保整个介质都置于确定的容器之中。因此，如果介质有可能自身调整为各处密度不同，我们也无法阻止。

B 点和 F 点代表介质的两种状态，其压力相同但密度相差甚远。全部介质可从状态 B 到达状态 F，不经过中间状态 $C\,D\,E$，而是直接由状态 B 一点一点地转化为状态 F。这样，整个介质的连续变化状态对应于图中直线 $B\,F$ 上的点，B 点代表介质完全处于稀薄状态，F 则代表介质完全处于凝聚状态。这就是气体或者蒸气在液化过程中发生的现象。

因此，一般情况下，恒定温度时压力与体积的关系对应于图中的折线 $A\;B\;F\;G$。不过，如果在液化过程中小心地避免介质与蒸气接触，就可能可以使介质保持在液态，并达到图中 E 点和 F 点之间的曲线部分所代表的状态。也有可能可以设计出某种阻止蒸气凝聚的方法从而达到曲线 $B\,C$ 上的点所代表的状态。

假设的曲线中从 C 点到 E 点的部分代表本质上不稳定的状态，因此是无法实现的。

Now let us suppose the medium to pass from *B* to *F* along the hypothetical curve *B C D E F* in a state always homogeneous, and to return along the straight line *F B* in the form of a mixture of liquid and vapour. Since the temperature has been constant throughout, no heat can have been transformed into work. Now the heat transformed into work is represented by the excess of the area *F D E* over *B C D*. Hence the condition which determines the maximum pressure of the vapour at given temperature is that the line *B F* cuts off equal areas from the curve above and below.

The higher the temperature, the greater the part of the pressure which depends on motion, as compared with that which depends on forces between the particles. Hence, as the temperature rises, the dip in the curve becomes less marked, and at a certain temperature the curve, instead of dipping, merely becomes horizontal at a certain point, and then slopes upward as before. This point is called the critical point. It has been determined for carbonic acid by the masterly researches of Andrews. It corresponds to a definite temperature, pressure and density.

At higher temperatures the curve slopes upwards throughout, and there is nothing corresponding to liquefaction in passing from the rarest to the densest state.

The molecular theory of the continuity of the liquid and gaseous states forms the subject of an exceedingly ingenious thesis by Mr. Johannes Diderik van der Waals*, a graduate of Leyden. There are certain points in which I think he has fallen into mathematical errors, and his final result is certainly not a complete expression for the interaction of real molecules, but his attack on this difficult question is so able and so brave, that it cannot fail to give a notable impulse to molecular science. It has certainly directed the attention of more than one inquirer to the study of the Low-Dutch language in which it is written.

The purely thermodynamical relations of the different states of matter do not belong to our subject, as they are independent of particular theories about molecules. I must not, however, omit to mention a most important American contribution to this part of thermodynamics by Prof. Willard Gibbs†, of Yale College, U.S., who has given us a remarkably simple and thoroughly satisfactory method of representing the relations of the different states of matter by means of a model. By means of this model, problems which had long resisted the efforts of myself and others may be solved at once.

(**11**, 357-359; 1875)

* Over de continuiteit van den gas en vloeistof toestand. Leiden: A. W. Sijthoff, 1873.

† "A method of geometrical representation of the thermodynamic properties of substances by means of surfaces." *Transactions of the Connecticut Academy of Arts and Sciences*, vol. II, Part 2.

现在，让我们设想介质始终以均匀的状态沿着假设的曲线 $B\ C\ D\ E\ F$ 从 B 变化到 F，并以蒸气和液体混合的形式沿着直线 $F\,B$ 返回。因为温度一直是恒定的，所以其间不会有热能转化为功，而转化为功的热能对应于区域 $F\,D\,E$ 与 $B\,C\,D$ 的面积差。因此，在给定温度下使蒸气压力达到最大的条件，就是直线 $B\,F$ 与上方和下方曲线围成的面积相等。

温度越高，由运动决定的那部分压力与由微粒间作用力决定的部分相比就越大。因此，随着温度升高，曲线的凹陷变得不明显，当温度达到某一个值时，曲线不再凹陷而在某一点达到水平线的位置之后曲线就像前面那样上升了。这个点称为临界点。安德鲁斯已经用巧妙的研究方法确定了碳酸的临界点。临界点对应于确定的温度、压力和密度。

在更高的温度下，曲线始终是向上倾斜的，在从最稀薄到最稠密的状态变化过程中不发生液化。

关于液态与气态连续性的分子理论构成了一篇极具才华的论文的主题，论文的作者是莱顿大学的毕业生约翰内斯·迪德里克·范德瓦尔斯先生*。尽管我觉得他在某些地方犯了一些数学错误，而且他的最终结果也确实不是对真实分子间相互作用的完整表达，但是他对这一难题所作的努力表现了他非凡的才华和勇气，以至于他的工作不可能不对分子科学产生重要的推动作用。确实有不止一位研究者受此驱使而去学习这篇文章撰写时所用的荷兰语。

物质不同状态间纯粹的热力学联系不属于我们的主题，因为它与这些关于分子的具体理论无关。不过，我决不能忘记提及一项极为重要的，由美国耶鲁大学的威拉德·吉布斯教授作出的热力学方面的贡献†，他利用一个模型给我们提供了一种极其简单又十分令人满意的表示不同物态间关联的方法。借助这个模型，那些长期以来阻碍我本人和其他人努力的难题可能立刻就会迎刃而解。

---

*《论液态和气态的连续性》，莱登赛特霍夫出版社，1873 年。

†《借助曲面表示物质热力学性质的几何描述法》，《康涅狄格艺术与科学学会学报》，第 2 卷，第 2 部分。

II

Let us now return to the case of a highly rarefied gas in which the pressure is due entirely to the motion of its particles. It is easy to calculate the mean square of the velocity of the particles from the equation of Clausius, since the volume, the pressure, and the mass are all measurable quantities. Supposing the velocity of every particle the same, the velocity of a molecule of oxygen would be 461 metres per second, of nitrogen 492, and of hydrogen 1,844, at the temperature 0°C.

The explanation of the pressure of a gas on the vessel which contains it by the impact of its particles on the surface of the vessel has been suggested at various times by various writers. The fact, however, that gases are not observed to disseminate themselves through the atmosphere with velocities at all approaching those just mentioned, remained unexplained, till Clausius, by a thorough study of the motions of an immense number of particles, developed the methods and ideas of modern molecular science.

To him we are indebted for the conception of the mean length of the path of a molecule of a gas between its successive encounters with other molecules. As soon as it was seen how each molecule, after describing an exceedingly short path, encounters another, and then describes a new path in a quite different direction, it became evident that the rate of diffusion of gases depends not merely on the velocity of the molecules, but on the distance they travel between each encounter.

I shall have more to say about the special contributions of Clausius to molecular science. The main fact, however, is, that he opened up a new field of mathematical physics by showing bow to deal mathematically with moving systems of innumerable molecules.

Clausius, in his earlier investigations at least, did not attempt to determine whether the velocities of all the molecules of the same gas are equal, or whether, if unequal, there is any law according to which they are distributed. He therefore, as a first hypothesis, seems to have assumed that the velocities are equal. But it is easy to see that if encounters take place among a great number of molecules, their velocities, even if originally equal, will become unequal, for, except under conditions which can be only rarely satisfied, two molecules having equal velocities before their encounter will acquire unequal velocities after the encounter. By distributing the molecules into groups according to their velocities, we may substitute for the impossible task of following every individual molecule through all its encounters, that of registering the increase or decrease of the number of molecules in the different groups.

By following this method, which is the only one available either experimentally or mathematically, we pass from the methods of strict dynamics to those of statistics and probability.

II

现在让我们回过头来看看极稀薄状态的气体，在这种状态下压力完全来源于气体微粒的运动。由于体积、压力和质量都是可观测量，因此很容易利用克劳修斯方程计算出微粒速率的均方。假设每种微粒具有相同的速率，则在 0℃时，氧分子的速率为每秒 461 米，氮分子为每秒 492 米，而氢分子为每秒 1,844 米。

已经有多位学者多次提出过用气体微粒对容器表面的撞击来解释气体对容器的压力。不过，我们一直不能解释为什么从未观测到气体以接近于上面所提到的速率散布到大气中的现象，直到克劳修斯通过对大量微粒的运动的全面研究发展了现代分子科学的方法和观念。

我们得益于他提出的一个概念，即一个气体分子在与其他分子相继发生两次碰撞之间所经过路程的平均长度。一旦看到每个气体分子是如何在经历了一段极短的路程后与另一个分子相撞，随后气体分子又沿着完全不同的方向开始新的旅程，那么很明显的就是：气体扩散的速率不仅依赖于分子的速度，还与分子在相继发生的碰撞之间途经的距离有关。

关于克劳修斯对分子科学的特殊贡献，我还有很多要说。最主要的是，他通过展示如何用数学方式处理含无数分子的运动系统从而开创了数学物理的新领域。

至少在早期的研究中，克劳修斯并没有试图去确定同一气体中的所有分子是否都具有相同的速率，或者如果它们的速率不等，其分布是否应遵循某种规律。他的第一个假说似乎假定所有分子的速率都是相等的。不过显而易见的是，如果碰撞发生在大量分子之间，那么即使它们的速率开始时是相等的，之后也会变得不相等，因为除了在某些几乎无法满足的条件下，两个碰撞前具有相同速率的分子在碰撞后总会获得不同的速率。通过将分子按其速率分组，我们就只需记录不同组内分子数量的增加或减少，而不必完成跟踪每一个分子的所有碰撞这项不可能完成的任务。

利用这种无论从实验角度还是数学角度来说都是唯一可行的方法，我们从严格的动力学方法转到了统计和概率的方法上。

When an encounter takes place between two molecules, they are transferred from one pair of groups to another, but by the time that a great many encounters have taken place, the number which enter each group is, on an average, neither more nor less than the number which leave it during the same time. When the system has reached this state, the numbers in each group must be distributed according to some definite law.

As soon as I became acquainted with the investigations of Clausius, I endeavoured to ascertain this law.

The result which I published in 1860 has since been subjected to a more strict investigation by Dr. Ludwig Boltzmann, who has also applied his method to the study of the motion of compound molecules. The mathematical investigation, though, like all parts of the science of probabilities and statistics, it is somewhat difficult, does not appear faulty. On the physical side, however, it leads to consequences, some of which, being manifestly true, seem to indicate that the hypotheses are well chosen, while others seem to be so irreconcilable with known experimental results, that we are compelled to admit that something essential to the complete statement of the physical theory of molecular encounters must have hitherto escaped us.

I must now attempt to give you some account of the present state of these investigations, without, however, entering into their mathematical demonstration.

I must begin by stating the general law of the distribution of velocity among molecules of the same kind.

If we take a fixed point in this diagram and draw from this point a line representing in direction and magnitude the velocity of a molecule, and make a dot at the end of the line, the position of the dot will indicate the state of motion of the molecule.

If we do the same for all the other molecules, the diagram will be dotted all over, the dots being more numerous in certain places than in others.

The law of distribution of the dots may be shown to be the same as that which prevails among errors of observation or of adjustment.

The dots in the diagram before you may be taken to represent the velocities of molecules, the different observations of the position of the same star, or the bullet-holes round the bull's eye of a target, all of which are distributed in the same manner.

The velocities of the molecules have values ranging from zero to infinity, so that in speaking of the average velocity of the molecules we must define what we mean.

当两个分子发生碰撞时，它们就会从原本所在组的一对转变成另一组的一对，而且在同时发生大量碰撞时，平均来看，同一时间段内进入某一个组的分子数量，不会多于或少于离开该组的分子数量。当系统达到这种状态时，分子在各组的分布一定符合某种确定的规律。

我在了解了克劳修斯的研究工作之后，便立即努力探求这一规律。

我在1860年发表的结果后来由路德维希·玻尔兹曼博士进行了更为严格的研究，他还将他的方法应用于对化合物分子运动的研究。数学研究，例如所有关于概率统计的科学部分，虽然有些困难，却不无裨益。从物理角度来说，数学方法的某些推论是非常正确的，似乎可以说明所选假说是正确的，而另一些却与已知的实验结果相违背，这使我们不得不承认，到目前为止，我们遗漏了关于分子碰撞物理理论完整表述中的某些核心部分。

现在我必须努力为各位介绍这方面研究的现状，不过，我们不会涉及其中的数学证明。

我必须从同种分子中速度分布的一般规律开始说起。

如果从图中选择一个固定的点，从这点出发画一条线来表示速度的方向和大小，并在线段末端画出端点，那么端点的位置就表示分子的运动状态。

如果我们对所有分子作同样处理，那么整张图上将画满了点，某些位置的点会比其他位置的点多。

可以看出，这些点的分布规律与观测或调节中经常出现的误差的分布规律是相同的。

你面前这张图中的点可以用来表示分子的速度，或者对同一颗恒星位置的不同观测结果，或者位于目标靶心周围的弹洞，上述这些具有相同的分布方式。

分子速率的取值范围是从零到无限大，因此说到分子的平均速率我们必须先定义我们指的是什么。

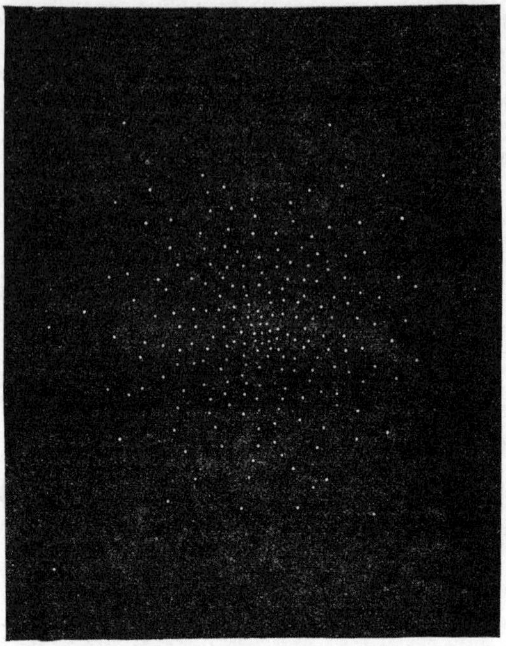

Fig. 2. Diagram of Velocities

The most useful quantity for purposes of comparison and calculation is called the "velocity of mean square". It is that velocity whose square is the average of the squares of the velocities of all the molecules.

This is the velocity given above as calculated from the properties of different gases. A molecule moving with the velocity of mean square has a kinetic energy equal to the average kinetic energy of all the molecules in the medium, and if a single mass equal to that of the whole quantity of gas were moving with this velocity, it would have the same kinetic energy as the gas actually has, only it would be in a visible form and directly available for doing work.

If in the same vessel there are different kinds of molecules, some of greater mass than others, it appears from this investigation that their velocities will be so distributed that the average kinetic energy of a molecule will be the same, whether its mass be great or small.

Here we have perhaps the most important application which has yet been made of dynamical methods to chemical science. For, suppose that we have two gases in the same vessel. The ultimate distribution of agitation among the molecules is such that the average kinetic energy of an individual molecule is the same in either gas. This ultimate state is also, as we know, a state of equal temperature. Hence the condition that two gases shall have the same temperature is that the average kinetic energy of a single molecule shall be the same in the two gases.

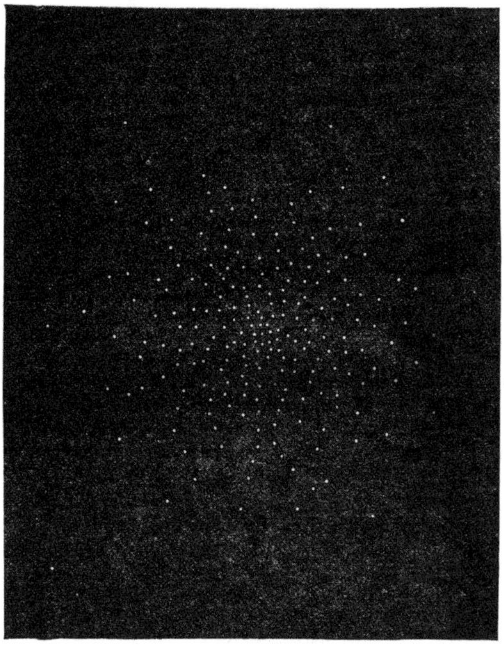

图 2. 速度图像

为了便于比较和计算，最有用的量叫作“均方速率”。这个量的平方是所有分子速率平方的平均值。

这就是上面给出的利用不同气体性质计算得到的速度。以均方速率运动的分子具有的动能等于介质内全体分子的平均动能，并且若以此速率运动的单个物体的质量等于气体的总质量，那么该物体与气体具有相同的动能，只是前者具有可见的形式并可直接用于做功。

如果同一容器中包含不同种类的分子，其中一些具有比较大的质量，那么根据这项研究，这些分子的速度分布方式应满足每种分子具有相同的平均动能，不管其质量是大还是小。

这里我们用到的可能是动力学方法最重要的应用，即将它应用于化学科学。比如，假定我们将两种气体置于同一容器中。分子运动的最终分布形式应满足两种气体中每一种的单个分子的平均动能相等。正如我们所知，这种最终态也是等温态。由此，两种气体具有相同温度的条件，就是两种气体中单个分子的平均动能相同。

Now, we have already shown that the pressure of a gas is two-thirds of the kinetic energy in unit of volume. Hence, if the pressure as well as the temperature be the same in the two gases, the kinetic energy per unit of volume is the same, as well as the kinetic energy molecule. There must, therefore, be the same number of molecules in unit of volume in the two gases.

This result coincides with the law of equivalent volumes established by Gay Lussac. This law, however, has hitherto tested on purely chemical evidence, the relative masses of the molecules of different substances having been deduced from the proportions in which the substances enter into chemical combination. It is now demonstrated on dynamical principles. The molecule is defined as that small portion of the substance which moves as one lump during the motion of agitation. This is a purely dynamical definition, independent of any experiments on combination.

The density of a gaseous medium, at standard temperature and pressure, is proportional to the mass of one of its molecules as thus defined.

We have thus a safe method of estimating the relative masses of molecules of different substances when in the gaseous state. This method is more to be depended on than those founded on electrolysis or on specific heat, because our knowledge of the conditions of the motion of agitation is more complete than our knowledge of electrolysis, or of the internal motions of the constituents of a molecule.

I must now say something about these internal motions, because the greatest difficulty which the kinetic theory of gases has yet encountered belongs to this part of the subject.

We have hitherto considered only the motion of the centre of mass of the molecule. We have now to consider the motion of the constituents of the molecule relative to the centre of mass.

If we suppose that the constituents of a molecule are atoms, and that each atom is what is called a material point, then each atom may move in three different and independent ways, corresponding to the three dimensions of space, so that the number of variables required to determine the position and configuration of all the atoms of the molecule is three times the number of atoms.

It is not essential, however, to the mathematical investigation to assume that the molecule is made up of atoms. All that is assumed is that the position and configuration of the molecule can be completely expressed by a certain number of variables.

Let us call this number n.

现在，我们已经指出，气体的压力是单位体积动能的2/3。因此，如果两种气体具有相同的压力和温度，那么它们每单位体积的动能以及每个分子的动能也是相同的。由此，两种气体单位体积内必定含有相同数量的分子。

这一结果与盖·吕萨克提出的等容定律是一致的。不过，到目前为止该定律只基于化学证据，不同物质的相对分子质量已经从物质参加化合反应时所占据的比例中推导出来了。现在已经用动力学原理证明了该定律。分子被定义为扰动运动中作为一个整体运动的物体的一小部分。这是一个纯粹的动力学定义，与任何化合实验无关。

在标准的温度和压力下，气态介质的密度正比于该气体按以上定义的一个分子的质量。

这样，我们就有一种可以估计气态时不同物质的相对分子质量的可靠方法。这种方法比那些建立在电解或者比热基础上的方法更可信赖，因为我们已经掌握的关于扰动运动条件方面的知识比关于电解或者分子结构内部运动的知识更完备。

现在我必须谈谈这些分子的内部运动，因为目前气体动力学理论遇到的最大困难就在这部分。

迄今为止我们考虑的只是分子质心的运动。现在我们不得不考虑分子组分相对于质心的运动。

如果我们假定分子是由原子构成的，并且每个原子都是所谓的质点，那么每个原子都可以在 3 个不同且独立的方向上运动，这对应于空间的 3 个维度，因此要确定分子中所有原子结构和位置所需的变量数就是原子个数的 3 倍。

不过，对数学研究来说，假定分子是由原子构成的并不是一个基本假设。全部的前提假设只是，分子的位置和结构可以用一定数量的变量完整地表达。

让我们将这个数设为 n。

Of these variables, three are required to determine the position of the centre of mass of the molecule, and the remaining n–3 to determine its configuration relative to its centre of mass.

To each of the n variables corresponds a different kind of motion.

The motion of translation of the centre of mass has three components.

The motions of the parts relative to the centre of mass have n–3 components.

The kinetic energy of the molecule may be regarded as made up of two parts—that of the mass of the molecule supposed to be concentrated at its centre of mass, and that of the motions of the parts relative to the centre of mass. The first part is called the energy of translation, the second that of rotation and vibration. The sum of these is the whole energy of motion of the molecule.

The pressure of the gas depends, as we have seen, on the energy of translation alone. The specific heat depends on the rate at which the whole energy, kinetic and potential, increases as the temperature rises.

Clausius had long ago pointed out that the ratio of increment of the whole energy to that of the energy of translation may be determined if we know by experiment the ratio of the specific heat at constant pressure to that at constant volume.

He did not, however, attempt to determine *à priori* the ratio of the two parts of the energy, though he suggested, as an extremely probable hypothesis, that the average values of the two parts of the energy in a given substance always adjust themselves to the same ratio. He left the numerical value of this ratio to be determined by experiment.

In 1860 I investigated the ratio of the two parts of the energy on the hypothesis that the molecules are elastic bodies of invariable form. I found, to my great surprise, that whatever be the shape of the molecules, provided they are not perfectly smooth and spherical, the ratio of the two parts of the energy must be always the same, the two parts being in fact equal.

This result is confirmed by the researches of Boltzmann, who has worked out the general case of a molecule having n variables.

He finds that while the average energy of translation is the same for molecules of all kinds at the same temperature, the whole energy of motion is to the energy of translation as n to 3.

For a rigid body n=6, which makes the whole energy of motion twice the energy of energy of translation.

在这些变量中，有 3 个是用来确定分子质心位置的，而其余 n–3 个则用来确定相对于质心的分子结构。

n 个变量的每一个都对应于一种不同的运动。

质心的平移运动有 3 个分量。

各部分相对于质心的运动则包含 n–3 个分量。

分子的动能可以看作是由两部分组成——其中一部分是假定整个分子的质量集中于其质心所产生的，另一部分是各部分相对质心运动所产生的。第一部分称为平动动能，第二部分称为转动和振动动能。上述量的总和就是分子运动的总能量。

正如我们所知，气体的压力只由平动动能决定。比热则依赖于温度升高时动能与势能的总增加量与温度增量的比值。

克劳修斯早就指出，如果我们能够通过实验获得恒压比热与恒容比热的比值，就能确定总能量增量与平动动能增量的比值。

但是，他并没有试图先验性地确定两部分能量的比值，尽管他提出了一个极有可能成立的假说：对于一种给定的物质，两部分能量的平均值总是具有相同的比值。他将这一比值的具体数值留待实验确定。

1860 年，我基于分子是不变形的弹性体的假说研究了两部分能量的比值。我惊讶地发现，不论分子是何种形状，只要它们不是绝对光滑和完美球状，那么两部分能量的比值就一定总是相同的，实际上这两部分能量是相等的。

这一结果被玻尔兹曼的研究工作所证实，他给出了具有 n 个变量的分子在一般情况下的结果。

他发现，相同温度下任何种类的分子都具有相同的平均平动动能，运动的总能量与平动动能之比为 $n:3$。

对于刚体来说 $n=6$，这使得运动的总能量为平动动能的 2 倍。

But if the molecule is capable of changing its form under the action of impressed forces, it must be capable of storing up potential energy, and if the forces are such as to ensure the stability of the molecule, the average potential energy will increase when the average energy of internal motion increases.

Hence, as the temperature rises, the increments of the energy of translation, the energy of internal motion, and the potential energy are as 3, (n–3), and e respectively, where e is a positive quantity of unknown value depending on the law of the force which binds together the constituents of the molecule.

When the volume of the substance is maintained constant, the effect of the application of heat is to increase the whole energy. We thus find for the specific heat of a gas at constant volume—

$$\frac{1}{2J}\frac{p_0V_0}{273^\circ}(n+e)$$

where p_0 and V_0 are the pressure and volume of unit of mass at zero centigrade, or 273° absolute temperature, and J is the dynamical equivalent of heat. The specific heat at constant pressure is

$$\frac{1}{2J}\frac{p_0V_0}{273^\circ}(n+2+e)$$

In gases whose molecules have the same degree of complexity the value of n is the same, and that of e *may* be the same.

If this is the case, the specific heat is inversely as the specific gravity, according to the law of Dulong and Petit, which is, to a certain degree of approximation, verified by experiment.

But if we take the actual values of the specific heat as found by Regnault and compare them with this formula, we find that $n + e$ for air and several other gases cannot be more than 4.9. For carbonic acid and steam it is greater. We obtain the same result if we compare the ratio of the calculated specific heats

$$\frac{2+n+e}{n+e}$$

with the ratio as determined by experiment for various gases, namely, 1.408.

And here we are brought face to face with the greatest difficulty which the molecular theory has yet encountered, namely, the interpretation of the equation $n + e = 4.9$.

If we suppose that the molecules are atoms—mere material points, incapable of rotatory energy or internal motion—then n is 3 and e is zero, and the ratio of the specific heats is 1.66, which is too great for any real gas.

但是，如果分子能够在外力作用下改变形状，它就一定可以储存势能，而如果外力是能保证分子的稳定性的那种，当内部运动的平均能量增加时，平均势能也会增加。

因此，随着温度升高，平动动能、内部运动能量和势能的增量分别为 3，$(n-3)$ 和 e，其中 e 是取值为正的未知量，其数值取决于将分子各组成部分束缚在一起的力的定律。

当物体的体积保持不变时，加热的效果是增加总能量。我们由此给出恒容气体的比热——

$$\frac{1}{2J}\frac{p_0V_0}{273^\circ}(n+e)$$

其中 p_0 和 V_0 是零摄氏度或绝对温度 273° 时单位质量的压力和体积，而 J 则是热的动力学当量。恒压条件下的比热为

$$\frac{1}{2J}\frac{p_0V_0}{273^\circ}(n+2+e)$$

对于那些有相同复杂程度的分子，n 值是相同的，e 值**可能**是相同的。

如果情况就是如此，那么根据杜隆–珀蒂定律，比热与比重成反比，这是一条在一定精度范围内已经被实验证明了的定律。

但是，如果我们采用勒尼奥所发现的实际比热值并将其与此公式进行比较就会发现，对空气和其他几种气体，$n+e$ 的值不会超过 4.9。而对碳酸和水蒸气来说这个值则大一些。如果将计算所得的比热之比值

$$\frac{2+n+e}{n+e}$$

与实验确定的比值相比，我们得到相同的结果，即 1.408。

此时，我们将面临分子理论中最大的困难，也就是，如何解释等式 $n+e=4.9$。

如果我们假设分子是不可分割的质点——没有转动能量或者内部运动—— 那么 n 为 3 而 e 为 0，比热的比值是 1.66，而这个值对于任何实际气体来说都太大了。

But we learn from the spectroscope that a molecule can execute vibrations of constant period. It cannot therefore be a mere material point, but a system capable of changing its form. Such a system cannot have less than six variables. This would make the greatest value of the ratio of the specific heats 1.33, which is too small for hydrogen, oxygen, nitrogen, carbonic oxide, nitrous oxide, and hydrochloric acid.

But the spectroscope tells us that some molecules can execute a great many different kinds of vibrations. They must therefore be systems of a very considerable degree of complexity, having far more than six variables. Now, every additional variable introduces an additional amount of capacity for internal motion without affecting the external pressure. Every additional variable, therefore, increases the specific heat, whether reckoned at constant pressure or at constant volume.

So does any capacity which the molecule may have for storing up energy in the potential form. But the calculated specific heat is already too great when we suppose the molecule to consist of two atoms only. Hence every additional degree of complexity which we attribute to the molecule can only increase the difficulty of reconciling the observed with the calculated value of the specific heat.

I have now put before you what I consider to be the greatest difficulty yet encountered by the molecular theory. Boltzmann has suggested that we are to look for the explanation in the mutual action between the molecules and the etherial medium which surrounds them. I am afraid, however, that if we call in the help of this medium, we shall only increase the calculated specific heat, which is already too great.

The theorem of Boltzmann may be applied not only to determine the distribution of velocity among the molecules, but to determine the distribution of the molecules themselves in a region in which they are acted on by external forces. It tells us that the density of distribution of the molecules at a point where the potential energy of a molecule is ψ, is proportional to $e^{-\frac{\psi}{\kappa\theta}}$ where θ is the absolute temperature, and κ is a constant for all gases. It follows from this, that if several gases in the same vessel are subject to an external force like that of gravity, the distribution of each gas is the same as if no other gas were present. This result agrees with the law assumed by Dalton, according to which the atmosphere may be regarded as consisting of two independent atmospheres, one of oxygen, and the other of nitrogen; the density of the oxygen diminishing faster than that of the nitrogen, as we ascend.

This would be the case if the atmosphere were never disturbed, but the effect of winds is to mix up the atmosphere and to render its composition more uniform than it would be if left at rest.

Another consequence of Boltzmann's theorem is, that the temperature tends to become equal throughout a vertical column of gas at rest.

但是，通过光谱仪我们了解到，分子可以进行周期固定的振动。因此分子不可能只是一个质点，而应是一个结构可以改变的体系。这一体系的变量不可能少于 6 个。这使得比热比值的最大值是 1.33，而这个结果对于氢气、氧气、氮气、碳氧化物、氮氧化物和氢氯酸来说都太小了。

但是光谱仪告诉我们某些分子可以进行多种不同类型的振动。因此它们必然是具有相当复杂程度的体系，具有的变量数远大于 6。那么，每增加一个变量，都会在不影响外部压力的前提下引入一些附加的内部运动的能力。因此，不论是在恒压还是恒容条件下计算，每增加一个变量都会使比热增加。

分子以势能形式储存能量的能力也是如此。但是，当我们假定分子仅由两个原子组成，计算得到的比热值就已经太大了。于是，分子的复杂程度每增加一点，都只会加大使比热的观测值与计算值相一致的难度。

现在我已将我认为的分子理论遇到的最大困难呈现在各位面前。玻尔兹曼曾经建议，应该从分子与围绕在其周围的以太介质的相互作用中寻求解释。不过，要是我们借助于这种介质的话，恐怕只会使已经过大的比热计算值变得更大。

玻尔兹曼定理不仅可以用于确定分子的速度分布，还可以用于确定在外力作用下分子自身在一个区域中的分布。它告诉我们，在分子势能为 ψ 的一点，分子的分布密度正比于 $e^{-\frac{\psi}{\kappa\theta}}$，其中 θ 为绝对温度，而 κ 对于所有气体都是常数。由此可知，如果同一容器中的几种气体受到外力（例如重力）作用，每一种气体的分布与没有其他气体存在时是一样的。这一结果与道尔顿提出的定律是一致的，根据这个定律，大气可以被视为是由两种独立的气体组成，一种为氧气，另一种为氮气；随着海拔升高，氧气的浓度比氮气的浓度减小得更快些。

以上是大气丝毫不受扰动的情况，而风的影响会将大气混匀，使其组成比保持静止时更均匀。

玻尔兹曼定理的另一推论是，静止状态的气体在垂直方向上温度趋于一致。

In the case of the atmosphere, the effect of wind is to cause the temperature to vary as that of a mass of air would do if it were carried vertically upwards, expanding and cooling as it ascends.

But besides these results, which I had already obtained by a less elegant method and published in 1866, Boltzmann's theorem seems to open up a path into a region more purely chemical. For if the gas consists of a number of similar systems, each of which may assume different states having different amounts of energy, the theorem tells us that the number in each state is proportional to $e^{-\frac{\psi}{\kappa\theta}}$ where ψ is the energy, θ the absolute temperature, and κ a constant.

It is easy to see that this result ought to be applied to the theory of the states of combination which occur in a mixture of different substances. But as it is only during the present week that I have made any attempt to do so, I shall not trouble you with my crude calculations.

I have confined my remarks to a very small part of the field of molecular investigation. I have said nothing about the molecular theory of the diffusion of matter, motion, and energy, for though the results, especially in the diffusion of matter and the transpiration of fluids are of great interest to many chemists, and though from them we deduce important molecular data, they belong to a part of our study the data of which, depending on the conditions of the encounter of two molecules, are necessarily very hypothetical. I have thought it better to exhibit the evidence that the parts of fluids are in motion, and to describe the manner in which that motion is distributed among molecules of different masses.

To show that all the molecules of the same substance are equal in mass, we may refer to the methods of dialysis introduced by Graham, by which two gases of different densities may be separated by percolation through a porous plug.

If in a single gas there were molecules of different masses, the same process of dialysis, repeated a sufficient number of times, would furnish us with two portions of the gas, in one of which the average mass of the molecules would be greater than in the other. The density and the combining weight of these two portions would be different. Now, it may be said that no one has carried out this experiment in a sufficiently elaborate manner for every chemical substance. But the processes of nature are continually carrying out experiments of the same kind; and if there were molecules of the same substance nearly alike, but differing slightly in mass, the greater molecules would be selected in preference to form one compound, and the smaller to form another. But hydrogen is of the same density, whether we obtain it from water or from a hydrocarbon, so that neither oxygen nor carbon can find in hydrogen molecules greater or smaller than the average.

对于大气来说，风的影响导致温度变化，例如，若一定质量的气体垂直向上运动，那么它在上升过程中会膨胀并会冷却。

然而，除了给出这些我已经于1866年就以不那么精巧的方法获得并发表的结果之外，玻尔兹曼的定理似乎还开辟了一条通向更纯粹的化学领域的道路。因为，如果气体是由大量相似的系统构成，每一个系统都可以假定为具有不同能量的不同状态，定理告诉我们每一状态中系统的数量正比于 $e^{-\frac{\psi}{\kappa\theta}}$，其中 ψ 表示能量，θ 表示绝对温度，而 κ 是一个常数。

容易看出，这一结果应当被用于发生在不同物质混合体中的组合的状态理论。不过，由于这项工作是我这个星期才刚刚开始的，我还不想以我粗糙的计算结果来搅扰诸位。

我一直将论题限制在分子研究领域的一个很小的范围内。我并没有提及关于物质扩散、运动以及能量的分子理论，因为，尽管这些结果，尤其是关于物质扩散和液体蒸发的结果，对很多化学家来说极有吸引力，并且从这些结果中我们可以推导出重要的分子数据，但这些数据依赖于两个分子碰撞的条件，因而属于我们研究工作中必然具有很多假设性的部分。我认为，还是展示一下能表明流体各部分运动的证据，并且描述一下此运动在不同质量的分子中的分布情况更好些。

为表明同种物质的所有分子都具有相同的质量，我们就要谈到格雷姆引入的透析法，用这种方法可以通过一个多孔塞进行过滤，将两种不同密度的气体分离。

如果在一种气体中存在不同质量的分子，将同样的透析过程足够多次地重复之后，我们就可以得到两部分气体，其中一部分的平均分子质量比另一部分的大。这两部分气体有不同的密度和化合量。现在，也许可以说，没有谁曾以足够精细的方式对每一种化学物质进行过这一实验。然而自然过程在持续不断地进行着这种类型的实验。如果同种物质具有十分相似但质量略有差别的分子，那么其中质量较大的分子将会被优先选择来形成一种化合物，而质量较小的分子则形成另一种。但是，无论是从水中还是从碳氢化合物中得到的氢都具有相同的密度，因此，氧和碳都不能在氢分子中找到大于或小于平均质量的个体。

The estimates which have been made of the actual size of molecules are founded on a comparison of the volumes of bodies in the liquid or solid state, with their volumes in the gaseous state. In the study of molecular volumes we meet with many difficulties, but at the same time there are a sufficient number of consistent results to make the study a hopeful one.

The theory of the possible vibrations of a molecule has not yet been studied as it ought, with the help of a continual comparison between the dynamical theory and the evidence of the spectroscope. An intelligent student, armed with the calculus and the spectroscope, can hardly fail to discover some important fact about the internal constitution of a molecule.

The observed transparency of gases may seem hardly consistent with the results of molecular investigations.

A model of the molecules of a gas consisting of marbles scattered at distances bearing the proper proportion to their diameters, would allow very little light to penetrate through a hundred feet.

But if we remember the small size of the molecules compared with the length of a wave of light, we may apply certain theoretical investigations of Lord Rayleigh's about the mutual action between waves and small spheres, which show that the transparency of the atmosphere, if affected only by the presence of molecules, would be far greater than we have any reason to believe it to be.

A much more difficult investigation, which has hardly yet been attempted, relates to the electric properties of gases. No one has yet explained why dense gases are such good insulators, and why, when rarefied or heated, they permit the discharge of electricity, whereas a perfect vacuum is the best of all insulators.

It is true that the diffusion of molecules goes on faster in a rarefied gas, because the mean path of a molecule is inversely as the density. But the electrical difference between dense and rare gas appears to be too great to be accounted for in this way.

But while I think it right to point out the hitherto unconquered difficulties of this molecular theory, I must not forget to remind you of the numerous facts which it satisfactorily explains. We have already mentioned the gaseous laws, as they are called, which express the relations between volume, pressure, and temperature, and Gay Lussac's very important law of equivalent volumes. The explanation of these may be regarded as complete. The law of molecular specific heats is less accurately verified by experiment, and its full explanation depends on a more perfect knowledge of the internal structure of a molecule than we as yet possess.

目前已有的对分子实际大小的估计都建立在将物体固态或液态时的体积与气态时的体积相比较的基础之上。在对分子体积的研究中我们遇到了很多困难，但同时也得到了很多可靠的结果，这使研究充满了希望。

对于分子可能的振动，相关理论的研究尚未开始，它本应在不断将动力学理论与光谱证据相比较的帮助下展开。用光谱工具和计算法武装起来的才智之士，一定会发现一些与分子内部结构有关的重要事实。

观测到的气体的透明度看来似乎很难与分子研究的结果一致。

一种认为气体是由间隔距离与其自身直径符合适当比例的分散小球组成的气体分子模型，只能允许极少量的光穿透 100 英尺的距离。

不过，要是我们还记得与光的波长相比分子的尺寸很微小，我们就会采用瑞利勋爵关于波与微小球体间相互作用的某些理论研究。他的研究表明，气体的透明度如果只受分子存在的影响，就会比我们有任何理由能够去相信的还要大得多。

一项几乎还没有人尝试过的更加困难的研究，是与气体的电学性质有关。还没有人能够解释为什么稠密的气体是非常良好的绝缘体，为什么气体在稀释或加热时可以放电，而完全的真空却是最好的绝缘体。

的确，稀薄气体中的分子扩散会进行得更快一些，因为分子运动的平均路程反比于密度。但是稠密气体与稀薄气体的电学差异似乎非常大以至于无法从这个角度来解释。

虽然我认为指出这种分子理论目前尚无法克服的困难是应该的，但我绝对不会忘记提醒诸位它已经令人满意地解释了大量问题。我们已经提到过气体定律，正如我们所说，它表达了气体的体积、压力与温度之间的关系，以及盖·吕萨克的极其重要的等容定律。可以认为关于这些定律的解释是完整的。分子比热定律已被实验相对粗略地验证了，而对它的完整解释则有赖于拥有比今天更完备的关于分子内部结构的知识。

But the most important result of these inquiries is a more distinct conception of thermal phenomena. In the first place, the temperature of the medium is measured by the average kinetic energy of translation of a single molecule of the medium. In two media placed in thermal communication, the temperature as thus measured tends to become equal.

In the next place, we learn how to distinguish that kind of motion which we call heat from other kinds of motion. The peculiarity of the motion called heat is that it is perfectly irregular; that is to say, that the direction and magnitude of the velocity of a molecule at a given time cannot be expressed as depending on the present position of the molecule and the time.

In the visible motion of a body, on the other hand, the velocity of the centre of mass of all the molecules in any visible portion of the body is the observed velocity of that portion, though the molecules may have also an irregular agitation on account of the body being hot.

In the transmission of sound, too, the different portions of the body have a motion which is generally too minute and too rapidly alternating to be directly observed. But in the motion which constitutes the physical phenomenon of sound, the velocity of each portion of the medium at any time can be expressed as depending on the position and the time elapsed; so that the motion of a medium during the passage of a sound-wave is regular, and must be distinguished from that which we call heat.

If, however, the sound-wave, instead of traveling onwards in an orderly manner and leaving the medium behind it at rest, meets with resistances which fritter away its motion into irregular agitations, this irregular molecular motion becomes no longer capable of being propagated swiftly in one direction as sound, but lingers in the medium in the form of heat till it is communicated to colder parts of the medium by the slow process of conduction.

The motion which we call light, though still more minute and rapidly alternating than that of sound, is, like that of sound, perfectly regular, and therefore is not heat. What was formerly called Radiant Heat is a phenomenon physically identical with light.

When the radiation arrives at a certain portion of the medium, it enters it and passes through it, emerging at the other side. As long as the medium is engaged in transmitting the radiation it is in a certain state of motion, but as soon as the radiation has passed through it, the medium returns to its former state, the motion being entirely transferred to a new portion of the medium.

Now, the motion which we call heat can never of itself pass from one body to another unless the first body is, during the whole process, hotter than the second. The motion of radiation, therefore, which passes entirely out of one portion of the medium and enters another, cannot be properly called heat.

不过，由这些探索得到的最重要的结果是关于热现象更加明确的概念。首先，介质的温度由介质中单个分子的平均平动动能来度量。在两种介质有热交换时，以此确定的温度有变得一致的倾向。

其次，我们知道了如何将所谓的热运动与其他类型的运动区分开。我们所谓的热运动的特性是完全无规则；也就是说，在任一给定时刻，一个分子的速度的方向和大小不可能用时间和现在分子所处的位置表示。

另一方面，在物体的可视运动中，物体的任一可视部分所含全部分子的质心速度，就是所观测到的该部分的速度。不过，分子也可能由于被加热而具有不规则的扰动。

同样地，在声音的传播过程中，物体不同部分的运动通常由于过于细微和过于快速变化而难以被直接观测。不过，在产生声音这一物理现象的运动过程中，介质中每一部分在任意时刻的速度都可以用位置和所经过的时间表示出来；因此在声波传播过程中介质的运动是规则的，必然不同于我们所说的热运动。

然而，如果声波不是以整齐有序的方式向前传播并使其所经之处的介质归于静止，而是遇到阻力从而将运动耗损在无规则的激发中，这种无规则的分子运动就无法再以声音的形式在一个方向上快速地传播，而只能以热的形式留在介质中，直到通过缓慢的传导过程流向介质较冷的部分。

被我们称为光的那种运动，尽管比声音运动更加细微和快速变化，但也和声音一样是非常规则的运动，因此不是热运动。以前被称为热辐射的物理现象在物理本质上与光是一样的。

辐射在到达介质的某一部分后，就会进入并穿过该部分，再从另一端出现。该部分介质在传导辐射时处于一种特定的运动状态，而一旦辐射穿过该部分介质，该部分介质便会回到先前的状态，而运动完全转移到了介质中新的部分。

现在，我们称为热的这种运动，不会自发地从一个物体进入另一个物体，除非在整个传导过程中，第一个物体总是比第二个物体的温度高一些。因此，从介质的一部分完全流出再进入另一部分的辐射运动，就不适宜被称作热了。

We may apply the molecular theory of gases to test those hypotheses about the luminiferous ether which assume it to consist of atoms or molecules.

Those who have ventured to describe the constitution of the luminiferous ether have sometimes assumed it to consist of atoms or molecules.

The application of the molecular theory to such hypotheses leads to rather startling results.

In the first place, a molecular ether would be neither more nor less than a gas. We may, if we please, assume that its molecules are each of them equal to the thousandth or the millionth part of a molecule of hydrogen, and that they can traverse freely the interspaces of all ordinary molecules. But, as we have seen, an equilibrium will establish itself between the agitation of the ordinary molecules and those of the ether. In other words, the ether and the bodies in it will tend to equality of temperature, and the ether will be subject to the ordinary gaseous laws as to pressure and temperature.

Among other properties of a gas, it will have that established by Dulong and Petit, so that the capacity for heat of unit of volume of the ether must be equal to that of unit of volume of any ordinary gas at the same pressure. Its presence, therefore, could not fail to be detected in our experiments on specific heat, and we may therefore assert that the constitution of the ether is not molecular.

(**11**, 374-377; 1875)

我们可以用气体分子理论去检验那些假定光以太是由原子或分子组成的假说。

那些敢于描绘光以太组成的人们也曾假定它是由原子或分子组成的。

将分子理论应用于这类假说会导致相当令人惊讶的结果。

首先，以太分子只会是气体本身。如果我们愿意的话，我们可以假定每个以太分子都是一个氢分子的千分之一或百万分之一，并且它们可以自由穿越任何常见分子的间隙。但是，正如我们已经看到的，普通分子的扰动与以太分子的扰动之间会自发地建立平衡。换句话说，以太与其中的物体会趋于温度相同的状态，那么以太就会服从诸如压力和温度等常见的气体定律。

以太除了具有气体的性质以外，根据杜隆和珀蒂建立的定律，相同压力下，单位体积以太的热容与单位体积任何普通气体的热容相等。因此，在比热实验中不可能检测不到以太的存在，从而我们可以断言以太不是由分子构成的。

（王耀杨 翻译；李芝芬 审稿）

The Law of Storms

J. J. Murphy

Editor's Note

Waterspout phenomena have been known since the time of ancient Greece, but their explanation was still being debated. Here Joseph John Murphy offers a new explanation for cyclones. He notes that air warms and expands whenever water vapour within it condenses into droplets, as the condensation releases latent heat. In a column of air, such a process may be self-amplifying: the expanding air lowers the pressure on air beneath it, causing that air to expand, cool and trigger further condensation. There is some basis in Murphy's idea, especially as waterspouts tend to occur where cold air exists over warmer bodies of water. But the dynamics causing waterspout formation, usually in connection with cumulus clouds high above, are considerably more complex.

I have to thank you for publishing, in *Nature* of Dec. 2, 1875, my letter in reply to M. Faye's theory of cyclones, and I have now to submit some remarks on his theory of waterspouts.

I understand him to maintain that the dark part of the waterspout, which we see, contains a core of transparent air, which is descending at the centre of a vortex, and that the dark visible external part is a cloud formed by an ascending counter-current.

All this is unproved, and I think baseless. No dynamical reason can be assigned why there should be a downward current at the centre of the vortex. If the waterspout is formed in a vortex, which I think probable, though I am not certain of it, the vortical motion will produce not a downward but an upward current at its centre, in consequence of the diminution of barometric pressure, due to the air being thrown to the circumference by the centrifugal force. We see such upward currents formed in the little dust-whirlwinds that form themselves over streets and roads in windy weather.

Further, if M. Faye's theory were true, and if the waterspout were transparent at the centre, it could not be so well defined and solid as it usually is, nor could it be formed so rapidly.

The true theory of waterspouts is expounded in Espy's "Philosophy of Storms", a work which, notwithstanding its great error of denying the rotation of cyclones, made an era in meteorology, and, so far as I am aware, is not yet superseded.

风暴定律

墨菲

编者按

在古希腊时代人们就知道海龙卷这种现象，不过关于该现象的解释一直存在争议。在这篇文章中，约瑟夫·约翰·墨菲对龙卷风提出了一种新的解释。他认为，当空气中的水汽凝结成小水滴时，空气就会变热并膨胀，因为凝结过程释放了潜热。在一个空气柱中，这种过程会自身放大：空气的膨胀使其下方的空气压强降低，导致下方的空气发生膨胀、冷却并进一步引起水汽的凝结。墨菲的观点有一定的事实根据，特别是海龙卷经常发生在那些较温暖的洋面上存在冷空气的地方。不过，海龙卷的发生通常与高空积云有关，其动力学成因是相当复杂的。

感谢你们在 1875 年 12 月 2 日的《自然》上发表了我对费伊的龙卷风理论的回信，现在我对他的海龙卷理论再作一些评论。

我的理解是他坚持以下两点：第一，我们看到的海龙卷的黑色部分包含一个在涡旋中心位置正在下沉的透明空气核；第二，可以看到的黑色部分的外围是由一股上升的空气流形成的云。

这些观点还都没有得到证实，而且，我觉得也没有事实根据。他的理论没能说明在涡旋中心会有一股向下运动的气流的动力学原因。如果海龙卷是涡旋式的（我认为这是很可能的，尽管还不能确定），那么涡旋运动就会在涡旋中心产生一个向上而不是向下运动的气流，因为在离心力作用下空气向外围运动从而导致中心气压降低。在有风的天气里，我们可以在街道上小规模的沙尘旋风中看到这种上升的气流。

此外，如果费伊的理论是正确的，并且海龙卷的中心是透明的，那么海龙卷就不会像它通常表现出的那样坚实而清晰，而且也不会那么迅速地形成。

埃斯皮的《风暴原理》一书阐述了关于海龙卷的准确理论。尽管该理论存在否认龙卷风旋转这一重大错误，但它却开创了气象学的新纪元，并且就我所知，到现在为止它依然是无可替代的。

When vapour is condensed into water, forming cloud, the latent heat of the vapour is liberated and expands the air. A simple calculation shows that, after deducting the destroyed volume of the condensed vapour, the increased volume of the air due to this expansion is between four and five times as great as the volume of the vapour before condensation. If, then, the air is nearly saturated with moisture, and the temperature in a state of convective equilibrium for dry air (that is to say, when the difference between the temperatures of any two strata is that due to the difference of their pressures), and condensation begins in any column of air, the effect of liberating this heat will be to make the air of that column warmer and lighter than the air at corresponding heights in the surrounding columns. What follows is from Espy's work, page 44:—

"It begins, by its diminished specific gravity, to rise, and then, if all circumstances are favourable, the cloud will increase as it ascends, and finally become of so great perpendicular depth, that by its less specific gravity the air below it, in consequence of diminished pressure, will so expand and cool by expansion, as to condense the vapour in it; and this process may go on so rapidly that the visible cone may appear to descend to the surface of the sea or earth from the place where it first appears, in about one or two seconds. The terms here employed must not be understood to mean that the cloud actually descends; it appears to the spectator to descend, but this is an optical deception, arising from new portions of invisible vapour constantly becoming condensed, while all the time the individual particles are in rapid motion upwards."

To this I will add as very probable, if not quite certain, that the rarefaction thus caused at the waterspout will produce an inflow of air from all sides, and this will produce a vortex at the centre; this again, by its centrifugal force, will increase the rarefaction, and thus will intensify the effect. But the commencement of the waterspout is in the way described by Espy in the above extract.

(**13**, 187; 1875)

Joseph John Murphy: Old Forge, Dunmurry, Co. Antrim, Dec. 12, 1875.

当水汽凝结为水滴而形成云时，水汽中的潜热就被释放出来并使空气膨胀。一个简单的计算说明：扣除水汽凝结造成的体积减小，潜热释放使空气膨胀而造成的空气体积增加大约是水汽凝结前空气体积的4~5倍。如果此时空气湿度接近饱和，而且温度正好能使干空气处于对流平衡状态（也就是说，此时任意两个气层之间的温度差都取决于它们之间的气压差），并且在某气柱中水汽发生了凝结，那么水汽凝结释放的潜热就会使该气柱中的空气比周围同高度的空气更暖更轻。接下来便会像埃斯皮的著作中第44页所说的：

“这部分空气由于比重减小而开始上升，如果此时环境条件都很适宜，那么随着空气的上升，云团会逐渐增大，最终其垂直厚度会变得非常大。这就会使得云团下比重较小的空气因为气压降低而膨胀变冷，从而使其中的水汽发生凝结。这个过程可能会进行得非常迅速，以至于可见的锥形云团在大概一两秒内就从其最初出现的地方下降到海面或者地面处。我们不能把这句话理解成云团真的在下降；在观察者看来云团似乎在下降，但这只是种视觉错觉，实际上是因为空气中的气体分子在快速上升的同时，不断有新的不可见水汽持续地发生凝结。”

对此，我想提出一个虽不是十分肯定但可能性很大的补充：海龙卷中的空气稀薄会使周围的空气从各个方向流入，使龙卷中心形成涡旋。在离心力作用下，龙卷中心的空气更加稀薄，从而加强了这种作用。不过，海龙卷最初的形成还是以上述摘录中埃斯皮描述的方式进行的。

（刘明 翻译；王鹏云 审稿）

On the Telephone, an Instrument for Transmitting Musical Notes by Means of Electricity

J. Munro

Editor's Note

Elisha Gray, *Nature* here reports, recently presented a paper describing a means for sending musical notes over long distances by electricity—in short, a telephone. Gray also showed how telegraphic messages could be transmitted this way. The device works by interrupting the electrical current at one end in a desired pattern, and then detecting a similar interruption at the other end, and linking this signal to a reed or box for producing sound. Gray demonstrated that as many as eight messages may be sent simultaneously. He invented his technique at nearly the same time as—some believe before—Alexander Graham Bell, who is widely credited as the inventor of the telephone.

MR. Elisha Gray recently read a paper before an American Society explaining his apparatus for transmitting musical notes by electricity. He showed experimentally how, by means of a current of electricity in a single wire, a number of notes could be reproduced simultaneously at a great distance, and how by this means also a number of telegraphic messages could be transmitted at once along a wire and separately received at the other end. One of Mr. Gray's apparatuses was exhibited in London at the last *soirée* of the Society of Telegraph Engineers by the president, Mr. Latimer Clark. The principle of the apparatus is as follows:—

A vibrating reed is caused to interrupt the electric current entering the wire a certain number of times per second and the current so interrupted at the sending end sets a similar reed vibrating at the distant end.

The sending reed is ingeniously maintained in constant vibration by a pair of intermittent electro-magnets which are magnetised and demagnetised by the vibrating reed itself.

Thus in Fig. 1 (which represents the transmitting part of the telephone and its connections for a single note), the current from the magnet battery flowing in the direction of the small arrow passes through the pair of electro-magnets A to the terminal *r* of the reed R, and thence by the spring contact *b* and the wire *bz* to the battery again, completing its circuit without passing through the other pair of electro-magnets B, which are not therefore magnetised. The reed R is consequently pulled over by the electro-magnets A. But on this taking place the spring contact *b* is broken and the circuit is no longer completed through *bz* but through the electro-magnets B, which are consequently magnetised, and

电话：一种利用电流传送音符的仪器

芒罗

编者按

《自然》的这篇文章报道的是伊莱沙·格雷最新提交的一篇论文，论文中描述了一种利用电流长距离传送音符的方法，简言之就是电话。另外，格雷还展示了如何通过这种方法传送电报信息。该装置的工作原理是：在一端以某种设计好的模式中断电流，然后在另一端探测到类似的中断电流信号，并将该信号连接到一个簧片或者盒子中来产生声音。格雷展示的这种装置可以同时传送多达 8 条信息。他差不多是在亚历山大·格雷厄姆·贝尔发明电话的同时发明了这种技术，甚至有些人认为格雷的发明比贝尔还早，不过，人们更普遍认为贝尔是电话的发明者。

最近，伊莱沙·格雷先生向美国某学会宣读了一篇论文，介绍了他发明的利用电流传送音符的设备，他用实验验证了如何利用单一导线中的电流使大量音符在很远的距离处能够同时重现，以及如何利用同样的方式使大量电报信息迅速通过一根导线传送而在接收端分别接收到信息。电报工程师协会会长拉蒂默·克拉克先生在该协会最近的一次晚宴（在伦敦）上展示了格雷先生的设备中的一种。这台设备的工作原理如下。

用一个振动簧片按每秒钟若干次中断进入电线的电流，那么这种在信号发送端的断续电流就会使得位于远程接收端的簧片产生类似的振动。

发送端的簧片通过一对间歇式电磁铁来精确地保持不断振动，而振动簧片本身又可以使电磁铁被磁化和退磁。

因此在图 1（示意了电话的发送部分及其产生单一音符所对应的连接）中，电流由磁铁电池出发，按照小箭头的方向经过一对电磁铁 A 流向簧片 R 的末端 r，而后，通过弹簧触点 b 沿 bz 又回到电池，没有经过另一对电磁铁 B 就完成了回路，因此 B 没有被磁化。结果簧片 R 被拉向磁化了的电磁铁 A。但是这将导致弹簧触点 b 断开，回路不再通过 bz 闭合，而是通过电磁铁 B 闭合，随之它被磁化并通过感应把簧片引向 B，因此簧片又被弹回到中间的位置，这样 b 点又被连接上，而电磁铁

tend by their induction on the reed to neutralise that of B. The reed therefore springs back to its intermediary position, but in so doing the contact at *b* is again made and the electro-magnets B again short-circuited and the reed pulled over (or rather *assisted* over, for it has its own resilience or spring) towards A; so this goes on keeping the reed in vibration between the electro-magnets and alternately making and breaking the spring contact *b* and also that of *a*, the number of contacts per second being dependant on the vibrating period of the reed.

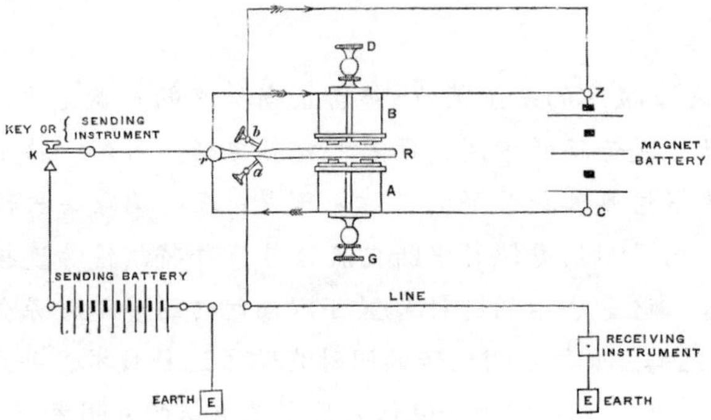

Fig. 1

While this is going on the reed of course emits its musical note. Two Leclanché or bichromate cells are sufficient to work the transmitter and give a good note. The spring contact *b* is to be adjusted by the screw there seen until the note emitted by the reed is both loud and pure. The magnets A and B are adjustable to or from the reed by the milled heads G and D.

The spring contact *a* just mentioned belongs properly to the line circuit. It is the intermittent contact which interrupts the current sent into the line. As will be seen from the diagram the circuit of the sending battery is made through the key K, the reed, and the spring contact *a*. On holding down the key K the current flows into the line, being interrupted, however, by the contact *a* as many times per second as the reed vibrates, and this intermittent current flowing to earth at the distant station, s made to elicit a corresponding note from the receiving apparatus there.

The receiving instruments are of two kinds, electro-magnetic and physiological.

In the first there is a plain double electro-magnet with a steel tongue having one end rigidly fixed to one pole, the other end being free to vibrate under the other pole. This stands over a wooden pipe closed at one end. Thus in Fig. 2 *t*T is the steel tongue fixed at *t* and free at T, while P is the sounding-pipe. The received current, coming from the line and passing through the electro-magnet M to earth, sets the tongue vibrating, and the pipe gives forth the same note as the reed at the sending station. Ten Daniell cells working through 1,000 ohms, give a good strong note, especially when the receiver is held in the hand close to the head. The screw *a*, Fig. 1, must be adjusted to give the best effect.

B 又被短路，弹簧重被拉向 A（或者更确切地说是因为簧片自身的弹力或弹性促使簧片偏向）；如此往复使得簧片 R 在一对电磁铁之间不停地振动，使弹簧触点 a 和 b 交替断开和接通，且每秒钟的接通次数取决于簧片 R 的振动周期。

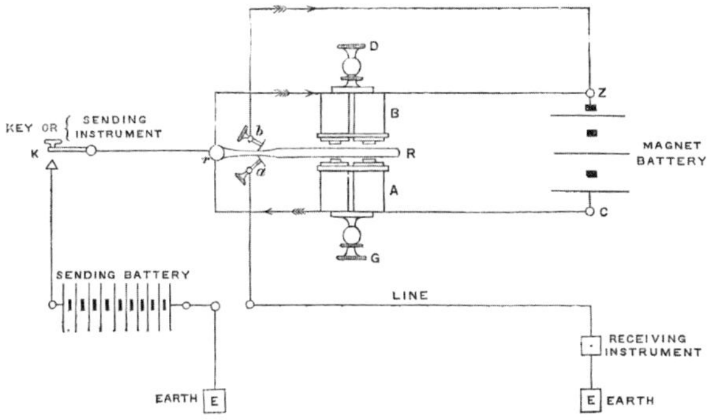

图 1

在上述的振动过程中簧片发出音符，两个勒克朗谢电池或者重铬酸盐电池足以启动信号发送器并能发出令人愉快的音符。通过螺丝可以调整弹簧触点 b 直到簧片发出的音符足够响亮和纯正。电磁铁 A 和 B 相对于簧片的位置可以用滚压了纹边的调节头 G 和 D 进行调整。

刚提到的弹簧触点 a 属于电话线路中的触点，它与簧片的间歇性接触不断地中断电话线路中的电流信号。正如图 1 中所示的，发送电池组的电路包括开关 K，簧片 R 和弹簧触点 a。接通开关按钮 K，电流流入电话线路，但随着簧片的振动，弹簧触点 a 会每秒多次断开而中断电流，这些间歇式的电流会流向远处的接地端，s 用于从接收设备引导出相应的信号。

信号接收设备分为两种，一种是电磁型的，另一种是生理型的。

第一类设备中有一个简单的双电磁铁，电磁铁上的钢舌一端被刚性地固定在一个电极上，另一端位于另一个电极下方，可以自由振动。这些竖立在一个一端封闭的木制管子上。如图 2 所示，tT 为钢舌，其中 t 为固定端，T 为自由端，P 为发音管。接收到的电流从电话线路流经电磁铁 M 最后到接地端，该电流使钢舌振动，于是发音管发出音符，这与发送端簧片发出的音符完全相同。10 个丹聂耳电池可以负载 1,000 欧姆的电阻，使设备发出强劲而优美的信号，尤其是当接收器被拿在手里靠近头部的时候。为了得到最好的效果，必须校正图 1 中的螺丝 a。

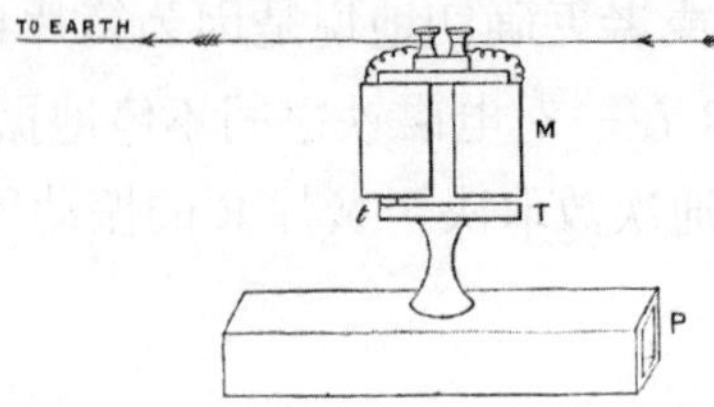

Fig. 2

The other receiving instrument is the most interesting of the two. It consists of a small induction coil used in conjunction with a peculiar sounding-box, as shown in Fig. 3.

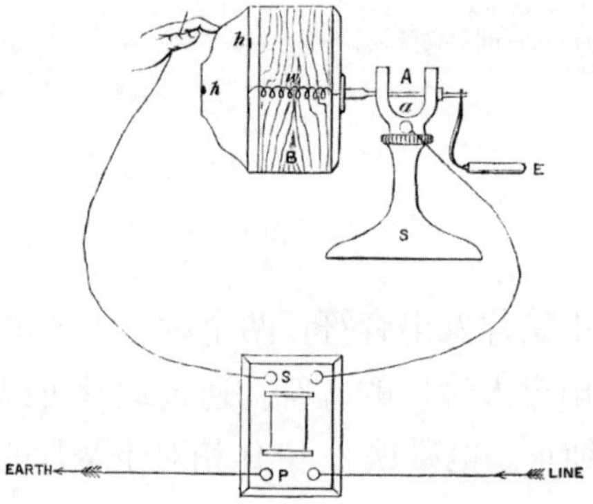

Fig. 3

Here the line-current is passed to earth through the primary circuit P of the small induction coil, and the induced current is led to the sounding-box. This consists of a flat hollow cylindrical wooden box B, covered by a convoluted face of sheet zinc with two air holes *hh*, perforated in it, this box is attached to a metal axle A, turning in forked iron bearings, insulated from but supported by an iron stand S. By this means the sounding-box can be revolved by the ebony handle E. The zinc face is connected across the empty interior of the box by a wire W to the metal bearings on the other side. One end of the secondary circuit of the induction coil is to be connected to the metal bearing by the terminal *a*, and the other to a short bare wire held in the left hand. On then striking a finger of the hand holding the wire smartly across the zinc face, the proper note is sounded by the box; or, what is more convenient, on turning the box by the insulated handle and keeping the point of the finger rubbing on its face, the note is heard. The rough under side of the finger pressed pretty hard on the bulging part of the face is best. The instant the current is put on by the sending key K, Fig. 1, the dry rasp of the skin on the zinc-surface becomes changed into a musical note.

These "sounders" can be made to receive indifferently a variety of notes. I have under my care at present a telephone with four transmitters tuned to give the four notes of the

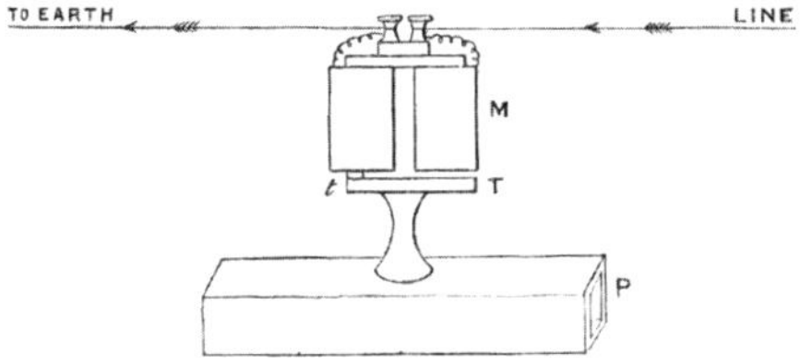

图 2

另一类接收设备是两类中最有意思的，它包括一个很小的感应线圈，这个感应线圈用来连接一个特殊的发音盒，如图 3 所示。

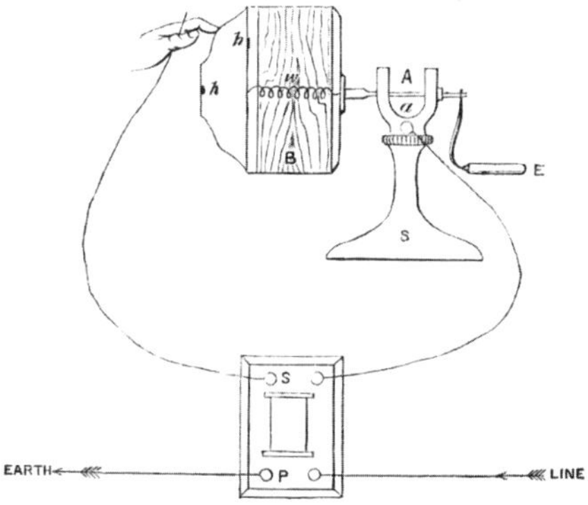

图 3

在这里，线路中的电流是通过带有小感应线圈的初始电路 P 后接地，感应电流通向发音盒。发音盒包括一个扁平的空心木质圆柱体 B，圆柱体表面缠绕着金属锌薄片，其中打有两个通气孔 *hh*，这个发音盒与金属轴 A 相连，A 可在铁制叉状支座上旋转，由铁架 S 支撑但与 S 之间是绝缘的。用这种方式可以通过黑檀木手柄 E 来转动发音盒，锌片表面通过金属丝 W 穿过中空的发音盒内部与另一端的金属支座连接。感应线圈次级电路的一端在接线点 *a* 与金属支架相连，另一端连着一根较短的裸线，用左手握住。接下来用握线的手的一个手指轻快地划过锌片表面，发音盒就会发出特定的音符；或者，更简便的方法是通过绝缘的手柄转动发音盒，使手指尖不停地摩擦锌片表面，这样也可以听到音符。被手指用力压住的锌片表面越粗糙，产生的声音越好。一旦通过发送端的开关 K（见图 1）接通电流，皮肤在锌片表面发出的单调声响就会转化成音符。

这些“发声器”可以制作成各种音符的接收器，目前我已经有一部具有 4 个发送器、能发出 4 种普通和弦音符的电话，以及两台能够准确解析任意一个音符或者

common chord, and two receivers, which interpret equally well any one of these notes or all together. But sounders are also made in the same way which will emit only one special note, and so are sensible only to the corresponding current. It is by their means that the telephone can be applied to multiplex telegraphy. As many as eight transmitters may be set to interrupt the line current according to the vibrations of eight different tuning-forks, and the resultant current can be made by means of eight special receivers to reproduce the same number of corresponding notes at the distant station. The current is controlled by eight keys at the sending end and sifted by eight sounders at the receiving end, each sounder being sensitive only to those portions of the current affected by its corresponding transmitter. The superimposed effect of the eight keys and transmitters on the line current can all be separately interpreted at the receiving end. Thus eight messages might be transmitted simultaneously along one wire in the same direction. It would seem hitherto, however, that this method of telegraphy by the telephone is inferior to the ordinary methods in point of speed of signalling, and in the length of circuit which can be worked by a given battery power.

(**14**, 30-32; 1876)

同时解析所有音符的接收器。但是发声器同样也可以做成只能发出一种特定音符，因此只对相应的电流敏感。这就意味着电话可以应用在多路电报上。最多可以有 8 个发送器，依据 8 个不同音叉的振动情况来中断各自电路中的电流，这个合成的电流可以通过远处接收端的 8 个特殊接收器来重现数量相同的对应音符。电流由发送端的 8 个开关控制，并通过接收端的 8 个发声器来过滤筛选，每一个发声器仅对与它相对应的发送器产生的电流敏感。发送端的 8 个开关及发送器对于线路电流所加的影响在接收端可以被分别解析出来。因此 8 种信息可以沿着一条电线向同一个方向同步传送。但是就目前来看，这种用电话发送电报的方式在信号传输速度方面和电池可以负载的线路长度方面不如普通的方法。

（胡雪兰 翻译；赵见高 审稿）

Maxwell's Plan for Measuring the Ether

Editor's Note

One of the most remarkable of Maxwell's scientific exploits was a scheme for telling the velocity of the Earth and the Solar System as a whole relative to the luminiferous ether, supposed at the time to be necessary for the propagation of electromagnetic waves. Maxwell's proposal to D. P. Todd, director of the Nautical Almanac office in Washington, D.C., was that accurate measurements of the rotation of Jupiter's satellites around their planet would allow this relative velocity through the ether to be derived. If Maxwell had been able to execute this plan (for which in Todd's opinion the data were not yet sufficiently accurate), he would have discovered that the ether is irrelevant to the propagation of electromagnetic waves and indeed does not exist—the foundation for Einstein's Theory of Special Relativity.

"On a Possible Mode of Detecting a Motion of the Solar System through the Luminiferous Ether". By the late Prof. J. Clerk Maxwell. In a letter to Mr. D. P. Todd, Director of the *Nautical Almanac* Office, Washington, U.S. Communicated by Prof. Stoke, Sec. R.S.

Mr. Todd has been so good as to communicate to me a copy of the subjoined letter, and has kindly permitted me to make any use of it.

As the notice referred to by Maxwell in the *Encyclopaedia Britannica* is very brief, being confined to a single sentence, and as the subject is one of great interest, I have thought it best to communicate the letter to the Royal Society.

From the researches of Mr. Huggins on the radial component of the relative velocity of our sun and certain stars, the coefficient of the inequality which we might expect as not unlikely, would be only something comparable with half a second of time. This, no doubt, would be a very delicate matter to determine. Still, for anything we know *à priori* to the contrary, the motion might be very much greater than what would correspond to this; and the idea has a value of its own, irrespective of the possibility of actually making the determination.

In his letter to me Mr. Todd remarks, "I regard the communication as one of extraordinary importance, although (as you will notice if you have access to the reply which I made) it is likely to be a long time before we shall have tables of the satellites of Jupiter sufficiently accurate to put the matter to a practical test."

麦克斯韦测量以太的计划

编者按

麦克斯韦最突出的科学成就之一是，他制定了地球和太阳系作为一个整体相对于以太的速度的测量计划。当时，以太被认为是电磁波传播的必要条件。麦克斯韦向华盛顿特区航海历书处的办公室主任托德建议，通过精确测量木星卫星围绕木星的公转可以得到太阳系整体相对于以太的速度。如果麦克斯韦当时真的实施了这个计划（托德认为该计划所需的数据还不够精确），那么他会发现，以太和电磁波的传播没有任何关系，而且实际上以太根本就不存在——这便是爱因斯坦的狭义相对论的基础。

《论一种探测太阳系在以太中运动的可能方式》，作者是已故的克拉克·麦克斯韦教授。这篇文章出现在麦克斯韦写给美国华盛顿**航海历书处**的办公室主任托德先生的一封信中，由皇家学会的秘书斯托克教授宣读。

托德先生十分慷慨地给了我一份附信的拷贝，并且大方地允许我使用它。

因为麦克斯韦提到的《大英百科全书》中的参考资料十分简略，仅仅只有一句话。而这个题目又十分吸引人，所以我觉得最好还是在皇家学会宣读这封信。

根据哈金斯先生对太阳和某些星体的相对速度径向分量的研究，系数上的差别（我们暂且认为结果可靠）大约只有半秒。毫无疑问，这对实验测量来说是一个很精细的问题。而且这和我们**先验的**观点相反，我们先验的观点中的运动比这里涉及的大很多。如果不考虑实际测量的可操作性，这个想法还是有它自身的价值的。

在托德先生给我的信中，他评论到，“虽然（如您将在我的答复中看到的那样）将这个计划付诸实施需要足够精确的木星卫星的数据表，而且这恐怕需要很长时间才能获得，但是我仍然认为这封信极其重要。”

I have not thought it expedient to delay the publication of the letter on the chance that something bearing on the subject might be found among Maxwell's papers.

(Copy)

Cavendish Laboratory,

Cambridge,

19th March, 1879

Sir,

I have received with much pleasure the tables of the satellites of Jupiter which you have been so kind as to send me, and I am encouraged by your interest in the Jovial system to ask you if you have made any special study of the apparent retardation of the eclipses as affected by the geocentric position of Jupiter.

I am told that observations of this kind have been somewhat put out of fashion by other methods of determining quantities related to the velocity of light, but they afford the *only* method, so far as I know, of getting any estimate of the direction and magnitude of the velocity of the sun with respect to the luminiferous medium. Even if we were sure of the theory of aberration, we can only get differences of position of stars, and in the terrestrial methods of determining the velocity of light, the light comes back along the same path again, so that the velocity of the earth with respect to the ether would alter the time of the double passage by a quantity depending on the square of the ratio of the earth's velocity to that of light, and this is quite too small to be observed.

But if J E is the distance of Jupiter from the earth, and l the geocentric longitude, and if l' is the longitude and λ the latitude of the direction in which the sun is moving through ether with velocity v, and if V is the velocity of light and t the time of transit from J to E,

$$\mathrm{JE} = [\,\mathrm{V} - v\cos\lambda\cos(l - l')\,]\,t$$

By a comparison of the values of t when Jupiter is in different signs of the zodiac, it would be possible to determine l' and $v\cos\lambda$.

I do not see how to determine λ, unless we had a planet with an orbit very much inclined to the ecliptic. It may be noticed that whereas the determination of V, the velocity of light, by this method depends on the differences of J E, that is, on the diameter of the earth's orbit, the determination of $v\cos\lambda$ depends on J E itself, a much larger quantity.

But no method can be made available without good tables of the motion of the satellites, and as I am not an astronomer, I do not know whether, in comparing the observations with the tables of Damoiseau, any attempt has been made to consider the term in $v\cos\lambda$.

将这封信推迟到可以在麦克斯韦的论文中找到相关内容的时候再发表，我认为并不适宜。

（原信的拷贝）

卡文迪什实验室，
剑桥，
1879 年 3 月 19 日

先生：

我非常高兴能收到您寄给我的木星卫星的数据表。受您对木星系统的兴趣的启发，我想问您有没有对由木星相对于地心的位置而造成的木星卫星蚀的明显延迟作过专门的研究。

我得知，在其他测量与光速有关的物理量的方法面前，这种观测已经有些过时了。但是，据我所知，这种观测给我们提供了估算太阳相对于以太介质的速度的方向和大小的**唯一**方法。即便我们坚信像差理论，我们得到的也只是恒星位置的差别。并且按照在地球上测量光速的方法，光按原路返回，地球相对于以太的速度会改变这一往返的时间。但是，这个时间上的改变量依赖于地球速度与光速之比的平方，因此，这个量太小，以至于无法观测。

但是，如果 JE 表示木星与地球之间的距离，l 表示地心经度，l' 和 λ 分别表示太阳在以太中运动方向的经度和纬度，v 表示太阳在以太中运动的速度，V 表示光速，t 表示光从木星到地球的传播时间，

$$\mathrm{JE} = [\,\mathrm{V} - v\cos\lambda\cos(l - l')\,]\,t$$

通过比较木星在黄道不同位置时的 t，就有可能确定 l' 和 $v\cos\lambda$。

除非有一个行星的轨道向黄道极大倾斜，不然我就无法知道怎样确定 λ。我们应该注意到，尽管用这种方法测量的光速 V 也与 JE 的变化有关，也就是说，依赖于地球轨道的直径，但是 $v\cos\lambda$ 却依赖于一个更大的量——JE 本身。

然而，没有精确的卫星运动数据就不会有任何行之有效的方法。我本人并不是天文学家，我不知道这些观测与达穆瓦素的数据表相比，是否曾经尝试过将 $v\cos\lambda$ 中的各项考虑在内。

I have, therefore, taken the liberty of writing to you, as the matter is beyond the reach of any one who has not made a special study of the satellites.

In the article E [ether] in the ninth edition of the "Encyclopaedia Britannica", I have collected all the facts I know about the relative motion of the ether and the bodies which move in it, and have shown that nothing can be inferred about this relative motion from any phenomena hitherto observed, except the eclipses, &c., of the satellites of a planet, the more distant the better.

If you know of any work done in this direction, either by yourself or others, I should esteem it a favour to be told of it.

Believe me,

Yours faithfully,

(Signed) J. Clerk Maxwell

(**21**, 314-315; 1880)

因此，我十分冒昧地给您写这封信。这个问题对没有专门研究过卫星的人来说，实在是勉为其难。

在《大英百科全书》第 9 版关于以太的文章中，我收集了所有我知道的关于以太以及在其中运动的物体的相对运动的资料，从这些资料中我发现，由目前观测到的实验现象，除了行星离我们越远，越有利于卫星蚀等现象以外，从其他任何已观测到的相关现象中都不能推断出有关这种相对运动的结论。

如果您知道有任何人，不论是您还是别人，做过这方面的工作，都请您告诉我，我将不胜感激。

相信我，

您忠实的，

（签名）克拉克·麦克斯韦

（王静 翻译；鲍重光 审稿）

Clerk Maxwell's Scientific Work

P. G. Tait

Editor's Note

James Clerk Maxwell died in November 1879. In this essay four months later, Scottish physicist Peter Guthrie Tait, who had known Maxwell from childhood, paid tribute to his accomplishments. Maxwell was producing influential, original work before the age of twenty. In 1864 he published a landmark paper giving the first complete statement of his theory of electricity and magnetism. It explained electromagnetic phenomena without recourse to action at a distance, and provided a unified view of what light is. Guthrie notes that the facility of Maxwell's thinking did not always translate into effective lectures. While the treatises he wrote were models of clarity, his extemporaneous lectures gave free rein to his imagination in a way that taxed his audiences.

AT the instance of Sir W. Thomson, Mr. Lockyer, and others I proceed to give an account of Clerk Maxwell's work, necessarily brief, but I hope sufficient to let even the non-mathematical reader see how very great were his contributions to modern science. I have the less hesitation in undertaking this work that I have been intimately acquainted with him since we were schoolboys together.

If the title of mathematician be restricted (as it too commonly is) to those who possess peculiarly ready mastery over symbols, whether they try to understand the significance of each step or no, Clerk Maxwell was not, and certainly never attempted to be, in the foremost rank of mathematicians. He was slow in "writing out", and avoided as far as he could the intricacies of analysis. He preferred always to have before him a geometrical or physical representation of the problem in which he was engaged, and to take all his steps with the aid of this: afterwards, when necessary, translating them into symbols. In the comparative paucity of symbols in many of his great papers, and in the way in which, when wanted, they seem to grow full-blown from pages of ordinary text, his writings resemble much those of Sir William Thomson, which in early life he had with great wisdom chosen as a model.

There can be no doubt that in this habit, of constructing a mental representation of every problem, lay one of the chief secrets of his wonderful success as an investigator. To this were added an extraordinary power of penetration, and an altogether unusual amount of patient determination. The clearness of his mental vision was quite on a par with that of Faraday; and in this (the true) sense of the word he was a mathematician of the highest order.

But the rapidity of his thinking, which he could not control, was such as to destroy, except for the very highest class of students, the value of his lectures. His books and his written

克拉克·麦克斯韦的科学工作

泰特

编者按

1879年11月，詹姆斯·克拉克·麦克斯韦逝世。在4个月后的这篇短文中，苏格兰物理学家彼得·格思里·泰特（他从小就认识麦克斯韦）热情称颂了麦克斯韦的成就。麦克斯韦在20岁之前就开始发表有影响力的原创研究论文。1864年，他发表了一篇首次完整阐述其电磁理论的论文。在这篇里程碑式的论文中，麦克斯韦抛开一定距离外的实际效应去解释电磁现象，并就光的本质提出了一个统一性的观点。格思里认为，麦克斯韦深刻的思考并没有通过他那些颇有影响的演讲全部体现出来。不过，与他那堪称逻辑清晰之典范的专著不同，他的即席演讲在某种程度上则更自由地展现了他那对听众来说过于跳跃的想象力。

应汤姆孙爵士、洛克耶先生以及其他一些人士的要求，我将对麦克斯韦的科学工作进行介绍。介绍必然是简略的，但我希望足以让即使没有数学背景的读者也能了解他对现代科学的伟大贡献。我与麦克斯韦在学生时代就已经熟识，因此，我毫不犹豫地接受了这个任务。

如果说数学家的头衔只属于那些不论是否试图理解每一步的意义，都对符号了如指掌的人（事实通常就是这样），那么克拉克·麦克斯韦就不是、也从来不试图成为这样的一流数学家。他总是不慌不忙地“完稿”，并且尽可能避免复杂的分析。他更喜欢以几何或物理的形式表示自己所研究的问题，并且总是借助于以下方式来完成下一步的研究：在必要时，将这些表示转化成符号的形式。在他众多的伟大著作中所出现的较少的符号，在需要的时候似乎又能拓展成为直接触及主题的成熟的篇章，从这些方面来看，他的著作与威廉·汤姆孙爵士的十分相似，他在早年就很有远见卓识地把爵士当作自己的榜样。

毫无疑问，为每个问题建立思维上的表示，这个习惯是麦克斯韦作为一名研究者能够获得巨大成功的主要秘诀之一。除此之外，他还拥有卓越的洞察力和非凡的意志力。他的思路清晰，堪比法拉第。从数学家一词的这种（真正的）意义来说，麦克斯韦就是一位最高层次的数学家。

然而，麦克斯韦的思维之敏捷，连他自己都无法控制，这使得他的讲座只能被极高层次的学生所接受，而其他人则很难从中受益。他的著作和演讲稿（通常是对

addresses (always gone over twice in MS.) are models of clear and precise exposition; but his *extempore* lectures exhibited in a manner most aggravating to the listener the extraordinary fertility of his imagination.

His original work was commenced at a very early age. His first printed paper, "*On the Description of Oval Curves, and those having a Plurality of Foci*", was communicated for him by Prof. Forbes to the Royal Society of Edinburgh, and inserted in the "*Proceedings*" for 1846, before he reached his fifteenth year. He had then been taught only a book or two of Euclid, and the merest elements of Algebra. Closely connected with this are three unprinted papers, of which I have copies (taken in the same year), on "*Descartes' Ovals*", "*The Meloid and Apioid*", and "*Trifocal Curves*". All of these, which are drawn up in strict geometrical form and divided into consecutive propositions, are devoted to the properties of plane curves whose equations are of the form

$$mr + nr' + pr'' + \cdots = constant$$

r, r', r'', &c., being the distances of a point on the curve from given fixed points, and m, n, p, &c., mere numbers. Maxwell gives a perfectly general method of tracing all such curves by means of a flexible and inextensible cord. When there are but two terms, if m and n have the same sign we have the ordinary Descartes' Ovals, if their signs be different we have what Maxwell called the Meloid and the Apioid. In each case a simple geometrical method is given for drawing a tangent at any point, and some of the other properties of the curves are elegantly treated.

Clerk Maxwell spent the years 1847–1850 at the University of Edinburgh, without keeping the regular course for a degree. He was allowed to work during this period, without assistance or supervision, in the Laboratories of Natural Philosophy and of Chemistry: and he thus experimentally taught himself much which other men have to learn with great difficulty from lectures or books. His reading was very extensive. The records of the University Library show that he carried home for study, during these years, such books as Fourier's *Théorie de la Chaleur*, Monge's *Géometrie Descriptive*, Newton's *Optics*, Willis' *Principles of Mechanism*, Cauchy's *Calcul Différentiel*, Taylor's *Scientific Memoirs*, and others of a very high order. These were *read through*, not merely consulted. Unfortunately no list is kept of the books consulted in the Library. One result of this period of steady work consists in two elaborate papers, printed in the *Transactions of the Royal Society of Edinburgh*. The first (dated 1849) "*On the Theory of Rolling Curves*", is a purely mathematical treatise, supplied with an immense collection of very elegant particular examples. The second (1850) is "*On the Equilibrium of Elastic Solids*". Considering the age of the writer at the time, this is one of the most remarkable of his investigations. Maxwell reproduces in it, by means of a special set of assumptions, the equations already given by Stokes. He applies them to a number of very interesting cases, such as the torsion of a cylinder, the formation of the large mirror of a reflecting telescope by means of a partial vacuum at the back of a glass plate, and the theory of Örsted's apparatus for the compression of water. But he

手稿的再一次重温）是表述清晰精确的典范；但是他的**即席**演讲却因为极富想象力的风格而使听众难以理解。

麦克斯韦最初的工作在他年轻时就已经着手展开。他发表的第一篇论文《论椭圆曲线及多焦点椭圆曲线》由福布斯教授代他在爱丁堡皇家学会的会议上宣读，并收录在《爱丁堡皇家学会会刊》中。当时是1846年，麦克斯韦尚不满15岁。那时的他只学过一两本欧几里德的书和最基本的代数基础。紧接着，他又写了另外3篇没有发表的文章，我有这几篇文章的拷贝，它们分别是：《笛卡尔椭圆》、《芜菁科昆虫形曲线和芹亚科植物形曲线》、《三焦点曲线》。这些建立在严格的几何形式上并且被分成论题连贯的文章，可以用来研究具有以下形式的方程所描述的平面曲线的性质：

$$mr + nr' + pr'' + \cdots = \text{常量}$$

r, r', r'' 等是从一个给定的固定点到曲线上某一点的距离，m, n, p 等仅仅是一些数字。麦克斯韦用容易弯曲但不能伸展的绳子，给出了一种理想的绘制这样曲线的一般方法。当方程中只有 m 和 n 两项时，如果两者同号，就得到普通的笛卡尔椭圆，如果两者异号，就得到麦克斯韦所谓的芜菁科昆虫形曲线和芹亚科植物形曲线。在每种情况下，麦克斯韦都给出了在任意一点作切线的简单几何方法，并且很好地处理了曲线其他方面的性质。

1847~1850年，克拉克·麦克斯韦就读于爱丁堡大学，在此他并不需要去学那些学位要求的常规课程，而是被允许在既无人帮助也无人指导的条件下在自然哲学与化学实验室工作。因此，他在实验中自学了很多东西，而这些是其他人很难从书本中或课堂上学到的。他的阅读非常广泛。大学图书馆的记录显示，他在这些年中借回家研读的书有傅立叶的《热力学理论》、蒙日的《几何学说明》、牛顿的《光学》、威利斯的《力学原理》、柯西的《微分计算》、泰勒的《科学回忆录》等高水平著作。这些书他全部**通读**，而不是仅仅翻阅一下。很可惜图书馆没有保存他在馆内阅览的书单。麦克斯韦这段时间持续学习的成果之一是发表在《爱丁堡皇家学会会报》上的两篇详细论文。第一篇《论曲线滚动理论》（1849年），这是一篇纯数学的论文，文章给出了大量简洁而恰当的例子。第二篇是《论弹性固体的平衡》（1850年）。考虑到作者当时的年龄，这可以被认为是他最不寻常的研究之一。麦克斯韦运用一系列特殊的假设，重新构造了已经由斯托克斯给出的方程。他将这些方程应用到许多非常有趣的情况中，比如圆柱体的扭曲、通过在玻璃板后形成局部真空的方法实现反射式望远镜的巨大镜面的构造、奥斯特的水压缩装置的理论。此外，他还将其方程应用于张力（向一个垂直穿过透明板的圆柱体施加力偶后在透明平板中产生的）

also applies his equations to the calculation of the strains produced in a transparent plate by applying couples to cylinders which pass through it at right angles, and the study (by polarised light) of the doubly-refracting structure thus produced. He expresses himself as unable to explain the permanence of this structure when once produced in isinglass, gutta percha, and other bodies. He recurred to the subject twenty years later, and in 1873 communicated to the Royal Society his very beautiful discovery of the *temporary* double refraction produced by shearing in viscous liquids.

During his undergraduateship in Cambridge he developed the germs of his future great work on "Electricity and Magnetism" (1873) in the form of a paper "On Faraday's Lines of Force", which was ultimately printed in 1856 in the "Trans. of the Cam. Phil. Soc." He showed me the MS. of the greater part of it in 1853. It is a paper of great interest in itself, but extremely important as indicating the first steps to such a splendid result. His idea of a fluid, incompressible and without mass, but subject to a species of friction in space, was confessedly adopted from the analogy pointed out by Thomson in 1843 between the steady flow of heat and the phenomena of statical electricity.

Other five papers on the same subject were communicated by him to the *Philosophical Magazine* in 1861–1862, under the title *Physical Lines of Force*. Then in 1864 appeared his great paper "*On a Dynamical Theory of the Electromagnetic Field*". This was inserted in the *Philosophical Transactions*, and may be looked upon as the first complete statement of the theory developed in the treatise on *Electricity and Magnetism*.

In recent years he came to the conclusion that such analogies as the conduction of heat, or the motion of the mass-less but incompressible fluid, depending as they do on Laplace's equation, were best symbolised by the quaternion notation with Hamilton's ∇ operator; and in consequence, in his work on electricity, he gives the expressions for all the more important physical quantities in their quaternion form, though without employing the calculus itself in their establishment. I have discussed in another place (*Nature*, vol. VII, p. 478) the various important discoveries in this remarkable work, which of itself is sufficient to secure for its author a foremost place among natural philosophers. I may here state that the main object of the work is to do away with "action at a distance," so far at least as electrical and magnetic forces are concerned, and to explain these by means of stresses and motions of the medium which is required to account for the phenomena of light. Maxwell has shown that, on this hypothesis, the velocity of light is the ratio of the electro-magnetic and electro-static units. Since this ratio, and the actual velocity of light, can be determined by absolutely independent experiments, the theory can be put at once to an exceedingly severe preliminary test. Neither quantity is yet fairly known within about 2 or 3 percent, and the most probable values of each certainly agree more closely than do the separate determinations of either. There can now be little doubt that Maxwell's theory of electrical phenomena rests upon foundations as secure as those of the undulatory theory of light. But the life-long work of its creator has left it still in its infancy, and it will probably require for its proper development the services of whole generations of mathematicians.

的计算，以及对同样产生的双折射结构的研究（用偏振光）。他无法解释为什么在云母、杜仲胶以及其他物体中一旦产生了这种结构就会持续存在。20 年以后，他又重新回到这个题目，并于 1873 年向皇家学会宣读了他非常美妙的发现——由黏性液体中的切变造成的**暂时**双折射。

在剑桥读本科时，麦克斯韦完成了最终于 1856 年发表在《剑桥哲学学会学报》上的论文《论法拉第力线》，这篇文章是他后来的伟大著作《电磁学》（1873 年）的雏形。1853 年，他曾给我看过这篇论文的大部分手稿。这篇论文本身就十分有趣，但更重要的是它显示了走向未来辉煌成就的第一步。麦克斯韦关于流体的观点是，流体不可压缩，没有质量，但是要克服空间中的某种摩擦力，这无疑采用了汤姆孙于 1843 年从稳定热流和静电现象之间类推出的结果。

关于这个主题还有另外 5 篇文章，以《论物理力线》为题发表在 1861~1862 年间的《哲学杂志》上。1864 年，他的伟大著作《电磁场的动力学理论》问世了。这篇文章被收录在《自然科学会报》上，可以将其看作对《电磁学》专著中所阐述理论的第一次完整论述。

近年，他得出了以下结论：对于这些遵循拉普拉斯方程的类似物理量，比如热传导或没有质量却不可压缩的流体运动来说，含有哈密顿算子 ∇ 的四维表示法是最佳的表示方式。接着，他在电学研究中给出了所有更重要的物理量的四维形式，然而他并没有在建立表示方法的同时进行计算。我在别处（《自然》，第 7 卷，第 478 页）讨论过这项伟大工作中的多个重要发现，这本身就足以使其作者跻身最伟大的科学家行列。我可以在这里说，这项工作的目标就是，至少在涉及电磁力时，弄清“超距作用”，并且要通过光现象所需的介质应力和运动来解释它们。麦克斯韦指出，在这种假设下，光速为电磁单元与静电单元之比。因为这个比值和光速都可以由完全独立的实验确定，所以上面的理论可以用极为严格的初级实验来检验。但是，目前这两个物理量都还没能在 2% ~ 3% 的误差范围内被清楚地认识，每个量的最可几数值当然比分散的数值更加集中。现在毫无疑问，麦克斯韦电现象理论建立的基础和光的波动学说的基础一样可信。但是，这个理论在其创建者的毕生努力下也还是处在初级阶段，它的合理发展可能还需要整整一代数学家的努力。

This was not the only work of importance to which he devoted the greater part of his time while an undergraduate at Cambridge. For he had barely obtained his degree before he read to the Cambridge Philosophical Society a remarkable paper *On the Transformation of Surfaces by Bending*, which appears in their *Transactions* with the date March 1854. The subject is one which had been elaborately treated by Gauss and other great mathematicians, but their methods left much to be desired from the point of view of simplicity. This Clerk Maxwell certainly supplied; and to such an extent that it is difficult to conceive that any subsequent investigator will be able to simplify the new mode of presentation as much as Maxwell simplified the old one. Many of his results, also, were real additions to the theory; especially his treatment of the *Lines of Bending*. But the whole matter is one which, except in its almost obvious elements, it is vain to attempt to popularise.

The next in point of date of Maxwell's greatest works is his "Essay on the Stability of the Motion of Saturn's Rings", which obtained the Adam's Prize in the University of Cambridge in 1857. This admirable investigation was published as a pamphlet in 1859. Laplace had shown in the *Mécanique Céleste* that a uniform solid ring cannot revolve permanently about a planet; for, even if its density were so adjusted as to prevent its splitting, a slight disturbance would inevitably cause it to fall in. Maxwell begins by finding what amount of *want* of uniformity would make a solid ring stable. He finds that this could be effected by a satellite rigidly attached to the ring, and of about $4\frac{1}{2}$ times its mass:—but that such an arrangement, while not agreeing with observation, would require extreme artificiality of adjustment of a kind not elsewhere observed. Not only so, but the materials, in order to prevent its behaving almost like a liquid under the great forces to which it is exposed, must have an amount of rigidity far exceeding that of any known substance.

He therefore dismisses the hypothesis of solid rings, and (commencing with that of a ring of equal and equidistant satellites) shows that a continuous liquid ring cannot be stable, but may become so when broken up into satellites. He traces in a masterly way the effects of the free and forced waves which must traverse the ring, under various assumptions as to its constitution; and he shows that the only system of rings which can dynamically exist must be composed of a very great number of separate masses, revolving round the planet with velocities depending on their distances from it. But even in this case the system of Saturn cannot be permanent, because of the mutual actions of the various rings. These mutual actions must lead to the gradual spreading out of the whole system, both inwards and outwards:—but if, as is probable, the outer ring is much denser than the inner ones, a very small increase of its external diameter would balance a large change in the inner rings. This is consistent with the progressive changes which have been observed since the discovery of the rings. An ingenious and simple mechanism is described, by which the motions of a ring composed of equal satellites can be easily demonstrated.

Another subject which he treated with great success, as well from the experimental as from the theoretical point of view, was the Perception of Colour, the Primary Colour Sensations, and the Nature of Colour Blindness. His earliest paper on these subjects bears

以上并不是麦克斯韦利用在剑桥读本科期间的大部分时间来完成的唯一重要的工作。因为直到他在剑桥哲学学会的会议上宣读了一篇著名的论文《论弯曲引起的表面变换》，并于1854年3月发表在《剑桥哲学学会学报》上，他才获得了学位。高斯和其他一些伟大的数学家都曾经详细研究过这个课题，但是从简洁的角度来看，他们的方法还有很多需要改进的地方。克拉克·麦克斯韦无疑完成了后续的简化工作，并且将其简化到了这样的程度：很难想象任何一个后继研究者在简化现有的新模型时，能达到像麦克斯韦简化旧模型时那样的程度。他的很多结果，是对原有理论的丰富，尤其是他对**弯曲线**所作的处理。但问题是，除了其中几乎显而易见的部分，这些结果都没有实现普及。

麦克斯韦的下一个伟大著作是《关于土星环运动稳定性的评论》，这篇文章在1857年获得了剑桥大学的亚当斯奖。这项令人称赞的研究在1859年被印成了小册子。拉普拉斯在他的《天体力学》中指出，均匀的固体环不可能持久地围绕行星转动，因为，即使环的密度调整到可以使其避免分裂的程度，一个微小的扰动也会不可避免地导致其塌陷。麦克斯韦从寻找使一个固体环稳定**所需的**均匀度开始。他发现，可以通过将一个卫星与这个环作刚性连接来实现环的稳定，其中卫星的质量等于环质量的$4\frac{1}{2}$倍。但是由于这种方法与观测不符，于是就需要极精巧的调节方式，而这种调节方式也没有在其他地方看到过。不但如此，为了避免环出现类似暴露在强力下的液体那样的表现，环的材料必须具有足够的硬度，而这种硬度远远超过了任何已知材料的硬度。

因此，麦克斯韦放弃了固体环的假设，开始设想一个由等距的相同行星组成的环。他指出，连续的液体环无法保持稳定，但是当它破裂成多个卫星的时候就可能达到稳定状态。他极为巧妙地论述了在各种假设的环结构中必须穿过环的或自由或受迫的波的影响，并且指出，唯一一种能够动态稳定存在的环形结构必须由大量分离的物质组成，这些物质围绕行星转动，其速度取决于它们到行星的距离。然而，即使在这种情况下，由于不同环之间的相互作用，土星系统也不可能持久稳定。因为这种相互作用必然会导致整个系统向内外两个方向扩散。但是，如果外环密度比内环密度大，外环直径很小的增大就能平衡内环直径很大的改变。这与环被发现以来所观察到的不断变化是一致的。麦克斯韦描述了一个独特而又简洁的机制，用它可以很容易地说明由相同卫星组成的环的运动。

麦克斯韦的另一个研究课题是颜色的感知、基础色觉以及色盲的本质。这项工作无论是从实验角度还是从理论角度来看都极为成功。他最早关于这些问题的文章诞生于1855年，第7篇则发表于1872年。“由于他在颜色组成方面的研究和其他光

date 1855, and the seventh has the date 1872. He received the Rumford Medal from the Royal Society in 1860, "For his Researches on the Composition of Colours and other optical papers". Though a triplicity about colour had long been known or suspected, which Young had (most probably correctly) attributed to the existence of three sensations, and Brewster had erroneously* supposed to be objective, Maxwell was the first to make colour-sensation the subject of actual measurement. He proved experimentally that any colour C (given in intensity of illumination as well as in character) may be expressed in terms of three arbitrarily chosen standard colours, X, Y, Z, by the formula

$$C = aX + bY + cZ$$

Here a, b, c are numerical coefficients, which may be positive or negative; the sign = means "matches", + means "superposed", and – directs the term to be taken to the other side of the equation.

These researches of Maxwell's are now so well known, in consequence especially of the amount of attention which has been called to the subject by Helmholtz' great work on Physiological Optics, that we need not farther discuss them here.

The last of his greatest investigations is the splendid Series on the Kinetic Theory of Gases, with the closely connected question of the sizes, and laws of mutual action, of the separate particles of bodies. The Kinetic Theory seems to have originated with D. Bernoulli; but his successors gradually reverted to statical theories of molecular attraction and repulsion, such as those of Boscovich. Herapath (in 1847) seems to have been the first to recall attention to the Kinetic Theory of gaseous pressure. Joule in 1848 calculated the average velocity of the particles of hydrogen and other gases. Krönig in 1856 (*Pogg. Ann.*) took up the question, but he does not seem to have advanced it farther than Joule had gone; except by the startling result that the weight of a mass of gas is only half that of its particles when at rest.

Shortly afterwards (in 1859) Clausius took a great step in advance, explaining, by means of the kinetic theory, the relations between the volume, temperature and pressure of a gas, its cooling by expansion, and the slowness of diffusion and conduction of heat in gases. He also investigated the relation between the length of the mean free path of a particle, the number of particles in a given space, and their least distance when in collision. The special merit of Clausius' work lies in his introduction of the processes of the theory of probabilities into the treatment of this question.

Then came Clerk Maxwell. His first papers are entitled "Illustrations of the Dynamical Theory of Gases", and appeared in the *Phil. Mag.* in 1860. By very simple processes he treats the collisions of a number of perfectly elastic spheres, first when all are of the same mass, secondly when there is a mixture of groups of different masses. He thus verifies

* All we can positively say to be erroneous is some of the principal arguments by which Brewster's view was maintained, for the subjective character of the triplicity has not been absolutely *demonstrated.*

学方面的论文”，麦克斯韦在1860年获得了皇家学会颁发的拉姆福德奖章。虽然人们知道或者猜测出颜色的三原色已经有很长一段时间了，例如杨曾经（很可能正确地）把三原色归因于存在三种主观色觉，而布鲁斯特错误地*猜测三原色是客观的，但是麦克斯韦却是第一个将实际测量引入色觉这个课题的人。他在实验中证明了，任何颜色C（以照度和特性的形式给出）都可以用三种任选的标准颜色X，Y，Z按照下面的公式表示出来：

$$C = aX + bY + cZ$$

其中，a，b，c是数值系数，可以取正也可以取负；等号表示“匹配”，加号表示“叠加”，减号表示把这一项移到方程的另一边。

现在麦克斯韦的这些研究已经是众所周知的了，特别是后来亥姆霍兹关于生理光学的伟大工作使得这个研究课题吸引了很多注意力，因此，我们就不需要在这里进一步详细讨论它们了。

麦克斯韦最后一项伟大的研究是他那套卓越的关于气体动力学的丛书。这套丛书的内容与物体离散粒子的大小和相互作用定律等问题密切相关。动力学理论似乎最早起源于伯努利，但是他的后继者逐渐回归到关于粒子吸引和排斥的统计理论上，就如博斯科维克所做的那样。赫拉帕斯（1847年）似乎是第一个重新注意到气体压力动力学理论的人。焦耳在1848年计算了氢气和其他气体微粒的平均速度。科隆尼格在1856年（《波根多夫年鉴》）考虑到了这个问题，但是除了得到气体的重量只有静止气体微粒的一半这一惊人结果之外，他似乎并没有进一步发展焦耳的结果。

在不久之后（1859年），克劳修斯取得了巨大的进展。他用动力学理论成功地解释了气体体积、温度、压强之间的关系，膨胀造成的冷却，以及气体中缓慢的热扩散和热传导。他还研究了气体微粒的平均自由程长度、给定空间中的粒子数以及粒子碰撞过程中的最小距离这三者之间的关系。克劳修斯工作的最大价值在于，他在处理这个问题时，引入了概率论的方法。

然后就是克拉克·麦克斯韦。他第一篇论文的题目是《气体动力学理论图示》，于1860年发表在《哲学杂志》上。他通过一个非常简单的过程来处理许多完全弹性小球的碰撞。他首先研究了所有小球的质量都相等的情况，然后研究了含有不同质

---

* 我们只能说布鲁斯特的观点所使用的主要论据是错误的，而三原色的主观性还没有完全被证实。

Gay-Lussac's law, that the number of particles per unit volume is the same in all gases at the same pressure and temperature. He explains gaseous friction by the transference to and fro of particles between contiguous strata of gas sliding over one another, and shows that the coefficient of viscosity is independent of the density of the gas. From Stokes' calculation of that coefficient he gave the first deduced approximate value of the mean length of the free path; which could not, for want of data, be obtained from the relation given by Clausius. He obtained a closely accordant value of the same quantity by comparing his results for the kinetic theory of diffusion with those of one of Graham's experiments. He also gives an estimate of the conducting power of air for heat; and he shows that the assumption of non-spherical particles, which during collision change part of their energy of translation into energy of rotation, is inconsistent with the known ratio of the two specific heats of air.

A few years later he made a series of valuable experimental determinations of the viscosity of air and other gases at different temperatures. These are described in *Phil. Trans.* 1866; and they led to his publishing (in the next volume) a modified theory, in which the gaseous particles are no longer regarded as perfectly elastic, but as repelling one another according to the law of the inverse fifth power of the distance. This paper contains some very powerful analysis, which enabled him to simplify the mathematical theory for many of its most important applications. Three specially important results are given in conclusion, and they are shown to be independent of the particular mode in which gaseous particles are supposed to act on one another. These are:—

1. In a mixture of particles of two kinds differing in amounts of mass, the average energy of translation of a particle must be the same for either kind. This is Gay Lussac's Law already referred to.
2. In a vertical column of mixed-gases, the density of each gas at any point is ultimately the same as if no other gas were present. This law was laid down by Dalton.
3. Throughout a vertical column of gas gravity has no effect in making one part hotter or colder than another; whence (by the dynamical theory of heat) the same must by true for all substances.

Maxwell has published in later years several additional papers on the Kinetic Theory, generally of a more abstruse character than the majority of those just described. His two latest papers (in the *Phil. Trans.* and *Camb. Phil. Trans.* of last year) are on this subject:— one is an extension and simplification of some of Boltzmann's valuable additions to the Kinetic Theory. The other is devoted to the explanation of the motion of the radiometer by means of this theory. Several years ago (*Nature*, vol. XII, p. 217), Prof. Dewar and the writer pointed out, and demonstrated experimentally, that the action of Mr. Crookes' very beautiful instrument was to be explained by taking account of the increased length of the mean free path in rarefied gases, while the then received opinions ascribed it either to evaporation or to a quasi-corpuscular theory of radiation. Stokes extended the explanation to the behaviour of disks with concave and convex surfaces, but the subject was not at all fully investigated from the theoretical point of view till Maxwell took it up. During the last ten years of his life he had no rival to claim concurrence with him in the whole wide domain

量小球的情况。他由此证明了盖·吕萨克定律，即在同温同压下，所有气体单位体积内的粒子数相等。他把气体的摩擦力解释为相邻气层之间微粒的相对运动，并且指出黏滞系数和气体的密度无关。他从斯托克斯对黏滞系数的计算出发，第一次推出了气体平均自由程的近似值，由于缺少数据，这个数值是无法从克劳修斯给出的关系中得到的。他把自己用扩散动力学理论计算出的结果和格雷厄姆一个实验的结果进行比较，发现二者能够很好地吻合。麦克斯韦还估算出了空气的热导率，并且指出，非球形粒子在碰撞中能将部分平动动能转化为转动动能的假设与已知的空气的两种比热之比不符。

几年后，他做了一系列很有价值的实验，来确定不同温度下空气和其他气体的黏性。这些工作的结果发表在 1866 年的《自然科学会报》上。在该杂志接下来的一卷上，他又发表了修正后的理论，此理论中不再把气体粒子视为完全弹性小球，而是认为粒子之间存在着与相互距离的五次方成反比的排斥作用。这篇论文包含一些非常有力的分析，这使得他可以为了理论的重要应用而进行数学理论上的简化。结论中给出了 3 个特别重要的结果，并且这些结果和气体粒子之间的相互作用模型无关。它们是：

1. 在由两种质量不同的粒子组成的混合物中，两种粒子的平均平动动能一定相等。这是盖·吕萨克定律已经提到过的。
2. 在装有混合气体的立柱容器中，每一种气体在任一点的密度最终都将相同，就好像没有其他气体存在一样。这个定律是道尔顿建立的。
3. 在整个装有气体的立柱容器中，重力并没有使某一部分的温度高于或低于另外一部分。因此（根据热动力学理论），这个规律应该适用于所有物质。

在之后的几年中，麦克斯韦发表了另外几篇关于动力学理论的文章，这些文章比上面提到的工作中的大多数都更加深奥。他最近的两篇论文发表在去年的《自然科学会报》和《剑桥哲学学报》上：一篇是对玻耳兹曼在动力学理论上的重要补充的推广和简化；另一篇文章中，他用这个理论解释了辐射计的运转。几年之后，杜瓦教授等人（《自然》，第 12 卷，第 217 页）指出并且通过实验证实了：考虑到稀薄气体中平均自由程的增加，无论是采用蒸气辐射理论还是准颗粒辐射理论都可以解释克鲁克斯先生非常精巧的仪器的作用。斯托克斯将其推广，用于解释具有凹凸表面的圆盘的行为，但是从理论的角度来看，这个课题在麦克斯韦着手之前并没有得到充分的研究。在麦克斯韦生命的最后十年里，他没有遇到能够在分子力学的广阔领域内和他平起平坐的对手，然而在更深奥的电学领域，倒是有两三个人能与他

of molecular forces, and but two or three in the still more recondite subject of electricity.

"Every one must have observed that when a slip of paper falls through the air, its motion, though undecided and wavering at first, sometimes becomes regular. Its general path is not in the vertical direction, but inclined to it at an angle which remains nearly constant, and its fluttering appearance will be found to be due to a rapid rotation round a horizontal axis. The direction of deviation from the vertical depends on the direction of rotation.... These effects are commonly attributed to some accidental peculiarity in the form of the paper...." So writes Maxwell in the *Cam. and Dub. Math. Jour.* (May, 1854), and proceeds to give an exceedingly simple and beautiful explanation of the phenomenon. The explanation is, of course, of a very general character, for the complete working out of such a problem appears to be, even yet, hopeless; but it is thoroughly characteristic of the man, that his mind could never bear to pass by any phenomenon without satisfying itself of at least its general nature and causes.

In the same volume of the *Math. Journal* there is an exceedingly elegant "problem" due to Maxwell, with his solution of it. In a note we are told that it was "suggested by the contemplation of the structure of the crystalline lens in fish". It is as follows:—

A transparent medium is such that the path of a ray of light within it is a given circle, the index of refraction being a function of the distance from a given point in the plane of the circle. Find the form of this function, and show that for light of the same refrangibility—

1. The path of *every ray within the medium* is a circle.
2. All the rays proceeding from any point in the medium will meet accurately in another point.
3. If rays diverge from a point without the medium and enter it through a spherical surface having that point for its centre, they will be made to converge accurately to a point within the medium.

Analytical treatment of this and connected questions, by a novel method, will be found in a paper by the present writer (*Trans. R.S.E.* 1865).

Optics was one of Clerk Maxwell's favourite subjects, but of his many papers on various branches of it, or subjects directly connected with it, we need mention only the following:—

"On the General Laws of Optical Instruments" (*Quart. Math. Jour.* 1858)
"On the Cyclide" (*Quart. Math. Journal*, 1868)
"On the best Arrangement for Producing a Pure Spectrum on a Screen" (*Proc. R.S.E.* 1868)
"On the Focal Lines of a Refracted Pencil" (*Math. Soc. Proc.* 1873)

A remarkable paper, for which he obtained the Keith Prize of the *Royal Society of Edinburgh*, is entitled "On Reciprocal Figures, Frames, and Diagrams of Forces." It is published in the *Transactions* of the Society for 1870. Portions of it had previously appeared in the *Phil. Mag* (1864).

相提并论。

“每个人都会注意到，一张纸片在空气中飘落时，虽然一开始摇摆不定，但是它的运动会趋于规则。它通常的路径并不是沿垂直方向，而是与垂直方向成一个角度，这个角度基本是一个常数。我们会发现纸片一开始的飘动是围绕一条水平轴快速转动。偏离垂直轴的方向取决于转动的方向……。这些结果通常被归因于纸张形状的某些偶然特性……”麦克斯韦在 1854 年 5 月发表于《剑桥与都柏林数学杂志》上的文章中这样写道，他想对这个现象作出非常简洁而漂亮的解释。当然，这个解释十分笼统，因为即使现在看来，完全解决这个问题也是希望渺茫。但这正是麦克斯韦的性格，他的思想决不容忍自己与任何连一般性质及成因都得不到满意解释的现象擦肩而过。

在《剑桥与都柏林数学杂志》的同一卷中，麦克斯韦提出了一个极其精彩的“问题”，并且自己作出了解答。我们从一则记录中得知，这个工作是“在思考鱼的晶状体结构时受到的启发”。内容如下：

所谓介质是透明的就是指光线在此介质中的传播路径是一个特定的圆，介质某一点的折射率是这一点到圆平面中给定点距离的函数。我找到了这个函数的形式，并且发现对于具有相同折射性质的光线来说——

1. **每一条光线在介质中的**路径都是一个圆。
2. 从介质中任意一点发出的光线，都会在另外一点精确相遇。
3. 如果光线在介质外的某一点发散，并且经由一个以此发散点为球心的球面进入介质，那么这些光线将精确地会聚到介质中的某一点上。

我本人在一篇文章（《爱丁堡皇家学会会报》，1865 年）中，用一种新颖的方法对这个问题及相关问题进行了分析。

光学是克拉克·麦克斯韦最喜欢的课题之一，但是在他关于光学不同分支或者直接与光学相关的大量文章中，我们只需提及下面这些：

《论光学仪器的普遍规律》（《数学季刊》，1858 年）
《论四次圆纹曲面》（《数学季刊》，1868 年）
《论在屏幕上生成纯光谱的最佳方案》（《爱丁堡皇家学会会刊》，1868 年）
《论折射光束的焦线》（《数学学会会刊》，1873 年）

麦克斯韦还有一篇意义重大的文章，题目是《论力的对应线图、框架和图解》，为此他获得了爱丁堡皇家学会的基思奖。文章于 1870 年发表在《爱丁堡皇家学会会报》上，文章中的一部分之前已经在《哲学杂志》（1864 年）上出现过。

The triangle and the polygon of forces, as well as the funicular polygon, had long been known; and also some corresponding elementary theorems connected with hydrostatic pressure on the faces of a polyhedron; but it is to Rankine that we owe the full principle of diagrams, and reciprocal diagrams, of frames and of forces. Maxwell has greatly simplified and extended Rankine's ideas: on the one hand facilitating their application to practical problems of construction, and on the other hand extending the principle to the general subject of stress in bodies. The paper concludes with a valuable extension to three dimensions of Sir George Airy's "Function of Stress".

His contributions to the *Proceedings of the London Mathematical Society* were numerous and valuable. I select as a typical specimen his paper on the forms of the stream-lines when a circular cylinder is moved in a straight line, perpendicular to its axis, through an infinitely extended, frictionless, incompressible fluid (vol. III, p. 224). He gives the complete solution of the problem; and, with his usual graphical skill, so prominent in his great work on Electricity, gives diagrams of the stream-lines, and of the paths of individual particles of the fluid. The results are both interesting and instructive in the highest degree.

In addition to those we have mentioned we cannot recall many pieces of *experimental* work on Maxwell's part:—with two grand exceptions. The first was connected with the determination of the British Association Unit of Electric Resistance, and the closely associated measurement of the ratio of the electrokinetic to the electrostatic unit. In this he was associated with Professors Balfour Stewart and Jenkin. The Reports of that Committee are among the most valuable physical papers of the age; and are now obtainable in a book-form, separately published. The second was the experimental verification of Ohm's law to an exceedingly close approximation, which was made by him at the Cavendish Laboratory with the assistance of Prof. Chrystal.

In his undergraduate days he made an experiment which, though to a certain extent physiological, was closely connected with physics. Its object was to determine why a cat always lights on its feet, however it may be let fall. He satisfied himself, by pitching a cat gently on a mattress stretched on the floor, giving it different initial amounts of rotation, that it instinctively made use of the conservation of Moment of Momentum, by stretching out its body if it were rotating so fast as otherwise to fall head foremost, and by drawing itself together if it were rotating too slowly.

I have given in this journal (vol. XVI, p. 119) a detailed account of his remarkable elementary treatise on "Matter and Motion", a work full of most valuable materials, and worthy of most attentive perusal not merely by students but by the foremost of scientific men.

His "Theory of Heat", which has already gone through several editions, is professedly elementary, but in many places is probably, in spite of its admirable definiteness, more difficult to follow than any other of his writings. In intrinsic importance it is of the same high order as his "Electricity", but as a whole it is *not* an elementary book. One of the

力的三角法则、多边形法则以及索状多边形法则早已为人们所熟知，一些和多面体表面静压有关的基础理论也是如此。但是兰金认为，我们缺少一套关于力的线图、框架、图解以及力本身的完整法则。麦克斯韦极大地简化并推广了兰金的观点：一方面，他使这个理论在建筑学实际问题上的应用更加方便；另一方面，他将这个理论推广到了物体中的压力这一一般主体。在文章结尾，麦克斯韦将乔治·艾里爵士的“压力函数”推广到了三维的情形，这个推广很有意义。

他为《伦敦数学学会会刊》贡献了大量有价值的稿件。我选取他的一篇文章作为其中的典范。这篇文章讨论了当一个圆柱体沿着垂直于轴的直线穿过一个不可压缩、无摩擦且可无限扩展的液体时液体中流线的形式（第 3 卷，第 224 页）。他给出了这个问题的完整解答，并且利用他在电学巨著中常用的图解技巧给出了流线和液体中单个粒子路径的图形。

除了上面提到的那些工作，麦克斯韦没有多少**实验性的**工作能被我们铭记，但是有两个工作是例外。第一个是关于确定电阻英制单位以及与其密切相关的电动力学单位与静电学单位之比的测量。这个工作是麦克斯韦和鲍尔弗·斯图尔特教授、詹金教授合作完成的。相关委员会的报告是这个时期最有价值的物理学论文之一，现在这些报告已经集结成册并且单独出版了。第二项工作是在实验上以极高的精度验证了欧姆定律。这项工作是他在卡文迪什实验室由克里斯托尔教授协助完成的。

麦克斯韦在读本科期间做了一个实验，虽然从某种程度上说这是一个生理学实验，但是它也和物理学密切相关。实验的目的是解释为什么猫在落地的时候总是能保持脚先着地。实验中他把猫轻轻地抛到毯子上，抛掷时，让猫具有不同的初始转动。麦克斯韦得到了满意的结果：猫在空中的时候，本能地利用了角动量守恒。如果给它的初始转动过快，它就会把身体伸展开，避免头先着地；相反地，如果给它的初始转动过慢，它就会缩成一团，最后总是能避免头先着地。

我曾经在贵刊上（第 16 卷，第 119 页）详细介绍了麦克斯韦著名的关于《物质和运动》的基础论文。他的这项工作有许多重要的结果，因此值得每一个人——不仅仅是学生，还包括一流的科学家——仔细研读。

他的《热学理论》已经出了好几版。尽管他自称这部书很基础，尽管书的思路的确非常清晰，但是其中许多地方可能比作者的其他任何著作都更难理解。这部书本身的重要性堪比他的《电学》，但是总体上却**不是**一本基础读物。克拉克·麦克斯

few knowable things which Clerk Maxwell did not know, was the distinction which most men readily perceive between what is easy and what is hard. What *he called* hard, others would be inclined to call altogether unintelligible. In the little book we are discussing there is matter enough to fill two or three large volumes without undue dilution (perhaps we should rather say, *with the necessary dilution*) of its varied contents. There is nothing flabby, so to speak, about anything Maxwell ever wrote: there is splendid muscle throughout, and an adequate bony structure to support it. "Strong meat for grown men" was one of his favourite expressions of commendation; and no man ever more happily exposed the true nature of the so-called "popular science" of modern times than he did when he wrote of "the forcible language and striking illustrations by which those who are past hope of being even beginners [in science] are prevented from becoming conscious of intellectual exhaustion before the hour has elapsed."

To the long list of works attached to Maxwell's name in the Royal Society's Catalogue of Scientific Papers may now be added his numerous contributions to the latest edition of the "Encyclopaedia Britannica"—Atom, Attraction, Capillarity, &c. Also the laborious task of preparing for the press, with copious and very valuable original notes, the "Electrical Researches of the Hon. Henry Cavendish." This work has appeared only within a month or two, and contains many singular and most unexpected revelations as to the early progress of the science of electricity. We hope shortly to give an account of it.

The works which we have mentioned would of themselves indicate extraordinary activity on the part of their author, but they form only a fragment of what he has published; and when we add to this the further statement, that Maxwell was always ready to assist those who sought advice or instruction from him, and that he has read over the proof-sheets of many works by his more intimate friends (enriching them by notes, always valuable and often of the quaintest character), we may well wonder how he found time to do so much.

Many of our readers must remember with pleasure the occasional appearance in our columns of remarkably pointed and epigrammatic verses, usually dealing with scientific subjects, and signed $\frac{dp}{dt}$*. The lines on Cayley's portrait, where determinants, roots of -1, space of n dimensions, the 27 lines on a cubic surface, &c., fall quite naturally into rhythmical English verse; the admirable synopsis of Dr. Ball's Treatise on Screws; the telegraphic love-letter with its strangely well-fitting *volts* and *ohms*; and specially the "Lecture to a Lady on Thomson's Reflecting Galvanometer", cannot fail to be remembered. No living man has shown a greater power of condensing the whole marrow of a question into a few clear and compact sentences than Maxwell shows in these verses. Always having a definite object, they often veiled the keenest satire under an air of charming innocence and *naïve* admiration. Here are a couple of stanzas from unpublished pieces of a similar kind:—first, some ghastly thoughts by an excited evolutionist—

---

* This *nom de plume* was suggested to him by me from the occurrence of his initials in the well-known expression of the second Law of Thermodynamics (for whose establishment on thoroughly valid grounds he did so much) $\frac{dp}{dt} = \text{J. C. M.}$

韦不知道的少数几个显而易见的事情之一就是，难与易的区别，而这是多数人都能分清的。**他所谓的**难事，其他人会认为是根本无法理解的。我们正在谈论的这本薄薄的书的内容不用过度展开（可能我们应该说，**经过必要的展开**），其内容就足以写满两三卷书。可以说，麦克斯韦写的东西从不松散拖沓：他的文章只有健美的肌肉和适量的用以支撑的骨架。“成人的强健肌肉”是麦克斯韦最喜欢的表达称赞的说法。“过来人总是希望自己保持那种［科学上的］初学者的状态，对他们来说，有力的语言和精彩的图示就是能在时间消逝之前避免灵感和智慧枯竭的灵药。”当麦克斯韦写出上面这段话时，恐怕没有人比他更乐于揭示现代所谓“流行科学”的真实本质了。

在皇家学会科学文献目录中，有麦克斯韦署名的工作已经可以列出一长串了。现在，应该还要加上他对《大英百科全书》中原子、吸引作用、毛细作用等方面所作的很多贡献。另外，他还为编写出版《亨利·卡文迪什电学研究》付出了辛勤的劳动，整理了丰富而珍贵的手稿。这项工作是在最近的一两个月才问世的，书中记录着电学早期发展历程带给我们的很多意想不到的非凡启迪。我们希望不久之后能介绍一下这项工作。

上面提到的工作已经显示了作者超常的科研能力，然而这些只是他已发表著作的冰山一角。如果我还补充说，麦克斯韦总是乐于帮助那些寻求建议或指引的人，而且他通读过很多好友的著作的校样（在上面所作的注释使其更加丰富，往往起到画龙点睛的作用），大家可能会觉得十分惊讶，他哪来那么多时间完成这么多事情。

很多读者一定还记得在我们的专栏中偶尔出现的那些非常尖锐的讽刺小诗，通常都是关于科学主题的，并且署名 $\frac{dp}{dt}$ *；在凯莱肖像画上的诗句中由行列式、–1 的根、n 维空间以及三次曲面上的 27 条线等很自然地组成的一首充满韵律的英文小诗；为鲍尔博士关于旋量的论文写的令人赞叹的简介；用出奇得体的**伏特**和**欧姆**组成的电报情书；特别是那篇《为女士所作的关于汤姆孙反射检流计的演讲》，所有这些都让人无法忘怀。当今世上没有人可以超越麦克斯韦在小诗中表现出的用几个清晰而简洁的句子就把问题的精髓概括出来的能力。这些小诗总是具有明确的目标，但是又把最尖锐的讽刺隐藏在迷人的纯真和纯朴的赞美之中。这里有两段没有发表的类似这种风格的片段：首先，是一个狂热的革命者的可怕想法——

---

* 这个笔名是我建议他取的，因为我在著名的热力学第二定律 $\frac{dp}{dt}$ = J. C. M. 中发现了他名字的首字母缩写（而他本人也为这个定律能够建立在坚实的基础之上做了许多工作）。

To follow my thoughts as they go on,
Electrodes I'd place in my brain;
Nay, I'd swallow a live entozöon,
New feelings of life to obtain—

next on the non-objectivity of Force—

Both Action and Reaction now are gone;
Just ere they vanished
Stress joined their hands in peace, and made them one,
Then they were banished.

It is to be hoped that these scattered gems may be collected and published, for they are of the very highest interest, as the work during leisure hours of one of the most piercing intellects of modern times. Every one of them contains evidence of close and accurate thought, and many are in the happiest form of epigram.

I cannot adequately express in words the extent of the loss which his early death has inflicted not merely on his personal friends, on the University of Cambridge, on the whole scientific world, but also, and most especially, on the cause of common sense, of true science, and of religion itself, in these days of much vain-babbling, pseudo-science, and materialism. But men of his stamp never live in vain; and in one sense at least they cannot die. The spirit of Clerk Maxwell still lives with us in his imperishable writings, and will speak to the next generation by the lips of those who have caught inspiration form his teachings and example.

(**21**, 317-321; 1880)

我跟随着自己的感觉，
我要把电极放进我的脑子里；
要不我就吞下活生生的寄生虫，
我的生命会有崭新的感受——

另一个，是关于力的非客观性——

作用力和反作用力都消失了；
就在他们消失之前，
压力让他们静静地携起手来，合二为一，
然后，他们被放逐天涯。

人们希望能够把这些散落的宝石结集出版，因为它们是现代最敏锐的智者中的一员闲暇时完成的作品，而又是如此有趣。每一首诗都证明了作者缜密的思维，并且很多都是以讽刺诗那种诙谐的手法写成的。

我实在无法用语言表达麦克斯韦的早逝是多么巨大的损失，受到损失的不仅仅是他的朋友、剑桥大学和整个科学界，特别是，在充满空谈、伪科学和物质主义的今天，人们对常识、真科学以及宗教本身的探究也会因此受到巨大的损失。然而，脚踏实地的人决不会生活在空谈中，至少从某种意义上讲，这样的人不会从世界上消失。麦克斯韦的精神会在他不朽的著作中与我们同在，并且这种精神会由那些受过他的教诲并以他为榜样的人，传承给下一代。

（王静 翻译；鲍重光 审稿）

Density of Nitrogen

Rayleigh

Editor's Note

Measurements of atomic weights of the elements—the weights, relative to hydrogen, of equal quantities—revealed that these were often close to whole numbers. William Prout suggested in 1815–1816 that hydrogen might thus be the building block of all atoms. In 1888 Lord Rayleigh at Cambridge determined the atomic weight of nitrogen, and found that the gas obtained from ammonia was slightly lighter, by one thousandth, than atmospheric nitrogen. Here Rayleigh appeals for an explanation of the discrepancy, which he suspects might be due to an impurity. Rayleigh later teamed up with William Ramsay of University College London, and in 1894 they announced a new, unreactive element in air, which they named argon after the Greek for "idle".

I am much puzzled by some recent results as to the density of *nitrogen*, and shall be obliged if any of your chemical readers can offer suggestions as to the cause. According to two methods of preparation I obtain quite distinct values. The relative difference, amounting to about 1/1,000 part, is small in itself; but it lies entirely outside the errors of experiment, and can only be attributed to a variation in the character of the gas.

In the first method the oxygen of atmospheric air is removed in the ordinary way by metallic copper, itself reduced by hydrogen from the oxide. The air, freed from CO_2 by potash, gives up its oxygen to copper heated in hard glass over a large Bunsen, and *then* passes over about a foot of red-hot copper in a furnace. This tube was used merely as an indicator, and the copper in it remained bright throughout. The gas then passed through a wash-bottle containing sulphuric acid, thence again through the furnace over *copper oxide*, and finally over sulphuric acid, potash, and phosphoric anhydride.

In the second method of preparation, suggested to me by Prof. Ramsay, everything remained unchanged, except that the *first* tube of hot copper was replaced by a wash-bottle containing liquid *ammonia*, through which the air was allowed to bubble. The ammonia method is very convenient, but the nitrogen obtained by means of it was 1/1,000 part *lighter* than the nitrogen of the first method. The question is, to what is the discrepancy due?

The first nitrogen would be too heavy, if it contained residual oxygen. But on this hypothesis something like 1 percent would be required. I could detect none whatever by means of alkaline pyrogallate. It may be remarked the density of this nitrogen agrees closely with that recently obtained by Leduc, using the same method of preparation.

氮气的密度

瑞利

编者按

元素原子量——相对于同等数量的氢的质量——的测定结果显示，各种元素原子量的数值通常都接近整数。1815~1816 年间，威廉·普劳特提出氢原子可能是组成其他原子的基本单元。1888 年，剑桥的瑞利勋爵测定了氮的原子量，结果发现通过氨得到的氮气比空气中的氮气轻千分之一。在这篇文章中，瑞利对这种差异提出了一种解释，他怀疑该差异可能是因为某种杂质造成的。后来，瑞利与伦敦大学学院的威廉姆·拉姆齐一起合作，并于 1894 年宣布发现空气中存在一种没有反应活性的新元素，他们根据希腊语中的“懒惰”一词将其命名为氩。

我对近来一些关于**氮气**密度的结果感到很困惑，如果你们读者中有熟悉化学的人可以提供相关原因的建议，我将不胜感激。用两种不同的制备方法，我得到了显然不同的数值。相对差别大约是 1/1,000，虽然这一差别本身并不大，但是这完全不属于实验误差，因此只能归因于气体性质的不同。

在第一种制备方法中，我们用普通的方法除去空气中的氧气，其中用到了通过氢气还原氧化铜制得的金属铜。首先是用钾碱除去空气中的 CO_2，然后使气体与硬质玻璃管中已经用本生灯加热的铜反应，以除去空气中的氧气，**接着**再使气体通过加热炉中的一块大约一英尺长的红热的铜。这个管道仅仅是用作指示剂，其中的铜应该始终保持亮红色。然后将气体通过一个盛有硫酸的洗气瓶，之后再使其通过加热炉中的**氧化铜**，最后依次通过硫酸、钾碱和磷酸酐。

第二种制备方法是拉姆齐教授向我建议的。与第一种方法相比，只有**第一个**加热铜的玻璃管被一个装有液**氨**的洗气瓶代替，这一洗气瓶允许空气在通过时产生气泡，除此之外再没有任何改变。这种氨法非常简便，但是通过这种方法得到的氮气比采用第一种方法制得的氮气**轻** 1/1,000。那么，产生这种差异的原因是什么呢？

用第一种方法制得的氮气中如果还有残余的氧气，那么其密度就会更大。但是按照这种假设，制得的氮气中就要有大约 1% 的氧气，但是我用碱性焦棓酸盐进行检测并没有发现任何氧气。值得注意的是，这种氮气的密度与最近勒迪克用同一种方法制得的氮气的密度非常接近。

On the other hand, can the ammonia-made nitrogen be too light from the presence of impurity? There are not many gases lighter than nitrogen, and the absence of hydrogen, ammonia, and water seems to be fully secured. On the whole it seemed the more probable supposition that the impurity was hydrogen, which in this degree of dilution escaped the action of the copper oxide. But a special experiment appears to exclude this explanation.

Into nitrogen prepared by the first method, but before its passage into the furnace tubes, one or two thousandths by volume of hydrogen were introduced. To effect this in a uniform manner the gas was made to bubble through a small hydrogen generator, which could be set in action under its own electromotive force by closing an external contact. The rate of hydrogen production was determined by a suitable galvanometer enclosed in the circuit. But the introduction of hydrogen had not the smallest effect upon the density, showing that the copper oxide was capable of performing the part desired of it.

Is it possible that the difference is independent of impurity, the nitrogen itself being to some extent in a different (dissociated) state?

I ought to have mentioned that during the fillings of the globe, the rate of passage of gas was very uniform, and about 2/3 litre per hour.

(**46**, 512-513; 1892)

Rayleigh: Terling Place, Witham, September 24.

另一方面，通过氨法制得的氮气会不会由于存在杂质而更轻呢？比氮气轻的气体并不是很多，而这种制备方法似乎可以完全保证得到的氮气中不含氢气、氨气和水蒸气。总的来看，最有可能的一种假设就是这种氮气中含有氢气，在此种稀释度下氢气逃过了与氧化铜的反应。但是，一个特殊的实验似乎又排除了这一解释。

在将第一种方法制得的氮气通入到炉管之前先向其中加入 1/1,000 ~ 2/1,000 体积的氢气。为了保证这一过程均匀稳定，我们使气体以冒泡的方式通过一个小型氢气发生器，可以通过切断一个外部接触来使这一氢气发生器在其自身的电动势下开始工作。通过一个连接在电路中的适当的检流计可以测定氢气产生的速率。但是，氢气的引入对氮气的密度没有丝毫的影响，这表明氧化铜能非常好地发挥除去氢气的作用。

那么氮气密度的这种差异会不会与其是否含有杂质无关，而是由于氮气本身在某种程度上就处于不同的状态（解离）呢？

另外，我还要说明一下，在气体注入球形容器的过程中，气体通过的速率非常平稳，大约是每小时 2/3 升。

（李世媛 翻译；李芝芬 审稿）

On a New Kind of Rays*

W. C. Röntgen

Editor's Note

This is an English translation of Wilhelm Conrad Röntgen's German report, in December 1895, of the discovery of X-rays. While experimenting with a cathode ray tube (also called a Crookes' or Lenard's tube, after earlier investigators), in which electrons or "kathode rays" are accelerated by electric fields, Röntgen found that the tube emits radiation that penetrates black paper and induces fluorescence in a screen on the other side. The rays also can penetrate matter and produce photographic images of "buried" objects such as bones. Röntgen deduces that these rays are not cathode rays, but seem instead to be akin to ultraviolet rays, yet with much greater penetrating power. He called them X-rays simply "for the sake of brevity".

(1) A discharge from a large induction coil is passed through a Hittorf's vacuum tube, or through a well-exhausted Crookes' or Lenard's tube. The tube is surrounded by a fairly close-fitting shield of black paper; it is then possible to see, in a completely darkened room, that paper covered on one side with barium platinocyanide lights up with brilliant fluorescence when brought into the neighbourhood of the tube, whether the painted side or the other be turned towards the tube. The fluorescence is still visible at two metres distance. It is easy to show that the origin of the fluorescence lies within the vacuum tube.

(2) It is seen, therefore, that some agent is capable of penetrating black cardboard which is quite opaque to ultra-violet light, sunlight, or arc-light. It is therefore of interest to investigate how far other bodies can be penetrated by the same agent. It is readily shown that all bodies possess this same transparency, but in very varying degrees. For example, paper is very transparent; the fluorescent screen will light up when placed behind a book of a thousand pages; printer's ink offers no marked resistance. Similarly the fluorescence shows behind two packs of cards; a single card does not visibly diminish the brilliancy of the light. So, again, a single thickness of tinfoil hardly casts a shadow on the screen; several have to be superposed to produce a marked effect. Thick blocks of wood are still transparent. Boards of pine two or three centimetres thick absorb only very little. A piece of sheet aluminium, 15 mm thick, still allowed the X-rays (as I will call the rays, for the sake of brevity) to pass, but greatly reduced the fluorescence. Glass plates of similar thickness behave similarly; lead glass is, however, much more opaque than glass free from lead. Ebonite several centimetres thick is transparent. If the hand be held before the fluorescent screen, the shadow shows the bones darkly, with only faint outlines of the surrounding tissues.

---

* By W. C. Röntgen. Translated by Arthur Stanton from the *Sitzungsberichte der Würzburger Physik-medic. Gesellschaft*, 1895.

论一种新型的射线*

伦琴

编者按

此文译自威廉·康拉德·伦琴 1895 年 12 月的一份关于发现了 X 射线的德文报告。当伦琴用阴极射线管（早期的科研人员也把它称作克鲁克斯管或莱纳德管，管中的电子或“阴极射线”被电场加速）进行实验时，他发现阴极射线管发射的射线能够穿透黑色的纸并在其另一侧的屏幕上显示出荧光。这种射线还可以穿透物质，人们可以利用它拍摄出像骨骼这样的“被遮挡的”物质的照片。伦琴推测这种射线不属于阴极射线，它似乎更接近紫外线，但具有更强的穿透力。“为了简便起见”，他把这类射线称为“X 射线”。

(1) 让大号感应线圈中产生的放电通过希托夫真空管，或者通过抽成真空的克鲁克斯管或莱纳德管。管子用黑纸包裹严实。在完全黑暗的房间里，将一面涂有铂氰酸钡的纸放在管子旁边，不论朝向管子的是涂有铂氰酸钡的一面还是没有涂的那面，纸都会被鲜艳的荧光照亮。在 2 米之外，这种荧光依然可见。很显然，荧光来源于真空管中。

(2) 由此可见，某些射线能够穿透这种紫外光、太阳光和电弧光都几乎不能透过的黑纸板。这就引起了人们研究这种射线到底能够多大程度地穿透其他物质的兴趣。很容易就能证明，所有物质对这种射线都是透明的，只是透明的程度大不相同。比如，纸是非常透明的，即使将荧光屏置于一本 1,000 页厚的书后面，我们仍然会在荧光屏上看到亮光，印刷油墨也不会造成明显的阻挡。类似地，单独一张卡片不会明显减弱光的强度，即使在两叠卡片后面我们也仍然能看到亮光。同样，单张锡箔纸的遮挡几乎不会使荧光屏上出现阴影，要产生明显的遮挡效果就必须重叠许多张锡箔纸。厚木块对于这种射线也是透明的。2~3 厘米厚的松木板的吸收效果非常微弱。15 毫米厚的铝板也能使 X 射线（为了简便起见，我将称这种射线为 X 射线）透过，但是能够大幅度地减弱荧光。玻璃板的作用与厚度相近的铝板类似，不过，含铅的玻璃对这种射线的阻挡效果比不含铅的玻璃更强。几厘米厚的硬质橡胶也是透明的。如果把手放在荧光屏前，屏幕上就会显示出骨骼的黑影，而周围组织则只有模糊的轮廓。

---

* 作者为伦琴。由阿瑟·斯坦顿译自 1895 年的《维尔茨堡物理–医学学会会刊》。

Water and several other fluids are very transparent. Hydrogen is not markedly more permeable than air. Plates of copper, silver, lead, gold, and platinum also allow the rays to pass, but only when the metal is thin. Platinum 0.2 mm thick allows some rays to pass; silver and copper are more transparent. Lead 1.5 mm thick is practically opaque. If a square rod of wood 20 mm in the side be painted on one face with white lead, it casts little shadow when it is so turned that the painted face is parallel to the X-rays, but a strong shadow if the rays have to pass through the painted side. The salts of the metals, either solid or in solution, behave generally as the metals themselves.

(3) The preceding experiments lead to the conclusion that the density of the bodies is the property whose variation mainly affects their permeability. At least no other property seems so marked in this connection. But that the density alone does not determine the transparency is shown by an experiment wherein plates of similar thickness of Iceland spar, glass, aluminium, and quartz were employed as screens. Then the Iceland spar showed itself much less transparent than the other bodies, though of approximately the same density. I have not remarked any strong fluorescence of Iceland spar compared with glass (see below, No. 4).

(4) Increasing thickness increases the hindrance offered to the rays by all bodies. A picture has been impressed on a photographic plate of a number of superposed layers of tinfoil, like steps, presenting thus a regularly increasing thickness. This is to be submitted to photometric processes when a suitable instrument is available.

(5) Pieces of platinum, lead, zinc, and aluminium foil were so arranged as to produce the same weakening of the effect. The annexed table shows the relative thickness and density of the equivalent sheets of metal.

	Thickness	Relative thickness	Density
Platinum	0.018 mm	1	21.5
Lead	0.050 mm	3	11.3
Zinc	0.100 mm	6	7.1
Aluminium	3.500 mm	200	2.6

From these values it is clear that in no case can we obtain the transparency of a body from the product of its density and thickness. The transparency increases much more rapidly than the product decreases.

(6) The fluorescence of barium platinocyanide is not the only noticeable action of the X-rays. It is to be observed that other bodies exhibit fluorescence, *e.g.* calcium sulphide, uranium glass, Iceland spar, rock-salt, &c.

水和其他几种液体对于这种射线都是非常透明的。氢气的透明度并没有明显强于空气。铜、银、铅、金和铂质的金属板只有在很薄的时候才能使这种射线透过。0.2 毫米的铂能使这种射线部分透过，银和铜则更透明一些。1.5 毫米厚的铅板基本上是不透明的。将一根边长为 20 毫米的方木棒的一个侧面涂上铅白，当木棒涂有铅白的面与射线平行时，几乎不会产生阴影，但是当射线必须穿过涂有铅白的一面时，就会产生明显的阴影。不论是固态的金属盐还是金属盐溶液，一般都能像金属本身一样阻挡该射线。

(3) 根据上述实验我们可以得出结论：物质的密度是这样一种性质，它的变化主要影响射线在该物质中的透过程度。至少其他性质的影响看起来都不如密度明显。不过，单是密度还不能完全决定物质对该射线的透明度。我用厚度相近的冰洲石板、玻璃板、铝板和石英板作为样品进行的实验表明，尽管这些物质具有近似相同的密度，但冰洲石对该射线的透明度却比其他物质小得多。在用冰洲石进行的实验中，我从来没有观察到像用玻璃进行的实验中出现的那样明显的荧光（见下文，第 4 部分）。

(4) 对于所有物体，增加厚度都会提高其对 X 射线的阻挡程度。我们已经在照相底片上对阶梯状叠放的多层锡箔进行了成像，得到的图像表现出了厚度的这种有规律的增加。如果根据此原理制成适当的仪器，则可以作为光度计使用。

(5) 为了得到对 X 射线相同的减弱效果，我将铂、铅、锌和铝分别制成如下规格的金属片。附表给出了具有相同减弱效果的各种金属片的密度和相对厚度。

	厚度	相对厚度	密度
铂	0.018 毫米	1	21.5
铅	0.050 毫米	3	11.3
锌	0.100 毫米	6	7.1
铝	3.500 毫米	200	2.6

从这些数据中可以清楚地看出，我们不可能根据金属密度与其厚度的乘积来确定其透明度。透明度增加的速度比该乘积减少的速度快很多。

(6) X 射线所产生的显著作用并不是只能使铂氰酸钡发出荧光。可以观测到，X 射线也能使其他一些物质发出荧光，例如硫化钙、铀玻璃、冰洲石和岩盐等。

Of special interest in this connection is the fact that photographic dry plates are sensitive to the X-rays. It is thus possible to exhibit the phenomena so as to exclude the danger of error. I have thus confirmed many observations originally made by eye observation with the fluorescent screen. Here the power of the X-rays to pass through wood or cardboard becomes useful. The photographic plate can be exposed to the action without removal of the shutter of the dark slide or other protecting case, so that the experiment need not be conducted in darkness. Manifestly, unexposed plates must not be left in their box near the vacuum tube.

It seems now questionable whether the impression on the plate is a direct effect of the X-rays, or a secondary result induced by the fluorescence of the material of the plate. Films can receive the impression as well as ordinary dry plates.

I have not been able to show experimentally that the X-rays give rise to any calorific effects. These, however, may be assumed, for the phenomena of fluorescence show that the X-rays are capable of transformation. It is also certain that all the X-rays falling on a body do not leave it as such.

The retina of the eye is quite insensitive to these rays: the eye placed close to the apparatus sees nothing. It is clear from the experiments that this is not due to want of permeability on the part of the structures of the eye.

(7) After my experiments on the transparency of increasing thicknesses of different media, I proceeded to investigate whether the X-rays could be deflected by a prism. Investigations with water and carbon bisulphide in mica prisms of 30° showed no deviation either on the photographic or the fluorescent plate. For comparison, light rays were allowed to fall on the prism as the apparatus was set up for the experiment. They were deviated 10 mm and 20 mm respectively in the case of the two prisms.

With prisms of ebonite and aluminium, I have obtained images on the photographic plate, which point to a possible deviation. It is, however, uncertain, and at most would point to a refractive index 1.05. No deviation can be observed by means of the fluorescent screen. Investigations with the heavier metals have not as yet led to any result, because of their small transparency and the consequent enfeebling of the transmitted rays.

On account of the importance of the question it is desirable to try in other ways whether the X-rays are susceptible of refraction. Finely powdered bodies allow in thick layers but little of the incident light to pass through, in consequence of refraction and reflection. In the case of the X-rays, however, such layers of powder are for equal masses of substance equally transparent with the coherent solid itself. Hence we cannot conclude any regular reflection or refraction of the X-rays. The research was conducted by the aid of finely-powdered rock-salt, fine electrolytic silver powder, and zinc dust already many times employed in chemical work. In all these cases the result, whether by the fluorescent screen

在这方面，特别让人感兴趣的是照相干版对 X 射线是敏感的。这就使我们可以将实验现象记录下来以避免出现错误。利用照相的方法，我已经确认了很多最初通过肉眼在荧光屏上观测得到的实验结果。X 射线穿透木块或纸板的能力很有用处。在对照相干版进行曝光时，可以不用除去遮光板或者其他保护盒，因此实验就不必在暗室中进行。当然，千万不要把装有未曝光照相干版的盒子放在真空管附近。

干版上留下的影像到底是 X 射线的直接效应，还是由干版材料发出的荧光引起的次级效应，现在看来还是一个令人疑惑的问题。和普通的干版一样，胶片也可以记录到影像。

我还没能通过实验证明 X 射线是否可以产生热效应，不过我们猜测它可以，因为荧光现象表明 X 射线可以引起能量转移。而且可以肯定的是，照射在物体上的 X 射线并没有全部以荧光形式离开物体。

人眼的视网膜对这种射线非常不敏感：即使眼睛离装置很近也看不见任何东西。实验结果清楚地表明，这并不是因为该射线在眼睛这部分结构中的透过程度不够。

(7) 在通过实验研究了不同介质随着厚度增加对该射线的透明度的变化之后，我又研究了 X 射线是否会被棱镜偏转。在顶角为 30° 的云母棱镜中分别装入水和二硫化碳进行实验，结果发现照相干版和荧光板上都没有显示出偏移。为了对照，我也用可见光在相同的实验装置上进行了实验，结果发现可见光在穿过上述两种棱镜时分别偏转了 10 毫米和 20 毫米。

在使用硬质橡胶和铝制成的棱镜进行实验时，我得到的照相干版上的影像显示射线可能发生了偏转，不过这一点还不能确定，而且偏转所对应的折射率最多也只有 1.05。使用荧光屏时则观测不到偏转。用较重金属进行的研究目前还没有任何结果，这是因为它们的透明度都很小，因而透射的射线非常微弱。

考虑到 X 射线能否被偏转这个问题的重要性，我们就有必要尝试用其他方法来研究 X 射线能否发生折射。由于反射和折射的原因，微细粉末形成的厚层几乎不能使入射光透过。不过，这种多层粉末对于 X 射线的透明度与同质量同组成的整块固体是一样的。因此我们不能得出 X 射线具有常规的反射或折射特性的结论。我又对细粉末状的岩盐、电解得到的细银粉和已经在化学实验中使用了多次的锌粉进行了研究。所有研究结果都表明，不论是用荧光屏还是用照相的方法，粉末与相应的固

or the photographic method, indicated no difference in transparency between the powder and the coherent solid.

It is, hence, obvious that lenses cannot be looked upon as capable of concentrating the X-rays; in effect, both an ebonite and a glass lens of large size prove to be without action. The shadow photograph of a round rod is darker in the middle than at the edge; the image of a cylinder filled with a body more transparent than its walls exhibits the middle brighter than the edge.

(8) The preceding experiments, and others which I pass over, point to the rays being incapable of regular reflection. It is, however, well to detail an observation which at first sight seemed to lead to an opposite conclusion.

I exposed a plate, protected by a black paper sheath, to the X-rays so that the glass side lay next to the vacuum tube. The sensitive film was partly covered with star-shaped pieces of platinum, lead, zinc, and aluminium. On the developed negative the star-shaped impression showed dark under platinum, lead, and, more markedly, under zinc; the aluminium gave no image. It seems, therefore, that these three metals can reflect the X-rays; as, however, another explanation is possible, I repeated the experiment with this only difference, that a film of thin aluminium foil was interposed between the sensitive film and the metal stars. Such an aluminium plate is opaque to ultra-violet rays, but transparent to X-rays. In the result the images appeared as before, this pointing still to the existence of reflection at metal surfaces.

If one considers this observation in connection with others, namely, on the transparency of powders, and on the state of the surface not being effective in altering the passage of the X-rays through a body, it leads to the probable conclusion that regular reflection does not exist, but that bodies behave to the X-rays as turbid media to light.

Since I have obtained no evidence of refraction at the surface of different media, it seems probable that the X-rays move with the same velocity in all bodies, and in a medium which penetrates everything, and in which the molecules of bodies are embedded. The molecules obstruct the X-rays, the more effectively as the density of the body concerned is greater.

(9) It seemed possible that the geometrical arrangement of the molecules might affect the action of a body upon the X-rays, so that, for example, Iceland spar might exhibit different phenomena according to the relation of the surface of the plate to the axis of the crystal. Experiments with quartz and Iceland spar on this point lead to a negative result.

(10) It is known that Lenard, in his investigations on kathode rays, has shown that they belong to the ether, and can pass through all bodies. Concerning the X-rays the same may be said.

体对 X 射线的透明度没有任何差别。

因此，很明显透镜是不能会聚 X 射线的。事实上，大尺寸的玻璃透镜和硬质橡胶透镜对 X 射线都没有会聚作用，这已经得到了证明。圆柱的透视影像显示，中间部分的阴影比边缘更深一些；如果在圆柱内部填入透明度比柱体材料更高的物质，那么在得到的影像中，中间部分会比边缘更亮一些。

(8) 上述实验和另外一些我未提及的实验，都表明这种射线不具备常规的反射能力。不过，我还是要详细介绍一个乍看上去似乎会使人们得出相反结论的实验。

实验中，我将一块用黑纸套保护起来的玻璃干版置于 X 射线中，使其玻璃面靠近真空管，并用铂、铅、锌和铝质的星形金属片部分地遮挡干版的感光膜。在显影后的负片上，铂片和铅片下方出现了黑色的星形影像，锌片下方的影像更加清晰，而铝片并没有产生阴影。由此看来，前三种金属可以反射 X 射线。不过也可能有另外的解释。我又重复了这一实验，这次唯一的不同之处是，我在感光膜和星形金属片之间插入了一块薄薄的铝箔。这块铝箔对紫外线是不透明的，但对 X 射线是透明的。结果出现了和以前一样的影像。这再次表明 X 射线在金属表面发生了反射。

如果综合考虑这一观测结果和其他一些结果，包括关于粉末透明度的结果以及关于表面状态不能有效改变 X 射线穿过物体的路径的结果，我们就会得出这样一个可能的结论：对于 X 射线来说，并不存在普通意义上的反射，物体对于 X 射线的作用，就像混浊介质对于可见光一样。

我还没有得到任何可以表明在不同介质表面 X 射线会发生折射的证据，这样看来，X 射线在所有物质中的传播速度可能都相同，而且在一种渗透一切物质、包容各种物质分子的介质中也是一样的。随着物质密度的增大，其分子对 X 射线的阻挡效果也变得更加明显。

(9) 分子的几何构型看起来可能会影响物质对 X 射线的阻挡作用，例如，对于冰洲石晶体而言，表面与晶轴之间相对取向的不同可能就会导致不同的现象。但是，为此而用石英和冰洲石进行的实验却得到了阴性的结果。

(10) 我们知道，莱纳德在对阴极射线的研究中已经指出，阴极射线属于以太，可以穿透任何物体。估计 X 射线可能也是这样的。

In his latest work, Lenard has investigated the absorption coefficients of various bodies for the kathode rays, including air at atmospheric pressure, which gives 4.10, 3.40, 3.10 for 1 cm, according to the degree of exhaustion of the gas in discharge tube. To judge from the nature of the discharge, I have worked at about the same pressure, but occasionally at greater or smaller pressures. I find, using a Weber's photometer, that the intensity of the fluorescent light varies nearly as the inverse square of the distance between screen and discharge tube. This result is obtained from three very consistent sets of observations at distances of 100 and 200 mm. Hence air absorbs the X-rays much less than the kathode rays. This result is in complete agreement with the previously described result, that the fluorescence of the screen can be still observed at 2 metres from the vacuum tube. In general, other bodies behave like air; they are more transparent for the X-rays than for the kathode rays.

(11) A further distinction, and a noteworthy one, results from the action of a magnet. I have not succeeded in observing any deviation of the X-rays even in very strong magnetic fields.

The deviation of kathode rays by the magnet is one of their peculiar characteristics; it has been observed by Hertz and Lenard, that several kinds of kathode rays exist, which differ by their power of exciting phosphorescence, their susceptibility of absorption, and their deviation by the magnet; but a notable deviation has been observed in all cases which have yet been investigated, and I think that such deviation affords a characteristic not to be set aside lightly.

(12) As the result of many researches, it appears that the place of most brilliant phosphorescence of the walls of the discharge-tube is the chief seat whence the X-rays originate and spread in all directions; that is, the X-rays proceed from the front where the kathode rays strike the glass. If one deviates the kathode rays within the tube by means of a magnet, it is seen that the X-rays proceed from a new point, *i.e.* again from the end of the kathode rays.

Also for this reason the X-rays, which are not deflected by a magnet, cannot be regarded as kathode rays which have passed through the glass, for that passage cannot, according to Lenard, be the cause of the different deflection of the rays. Hence I conclude that the X-rays are not identical with the kathode rays, but are produced from the kathode rays at the glass surface of the tube.

(13) The rays are generated not only in glass. I have obtained them in an apparatus closed by an aluminium plate 2 mm thick. I purpose later to investigate the behaviour of other substances.

(14) The justification of the term "rays", applied to the phenomena, lies partly in the regular shadow pictures produced by the interposition of a more or less permeable body between the source and a photographic plate or fluorescent screen.

莱纳德在最近的工作中研究了各种物体对阴极射线的吸收系数，比如一个大气压下的空气的吸收系数，根据放电管抽真空程度的不同，每一厘米对应的吸收系数分别是 4.10、3.40 和 3.10。为了根据放电的本质来作出判断，我在基本相同的压强下进行了研究，不过偶尔也会用更高一点或更低一点的压强。利用韦伯光度计，我发现荧光的强度近似与屏幕到放电管距离的平方成反比。这个结论是根据三组非常一致的观测结果得到的，其观测距离分别为 100 毫米和 200 毫米。因此，空气对 X 射线的吸收比对阴极射线的吸收低很多。这一结果与前述的在距真空管 2 米处的屏幕上仍会出现荧光的结果是完全一致的。大体上，其他物质的性质与空气类似，它们对 X 射线比对阴极射线更加透明。

(11) 另一个更明显也更值得关注的区别是磁场的作用。即使是在非常强的磁场中，我也没有观测到 X 射线的任何偏转。

阴极射线在磁场作用下会发生偏转，这是它的独特性质之一。赫兹和莱纳德曾经观测到存在好几种阴极射线，它们的区别在于激发磷光的能力不同、被吸收的容易程度不同以及在磁场作用下的偏转不同。但是对于所有已经被研究过的阴极射线，人们都观测到了显著的偏转，我认为这种偏转代表了阴极射线的一种绝不该被忽视的特性。

(12) 很多研究结果表明，放电管管壁上磷光最强的位置是在 X 射线产生并向四周各个方向发散的那个源头处，也就是说，X 射线产生于阴极射线轰击玻璃的前沿位置。如果利用磁场使管中的阴极射线偏转，就会看到 X 射线从另一个位置上产生，但仍然是在阴极射线的终端位置。

基于这一原因，我们不能把在磁场作用下并不偏转的 X 射线看作是已经穿透玻璃的阴极射线，因为按照莱纳德的说法，这条通道不可能是由阴极射线的不同偏转造成的。由此我断定，X 射线与阴极射线是不同的，它是阴极射线作用于真空管的玻璃表面而产生的。

(13) 并不是只有用玻璃才能产生 X 射线。我曾利用一种被 2 毫米厚的铝板包裹起来的装置得到了 X 射线。以后我将研究其他物质是否也能产生 X 射线。

(14) 在描述这种现象时我使用了“射线”这个词，这在一定程度上是因为，将不太透明的物体插入到源和照相干版或荧光屏之间时会产生规则的阴影。

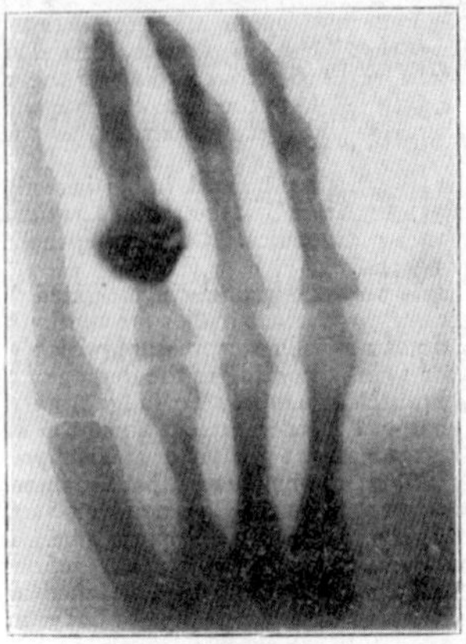

Fig. 1. Photograph of the bones in the fingers of a living human hand.
The third finger has a ring upon it.

I have observed and photographed many such shadow pictures. Thus, I have an outline of part of a door covered with lead paint; the image was produced by placing the discharge-tube on one side of the door, and the sensitive plate on the other. I have also a shadow of the bones of the hand (Fig. 1), of a wire wound upon a bobbin, of a set of weights in a box, of a compass card and needle completely enclosed in a metal case (Fig. 2), of a piece of metal where the X-rays show the want of homogeneity, and of other things.

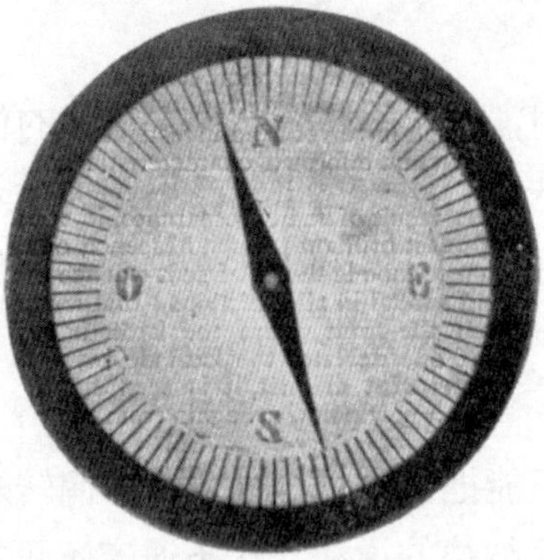

Fig. 2. Photograph of a compass card and needle completely enclosed in a metal case

For the rectilinear propagation of the rays, I have a pin-hole photograph of the discharge apparatus covered with black paper. It is faint but unmistakable.

(15) I have sought for interference effects of the X-rays, but possibly, in consequence of their small intensity, without result.

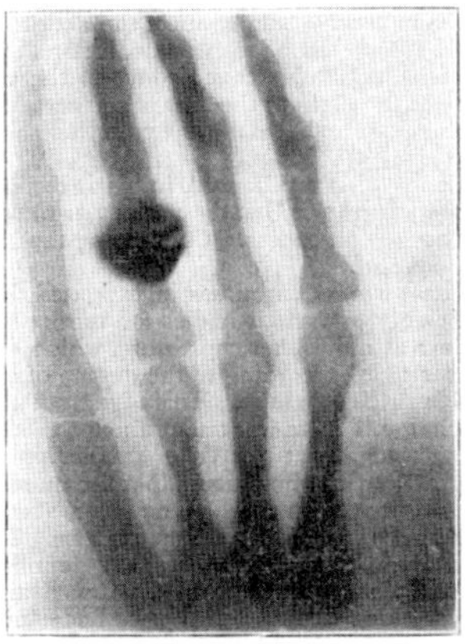

图 1. 活人手指（第三指上戴着一枚戒指）骨骼的影像

我已经观察并用照相记录了很多这样的阴影。由此，我记录下了门的局部轮廓。我是用含铅涂料刷了门的轮廓，然后把放电管放置在门的一侧，而把光敏照相干版放置在另一侧，这样就得到了门的局部轮廓的影像。我还记录了其他许多物体的阴影，这包括手掌骨骼（图 1）、缠在绕线筒上的导线、一套装在盒子里的砝码、完全密封于金属盒子中的罗经刻度盘和指针（图 2）、一块在 X 射线下显示出具有不均匀缺陷的金属片，以及其他一些物品。

图 2. 完全密封于金属盒子中的罗经刻度盘和指针的影像

为了说明射线的直线传播，我用针孔照相的方法拍摄了用黑纸覆盖的放电装置。照片虽然有些模糊，但却可以明白无误地分辨出装置。

(15) 我曾经试图寻找 X 射线的干涉效应，但并没有检测到，这可能是由于强度太低的缘故。

(16) Researches to investigate whether electrostatic forces act on the X-rays are begun but not yet concluded.

(17) If one asks, what then are these X-rays; since they are not kathode rays, one might suppose, from their power of exciting fluorescence and chemical action, them to be due to ultra-violet light. In opposition to this view a weighty set of considerations presents itself. If X-rays be indeed ultra-violet light, then that light must possess the following properties.

(*a*) It is not refracted in passing from air into water, carbon bisulphide, aluminium, rock-salt, glass or zinc.
(*b*) It is incapable of regular reflection at the surfaces of the above bodies.
(*c*) It cannot be polarised by any ordinary polarising media.
(*d*) The absorption by various bodies must depend chiefly on their density.

That is to say, these ultra-violet rays must behave quite differently from the visible, infra-red, and hitherto known ultra-violet rays.

These things appear so unlikely that I have sought for another hypothesis.

A kind of relationship between the new rays and light rays appears to exist; at least the formation of shadows, fluorescence, and the production of chemical action point in this direction. Now it has been known for a long time, that besides the transverse vibrations which account for the phenomena of light, it is possible that longitudinal vibrations should exist in the ether, and, according to the view of some physicists, must exist. It is granted that their existence has not yet been made clear, and their properties are not experimentally demonstrated. Should not the new rays be ascribed to longitudinal waves in the ether?

I must confess that I have in the course of this research made myself more and more familiar with this thought, and venture to put the opinion forward, while I am quite conscious that the hypothesis advanced still requires a more solid foundation.

(**53**, 274-276; 1896)

(16) 关于静电力对 X 射线是否有作用的研究工作正在进行，但目前尚无结论。

(17) 也许有人会问 X 射线到底是什么。既然这种射线不是阴极射线，有人可能就会根据其激发荧光和引发化学反应的能力猜想它是紫外光。然而，一系列认真的思考都是反对这种观点的。如果 X 射线真是一种紫外光，那么这种紫外光就必须具有如下性质：

(*a*) 它在由空气进入水、二硫化碳、铝、岩盐、玻璃或锌时，不会发生折射。

(*b*) 它在上述物质的表面不会发生常规的反射。

(*c*) 任何普通的偏振介质都不能使它偏振。

(*d*) 不同物质对它的吸收主要取决于该物质的密度。

也就是说，这种紫外线必须具有与可见光、红外线以及迄今为止已知的紫外线都十分不同的性质。

看起来这些是很难成立的，因此我想到了另一种假说。

这种新型的射线与普通光之间看起来应该存在着某种关联，至少在形成阴影、激发荧光以及引发化学反应这些方面都是相似的。长期以来我们都知道，除了能够解释光现象的横向振动外，在以太中可能存在纵向振动，某些物理学家甚至认为纵向振动是必定存在的。尽管目前人们还不完全清楚纵向振动是否存在，也没有通过实验论证这种纵向振动的性质，但是，难道我们就不能认为这种新型的射线属于以太中的纵波吗？

我必须要承认的是，在研究过程中我越来越倾向于这一观点。另外我也十分清楚这一新假说还需要更为可靠的证据，我承认在目前的情况下抛出这一观点是比较冒昧的。

（王耀杨 翻译；江丕栋 审稿）

Professor Röntgen's Discovery

A. A. C. Swinton

Editor's Note

One of the most sobering things about this verification of Röntgen's discovery of X-rays, less than a month after they were first reported, is that it shows how tepid the reception of great discoveries can be among scientific peers. Campbell-Swinton hints that the newspapers have been getting excited over a phenomenon that is not "entirely novel". But that is because he somewhat misinterprets Röntgen's results. Swinton insists on regarding the X-rays as "some portion of the kathode radiations", and points out that cathode rays are already known to produce photographic images—missing Röntgen's claim that his X-rays are not cathode rays at all.

THE newspaper reports of Prof. Röntgen's experiments have, during the past few days, excited considerable interest. The discovery does not appear, however, to be entirely novel, as it was noted by Hertz that metallic films are transparent to the kathode rays from a Crookes or Hittorf tube, and in Lenard's researches, published about two years ago, it is distinctly pointed out that such rays will produce photographic impressions. Indeed, Lenard, employing a tube with an aluminium window, through which the kathode rays passed out with comparative ease, obtained photographic shadow images almost identical with those of Röntgen, through pieces of cardboard and aluminium interposed between the window and the photographic plate.

Prof. Röntgen has, however, shown that this aluminium window is unnecessary, as some portion of the kathode radiations that are photographically active will pass through the glass walls of the tube. Further, he has extended the results obtained by Lenard in a manner that has impressed the popular imagination, while, perhaps most important of all, he has discovered the exceedingly curious fact that bone is so much less transparent to these radiations than flesh and muscle, that if a living human hand be interposed between a Crookes tube and a photographic plate, a shadow photograph can be obtained which shows all the outlines and joints of the bones most distinctly.

Working upon the lines indicated in the telegrams from Vienna, recently published in the daily papers, I have, with the assistance of Mr. J. C. M. Stanton, repeated many of Prof. Röntgen's experiments with entire success. According to one of our first experiments, an ordinary gelatinous bromide dry photographic plate was placed in an ordinary camera back. The wooden shutter of the back was kept closed, and upon it were placed miscellaneous articles such as coins, pieces of wood, carbon, ebonite, vulcanised fibre, aluminium, &c., all being quite opaque to ordinary light. Above was supported a Crookes tube, which was excited for some minutes. On development, shadows of all the articles

伦琴教授的发现

斯温顿

编者按

距离伦琴最初宣布发现X射线还不到一个月，就有了这篇对伦琴的发现的查证，这是一个应该引起人们警醒的事例，它表明科学界同行对重大发现的态度也会是冷淡的。坎贝尔–斯温顿含蓄地指出，报业为之感到兴奋的现象实际上并不是一个“全新的”现象。但他之所以这样说是因为他在某种程度上误会了伦琴得到的结论。斯温顿坚持把X射线看作是“阴极射线的一部分”，并指出人们早就知道阴极射线能够用于拍摄照片，可他没有注意到伦琴所称的X射线根本就不是阴极射线。

前一段时间，关于伦琴教授的实验的新闻报道引起了相当广泛的关注。不过，这一发现似乎并不是全新的，因为赫兹就曾注意到从克鲁克斯管或希托夫管中发射出来的阴极射线能够穿透金属薄片，而大约两年前莱纳德就在其发表的研究报告中明确地指出这种射线可以产生影像。莱纳德使用了一个带铝窗的管子，阴极射线可以比较容易地从此窗中穿出，伦琴的实验中则是射线穿过了插在管窗与照相干版之间的纸板和铝片。实际上，莱纳德得到了与伦琴的结果几乎完全一样的阴影图像。

不过，伦琴教授已经阐明这个铝窗并不是必需的，因为一部分能够引起成像的阴极辐射是从玻璃管壁中穿出的。此外，他还以一种能给人们留下深刻印象的方式推广了莱纳德得到的结果，而也许最为重要的是，他发现了一个极为新奇的现象，即这种辐射穿透骨骼的能力比穿透肌肉的能力差很多，如果将活人的手置于克鲁克斯管与照相干版之间，就能得到一张非常清晰地显示出骨骼关节轮廓的阴影图像。

最近，许多日报都刊载了来自维也纳的电报，根据其中提供的线索，在斯坦顿先生的协助下，我已经完全成功地重复了伦琴教授的很多实验。在最初的一次实验中，我们将一张普通的凝胶溴化物照相干版放置在普通相机后面。背面的木质快门始终保持关闭状态，并紧接着放置各种物品，诸如硬币、木块、炭、硬质橡胶、硬化纤维和铝等，所有这些物品对于普通的可见光都是完全不透明的。在这些物品上方固定一个已经激发了几分钟的克鲁克斯管。显影后，放置的所有物品的阴影都清晰可

placed on the slide were clearly visible, some being more opaque than others. Further experiments were tried with thin plates of aluminium or of black vulcanised fibre interposed between the objects to be photographed and the sensitive surface, this thin plate being used in place of the wood of the camera back. In this manner sharper shadow pictures were obtained. While most thick metal sheets appear to be entirely opaque to the radiations, aluminium appears to be relatively transparent. Ebonite, vulcanised fibre, carbon, wood, cardboard, leather and slate are all very transparent, while, on the other hand, glass is exceedingly opaque. Thin metal foils are moderately opaque, but not altogether so.

As tending to the view that the radiations are more akin to ultraviolet than to infra-red light, it may be mentioned that a solution of alum in water is distinctly more transparent to them than a solution of iodine in bisulphide of carbon.

So far as our own experiments go, it appears that, at any rate without very long exposures, a sufficiently active excitation of the Crookes tube is not obtained by direct connection to an ordinary Rhumkorff induction coil, even of a large size. So-called high frequency currents, however, appear to give good results, and our own experiments have been made with the tube excited by current obtained from the secondary circuit of a Tesla oil coil, through the primary of which were continuously discharged twelve half-gallon Leyden jars, charged by an alternating current of about 20,000 volts pressure, produced by a transformer with a spark-gap across its high-pressure terminals.

For obtaining shadow photographs of inanimate objects, and for testing the relative transparency of different substances, the particular form of Crookes tube employed does not appear to greatly signify, though some forms are, we find, better than others. When, however, the human hand is to be photographed, and it is important to obtain sharp shadows of the bones, the particular form of tube used and its position relative to the hand and sensitive plate appear to be of great importance. So far, owing to the frequent destruction of the tubes, due to overheating of the terminals, we have not been able to ascertain exactly the best form and arrangement for this purpose, except that it appears desirable that the electrodes in the tube should consist of flat and not curved plates, and that these plates should be of small dimensions.

The accompanying photograph of a living human hand (Fig. 1) was exposed for twenty minutes through an aluminium sheet 0.0075 in thickness, the Crookes tube, which was one of the kind containing some white phosphorescent material (probably sulphide of barium), being held vertically upside down, with its lowest point about two inches above the centre of the hand.

见，其中一些物品的阴影比其他的更加明显。在后来的实验中，我们尝试着将薄铝板或黑色硬化纤维薄板插入待成像的物体与感光表面之间，这一薄板用来代替相机后的木质快门。用这种方式我们得到了更加清晰的阴影图像。大部分厚金属板对于这种辐射似乎都是完全不透明的，而铝板则似乎比较透明。硬质橡胶、硬化纤维、炭、木块、纸板、皮革以及石板都是非常透明的，相反，玻璃则是非常不透明的。薄金属板是中等透明的，不过也不全是这样。

为了支持该辐射更类似于紫外线而不是红外线的观点，我们要说明一下，与碘的二硫化碳溶液相比，明矾的水溶液对该辐射的透明度明显好得多。

就我们的实验情况来说，如果不进行很长时间的曝光，单靠将一个普通的拉姆科夫感应线圈与克鲁克斯管直接相连，即使是用大号的线圈，看起来似乎也无法使克鲁克斯管产生足够引起成像活性的激发辐射。不过，我们常说的高频电流看来能够给出好的结果。我们在实验中使用的克鲁克斯管，是由特斯拉油线圈的次级电路产生的电流来激发的，通过其初级电路的是连续放电的 12 个半加仑莱顿瓶，这些莱顿瓶通过电压约为 20,000 伏特的交流电进行充电，此交流电由一高压端带有放电间隙的变压器产生。

要获得无生命物体的阴影或检验不同物质对此种辐射的相对透明度，使用哪种结构的克鲁克斯管看起来并不是非常要紧，尽管我们发现某些结构的管子比另外一些好一点。但是，在获取人手的影像时，重要的是得到骨骼的清晰阴影，那么所使用的管子的特殊结构以及管子相对于人手和感光干版的位置就显得至关重要了。到目前为止，由于管子经常因其末端过热而毁坏，我们还没能弄清楚对于上述目的来说什么样的管子结构和摆放位置是最好的，不过能够确定的是，管子中的电极应该采用平板电极而不是曲面电极，并且应该用尺寸较小的电极板。

本文所附的活人手掌的影像（图 1）是该辐射穿过厚度为 0.0075 的铝片持续曝光 20 分钟而得到的，实验中使用的克鲁克斯管中包含某种白色磷光物质（可能是硫化钡），管子颠倒后垂直放置，其最低点位于掌心上方大约 2 英寸处。

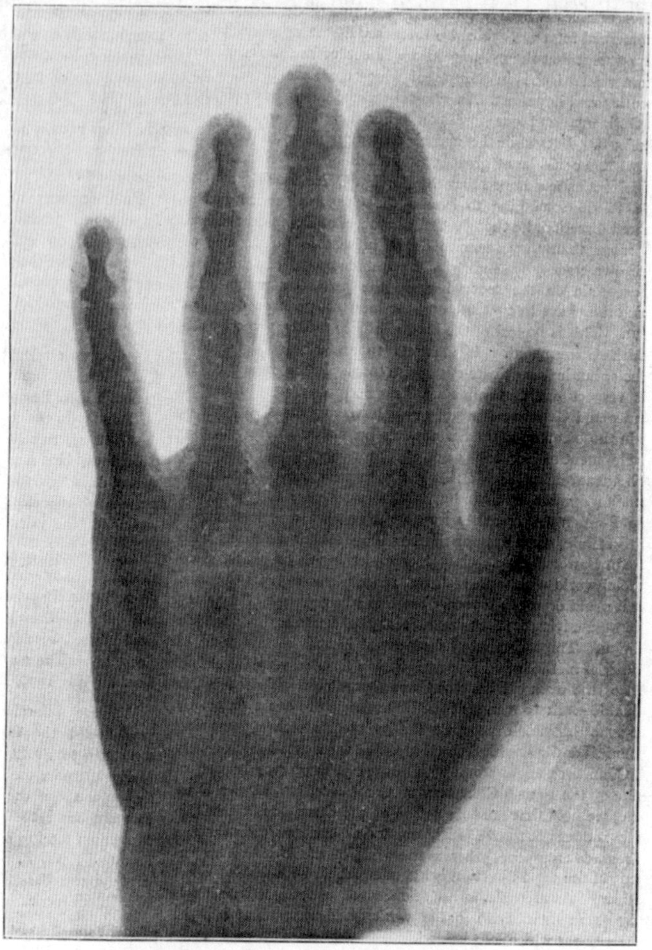

Fig. 1. Photograph of a living human hand

By substituting a thin sheet of black vulcanised fibre for the aluminium plate, we have since been able to reduce the exposure required to four minutes. Indeed with the aluminium plate, the twenty minutes' exposure appears to have been longer than was necessary. Further, having regard to the great opacity of glass, it seems probable that where ordinary Crookes tubes are employed, a large proportion of the active radiations must be absorbed by the glass of the tube itself. If this is so, by the employment of a tube partly constructed of aluminium, as used by Lenard, the necessary length of exposure could be much reduced.

(**53**, 276-277; 1896)

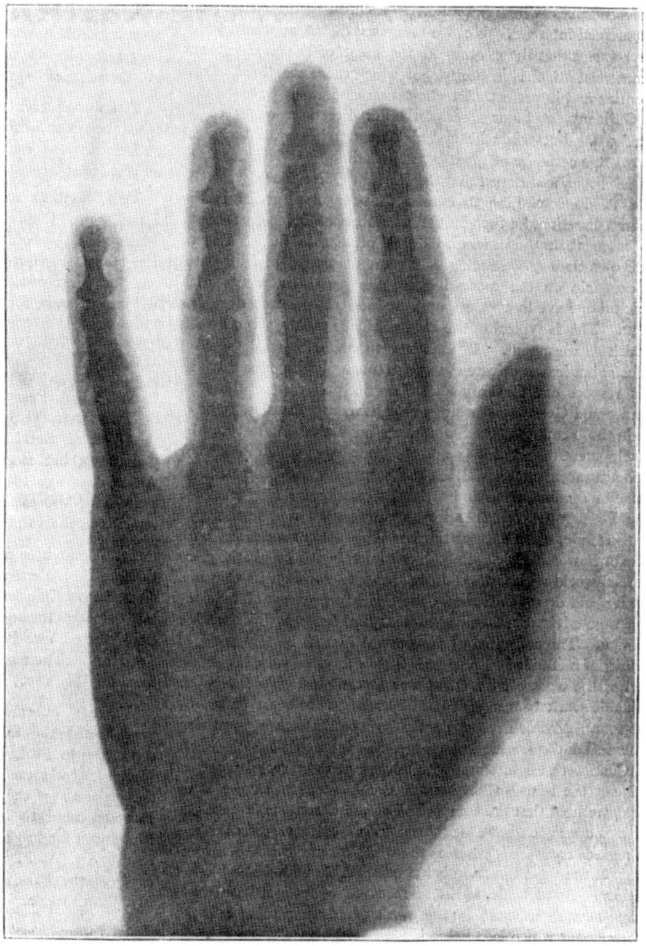

图 1. 活人手掌的影像

用一块黑色硬化纤维薄板代替铝板后，我们发现可以把曝光时间缩短到 4 分钟。实际上，在使用铝板的情况下，20 分钟的曝光时间似乎也比必需的曝光时间长一些。此外，考虑到玻璃很显著的不透明性，看来在使用普通克鲁克斯管时，大部分能够形成影像的辐射一定被管子自身的玻璃壁吸收了。如果确实如此，那么使用部分为铝质材料的管子（如同莱纳德所用的那样）的话，必需的曝光时间应该会大大缩短。

（王耀杨 翻译；江丕栋 审稿）

New Experiments on the Kathode Rays*

M. J. Perrin

Editor's Note

Physicists were puzzled by cathode rays, which carried energy from a negative electrode (cathode) toward a positive electrode inside a vacuum tube. Experimenters had determined that they originated at the cathode, but could not say what they were. In this classic paper the French physicist Jean Perrin reports experiments that helped to clarify the mystery. He placed into a cathode ray tube a metal cylinder linked to an electroscope, which would measure any charge deposited into it. When the cathode rays were directed into the tube, Perrin detected a significant negative charge. Perrin speculated that the charge carriers were negative ions created near the cathode. In fact they were electrons, as J. J. Thomson discovered one year later.

(1) Two hypotheses have been propounded to explain the properties of the kathode rays.

Some physicists think with Goldstein, Hertz, and Lenard, that this phenomenon is like light, due to vibrations of the ether †, or even that it is light of short wavelength. It is easily understood that such rays may have a rectilinear path, excite phosphorescence, and affect photographic plates.

Others think, with Crookes and J. J. Thomson, that these rays are formed by matter which is negatively charged and moving with great velocity, and on this hypothesis their mechanical properties, as well as the manner in which they become curved in a magnetic field, are readily explicable.

This latter hypothesis has suggested to me some experiments which I will now briefly describe, without for the moment pausing to inquire whether the hypothesis suffices to explain all the facts at present known, and whether it is the only hypothesis that can do so. Its adherents suppose that the kathode rays are negatively charged; so far as I know, this electrification has not been established, and I first attempted to determine whether it exists or not.

(2) For that purpose I had recourse to the laws of induction, by means of which it is possible to detect the introduction of electric charges into the interior of a closed electric conductor, and to measure them. I therefore caused the kathode rays to pass into a

* Translation of a paper by M. Jean Perrin, read before the Paris Academy of Sciences on December 30, 1895.

† These vibrations might be something different from light; recently M. Jaumann, whose hypotheses have since been criticised by M. H. Poincaré, supposed them to be longitudinal.

关于阴极射线的新实验 *

佩兰

编者按

物理学家们对在真空管中把能量从负极（阴极）带到正极的阴极射线感到迷惑不解。实验可以证明它们来自阴极，但不能说明它们到底是什么。法国物理学家让·佩兰在这篇经典论文中用实验揭开了阴极射线的神秘面纱。他在阴极射线管内放置了一个与验电器相连的金属圆柱体，验电器可以测量进入其中的电荷。当阴极射线进入真空管时，佩兰检测到了大量的负电荷。佩兰推测这些载流子是在阴极附近产生的负离子。实际上它们就是一年之后汤姆逊发现的电子。

(1) 现在可以解释阴极射线性质的假说有两种。

一部分物理学家和戈尔德施泰因、赫兹、莱纳德的意见一致，认为这种现象和光一样，是由以太的振动引起的†，或者它就是一种短波长的光。很容易就可以理解，这种射线可能是沿直线传播的，能激发磷光，而且可以使照相干版感光。

另一部分人则与克鲁克斯、汤姆逊持相同的观点，认为这种射线是由带负电的物质组成，并以极快的速度运动。用这种假说可以解释它们的力学性质，也可以很容易地说明为什么它们在磁场中的路径会弯曲。

后一种假说启发我进行了一些实验，我将在这里简单地描述这些实验，暂时先不管这个假说是否能解释目前所有已知的现象，或者是否只有这一种假说可以解释这些现象。它的支持者认为阴极射线是带负电的，而据我所知，这种带电性还没有被确认，我首先要确定它是否带电。

(2) 为了达到这个目的，我将借助电磁感应定律，用这个定律，我们可以检测引入闭合导电体内部的电量，并且进行定量测量。因此，我让阴极射线通过法拉第

---

* 这篇文章翻译自让·佩兰在1895年12月30日向巴黎科学院宣读的论文。

† 这种振动可能与光有些不同。最近，尧曼（其猜想曾经被普安卡雷批判过）提出这种振动可能是纵向的。

Faraday's cylinder. For this purpose I employed the vacuum tube represented in Fig. 1. A B C D is a tube with an opening α in the centre of the face B C. It is this tube which plays the part of a Faraday's cylinder. A metal thread soldered at S to the wall of the tube connects this cylinder with an electroscope.

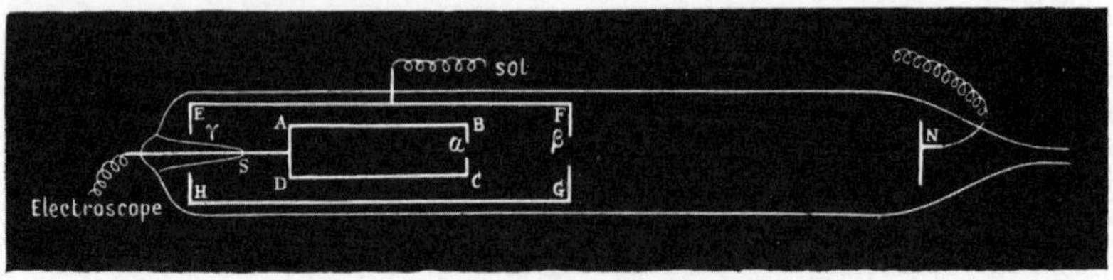

Fig. 1

E F G H is a second cylinder in permanent communication with the earth, and pierced by two small openings at β and γ; it protects the Faraday's cylinder from all external influence. Finally, at a distance of about 0.10 m in front of F G , was placed an electrode N. The electrode N served as kathode; the anode was formed by the protecting cylinder E F G H; thus a pencil of kathode rays passed into the Faraday's cylinder. This cylinder invariably became charged with negative electricity.

The vacuum tube could be placed between the poles of an electro-magnet. When this was excited, the kathode rays, becoming deflected, no longer passed into the Faraday's cylinder, and this cylinder was then not charged; it, however, became charged immediately the electromagnet ceased to be excited.

In short, the Faraday's cylinder became negatively charged when the kathode rays entered it, and only when they entered it; *the kathode rays are then charged with negative electricity*.

The quantity of electricity which these rays carry can be measured. I have not finished this investigation, but I shall give an idea of the order of magnitude of the charges obtained when I say that for one of my tubes, at a pressure of 20 microns of mercury, and for a single interruption of the primary of the coil, the Faraday's cylinder received a charge of electricity sufficient to raise a capacity of 600 C. G. S. units to 300 volts.

(3) The kathode rays being negatively charged, the principle of the conservation of electricity drives us to seek somewhere the corresponding positive charges. I believe that I have found them in the very region where the kathode rays are formed, and that I have established the fact that they travel in the opposite direction, and fall upon the kathode. In order to verify this hypothesis, it is sufficient to use a hollow kathode pierced with a small opening by which a portion of the attracted positive electricity might enter. This electricity could then act upon a Faraday's cylinder inside the kathode.

圆筒。为此我设计了如图 1 所示的真空管。A B C D 是一根管子，在 B C 面中心的 α 处有一个小孔。正是这根管子起到了法拉第圆筒的作用。一根焊接在管壁 S 处的金属线将圆筒和外部的验电器连接起来。

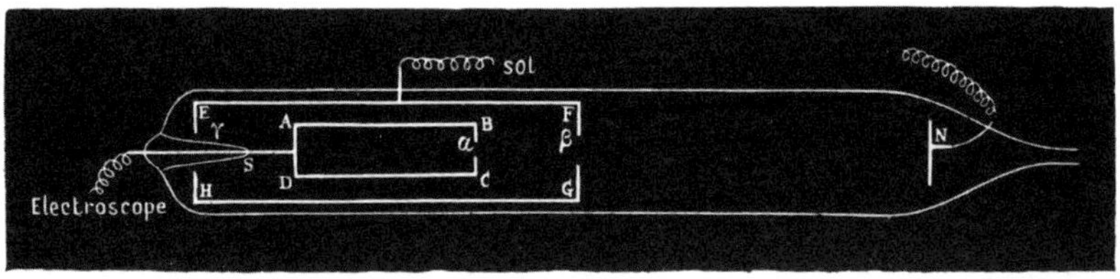

图 1

E F G H 是另一个圆筒，永久接地，并且在 β、γ 处穿有两个小孔；这个圆筒可以屏蔽外界对法拉第圆筒的干扰。最后，在 FG 前面大约 0.1 米的地方有一个电极 N。电极 N 作为阴极，屏蔽圆筒 E F G H 作为阳极。在这样的条件下，将一束阴极射线通入法拉第圆筒，这个圆筒将一直带负电。

真空管可以放置在电磁铁的两极之间。当电磁铁通电时，阴极射线将发生偏转，不能再通入法拉第圆筒，这个圆筒也将不再带电，而当电磁铁断电之后，法拉第圆筒马上又带电了。

简单地说就是，当且仅当有阴极射线进入时，法拉第圆筒带负电，**所以阴极射线一定带负电**。

射线所带的电量可以被测量出来。我还没有完成这项研究，但是我可以给出一个有关所获电量的数量级的概念，对于一个压力为 20 微米汞柱的真空管，将初级线圈截断，法拉第圆筒接收的电量足以使 600 单位（厘米克秒制）的电容器的电势差提高到 300 伏特。

（3）阴极射线带负电，根据电荷守恒定律，我们应该能在某处找到相应的正电荷。我确信我已经在阴极射线产生的地方找到了正电荷，我认为它们向相反的方向运动，而后撞在了阴极上。为了证明这种说法，只要用一个中空的阴极就行，在阴极上穿一个小孔，以使一部分被吸引过来的正电荷可以由此通过。进入阴极的正电荷会影响阴极内部的法拉第圆筒。

The protecting cylinder E F G H with its opening β fulfilled these conditions, and this time I therefore employed it as the kathode, the electrode N being the anode. The Faraday's cylinder is then invariably charged with *positive electricity*. The positive charges were of the order of magnitude of the negative charges previously obtained.

Thus, at the same time as negative electricity is *radiated* from the kathode, positive electricity travels towards that kathode.

I endeavoured to determine whether this positive flux formed a second system of rays absolutely symmetrical to the first.

(4) For that purpose I constructed a tube (Fig. 2) similar to the preceding, except that between the Faraday's cylinder and the opening β was placed a metal diaphragm pierced with an opening β', so that the positive electricity which entered by β could only affect the Faraday's cylinder if it also traversed the diaphragm β'. Then I repeated the preceding experiments.

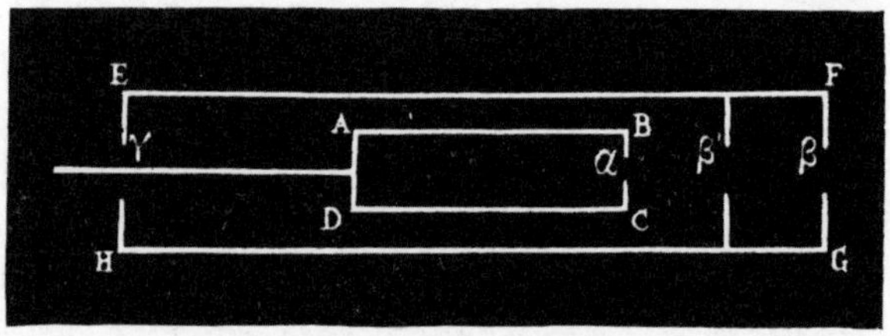

Fig. 2

When N was the kathode, the rays emitted from the kathode passed through the two openings β and β' without difficulty, and caused a strong divergence of the leaves of the electroscope. But when the protecting cylinder was the kathode, the positive flux, which, according to the preceding experiment, entered at β, did not succeed in separating the gold leaves except at very low pressures. When an electrometer was substituted for the electroscope, it was found that the action of the positive flux was real but very feeble, and increased as the pressure decreased. In a series of experiments at a pressure of 20 microns, it raised a capacity of 2,000 C. G .S. units to 10 volts; and at a pressure of 3 microns, during the same time, it raised the potential to 60 volts.*

By means of a magnet this action could be entirely suppressed.

(5) These results as a whole do not appear capable of being easily reconciled with the theory which regards the kathode rays as an ultra-violet light. On the other hand, they agree well with the theory which regards them as a material radiation, and which, as it appears to me, might be thus enunciated.

* The breaking of the tube has temporarily prevented me from studying the phenomenon at lower pressures.

带有小孔 β 的屏蔽圆筒 E F G H 满足以上条件，因此这一回我用 E F G H 作阴极，电极 N 作阳极。这样，法拉第圆筒就会一直带**正电**。其所带正电荷和前面所测的负电荷的数量级相同。

这说明，在阴极**发射**负电荷的同时，正电荷也在向阴极运动。

我下决心要确定这种正电流是否能形成另一个和阴极射线完全对称的射线系统。

(4) 为此我构造了一个和前面类似的管子（见图 2），与前面管子唯一的不同之处是，在法拉第圆筒和小孔 β 之间放置了一个金属膜片，膜片上有一个小孔 β′，这样，从 β 进入的正电荷只有在也通过 β′ 的情况下才能作用于法拉第圆筒。我用这个装置重复了上面的实验。

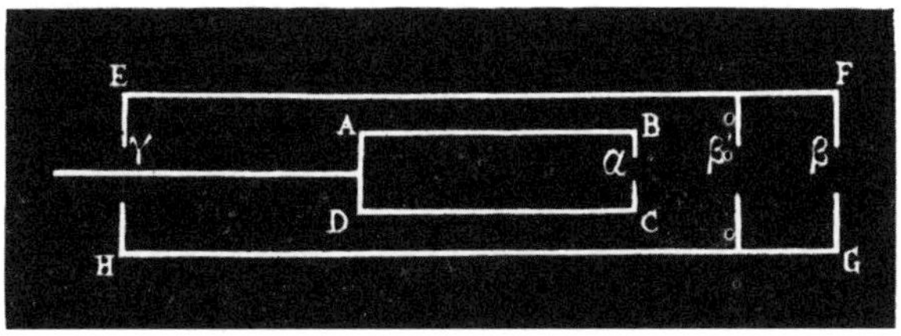

图 2

当 N 作为阴极时，由阴极发射的射线可以顺利地通过 β 和 β′ 两个孔，使验电器的两个叶片张得很大。但是当屏蔽圆筒作为阴极时，根据前述的实验，通过 β 进入的正电流并没有使验电器的金箔张开，除非是在压力很低的情况下。当用静电计代替验电器时，可以看到正电流的确存在但非常微弱，并随着压力的减小而增大。在 20 微米汞柱条件下进行的一系列实验中，这一正电流可以把 2,000 单位（厘米克秒制）的电容器的电势差提高到 10 伏特；在压力为 3 微米汞柱时，同样时间内电势差被提高到了 60 伏特。*

利用磁铁可以完全地抑制这种作用。

(5) 总的来说，把阴极射线看作是紫外线的假说似乎不太容易解释这些实验结果。另一方面，这些实验结果与认为阴极射线是一种物质辐射的假说符合得很好，在我看来，这些实验结果恐怕只能这么解释。

* 真空管的爆裂使我暂时无法再在低压条件下研究这一现象。

In the neighbourhood of the kathode, the electric field is sufficiently intense to break into pieces (*into ions*) certain of the molecules of the residual gas. The negative ions move towards the region where the potential is increasing, acquire a considerable speed, and form the kathode rays; their electric charge, and consequently their mass (at the rate of one valence-gramme for 100,000 Coulombs) is easily measurable. The positive ions move in the opposite direction; they form a diffused brush, sensitive to the magnet, and not a radiation in the correct sense of the word.*

(**53**, 298-299; 1896)

* This work has been carried out in the laboratory of the Normal School, and in that of M. Pellat at the Sorbonne.

在阴极附近，电场强度强到足以把一定量的残余气体分子打成碎片（**变成离子**）。负离子向着电势增加的方向运动，速度很大，这形成了阴极射线，它们的电量很容易被测定，从而其质量（100,000 库仑对应 1 克当量）也很容易得到。正离子向相反方向运动，形成一个发散的尾巴，对磁场非常敏感，准确地说这就不是辐射了。*

（王锋 翻译；江丕栋 审稿）

* 这项研究是在师范学院的实验室和索邦大学佩拉的实验室中进行的。

The Effect of Magnetisation on the Nature of Light Emitted by a Substance*

P. Zeeman

Editor's Note

Does a magnetic field influence the light emitted by an atom? Here Pieter Zeeman reports the first evidence that it does. Zeeman heated sulphur in a ceramic chamber with transparent ends, and placed the chamber in a magnetic field. With the light of an arc lamp, he then measured the absorption spectrum and found a broadening of certain lines, attributing this to a change in the frequency of the absorbed light. Zeeman noted that the polarization of light emitted in the presence of a field behaves as predicted by Lorentz, owing to the circular motion of charged particles within the atom. He estimates the charge/mass ratio for these particles as being about 10^7.

IN consequence of my measurements of Kerr's magneto-optical phenomena, the thought occurred to me whether the period of the light emitted by a flame might be altered when the flame was acted upon by magnetic force. It has turned out that such an action really occurs. I introduced into an oxyhydrogen flame, placed between the poles of a Ruhmkorff's electromagnet, a filament of asbestos soaked in common salt. The light of the flame was examined with a Rowland's grating. Whenever the circuit was closed both D lines were seen to widen.

Since one might attribute the widening to the known effects of the magnetic field upon the flame, which would cause an alteration in the density and temperature of the sodium vapour, I had resort to a method of experimentation which is much more free from objection.

Sodium was strongly heated in a tube of biscuit porcelain, such as Pringsheim used in his interesting investigations upon the radiations of gases. The tube was closed at both ends by plane parallel glass plates, whose effective area was 1 cm. The tube was placed horizontally between the poles, at right angles to the lines of force. The light of an arc lamp was sent through. The absorption spectrum showed both D lines. The tube was continuously rotated round its axis to avoid temperature variations. Excitation of the magnet caused immediate widening of the lines. It thus appears very probable that the period of sodium light is altered in the magnetic field. It is remarkable that Faraday, as early as 1862, had made the first recorded experiment in this direction, with the incomplete resources of that period, but with a negative result (Maxwell, "Collected Works", vol. II, p. 790).

---

* Translated by Arthur Stanton from the *Proceedings of the Physical Society of Berlin*.

磁化对物质发射的光的性质的影响*

塞曼

编者按

磁场会影响原子发射的光吗？彼得·塞曼在这篇报告中首次证明这种效应是存在的。塞曼在两端透明的陶瓷真空室中加热硫黄，并把这个真空室放入磁场中。在弧光灯的照射下，他测量了吸收光谱并发现某些特定的谱线出现了加宽的现象，他把这归因于被吸收光线的频率的改变。塞曼特别提到，有场存在时发射出来的光的偏振与洛伦兹预言的一样，是由原子内带电粒子的圆周运动产生的。他估计这些粒子的荷质比约为 10^7。

我在对克尔磁光效应进行测量时突然产生了这样的想法：当磁力作用于火焰时，火焰发射出的光的周期是否会发生变化。结果证实这样的作用确实存在。我把浸泡在普通食盐中的石棉丝放在置于鲁姆科夫电磁体两极之间的氢氧焰中。火焰光用罗兰光栅检验。每当电路接通时都能看到两条 D 线的加宽。

鉴于也许有人会将谱线加宽归因于磁场对火焰的某种已知作用使钠蒸气的密度和温度发生了变化，我已采用了更加没有异议的实验方法进行了确证。

我们在素瓷管（与普林斯海姆在他著名的气体辐射实验中所用的一样）内对钠进行高温加热。管的两端用两块相互平行的平玻璃板密封，其有效区域为 1 厘米。该管被水平地置于两极之间，与磁力线垂直。弧光灯的光穿过其中，吸收光谱中显示出两条 D 线。管子不停地绕着它的轴自转以保持各处温度均衡，磁作用使谱线迅速加宽。很可能是因为钠光的周期在磁场中发生了变化。值得注意的是，这方面第一个有记录的实验是法拉第早在 1862 年进行的，那时的资源并不完备，得到的是阴性的结果（麦克斯韦，《文集》，第 2 卷，第 790 页）。

---

* 由阿瑟·斯坦顿翻译自《柏林物理学会会刊》。

It has been already stated what, in general, was the origin of my own research on the magnetisation of the lines in the spectrum. The possibility of an alteration of period was first suggested to me by the consideration of the accelerating and retarding forces between the atoms and Maxwell's molecular vortices; later came an example suggested by Lord Kelvin, of the combination of a quickly rotating system and a double pendulum. However, a true explanation appears to me to be afforded by the theory of electric phenomena propounded by Prof. Lorentz.

In this theory, it is considered that, in all bodies, there occur small molecular elements charged with electricity, and that all electrical processes are to be referred to the equilibrium or motion of these "ions". It seems to me that in the magnetic field the forces directly acting on the ions suffice for the explanation of the phenomena.

Prof. Lorentz, to whom I communicated my idea, was good enough to show me how the motion of the ions might be calculated, and further suggested that if my application of the theory be correct there would follow these further consequences: that the light from the edges of the widened lines should be circularly polarised when the direction of vision lay along the lines of force; further, that the magnitude of the effect would lead to the determination of the ratio of the electric charge the ion bears to its mass. We may designate the ratio e/m. I have since found by means of a quarter-wave length plate and an analyser, that the edges of the magnetically-widened lines are really circularly polarised when the line of sight coincides in direction with the lines of force. An altogether rough measurement gives 10^7 as the order of magnitude of the ratio e/m when e is expressed in electromagnetic units.

On the contrary, if one looks at the flame in a direction at right angles to the lines of force, then the edges of the broadened sodium lines appear plane polarised, in accordance with theory. Thus there is here direct evidence of the existence of ions.

This investigation was conducted in the Physical Institute of Leyden University, and will shortly appear in the "Communications of the Leyden University".

I return my best thanks to Prof. K. Onnes for the interest he has shown in my work.

(**55**, 347; 1897)

P. Zeeman: Amsterdam.

前面已经介绍了我对谱线磁化进行研究的起因。周期变化的可能性使我首先想到的是原子与麦克斯韦分子涡旋之间的加速和减速作用力；然后想到的是开尔文勋爵提出的一个快速旋转体系与双摆复合体的例子。然而，使我受到启发并最终得出正确结论的是洛伦兹教授提出的关于电现象的理论。

这个理论认为：在所有物体中，都存在小的、带电的分子单元，所有电的过程都与这些“离子”的平衡或运动有关。在我看来，只要认为在磁场中力直接作用于这些离子上，就足以解释这些现象。

我向洛伦兹教授阐述了我的观点，他友好地告诉我离子如何运动也许是可以计算的，并进一步建议说，如果我对该理论的应用是正确的，那么就会出现以下结果：当沿磁力线方向观察时，从加宽谱线边缘发出的光应该是圆偏振光；此外，这个效应的大小将能使人们测定离子所带电荷与其质量的比值。我们可以用 e/m 表示这个比值。后来我用四分之一波片和检偏器测量发现，当观测方向与磁力线一致时，磁场加宽谱线的边缘果然是圆偏振的。粗略的测定表明，如果用 e 来表示电磁单位，e/m 这一比值的数量级大约为 10^7。

反之，如果观察火焰的方向与磁力线垂直，加宽的钠线边缘出现的是平面偏振光，这与理论相符。这些都是离子存在的直接证据。

这项研究是在莱顿大学物理研究所进行的，不久之后研究报告将刊登在《莱顿大学学报》上。

非常感谢昂内斯教授对我的工作的重视。

（沈乃澂 翻译；赵见高 审稿）

An Undiscovered Gas

W. Ramsay

Editor's Note

Sir William Ramsay was professor of chemistry at University College London and was awarded a Nobel prize in 1904 for his discovery of four of the rare gases, helium (in concert with Lord Rayleigh), neon, argon and xenon. Ramsay's address to the Chemistry Section of the British Association for the Advancement of Science in 1897 was principally about his reasons for believing that neon would exist, but is also interesting because it illustrates how tentative were ideas about the periodic table of the elements.

A sectional address to members of the British Association falls under one of three heads. It may be historical, or actual, or prophetic; it may refer to the past, the present, or the future. In many cases, indeed in all, this classification overlaps. Your former presidents have given sometimes a historical introduction, followed by an account of the actual state of some branch of our science, and, though rarely, concluding with prophetic remarks. To those who have an affection for the past, the historical side appeals forcibly; to the practical man, and to the investigator engaged in research, the actual, perhaps, presents more charm; while to the general public, to whom novelty is often more of an attraction than truth, the prophetic aspect excites most interest. In this address I must endeavour to tickle all palates; and perhaps I may be excused if I take this opportunity of indulging in the dangerous luxury of prophecy, a luxury which the managers of scientific journals do not often permit their readers to taste.

The subject of my remarks today is a new gas. I shall describe to you later its curious properties; but it would be unfair not to put you at once in possession of the knowledge of its most remarkable property—it has not yet been discovered. As it is still unborn, it has not yet been named. The naming of a new element is no easy matter. For there are only twenty-six letters in our alphabet, and there are already over seventy elements. To select a name expressible by a symbol which has not already been claimed for one of the known elements is difficult, and the difficulty is enhanced when it is at the same time required to select a name which shall be descriptive of the properties (or want of properties) of the element.

It is now my task to bring before you the evidence for the existence of this undiscovered element.

It was noticed by Döbereiner, as long ago as 1817, that certain elements could be arranged in groups of three. The choice of the elements selected to form these triads was made on account of their analogous properties, and on the sequence of their atomic

一种尚未发现的气体

拉姆齐

编者按

威廉·拉姆齐爵士是伦敦大学学院的化学教授，他曾因发现了氦（与瑞利勋爵合作）、氖、氩和氙这4种稀有气体而荣获1904年的诺贝尔奖。拉姆齐于1897年向英国科学促进会化学分部作了报告，主要阐述了他认为氖存在的观点及理由，报告之所以吸引人还在于它表明了当时关于元素周期表的许多观点还是非常不确定的。

向英国科学促进会成员所作的部门性演讲，大体可归入以下三类中的一种。它可以是历史性的，现实性的或是预见性的；它可能会涉及过去、现在或将来。在很多情况下，实际上是在所有情况下，这些分类间都是有交集的。前任主席有时会先给出一段历史介绍，随后是一段关于某些分支学科现状的说明，接着，虽然很少见，还是会以预见性的评述作为结语。对那些热衷于过去的听众来说，历史性的一面似乎更吸引人；对于实干家以及研究人员来说，可能现实性的介绍更有吸引力；而对于广大民众来说，新奇事物比科学真理更有诱惑力，预言性的内容将引起更多关注。在这段演讲中我将竭尽全力满足所有人的需求；如果我有凭借诸位给予我的宽容而作出危险的狂妄预言，敬请大家原谅，这种狂妄是科学杂志的管理者们不常让读者们体验的。

今天我要谈的主题是一种新的气体。随后我将向各位介绍它的奇妙性质；但是，为了公平起见，我要让你们马上了解到它最不同寻常的性质——它至今尚未被发现。由于尚未问世，它也还没有名字。一种新元素的命名可不是件简单的事。因为在我们的字母表中只有26个字母，而目前已有70多种元素。选择一个从未被其他已知元素用过的符号作为名字是很困难的，同时还要使所选择的名字能描述该元素性质（或者是希望它所具备的性质），那就更困难了。

现在，我要将这种尚未被发现的元素存在的证据呈现给你们。

早在1817年，德贝赖纳就注意到，某些元素可以按照三个一组的方式进行排列。选择元素构成这样的三元素组，是以它们具有的相似的性质，以及它们的原子量大

weights, which had at that time only recently been discovered. Thus calcium, strontium, and barium formed such a group; their oxides, lime, strontia, and baryta are all easily slaked, combining with water to form soluble lime-water, strontia-water, and baryta-water. Their sulphates are all sparingly soluble, and resemblance had been noticed between their respective chlorides and between their nitrates. Regularity was also displayed by their atomic weights. The numbers then accepted were 20, 42.5, and 65; and the atomic weight of strontium, 42.5, is the arithmetical mean of those of the other two elements, for (65+20)/2=42.5. The existence of other similar groups of three was pointed out by Döbereiner, and such groups became known as "Döbereiner's triads".

Another method of classifying the elements, also depending on their atomic weights, was suggested by Pettenkofer, and afterwards elaborated by Kremers, Gladstone, and Cooke. It consisted in seeking for some expression which would represent the differences between the atomic weights of certain allied elements. Thus, the difference between the atomic weight of lithium, 7, and sodium, 23, is 16; and between that of sodium and of potassium, 39, is also 16. The regularity is not always so conspicuous; Dumas, in 1857, contrived a somewhat complicated expression which, to some extent, exhibited regularity in the atomic weights of fluorine, chlorine, bromine, and iodine; and also of nitrogen, phosphorus, arsenic, antimony and bismuth.

The upshot of these efforts to discover regularity was that, in 1864, Mr. John Newlands, having arranged the elements in eight groups, found that when placed in the order of their atomic weights, "the eighth element, starting from a given one, is a kind of repetition of the first, like the eighth note of an octave in music." To this regularity he gave the name "The Law of Octaves".

The development of this idea, as all chemists know, was due to the late Prof. Lothar Meyer, of Tübingen, and to Prof. Mendeléeff, of St. Petersburg. It is generally known as the "Periodic Law". One of the simplest methods of showing this arrangement is by means of a cylinder divided into eight segments by lines drawn parallel to its axis; a spiral line is then traced round the cylinder, which will, of course, be cut by these lines eight times at each revolution. Holding the cylinder vertically, the name and atomic weight of an element is written at each intersection of the spiral with a vertical line, following the numerical order of the atomic weights. It will be found, according to Lothar Meyer and Mendeléeff, that the elements grouped down each of the vertical lines form a natural class; they possess similar properties, form similar compounds, and exhibit a graded relationship between their densities, melting-points, and many of their other properties. One of these vertical columns, however, differs from the others, inasmuch as on it there are three groups, each consisting of three elements with approximately equal atomic weights. The elements in question are iron, cobalt, and nickel; palladium, rhodium, and ruthenium; and platinum, iridium, and osmium. There is apparently room for a fourth group of three elements in this column, and it may be a fifth. And the discovery of such a group is not unlikely, for when this table was first drawn up Prof. Mendeléeff drew attention to certain gaps, which have since been filled up by the discovery of gallium, germanium, and others.

小顺序（当时刚刚发现的）为依据的。据此，钙、锶和钡形成一个三元素组；它们的氧化物，石灰、锶土和重土都容易熟化，即与水结合形成可溶的石灰水、锶土水和重土水。它们的硫酸盐的溶解度都很小，而且它们各自的氯化物以及硝酸盐也都具有相似之处。它们的原子量也体现出了规律性。当时公认的数值分别为 20、42.5 和 65；而锶的原子量 42.5 是另外两种元素原子量的算术平均值，因为 (65+20)/2=42.5。德贝赖纳还指出了其他一些类似的三元素组的存在，它们后来被称为“德贝赖纳三元素组”。

另外一种根据原子量进行元素分类的方法是由佩滕科费尔提出的，后来克雷默斯、格拉德斯通和库克进行了详细阐述。这种方法是寻找某种表达式以描述某些相关元素的原子量之差。比如，锂的原子量是 7，它和钠的原子量 23 之间的差是 16；而钠和钾(原子量 39)的原子量之差也是 16。规律性并非总是这样显而易见；1857 年，杜马设计出一个稍显复杂的表达式，从某种程度上体现了氟、氯、溴和碘以及氮、磷、砷、锑和铋的原子量之间的规律性。

这些寻找规律性的努力的最终结果是，在 1864 年约翰 · 纽兰兹先生将元素分为 8 组，他发现当元素按照原子量的顺序排列时，“以某一特定元素为起始的 8 个元素形成一种周而复始的循环，犹如音乐中八度音阶中的 8 个音符。”他将这种规律性称为“八音律”。

正如所有化学家都知道的那样，这一观点的发展应当归功于图宾根大学已故的洛塔尔·迈耶尔教授和圣彼得堡大学的门捷列夫教授。这种规律通常被称为“周期律”。展示这种排列的最简单的方法之一，是利用一个被若干条与轴平行的直线分割为 8 部分的圆柱面；一条螺旋线环绕柱面，当然，它会在每一次环绕中被这些直线切割 8 次。将柱面垂直放置，元素的名称和原子量就写在螺旋线与垂线的每个交点上，前面还写着按原子量排列的序数。根据洛塔尔 · 迈耶尔和门捷列夫的观点，每条垂线上的元素组成一个自然类；它们表现出相似的性质，形成类似的化合物，并在密度、熔点和许多其他性质上体现出递变性。不过，在这些垂直列中有一个是与众不同的，因为其中包含 3 个组，每一组都由原子量几乎相同的元素组成。这些元素是铁、钴和镍；钯、铑和钌；铂、铱和锇。在这一纵列中显然还为第 4 个三元素组留有余地，可能还会有第 5 个。发现这样一个组并不是不可能的，因为早在这张表刚刚被草拟出来时，门捷列夫教授就注意到其中特定位置的空缺，这些空缺逐渐被后来发现的镓、锗等其他元素填充。

The discovery of argon at once raised the curiosity of Lord Rayleigh and myself as to its position in this table. With a density of nearly 20, if a diatomic gas, like oxygen and nitrogen, it would follow fluorine in the periodic table; and our first idea was that argon was probably a mixture of three gases, all of which possessed nearly the same atomic weights, like iron, cobalt, and nickel. Indeed, their names were suggested, on this supposition, with patriotic bias, as Anglium, Scotium, and Hibernium! But when the ratio of its specific heats had, at least in our opinion, unmistakably shown that it was molecularly monatomic, and not diatomic, as at first conjectured, it was necessary to believe that its atomic weight was 40, and not 20, and that it followed chlorine in the atomic table, and not fluorine. But here arises a difficulty. The atomic weight of chlorine is 35.5, and that of potassium, the next element in order in the table, is 39.1; and that of argon, 40, follows, and does not precede, that of potassium, as it might be expected to do. It still remains possible that argon, instead of consisting wholly of monatomic molecules, may contain a small percentage of diatomic molecules; but the evidence in favour of this supposition is, in my opinion, far from strong. Another possibility is that argon, as at first conjectured, may consist of a mixture of more than one element; but, unless the atomic weight of one of the elements in the supposed mixture is very high, say 82, the case is not bettered, for one of the elements in the supposed trio would still have a higher atomic weight than potassium. And very careful experiments, carried out by Dr. Norman Collie and myself, on the fractional diffusion of argon, have disproved the existence of any such element with high atomic weight in argon, and, indeed, have practically demonstrated that argon is a simple substance, and not a mixture.

The discovery of helium has thrown a new light on this subject. Helium, it will be remembered, is evolved on heating certain minerals, notably those containing uranium; although it appears to be contained in others in which uranium is not present, except in traces. Among these minerals are clèveite, monazite, fergusonite, and a host of similar complex mixtures, all containing rare elements, such as niobium, tantalum, yttrium, cerium, &c. The spectrum of helium is characterised by a remarkably brilliant yellow line, which had been observed as long ago as 1868 by Profs. Frankland and Lockyer in the spectrum of the sun's chromosphere, and named "helium" at that early date.

The density of helium proved to be very close to 2.0, and, like argon, the ratio of its specific heat showed that it, too, was a monatomic gas. Its atomic weight therefore is identical with its molecular weight, viz. 4.0, and its place in the periodic table is between hydrogen and lithium, the atomic weight of which is 7.0.

The difference between the atomic weights of helium and argon is thus 36, or 40 – 4. Now there are several cases of such a difference. For instance, in the group the first member of which is fluorine we have—

<table>
<tr><td>Fluorine</td><td>19</td><td rowspan="2">16.5</td></tr>
<tr><td>Chlorine</td><td>35.5</td></tr>
<tr><td>Manganese</td><td>55</td><td>19.5</td></tr>
</table>

氩的发现立即激起了瑞利勋爵和我本人对它在周期表中位置的好奇心。它的密度大约是 20，如果是像氧气和氮气一样的双原子气体，那么在周期表中它应该是紧随氟元素之后；我们最初的想法是，氩很可能是 3 种气体的混合物，它们具有几乎相同的原子量，就像铁、钴和镍一样。实际上，基于这种假定，按照具有爱国主义倾向的方式来命名，应该把它们分别称作 Anglium，Scotiun 和 Hibernium。但是，当比热比准确无误地表明它是单原子分子而非最初所猜测的双原子分子时，我们只能相信它的原子量是 40 而不是 20，而它在周期表中的位置也应是位于氯之后，而不是氟之后——至少在我们看来是这样。不过这又产生了新的困难。氯的原子量是 35.5，周期表中下一个元素钾的原子量是 39.1；而氩的原子量是 40，并不像我们期待的那样比钾的小，而是比它大。还有可能氩并不完全是由单原子分子组成，而是包含一小部分的双原子分子；不过在我看来，支持这一假设的证据非常不足。另外一种可能性就是氩可能是多种元素的混合物，正如最初猜测的那样；但是，除非这个假定的混合物中某一种元素的原子量非常高，比如说有 82，否则情况也并不理想，因为在这个假定的三元素组中要有一种元素的原子量比钾的高。由诺曼 · 科利和我本人所做的非常严谨的关于氩分馏扩散的实验已经证明，在氩中不存在任何高原子量的元素，事实上，氩是一种纯净物而不是混合物。

氦的发现给这一问题带来了转机。我们应该记得，加热某些矿物，特别是那些含铀的矿物时，会释放出氦；尽管氦似乎也存在于另外一些并不含铀的矿物中，但是含量非常低。这些矿物包括钇铀矿、独居石、褐钇铌（钽）矿以及很多类似的复杂混合物，它们都含有诸如铌、钽、钇和铈等稀有元素。氦光谱的特征是有一条显著而明亮的黄线，弗兰克兰和洛克耶教授早在 1868 年就在太阳色球的光谱中观察到了这一点，并从那时起称它为“氦”。

经证实，氦的密度非常接近 2.0，而且，与氩一样，它的比热比表明它也是一种单原子气体。因此，它的原子量与分子量是一样的，也就是 4.0，而它在周期表中的位置则介于氢和原子量为 7.0 的锂之间。

因此，氦与氩的原子量之差为 40－4，也就是 36。这里还有几个差值与此相同的例子。例如，在以氟作为第一个成员的那组元素中，我们看到——

氟	19	16.5
氯	35.5	
锰	55	19.5

In the oxygen group—

Oxygen	16	16
Sulphur	32	
Chromium	52.3	20.3

In the nitrogen group—

Nitrogen	14	17
Phosphorus	31	
Vanadium	51.4	20.4

And in the carbon group—

Carbon	12	16.3
Silicon	28.3	
Titanium	48.1	19.8

These instances suffice to show that approximately the differences are 16 and 20 between consecutive members of the corresponding groups of elements. The total differences between the extreme members of the short series mentioned are—

Manganses – Fluorine	36
Chromium – Oxygen	36.3
Vanadium – Nitrogen	37.4
Titanium – Carbon	36.1

This is approximately the difference between the atomic weights of helium and argon, 36.

There should, therefore, be an undiscovered element between helium and argon, with an atomic weight 16 units higher than than of helium, and 20 units lower than that of argon, namely 20. And if this unknown element, like helium and argon, should prove to consist of monatomic molecules, then its density should be half its atomic weight, 10. And pushing the analogy still further, it is to be expected that this element should be as indifferent to union with other elements as the two allied elements.

My assistant, Mr. Morris Travers, has indefatigably aided me in a search for this unknown gas. There is a proverb about looking for a needle in a haystack; modern science, with the aid of suitable magnetic appliances, would, if the reward were sufficient, make short work of that proverbial needle. But here is a supposed unknown gas, endowed no doubt with negative properties, and the whole world to find it in. Still, the attempt had to be made.

We first directed our attention to the sources of helium—minerals. Almost every mineral which we could obtain was heated in a vacuum, and the gas which was evolved examined. The results are interesting. Most minerals give off gas when heated, and the gas contains, as a rule, a considerable amount of hydrogen, mixed with carbonic acid, questionable traces of nitrogen, and carbonic oxide. Many of the minerals, in addition, gave helium, which proved to be widely distributed, though only in minute proportion. One mineral—malacone—gave appreciable quantities of argon; and it is noteworthy that argon was not

在氧这一组中——

氧	16	16
硫	32	
铬	52.3	20.3

在氮这一组中——

氮	14	17
磷	31	
钒	51.4	20.4

而在碳这一组中——

碳	12	16.3
硅	28.3	
钛	48.1	19.8

这些实例足以表明，各组元素的连续成员之间的原子量之差大约都是 16 和 20。以上各组中两端元素的总的原子量之差为——

锰－氟	36
铬－氧	36.3
钒－氮	37.4
钛－碳	36.1

这与氦和氩的原子量之差 36 也是近似一致的。

可见，在氦与氩之间应该有一种尚未发现的元素，它的原子量比氦的原子量大 16 而比氩的原子量小 20，也就是 20。而且，如果这种未知元素像氦和氩一样被证实由单原子分子构成，那么它的密度应该是其原子量的一半——10。进一步类比，可以预期这种元素像它的两种同类元素一样不易与其他元素结合。

我的助手莫里斯 · 特拉弗斯先生一直毫不厌倦地帮助我研究这种未知气体。有句俗语叫大海捞针；如果报酬足够丰厚，那么在有合适的磁力仪器帮助下的现代科学将大大简化大海捞针的工作。但现在是要在全世界寻找一种假想的未知气体，而且无疑它还具有不利的性质。所以，仍然需要努力尝试。

我们首先将注意力集中在氦的来源——矿物上。我们几乎将每一种可能得到的矿物都在真空中加热，并检验释放出来的气体。结果很有趣，大多数矿物在加热时释放出气体，通常有大量的氢气，另外还混杂着碳酸，不太确定的痕量氮气，还有一氧化碳。此外，有很多种矿物释放出氦气，这证明它的分布很广，不过只占很小的比例。一种叫做水锆石的矿物能释放出大量的氩气；值得注意的是，除了这种矿石和一份陨铁标本以外，在其他物质中并没有发现氩，而且奇怪的是，在水锆石中

found except in it (and, curiously, in much larger amount than helium), and in a specimen of meteoric iron. Other specimens of meteoric iron were examined, but were found to contain mainly hydrogen, with no trace of either argon or helium. It is probable that the sources of meteorites might be traced in this manner, and that each could be relegated to its particular swarm.

Among the minerals examined was one to which our attention had been directed by Prof. Lockyer, named eliasite, from which he said that he had extracted a gas in which he had observed spectrum lines foreign to helium. He was kind enough to furnish us with a specimen of this mineral, which is exceedingly rare, but the sample which we tested contained nothing but undoubted helium.

During a trip to Iceland in 1895, I collected some gas from the boiling springs there; it consisted, for the most part, of air, but contained somewhat more argon than is usually dissolved when air is shaken with water. In the spring of 1896 Mr. Travers and I made a trip to the Pyrenees to collect gas from the mineral springs of Cauterets, to which our attention had been directed by Dr. Bouchard, who pointed out that these gases are rich in helium. We examined a number of samples from the various springs, and confirmed Dr. Bouchard's results, but there was no sign of any unknown lines in the spectrum of these gases. Our quest was in vain.

We must now turn to another aspect of the subject. Shortly after the discovery of helium, its spectrum was very carefully examined by Profs. Runge and Paschen, the renowned spectroscopists. The spectrum was photographed, special attention being paid to the invisible portions, termed the "ultra-violet" and "infra-red". The lines thus registered were found to have a harmonic relation to each other. They admitted of division into two sets, each complete in itself. Now, a similar process had been applied to the spectrum of lithium and to that of sodium, and the spectra of these elements gave only one series each. Hence, Profs. Runge and Paschen concluded that the gas, to which the provisional name of helium had been given, was, in reality, a mixture of two gases, closely resembling each other in properties. As we know no other elements with atomic weights between those of hydrogen and lithium, there is no chemical evidence either for or against this supposition. Prof. Runge supposed that he had obtained evidence of the separation of these imagined elements from each other by means of diffusion; but Mr. Travers and I pointed out that the same alteration of spectrum, which was apparently produced by diffusion, could also be caused by altering the pressure of the gas in the vacuum tube; and shortly after Prof. Runge acknowledged his mistake.

These considerations, however, made it desirable to subject helium to systematic diffusion, in the same way as argon had been tried. The experiments were carried out in the summer of 1896 by Dr. Collie and myself. The result was encouraging. It was found possible to separate helium into two portions of different rates of diffusion, and consequently of different density by this means. The limits of separation, however, were not very great. On the one hand, we obtained gas of a density close on 2.0; and on the other, a sample of

氩含量比氦高很多。在检验其他陨铁标本时，却发现它们主要含氢，没有一丁点氩或氦。也许利用这种方式能够追溯陨星的来源，并将每一个都归入其所在的特定门类中。

我们的注意力被洛克耶教授拉到一种被检验过的名叫脂铅铀矿的矿物上，洛克耶教授说他已经从中提取出一种具有和氦不同的光谱线的气体。他还非常友好地为我们提供了这种极为稀有的矿物的一份样本，但是我们检验的样本中只含有氦。

1895 年在冰岛旅行期间，我从那里的沸泉中收集了一些气体；这些气体主要就是空气，但是其中所含的氩却比通常把水放在空气中振荡时水中所溶解的氩多一些。1896 年春，特拉弗斯先生和我前往比利牛斯山收集科特雷矿泉水中的气体；这一次我们接受了布沙尔博士的指点，他指出这种气体中富含氦气。我们检验了来自不同泉水的大量样本，证实了布沙尔博士的结论，但是这些气体的光谱中没有任何未知谱线的迹象。我们的探索仍然一无所获。

现在我们必须转向问题的另一个方面。在发现氦之后不久，著名光谱学家龙格教授和帕邢教授就非常细致地检验了它的光谱。他们对光谱照了相，并特别注意到其中的不可见部分，人们称之为“紫外区”和“红外区”。他们发现这样记录的谱线彼此间具有谐波关系。它们可以被分为两个子集，各自是一个完整的体系。现在，对锂和钠的光谱也进行了类似的研究，但这两种元素的谱图中每一种只出现了一个谱线系列。由此，龙格教授和帕邢教授得出结论认为，被我们称为氦的这种气体实际上是两种性质极为相似的气体的混合物。正如我们所知，没有其他元素的原子量介于氢和锂之间，因而没有化学上的证据支持或反对这一假设。龙格教授认为他已经得到了通过扩散方法分离这些假想元素的证据；但是特拉弗斯先生和我认为，看似由扩散产生的谱图变化，也可能是由真空管中气体压强的变化引起的；不久之后龙格教授承认了他的错误。

不过，这些研究使人们想到可以用系统扩散的方法研究氦，就像曾经研究氩时所尝试过的一样。1896 年夏天，科利博士和我进行了实验，结果是令人鼓舞的。我们发现，这种方法可以将氦分离为扩散速率不同从而具有不同密度的两个部分。不过，分离的极限还不是很大。我们一方面得到了一种密度接近于 2.0 的气体，另一方面

density 2.4 or thereabouts. The difficulty was increased by the curious behaviour, which we have often had occasion to confirm, that helium possesses a rate of diffusion too rapid for its density. Thus, the density of the lightest portion of the diffused gas, calculated from its rate of diffusion, was 1.874; but this corresponds to a real density of about 2.0. After our paper, giving an account of these experiments, had been published, a German investigator, Herr A. Hagenbach, repeated our work and confirmed our results.

The two samples of gas of different density differ also in other properties. Different transparent substances differ in the rate at which they allow light to pass through them. Thus, light travels through water at a much slower rate than through air, and at a slower rate through air than through hydrogen. Now Lord Rayleigh found that helium offers less opposition to the passage of light than any other substance does, and the heavier of the two portions into which helium had been split offered more opposition than the lighter portion. And the retardation of the light, unlike what has usually been observed, was nearly proportional to the densities of the samples. The spectrum of these two samples did not differ in the minutest particular; therefore it did not appear quite out of the question to hazard the speculation that the process of diffusion was instrumental, not necessarily in separating two kinds of gas from each other, but actually in removing light molecules of the same kind from heavy molecules. This idea is not new. It had been advanced by Prof. Schützenberger (whose recent death all chemists have to deplore), and later, by Mr. Crookes, that what we term the atomic weight of an element is a mean; that when we say the atomic weight of oxygen is 16, we merely state that the average atomic weight is 16; and it is not inconceivable that a certain number of molecules have a weight somewhat higher than 32, while a certain number have a lower weight.

We therefore thought it necessary to test this question by direct experiment with some known gas; and we chose nitrogen, as a good material with which to test the point. A much larger and more convenient apparatus for diffusing gases was built by Mr. Travers and myself, and a set of systematic diffusions of nitrogen was carried out. After thirty rounds, corresponding to 180 diffusions, the density of the nitrogen was unaltered, and that of the portion which should have diffused most slowly, had there been any difference in rate, was identical with that of the most quickly diffusing portion—*i.e.* with that of the portion which passed first through the porous plug. This attempt, therefore, was unsuccessful; but it was worth carrying out, for it is now certain that it is not possible to separate a gas of undoubted chemical unity into portions of different density by diffusion. And these experiments rendered it exceedingly improbable that the difference in density of the two fractions of helium was due to separation of light molecules of helium from heavy molecules.

The apparatus used for diffusion had a capacity of about two litres. It was filled with helium, and the operation of diffusion was carried through thirty times. There were six reservoirs each full of gas, and each was separated into two by diffusion. To the heavier portion of one lot, the lighter portion of the next was added, and in this manner all six reservoirs were successfully passed through the diffusion apparatus. This process was

还得到一种密度在 2.4 左右的样品。我们常常发现氦的扩散速率相对于它的密度来说实在太快了，这种奇怪的现象也增加了研究的难度。由此，利用扩散速率计算出扩散所得气体中最轻部分的密度为 1.874，但对应的真实密度却大约是 2.0。在我们介绍上述实验的论文发表之后，一位德国研究者，哈根巴赫先生重复了我们的工作并确认了这一结果。

这两种密度不同的气体样品在其他性质上也不相同。光穿过不同透明物体的速率是不同的。因此，光在水中的传播速率比在空气中低很多，而在空气中的传播速率又比在氢气中低。瑞利勋爵发现，氦对光透过的阻碍比其他任何物质都小，而通过分离氦所得到的两个组分中，较重的部分对光透过的阻碍比较轻的部分大一些。与通常所观测到的不同，对光的阻碍几乎正比于样品的密度。两种样品的光谱在最微小的细节上也毫无区别；由此，看起来我们可以大胆地认为，扩散过程是起作用的，但不一定有助于将两种气体彼此分开，而实际上是对将同一种气体中较轻的分子从较重的分子中分离出去有帮助。这并不是一个全新的观点。最先由舒岑贝热教授（最近他的去世令所有化学家感到痛惜），还有后来的克鲁克斯先生提出，我们所谓的某种元素的原子量是平均值；当我们说氧的原子量是 16 时，我们只是指出氧的平均原子量是 16；不难想象，有一定数量分子的重量比 32 稍高，同时还有一定数量分子的重量略低一些。

由此，我们认为，必须用某些已知气体直接进行实验来验证这一问题；我们选择了氮气这种合适的物质来检验这一观点。特拉弗斯和我建立了更大、更方便的用于气体扩散的装置，并对氮气进行了一组系统扩散。经过 30 轮相当于 180 次扩散之后，氮气的密度没有变化，如果扩散速率确实有所不同的话，那么扩散速率最慢的组分与扩散速率最快的组分——即最先通过多孔塞的部分——密度是一样的。可见，这次的努力是不成功的；不过值得尝试，因为现在可以确定，不可能通过扩散将化学上确实均一的一种气体分离为几个密度不同的部分。这些实验证明，氦的两个组分的密度差异不可能是由较轻的氦分子与较重的分子分离引起的。

用于扩散实验的装置的容积大约是 2 升。向装置中充满氦气，进行 30 次扩散操作。共有 6 个充满气体的储气槽，每个储气槽中的气体通过扩散分离为两部分。一个储气槽中较轻的气体被加入到前一个储气槽中较重的气体中，全部 6 个储气槽中的气体以这种方式连续地通过扩散装置。将这一过程进行 30 次，每一次每个储气槽中的

carried out thirty times, each of the six reservoirs having had its gas diffused each time, thus involving 180 diffusions. After this process, the density of the more quickly diffusing gas was reduced to 2.02, while that of the less quickly diffusing had increased to 2.27. The light portion on re-diffusion hardly altered in density, while the heavier portion, when divided into three portions by diffusion, showed a considerable difference in density between the first third and the last third. A similar set of operations was carried out with a fresh quantity of helium, in order to accumulate enough gas to obtain a sufficient quantity for a second series of diffusions. The more quickly diffusing portions of both gases were mixed and re-diffused. The density of the lightest portion of these gases was 1.98; and after other 15 diffusions, the density of the lightest portion had not decreased. The end had been reached; it was not possible to obtain a lighter portion by diffusion. The density of the main body of this gas is therefore 1.98; and its refractivity, air being taken as unity, is 0.1245. The spectrum of this portion does not differ in any respect from the usual spectrum of helium.

As re-diffusion does not alter the density or the refractivity of this gas, it is right to suppose that either one definite element has now been isolated; or that if there are more elements than one present, they possess the same, or very nearly the same, density and refractivity. There may be a group of elements, say three, like iron, cobalt, and nickel; but there is no proof that this idea is correct, and the simplicity of the spectrum would be an argument against such a supposition. This substance, forming by far the larger part of the whole amount of the gas, must, in the present state of our knowledge, be regarded as pure helium.

On the other hand, the heavier residue is easily altered in density by re-diffusion, and this would imply that it consists of a small quantity of a heavy gas mixed with a large quantity of the light gas. Repeated re-diffusion convinced us that there was only a very small amount of the heavy gas present in the mixture. The portion which contained the largest amount of heavy gas was found to have the density 2.275, and its refractive index was found to be 0.1333. On re-diffusing this portion of gas until only a trace sufficient to fill a Plücker's tube was left, and then examining the spectrum, no unknown lines could be detected, but, on interposing a jar and spark gap, the well-known blue lines of argon became visible; and even without the jar the red lines of argon, and the two green groups were distinctly visible. The amount of argon present, calculated from the density, was 1.64 percent, and from the refractivity 1.14 percent. The conclusion had therefore to be drawn that the heavy constituent of helium, as it comes off the minerals containing it, is nothing new, but, so far as can be made out, merely a small amount of argon.

If, then, there is a new gas in what is generally termed helium, it is mixed with argon, and it must be present in extremely minute traces. As neither helium nor argon has been induced to from compounds, there does not appear to be any method, other than diffusion, for isolating such a gas, if it exists, and that method has failed in our hands to give any evidence of the existence of such a gas. It by no means follows that the gas does not exist; the only conclusion to be drawn is that we have not yet stumbled on the material

气体都进行扩散，一共进行 180 次扩散。经过上述过程后，扩散速率较快的气体密度下降到 2.02,而扩散较慢的则升高到 2.27。重复扩散所得的较轻组分密度很难改变，而较重组分通过扩散分成了三部分，第一部分与第三部分的密度表现出相当大的差异。为了收集足够多的气体以进行第二系列的扩散，又对一些新制的氦气进行了一系列类似的操作。最后，两批气体中扩散较快的组分被混在一起并进行了重复扩散。其中最轻组分的气体密度为 1.98；再进行 15 次扩散后，最轻组分的气体密度也没有减小。这已经达到了极限；不可能通过扩散得到更轻的组分了。因此这种气体的主要部分的密度就是 1.98；而若将气体视为均一的话，它的折射率就是 0.1245。这个组分的光谱与通常的氦光谱从任何方面看都没有差别。

重复扩散并没有改变这种气体的密度和折射率，由此可知，不管是现在已分离出一种确定的元素，还是其中含有更多的元素，它们都有相同或近乎相同的密度和折射率。也许会有一组元素，比如像铁、钴和镍这样的三种元素；但是没有证据表明这种观点是正确的，而光谱的简单性也不支持这种假设。从我们目前的认识水平来看，占全部气体中一大部分的这种物质应该是纯净的氦气。

另一方面，这种较重的残余物很容易通过重复扩散改变其密度，这意味着它是由少量的重气体和大量的轻气体组成的混合物。不断重复扩散后得到的结果使我们确信混合物中只存在极少量的重气体。含有最多重气体部分的密度是 2.275，其折射率为 0.1333。将这部分气体继续进行重复扩散，直到剩余的量只够充满一个普吕克管为止，接着检验其光谱，没有检测到未知谱线，不过将其装入广口瓶中再放入放电器，就可以看到众所周知的氩特有的蓝色谱线；即使没有装入广口瓶，氩的红色谱线以及两组绿色谱线也是明显可见的。由密度计算出的氩的含量为 1.64%，而由折射率计算得到的是 1.14%。由此可以得出结论，在我们目前的认识水平下，从含氦矿物中释放出来的氦气中的较重组分并不是什么新物质，而只是少量的氩气。

那么，如果在我们通常所称的氦气中存在新的气体，那它应该是混在氩气中的，而且是痕量的。如果这种气体存在的话，由于尚不能使氦气或氩气形成化合物，看起来除了扩散之外也没有任何其他方法能够分离出这种气体，而扩散的方法又无法给出任何关于这种气体存在的证据。但这决不能说明该气体不存在；唯一可以得出

which contains it. In fact, the haystack is too large and the needle too inconspicuous. Reference to the periodic table will show that between the elements aluminium and indium there occurs gallium, a substance occurring only in the minutest amount on the earth's surface; and following silicon, and preceding tin, appears the element germanium, a body which has as yet been recognised only in one of the rarest of minerals, argyrodite. Now, the amount of helium in fergusonite, one of the minerals which yields it in reasonable quantity, is only 33 parts by weight in 100,000 of the mineral; and it is not improbable that some other mineral may contain the new gas in even more minute proportion. If, however, it is accompanied in its still undiscovered source by argon and helium, it will be a work of extreme difficulty to effect a separation from these gases.

In these remarks it has been assumed that the new gas will resemble argon and helium in being indifferent to the action of reagents, and in not forming compounds. This supposition is worth examining. In considering it, the analogy with other elements is all that we have to guide us.

We have already paid some attention to several triads of elements. We have seen that the differences in atomic weights between the elements fluorine and manganese, oxygen and chromium, nitrogen and vanadium, carbon and titanium, is in each case approximately the same as that between helium and argon, viz. 36. If elements further back in the periodic table be examined, it is to be noticed that the differences grow less, the smaller the atomic weights. Thus, between boron and scandium, the difference is 33; between beryllium (glucinum) and calcium, 31; and between lithium and potassium, 32. At the same time, we may remark that the elements grow liker each other, the lower the atomic weights. Now, helium and argon are very like each other in physical properties. It may be fairly concluded, I think, that in so far they justify their position. Moreover, the pair of elements which show the smallest difference between their atomic weights is beryllium and calcium; there is a somewhat greater difference between lithium and potassium. And it is in accordance with this fragment of regularity that helium and argon show a greater difference. Then again, sodium, the middle element of the lithium triad, is very similar in properties both to lithium and potassium; and we might, therefore, expect that the unknown element of the helium series should closely resemble both helium and argon.

Leaving now the consideration of the new element, let us turn our attention to the more general question of the atomic weight of argon, and its anomalous position in the periodic scheme of the elements. The apparent difficulty is this: The atomic weight of argon is 40; it has no power to form compounds, and thus possesses no valency; it must follow chlorine in the periodic table, and precede potassium; but its atomic weight is greater than that of potassium, whereas it is generally contended that the elements should follow each other in the order of their atomic weights. If this contention is correct, argon should have an atomic weight smaller than 40.

Let us examine this contention. Taking the first row of elements, we have:

的结论是，我们还没有碰到过包含它的物质。事实上，大海太宽广了，而针则太不显眼了。周期表显示在铝和铟元素之间还应该存在镓，它是一种在地壳中含量极少的物质；而位于硅元素之后锡元素之前的是锗——一种至今只在硫银锗矿（最稀有的矿物之一）中发现过的元素。褐钇铌（钽）矿是可以产生适量氦气的矿物之一，当这种矿物的总重量为100,000时，氦只占其中的33；其他一些矿物中所含新气体的比例可能更低。然而，如果在尚未发现的来源中新气体也是与氩和氦共存的话，那么把它从这些气体中分离出来将是一项极其艰巨的任务。

在上面的评述中我们假定，新气体与氩和氦类似，对于试剂反应呈惰性，不形成化合物。这一假定尚有待检验。在考虑这一问题时，与其他元素的类比是我们唯一的指导思想。

我们已经关注了若干个三元素组。我们看到，氟与锰、氧与铬、氮与钒、碳与钛这些元素之间的原子量差值和氦与氩之间的差值是近似相等的，大约都是36。如果进一步检验周期表中靠后的元素，可以看到原子量越小，其差值越小。硼与钪之间的原子量差值为33；铍与钙之间的差值为31；而锂与钾之间的差值则是32。同时，我们也可以说，元素之间越相似，其原子量的差值越小。在物理性质上，氦和氩非常相似。我认为，在此范围内可以适当地得出结论，它们的性质证实了它们位置的正确性。此外，原子量差值最小的元素对是铍和钙；锂和钾之间的差值则更大一些。与这一规律相吻合的是，氦和氩之间存在更大的原子量差值。而且，锂所在的三元素组中的中间元素钠的性质与锂和钾颇为相似；由此，我们就可以预期，位于氦所在系列中的这一未知元素的性质与氦和氩应该非常相似。

暂时不考虑新元素的问题，让我们将注意力转向更一般性的问题，即氩的原子量以及它在元素周期表中的反常位置。我们面临的困难显然是这样的：氩的原子量是40；它没有形成化合物的能力，由此表现为没有化合价；在周期表中它必须紧随氯元素之后而位于钾元素之前；但是它的原子量比钾的原子量大，而一般认为元素应该按照其原子量的顺序依次排列。如果这一观点是正确的，氩的原子量就应该小于40。

让我们检验一下这种观点。考察第一行元素，我们得到：

Li = 7, Be = 9.8, B = 11, C = 12, N = 14, O = 16, F = 19, ? = 20.

The differences are:

2.8, 1.2, 1.0, 2.0, 2.0, 3.0, 1.0.

It is obvious that they are irregular. The next row shows similar irregularities. Thus:

(? = 20), Na = 23, Mg = 24.3, Al = 27, Si = 28, P = 31, S = 32,
Cl = 35.5, A = 40.

And the differences:

3.0, 1.3, 2.7, 1.0, 3.0, 1.0, 3.5, 4.5.

The same irregularity might be illustrated by a consideration of each succeeding row. Between argon and the next in order, potassium, there is a difference of −0.9; that is to say, argon has a higher atomic weight than potassium by 0.9 unit; whereas it might be expected to have a lower one, seeing that potassium follows argon in the table. Further on in the table there is a similar discrepancy. The row is as follows:

Ag = 108, Cd = 112, In = 114, Sn = 119, Sb = 120.5,
Te = 127.7, I = 127.

The differences are:

4.0, 2.0, 5.0, 1.5, 7.2, −0.7.

Here, again, there is a negative difference between tellurium and iodine. And this apparent discrepancy has led to many and careful redeterminations of the atomic weight of tellurium. Prof. Brauner, indeed, has submitted tellurium to methodical fractionation, with no positive results. All the recent determinations of its atomic weight give practically the same number, 127.7.

Again, there have been almost innumerable attempts to reduce the differences between the atomic weights to regularity, by contriving some formula which will express the numbers which represent the atomic weights, with all their irregularities. Needless to say, such attempts have in no case been successful. Apparent success is always attained at the expense of accuracy, and the numbers reproduced are not those accepted as the true atomic weights. Such attempts, in my opinion, are futile. Still, the human mind does not rest contented in merely chronicling such an irregularity; it strives to understand why such an irregularity should exist. And, in connection with this, there are two matters which call for our consideration. These are: Does some circumstance modify these "combining proportions" which we term "atomic weights"? And is there any reason to suppose that we can modify them at our will? Are they true "constants of nature", unchangeable, and once for all determined? Or are they constant merely so long as other circumstances, a change in which would modify them, remain unchanged?

Li = 7，Be = 9.8，B = 11，C = 12，N = 14，O = 16，F = 19，？ = 20。

其差值为：

2.8，1.2，1.0，2.0，2.0，3.0，1.0。

很显然，这是不规则的。下一行显示出类似的不规则性。它们是：

（？ = 20），Na = 23，Mg = 24.3，Al = 27，Si = 28，P = 31，S = 32，
Cl = 35.5，A = 40。

而其差值为：

3.0，1.3，2.7，1.0，3.0，1.0，3.5，4.5。

考察后面每一行都可以发现同样的不规则性。氩与紧随其后的钾之间的原子量差值为 –0.9；也就是说，氩的原子量比钾大 0.9 个单位；但鉴于在周期表中钾位于氩之后，氩的原子量应该低于钾的原子量。在表中还有类似的矛盾。如此行所示：

Ag = 108，Cd = 112，In = 114，Sn = 119，Sb = 120.5，
Te = 127.7，I = 127。

其差值为：

4.0，2.0，5.0，1.5，7.2，–0.7。

这里，在碲与碘之间再次出现了负的原子量差值。这一明显的矛盾引发了很多谨慎地重测碲原子量的工作。确实，布劳纳教授已经对碲进行了系统的分馏，但是没有得到预期的结果。近来所有对碲的原子量进行测定的工作都得出了同样的结果——127.7。

另外，为了归纳出原子量差值所具有的规律性，人们进行了几乎无数次的尝试工作，设计某种可以反映原子量数值的不规律性的公式。不用说，这些尝试都没有成功。一些表面上的成功总是通过牺牲精确性来获得的，公式中得出的数值并不是我们公认的原子量的真实值。在我看来，这些努力是徒劳的。当然，人类的头脑不会仅仅满足于记述这一不规律性，而是要努力理解这种不规律性为什么会存在。并且，有两件与此相关的事需要考虑，即，是否有某种条件可以改变被我们称为“原子量”的“组合比例”？以及是否有理由使我们相信我们可以按照自己的意愿改变它们？它们是被一次性确定了的真正不可改变的“自然常量”，还是只在其他条件不变时才保持恒定，而其中任一条件的改变都将使它们发生变化呢？

In order to understand the real scope of such questions, it is necessary to consider the relation of the "atomic weights" to other magnitudes, and especially to the important quantity termed "energy".

It is known that energy manifests itself under different forms, and that one form of energy is quantitatively convertible into another form, without loss. It is also known that each form of energy is expressible as the product of two factors, one of which has been termed the "intensity factor", and the other the "capacity factor". Prof. Ostwald, in the last edition of his "Allgemeine Chemie", classifies some of these forms of energy as follows:

Kinetic energy	is the product of	Mass into the square of velocity.
Linear	... ...	Length into force.
Surface	... ...	Surface into surface tension.
Volume	... ...	Volume into pressure.
Heat	... ...	Heat capacity (entropy) into temperature.
Electrical	... ...	Electrical capacity into potential.
Chemical	... ...	"Atomic weight" into affinity.

In each statement of factors, the "capacity factor" is placed first, and the "intensity factor" second.

In considering the "capacity factors", it is noticeable that they may be divided into two classes. The two first kinds of energy, kinetic and linear, are *independent of the nature of the material* which is subject to the energy. A mass of lead offers as much resistance to a given force, or, in other words, possesses as great inertia as an equal mass of hydrogen. A mass of iridium, the densest solid, counterbalances an equal mass of lithium, the lightest known solid. On the other hand, surface energy deals with molecules, and not with masses. So does volume energy. The volume energy of two grammes of hydrogen, contained in a vessel of one litre capacity, is equal to that of thirty-two grammes of oxygen at the same temperature, and contained in a vessel of equal size. Equal masses of tin and lead have not equal capacity for heat; but 119 grammes of tin has the same capacity as 207 grammes of lead; that is, equal atomic masses have the same heat capacity. The quantity of electricity conveyed through an electrolyte under equal difference of potential is proportional, not to the mass of the dissolved body, but to its equivalent; that is, to some simple fraction of its atomic weight. And the capacity factor of chemical energy is the atomic weight of the substance subjected to the energy. We see, therefore, that while mass or inertia are important adjuncts of kinetic and linear energies, all other kinds of energy are connected with atomic weights, either directly or indirectly.

Such considerations draw attention to the fact that quantity of matter (assuming that there exists such a carrier of properties as we term "matter") need not necessarily be measured by its inertia, or by gravitational attraction. In fact the word "mass" has two totally distinct significations. Because we adopt the convention to measure quantity of matter by its mass, the word "mass" has come to denote "quantity of matter." But it is open to any one to

为了理解这些问题的真正内涵，有必要考虑“原子量”与其他参量之间的关联，尤其是与我们称为“能量”的这个重要参量之间的关系。

我们知道能量以不同形式存在，并且一种形式的能量可以毫无损耗地转化为另一种形式的能量。我们还知道，每一种形式的能量都可以表示为两个因子的乘积，其中一个称为“强度因子”，另一个称为“容量因子”。奥斯特瓦尔德教授在他最新版的《普通化学》中，对几种形式的能量进行了如下分类：

动能	是	质量与速率平方的	乘积。
线性能	是	长度与力的	乘积。
表面能	是	表面与表面张力的	乘积。
体积能	是	体积与压力的	乘积。
热能	是	热容量（熵）与温度的	乘积。
电能	是	电容与电势的	乘积。
化学能	是	“原子量”与亲合力的	乘积。

在对各因子的每一组表述中，“容量因子”放在前面，而“强度因子”放在后面。

考察一下“容量因子”，不难看出它们可以分为两类。前两种能量，动能和线性能，**不依赖于物质的属性**（它是受能量支配的）。一定质量的铅与等质量的氢对给定的力产生相同的反作用力，换言之，它们具有同样大的惯性。一定质量的铱（密度最大的固体）与等质量的锂（已知最轻的固体）也能相平衡。另一方面，表面能所涉及的是分子而不是质量，体积能也是如此。放置在一个容量为 1 升的容器中的 2 克氢，与相同温度下放置在相同体积的容器中的 32 克氧具有相同的体积能。等质量的锡和铅的热容量并不相同；但 119 克锡与 207 克铅的热容量是相同的；也就是说，相同的原子数量对应相同的热容量。同一电势差，在电解质溶液中传输的电量正比于溶质的当量而不是其质量；也就是正比于其原子量的某个简单百分比。化学能的容量因子则是该能量所支配物质的原子量。由此我们看到，因为质量或惯性是动能与线性能的重要附属条件，所以其他各种类型的能量都直接或间接地与原子量相关联。

这些考虑使我们注意到物质的量（假定对于我们所说的“物质”的性质确实存在这样一个载体）不一定非要通过其惯性或重力来测定。实际上，“质量”这个词具有两方面完全不同的含义。由于我们习惯于通过质量测量物质的多少，“质量”这个词逐渐用以表示“物质的量”。但是，任何其他的能量因子也都可以用来测定物质

measure a quantity of matter by any other of its energy factors. I may, if I choose, state that those quantities of matter which possess equal capacities for heat are equal; or that "equal numbers of atoms" represent equal quantities of matter. Indeed, we regard the value of material as due rather to what it can do, than to its mass; and we buy food, in the main, on an atomic, or perhaps, a molecular basis, according to its content of albumen. And most articles depend for their value on the amount of food required by the producer or the manufacturer.

The various forms of energy may therefore be classified as those which can be referred to an "atomic" factor, and those which possess a "mass" factor. The former are in the majority. And the periodic law is the bridge between them; and yet, an imperfect connection. For the atomic factors, arranged in the order of their masses, display only a partial regularity. It is undoubtedly one of the main problems of physics and chemistry to solve this mystery. What the solution will be is beyond my power of prophecy; whether it is to be found in the influence of some circumstance on the atomic weights, hitherto regarded as among the most certain "constants of nature"; or whether it will turn out that mass and gravitational attraction are influenced by temperature, or by electrical charge, I cannot tell. But that some means will ultimately be found of reconciling these apparent discrepancies, I firmly believe. Such a reconciliation is necessary, whatever view be taken of the nature of the universe and of its mode of action; whatever units we may choose to regard as fundamental among those which lie at our disposal.

In this address I have endeavoured to fulfil my promise to combine a little history, a little actuality, and a little prophecy. The history belongs to the Old World; I have endeavoured to share passing events with the New; and I will ask you to join with me in the hope that much of the prophecy may meet with its fulfilment on this side of the ocean.

(**56**, 378-382; 1897)

的量。如果让我选择，我会说，那些具有相同热容量的物质的量是相同的；或者说，“相同数量的原子”代表相同的物质的量。实际上，我们衡量物质的价值是看它的能力而不是它的质量；我们购买食物时，大体上在原子或者也许是分子的基础上来讲，是根据其中的蛋白质含量。而大多数商品的价格都取决于生产者或制造商所需食物的量。

由此，各种形式的能量可以分为两类，即具有可以称之为“原子”因子的能量，和具有“质量”因子的能量。前者是占大多数的。周期律则是两者之间的桥梁；只是目前尚不够完善。因为按照其质量顺序排列的原子因子只表现出部分规律性。如何揭开这个秘密无疑是物理和化学领域的主要难题之一。我没有能力预言答案会是什么。是否会发现某些条件可以影响迄今仍被视为最为确定的“自然常量”之一的原子量，或者是否会发现质量和重力作用受温度或者电荷的影响，我都无法预言。但是我绝对相信，我们终将找到调和这些表面矛盾的方法。无论对宇宙的本质及其运动模式有何看法，无论我们将选择什么作为基本单位，这样一种调和都是必需的。

在这篇报告中，我努力实践了我开始的承诺——将一些历史、现实和预言结合起来。历史属于过去的世界；我一直努力与新世界分享过去的事；让我们共同期望，很多预言将会在大洋此岸得到证实。

（王耀杨 翻译；汪长征 审稿）

Distant Electric Vision

A. A. Campbell-Swinton

Editor's Note

Here the electrical engineer Alan Archibald Campbell-Swinton comments on the recent discussion of a Mr. Bidwell on the technical obstacles to transmitting visual signals electrically over long distances—that is, to television. The need for 160,000 synchronized operations per second, he notes, might be achieved with separate cathode rays at the transmitting and receiving ends, which could be swept over a display surface in less than the one-tenth of a second necessary for visual persistence. The more demanding challenge lay in finding a means for transmitting the high-frequency signals required for visual fields over long distances. Campbell-Swinton is today seen as the first man to have clearly envisaged how television might work.

REFERRING to Mr. Shelford Bidwell's illuminating communication on this subject published in *Nature* of June 4, may I point out that though, as stated by Mr. Bidwell, it is wildly impracticable to effect even 160,000 synchronised operations per second by ordinary mechanical means, this part of the problem of obtaining distant electric vision can probably be solved by the employment of two beams of kathode rays (one at the transmitting and one at the receiving station) synchronously deflected by the varying fields of two electromagnets placed at right angles to one another and energised by two alternating electric currents of widely different frequencies, so that the moving extremities of the two beams are caused to sweep synchronously over the whole of the required surfaces within the one-tenth of a second necessary to take advantage of visual persistence.

Indeed, so far as the receiving apparatus is concerned, the moving kathode beam has only to be arranged to impinge on a sufficiently sensitive fluorescent screen, and given suitable variations in its intensity, to obtain the desired result.

The real difficulties lie in devising an efficient transmitter which, under the influence of light and shade, shall sufficiently vary the transmitted electric current so as to produce the necessary alterations in the intensity of the kathode beam of the receiver, and further in making this transmitter sufficiently rapid in its action to respond to the 160,000 variations per second that are necessary as a minimum.

Possibly no photoelectric phenomenon at present known will provide what is required in this respect, but should something suitable be discovered, distant electric vision will, I think, come within the region of possibility.

(**78**, 151; 1908)

A. A. Campbell-Swinton: 66 Victoria Street, London, S.W., June 12.

远程电视系统

坎贝尔–斯温顿

编者按

电气工程师艾伦·阿奇博尔德·坎贝尔–斯温顿在这篇文章中评价了比德韦尔先生最近关于视觉信号在通过电力远距离传输（即电视）时遇到技术障碍的论述。他特别提到，每秒所需的160,000次同步操作也许可以通过分别位于发送端和接收端的阴极射线来完成，这样才能在显示区域内以低于1/10秒的时间进行扫描以满足视觉暂留的需要。找到一种传输高频信号的手段以满足远程视场的需要对我们来说是一个更高层次的挑战。在今天看来，坎贝尔–斯温顿是第一个明确提出电视工作原理的人。

谢尔福德·比德韦尔先生在6月4日的《自然》上就远程电视系统这一主题发表了一篇颇具启发性的通讯文章，他在该文章中指出，通过普通的力学方法想要在每秒内有效执行160,000次同步操作是完全不可能实现的，但我想说的是，对于实现远程电视系统所面临的这部分问题，通过引入两束阴极射线（一束在发送站，一束在接收站）就很有可能得到解决。用两个互相垂直放置的电磁铁产生的交变磁场来使这两束射线发生同步偏转，用两个频率迥异的交流电流来驱动电磁铁，这样就能使两束射线变化中的极限状态在1/10秒的时间内同步地扫过整个屏幕表面，而在1/10秒内完成这些是达到视觉暂留效果所必需的。

对于接收装置来说，实际上只要使变化的阴极射线在撞击足够敏感的荧光屏后能在其上形成影像，并使不同强度射线产生的影像有合适的差异即可，这样就能获得期望的结果了。

真正的困难在于发射装置的设计。首先，这个发射装置要有能力在光线和阴影的作用下发射出变化细节足够丰富的电流，以便能使接收端阴极射线的强度产生必要的变化。其次，这个发射装置的速度要足够快，每秒最少要能对160,000次变化作出响应。

可能现在还没有任何光电仪器能够达到这些要求，但人们肯定能发明出满足要求的东西。我相信远程电视系统最终可以实现。

（刘东亮 翻译；李军刚 审稿）

Intra-Atomic Charge

F. Soddy

Editor's Note

What was the internal structure of an atom? While Ernest Rutherford's experiments in 1911 had convinced him that the atom contained a dense, positively charged nucleus, others were not so sure. Here Rutherford's sometime collaborator Frederick Soddy suggested that the nucleus must also contain negative charges, expelled during so-called radioactive beta decay. Soddy introduces the term "isotope": atoms essentially identical in their chemical properties but with differing nuclei. For any given nuclear charge, he asserted, an atom may have any number of electrons in an "outer ring system". Changes in this number are a consequence of chemical action, with no effect on the nucleus. Clarification of this view awaited the discovery of the proton and neutron.

THAT the intra-atomic charge of an element is determined by its place in the periodic table rather than by its atomic weight, as concluded by A. van der Broek (*Nature*, November 27, p. 372), is strongly supported by the recent generalisation as to the radio-elements and the periodic law. The successive expulsion of one α and two β particles in three radio-active changes in any order brings the intra-atomic charge of the element back to its initial value, and the element back to its original place in the table, though its atomic mass is reduced by four units. We have recently obtained something like a direct proof of van der Broek's view that the intra-atomic charge of the nucleus of an atom is not a purely positive charge, as on Rutherford's tentative theory, but is the difference between a positive and a smaller negative charge.

Fajans, in his paper on the periodic law generalisation (*Physikal. Zeitsch.*, 1913, vol. XIV, p. 131), directed attention to the fact that the changes of chemical nature consequent upon the expulsion of α and β particles are precisely of the same kind as in ordinary electrochemical changes of valency. He drew from this the conclusion that radio-active changes must occur in the same region of atomic structure as ordinary chemical changes, rather than with a distinct inner region of structure, or "nucleus", as hitherto supposed. In my paper on the same generalisation, published immediately after that of Fajans (*Chem. News*, February 28), I laid stress on the absolute identity of chemical properties of different elements occupying the same place in the periodic table.

A simple deduction from this view supplied me with a means of testing the correctness of Fajans's conclusion that radio-changes and chemical changes are concerned with the same region of atomic structure. On my view his conclusion would involve nothing else than that, for example, uranium in its tetravalent uranous compounds must be chemically

原子内的电荷

索迪

编者按

原子的内部结构是怎样的？1911 年，当欧内斯特·卢瑟福用实验验证了原子包含一个致密的带正电的核时，其他人并没有表示十分肯定。在这篇文章中，曾经与卢瑟福合作过的弗雷德里克·索迪提出原子核中必须同时也包含负电荷，这些负电荷会在放射性 β 衰变时发射出去。索迪引入了“同位素”的概念，即原子核结构不同但化学性质基本一致的原子。他认为，对于任意给定的核电荷，原子“外层系统”可以排布任意的电子数量。化学反应能引起外层电子数的变化，但对原子核没有影响。后来质子和中子的发现证实了他的观点。

范德布鲁克断言（《自然》，11 月 27 日，第 372 页），一种元素原子内的电荷是由它在周期表中的位置而不是它的原子量确定的，这一论断受到最近一些关于放射性元素与周期律的结论的强力支持。如果在 3 次放射性变化中相继发射出 1 个 α 粒子和 2 个 β 粒子，那么不管这 3 次衰变的次序如何，都会使该元素的原子内电荷回到初始数值，元素也回到了它在周期表中的原始位置，但是它的原子质量却减少了 4 个单位。最近我们获得了一些证据，能够直接支持范德布鲁克的观点：就像卢瑟福的初步理论所指出的，原子核内的电荷并不是单纯的正电荷，而是正电荷与较小的负电荷的差值。

法扬斯在他那篇关于周期律的一般法则的论文(《物理学杂志》,1913 年,第 14 卷,第 131 页）中，特别指出了如下事实：由于 α 粒子和 β 粒子的发射而引起的元素化学性质的改变，与普通的会发生价态变动的由电化学变化引起的物质化学性质的改变完全属于同一类型。他由此得出的结论是，放射性变化必定与普通的化学变化一样发生在原子结构的同一区域，而不是像我们目前所假设的——放射性变化发生在被称为“核”的一个完全不同的内部区域。在法扬斯的论文发表后不久，我也针对同一主题发表了一篇论文（《化学新闻》，2 月 28 日），文中我将重点放在了周期表中处于同一位置的不同元素具有完全相同的化学性质这一点上。

此观点的一个简单推论为我提供了一种检验法扬斯认为的放射性变化与化学变化发生于原子结构中同一区域这一观点是否正确的方法。我认为，他的结论其实就是下面的意思：举例来说，铀的四价化合物中的铀元素必定与钍化合物中的钍元

identical with and non-separable from thorium compounds. For uranium X, formed from uranium I by expulsion of an α particle, is chemically identical with thorium, as also is ionium formed in the same way from uranium II. Uranium X loses two β particles and passes back into uranium II, chemically identical with uranium. Uranous salts also lose two electrons and pass into the more common hexavalent uranyl compounds. If these electrons come from the same region of the atom uranous salts should be chemically non-separable from thorium salts. But they are not.

There is a strong resemblance in chemical character between uranous and thorium salts, and I asked Mr. Fleck to examine whether they could be separated by chemical methods when mixed, the uranium being kept unchanged throughout in the uranous or tetravalent condition. Mr. Fleck will publish the experiments separately, and I am indebted to him for the result that the two classes of compounds can readily be separated by fractionation methods.

This, I think, amounts to a proof that the electrons expelled as β rays come from a nucleus not capable of supplying electrons to or withdrawing them from the ring, though this ring is capable of gaining or losing electrons from the exterior during ordinary electro-chemical changes of valency.

I regard van der Broek's view, that the number representing the net positive charge of the nucleus is the number of the place which the element occupies in the periodic table when all the possible places from hydrogen to uranium are arranged in sequence, as practically proved so far as the relative value of the charge for the members of the end of the sequence, from thallium to uranium, is concerned. We are left uncertain as to the absolute value of the charge, because of the doubt regarding the exact number of rare-earth elements that exist. If we assume that all of these are known, the value for the positive charge of the nucleus of the uranium atom is about 90. Whereas if we make the more doubtful assumption that the periodic table runs regularly, as regards numbers of places, through the rare-earth group, and that between barium and radium, for example, two complete long periods exist, the number is 96. In either case it is appreciably less than 120, the number were the charge equal to one-half the atomic weight, as it would be if the nucleus were made out of α particles only. Six nuclear electrons are known to exist in the uranium atom, which expels in its changes six β rays. Were the nucleus made up of α particles there must be thirty or twenty-four respectively nuclear electrons, compared with ninety-six or 102 respectively in the ring. If, as has been suggested, hydrogen is a second component of atomic structure, there must be more than this. But there can be no doubt that there must be some, and that the central charge of the atom on Rutherford's theory cannot be a pure positive charge, but must contain electrons, as van der Broek concludes.

素具有完全相同的化学性质且不可区分。因为，由铀 I 发射 1 个 α 粒子而形成的铀 X 与钍，以及由铀 II 以同样方式形成的“锾”，在化学性质上是一致的。铀 X 发射 2 个 β 粒子就又回到铀 II，铀 II 与铀的化学性质是一致的。亚铀盐也可以失去 2 个电子并形成更为常见的六价铀的化合物。如果这些电子来自原子中的同一区域，那么亚铀盐和钍盐应该具有相同的化学性质并且不可区分。然而事实并非如此。

亚铀盐与钍盐的化学性质非常相似，我已经请弗莱克先生研究是否可以在始终保持铀元素的四价或六价状态不发生改变的前提下用化学方法将铀盐与钍盐的混合物中的铀和钍分离开来。弗莱克先生将独立发表他的实验，我很感激能够使用他取得的结果，即通过分馏的方法可以很顺利地分离这两类化合物。

我认为这足以证明，以 β 射线形式发射出的电子来自核，尽管核无法为核外圈层提供电子或从中取走电子，但核外圈层可以在普通的会发生价态变动的电化学变化中从外部获取电子或失去电子。

在研究从铊到铀这些处在周期表序列尾部的成员的电荷相对值时，我把范德布鲁克的观点当作是已经被证实的，即从氢到铀按顺序排好周期表中所有可能的位置后，代表核所具有的净正电荷的数值就正好是元素在周期表中所处位置对应的数值。不过我们仍旧不能确定电荷的绝对值，因为对到底存在多少种稀土元素还存在疑问。如果我们假定这些都是已知的，那么铀原子核中正电荷的数值大约是 90。如果我们采取另一个更不确定的假设，即从周期表中位置对应的数值的角度来看，包括全部稀土元素以及介于钡与镭之间的元素在内的周期表是规则排布的，整个周期表有两个完整的长周期，这样得到的数值就会是 96。不管是哪种情况，该数值都明显小于 120，即便假设其电荷数值就是 120，那也只是其原子量的一半，如果真是这样，那核可能就只有 α 粒子了。已知铀原子中存在 6 个核电子，在放射性变化中这些核电子以 6 次 β 射线的形式发射出来。如果核是由 α 粒子构成的话，那么与核外圈层中具有 96 个或者 102 个电子相对应，就必须有 30 个或者 24 个核电子。另外已经有人提出氢是原子结构的另一种构件，如果考虑上这一点的话，那么其核电子数就不止于此了。但是有一点毋庸置疑，那就是必定存在一些核电子，而且卢瑟福理论中原子的中心电荷也不是单纯的正电荷，而是必定包含负电子，就像范德布鲁克断定的那样。

So far as I personally am concerned, this has resulted in a great clarification of my ideas, and it may be helpful to others, though no doubt there is little originality in it. The same algebraic sum of the positive and negative charges in the nucleus, when the arithmetical sum is different, gives what I call "isotopes" or "isotopic elements", because they occupy the same place in the periodic table. They are chemically identical, and save only as regards the relatively few physical properties which depend upon atomic mass directly, physically identical also. Unit changes of this nuclear charge, so reckoned algebraically, give the successive places in the periodic table. For any one "place," or any one nuclear charge, more than one number of electrons in the outer-ring system may exist, and in such a case the element exhibits variable valency. But such changes of number, or of valency, concern only the ring and its external environment. There is no in- and out-going of electrons between ring and nucleus.

(**92**, 399-400; 1913)

Frederick Soddy: Physical Chemistry Laboratory, University of Glasgow.

就目前我个人的思考结果来说，以上是对我的观点的一个很明晰的解释。虽然很明显其中并无多少创见，但也许会对其他人有所帮助吧。根据核内正负电荷的代数和相同而算数和不同的现象，我提出了“同位素”或“同位置元素”的概念，因为它们在周期表中处于相同的位置。它们在化学性质上是完全相同的，除了很有限的一些直接决定于原子量的物理性质外，其他大部分物理性质也是完全相同的。从数学角度来看，核电荷数会单位递增地发生变化，这使得周期表中的位置也连续地变化。对于周期表中任意一个位置，或者说任意一个确定的核电荷数，可以存在不止一种外层电子数，在这种情况下元素就表现出了不同的价态。不过，这种电子数或者价态的变化只是考虑了核外圈层及其外部环境，并没有把核外圈层与核之间的电子进出过程考虑进去。

（王耀杨 翻译；汪长征 审稿）

The Structure of the Atom

E. Rutherford

Editor's Note

Responding to a comment by Frederick Soddy, Ernest Rutherford here clarifies his view on the structure of the atomic nucleus. Soddy had suggested that Rutherford believed the nucleus to contain positive charges only. On the contrary, Rutherford insists, he believes only that the atomic nucleus is small, dense, and positively charged overall. Moreover, he thinks that two of the key products of radioactive decay—alpha and beta particles—might both originate from the nucleus. Rutherford supports a recent suggestion that the charge on the atomic nucleus is equal to the atomic number, and not to half the atomic weight. This observation prefigured the revelation that the atomic number is the equal to the number of protons in the nucleus.

IN a letter to this journal last week, Mr. Soddy has discussed the bearing of my theory of the nucleus atom on radio-active phenomena, and seems to be under the impression that I hold the view that the nucleus must consist entirely of positive electricity. As a matter of fact, I have not discussed in any detail the question of the constitution of the nucleus beyond the statement that it must have a resultant positive charge. There appears to me no doubt that the α particle does arise from the nucleus, and I have thought for some time that the evidence points to the conclusion that the β particle has a similar origin. This point has been discussed in some detail in a recent paper by Bohr (*Phil. Mag.*, September, 1913). The strongest evidence in support of this view is, to my mind, (1) that the β ray, like the α ray, transformations are independent of physical and chemical conditions, and (2) that the energy emitted in the form of β and γ rays by the transformation of an atom of radium C is much greater than could be expected to be stored up in the external electronic system. At the same time, I think it very likely that a considerable fraction of the β rays which are expelled from radio-active substances arise from the external electrons. This, however, is probably a secondary effect resulting from the primary expulsion of a β particle from the nucleus.

The original suggestion of van der Broek that the charge on the nucleus is equal to the atomic number and not to half the atomic weight seems to me very promising. This idea has already been used by Bohr in his theory of the constitution of atoms. The strongest and most convincing evidence in support of this hypothesis will be found in a paper by Moseley in *The Philosophical Magazine* of this month. He there shows that the frequency of the X radiations from a number of elements can be simply explained if the number of unit charges on the nucleus is equal to the atomic number. It would appear that the charge on the nucleus is the fundamental constant which determines the physical and chemical properties of the atom, while the atomic weight, although it approximately follows the order of the nucleus charge, is probably a complicated function of the latter depending on the detailed structure of the nucleus.

(**92**, 423; 1913)

E. Rutherford: Manchester, December 6, 1913.

原子结构

卢瑟福

编者按

为了回应弗雷德里克·索迪的意见，欧内斯特·卢瑟福在这篇文章中进一步解释了他对原子核结构的观点。索迪曾指出卢瑟福认为原子核只包含正电荷，而卢瑟福却强调说他只承认原子核很小、很致密，以及整体带正电。此外，卢瑟福认为，放射性衰变的两个主要产物，即 α 粒子和 β 粒子，可能都源自原子核。卢瑟福同意最近有人提出的关于核电荷等于原子序数而非原子量的一半的观点。这个结果预示了原子序数与原子核中的质子数相等这一关系。

在上周致贵刊的一封信中，索迪先生对我关于放射性现象中原子核的理论进行了相关讨论，他似乎以为，我认定原子核必须完全由带正电荷的粒子构成。事实上，我只是认为原子核必定具有总和为正的电荷，而对它的具体构成并未发表看法。在我看来，α 粒子无疑是产生于核的，而且经过一段时间的思考，我认为有证据表明 β 粒子也源自核。玻尔在最近的一篇论文（《哲学杂志》，1913 年 9 月）中对这一点进行了较为详细的讨论。支持这一观点的最强有力的证据是：(1) 与 α 射线一样，β 射线的衰变也是与物理和化学条件无关的；(2) 镭原子发生衰变时以 β 射线或 γ 射线的形式放出的能量 C，比预想的外部电子系统所存储的能量大得多。不过，放射性物质发射出的 β 射线可能有相当一部分来源于外部电子。这也许是原子核中的 β 粒子的初级辐射引起的一种次级效应。

范德布鲁克最先提出，核所带的基本电荷数应等于原子序数而不是原子量的一半，我认为这是非常有可能的。玻尔在他的原子结构理论中已经应用了这种观点。在莫塞莱本月发表于《哲学杂志》上的文章中，可以找到支持这个假说的最强有力也最令人信服的证据。文中表明，如果核所带的基本电荷数等于原子序数，就能方便地对多种元素所发出的 X 辐射的频率进行解释。看起来，核所带的基本电荷数是确定原子物理和化学性质的基本常数，尽管原子量的顺序与相应原子核电荷数的顺序基本一致，但原子量并非如人们以前所设想的那样是其核电荷数的两倍，可能是核电荷数的复杂函数，其函数关系与核的具体结构有关。

（王耀杨 翻译；鲍重光 审稿）

The Reflection of X-Rays

Editor's Note

The German physicist Max von Laue demonstrated the phenomenon of X-ray diffraction in 1912. This high-energy form of light, with wavelengths comparable to the spacing between molecules in crystalline solids, could be used to reveal crystal structures. Other physicists had deduced the relationship between the lattice spacing and the angles at which bright diffraction spots should occur. Here Maurice de Broglie, the brother of physicist Louis de Broglie, introduces what came to be known as the rotating-crystal method for recording X-ray diffraction from a single crystal. The technique detects X-rays reflected along the surface of a series of so-called "Laue cones". It became the standard method of X-ray diffraction for many years.

IN view of the great interest of Prof. Bragg's and Messrs. Moseley and Darwin's researches on the distribution of the intensity of the primary radiation from X-ray tubes, it may be of interest to describe an alternate method which I have found very convenient (*Comptes rendus*, November 17, 1913).

As we know, the wavelength of the reflected ray is defined by the equation $n\lambda = 2d\sin\theta$, where n is a whole number, d the distance of two parallel planes, and θ the glancing angle. If one mounts a crystal with one face in the axis of an instrument that turns slowly and regularly, such as, for instance, a registering barometer, the angle changes gradually and continuously.

If, therefore, one lets a pencil of X-rays, emerging from a slit, be reflected from this face on to a photographic plate, one finds the true spectrum of the X-rays on the plate, supposing intensity of the primary beam to have remained constant. (This can be tested by moving another plate slowly before the primary beam during the exposure.)

The spectra thus obtained are exactly analogous to those obtained with a diffraction grating, and remind one strongly of the usual visual spectra containing continuous parts, bands, and lines.

So far I have only identified the doublet, $11°17'$ and $11°38'$, described by Messrs. Moseley and Darwin. The spectra contain also a number of bright lines about two octaves shorter than these, and the continuous spectrum is contained within about the same limits. These numbers may be used in the interpretations of diffraction Röntgen patterns, as they were obtained with tubes of the same hardness as those used for producing these latter.

X 射线的反射

编者按

1912 年，德国物理学家马克斯·冯·劳厄证明了 X 射线的衍射现象。这种形式的光能量很高，其波长与晶格中分子之间的距离相近，因而可以用于研究晶体的结构。其他物理学家推算出了晶格间距与预计会出现明亮衍射斑点的角度之间的关系。在这篇文章中，物理学家路易斯·德布罗意的哥哥莫里斯·德布罗意介绍了记录单晶 X 射线衍射的方法，后来被称作旋转晶体法。该技术探测沿着一组被称作“劳厄锥”的晶面反射的 X 射线。许多年来这种方法一直是人们研究 X 射线衍射的标准方法。

鉴于大家对布拉格教授、莫塞莱先生以及达尔文先生关于从 X 射线管发射的初级辐射强度分布的研究有极大的兴趣，我已发现的另一种很方便的方法可能也会引起大家的兴趣（《法国科学院院刊》，1913 年 11 月 17 日）。

如我们所知，反射射线的波长由方程 $n\lambda = 2d\sin\theta$ 确定，式中 n 是一个整数，d 是两个平行平面间的距离，θ 是掠射角。如果我们将一块晶体放在一台缓慢而有规律旋转的仪器（例如，记录式气压计）上，使晶体的一个表面沿仪器的轴向，那么反射的角度将连续不断地发生变化。

如果我们使一束 X 射线从狭缝中射出，并被此晶面反射到照相干版上，假定初级光束的强度保持不变，即可在干版上得到 X 射线的真实光谱。（可以通过曝光时在初级光束前方缓慢移动另一块照相干版来检验光束强度是否保持不变。）

这样得到的光谱与用衍射光栅得到的光谱非常类似，而且很容易使人想起那些含有连续区、谱带和谱线的普通可见光谱。

至今，我仅确认了莫塞莱先生和达尔文先生所描述的双线：11° 17′ 和 11° 38′。这个光谱还包含比双线短大约两个倍频程的许多亮线，还有在大致相同范围内的连续谱。这些光谱线也许可用于解释伦琴的衍射图样，因为产生伦琴的衍射图样时使用的管子与产生这些谱线时使用的管子具有相同的硬度。

The arrangement described above enables us to distinguish easily the spectra of different orders, as the interposition of an absorbing layer cuts out the soft rays, but does not weaken appreciably the hard rays of the second and higher orders.

It is convenient also for absorption experiments; thus a piece of platinum foil of 0.2 mm thickness showed transparent bands. The exact measurements will be published shortly, as well as the result of some experiments I am engaged upon at present upon the effect of changing the temperature of the crystal.

Maurice de Broglie

* * *

As W. L. Bragg first showed, when a beam of soft X-rays is incident on a cleavage plane of mica, a well-defined proportion of the beam suffers a reflection strictly in accordance with optical laws. In addition to this generally reflected beam, Bragg has shown that for certain angles of incidence, there occurs a kind of selective reflection due to reinforcement between beams incident at these angles on successive parallel layers of atoms.

Experiments I am completing seem to show that a generally reflected beam of rays on incidence at a second crystal surface again suffers optical reflection; but the degree of reflection is dependent on the orientation of this second reflector relative to the first.

The method is a photographic one. The second reflector is mounted on a suitably adapted goniometer, and the photographic plate is mounted immediately behind the crystal. The beam is a pencil 1.5 mm in diameter. When the two reflectors are parallel the impression on the plate, due to the two reflections, is clear. But as the second reflector is rotated about an axis given by the reflected beam from the first and fixed reflector, the optically reflected radiation from the second reflector—other conditions remaining constant—diminishes very appreciably. As the angle between the reflectors is increased from 0° to 90°, the impression recorded on the photographic plate diminishes in intensity. For an angle of 20° it is still clear; for angles in the neighbourhood of 50° it is not always detectable; and for an angle of 90° it is very rarely detectable in the first stages of developing, and is then so faint that it never appears on the finished print.

These results, then, would show that the generally reflected beam of X-rays is appreciably polarised in a way exactly analogous to that of ordinary light. Owing to the rapidity with which the intensity of the generally reflected beam falls off with the angle of incidence of the primary beam, it has not been possible to work with any definiteness with angles of incidence greater than about 78°, and this is unfortunately a considerably larger angle than the probable polarising angle. Experiments with incidence in the neighbourhood of 45° should prove peculiarly decisive, for whereas ordinary light cannot as a rule be completely polarised by reflection, the reflection of X-rays, which occurs at planes of atoms, is

上述装置能使我们很容易地区别不同级的谱线，因为插入吸收层可以截断软射线，但并不会太明显地减弱次级和更高级的硬射线。

进行吸收实验也是很方便的。用这样的方法，一片 0.2 毫米厚的铂箔会显示出透射带。至于确切的测量方法以及目前我在改变晶体温度方面所做的实验得到的结果，都将在我即将发表的文章中进行介绍。

莫里斯·德布罗意

*　*　*

正如布拉格首先指出的，当一束软 X 射线入射到云母的一个解理面上时，光束反射部分所占的比例严格遵照光学定律。除了这种普通的反射光束，布拉格指出，对于特定的入射角，由于以这些角度入射到原子中连续的平行层上的光束彼此相互增强，因而会产生一种选择性的反射。

我正在进行的实验似乎表明，当一个普通的反射束入射到下一个晶体表面上时，还会再次产生光学反射，但是，反射的程度取决于第二个反射面与第一个反射面之间的夹角。

我用的是照相记录法。第二个反射面被安装在调整好的测角仪上，照相干版紧贴在晶体后面。光束直径为 1.5 毫米。当两个反射面平行时，两次反射在照相干版上产生的影像是清晰的。但是，当以第一个固定反射面上反射的射线为轴旋转第二个反射面时，假如其他条件不变，则从第二个反射面上反射的光学辐射明显减弱。当反射面之间的夹角从 0° 增加到 90° 时，照相干版上记录的影像的清晰度不断降低。在角度为 20° 时，影像仍然很清晰；角度为 50° 左右时，经常检测不到影像；角度为 90° 时，由于影像非常微弱，在第一步显影阶段已经很难看到，而在最终冲洗出的照片上则从来没有出现过。

以上结果表明，X 射线普通反射束的偏振特性在某种程度上完全类似于普通光的偏振。由于普通反射束的强度随着初级光束入射角的增加而迅速下降，当入射角超过 78° 后就很难得到确定无疑的影像了。遗憾的是，这个角度比可能的偏振角大很多。入射角在 45° 附近的实验尤其具有决定意义，虽然通常不能通过反射使普通光完全偏振，但在原子平面上 X 射线的反射并不受被辐照的晶体表面上任何污染的影响，一旦出现偏振，那么偏振角处被反射的辐射将是完全偏振的。被选择性反射

independent of any contamination of the exposed crystal surface, and polarisation, once established, should prove complete for radiation reflected at the polarising angle. The selectively reflected X-rays seem to show the same effects as does the generally reflected beam. Selectively reflected radiation is always detectable after the second reflection, but this seems due to the selectively reflected radiation produced at the second reflector by the unpolarised portion of the beam generally reflected at the first reflector.

The application of a theory of polarisation to explain the above results is interestingly supported by the fact that in the case of two reflections by parallel reflectors, the proportion of X-rays reflected at the second reflector is invariably greater than the proportion of rays reflected at the first; that is, the ratio of reflected radiation to incident radiation at the second reflector is always greater than the same ratio at the first reflector. This might be expected if vibrations perpendicular to the plane of incidence are to be reflected to a greater extent than those in the plane of incidence. The proportion of such vibrations is larger in the beam incident on the second reflector than in the original beam, and a greater proportion of radiation would be reflected at the second reflector than could be at the first. For the case of parallel reflectors and incidence of a primary beam on the first at the polarising angle, the reflection at the second should be complete.

E. Jacot

(**92**, 423-424; 1913)

Maurice de Broglie: 29, Rue Chateaubriand, Paris, December 1.
E. Jacot: South African College, Cape Town, November 14.

的 X 射线与普通的反射有相同的效应。在第二次反射后总能检测到被选择性反射的辐射，但这似乎是由第一个反射面处被反射的那部分非偏振辐射在第二个反射面上发生选择性反射而造成的。

偏振理论可以用来解释上述结果，这得到了以下事实的强烈支持，在由两个平行反射面引起的两次反射中，第二个反射面上被反射的 X 射线总是多于第一个反射面上被反射的 X 射线；即在第二个反射面上被反射的辐射相对于入射辐射的比例，总是大于第一个反射面。这就可以预期，垂直于入射平面的振动被反射的量比在入射平面内的振动更大。与原光束相比，入射到第二个反射面上的光束中垂直于入射平面的振动所占比例更大一些，因此在第二个反射面上被反射的辐射的比例大于第一个反射面。在两个反射面相互平行的情况下，如果初始光束以偏振角入射到第一个反射面上，则在第二个反射面上它将被完全反射。

贾科

（沈乃澂 翻译；江丕栋 审稿）

The Constitution of the Elements

F. W. Aston

Editor's Note

Francis Aston was a physicist at the Cavendish Laboratory of the University of Cambridge who invented the instrument now called the mass spectrometer, which could in principle measure the masses of individual atoms. The principle was to electrify the atoms by using an electric field to remove one or more of their electrons, accelerate them in the same electric field, and deflect them by means of a magnetic field. A beam of atoms thus ionised would travel in a curved path onto a screen and the mass of the atom related to some standard could then be inferred from the deflection of its path. At the beginning of what proved to be his life's work, Aston had come to the conclusion that several familiar atoms consisted of isotopes with different masses but the same overall electric charge. In this letter, he drew attention to the fact that the measured masses of atoms appeared to be very near to integral multiples of the mass of the hydrogen atom, and suggested that this may "do much to elucidate the ultimate structure of matter".

IT will doubtless interest readers of *Nature* to know that other elements besides neon (see *Nature* for November 27, p. 334) have now been analysed in the positive-ray spectrograph with remarkable results. So far oxygen, methane, carbon monoxide, carbon dioxide, neon, hydrochloric acid, and phosgene have been admitted to the bulb, in which, in addition, there are usually present other hydrocarbons (from wax, etc,) and mercury.

Of the elements involved hydrogen has yet to be investigated; carbon and oxygen appear, to use the terms suggested by Paneth, perfectly "pure"; neon, chlorine, and mercury are unquestionably "mixed". Neon, as has been already pointed out, consists of isotopic elements of atomic weights 20 and 22. The mass-spectra obtained when chlorine is present cannot be treated in detail here, but they appear to prove conclusively that this element consists of at least two isotopes of atomic weights 35 and 37. Their elemental nature is confirmed by lines corresponding to double charges at 17.50 and 18.50, and further supported by lines corresponding to two compounds HCl at 36 and 38, and in the case of phosgene to two compounds COCl at 63 and 65. In each of these pairs the line corresponding to the smaller mass has three or four times the greater intensity.

Mercury, the parabola of which was used as a standard of mass in the earlier experiments, now proves to be a mixture of at least three or four isotopes grouped in the region corresponding to 200. Several, if not all, of these are capable of carrying three, four, five, or even more charges. Accurate values of their atomic weights cannot yet be given.

元素的组成

阿斯顿

编者按

剑桥大学卡文迪什实验室的物理学家弗郎西斯·阿斯顿发明了我们今天称为质谱仪的仪器，这种仪器原则上可以测定单个原子的质量。其原理是，利用电场移除原子中的一个或多个电子而将原子电离，使原子在该电场中加速，再利用磁场使之偏转。电离了的原子束经由弯曲轨迹到达接收屏，通过其轨迹的偏转程度可以推断此原子相对于某种标准的质量。在这项堪称终身成就工作的早期，阿斯顿断言几种常见元素是由质量不同但总电荷数相同的同位素组成。在这封快报中，他注意到已测得的原子质量似乎非常接近于氢原子质量的整数倍这一事实，并指出这将“大大有助于阐明物质的基本结构”。

《自然》的读者们一定会有兴趣知道，用阳极射线谱仪分析除氖之外的其他一些元素（参见《自然》，11 月 27 日，第 334 页）后得到了不同寻常的结果。目前已经用此仪器对氧、甲烷、一氧化碳、二氧化碳、氖、氢氯酸和光气进行了分析，不过仪器中通常还存在其他烃类（来自蜡等）和汞。

相关元素中，氢还需要被研究。按照帕内特的说法，碳和氧似乎是绝对“纯的”。而氖、氯和汞则无疑是“混合的”。如同已经指出的那样，氖由原子量分别为 20 和 22 的同位素组成。这里无法详细讨论有氯存在时得到的质谱，但似乎可以确认这种元素至少由原子量为 35 和 37 的两种同位素组成。根据位于 17.50 和 18.50 的与双电荷微粒有关的谱线，可以证实它们的元素性质。与两种 HCl 相对应的位于 36 和 38 处的两条谱线以及在研究光气时由两种 COCl 所形成的位于 63 和 65 处的两条谱线进一步支持了这样的元素性质。在上述的每组谱线对中，较小质量对应的谱线强度是较大质量对应谱线强度的 3~4 倍。

在早期实验中汞的抛物线曾被用作质量标准，现在证实汞至少是由 3~4 种原子量集中在 200 左右的同位素混合而成。其中几种，如果不是全部的话，可以携带 3 个、4 个、5 个甚至更多个电荷。目前尚无法给出它们的原子量的准确数值。

A fact of the greatest theoretical interest appears to underlie these results, namely, that of more than forty different values of atomic and molecular mass so far measured, all, without a single exception, fall on whole numbers, carbon and oxygen being taken as 12 and 16 exactly, and due allowance being made for multiple charges.

Should this integer relation prove general, it should do much to elucidate the ultimate structure of matter. On the other hand, it seems likely to make a satisfactory distinction between the different atomic and molecular particles which may give rise to the same line on a mass-spectrum a matter of considerable difficulty.

(**104**, 393; 1919)

F. W. Aston: Cavendish Laboratory, December 6.

这些结果之下似乎存在一个极具理论价值的事实，即目前已测得的超过 40 个原子或分子的质量数据，无一例外都是整数，这里以碳原子量和氧原子量正好为 12 和 16 为准确值，并适当考虑带多个电荷造成的影响。

如果这一整数关系被证明是普遍存在的，它将大大有助于阐明物质的基本结构。另一方面，要对质谱中给出相同谱线的不同原子和分子进行令人满意的区分，看来还是一件相当困难的事情。

（王耀杨 翻译；李芝芬 审稿）

Einstein's Relativity Theory of Gravitation

E. Cunningham

Editor's Note

Arthur Eddington had recently announced his measurements of the deflection of starlight during a solar eclipse, in apparent agreement with Einstein's general theory of relativity. Here Ebenezer Cunningham surveys Einstein's ideas. Einstein's 1905 work had established a link between inertia and energy, and his new work pursued the question of whether gravity too might be linked to energy. What emerges from the theory, Cunningham argues, is a view in which there is no ultimate criterion for the equality of space or time intervals, but only the equivalence of an infinite number of ways of mapping out physical events. All this has been made possible, he notes, by Einstein adopting mathematics already developed by Riemann, Levi-Civita and others.

I

THE results of the Solar Eclipse Expeditions announced at the joint meeting of the Royal Society and Royal Astronomical Society on November 6 brought for the first time to the notice of the general public the consummation of Einstein's new theory of gravitation. The theory was already in being before the war; it is one of the few pieces of pure scientific knowledge which have not been set aside in the emergency; preparations for this expedition were in progress before the war had ceased.

Before attempting to understand the theory which, if we are to believe the daily Press, has dimmed the fame of Newton, it may be worth while to recall what it was that he did. It was not so much that he, first among men, used the differential calculus. That claim was disputed by Leibniz. Nor did he first conceive the exact relations of inertia and force. Of these, Galileo certainly had an inkling. Kepler, long before, had a vague suspicion of a universal gravitation, and the law of the inverse square had, at any rate, been mooted by Hooke before the "Principia" saw the light. The outstanding feature of Newton's work was that it drew together so many loose threads. It unified phenomena so diverse as the planetary motions, exactly described by Kepler, the everyday facts of falling bodies, the rise and fall of the tides, the top-like motion of the earth's axis, besides many minor irregularities in lunar and planetary motions. With all these drawn into such a simple scheme as the three laws of motion combined with the compact law of the inverse square, it is no wonder that flights of speculation ceased for a time. The universe seemed simple and satisfying. For a century at least there was little to do but formal development of Newton's dynamics. In the mid-eighteenth century Maupertuis hinted at a new physical doctrine. He was not content to think of the universe as a great clock the wheels of which turned inevitably and irrevocably according to a fixed rule. Surely there must be some purpose, some divine economy in all its motions. So he propounded a principle of least action. But it soon appeared that this was only Newton's laws in a new guise; and so the eighteenth century closed.

爱因斯坦关于万有引力的相对论

坎宁安

编者按

阿瑟·爱丁顿最近宣布，他在日食期间对星光偏转的测量明确地证实了爱因斯坦的广义相对论。在这篇文章中，埃比尼泽·坎宁安简单描述了爱因斯坦的理论。爱因斯坦在1905年就已经建立了惯性和能量之间的联系，他现在的工作主要是考查重力是否也有可能与能量相关联。坎宁安指出，这个理论说明了这样一个观点：虽然不存在衡量空间间隔和时间间隔完全相等的绝对标准，但可以设计无限多种方式以保证多个物理事件的等同性。他说，爱因斯坦利用黎曼、列维齐维塔和其他人的数学理论已经验证了这些结论。

I

在11月6日皇家学会和皇家天文学会共同举办的会议上宣布的日食观测结果，使得爱因斯坦关于万有引力的新理论受到了公众的广泛关注。这个理论在战争之前就已经被提出来了；它是在战争中极少数没有被丢到一边的纯科学工作之一；日食观测的准备工作在战争结束之前就已经着手进行了。

如果在试图理解这个新理论之前，我们就已经像日报上说的那样，认为牛顿在其面前也会黯然失色，那么最好首先回顾一下牛顿所做的工作。并不能说牛顿是第一个使用微分计算的人。有些议论认为这是莱布尼茨的首创。牛顿也不是第一个认识到惯性和力之间关系的人。这方面，肯定是伽利略首先对此进行了初步的设想。开普勒在很久以前就对万有引力有过模糊的猜想，但是，引力与距离平方成反比的规律是在该“原理”建立之前，由胡克首先发现的。牛顿所做工作的杰出之处在于，他把这么多松散的线索整合到了一起。行星运动（已被开普勒精确描述过）、日常生活中的落体运动、潮汐的涨落、地轴的进动以及在月球和行星运动中出现的许多小的不规则性——牛顿将所有这些现象都统一了起来。牛顿把所有这些都归入了一个包括力学三大定律和简洁的平方反比关系的基本框架中，也难怪在后来很长一段时间内，科学上的思索都停滞了。整个宇宙都看似简单而圆满。在后来至少一个世纪的时间里，除了在形式上发展一下牛顿力学，没有其他工作可做。在18世纪中叶，莫佩尔蒂暗示了一种新的物理学说。他不满足于认为宇宙是一个在某种确定的法则下永不停止、永不倒退的钟表。宇宙的运动一定是有目的的，一定有一种神圣的力量在支配着它的运动。因此，他提出了最小作用原理。但是人们很快发现这只不过是牛顿定律的一种新的外在形式而已，然后18世纪就这样结束了。

The nineteenth saw great changes. When it closed, the age of electricity had come. Men were peering into the secrets of the atom. Space was no longer a mighty vacuum in the cold emptiness of which rolled the planets. It was filled in every part with restless energy. Ether, not matter, was the last reality. Mass and matter were electrical at bottom. A great problem was set for the present generation: to reconcile one with the other the new laws of electricity and the classical dynamics of Newton. At this point the principle of least action began to assume greater importance; for the old and the new schemes of the universe had this in common, that in each of them the time average of the difference between the kinetic and the potential energies appears to be a minimum.

One of the main difficulties encountered by the electrical theory of matter has been the obstinate refusal of gravitation to come within its scope. Quietly obeying the law of the inverse square, it heeded not the bustle and excitement of the new physics of the atom, but remained, independent and inevitable, a constant challenge to rash claimants to the key of the universe. The electrical theory seemed on the way to explain every property of matter yet known, except the one most universal of them all. It could trace to its origins the difference between copper and glass, but not the common fact of their weight; and now the ether began silently to steal away.

One matter that has seriously troubled men in Newton's picture of the universe is its failure to accord with the philosophic doctrine of the relativity of space and time. The vital quantity in dynamics is the acceleration, the change of motion of a body. This does not mean that Newton assumed the existence of some ultimate framework in space relative to which the actual velocity of a body can be uniquely specified, for no difference is made to his laws if any arbitrary constant velocity is added to the velocity of every particle of matter at all time. The serious matter is that the laws cannot possibly have the same simplicity of form relative to two frameworks of which one is in rotation or non-uniform motion relative to the other. It seems, for instance, that if Newton were right, the term "fixed direction" in space means something, but "fixed position" means nothing. It seems as if the two must stand or fall together. And yet the physical relations certainly make a distinction. Why this should be so has not yet been made known to us. Whatever new theory we adopt must take account of the fact.

It was with some feeling of relief that men hailed the advent of the ether as a substitute for empty space, though we may note in passing that some philosophers—Comte, for example—have held that the concept of an ether, infinite and intangible, is as illogical as that of an absolute space. But, jumping at the notion, physicists proposed to measure all velocities and rotations relative to it. Alas! the ether refused to disclose the measurements. Explanations were soon forthcoming to account for its reluctance; but these were so far-reaching that they explained away the ether itself in the sense in which it was commonly understood. At any rate, they proved that this creature of the scientific imagination was not one, but many. It quite failed to satisfy the cravings for a permanent standard against which motion might be measured. The problem was left exactly where it was before. This was prewar relativity, summarised by Einstein in 1905. The physicists complained loudly that he was taking away their ether.

19世纪发生了重大的变化。在这个世纪末，电的时代到来了。人类要揭示原子中的秘密。太空也不再是有行星在其中运行的寒冷而空洞的广袤真空。宇宙中的每个部分都注满了运动着的能量。以太，而不是物质，才是最终的存在形式。物质实际上都是带电的。一个重大的问题摆在了当代人的面前：如何将新的电学理论和经典的牛顿力学原理统一起来。这样，最小作用原理就变得重要起来；因为它是新旧两种宇宙观相通的部分，在两者中，动能和势能之差的时间平均都应该取最小值。

物质的电学理论遇到的一个最主要的问题是，无法把万有引力引入到这个理论框架之中。在电学理论遵从平方反比定律的情况下，新兴的原子物理的蓬勃发展却被忽视掉了，电学独自向破解宇宙之谜的目标发起了挑战。电学理论力图说明物质的所有其他性质，而唯独将人们最为熟知的属性排除在外。它可以说清铜和玻璃之间存在差别的原因，但是不能解释它们共同的特性：重量。至此，以太学说也开始默默地销声匿迹了。

在牛顿的宇宙框架中最令人困扰的一点是：它无法与时空相对性的哲学学说达成一致。加速度在力学中是一个关键的量，它反映了物体运动状态的变化。但这并不意味着牛顿假设了空间中存在一个终极的参考系，一个物体的实际速度相对于这个参考系是唯一确定的，因为在他的理论中，任意一个质点的运动速度总可以加上一个常数速度。这种处理方法的严重缺陷是，当处理两个参考系的问题时，如果其中一个参考系相对于另一个做转动或者非匀速运动，那么前面所说的法则就不可能保持如此简单的形式。这样看来，如果牛顿是正确的，那么在空间中，“固定方向”是有意义的，但“固定位置”却没有任何意义。两者看似应该同时成立或者同时不成立，但是物理上的关系显然是有区别的。为什么会这样呢？我们还不知道。无论我们采用什么样的新理论，都要考虑到上面的问题。

当我们引入了以太的概念，用它来代替空无一物的空间时，问题看似得到了解决，尽管我们也许注意到，在传播这一概念的时候，一些哲学家，比如孔德，认为无限且无形的以太和绝对空间一样不合逻辑。但让我们先把这些看法抛在一边，物理学家们提出要测量所有相对于以太的平动和转动。唉！可惜以太却拒绝我们的测量。很快就出现了对这一难题的解释，但是这与能够用一种大家可以接受的方式来为以太辩解还有非常遥远的距离。无论如何，它们证明了这种科学想象的创造并不是唯一的，而是有很多种。我们渴望找到一种可以用来测量运动的永久标准，而以太的概念是非常失败的。问题和从前一样没有得到解决。还是战争之前由爱因斯坦于1905年总结出来的相对论解决了这个问题。物理学家们则强烈地抱怨爱因斯坦摒弃了他们的以太。

Let it not be thought, however, that the results of the hypothesis then advanced were purely negative. They showed quite clearly that many current ideas must be modified, and in what direction this must be done. Most notably it emphasised the fact that inertia is not a fundamental and invariable property of matter; rather it must be supposed that it is consequent upon the property of energy. And, again, energy is a relative term. One absolute quantity alone remained; one only stood independent of the taste or fancy of the observer, and that was "action". While the ether and the associated system of measurement could be selected as any one of a legion, the principle of least action was satisfied in each of them, and the magnitude of the action was the same in all.

But, still, gravitation had to be left out; and the question from which Einstein began the great advance now consummated in success was this. If energy and inertia are inseparable, may not gravitation, too, be rooted in energy? If the energy in a beam of light has momentum, may it not also have weight?

The mere thought was revolutionary, crude though it be. For if at all possible it means reconsidering the hypothesis of the constancy and universality of the velocity of light. This hypothesis was essential to the yet infant principle of relativity. But if called in question, if the velocity of light is only approximately constant because of our ordinary ways of measuring, the principle of relativity, general as it is, becomes itself an approximation. But to what? It can only be to something more general still. Is it possible to maintain anything at all of the principle with that essential limitation removed?

Here was exactly the point at which philosophers had criticised the original work of Einstein. For the physicist it did too much. For the philosopher it was not nearly drastic enough. He asked for an out-and-out relativity of space and time. He would have it that there is no ultimate criterion of the equality of space intervals or time intervals, save complete coincidence. All that is asked is that the order in which an observer perceives occurrences to happen and objects to be arranged shall not be disturbed. Subject to this, any way of measuring will do. The globe may be mapped on a Mercator projection, a gnomonic, a stereographic, or any other projection; but no one can say that one is a truer map than another. Each is a safe guide to the mariner or the aviator. So there are many ways of mapping out the sequences of events in space and time, all of which are equally true pictures and equally faithful servants.

This, then, was the mathematical problem presented to Einstein and solved. The pure mathematics required was already in existence. An absolute differential calculus, the theory of differential invariants, was already known. In pages of pure mathematics that the majority must always take as read, Riemann, Christoffel, Ricci, and Levi-Civita supplied him with the necessary machinery. It remained out of their equations and expressions to select some which had the nearest kinship to those of mathematical physics and to see what could be done with them.

(**104**, 354-356; 1919)

但是，我们不要以为后来提出的假说都是不正确的。这些假说明确地表明现有的许多观点需要修正，并且说明了修正的方向应该在哪里。尤其是它强调了一个事实：惯性不是物质的一个不会发生变化的基本属性，而应该被看作是随着能量的变化而变化的。我们要再次强调，能量是一个相对量。一个绝对量是独立不变的；它不依赖于观察者的体验和想象，这就是“作用”。当以太及与它相关的测量系统被选择作为大量作用中的任意一员时，它们都满足最小作用原理，并且作用量的总和不变。

但是，万有引力仍然没有被考虑进去，爱因斯坦正是从这个问题出发，现在已成功地获得了巨大的进展。如果能量和惯性是不可分割的，那么重力难道就不能建立在能量的基础之上吗？如果能量是一束具有动量的光束，那么它为什么不能有重量呢？

这样的想法是革命性的，尽管它还不够成熟。因为，如果这是可能的，那么它就意味着需要重新考虑光速不变性和普适性的假说。这个假说是尚不成熟的相对论的基本原则。但是如果它被质疑，如果光速只是因为我们通常的测量方法不够精确才大致不变，那么，被大家普遍接受的相对论法则就只是一种近似，就像它本身的系统一样。但这是对什么的近似呢？只能是对一种更加普遍的原理的近似。当消除了那种根本上的限制以后，原来的法则中还有没有什么东西可以保留下来呢？

就是在这个问题上，哲学家们批判了爱因斯坦早期的工作。对物理学家来说，这样的批评太偏激了。对哲学家来说，这种批评还远算不上严厉。爱因斯坦开始寻找一种彻底的时空相对论。他认为，除了完全重合之外，不存在衡量空间间隔和时间间隔完全相等的绝对标准。他只要求观察者观察到的事件的发生顺序和物体摆放的顺序不被打乱。在这个前提下，任何测量方法都将是可行的。这就好像我们可以用墨卡托投影法、心射切面投影法、立体投影法或者任何一种其他的方法去画地球，而没有人会说其中哪一种地图较之其他地图更准确。飞行员和海员使用任何一种地图都是安全可靠的。因此，也有很多方法可以标定时空中事件发生的顺序，每一种都描述了真实的情况，每一种都同样可信。

这样，爱因斯坦接下来就只需要解决那些数学上的问题了。他所需要的纯数学方法已经存在。绝对微分、微分不变量理论，这些都是已知的。大多数人经常研读的纯数学著作，如黎曼、克里斯托弗尔、里奇和列维齐维塔的著作，都为爱因斯坦提供了必要的数学工具。剩下的工作就是从那些方程和表达式中选出最接近数学物理的部分，并想办法把它们解出来。

II. The Nature of the Theory

In the first article an attempt was made to show the roads which led to Einstein's adventure of thought. On the physical side briefly it was this. Newton associated gravitation definitely with mass. Electromagnetic theory showed that the mass of a body is not a definite and invariable quantity inherent in matter alone. The energy of light and heat certainly has inertia. Is it, then, also susceptible to gravitation, and, if so, exactly in what manner? The very precise experiments of Eötvös rather indicated that the mass of a body, as indicated by its inertia, is the same as that which is affected by gravitation.

Also, how must the expression of Newton's law of gravitation be modified to meet the new view of mass? How, also, must the electromagnetic theory and the related pre-war relativity be adapted to allow of the effect of gravitation? With the relaxation of the stipulation that the velocity of light shall be constant, will the principle of relativity become more general and acceptable to the philosophic doctrine of relativity, or will it, on the other hand, become completely impossible?

One point arises immediately. The out-and-out relativist will not admit an absolute measure of acceleration any more than of velocity. The effect, however, of an accelerated motion is to produce an apparent change in gravitation; the measure of gravitation at any place must therefore be a relative quantity depending upon the choice which the observer makes as to the way in which he will measure velocities and accelerations. This is one of Einstein's fundamental points. It has been customary in expositions of mechanics to distinguish between so-called "centrifugal force" and "gravitational force". The former is said to be fictitious, being simply a manifestation of the desire of a body to travel uniformly in a straight line. On the other hand, gravitation has been called a real force because associated with a cause external to the body on which it acts.

Einstein asks us to consider the result of supposing that the distinction is not essential. This was his so-called "principle of equivalence". It led at once to the idea of a ray of light being deviated as it passes through a field of gravitational force. An observer near the surface of the earth notes objects falling away from him towards the earth. Ordinarily, he attributes this to the earth's attraction. If he falls with them, his sense of gravitation is lost. His watch ceases to press on the bottom of his pocket; his feet no longer press on his boots. To this falling observer there is no gravitation. If he had time to think or make observations of the propagation of light, according to the principle of equivalence he would now find nothing gravitational to disturb the rectilinear motion of light. In other words, a ray of light propagated horizontally would share in his vertical motion. To an observer not falling, and, therefore, cognisant of a gravitational field, the path of the ray would therefore be bending downward towards the earth.

The systematic working out of this idea requires, as has been remarked, considerable mathematics. All that can be attempted here is to give a faint indication of the line of attack, mainly by way of analogy.

II. 理论的本质

第一篇文章旨在说明爱因斯坦的思考方法，从物理学的角度来看，简要的说明就是这样。牛顿明确地将万有引力和质量联系起来。电磁学理论表明，一个物体的质量并不是物质确定不变的内在属性。光能和热能当然都具有惯性，那么，它也会受到万有引力的影响吗？如果确实如此，确切的作用方式又是怎样的呢？厄缶的精确实验更表明了由其惯性所表示的物体质量同样受到万有引力的影响。

牛顿万有引力定律的描述要怎样修正才能符合关于质量的新观点呢？电磁学理论和战争前提出的相对论要怎样调整才能允许万有引力效应的存在呢？在解除了光速不变的约束之后，相对论原理会不会成为一种更加普遍且被相对性的哲学理论所接受的法则呢？或者说，另一方面，它会不会被证明完全不可行呢？

这里马上就引出了一个观点。彻底的相对论者只能对速度进行绝对测量，却不能对加速度进行绝对测量。然而，加速运动在万有引力场中会发生明显的变化；因此在任意地点，万有引力的测量值肯定都是相对的，它取决于测量者所选择的测量速度和加速度的方式。这是爱因斯坦的主要观点之一。在力学上，对所谓“离心力”和“万有引力”的解释通常是有区别的。前者是一个虚拟的力，仅仅表现了物体要做匀速直线运动的趋势。另一方面，万有引力被认为是一个真实的力，因为它和作用于物体的外界因素有关。

爱因斯坦让我们考虑，如果这种区别不是本质的，结果会怎样。这就是他所说的“等效原理”。它立刻就引出了这样的设想：一束光穿过万有引力场时会发生弯曲。一个在地球表面附近的观察者会看到物体远离自己落向地球。一般来说，他会把这归结为地球的引力。如果他和物体一起下落，那么他对万有引力的感觉就会消失。他的怀表不再压在衣袋底部，他的脚也不再压在靴子上。对于这位正在下落的观察者来说，他是观察不到万有引力存在的。如果他有时间观察和思考光的传播，那么按照等效原理，他将看不到光线的直线运动被万有引力所干扰。换句话说，一束沿水平方向传播的光线，将与观察者一起同时做垂直运动。因此，一个没有下落的观察者可以感觉到万有引力的存在，所以光线在向地球运动的过程中会发生弯曲。

就像前面提到的那样，要把这样一个想法系统地求解出来，需要大量的数学运算。这里我们所能做的只是用模糊的示意来说明这个原理，主要通过类比法。

It is no new discovery to speak of time as a fourth dimension. Every human mind has the power in some degree of looking upon a period of the history of the world as a whole. In doing this, little difference is made between intervals of time and intervals of space. The whole is laid out before him to comprehend in one glance. He can at the same time contemplate a succession of events in time, and the spatial relations of those events. He can, for instance, think simultaneously of the growth of the British Empire chronologically and territorially. He can, so to speak, draw a map, a four-dimensional map, incapable of being drawn on paper, but none the less a picture of a domain of events.

Let us pursue the map analogy in the familiar two-dimensional sense. Imagine that a map of some region of the globe is drawn on some material capable of extension and distortion without physical restriction save that of the preservation of its continuity. No matter what distortion takes place, a continuous line marking a sequence of places remains continuous, and the places remain in the same order along that line. The map ceases to be any good as a record of distance travelled, but it invariably records certain facts, as, for example, that a place called London is in a region called England, and that another place called Paris cannot be reached from London without crossing a region of water. But the common characteristic of maps of correctly recording the shape of any small area is lost.

The shortest path from any place on the earth's surface to any other place is along a great circle; on all the common maps, one series of great circles, the meridians, is mapped as a series of straight lines. It might seem at first sight that our extensible map might be so strained that all great circles on the earth's surface might be represented by straight lines. But, as a matter of fact, this is not so. We might represent the meridians and the great circles through a second diameter of the earth as two sets of straight lines, but then every other great circle would be represented as a curve.

The extension of this to four dimensions gives a fair idea of Einstein's basic conception. In a world free from gravitation we ordinarily conceive of free particles as being permanently at rest or moving uniformly in straight lines. We may imagine a four-dimensional map in which the history of such a particle is recorded as a straight line. If the particle is at rest, the straight line is parallel to the time axis; otherwise it is inclined to it. Now if this map be strained in any manner, the paths of particles are no longer represented as straight lines. Any person who accepts the strained map as a picture of the facts may interpret the bent paths as evidence of a "gravitational field", but this field can be explained right away as due to his particular representation, for the paths can all be made straight.

But our two-dimensional analogy shows that we may conceive of cases where no amount of straining will make all the lines that record the history of free particles simultaneously straight; pure mathematics can show the precise geometrical significance of this, and can write down expressions which may serve as a measure of the deviations that cannot be removed. The necessary calculus we owe to the genius of Riemann and Christoffel.

把时间作为第四维并不是一个新的发现。每一个人在某种程度上都会把世界历史的一段时期当作一个整体来看待。在这样做的时候，时间段和空间段没有什么区别。在匆匆一瞥之中所有的东西都呈现到他面前要他去了解。他在仔细考虑一系列事件发生时间的同时，还要将它们和发生地点联系起来。比如，他能同时从时间顺序和疆域范围两方面来考虑大英帝国的扩张。所以可以这样说，他可以画一个地图，一个四维的地图，尽管不能画在纸上，但依然可以描述一系列事件。

让我们用我们所熟悉的二维地图进行类比。我们可以想象有一个描述世界上某个区域的地图，用来制作这个地图的材料可以不受物理限制，随意延展和扭曲以保持它的连续性。不管这种材料如何扭曲变形，它上面表示地点次序的连续直线依然保持连续，沿着这条直线各个地点的排列顺序不变。这种地图无法记录旅行的距离，但是它可以忠实地记录某些特定的事实，比如，伦敦位于英国境内，从伦敦到另一个地方——巴黎，不可能不跨越海洋。但是一般地图所具有的记录任意一小块地方的功能在这种地图中完全丧失了。

在地球表面从一地到另一地的最短路径是沿着大圆的路径； 在普通的地图上，一系列的大圆，即经线，是用一系列的直线来表示的。初看起来，我们的可伸缩地图可以被拉伸开，这样，地球表面的所有大圆都可以呈直线。可事实并不是这样。我们可以把经线和另外一组由地球的另外一个直径确定的大圆表示成两组直线，但是这样做之后，所有其他的大圆就只能表示为曲线了。

将上述观点扩展到四维时空，就构成了爱因斯坦的基本概念。我们通常认为在没有万有引力的世界里，自由粒子将永远静止或者做匀速直线运动。我们可以想象一下，在四维的地图中，粒子的历史被记录为一条直线。如果这个粒子是静止的，那么这条直线就平行于时间轴，否则就是倾斜的。现在，如果地图以任意方式被拉伸，那么粒子的路径就不再被表示为直线。如果我们能接受可伸缩地图作为表征事实的方式，就可以把这些弯曲的路径看作是“万有引力场”的证据，但是，这种引力场也可以马上被解释成是由这种特殊的表示方法造成的，因为所有路径都可以变成直线。

但是类似的二维地图告诉我们，没有任何一种变形方式可以使所有记录自由粒子历史的线都同时呈直线；关于这一点，纯数学可以给出它在几何学上的精确证明，还可以给出表达式，用来度量那些不可消除的弯曲。天才的黎曼和克里斯托弗尔给出了我们所需的微积分算法。

Einstein now identifies the presence of curvatures that cannot be smoothed out with the presence of matter. This means that the vanishing of certain mathematical expressions indicates the absence of matter. Thus he writes down the laws of the gravitational field in free space. On the other hand, if the expressions do not vanish, they must be equal to quantities characteristic of matter and its motion. These equalities form the expression of his law of gravitation at points where matter exists.

The reader will ask: What are the quantities which enter into these equations? To this only a very insufficient answer can here be given. If, in the four-dimensional map, two neighbouring points be taken, representing what may be called two neighbouring occurrences, the actual distance between them measured in the ordinary geometrical sense has no physical meaning. If the map be strained, it will be altered, and therefore to the relativist it represents something which is not in the external world of events apart from the observer's caprice of measurement. But Einstein assumes that there is a quantity depending on the relation of the points one to the other which is invariant—that is, independent of the particular map of events. Comparing one map with another, thinking of one being strained into the other, the relative positions of the two events are altered as the strain is altered. It is assumed that the strain at any point may be specified by a number of quantities (commonly denoted g_{rs}), and the invariable quantity is a function of these and of the relative positions of the points.

It is these quantities g_{rs} which characterise the gravitational field and enter into the differential equations which constitute the new law of gravitation.

It is, of course, impossible to convey a precise impression of the mathematical basis of this theory in non-mathematical terms. But the main purpose of this article is to indicate its very general nature. It differs from many theories in that it is not devised to meet newly observed phenomena. It is put together to satisfy a mental craving and an obstinate philosophic questioning. It is essentially pure mathematics. The first impression on the problem being stated is that it is incapable of solution; the second of amazement that it has been carried through; and the third of surprise that it should suggest phenomena capable of experimental investigation. This last aspect and the confirmation of its anticipations will form the subject of the next article.

(**104**, 374-376; 1919)

III. The Crucial Phenomena

In the article last week an attempt was made to indicate the attitude of the complete relativist to the laws which must be obeyed by gravitational matter. The present article deals with particular conclusions.

现在，爱因斯坦认为，无法消除的弯曲表示物质的存在。这意味着，如果某个数学表达式为零则表示没有物质存在。于是他写出了自由空间中万有引力场的定律。另一方面，如果表达式不为零，那么它们一定等于描述物质及其运动的物理量。这些方程就是爱因斯坦在有物质存在的点上构筑的万有引力定律表达式。

读者可能会问：这些方程中都有哪些物理量？在这里我们只能给出一个非常不充分的回答。如果在四维地图中，两个相邻的点被认为代表两个相邻的事件，那么用普通几何方法测量的两点之间的实际距离将没有任何物理意义。如果这个地图发生变形，它就会被改变，所以对于相对论者来说，撇开观察者反复无常的测量结果，它代表了某种不存在于由事件组成的外部世界中的东西。但是爱因斯坦认为有一个物理量依赖于两个点之间的关系，具有不变性，也就是说，它不依赖于某种事件地图。比较一个地图和另一个地图，设想其中一个发生变形而成为另一个，代表两个事件的点的相对位置会随着变形方式的变化而变化。可以假设，任意点的变形可以用一些物理量来表示（一般记做 g_{rs}），而不变量是这些物理量和事件点相对位置的函数。

这个表征万有引力场的物理量 g_{rs} 被引入构成万有引力新定律的微分方程中。

当然，我们不可能用非数学语言将这个理论用数字精确地表达出来。但是这篇文章的主要目的是说明它的普遍特征。这个理论与许多其他理论的不同之处在于，它不是为了解释某个新发现而被构建的。构建它的目的是为了满足精神上的渴望和应对哲学上的质疑。它在本质上是纯数学问题。这个问题给我们的第一印象是，它是不可能被攻破的；第二点出人意料的是，它居然被攻破了；第三个令人惊奇的是，它竟然预言了可以用实验研究的现象。关于最后一方面以及对其预言的证实将是下一篇文章的主题。

III. 关键的现象

上周的文章旨在说明一个完全的相对论者对万有引力物质必须遵循的法则的态度。而本文是要介绍一些特定的结论。

As Minkowski remarked in reference to Einstein's early restricted principle of relativity: "From henceforth, space by itself and time by itself do not exist; there remains only a blend of the two" ("Raum und Zeit", 1908). In this four-dimensional world that portrays all history let (x_1, x_2, x_3, x_4) be a set of coordinates. Any particular set of values attached to these coordinates marks an event. If an observer notes two events at neighbouring places at slightly different times, the corresponding points of the four-dimensional map have coordinates slightly differing one from the other. Let the differences be called (dx_1, dx_2, dx_3, dx_4). Einstein's fundamental hypothesis is this: there exists a set of quantities g_{rs} such that

$$g_{11}dx_1^2 + 2g_{12}dx_1dx_2 + \cdots + g_{44}dx_4^2$$

has the same value, no matter how the four-dimensional map is strained. In any strain g_{rs} is, of course, changed, as are also the differences *dx*.*

If the above expression be denoted by $(ds)^2$, *ds* may conveniently be called the *interval* between two events (not, of course, in the sense of time interval). In the case of a field in which there is no gravitation at all, if dx_4 is taken to be *dt*, it is supposed that ds^2 reduces to the expression $dx_1^2 + dx_2^2 + dx_3^2 - c^2dt^2$, where *c* is the velocity of light. If this is put equal to zero, it simply expresses the condition that the neighbouring events correspond to two events in the history of a point travelling with the velocity of light.

Einstein is now able to write down differential equations connecting the quantities g_{rs} with the coordinates (x_1, x_2, x_3, x_4), which are in complete accord with the requirement of complete relativity. † These equations are assumed to hold at all points of space unoccupied by matter, and they constitute Einstein's law of gravitation.

Planetary Motion

The next step is to find a solution of the equations when there is just one point in space at which matter is supposed to exist, one point which is a singularity of the solution. This can be effected completely ‡ : that is, a unique expression is obtained for the interval between two neighbouring events in the gravitational field of a single mass. This mass is now taken to be the sun.

It is next assumed that in the four-dimensional map (which, by the way, has now a bad twist in it, that cannot be strained out, all along the line of points corresponding to the

* The gravitational field is specified by the set of quantities g_{rs}. When the gravitational field is small, these are all zero, except for g_{44}, which is approximately the ordinary Newtonian gravitational potential.

† These equations take the place of the old Laplace equation $\nabla^2 V=0$. Just as that equation is the only differential equation of the second order which is entirely independent of any change of ordinary space coordinates, so Einstein equations are uniquely determined by the condition of relativity.

‡ The result is that the invariant interval *ds* is given by $ds^2 = (1 - 2m/r)(dt^2 - dr^2) - r^2(d\theta^2 + \sin^2\theta d\phi^2)$, the four coordinates being now interpreted as time and ordinary spherical polar coordinates.

闵可夫斯基这样评论爱因斯坦早期的狭义相对论："从今以后，单独的空间和单独的时间都将不存在；二者只能作为一个复合体而存在"（《空间和时间》，1908 年）。在这个描述了所有历史事件的四维空间中，(x_1, x_2, x_3, x_4) 被视为一组坐标。把任意一组特定数值代入坐标中，都能表示一个事件。如果一个观察者观察到发生地点和时间都很接近的两个事件，那么四维地图上相应两点的坐标也区别不大。我们把它们之间的差别表示为 (dx_1, dx_2, dx_3, dx_4)。爱因斯坦的基本假设是这样的：存在一组 g_{rs}，无论四维地图发生什么样的变形，

$$g_{11}dx_1^2 + 2g_{12}dx_1dx_2 + \cdots + g_{44}dx_4^2$$

都具有相同的值。在发生变形时，g_{rs} 的值当然会发生变化，差值 dx 也同样会改变。*

如果上面的表达式被记作 $(ds)^2$，为方便起见，我们可以把 ds 称作两个事件的**间隔**（当然不是一般观念中的时间间隔）。在万有引力场不存在的情况下，如果 dx_4 用 dt 来代替，那么 ds^2 就会退化成表达式 $dx_1^2 + dx_2^2 + dx_3^2 - c^2dt^2$，这里 c 是光速。如果这个表达式等于零，则表示一个以光速运动的点所经历的两个相邻的事件。

现在，爱因斯坦就可以写出将物理量 g_{rs} 与坐标 (x_1, x_2, x_3, x_4) 相联系的微分方程了，这与完善相对论的要求完全一致。† 这些方程包含了空间中所有未被物质占据的点，它们构成了爱因斯坦的万有引力定律。

行星的运动

下一步的任务是找到一个空间中只有一个点被物质占据的方程解，而这个点是解的奇点。这完全可以做到‡：也就是说，对于单一质点在引力场中的两个相邻事件的间隔，我们可以得到唯一的表达式。太阳可以当作这样的一个质点。

接下来假设在四维地图（现在这个地图严重扭曲，不能把对应于太阳每个时刻位置的点组成的线拉伸开）中，在太阳引力场中运动的质点的路径将是图上任意两

* 万有引力场由一组物理量 g_{rs} 来说明，当万有引力场很小的时候，这些值除了 g_{44} 以外都为零，这就近似成为牛顿的万有引力势场。

† 这些方程代替了旧的拉普拉斯方程 $\nabla^2V=0$。正如拉普拉斯方程是唯一一个完全不受普通空间坐标体系变化影响的二阶微分方程一样，爱因斯坦方程是唯一一个由相对论条件确定的方程。

‡ 结果是：不变的间隔 ds 可以由下式确定，$ds^2 = (1 - 2m/r)(dt^2 - dr^2) - r^2(d\theta^2 + \sin^2\theta d\phi^2)$，这四个坐标可以解释为时间和普通的球面极坐标。

positions of the sun at every instant of time) the path of a particle moving under the gravitation of the sun will be the most direct line between any two points on it, in the sense that the sum of all the intervals corresponding to all the elements of its path is the least possible.* Thus the equations of motion are written down. The result is this:

The motion of a particle differs only from that given by the Newtonian theory by the presence of an additional acceleration towards the sun equal to three times the mass of the sun (in gravitational units) multiplied by the square of the angular velocity of the planet about the sun.

In the case of the planet Mercury, this new acceleration is of the order of 10^{-8} times the Newtonian acceleration. Thus up to this order of accuracy Einstein's theory actually arrives at Newton's laws: surely no dethronement of Newton.

The effect of the additional acceleration can easily be expressed as a perturbation of the Newtonian elliptic orbit of the planet. It leads to the result that the major axis of the orbit must rotate in the plane of the orbit at the rate of 42.9″ per century.

Now it has long been known that the perihelion of Mercury does actually rotate at the rate of about 40″ per century, and Newtonian theory has never succeeded in explaining this, except by *ad hoc* assumptions of disturbing matter not otherwise known.

Thus Einstein's theory almost exactly accounts for the one outstanding failure of Newton's scheme, and, we may note, does not introduce any discrepancy where hitherto there was agreement.

The Deflection of Light by Gravitation

The new theory having justified itself so far, it was thought worth while for British astronomers to devote their main energies at the recent solar eclipse to testing its prediction of an entirely new phenomenon.

As was remarked above, the propagation of light in the ordinary case of freedom from gravitational effect is represented by the equation $ds = 0$.

This Einstein boldly transfer to his generalised theory. After all, it is quite a natural assumption. The propagation of light is a purely objective phenomenon. The emission of a disturbance from one point at one moment, and its arrival at another point at another moment, are events distinct and independent of the existence of an observer. Any law that connects them must be one which is independent of the map the observer uses; ds being an invariant quantity, $ds = 0$ expresses such an invariant law.

* This corresponds to the fact that in a field where there is no acceleration at all the path of a particle is the shortest distance between two points.

点之间最直的线，即与路径上所有组成部分对应的所有间隔之和尽可能最小。* 这样，就可以写出运动方程。结果如下：

一个质点的运动与牛顿理论给出的运动的不同之处在于多出了一个朝向太阳的加速度，这个加速度的值等于三倍的太阳质量（万有引力单位）乘以行星绕太阳运动的角速度的平方。

对于水星，这个新加速度的量级是牛顿加速度的 10^{-8}。这样，低于这个精确度，爱因斯坦理论就还原成了牛顿理论：牛顿理论当然不会失效。

这个多出来的加速度可以被看作是牛顿椭圆行星轨道的一种扰动。这就造成了轨道主轴在轨道平面中以每世纪 42.9 角秒的速度进动。

很久以前我们就知道，水星的近日点的确在以每世纪约 40 角秒的速度进动，牛顿理论从来没有成功地解释过这个现象，除非特意假设有一个在其他情况下未曾出现的干扰物体。

这样，爱因斯坦的理论就彻底解决了牛顿理论框架中的一个重要不足，并且，我们注意到，爱因斯坦理论和牛顿理论没有矛盾，至今它们仍然是统一的。

光在万有引力作用下的偏转

到目前为止，这个新理论的正确性已经得到了证明，英国天文学家们正将他们的主要精力放在近期的日食上，为的是检验此理论所预言的一个全新的现象，大家都认为这是一件值得做的事情。

正如上面所说的那样，在没有万有引力效应的情况下，光的传播可以表示为方程 $ds = 0$。

爱因斯坦大胆地将它移植到了他的广义理论中。毕竟，这是一个很自然的假设。光的传播是一个纯客观的现象。在某一时刻某一点发生的干扰，以及这个干扰在另一时刻到达另一点，这两者是不同的事件，与观察者是否存在无关。任何将它们联系起来的法则都一定不依赖于观察者所使用的地图；ds 是一个不变的量，$ds = 0$ 表示出了这样一个不变的法则。

---

* 它对应于这样一个事实：在一个没有加速度的场中，粒子运动的路径是两点之间的最短距离。

This leads at once to a law of variation of the velocity of light in the gravitational field of the sun.

$$v = c(1 - 2m/r)$$

Here m, as before, is the mass of the sun in gravitational units, and is equal to 1.47 kilometres, while c is the velocity of light at a great distance from the sun. Thus the path of a ray is the same as that if, on the ordinary view, it were travelling in a medium the refractive index of which was $(1 - 2m/r)^{-1}$. In this medium the refractive index would increase in approaching the sun, so that the rays would be bent round towards the sun in passing through it. The total amount of the deflection for a ray which just grazes the sun's surface works out to be 1.75", falling off as the inverse of the distance of nearest approach.

The apparent position of a star near to the sun is thus further from the sun's centre than the true position. On the photographic plate in the actual observations made by the Eclipse Expedition the displacement of the star image is of the order of a thousandth of an inch. The measurements show without doubt such a displacement. The stars observed were, of course, not exactly at the edge of the sun's disc; but on reduction, allowing for the variation inversely as the distance, they give for the bend of a ray just grazing the sun the value 1.98", with a probable error of 6 percent, in the case of the Sobral expedition, and of 1.64" in the Principe expedition.

The agreement with the theory is close enough, but, of course, alternative possible causes of the shift have to be considered. Naturally, the suggestion of an actual refracting atmosphere surrounding the sun has been made. The existence of this, however, seems to be negatived by the fact that an atmosphere sufficiently dense to produce the refraction in question would extinguish the light altogether, as the rays would have to travel a million miles or so through it. The second suggestion, made by Prof. Anderson in *Nature* of December 4, that the observed displacement might be due to a refraction of the ray in travelling through the earth's atmosphere in consequence of a temperature gradient within the shadow cone of the moon, seems also to be negatived. Prof. Eddington estimates that it would require a change of temperature of about 20°C per minute at the observing station to produce the observed effect. Certainly no such temperature change as this has ever been noted; and, in fact, in Principe, at which the Cambridge expedition made its observations, there was practically no fall of temperature.

Gravitation and the Solar Spectrum

It was suggested by Einstein that a further consequence of his theory would be an apparent discrepancy of period between the vibrations of an atom in the intense gravitational field of the sun and the vibrations of a similar atom in the much weaker field of the earth. This is arrived at thus. An observer would not be able to infer the intensity of the gravitational field in which he was placed from any observations of atomic vibrations in the same field: that is, an observer on the sun would estimate the period of vibration of

这立刻就引出了在太阳引力场中光速不变的定律。

$$v = c(1-2m/r)$$

这里的 m 和前面一样，是太阳在万有引力单位下的质量，它相当于 1.47 千米，c 是远离太阳处的光速。这样，从通常的观点上看，一束光的行进路线和它在折射率为 $(1-2m/r)^{-1}$ 的介质中传播一样。在这个介质中，越接近太阳折射率就越大，所以光线在经过太阳的时候就会发生弯曲。一束刚好掠过太阳表面的光的偏转角度是 1.75 角秒，偏转角度会随着光线与太阳之间最小距离的倒数的减小而减小。

靠近太阳的恒星的视位置比它的真实位置离日心更远。在观测日食时所拍的照相底片上，恒星图像位移的数量级为千分之一英尺。测量结果毫无疑问地显示了这样的一个位移。当然，观测的恒星并非恰好在日面的边缘；但是可以利用偏转量和距离成反比的关系进行化规，在索布拉尔的观测队测算出一束刚好掠过太阳表面的光线的偏转角度是 1.98 角秒，允许的误差范围是 6%；在普林西比岛的观测队得到的结果是 1.64 角秒。

实验和理论已经足够吻合，但是，也必须考虑到引起位移的其他可能原因。很自然的，有人怀疑在太阳周围有一个能发生真正的折射效应的大气层。但是，它的存在却可以被以下因素否定，即产生这样的折射作用要求大气层足够厚，而这么厚的大气层会使光线消失，因为光线需要经过 100 万公里左右才能穿过去。第二种可能性是安德森教授在 12 月 4 日的《自然》上提出的。他认为，观察到的位移可能是由光线穿过地球大气层时的折射造成的，因为在月影锥内存在温度梯度，温度梯度可以引发折射现象，这个猜测也不成立。爱丁顿教授估算过，要产生我们观测到的效应，观测站的温度变化应达到每分钟 20℃之多。这样的温度变化当然从未有过；事实上，在剑桥考察队进行观测的普林西比岛，温度并未下降过。

万有引力和太阳光谱

爱因斯坦指出，他的理论还会进一步导出这样的结果：原子在太阳的强引力场中的振动周期和其在弱得多的地球引力场中的振动周期有明显的差别。这个结果是这样得到的：当一个观察者与他所观察的振动原子处在同一个引力场中时，他不可能通过观察原子的振动来判断这个引力场的强度。也就是说，一个在太阳上的观察者测量到的原子振动周期与一个相似原子在地球上的振动周期相同，前提是他本人

an atom there to be the same that he would find for a similar atom in the earth's field if he transported himself thither. But on transferring himself he automatically changes his scale of time; in the new scale of time the solar atom vibrates differently, and, therefore, is not synchronous with the terrestrial atom.

Observations of the solar spectrum so far are adverse to the existence of such an effect. What, then, is to be said? Is the theory wrong at this point? If so, it must be given up, in spite of its extraordinary success in respect of the other two phenomena.

Sir Joseph Larmor, however, is of opinion that Einstein's theory itself does not in reality predict the displacement at all. The present writer shares his opinion. Imagine, in fact, two identical atoms originally at a great distance from both sun and earth. They have the same period. Let an observer A accompany one of these into the gravitational field of the sun, and an observer B accompany the other into the field of the earth. In consequence of A and B having moved into different gravitational fields, they make different changes in their scales of time, so that actually the solar observer A will find a different period for the solar atom from that which B, on the earth, attributes to his atom. It is only when the two observers choose so to measure space and time that they consider themselves to be in identical gravitational fields that they will estimate the periods of the atoms alike. This is exactly what would happen if B transferred himself to the same position as A. Thus, though an important point remains to be cleared up, it cannot be said that it is one which at present weighs against Einstein's theory.

(**104**, 394-395; 1919)

来到地球。但是在他转换观测地点的时候，他会自动调整时间尺度；在新的时间尺度中，太阳引力场中的原子振动周期将发生变化，所以就与地球上的原子不同步了。

迄今为止，太阳光谱的观测结果并不支持这种效应的存在。接下来我们该说些什么呢？这个理论在这一点上是否错了？如果是这样，它必须被放弃，尽管它成功地解释了另外两个现象。

然而，约瑟夫·拉莫尔爵士认为，事实上爱因斯坦的理论本身并没有预言过这种移位效应。本文作者也同意他的观点。事实上，我们可以想象，两个相同的原子最初处于既远离太阳又远离地球的某地。它们具有相同的周期。让观察者A伴随着其中一个原子来到太阳引力场中，让观察者B伴随着另一个原子来到地球引力场中。由于A和B进入的引力场不同，他们的时间尺度发生的变化也不同，因此，太阳上的观察者A将发现太阳上原子的振动周期与地球上的观察者B看到的地球上原子的振动周期不同。只有当两个观察者在同一个引力场中来测量时空时，他们才会判断出两原子的周期相同。如果B来到A的位置，就会发生以上所说的情况。所以，尽管还有一个要点有待澄清，但我们目前还不能说这是一个与爱因斯坦理论相悖的现象。

（王静 翻译；鲍重光 审稿）

A Brief Outline of the Development of the Theory of Relativity*

A. Einstein

Editor's Note

By 1921, Einstein's theory of relativity was widely accepted. Here he describes the theory's historical development, starting from the aim to rid physics of reliance on action at a distance. Maxwell had achieved this for electricity and magnetism, and his mathematical formulation led others to suppose that all space is filled with an ether that carried the electric and magnetic fields. This led to difficulties, which Einstein overcame by abandoning the belief that events may be simultaneous regardless of an observer's motion. But this new understanding didn't encompass gravity. The supposition that gravity and inertia are identical prompted the general theory of relativity. Einstein wonders if gravitational and electrical phenomena might be unified in a theory of all nature's forces—a theory physicists still seek today.

THERE is something attractive in presenting the evolution of a sequence of ideas in as brief a form as possible, and yet with a completeness sufficient to preserve throughout the continuity of development. We shall endeavour to do this for the Theory of Relativity, and to show that the whole ascent is composed of small, almost self-evident steps of thought.

The entire development starts off from, and is dominated by, the idea of Faraday and Maxwell, according to which all physical processes involve a continuity of action (as opposed to action at a distance), or, in the language of mathematics, they are expressed by partial differential equations. Maxwell succeeded in doing this for electro-magnetic processes in bodies at rest by means of the conception of the magnetic effect of the vacuum-displacement-current, together with the postulate of the identity of the nature of electro-dynamic fields produced by induction, and the electro-static field.

The extension of electro-dynamics to the case of moving bodies fell to the lot of Maxwell's successors. H. Hertz attempted to solve the problem by ascribing to empty space (the ether) quite similar physical properties to those possessed by ponderable matter; in particular, like ponderable matter, the ether ought to have at every point a definite velocity. As in bodies at rest, electro-magnetic or magneto-electric induction ought to be determined by the rate of change of the electric or magnetic flow respectively, provided that these velocities of alteration are referred to surface elements moving with the body. But the theory of Hertz was opposed to the fundamental experiment of Fizeau on the

---

* Translated by Dr. Robert W. Lawson.

相对论发展概述*

爱因斯坦

编者按

爱因斯坦的相对论在1921年得到了大家的广泛认可。在这篇演讲稿中，爱因斯坦从物理学力图摆脱超距作用的影响开始，对这个理论的历史沿革进行了回顾。麦克斯韦的电磁理论完成了这一使命，其他物理学家根据他的数学公式提出了以太假说，即认为所有的空间中都充满了作为电场和磁场媒介的以太。这使相对性原理遇到了困难，而爱因斯坦通过放弃事件的同时性可能与观测者的运动无关的观点摆脱了这个困境。但是这种新的理解没有考虑到重力。爱因斯坦猜测重力和惯性可能具有同一性，这一构想促使他提出了广义相对论。爱因斯坦设想重力和电现象或许也可以用一种适用于所有自然力的理论统一起来，这也是今天的物理学家们正在寻找的理论。

用尽可能简练的语言来阐述一系列观念的演变，但仍充分完整地把这种演变的连续性保留下来，这是一件很吸引人的事。在讲述相对论的发展时，我们将尽力做到这一点，并说明其整个发展过程是由一系列细微而又不言而喻的思维过程构成的。

整个发展历程始于并受制于法拉第和麦克斯韦的观念，按照他们的观念，所有的物理过程都包含连续作用（与超距作用相反），或者用数学语言来表示就是利用偏微分方程来描述物理过程。麦克斯韦利用真空位移电流的磁效应概念以及感生电动力场和静电场在本质上完全相同这一假定，成功地构筑了描述静止介质中电磁过程的偏微分方程。

把电动力学理论推广到运动物体的重任落在了麦克斯韦的后继者的身上。赫兹试图通过赋予虚空（以太）与一般有重物质颇类似的物理性质来解决这个问题。特别是，与有重物质一样，以太在空间的每一点上都应该有确定的速度。正如静止物体那样，如果电流或磁流的变化速度是以随物体一起运动的曲面元作为参考的话，那么电磁感应或者磁电感应应当分别由电流或磁流的变化率决定。但是赫兹的理论与斐索有关光在流动液体中传播的基本实验相矛盾。就是说，麦克斯韦理论对运动

* 由罗伯特·劳森博士翻译。

propagation of light in flowing liquids. The most obvious extension of Maxwell's theory to the case of moving bodies was incompatible with the results of experiment.

At this point, H. A. Lorentz came to the rescue. In view of his unqualified adherence to the atomic theory of matter, Lorentz felt unable to regard the latter as the seat of continuous electro-magnetic fields. He thus conceived of these fields as being conditions of the ether, which was regarded as continuous. Lorentz considered the ether to be intrinsically independent of matter, both from a mechanical and a physical point of view. The ether did not take part in the motions of matter, and a reciprocity between ether and matter could be assumed only in so far as the latter was considered to be the carrier of attached electrical charges. The great value of the theory of Lorentz lay in the fact that the entire electro-dynamics of bodies at rest and of bodies in motion was led back to Maxwell's equations of empty space. Not only did this theory surpass that of Hertz from the point of view of method, but with its aid H. A. Lorentz was also pre-eminently successful in explaining the experimental facts.

The theory appeared to be unsatisfactory only in *one* point of fundamental importance. It appeared to give preference to one system of coordinates of a particular state of motion (at rest relative to the ether) as against all other systems of coordinates in motion with respect to this one. In this point the theory seemed to stand in direct opposition to classical mechanics, in which all inertial systems which are in uniform motion with respect to each other are equally justifiable as systems of coordinates (Special Principle of Relativity). In this connection, all experience also in the realm of electro-dynamics (in particular Michelson's experiment) supported the idea of the equivalence of all inertial systems, *i.e.* was in favour of the special principle of relativity.

The Special Theory of Relativity owes its origin to this difficulty, which, because of its fundamental nature, was felt to be intolerable. This theory originated as the answer to the question: Is the special principle of relativity really contradictory to the field equations of Maxwell for empty space? The answer to this question appeared to be in the affirmative. For if those equations are valid with reference to a system of coordinates K, and we introduce a new system of coordinates K′ in conformity with the—to all appearances readily establishable—equations of transformation

$$\left.\begin{aligned} x' &= x - vt \\ y' &= y \\ z' &= z \\ t' &= t \end{aligned}\right\} \text{(Galileo transformation)},$$

then Maxwell's field equations are no longer valid in the new coordinates (x', y', z', t'). But appearances are deceptive. A more searching analysis of the physical significance of space and time rendered it evident that the Galileo transformation is founded on arbitrary assumptions, and in particular on the assumption that the statement of simultaneity has a meaning which is independent of the state of motion of the system of coordinates used. It was shown that the field equations for *vacuo* satisfy the special principle of relativity,

物体的这种最直接的推广与实验结果不符。

正在这时，洛伦兹进行了补救。由于洛伦兹是物质原子理论的忠实支持者，所以他觉得不能把物质看成是连续电磁场的所在地。因此，他设想这些场是连续的以太的某种状态。洛伦兹认为，从力学和物理学两方面的观点来看，以太在本质上与物质无关。以太不参与物质的运动，以太和物质之间的相互关系仅在于，物质被看成是所附电荷的载体。洛伦兹理论的重要价值在于，它使包括静止物体和运动物体在内的整个电动力学回归到了真空中的麦克斯韦方程。该理论不仅在方法论上超越了赫兹的理论，而且洛伦兹还利用它非常成功地解释了许多实验事实。

这个理论似乎仅仅在**一个**重要的基本点上不能令人满意。这就是，似乎某个具有特殊运动状态的坐标系（它相对于以太是静止的）要比相对于这个坐标系运动的所有其他坐标系更加优越。从这一点上来看，这个理论好像违背了经典力学，因为在经典力学中，所有相互间做匀速运动的惯性系都同样有理由被用来当作坐标系（狭义相对性原理）。在这一点上，包括电动力学领域在内的所有经验（尤其是迈克尔逊实验）都支持所有惯性系均等价这一观点，即都支持狭义相对性原理。

狭义相对论就是为解决这一困难而诞生的，这个困难由于它具有的根本性而无法让人容忍。狭义相对论最初被用于解答下述问题：狭义相对性原理真的与真空中的麦克斯韦场方程矛盾吗？答案似乎是肯定的。因为，如果某些方程对于坐标系 K 是成立的，而且我们引进一个新的坐标系 K′，使它符合于（显然容易做到）如下的变换方程：

$$\left.\begin{aligned} x' &= x - vt \\ y' &= y \\ z' &= z \\ t' &= t \end{aligned}\right\}\text{（伽利略变换），}$$

那么麦克斯韦场方程组在新的坐标系 $(x',\ y',\ z',\ t')$ 中不再成立。但表面现象是靠不住的，在更透彻地分析时间和空间的物理意义后发现，伽利略变换是建立在几个相当任意的假设上面的，尤其是假设同时性的陈述与所使用的坐标系的运动状态无关。研究表明，如果我们利用下面的变换方程，则真空中的场方程可以满足狭义

provided we make use of the equations of transformation stated below:

$$\left.\begin{aligned} x' &= \frac{x - vt}{\sqrt{1 - v^2/c^2}} \\ y' &= y \\ z' &= z \\ t' &= \frac{t - vx/c^2}{\sqrt{1 - v^2/c^2}} \end{aligned}\right\} \text{(Lorentz transformation)}$$

In these equations x, y, z represent the coordinates measured with measuring-rods which are at rest with reference to the system of coordinates, and t represents the time measured with suitably adjusted clocks of identical construction, which are in a state of rest.

Now in order that the special principle of relativity may hold, it is necessary that all the equations of physics do not alter their form in the transition from one inertial system to another, when we make use of the Lorentz transformation for the calculation of this change. In the language of mathematics, all systems of equations that express physical laws must be co-variant with respect to the Lorentz transformation. Thus, from the point of view of method, the special principle of relativity is comparable to Carnot's principle of the impossibility of perpetual motion of the second kind, for, like the latter, it supplies us with a general condition which all natural laws must satisfy.

Later, H. Minkowski found a particularly elegant and suggestive expression for this condition of co-variance, one which reveals a formal relationship between Euclidean geometry of three dimensions and the space-time continuum of physics.

Euclidean Geometry of Three Dimensions.	*Special Theory of Relativity.*
Corresponding to two neighbouring points in space, there exists a numerical measure (distance ds) which conforms to the equation $ds^2 = dx_1^2 + dx_2^2 + dx_3^2$	Corresponding to two neighbouring points in space-time (point events), there exists a numerical measure (distance ds) which conforms to the equation $ds^2 = dx_1^2 + dx_2^2 + dx_3^2 + dx_4^2$
It is independent of the system of coordinates chosen, and can be measured with the unit measuring-rod.	It is independent of the inertial system chosen, and can be measured with the unit measuring-rod and a standard clock. x_1, x_2, x_3 are here rectangular coordinates, whilst $x_4 = \sqrt{-1}ct$ is the time multiplied by the imaginary unit and by the velocity of light.
The permissible transformations are of such a character that the expression for ds^2 is invariant, *i.e.* the linear orthogonal transformations are permissible.	The permissible transformations are of such a character that the expression for ds^2 is invariant, *i.e.* those linear orthogonal substitutions are permissible which maintain the semblance of reality of x_1, x_2, x_3, x_4. These substitutions are the Lorentz transformations.
With respect to these transformations, the laws of Euclidean geometry are invariant.	With respect to these transformations, the laws of physics are invariant.

From this it follows that, in respect of its *rôle* in the equations of physics, though not with regard to its physical significance, time is equivalent to the space coordinates (apart from the relations of reality). From this point of view, physics is, as it were, a Euclidean geometry of four dimensions, or, more correctly, a statics in a four-dimensional Euclidean continuum.

相对性原理：

$$\left.\begin{aligned} x' &= \frac{x - vt}{\sqrt{1 - v^2/c^2}} \\ y' &= y \\ z' &= z \\ t' &= \frac{t - vx/c^2}{\sqrt{1 - v^2/c^2}} \end{aligned}\right\}\text{（洛伦兹变换）}$$

在上述方程中 x、y、z 表示位置坐标，用相对于坐标系静止的量尺来测量；t 表示时间，用处于静止状态的经过适当校准并且具有相同构造的时钟来测量。

现在，为了使狭义相对性原理成立，要求所有的物理方程在使用洛伦兹变换来计算它们从一个惯性系到另一个惯性系的转换时其形式保持不变。用数学语言来描述就是，所有描述物理定律的方程相对于洛伦兹变换必须是协变的。因此，从方法论的角度来看，狭义相对性原理可以与第二种永恒运动不能实现的卡诺定理相比拟，因为正如卡诺定理一样，狭义相对性原理为我们提供了所有自然规律必须遵守的一般法则。

随后，闵可夫斯基找到了一个特别简洁而又极具启发性的方式来表述这个协变条件，揭示了三维欧几里德几何学与物理学中时空连续统之间的对应关系。

三维欧几里德几何学	狭义相对论
对应于空间中两个相邻的点，存在一种按如下方程计算的数值度量（距离 ds）： $ds^2 = dx_1^2 + dx_2^2 + dx_3^2$	对应于时空中两个相邻的点（点事件），存在一种按如下方程计算的数值度量（距离 ds）： $ds^2 = dx_1^2 + dx_2^2 + dx_3^2 + dx_4^2$
该距离与所选的坐标系无关，并可以用单位量尺测量。	该距离与所选择的惯性系无关，并可以用单位量尺和标准时钟测量。这里的 x_1，x_2，x_3 是直角坐标系中的坐标，而 $x_4 = \sqrt{-1}ct$ 是时间和虚数单位以及光速的乘积。
保持 ds^2 表达式不变的变换是允许的，即线性正交变换是允许的。	保持 ds^2 表达式不变的变换是允许的，即那些线性正交变换是允许的，它们保持了 x_1，x_2，x_3，x_4 表面上的实数性，这些变换就是洛伦兹变换。
相对于这些变换，欧几里德几何学中的定律保持不变。	相对于这些变换，物理学中的定律保持不变。

由此可以看出，时间坐标在物理方程中的**作用**与空间坐标等价（除了实数性之外），虽然不是就其物理意义而言。按照这种观点，物理学过去和现在都是一种四维

The development of the special theory of relativity consists of two main steps, namely, the adaptation of the space-time "metrics" to Maxwell's electro-dynamics, and an adaptation of the rest of physics to that altered space-time "metrics". The first of these processes yields the relativity of simultaneity, the influence of motion on measuring-rods and clocks, a modification of kinematics, and in particular a new theorem of addition of velocities. The second process supplies us with a modification of Newton's law of motion for large velocities, together with information of fundamental importance on the nature of inertial mass.

It was found that inertia is not a fundamental property of matter, nor, indeed, an irreducible magnitude, but a property of energy. If an amount of energy E be given to a body, the inertial mass of the body increases by an amount E/c^2, where c is the velocity of light *in vacuo*. On the other hand, a body of mass m is to be regarded as a store of energy of magnitude mc^2.

Furthermore, it was soon found impossible to link up the science of gravitation with the special theory of relativity in a natural manner. In this connection I was struck by the fact that the force of gravitation possesses a fundamental property, which distinguishes it from electro-magnetic forces. All bodies fall in a gravitational field with the same acceleration, or—what is only another formulation of the same fact—the gravitational and inertial masses of a body are numerically equal to each other. This numerical equality suggests identity in character. Can gravitation and inertia be identical? This question leads directly to the General Theory of Relativity. Is it not possible for me to regard the earth as free from rotation, if I conceive of the centrifugal force, which acts on all bodies at rest relatively to the earth, as being a "real" field of gravitation, or part of such a field? If this idea can be carried out, then we shall have proved in very truth the identity of gravitation and inertia. For the same property which is regarded as *inertia* from the point of view of a system not taking part in the rotation can be interpreted as *gravitation* when considered with respect to a system that shares the rotation. According to Newton, this interpretation is impossible, because by Newton's law the centrifugal field cannot be regarded as being produced by matter, and because in Newton's theory there is no place for a "real" field of the "Koriolis-field" type. But perhaps Newton's law of field could be replaced by another that fits in with the field which holds with respect to a "rotating" system of coordinates? My conviction of the identity of inertial and gravitational mass aroused within me the feeling of absolute confidence in the correctness of this interpretation. In this connection I gained encouragement from the following idea. We are familiar with the "apparent" fields which are valid relatively to systems of coordinates possessing arbitrary motion with respect to an inertial system. With the aid of these special fields we should be able to study the law which is satisfied in general by gravitational fields. In this connection we shall have to take account of the fact that the ponderable masses will be the determining factor in producing the field, or, according to the fundamental result of the special theory of relativity, the energy density—a magnitude having the transformational character of a tensor.

欧几里德几何学，或者，更确切地说，是四维欧几里德连续统中的一种静力学。

狭义相对论的发展经历了两个主要阶段，这就是使时空“度规”适合于麦克斯韦电动力学，以及使物理学中的其余部分适合于这个新的时空“度规”。第一个阶段的成果有同时性的相对性、运动对量尺及时钟的影响、运动学的修正，特别是还有新的速度相加定理。第二个阶段给我们提供了在高速情况下对牛顿运动定律的修正，以及关于惯性质量本质的具有基本重要性的知识。

研究表明惯性不是物质的基本属性，也不是一个不能分解的基本量，而只是能量的一个属性。如果我们赋予一个物体大小为 E 的能量，则此物体的惯性质量将增加 E/c^2，这里 c 是光在真空中的传播速度。同样，一个质量为 m 的物体将被认为具有 mc^2 的能量。

此外，我们很快发现很难把引力科学同狭义相对论以自然的方式联系起来。这种情况使我意识到引力具有一种不同于电磁力的基本性质。在引力场中，所有物体都以相同的加速度下落，或者说，一个物体的引力质量和惯性质量在数值上是相等的（这只不过是同一个事实的另一种表达方式）。这种数值上的相同暗示着两者本质上的等同。引力和惯性能够等同吗？这个问题直接导致了广义相对论的产生。如果我把作用于所有相对于地球静止的物体上的离心力想象成是一个“真实的”引力场，或者是这种引力场的一部分，那我难道不能认为地球是不转动的吗？如果这个想法能够实现，那么我们已经真正证明了引力和惯性的等同性。在不随地球转动的参考系里看来是**惯性**的这一特性，在随地球一起转动的参考系里可以被解释为**引力**。按照牛顿的观点，这样的解释是说不通的，因为牛顿定律告诉我们，离心力场不能被看作是由物质产生的，而且因为在牛顿的理论中，没有把“科里奥利场”这种类型的场当成是“真实的”场。但是，或许牛顿的有关场的定律可以用场的另外一种定律取代，这种定律既适合于这种场又在“转动”坐标系中成立？我坚信惯性质量与引力质量的等同性，这使我有绝对的信心认为上述解释是正确的。就这一点来说，我从以下观点中受到了鼓舞。我们熟悉“表观的”场，这些场在那些相对于一个惯性系做任意运动的坐标系中是有效的。借助于这些特殊的场，我们应当有可能研究引力场通常所满足的定律。关于这一点，我们不得不考虑这样的事实，即有重物质是产生场的决定性因素。或者可以这样表达，按照狭义相对论的基本结果，能量密度这个具有张量变换特性的物理量是产生场的决定因素。

On the other hand, considerations based on the metrical results of the special theory of relativity led to the result that Euclidean metrics can no longer be valid with respect to accelerated systems of coordinates. Although it retarded the progress of the theory several years, this enormous difficulty was mitigated by our knowledge that Euclidean metrics holds for small domains. As a consequence, the magnitude *ds*, which was physically defined in the special theory of relativity hitherto, retained its significance also in the general theory of relativity. But the coordinates themselves lost their direct significance, and degenerated simply into numbers with no physical meaning, the sole purpose of which was the numbering of the space-time points. Thus in the general theory of relativity the coordinates perform the same function as the Gaussian coordinates in the theory of surfaces. A necessary consequence of the preceding is that in such general coordinates the measurable magnitude *ds* must be capable of representation in the form

$$ds^2 = \sum_{uv} g_{uv}\, dx_u\, dx_v \ ,$$

where the symbols g_{uv} are functions of the space-time coordinates. From the above it also follows that the nature of the space-time variation of the factors g_{uv} determines, on one hand the space-time metrics, and on the other the gravitational field which governs the mechanical behaviour of material points.

The law of the gravitational field is determined mainly by the following conditions: First, it shall be valid for an arbitrary choice of the system of coordinates; secondly, it shall be determined by the energy tensor of matter; and thirdly, it shall contain no higher differential coefficients of the factors g_{uv} than the second, and must be linear in these. In this way a law was obtained which, although fundamentally different from Newton's law, corresponded so exactly to the latter in the deductions derivable from it that only very few criteria were to be found on which the theory could be decisively tested by experiment.

The following are some of the important questions which are awaiting solution at the present time. Are electrical and gravitational fields really so different in character that there is no formal unit to which they can be reduced? Do gravitational fields play a part in the constitution of matter, and is the continuum within the atomic nucleus to be regarded as appreciably non-Euclidean? A final question has reference to the cosmological problem. Is inertia to be traced to mutual action with distant masses? And connected with the latter: Is the spatial extent of the universe finite? It is here that my opinion differs from that of Eddington. With Mach, I feel that an affirmative answer is imperative, but for the time being nothing can be proved. Not until a dynamical investigation of the large systems of fixed stars has been performed from the point of view of the limits of validity of the Newtonian law of gravitation for immense regions of space will it perhaps be possible to obtain eventually an exact basis for the solution of this fascinating question.

(**106**, 782-784; 1921)

另一方面，基于对狭义相对论度规结果的考虑导致了这样的结论，即欧几里德度规不再适用于加速参考系。尽管它使理论的进程延迟了几年，但是这个巨大的困难在我们认识到欧几里德度规对于小的区域依然适用之后就变得比较容易解决了。结果，ds 这个迄今在狭义相对论中定义的物理量，在广义相对论中其物理含义仍保持不变。但是坐标本身失去了直接的意义，而完全退化成没有物理意义的数字，它们的唯一用途是标记时空点，因此广义相对论中的坐标与曲面论中高斯坐标的作用相同。以上的叙述必然给出这样一个结论：可测量的量 ds 必定可以用这种广义坐标表达为以下形式：

$$ds^2 = \sum_{uv} g_{uv}\, dx_u\, dx_v \text{ ,}$$

这里符号 g_{uv} 是时空坐标的函数。以上的分析还表明，因子 g_{uv} 随时空变化的特性，一方面决定了时空度规，另一方面还决定了支配质点力学行为的引力场。

引力场的定律主要取决于以下几个条件：第一，它应该在任意选择的坐标系中都是有效的；第二，它应当由物质的能量张量决定；第三，它所包含的因子 g_{uv} 的微分系数最高不超过二阶，并且它相对于这些微分都是线性的。这样我们就得到了一个定律，虽然它在本质上不同于牛顿定律，但是在由它给出的推论中它与牛顿定律吻合得很好，以至于发现只有很少几个判据能够用来对它进行决定性的实验检验。

下面是一些目前尚待解决的重要问题。电场和引力场在特性上真的有那么不同以至于不能用一个形式上的统一体把它们都包含进去吗？引力场对物质的结构起一定作用吗？原子核内部的连续统在相当程度上可以被看作是非欧氏的吗？最后一个问题涉及宇宙论，即惯性是否源自于远距离物质之间的相互作用？与后者相关的是： 宇宙的空间范围是有限的吗？在这一点上我的观点和爱丁顿的观点相悖。与马赫一样，我感到迫切需要一个肯定的答案，但暂时还找不到证据证明。只有从这样的观点（即认为在广袤宇宙空间使用牛顿引力定律具有局限性的观点）出发来对庞大恒星系的动力学进行研究之后，才有可能最终获得解决这一令人困惑的问题的确切依据。

（王锋 翻译；张元仲 审稿）

Atomic Structure

N. Bohr

Editor's Note

Here Danish physicist Niels Bohr ponders the possibility of explaining atomic properties on the basis of the new quantum theory. Physicists had conjectured about how the grouping of elements in the periodic table might reflect specific patterns of electrons arranged around the nucleus, yet without explaining how such patterns arise. As Bohr notes, his model of the hydrogen atom suggested that electrons may fall into distinct shells. Much of the periodic table, says Bohr, can be understood as the successive filling of these shells. The stability of filled shells would explain the chemical inactivity of inert gases such as helium and argon. Bohr's model was supplanted by a more mathematically rigorous quantum theory, yet much of his qualitative picture remains intact today.

IN a letter to *Nature* of November 25 last Dr. Norman Campbell discusses the problem of the possible consistency of the assumptions about the motion and arrangement of electrons in the atom underlying the interpretation of the series spectra of the elements based on the application of the quantum theory to the nuclear theory of atomic structure, and the apparently widely different assumptions which have been introduced in various recent attempts to develop a theory of atomic constitution capable of accounting for other physical and chemical properties of the elements. Dr. Campbell puts forward the interesting suggestion that the apparent inconsistency under consideration may not be real, but rather appear as a consequence of the formal character of the principles of the quantum theory, which might involve that the pictures of atomic constitution used in explanations of different phenomena may have a totally different aspect, and nevertheless refer to the same reality. In this connection he directs attention especially to the so-called "principle of correspondence", by the establishment of which it has been possible—notwithstanding the fundamental difference between the ordinary theory of electromagnetic radiation and the ideas of the quantum theory—to complete certain deductions based on the quantum theory by other deductions based on the classical theory of radiation.

In so far as it must be confessed that we do not possess a complete theory which enables us to describe in detail the mechanism of emission and absorption of radiation by atomic systems, I naturally agree that the principle of correspondence, like all other notions of the quantum theory, is of a somewhat formal character. But, on the other hand, the fact that it has been possible to establish an intimate connection between the spectrum emitted by an atomic system—deduced according to the quantum theory on the assumption of a certain type of motion of the particles of the atom—and the constitution of the radiation, which, according to the ordinary theory of electromagnetism, would result from the same type of motion, appears to me do afford an argument in favour of the reality of the assumptions of the spectral theory of a kind scarcely compatible with Dr. Campbell's suggestion. On

原子结构

玻尔

编者按

在这篇文章中丹麦物理学家尼尔斯·玻尔反复思索是否可以用新的量子理论来解释原子的性质。物理学家们猜测周期表中的元素分组可能在某种程度上反映了核外电子的排布情况，但没有说明这种排布模式是如何形成的。玻尔指出，他的氢原子模型可以说明电子为什么会填入不同的壳层。玻尔说，周期表中大部分元素的核外电子都可以被认为是连续填充这些壳层的。充满电子的壳层结构比较稳定，因而表现出化学反应上的惰性，如氦、氩等惰性气体。虽然玻尔的模型已经被数学上更为严密的量子理论取代，但他对核外电子排布的大部分定性描述至今仍然有效。

诺曼·坎贝尔博士在去年 11 月 25 日写给《自然》的信中，谈到有关原子中电子运动和排布的假说是否可能保持前后一致的问题，这关系到量子理论能否作为原子结构的核心理论对一系列元素的光谱进行解释的问题，为了建立一个能够解释元素其他物理化学性质的原子组成理论，人们在最近的研究中提出了许多差异很大的假设。坎贝尔博士的观点令人振奋，他提出，表面上的矛盾可能不是真的，而是因为量子理论的原理具有表观化的特征，其对原子构成的描述用于解释不同的现象时可能使用完全不同的形式。他把人们对这个问题的注意力转向了所谓的“对应原理”，尽管普通电磁辐射理论与量子理论存在本质上的不同，但在建立了“对应原理”之后，就可能可以用经典辐射理论的结论推导出量子力学理论中的某些推论。

必须承认，我们至今尚未建立起一个完整的理论来详细地描述原子系统发射辐射和吸收辐射的机制，我当然同意“对应原理”像所有其他的量子理论概念一样，在某种意义上也带有形式化的特征。但是，从另一个角度上说，现在已经有可能在一个原子系统——根据基于假设原子中的粒子做某种形式的运动的量子理论推导得到——的发射光谱和由同样类型的运动产生的放射物的构成之间建立一种紧密的关系，根据普通的电磁学理论，我认为它提供了一个论点，这个论点支持一种与坎贝尔博士的建议几乎完全不相符的光谱理论假说。相反，如果我们承认用量子理论解

the contrary, if we admit the soundness of the quantum theory of spectra, the principle of correspondence would seem to afford perhaps the strongest inducement to seek an interpretation of the other physical and chemical properties of the elements on the same lines as the interpretation of their series spectra; and in this letter I should like briefly to indicate how it seems possible by an extended use of this principle to overcome certain fundamental difficulties hitherto involved in the attempts to develop a general theory of atomic constitution based on the application of the quantum theory to the nucleus atom.

The common character of theories of atomic constitution has been the endeavour to find configurations and motions of the electrons which would seem to offer an interpretation of the variations of the chemical properties of the elements with the atomic number as they are so clearly exhibited in the well-known periodic law. A consideration of this law leads directly to the view that the electrons in the atom are arranged in distinctly separate groups, each containing a number of electrons equal to one of the periods in the sequence of the elements, arranged according to increasing atomic number. In the first attempts to obtain a definite picture of the configuration and motion of the electrons in these groups it was assumed that the electrons within each group at any moment were placed at equal angular intervals on a circular orbit with the nucleus at the centre, while in later theories this simple assumption has been replaced by the assumptions that the configurations of electrons within the various groups do not possess such simple axial symmetry, but exhibit a higher degree of symmetry in space, it being assumed, for instance, that the configuration of the electrons at any moment during their motions possesses polyhedral symmetry. All such theories involve, however, the fundamental difficulty that no interpretation is given why these configurations actually appear during the formation of the atom through a process of binding of the electrons by the nucleus, and why the constitution of the atom is essentially stable in the sense that the original configuration is reorganized if it be temporarily disturbed by external agencies. If we reckon with no other forces between the particles except the attraction and repulsion due to their electric charges, such an interpretation claims clearly that there must exist an intimate interaction or "coupling" between the various groups of electrons in the atom which is essentially different from that which might be expected if the electrons in different groups are assumed to move in orbits quite outside each other in such a way that each group may be said to form a "shell" of the atom, the effect of which on the constitution of the outer shells would arise mainly from the compensation of a part of the attraction from the nucleus due to the charge of the electrons.

These considerations are seen to refer to essential features of the nucleus atom, and so far to have no special relation to the character of the quantum theory, which was originally introduced in atomic problems in the hope of obtaining a rational interpretation of the stability of the atom. According to this theory an atomic system possesses a number of distinctive states, the co-called "stationary states", in which the motion can be described by ordinary mechanics, and in which the atom can exist, at any rate for a time, without emission of energy radiation. The characteristic radiation from the atom is emitted only during a transition between two such states, and this process of transition cannot be described by ordinary mechanics, any more than the character of the emitted radiation

释光谱是合理的，那么“对应原理”在解释元素的系列光谱以及元素的其他物理化学性质方面也许能够得出最可信的推论；在这封信中，我将简单地介绍当我们在试图把量子理论应用于原子核系统以构建一个普适的原子组成理论时，是如何广泛地运用这一原则来克服目前遇到的主要难题的。

一系列原子组成理论的共性是都在极力寻找电子的排列和运动规律以解释元素的化学性质为什么会随原子序数的增加而发生周期性变化，正如众所周知的周期率明确指出的那样。周期率使人们猜想，原子中的电子可以被分为不同的组，每一组包括的电子数等于元素序列的一个周期。在早先试图明确描述各组中电子排布和运动的理论中，人们假设每一组中的电子在任一时刻都以相等的角间距排布在以原子核为中心的圆形轨道上，而在后来的理论中，这一简单的假设被新的假设取代，即认为各组中的电子排布不具有这种单一的轴对称，但在空间上则表现出更高程度的对称。例如，假设在电子运动的任一时刻它们的排布呈多面体对称。但是，所有这些理论都难以解释为什么通过原子核束缚电子而生成原子的过程中会出现这样的排布，也不能解释当外部介质暂时侵入时为什么原子的结构在原始排列重组后依然保持基本稳定。如果我们设想，除了由于所带电荷产生的吸引力或排斥力之外，粒子之间再无其他作用力，那么这样的解释就明确要求在原子中不同组的电子之间必须存在一种紧密的相互作用或“耦合”，这完全不同于认为各组电子运动轨道之间的距离也许远到足以使每一组电子单独形成原子的一个“壳层”的假设，耦合效应对外层电子的影响主要是部分地抵消由于电子带负电荷而受到的来自原子核的吸引力。

上述论点被认为涉及核原子的基本特征，与量子理论的特殊性没有关系，人们把量子理论引入原子体系的初衷是希望它能够对原子的稳定性给出合理的解释。根据量子理论，原子体系中存在若干种不同的状态，即所谓的“定态”，电子在定态中的运动不能用一般理论来解释，在某一段时间内，原子可以以任意速率运动，但不会以辐射方式释放能量。只有在两个定态之间发生跃迁时原子才会发射特征辐射，而这种跃迁过程不能用一般理论来描述，更不用说利用普通的电磁理论根据运动规律计算发射的特征辐射了。与普通电磁理论形成鲜明的对照，量子理论假设跃迁通

can be calculated from the motion by the ordinary theory of electro-magnetism, it being, in striking contrast to this theory, assumed that the transition is always followed by an emission of monochromatic radiation the frequency of which is determined simply from the difference of energy in the two states. The application of the quantum theory to atomic problems—which took its starting point from the interpretation of the simple spectrum of hydrogen, for which no *a priori* fixation of the stationary states of the atoms was needed—has in recent years been largely extended by the development of systematic methods for fixing the stationary states corresponding to certain general classes of mechanical motions. While in this way a detailed interpretation of spectroscopic results of a very different kind has been obtained, so far as phenomena which depend essentially on the motion of one electron in the atom were concerned, no definite elucidation has been obtained with regard to the constitution of atoms containing several electrons, due to the circumstance that the methods of fixing stationary states were not able to remove the arbitrariness in the choice of the number and configurations of the electrons in the various groups, or shells, of the atom. In fact, the only immediate consequence to which they lead is that the motion of every electron in the atom will on a first approximation correspond to one of the stationary states of a system consisting of a particle moving in a central field of force, which in their limit are represented by the various circular or elliptical stationary orbits which appear in Sommerfeld's theory of the fine structure of the hydrogen lines. A way to remove the arbitrariness in question is opened, however, by the introduction of the correspondence principle, which gives expression to the tendency in the quantum theory to see not merely a set of formal rules for fixing the stationary states of atomic systems and the frequency of the radiation emitted by the transitions between these states, but rather an attempt to obtain a rational generalization of the electromagnetic theory of radiation which exhibits the discontinuous character necessary to account for the essential stability of atoms.

Without entering here on a detailed formulation of the correspondence principle, it may be sufficient for the present purpose to say that it establishes an intimate connection between the character of the motion in the stationary states of an atomic system and the possibility of a transition between two of these states, and therefore offers a basis for a theoretical examination of the process which may be expected to take place during the formation and reorganisation of an atom. For instance, we are led by this principle directly to the conclusion that we cannot expect in actual atoms configurations of the type in which the electrons within each group are arranged in rings or configurations of polyhedral symmetry, because the formation of such configurations would claim that all the electrons within each group should be originally bound by the atom at the same time. On the contrary, it seems necessary to seek the configurations of the electrons in the atoms among such configurations as may be formed by the successive binding of the electrons one by one, a process the last stages of which we may assume to witness in the emission of the series spectra of the elements. Now on the correspondence principle we are actually led to a picture of such a process which not only affords a detailed insight into the structure of these spectra, but also suggests a definite arrangement of the electrons in the atom of a type which seems suitable to interpret the high-frequency spectra and the chemical properties of the elements. Thus from a consideration of the possible transitions

常伴随着发出单色辐射的过程，频率只取决于两个状态之间的能量差。最初应用量子理论解决原子中的问题是从解释简单的氢原子光谱开始的，因为氢原子不需要先验地确定定态，现在人们根据特定力学运动的类别已经得到了确定定态的系统化方法，因而量子力学理论已经被广泛地用于解决原子中的问题。虽然通过这种方式我们已经获得了许多差异很大的对光谱结果的详细解释，但只局限于研究原子中与单电子运动有关的现象。对于包括几个电子的原子，则由于确定定态的方法不能够排除在原子内不同组（或称壳层）中选择电子数量及其排布时的随意性，而无法得到明确的解释。实际上，这导致的唯一的直接后果是：原子中每个电子的运动情况都可以用一个在中心力场中运动的粒子的一个定态来近似，这种近似只限于用索末菲关于氢线精细结构理论中的圆形或椭圆形轨道来描述定态。但在引入了“对应原理”之后，人们就有了消除随意性的方法，这表明量子理论的发展趋势不仅在于建立一套正式的规则以确定原子体系中的定态和在这些定态之间发生跃迁时发射的辐射频率，而且还要努力总结出合理的辐射电磁理论法则，这一法则对不连续特性的解释必须能够说明原子是稳定的。

在这里不用详述“对应原理”的具体内容，为了解决现在的矛盾，也许可以认为它足以在原子体系不同定态的运动特征与两个定态发生跃迁的可能性之间建立很强的相关性，因而为用理论检验一个原子形成或重组时可能发生的过程提供了依据。例如，根据这个原理我们得出的直接结论是：在实际原子中每一组电子的排布都不可能是环形也都不是多边形对称，因为形成这样的排布要求各组中所有的电子都必须在开始的时候同时被原子束缚住。恰恰相反，我们有必要在原子中寻找使电子有可能是一个一个连续地被原子束缚住的排布方式，我们也许能在元素系列发射光谱中观察到这个过程的最后阶段。现在利用“对应原理”，我们对该过程的描述不仅能够精确地解析这些光谱的结构，还能明确地提出原子中电子的排布方式，既适合解释高频光谱，又能说明元素的化学性质。因此从考虑定态之间可能出现的跃迁入手，根据束缚每一个电子的不同步骤，我们首先假设只有最开始的 2 个电子在可以被称为 1–量子的轨道上运动，1–量子轨道近似于一个中心体系的定态，即一个电子绕一个原子核旋转的体系的基准状态。在最开始的 2 个电子之后被束缚的电子将不能通

between stationary states, corresponding to the various steps of the binding of each of the electrons, we are led in the first place to assume that only the two first electrons move in what may be called one-quantum orbits, which are analogous to that stationary state of a central system which corresponds to the normal state of a system consisting of one electron rotating round a nucleus. The electrons bound after the first two will not be able by a transition between two stationary states to procure a position in the atom equivalent to that of these two electrons, but will move in what may be called multiple-quanta orbits, which correspond to other stationary states of a central system.

The assumption of the presence in the normal state of the atom of such multiple-quanta orbits has already been introduced in various recent theories, as, for instance, in Sommerfeld's work on the high-frequency spectra and in that of Landé on atomic dimensions and crystal structure; but the application of the correspondence principle seems to offer for the first time a rational theoretical basis for these conclusions and for the discussion of the arrangement of the orbits of the electrons bound after the first two. Thus by means of a closer examination of the progress of the binding process this principle offers a simple argument for concluding that these electrons are arranged in groups in a way which reflects the periods exhibited by the chemical properties of the elements within the sequence of increasing atomic numbers. In fact, if we consider the binding of a large number of electrons by a nucleus of high positive charge, this argument suggests that after the first two electrons are bound in one-quantum orbits, the next eight electrons will be bound in two-quanta orbits, the next eighteen in three-quanta orbits, and the next thirty-two in four-quanta orbits.

Although the arrangements of the orbits of the electrons within these groups will exhibit a remarkable degree of spatial symmetry, the groups cannot be said to form simple shells in the sense in which this expression is generally used as regards atomic constitution. In the first place, the argument involves that the electrons within each group do not all play equivalent parts, but are divided into sub-groups corresponding to the different types of multiple-quanta orbits of the same total number of quanta, which represents the various stationary states of an electron moving in a central field. Thus, corresponding to the fact that in such a system there exist two types of two-quanta orbits, three types of three-quanta orbits, and so on, we are led to the view that the above-mentioned group of eight electrons consists of two sub-groups of four electrons each, the group of eighteen electrons of three sub-groups of six electrons each, and the group of thirty-two electrons of four sub-groups of eight electrons each.

Another essential feature of the constitution described lies in the configuration of the orbits of the electrons in the different groups relative to each other. Thus for each group the electrons within certain sub-groups will penetrate during their revolution into regions which are closer to the nucleus than the mean distances of the electrons belonging to groups of fewer-quanta orbits. This circumstance, which is intimately connected with the essential features of the processes of successive binding, gives just that expression for the "coupling" between the different groups which is a necessary condition for the

过两个定态之间的跃迁进入原子中与最开始的 2 个电子等同的位置，但将在可以被称作多量子的轨道上运动，相当于一个中心体系中的其他定态。

最近的一些理论已经假定过这样的多量子轨道存在于正常状态的原子中，如在索末菲关于高频光谱的论文中，以及在朗代关于原子大小和晶体结构的著作中；但正是对应原理第一次为这些推论提供了合理的理论依据，也为在最开始的 2 个电子之后被束缚的电子轨道如何分配这一问题提供了解答。因此通过对束缚过程的进一步研究，对应原理提出了一个简单的规则，认为这些电子分组排列的方式也是元素化学性质随原子序数递增而表现出周期性的反映。实际上，如果我们考虑的是大量电子被一个带较多正电荷的原子核束缚，该规则指出：最开始的 2 个电子位于 1–量子轨道，后 8 个电子被束缚于 2–量子轨道，然后 18 个电子在 3–量子轨道，再后面 32 个电子位于 4–量子轨道。

尽管各组中电子轨道的排列呈现出惊人的空间对称性，但不能因此而认为这些组在原子构造上显示出大家普遍接受的简单壳层结构。首先，该规则认为，同一壳层中的电子并不都扮演同样的角色，总量子数相等的多量子轨道有不同的类型，对应于这些不同的类型，电子又被归入不同的子壳层，子壳层也是描绘电子在中心力场中运动的定态。因此，根据一个体系中存在 2 类 2–量子轨道，3 类 3–量子轨道，以此类推，我们可以得到这样的结论——上面提到的 8 电子壳层由 2 个 4 电子子壳层构成，18 电子壳层由 3 个 6 电子子壳层构成，32 电子壳层由 4 个 8 电子子壳层构成。

原子组成的另一个基本特征是：不同壳层中电子轨道的构型相互关联。因而对每一个壳层来说，其中某些子壳层中的电子在绕核旋转过程中会进入离核距离比量子数较低轨道（即更内层的轨道）中电子离核的平均距离更近的区域。该现象与连续成键过程的基本特征联系紧密，说明不同壳层之间存在“耦合”现象，这种耦合正是使原子构型保持稳定的必要条件。事实上，这样的耦合是整个理论的主要特征，

stability of atomic configurations. In fact, this coupling is the predominant feature of the whole picture, and is to be taken as a guide for the interpretation of all details as regards the formation of the different groups and their various sub-groups. Further, the stability of the whole configuration is of such a character that if any one of the electrons is removed from the atom by external agencies not only may the previous configuration be reorganised by a successive displacement of the electrons within the sequence in which they were originally bound by the atom, but also the place of the removed electron may be taken by any one of the electrons belonging to more loosely bound groups or sub-groups through a process of direct transition between two stationary states, accompanied by an emission of a monochromatic radiation. This circumstance—which offers a basis for a detailed interpretation of the characteristic structure of the high-frequency spectra of the elements—is intimately connected with the fact that the electrons in the various sub-groups, although they may be said to play equivalent parts in the harmony of the inter-atomic motions, are not at every moment arranged in configurations of simple axial or polyhedral symmetry as in Sommerfeld's or Landé's work, but that their motions are, on the contrary, linked to each other in such a way that it is possible to remove any one of the electrons from the group by a process whereby the orbits of the remaining electrons are altered in a continuous manner.

These general remarks apply to the constitution and stability of all the groups of electrons in the atom. On the other hand, the simple variations indicated above of the number of electrons in the groups and sub-groups of successive shells hold only for that region in the atom where the attraction from the nucleus compared with the repulsion from the electrons possesses a preponderant influence on the motion of each electron. As regards the arrangements of the electrons bound by the atom at a moment when the charges of the previously bound electrons begin to compensate the greater part of the positive charge of the nucleus, we meet with new features, and a consideration of the conditions for the binding process forces us to assume that new, added electrons are bound in orbits of a number of quanta equal to, or fewer than, that of the electrons in groups previously bound, although during the greater part of their revolution they will move outside the electrons in these groups. Such a stop in the increase, or even decrease, in the number of quanta characterising the orbits corresponding to the motion of the electrons in successive shells takes place, in general, when somewhat more than half the total number of electrons is bound. During the progress of the binding process the electrons will at first still be arranged in groups of the indicated constitution, so that groups of three-quanta orbits will again contain eighteen electrons and those of two-quanta orbits eight electrons. In the neutral atom, however, the electrons bound last and most loosely will, in general, not be able to arrange themselves in such a regular way. In fact, on the surface of the atom we meet with groups of the described constitution only in the elements which belong to the family of inactive gases, the members of which from many points of view have also been acknowledged to be a sort of landmark within the natural system of the elements. For the atoms of these elements we must expect the constitutions indicated by the following symbols:

Helium	(2_1),	Krypton	$(2_1 8_2 18_3 8_2)$,
Neon	$(2_1 8_2)$	Xenon	$(2_1 8_2 18_3 18_3 8_2)$,
Argon	$(2_1 8_2 8_2)$,	Niton*	$(2_1 8_2 18_3 32_4 18_3 8_2)$,

* "Niton" was the provisional name in 1921 for the radioactive gas now called "radon".

也是理解不同壳层及其子壳层结构所有细节的基础。另外，整个原子结构具有保持稳定的特点——如果原子中的任何一个电子由于外界因素被移走，不仅被原子束缚的电子会在以前原子构型的基础上通过连续位移而进行重新组合，而且那些受束缚较弱的壳层或子壳层中的电子可以通过两个定态之间的直接跃迁而占据被移走的电子的位置，同时发出单色辐射。这个能为详细解释元素高频光谱特征结构提供依据的现象与以下的事实密切相关：尽管我们认为不同子壳层中的电子在原子内的简谐运动中起着同样的作用，但它们并非每时每刻都在外形上保持索末菲和朗代的著作中所说的简单轴对称或者多面体对称，相反，这些电子的行为是通过这样的方式相互联系的——当外力从壳层中移走任何一个电子后，其余电子的轨道将发生连续的变化。

这些观点可用于解释原子中所有壳层电子的组成方式和稳定性。另外，在连续的壳层和子壳层中，上述电子数目的简单变化只会发生在原子中的某个区域，在这个区域内，原子核的吸引力对每个电子运动的影响远远大于电子之间的排斥力。考虑到之前被原子核束缚的电子的电荷开始抵消掉原子核的大部分正电荷时的电子排布情况，我们得到了一些新的观点，成键过程要求的条件使我们不得不假设新增电子所在轨道的量子数等于或小于以前该壳层被束缚的电子，尽管它们在绕原子核旋转的大部分时间内都在该壳层中其他电子的外侧运动。一般说来，表征连续壳层中电子运动状态的轨道量子数或许在一半以上的电子被束缚后就不再增减了。在成键过程中，电子仍会首先排布在确定结构的壳层中，因此 3–量子轨道壳层还会有 18 个电子，而 2–量子轨道壳层还会有 8 个电子。然而，在中性原子中，最后成键且受束缚最弱的电子通常不会按照这样的规则排布。事实上，在原子表层，我们仅在惰性气体元素中观察到上面描述的壳层结构，因此，从各方面看，惰性气体家族都可以称得上是人们认识天然元素体系的里程碑。我们料想这些元素的原子组成可以用符号表示如下：

氦 (2_1),	氪 $(2_1 8_2 18_3 8_2)$,
氖 $(2_1 8_2)$,	氙 $(2_1 8_2 18_3 18_3 8_2)$,
氩 $(2_1 8_2 8_2)$,	氡 *$(2_1 8_2 18_3 32_4 18_3 8_2)$,

* "Niton" 是现在被称为 "radon" 的放射性气体在 1921 年的临时名称。

where the large figures denote the number of electrons in the groups starting from the innermost one, and the small figures the total number of quanta characterising the orbits of electrons within each group.

These configurations are distinguished by an inherent stability in the sense that it is especially difficult to remove any of the electrons from such atoms so as to form positive ions, and that there will be no tendency for an electron to attach itself to the atom and to form a negative ion. The first effect is due to the large number of electrons in the outermost group; hence the attraction from the nucleus is not compensated to the same extent as in configurations where the outer group consists only of a few electrons, as is the case in those families of elements which in the periodic table follow immediately after the elements of the family of the inactive gases, and, as is well known, possess a distinct electro-positive character. The second effect is due to the regular constitution of the outermost group, which prevents a new electron from entering as a further member of this group. In the elements belonging to the families which in the periodic table precede the family of the inactive gases we meet in the neutral atom with configurations of the outermost group of electrons which, on the other hand, exhibit a great tendency to complete themselves by the binding of further electrons, resulting in the formation of negative ions.

The general lines of the latter considerations are known from various recent theories of atomic constitution, such as those of A. Kossel and G. Lewis, based on a systematic discussion of chemical evidence. In these theories the electro-positive and electro-negative characters of these families in the periodic table are interpreted by the assumption that the outer electrons in the atoms of the inactive gases are arranged in especially regular and stable configurations, without, however, any attempt to give a detailed picture of the constitution and formation of these groups. In this connection it may be of interest to direct attention to the fundamental difference between the picture of atomic constitution indicated in this letter and that developed by Langmuir on the basis of the assumption of stationary or oscillating electrons in the atom, referred to in Dr. Campbell's letter. Quite apart from the fact that in Langmuir's theory the stability of the configuration of the electrons is considered rather as a postulated property of the atom, for which no detailed *a priori* interpretation is offered, this difference discloses itself clearly by the fact that in Langmuir's theory a constitution of the atoms of the inactive gases is assumed in which the number of electrons is always largest in the outermost shell. Thus the sequence of the number of electrons within the groups of a niton atom is, instead of that indicated above, assumed to be 2, 8, 18, 18, 32, such as the appearance of the periods in the sequence of the elements might seem to claim at first sight.

The assumption of the presence of the larger groups in the interior of the atom, which is an immediate consequence of the argument underlying the present theory, appears, however, to offer not merely a more suitable basis for the interpretation of the general properties of the elements, but especially an immediate interpretation of the appearance of such families of elements within the periodic table, where the chemical properties of

其中较大的数字代表从最内层开始每一壳层上的电子数，较小的数字代表每一壳层中电子轨道的总量子数。

惰性气体的原子结构具有很强的内在稳定性，从某种意义上说，很难从这样的原子中移走任何一个电子使其变成正离子；同样，一个电子与这类原子结合而使其变成负离子也是不可能的。前者是由于最外层有大量的电子，因此对原子核吸引力的抵消程度大于那些外层只有较少电子的原子，如元素周期表中紧跟在惰性气体之后的那些元素，正如大家所知道的，这种结构具有明显的正电特性。后者是由于最外层电子的规则排列阻止了一个外来电子加入该壳层成为新成员。另一方面，我们发现在元素周期表中排在惰性气体之前的那些族的元素，它们的中性原子的最外层电子结构非常倾向于吸引外来电子以填满该壳层从而形成负离子。

后面提到的种种看法都是根据最近关于原子组成的各种理论得到的，如科塞尔和刘易斯建立在化学实验系统分析基础上的理论。这些理论认为，如果假定惰性气体原子的外层电子结构非常规则和稳定，就可以理解为什么元素周期表中某些族的元素带有正电特性而另一些族的元素带有负电特性，但没有人试图对原子壳层的结构和成因给予详细解释。关于这一点也许更值得关注的是：本文中所阐述的原子结构理论与坎贝尔博士在来信中提到的朗缪尔的理论有本质的不同，朗缪尔假设电子在原子中处于定态或振荡态。本文提到的理论与朗缪尔的理论的不同之处在于：电子排布的稳定性被视为是原子的必要属性，但关于这一点没有给出详细的**先验的**解释，这种不同本身很明确地表明，朗缪尔的理论认为惰性气体原子的最外层电子数总是最大的。因此，氡原子每一壳层上的电子数不再是上文提到的，而是 2、8、18、18、32，乍一看似乎在表面上仍然保留了元素的周期性。

然而，作为当前理论的一个直接推论，原子内部存在较大壳层的假设不仅更适合于解释各种元素的一般性质，而且能马上解释元素周期表中为什么存在相邻元素的化学性质相差非常小的族。事实上，这些族的存在是因为随着原子序数的递增，增加的电子填充到了原子内部的可以容纳大量电子的壳层中。如此说来，我们也许

successive elements differ only very slightly from each other. The existence of such families appears, in fact, as a direct consequence of the formation of groups containing a larger number of electrons in the interior of the atom when proceeding through the sequence of the elements. Thus in the family of the rare earths we may be assumed to be witnessing the successive formation of an inner group of thirty-two electrons at that place in the atom where formerly the corresponding group possessed only eighteen electrons. In a similar way we may suppose the appearance of the iron, palladium, and platinum families to be witnessing stages of the formation of groups of eighteen electrons. Compared with the appearance of the family of the rare earths, however, the conditions are here somewhat more complicated, because we have to do with the formation of a group which lies closer to the surface of the atom, and where, therefore, the rapid increase in the compensation of the nuclear charge during the progress of the binding process plays a greater part. In fact, we have to do in the cases in question, not, as in the rare earths, with a transformation which in its effects keeps inside one and the same group, and where, therefore, the increase in the number in this group is simply reflected in the number of the elements within the family under consideration, but we are witnesses of a transformation which is accompanied by a confluence of several outer groups of electrons.

In a fuller account which will be published soon the questions here discussed will be treated in greater detail. In this letter it is my intention only to direct attention to the possibilities which the elaboration of the principles underlying the spectral applications of the quantum theory seems to open for the interpretation of other properties of the elements. In this connection I should also like to mention that it seems possible, from the examination of the change of the spectra of the elements in the presence of magnetic fields, to develop an argument which promises to throw light on the difficulties which have hitherto been involved in the explanation of the characteristic magnetic properties of the elements, and have been discussed in various recent letters in *Nature*.

(**107**, 104-107; 1921)

N. Bohr: Copenhagen, February 14.

可以认为在稀土族元素中，电子在最多可以容纳 32 个电子而原来仅含有 18 个电子的内壳层中连续地填充。我们同样可以猜测到铁、钯、铂族元素是在逐步填充最多可容纳 18 个电子的壳层。然而，与稀土族相比，铁、钯、铂族元素的情况更复杂一些，因为在靠近原子表面的壳层填充电子时我们不得不面临的难题是，在成键过程中对核电荷的补偿的增加速度非常快。事实上，我们还必须用一种在同一壳层内部发生的转变来解决这些在稀土元素中不存在的难题，因此该壳层中电子数目的增加只是反映了被考虑的族中原子序数的增加，但是我们发现了伴随着几个外层电子汇合的转化过程。

我在即将发表的报告中将更详细地说明这里提到的问题。我写这封信的目的只是让大家注意，应用量子理论解释光谱的那些基本原理的详细阐述，看起来也有可能同时解释了元素的其他性质。就此而言，我还想提醒大家，通过研究磁场中元素光谱的变化，也许可以提出一种有望解决迄今为止人们在解释元素磁特性时遇到的困难的观点，《自然》近期的几篇快报已经对此问题进行了讨论。

（王锋 翻译；李淼 审稿）

The Dimensions of Atoms and Molecules

W. L. Bragg and H. Bell

Editor's Note

In the early 1920s, physicists were struggling to understand the physical structure of atoms using a model introduced by Bohr in 1913, in which electrons filled up shells around the nucleus. Bohr's arguments predicted that elements within any particular period (row) of the periodic table should have roughly the same atomic size, but that atomic size should jump markedly when moving from one period to the next. Here Lawrence Bragg and H. Bell report data on the dimensions of atoms that support Bohr's ideas. Estimating these dimensions from crystal densities and liquid viscosity, they find that elements near the end of rows in the table have almost identical dimensions, while a definite increase happens from one period to the next.

CERTAIN relations which are to be traced between the distances separating atoms in a crystal make it possible to estimate the distance between their centres when linked together in chemical combination. On the Lewis-Langmuir theory of atomic constitution, two electro-negative elements when combined hold one or more pairs of electrons in common, so that the outer electron shell of one atom may be regarded as coincident with that of the other at the point where the atoms are linked together. From this point of view, estimates may be made (W. L. Bragg, *Phil. Mag.*, vol. XI, August, 1920) from crystal data of the diameters of these outer shells. The outer shell of neon, for example, was estimated from the apparent diameters of the carbon, nitrogen, oxygen, and fluorine atoms, which show a gradual approximation to a minimum value of 1.30×10^{-8} cm. The diameters of the inert gases as found in this way are given in the second column of the following table:

Gas	Diameter 2σ (Crystals)	Diameter $2\sigma'$ (Viscosity)	Difference $2\sigma'-2\sigma$
Helium	—	1.89	—
Neon	1.30	2.35	1.05
Argon	2.05	2.87	0.82
Krypton	2.35	3.19	0.84
Xenon	2.70	3.51	0.81

In the third column are given Rankine's values (A. O. Rankine, *Proc. Roy. Soc.*, A, vol. XCVIII, 693, pp. 360–374, February, 1921) for the diameters of the inert gases calculated from their viscosities by Chapman's formula (S. Chapman, *Phil. Trans. Roy. Soc.*, A, vol. CCXVI, pp. 279–348, December, 1915). These are considerably greater than the diameters calculated from crystals, but this is not surprising in view of our ignorance both of the field of force surrounding the outer electron shells and of the nature of the

原子和分子的尺度

布拉格，贝尔

编者按

20世纪20年代初期的物理学家总想应用1913年玻尔提出的模型来努力理解原子的物理结构，在玻尔的模型中，电子填充原子核外的壳层。玻尔的理论预言，在周期表中任意一个周期（一行）的元素应该具有大致相同的原子尺寸，但是从一个周期变化到下一个周期时，原子的大小将发生显著的变化。在这篇文章中，劳伦斯·布拉格和贝尔列举了各种原子的尺寸数据，这些数据支持玻尔的理论。根据这些由晶体密度和液体黏度估算得到的数据，他们发现周期表中每行靠近行尾的元素具有几乎相同的尺寸，而当元素从一个周期过渡到下一个周期时，原子尺寸出现了一定的增加。

通过探索晶体中原子间距离之间的特定关系，可以估算出它们在化学结合中彼此连接时两者中心之间的距离。根据原子结构的刘易斯–朗缪尔理论，两种负电性的元素在结合时共用一对或若干对电子，这可以看作是一个原子的外层电子与另一个原子的外层电子在两个原子连接处重合。从这种观点出发，我们可以利用某些外壳直径的晶体数据作出估计（布拉格，《哲学杂志》，第11卷，1920年8月）。例如，氖的外壳直径就是根据碳、氮、氧和氟原子的表观直径估计得到的——估算逐渐逼近于一个最小值 1.30×10^{-8} cm。下表中的第二列给出了以上述方式得到的惰性气体的直径：

气体	直径2σ（晶体）	直径$2\sigma'$（黏度）	差值 $2\sigma'-2\sigma$
氦	—	1.89	—
氖	1.30	2.35	1.05
氩	2.05	2.87	0.82
氪	2.35	3.19	0.84
氙	2.70	3.51	0.81

第三列中给出了兰金的值（兰金，《皇家学会学报》，A辑，第98卷，第693期，第360~374页，1921年2月），这是根据查普曼公式由黏度计算出来的惰性气体的直径（查普曼，《皇家学会自然科学会报》，A辑，第216卷，第279~348页，1915年12月），它们明显大于利用晶体数据计算出的直径。但是，考虑到我们忽略了外部电子层周围的力场以及将原子连接在一起的共用电子的性质，这就不足为奇了，

electron-sharing which links the atoms together, for it is quite possible that their structures might coalesce to a considerable extent. The constancy of the differences between the two estimates given in the fourth column shows that the *increase* in the size of the atom as each successive electron shell is added is nearly the same (except in the case of neon), whether measured by viscosity or by the crystal data. Further, Rankine has shown that the molecule Cl_2 behaves as regards its viscosity like two argon atoms with a distance between their centres very closely equal to that calculated from crystals, and that the same is true for the pairs Br_2 and krypton, I_2 and xenon.

We see, therefore, that the evidence both of crystals and viscosity measurements indicates that (*a*) the elements at the end of any one period in the periodic table are very nearly identical as regards the diameters of their outer electron shells, and (*b*) in passing from one period to the next there is a definite increase in the dimensions of the outer electron shell, the absolute amount of this increase estimated by viscosity agreeing closely with that determined from crystal measurements.

A further check on these measurements is afforded by the infra-red absorption spectra of HF, HCl, and HBr. The wave-number difference δv between successive absorption lines determines the moment of inertia I of the molecule in each case, the formula being

$$\delta v = \frac{h}{4\pi^2 c\mathrm{I}},$$

where h is Planck's constant and c the velocity of light.

It is therefore possible to calculate the distances between the centres of the nuclei in each molecule, for

$$s^2 = \frac{m + m'}{mm'} \cdot \frac{h}{4\pi^2 c m_{\mathrm{H}} \delta v},$$

where m and m' are the atomic weights relative to hydrogen and m_{H} the mass of the hydrogen atom. The following table gives these distances (E. S. Imes, *Astroph. Journal*, vol. 1, p. 251, 1919). It will be seen that there are again increases in passing from F to Cl and Cl to Br, which agree closely with the increases in the radii σ of the electron shells given by the crystal and viscosity data.

	$s \times 10^8$				$\sigma \times 10^8$ (Crystals)		$\sigma' \times 10^8$ (Viscosity)	
HF	0.93		Neon	(=F)	0.65		1.17	
		0.35				*0.37*		*0.26*
HCl	1.28		Argon	(=Cl)	1.02		1.43	
		0.15				*0.15*		*0.15*
HBr	1.43		Krypton	(= Br)	1.17		1.58	
						0.18		*0.17*
HI	—		Xenon	(=I)	1.35		1.75	

The increase from fluorine to chlorine of 0.35×10^{-8} cm confirms the estimate given by crystals of 0.37×10^{-8} cm, as against the estimate 0.26×10^{-8} cm given by viscosity data.

因为很有可能它们的结构会有一定程度的重叠。第四列显示了两种估计值之间差值的恒定性，这表明，不论结果是通过黏度还是通过晶体数据测得的，在相继加入各个电子层时，原子尺寸的**增加**几乎是一样的（氖的情况例外）。兰金还进一步指出，Cl_2 在黏度性质方面的表现如同两个氩原子，其中心间距与根据晶体数据计算出来的几乎相等，对于 Br_2 与氪以及 I_2 与氙来说也是如此。

由此我们看到，晶体和黏度的测量结果都指出：(*a*) 就其外部电子层的直径来说，周期表中任一周期末尾的元素几乎是一样的；(*b*) 从一个周期过渡到下一个周期时，外部电子层尺寸会有一个确定的增加量，通过黏度估计出的这一增量的绝对数值与利用晶体数据确定的结果非常接近。

HF、HCl 和 HBr 的红外吸收光谱可以进一步证实上述测量结果。相继的吸收谱线的波数差 δv 决定了各种情况下分子的转动惯量 I，公式为

$$\delta v = \frac{h}{4\pi^2 c\mathrm{I}},$$

其中，h 为普朗克常数，c 为光速。

由此就有可能计算出每个分子中核心间的距离，因为

$$s^2 = \frac{m + m'}{mm'} \cdot \frac{h}{4\pi^2 c m_{\mathrm{H}} \delta v},$$

其中，m 和 m' 为相对于氢的原子量，m_{H} 为氢原子的质量。下面的表格给出了这些距离（艾姆斯，《天体物理学杂志》，第 1 卷，第 251 页，1919 年）。我们将看到，从 F 过渡到 Cl 以及从 Cl 过渡到 Br，该距离都有所增加，并与通过晶体数据和黏度给出的电子壳层半径 σ 的增量非常接近。

	$s \times 10^8$				$\sigma \times 10^8$（晶体）		$\sigma' \times 10^8$（黏度）	
HF	0.93		氖	(=F)	0.65		1.17	
		0.35				*0.37*		*0.26*
HCl	1.28		氩	(=Cl)	1.02		1.43	
		0.15				*0.15*		*0.15*
HBr	1.43		氪	(=Br)	1.17		1.58	
						0.18		*0.17*
HI	—		氙	(=I)	1.35		1.75	

从氟到氯的增量 0.35×10^{-8}cm 肯定了晶体测量给出的估计值 0.37×10^{-8} cm，但与黏度数据给出的估计值 0.26×10^{-8} cm 不一致。根据上述结果可知，要得到氢原子

It follows from the above that the distance between the hydrogen nucleus and the centre of an electro-negative atom to which it is attached is obtained by adding 0.26×10^{-8} cm to the radius of the electro-negative atom as given by crystal structures. The radius of the inner electron orbit, according to Bohr's theory, is 0.53×10^{-8} cm, double this value. The crystal data, therefore, predict the value $\delta v = 13.0\ \text{cm}^{-1}$ for the HI molecule, corresponding to a distance 1.61×10^{-8} cm between their atomic centres.

This evidence is interesting as indicating that the forces binding the atoms together are localised at that part of the electron shell where linking takes place.

(**107**, 107; 1921)

W. L. Bragg, H. Bell: Manchester University, March 16.

核与附着其上的负电性原子中心之间的距离，只需将该原子根据晶体结构得到的半径加上 0.26×10^{-8} cm 即可。根据玻尔理论，内部电子轨道半径为 0.53×10^{-8} cm，是这一数值的两倍。因此，晶体数据预言 HI 分子的 $\delta v = 13.0\ \text{cm}^{-1}$，与其原子中心间的距离 1.61×10^{-8} cm 相符。

这个证据是引人关注的，因为它表明，将原子束缚在一起的力就位于电子层中发生连接的地方。

（王耀杨 翻译；李芝芬 审稿）

Waves and Quanta

L. de Broglie

Editor's Note

By 1923, physicists were facing up to the implications of Planck's and Einstein's discoveries about the quantized nature of light. Although a wave phenomenon, light also seemed particulate. Bohr's model of the atom had exploited the quantization principle for electrons. Here Louis de Broglie suggests that particles with mass, such as electrons, may also have associated waves, and that this idea could put Bohr's view on firmer ground. De Broglie says that Bohr's results can be obtained by demanding that an integral number of such electron waves must fit into its orbit around the nucleus. De Broglie's suggestion was confirmed in dramatic fashion by the discovery in 1927 of electron diffraction by crystals.

THE quantum relation, energy = $h \times$ frequency, leads one to associate a periodical phenomenon with any isolated portion of matter or energy. An observer bound to the portion of matter will associate with it a frequency determined by its internal energy, namely, by its "mass at rest." An observer for whom a portion of matter is in steady motion with velocity βc, will see this frequency lower in consequence of the Lorentz–Einstein time transformation. I have been able to show (*Comptes rendus*, September 10 and 24, of the Paris Academy of Sciences) that the fixed observer will constantly see the internal periodical phenomenon in phase with a wave the frequency of which $\nu = \frac{m_0 c^2}{h\sqrt{1-\beta^2}}$ is determined by the quantum relation using the whole energy of the moving body—provided it is assumed that the wave spreads with the velocity c/β. This wave, the velocity of which is greater than c, cannot carry energy.

A radiation of frequency ν has to be considered as divided into atoms of light of very small internal mass ($<10^{-50}$ gm) which move with a velocity very nearly equal to c given by $\frac{m_0 c^2}{\sqrt{1-\beta^2}} = h\nu$. The atom of light slides slowly upon the non-material wave the frequency of which is ν and velocity c/β, very little higher than c.

The "phase wave" has a very great importance in determining the motion of any moving body, and I have been able to show that the stability conditions of the trajectories in Bohr's atom express that the wave is tuned with the length of the closed path.

The path of a luminous atom is no longer straight when this atom crosses a narrow opening; that is, diffraction. It is then *necessary* to give up the inertia principle, and we must suppose that any moving body follows always the ray of its "phase wave"; its path will then bend by passing through a sufficiently small aperture. Dynamics must undergo

波与量子

德布罗意

编者按

物理学家们在1923年的主要工作是探讨普朗克和爱因斯坦关于光的量子性质的发现。虽然光是一种波，但它似乎也具有粒子的特性。玻尔的原子模型已经用到了电子的量子化原则。路易斯·德布罗意在本文中提出像电子这样有质量的粒子可能也有相对应的波，这一观点为玻尔的理论奠定了更坚实的基础。德布罗意认为，玻尔理论中的结果可以由电子波波长取整数以适合核外电子轨道得到。1927年，德布罗意的假设就被电子在晶体中的衍射实验成功证实了。

量子关系式，即能量 = 普朗克常数 h × 频率，使人们可以把一个周期现象与任一孤立的物质或能量联系起来。与物体一起运动的观察者观测到的周期现象的频率将由物体的内部能量，即“静止质量”决定。当物体相对于观察者以 βc 的速度匀速运动时，由洛伦兹–爱因斯坦时间变换公式，观察者观测到的频率变低。我已经指出（巴黎科学院的《法国科学院院刊》，9月10日和24日），这位固定不动的观察者总能通过波的相位看到频率为 $\nu = \dfrac{m_0 c^2}{h\sqrt{1-\beta^2}}$ 的内禀周期性现象，该公式是依据量子关系式推导出来的，利用了运动物体的总能量，并假设波以 c/β 的速度传播。这个速度大于 c 的波不能携带能量。

频率为 ν 的辐射必须被看成由内部质量极小（小于 10^{-50} 克）的光原子组成，其运动速度（由公式 $\dfrac{m_0 c^2}{\sqrt{1-\beta^2}} = h\nu$ 决定）很接近于 c。这些光原子沿着频率为 ν，速度为 c/β（仅比 c 略高一点）的非实物波缓慢行进。

“相位波”对于确定任何物体的运动都是至关重要的。我已经指出，玻尔原子轨道的稳定条件表明相位波的波长应与闭合轨道的长度相匹配。

当光原子穿过一个小孔时，它的路径就不再是一条直线，即产生了衍射现象。因此**必须**摒弃惯性原理，我们必须假定，任何运动物体总沿着它的“相位波”的放

the same evolution that optics has undergone when undulations took the place of purely geometrical optics. Hypotheses based upon those of the wave theory allowed us to explain interferences and diffraction fringes. By means of these new ideas, it will probably be possible to reconcile also diffusion and dispersion with the discontinuity of light, and to solve almost all the problems brought up by quanta.

(**112**, 540; 1923)

Louis de Broglie: Paris, September 12.

射路径行进。因此，当它经过一个足够小的孔时，其轨迹会发生弯曲。如同几何光学被波动光学取代一样，实物粒子的动力学也应经历相应的变革。我们可以利用基于波动理论的假设解释干涉和衍射产生的条纹。借助这些新思想，还有可能把漫射和散射现象与光的不连续性联系起来，解决由量子引出的几乎所有问题。

（王锋 翻译；刘纯 审稿）

Australopithecus africanus: the Man-Ape of South Africa

R. A. Dart

Editor's Note

Raymond Dart's discovery of the face and brain cast of a juvenile ape-like creature in South Africa, reported here, can be marked as the beginning of the modern era of the study of fossil man. Until that date, all members of the fossil human family were either definitely apes, such as Dryopithecus, or clearly close to humans, such as Neanderthal Man or Pithecanthropus (nowadays *Homo erectus*). Having something so clearly transitional raised challenging questions about the course of human evolution. One was the very human-like teeth associated with a small, ape-like brain, at complete variance with the then-current dogma that human ancestors evolved bigger brains before human-like teeth—amplified by Piltdown Man, now known to have been a hoax.

TOWARDS the close of 1924, Miss Josephine Salmons, student demonstrator of anatomy in the University of the Witwatersrand, brought to me the fossilised skull of a cercopithecid monkey which, through her instrumentality, was very generously loaned to the Department for description by its owner, Mr. E. G. Izod, of the Rand Mines Limited. I learned that this valuable fossil had been blasted out of the limestone cliff formation—at a vertical depth of 50 feet and a horizontal depth of 200 feet—at Taungs, which lies 80 miles north of Kimberley on the main line to Rhodesia, in Bechuanaland, by operatives of the Northern Lime Company. Important stratigraphical evidence has been forthcoming recently from this district concerning the succession of stone ages in South Africa (Neville Jones, *Jour. Roy. Anthrop. Inst.*, 1920), and the feeling was entertained that this lime deposit, like that of Broken Hill in Rhodesia, might contain fossil remains of primitive man.

I immediately consulted Dr. R. B. Young, professor of geology in the University of the Witwatersrand, about the discovery, and he, by a fortunate coincidence, was called down to Taungs almost synchronously to investigate geologically the lime deposits of an adjacent farm. During his visit to Taungs, Prof. Young was enabled, through the courtesy of Mr. A. F. Campbell, general manager of the Northern Lime Company, to inspect the site of the discovery and to select further samples of fossil material for me from the same formation. These included a natural cercopithecid endocranial cast, a second and larger cast, and some rock fragments disclosing portions of bone. Finally, Dr. Cordon D. Laing, senior lecturer in anatomy, obtained news, through his friend Mr. Ridley Hendry, of another primate skull from the same cliff. This cercopithecid skull, the possession of Mr. De Wet, of the Langlaagte Deep Mine, has also been liberally entrusted by him to the Department for scientific investigation.

南方古猿非洲种：南非的人猿

达特

编者按

这篇文章报道的是雷蒙德·达特在南非发现了一件幼年类人猿的头骨，这被认为是现代人类化石研究的开端。在此之前，所有的人类化石要么确定无疑地属于猿类，比如森林古猿，要么非常接近人类，比如尼安德特人和爪哇猿人（现在称为直立人）。这块化石具有非常明确的过渡性特征，这对人类的进化历程提出了质疑。这块化石中的牙齿与人类的牙齿非常相似，脑比较小，类似于古猿，这与当时的主流观点是完全相悖的。由于皮尔当人的发现，当时人们广泛接受的主流观点认为，人类祖先脑量的增大先于类似现代人牙齿的出现，不过现在看来皮尔当人不过是场骗人的闹剧。

在 1924 年岁末年终之际，威特沃特斯兰德大学解剖学专业的学生助教约瑟芬·萨蒙斯小姐给我带来了一件猴的头骨化石。因为她的关系，这件化石的主人兰德矿业有限公司的伊佐德先生才非常慷慨地把头骨化石借给学院用于描述研究。我得知这块珍贵的化石是从垂直高 50 英尺，水平宽 200 英尺的一个石灰岩悬崖中炸出来的，地点是在汤恩。汤恩位于金伯利北部 80 英里，在通往贝专纳兰的罗得西亚市的主干线上，为北方石灰公司所有。最近出现的重要的地层学证据表明，这一地区与南非岩层年代的连续性有关（内维尔·琼斯，《皇家人类学研究院院刊》，1920 年），我想这个石灰岩沉积层可能像罗得西亚的布罗肯希尔山一样，可能包含原始人类的化石遗迹。

我立即与威特沃特斯兰德大学的地质学教授扬博士讨论了相关发现。非常巧合的是，几乎同时，他被派遣到汤恩附近的一个农场去调查石灰岩沉积层的地质情况。在汤恩，经北方石灰公司总经理坎贝尔先生的首肯，扬教授获准探查化石发现地，并从同一形成层中给我挑选了更多的化石标本。这些标本包括一个天然的猴颅内模，另一个更大的颅内模，和一些漏出部分骨头的岩石碎块。后来，一位年长的解剖学教师科登·莱恩博士通过他的朋友里德利·亨德里先生获知，同一悬崖中又发现了一件灵长类头骨。这块头骨来自兰拉格特深矿，已经由他的拥有者德威特先生委托给学院作科研之用。

Fig. 1. Norma facialis of *Australopithecus africanus* aligned on the Frankfort horizontal

The cercopithecid remains placed at our disposal certainly represent more than one species of catarrhine ape. The discovery of Cercopithecidae in this area is not novel, for I have been informed that Mr. S. Haughton has in the press a paper discussing at least one species of baboon from this same spot (Royal Society of South Africa). It is of importance that, outside of the famous Fayüm area, primate deposits have been found on the African mainland at Oldaway (Hans Reck, *Silsungsbericht der Gesellsch. Naturforsch. Freunde*, 1914), on the shores of Victoria Nyanza (C. W. Andrews, *Ann. Mag. Nat. Hist.*, 1916), and in Bechuanaland, for these discoveries lend promise to the expectation that a tolerably complete story of higher primate evolution in Africa will yet be wrested from our rocks.

In manipulating the pieces of rock brought back by Prof. Young, I found that the larger natural endocranial cast articulated exactly by its fractured frontal extremity with another piece of rock in which the broken lower and posterior margin of the left side of a mandible was visible. After cleaning the rock mass, the outline of the hinder and lower part of the facial skeleton came into view. Careful development of the solid limestone in which it was embedded finally revealed the almost entire face depicted in the accompanying photographs.

It was apparent when the larger endocranial cast was first observed that it was specially important, for its size and sulcal pattern revealed sufficient similarity with those of the chimpanzee and gorilla to demonstrate that one was handling in this instance an anthropoid and not a cercopithecid ape. Fossil anthropoids have not hitherto been recorded south of the Fayüm in Egypt, and living anthropoids have not been discovered in recent times south of Lake Kivu region in Belgian Congo, nearly 2,000 miles to the north, as the crow flies.

All fossil anthropoids found hitherto have been known only from mandibular or maxillary fragments, so far as crania are concerned, and so the general appearance of the types they

图 1. 南方古猿非洲种的前面观（已在眼耳平面上对齐）

我们手里有的这些猴化石绝不仅仅只代表狭鼻猴的一个种。在这一地区发现猴类物种并不是什么新闻，就我所知，霍顿先生的一篇已投稿的文章中讨论了至少一种来自这一地区的狒狒（南非皇家学会）。更重要的是，在著名的法尤姆地区之外，非洲大陆的奥德威（汉斯·雷克，《研究者协会会刊》，1914 年），维多利亚–尼亚萨湖岸（安德鲁斯，《自然史年鉴》，1916 年），以及贝专纳兰都有灵长类化石发现，这些发现使我有希望从我们这些化石研究中获得一个关于非洲高等灵长类动物进化的相对完整的故事，当然这需要费一番辛苦。

在处理扬教授带回的这些化石岩块时，我发现较大的那块颅内模在额骨前端破裂了，但它正好与另一块相连，其中可以清楚地看到下颌骨左侧的后下缘。清理完这些岩块，面部骨骼的后下部分轮廓就呈现出来了。然后，经过小心处理包埋着头骨的坚硬的石灰石，最终一张几乎完整的面部出现了（如图所示）。

很明显，这个大的颅内模的首次发现相当重要，因为它的尺寸和沟回形状与黑猩猩和大猩猩非常相似，这表明我们手中的并非猴类化石，而是一个类人猿。到目前为止在埃及法尤姆以南至今也没有发现类人猿化石，比属刚果的基伍湖（按照直线距离计算距北部将近 2,000 英里）以南的地区也没有关于现存类人猿的记录。

到目前为止，所有关于类人猿化石的知识，就头骨而言，仅仅来源于上颌骨或者下颌骨，这些化石所代表的类型也都不清楚。因此事实上，长满牙齿的完整面部

represented has been unknown; consequently, a condition of affairs where virtually the whole face and lower jaw, replete with teeth, together with the major portion of the brain pattern, have been preserved, constitutes a specimen of unusual value in fossil anthropoid discovery. Here, as in *Homo rhodesiensis*, Southern Africa has provided documents of higher primate evolution that are amongst the most complete extant.

Apart from this evidential completeness, the specimen is of importance because it exhibits an extinct race of apes *intermediate between living anthropoids and man*.

In the first place, the whole cranium displays *humanoid* rather than anthropoid lineaments. It is markedly dolichocephalic and leptoprosopic, and manifests in a striking degree the *harmonious relation* of calvaria to face emphasised by Pruner-Bey. As Topinard says, "A cranium elongated from before backwards, and at the same time elevated, is already in harmony by itself; but if the face, on the other hand, is elongated from above downwards, and narrows, the harmony is complete." I have assessed roughly the difference in the relationship of the glabella-gnathion facial length to the glabella-inion calvarial length in recent African anthropoids of an age comparable with that of this specimen (depicted in Duckworth's "Anthropology and Morphology", second edition, vol. I), and find that, if the glabella-inion length be regarded in all three as 100, then the glabella-gnathion length in the young chimpanzee is approximately 88, in the young gorilla 80, and in this fossil 70, which proportion suitably demonstrates the enhanced relationship of cerebral length to facial length in the fossil (Fig. 2).

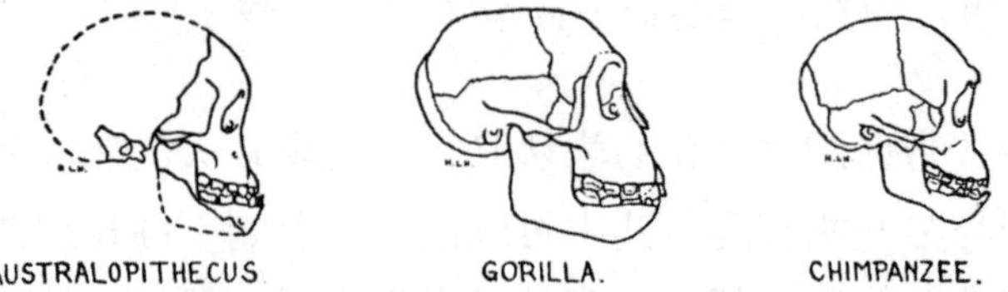

Fig. 2. Cranial form in living anthropoids of similar age (after Duckworth) and in the new fossil. For this comparison, the fossil is regarded as having the same calvarial length as the gorilla.

The glabella is tolerably pronounced, but any traces of the salient supra-orbital ridges, which are present even in immature living anthropoids, are here entirely absent. Thus the relatively increased glabella-inion measurement is due to brain and not to bone. Allowing 4 mm for the bone thickness in the inion region, that measurement in the fossil is 127 mm; *i.e.* 4 mm less than the same measurement in an adult chimpanzee in the Anatomy Museum at the University of the Witwatersrand. The orbits are not in any sense detached from the forehead, which rises steadily from their margins in a fashion amazingly human. The interorbital width is very small (13 mm) and the ethmoids are not blown out laterally as in modern African anthropoids. This lack of ethmoidal expansion causes the lacrimal fossae to face posteriorly and to lie relatively far back in the orbits, as in man. The orbits, instead of being subquadrate as in anthropoids, are almost circular, furnishing an orbital index of 100, which is well within the range of human variation (Topinard,

和下颌，连同脑结构的主要部分，被一起保留了下来，这构成了类人猿发现史上一件不寻常的宝贵的标本。在这里，如同罗德西亚人，南非提供了有关高等灵长类动物进化的现存最完整的资料。

抛开证据的完整性不谈，这个标本的重要性在于它揭示了一种已经灭绝的猿类，**介于现存类人猿与人类之间**的中间类型。

首先，整个头骨轮廓显示的是**人类的**特征，而非类人猿的。其显著特征：颅长，面窄，这些都非常符合普瑞纳贝（人类学家）强调的颅顶与脸的**和谐关系**。如托皮纳尔所说："头盖骨从前向后延伸，同时抬高，本身已经是和谐的；但另一方面如果面部 也从下向上延伸，并且变窄，就和谐完整了。"在粗略估计了这个标本和与之年龄相近的近代非洲类人猿的面长（眉间至颔下点）与颅长（眉间至枕骨隆突）比例的差异后（详见迪克沃斯的《人类学与形态学》，第 2 版，第 1 卷），我发现，如果三个物种的颅长都为 100，那么年轻黑猩猩的面长为 88，年轻大猩猩的面长为 80，这个化石标本的面长则为 70。这一比例很好地证明了化石中颅长与面长之比不断增加的关系（图 2）。

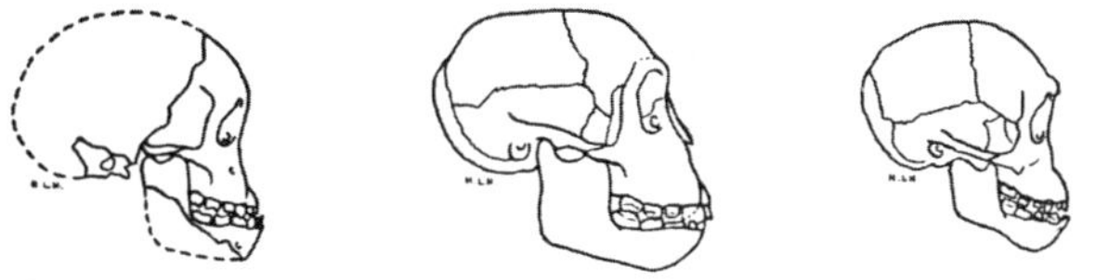

图 2. 新发现化石及年岁接近的现存类人猿（根据迪克沃斯的著作）的颅骨形状。通过这个比较可以看到，新发现化石的颅盖骨与大猩猩的一样长。

在这个化石标本中，眉间还算突出，然而突出的眶上脊却完全缺失，这一特征即使在现存的幼年类人猿中也是存在的。这就是说相对增长的眉间至枕外隆突之间的距离是由于脑量增长而并非骨骼生长所致。留出 4 毫米作为枕外隆突区域的骨骼厚度，这个化石的颅长为 127 毫米，也只比威特沃特斯兰德大学解剖学博物馆中成年黑猩猩的颅长少 4 毫米。其眼眶没有任何从前额分离的迹象，而是从边缘平稳突出，与人类的眼眶模式极其相似。眼间宽度很小（13 毫米），而筛骨也没有像现代非洲类人猿一样从侧面鼓起。由于筛骨没有膨大，使得其泪腺沟朝后，像人一样位于眼眶相对较后的位置。眼眶几乎是圆形的，不像类人猿的方形。设定眼眶指数为 100，则它完全位于人类变化范围内（托皮纳尔，《人类学》）。颧骨，颧弓，上颌骨和下颌骨，所有这些显示的是精巧的人类的特征。面部突颌度相对轻微，弗劳尔颌指数

"Anthropology"). The malars, zygomatic arches, maxillae, and mandible all betray a delicate and humanoid character. The facial prognathism is relatively slight, the gnathic index of Flower giving a value of 109, which is scarcely greater than that of certain Bushmen (Strandloopers) examined by Shrubsall. The nasal bones are not prolonged below the level of the lower orbital margins, as in anthropoids, but end above these, as in man, and are incompletely fused together in their lower half. Their maximum length (17 mm) is not so great as that of the nasals in *Eoanthropus dawsoni*. They are depressed in the median line, as in the chimpanzee, in their lower half, but it seems probable that this depression has occurred post-mortem, for the upper half of each bone is arched forwards (Fig. 1). The nasal aperture is small and is just wider than it is high (17 mm × 16 mm). There is no nasal spine, the floor of the nasal cavity being continuous with the anterior aspect of the alveolar portions of the maxillae, after the fashion of the chimpanzee and of certain New Caledonians and negroes (Topinard, *loc. cit.*).

In the second place, the dentition is *humanoid* rather than anthropoid. The specimen is juvenile, for the first permanent molar tooth only has erupted in both jaws on both sides of the face; *i.e.* it corresponds anatomically with a human child of six years of age. Observations upon the milk dentition of living primates are few, and only one molar tooth of the deciduous dentition in one fossil anthropoid is known (Gregory, "The Origin and Evolution of the Human Dentition", 1920). Hence the data for the necessary comparisons are meagre, but certain striking features of the milk dentition of this creature may be mentioned. The tips of the canine teeth transgress very slightly (0.5–0.75 mm) the general margin of the teeth in each jaw, *i.e.* very little more than does the human milk canine. There is no diastema whatever between the premolars and canines on either side of the lower jaw, such as is present in the deciduous dentition of living anthropoids; but the canines in this jaw come, as in the human jaw, into alignment with the incisors (Gregory, *loc. cit.*). There is a diastema (2 mm on the right side, and 3 mm on the left side) between the canines and lateral incisors of the upper jaw; but seeing, first, that the incisors are narrow, and, secondly, that diastemata (1 mm–1.5 mm) occur between the central incisors of the upper jaw and between the medial and lateral incisors of both sides in the lower jaw, and, thirdly, that some separation of the milk teeth takes place even in mankind (Tomes, "Dental Anatomy", seventh edition) during the establishment of the permanent dentition, it is evident that the diastemata which occur in the upper jaw are small. The lower canines, nevertheless, show wearing facets both for the upper canines and for the upper lateral incisors.

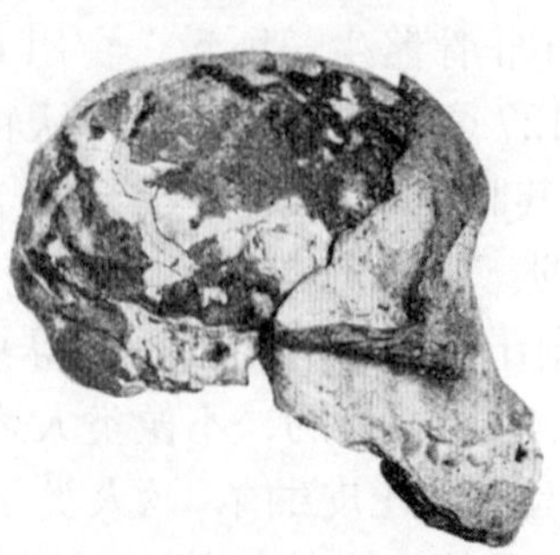

Fig. 3. Norma lateralis of *Australopithecus africanus* aligned on the Frankfort horizontal.

为 109，这一指数几乎不大于舒本萨尔检测过的布希曼人。鼻骨向下延伸不低于眶下沿水平，这点像类人猿，而止于眶下沿之上，这点又像人类，并且下半部分不完全地融合在一起。鼻骨的最大长度为 17 毫米，不像道森曙人的那么大。鼻骨下半部分从中线部位下陷，这点像黑猩猩，然而这种下陷也可能是发生在死后，因为其上半部分的每块骨头都向前拱起（图 1）。鼻孔小，其宽度刚好大于其高度（17 毫米 ×16 毫米）。没有鼻棘，鼻腔底面与上颌齿槽部的前面相连，这与黑猩猩和一些新苏格兰人及黑人的样式相仿（如前面托皮纳尔所述）。

其次，其齿系是**人类的**而非类人猿的。从两侧第一恒臼齿刚刚萌出可以判断这一标本还是幼年，从解剖学上说大约相当于人类的 6 岁儿童。对现存灵长类乳齿系的研究还很少，而化石类人猿中也只有一个乳臼齿的记录（格雷戈里，《人类齿系的起源与进化》，1920 年）。因此对乳齿作必要比较的数据太少，但是这个物种乳齿系显著的特征还是值得一提的。在两颌中，犬齿尖端略微超出整个齿列（0.5 ~ 0.75 毫米），例如，只比人类的犬齿突出一点点。下颌两侧的前臼齿和犬齿之间没有齿隙，现存类人猿的乳齿也是如此；然而其犬齿与门齿排列在一起，这点与人类一样（格雷戈里，见上述引文）。上颌犬齿与侧门齿之间有齿隙（右侧为 2 毫米，左侧为 3 毫米），但应注意到：（1）门齿狭窄；（2）上颌中门齿之间以及下颌两侧中门齿与侧门齿之间都有间隙裂（1~1.5 毫米）；（3）在恒齿形成过程中，即使在人类中，乳齿分开的现象也有发生（托姆斯，《牙齿解剖学》，第 7 版）。显然上颌的牙间隙比较窄。然而，下犬齿有相对于上犬齿以及上侧门齿的磨损面。

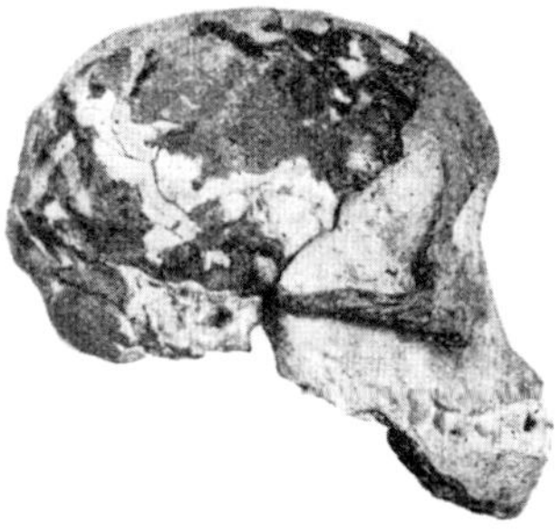

图 3. 南方古猿非洲种的侧面观（已在眼耳平面上对齐）

The incisors as a group are irregular in size, tend to overlap one another, and are almost vertical, as in man; they are not symmetrical and well spaced, and do not project forwards markedly, as in anthropoids. The upper lateral incisors do project forwards to some extent and perhaps also do the upper central incisors very slightly, but the lateral lower incisors betray no evidence of forward projection, and the central lower incisors are not even vertical as in most races of mankind, but are directed slightly backwards, as *sometimes* occurs in man. Owing to these remarkably human characters displayed by the deciduous dentition, when contour tracings of the upper jaw are made, it is found that the jaw and the teeth, as a whole, take up a parabolic arrangement comparable only with that presented by mankind amongst the higher primates. These facts, together with the more minute anatomy of the teeth, will be illustrated and discussed in the memoir which is in the process of elaboration concerning the fossil remains.

In the third place, the mandible itself is *humanoid* rather than anthropoid. Its ramus is, on the whole, short and slender as compared with that of anthropoids, but the bone itself is more massive than that of a human being of the same age. Its symphyseal region is virtually complete and reveals anteriorly a more vertical outline than is found in anthropoids or even in the jaw of Piltdown man. The anterior symphyseal surface is scarcely less vertical than that of Heidelberg man. The posterior symphyseal surface in living anthropoids differs from that of modern man in possessing a pronounced posterior prolongation of the lower border, which joins together the two halves of the mandible, and so forms the well-known *simian shelf* and above it a deep genial impression for the attachment of the tongue musculature. In this character, *Eoanthropus dawsoni* scarcely differs from the anthropoids, especially the chimpanzee; but this new fossil betrays no evidence of such a shelf, the lower border of the mandible having been massive and rounded after the fashion of the mandible of *Homo heidelbergensis*.

Fig. 4. Norma basalis of *Australopithecus africanus* aligned on the Frankfort horizontal.

其门齿大小不规则，倾向于彼此重叠，几乎垂直，这点像人类；它们呈不对称分布，空间分布也不甚合理，没有显著向前突出，这点像类人猿。上侧门齿的确有一定程度的向前突出，上中门齿似乎也略为有点向前突出；然而下侧门齿丝毫没有向前突出的迹象，并且下中门齿不是像大多数人种中那样整齐地垂直向上，而是像人类中**有时**发生的那样略为向后倾斜。基于乳齿系所显示的这些显著的人类特征，当绘出上颌的轮廓线之后，发现颌与牙齿在整体上呈抛物线的排列方式，仅能与高等灵长类动物中的人类相比拟。这些事实，以及更多的牙齿微细解剖特征将在即将发表的有关化石的论文中阐述和讨论。

第三，下颌骨本身是**人类的**而非类人猿的。从整体上看，下颌支短而纤细，与类人猿相仿，但骨头本身比同龄的人类的大。其联合区近乎完整，并且从前面显示比类人猿甚至比皮尔当人更为垂直的轮廓。其前端联合面几乎与海德堡人的一样垂直。现存类人猿下颌骨后联合面与现代人类的区别在于其下沿明显地向后延伸，将下颌骨后部连接在一起形成著名的**猿板**结构，这一结构上的深印迹在于舌头肌肉组织的附着处。在这一特征上，道森曙人与类人猿，尤其是黑猩猩几乎没有差别，但在这一新化石标本中没有找到这一结构，其下颌骨下沿粗壮而圆隆，类似于海德堡人的下颌骨。

图 4. 南方古猿非洲种的底面观（已在眼耳平面上对齐）

That hominid characters were not restricted to the face in this extinct primate group is borne out by the relatively forward situation of the foramen magnum. The position of the basion can be assessed within a few millimetres of error, because a portion of the right exoccipital is present alongside the cast of the basal aspect of the cerebellum. Its position is such that the basi-prosthion measurement is 89 mm, while the basi-inion measurement is at least 54 mm. This relationship may be expressed in the form of a "head-balancing" index of 60.7. The same index in a baboon provides a value of 41.3, in an adult chimpanzee 50.7, in Rhodesian man 83.7, in a dolichocephalic European 90.9, and in a brachycephalic European 105.8. It is significant that this index, which indicates in a measure the poise of the skull upon the vertebral column, points to the assumption by this fossil group of an attitude appreciably more erect than that of modern anthropoids. The improved poise of the head, and the better posture of the whole body framework which accompanied this alteration in the angle at which its dominant member was supported, is of great significance. It means that a greater reliance was being placed by this group upon the feet as organs of progression, and that the hands were being freed from their more primitive function of accessory organs of locomotion. Bipedal animals, their hands were assuming a higher evolutionary role not only as delicate tactual, examining organs which were adding copiously to the animal's knowledge of its physical environment, but also as instruments of the growing intelligence in carrying out more elaborate, purposeful, and skilled movements, and as organs of offence and defence. The latter is rendered the more probable, in view, first, of their failure to develop massive canines and hideous features, and, secondly, of the fact that even living baboons and anthropoid apes can and do use sticks and stones as implements and as weapons of offence ("Descent of Man", p. 81 *et seq.*).

Lastly, there remains a consideration of the endocranial cast which was responsible for the discovery of the face. The cast comprises the right cerebral and cerebellar hemispheres (both of which fortunately meet the median line throughout their entire dorsal length) and the anterior portion of the left cerebral hemisphere. The remainder of the cranial cavity seems to have been empty, for the left face of the cast is clothed with a picturesque lime crystal deposit; the vacuity in the left half of the cranial cavity was probably responsible for the fragmentation of the specimen during the blasting. The cranial capacity of the specimen may best be appreciated by the statement that the length of the cavity could not have been less than 114 mm, which is 3 mm greater than that of an adult chimpanzee in the Museum of the Anatomy Department in the University of the Witwatersrand, and only 14 mm less than the greatest length of the cast of the endocranium of a gorilla chosen for casting on account of its great size. Few data are available concerning the expansion of brain matter which takes place in the living anthropoid brain between the time of eruption of the first permanent molars and the time of their becoming adult. So far as man is concerned, Owen ("Anatomy of Vertebrates", vol. III) tells us that "The brain has advanced to near its term of size at about ten years, but it does not usually obtain its full development till between twenty and thirty years of age." R. Boyd (1860) discovered an increase in weight of nearly 250 grams in the brains of male human beings after they had reached the age of seven years. It is therefore reasonable to believe that the adult forms typified by our present specimen possessed brains which were larger than

在这一灭绝的灵长类中这样的人类特征不仅仅局限于面部，其枕骨大孔处于相对朝前的位置也是一个很好的例证。由于沿小脑基部轮廓的外侧处的一部分右枕骨保存了下来，对颅底点的测量可以控制在几毫米的误差之内。它的位置可以参考两个位置点：即颅底点距上颌齿槽前缘点 89 毫米，颅底点至枕外隆突至少 54 毫米。这可以用头“平衡指数”的方式表达为 60.7。狒狒这一指数为 41.3，成年黑猩猩为 50.7，罗得西亚人为 83.7，长头欧洲人为 90.9，短头欧洲人为 105.8。很重要的一点是，这一指数反映的是头骨在脊柱上的姿态，从这一指数可以推论，化石标本所代表的类群表现出比现代类人猿更直立的一种姿态。头部姿势的提升，以及伴随这一角度改变而来的整个身体框架的姿态的改进，对于生物体本身来说太重要了。这意味着这一类群的生物更多地依赖于脚作为身体行进的器官，而手则被解放出来，其功能不再只是原始的移动器官的附属品。两足动物的手被认为是一种高等的进化，因为手已经不仅仅只是一个精妙的触觉上的感知器官，使动物获得更丰富的物理环境知识，更重要的是手已经成为提高智力的一种工具，能够承担更精细的，更有目的性的，更有技巧性的运动，并作为防御以及进攻的器官。而后者的可能性居多，这些体现在，首先，没有形成巨大的犬齿和丑陋的面貌特征，其次，一个不容忽略的事实是，即使现存的狒狒和类人猿也能够并确实在使用树枝和石头作为工具以及攻击的武器（《人类的由来》，第 81 页）。

最后，除了考虑面部特征外，对于颅内模型还有一些新的考虑。这个标本包含右大脑半球和小脑半球（很幸运两者都贯穿整个背侧长度达到中线位置）以及左大脑半球的前面部分。化石标本的左面部包裹了一层别致的石灰石晶体沉积物，由此推断颅腔的其余部分可能是空的；而颅腔左半部分的中空可能是因为爆破中标本碎裂了。标本的颅容量可以通过腔体长度计算，这一标本的颅腔长度不小于 114 毫米，这比威特沃特斯兰德大学解剖学博物馆中成年黑猩猩的颅腔长度长 3 毫米，而比一只由于其体形巨大被挑选出来用于制作颅内模型的大猩猩的颅腔最大长度仅仅少 14 毫米。目前，在长出第一恒臼齿到成为成体这段时间，有关现存类人猿脑量扩张的数据还基本没有。就人类而言，欧文（《脊椎动物解剖学》，第 3 卷）指出，“在大约 10 岁时脑能够发育到接近成年的体积大小，而通常要到 20~30 岁才能发育完全。”博伊德（1860 年）发现 7 岁以后人类男性个体的脑重量将增长近 250 克。因此有理由相信，目前我们手头的这个化石标本代表的成年脑量，应该比这个幼年标本的大，如果不超过的话，也应该等于一个发育完全的成年大猩猩的脑量。

that of this juvenile specimen, and equalled, if they did not actually supersede, that of the gorilla in absolute size.

Whatever the total dimensions of the adult brain may have been, there are not lacking evidences that the brain in this group of fossil forms was distinctive in type and was an instrument of greater intelligence that that of living anthropoids. The face of the endocranial cast is scarred unfortunately in several places (cross-hatched in the dioptographic tracing—see Fig. 5). It is evident that the relative proportion of cerebral to cerebellar matter in this brain was greater than in the gorilla's. The brain does not show that general pre- and post-Rolandic flattening characteristic of the living anthropoids, but presents a rounded and well-filled-out contour, which points to a symmetrical and balanced development of the faculties of associative memory and intelligent activity. The pithecoid type of parallel sulcus is preserved, but the sulcus lunatus has been thrust backwards towards the occipital pole by a pronounced general bulging of the parieto-temporo-occipital association areas.

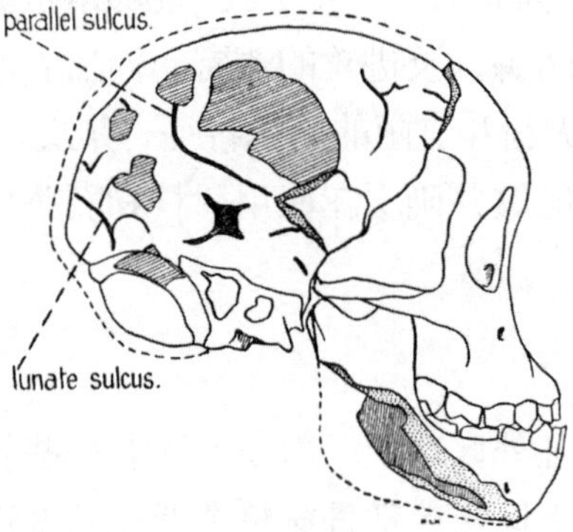

Fig. 5. Dioptographic tracing of *Australopithecus africanus* (right side), $\times\frac{1}{3}$.

To emphasise this matter, I have reproduced (Fig. 6) superimposed coronal contour tracings taken at the widest part of the parietal region in the gorilla endocranial cast and in this fossil. Nothing could illustrate better the mental gap that exists between living anthropoid apes and the group of creatures which the fossil represents than the flattened atrophic appearance of the parietal region of the brain (which lies between the visual field on one hand, and the tactile and auditory fields on the other) in the former and its surgent vertical and dorso-lateral expansion in the latter. The expansion in this area of the brain is the more significant in that it explains the posterior *humanoid* situation of the sulcus lunatus. It indicates (together with the narrow interorbital interval and human characters of the orbit) the fact that this group of beings, having acquired the faculty of stereoscopic vision, had profited beyond living anthropoids by setting aside a relatively much larger area of the cerebral cortex to serve as a storehouse of information concerning their objective environment as its details were simultaneously revealed to the senses of vision and touch, and also of hearing. They possessed to a degree unappreciated by living anthropoids the use of their hands and ears and the consequent faculty of associating with the colour,

无论成体脑的大小如何，我们都不难发现形成这种化石的种群的脑的类型，与现存类人猿相比不仅是明显不同的，而且是更加高级的。不幸的是颅腔标本表面有多处破损（图 5，用交叉平行线画出的阴影部分）。显然，这个颅腔中大脑与小脑的相对比例高于大猩猩。这个脑没有表现出一般现存类人猿的罗蓝氏区前后扁平的模式特征，而是显示出一个变圆的、更充盈的轮廓，表明与记忆和智力活动相关的官能得到了对称的、平衡的发展。类人猿类型的平行沟得以保存下来，然而由于顶骨–颞骨–枕骨联合区域显著的整体突出使得月状沟被向后推向枕侧。

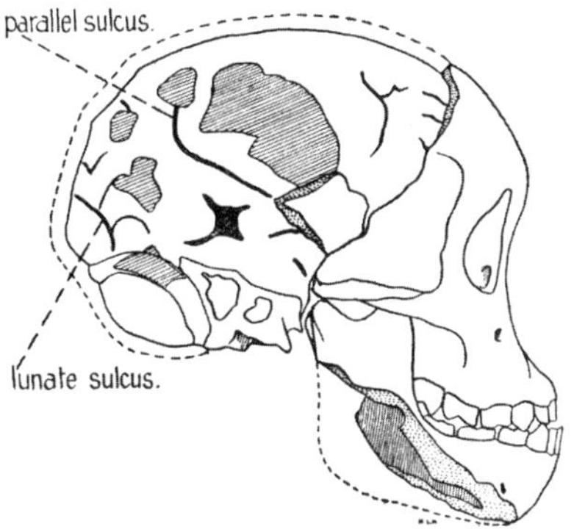

图 5. 南方古猿非洲种透视素描图（右侧），× $\frac{1}{3}$

为了强调这一点，我绘制了大猩猩和这个化石标本的颅内模顶区最宽处的轮廓叠加线（图 6）。大猩猩脑顶区（处于视觉区与触觉区及听觉区之间）扁平、萎缩，而化石标本相应区域陡立并向背部膨大，最好地说明了现存类人猿与化石标本代表的生物类群之间的智力差异。更重要的是这一脑区的膨大揭示了月状沟靠后这种**人类化**情况的原因。这一特征，加上窄的眶间隔，具有人类特征的眼眶，表明这一类群的生物已经获得了立体视觉的能力，这一超越现存类人猿的能力得益于将一块相对较大的大脑皮层区域设置为同时向视觉、触觉以及听觉传递有关客观环境细节的信息储藏库。一定程度上它们拥有类人猿不能相比的支配它们手、耳的使用的能力，并随之而来获知有关颜色、形状和物体的总体面貌、重量、质地、弹性和柔韧性，以及物体发出的声音所代表的意义的能力。换言之，与近代的猿类相应的器官相比，更有意识性和目的性地用它们的眼睛看，用它们的耳朵听，用它们的手进行操作。

form, and general appearance of objects, their weight, texture, resilience, and flexibility, as well as the significance of sounds emitted by them. In other words, their eyes saw, their ears heard, and their hands handled objects with greater meaning and to fuller purpose than the corresponding organs in recent apes. They had laid down the foundations of that discriminative knowledge of the appearance, feeling, and sound of things that was a necessary milestone in the acquisition of articulate speech.

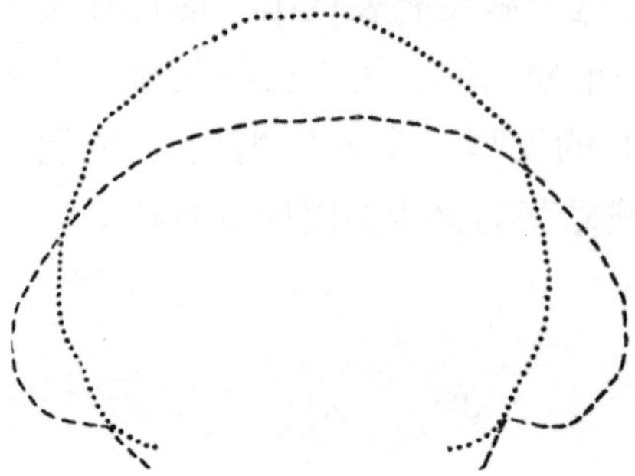

Fig. 6. Contour tracings of coronal sections through the widest part of the parietal region of the endocranial casts in Australopithecus ... and in a gorilla....

There is, therefore, an ultra-simian quality of the brain depicted in this immature endocranial cast which harmonises with the ultra-simian features revealed by the entire cranial topography and corroborates the various inferences drawn therefrom. The two thousand miles of territory which separate this creature from its nearest living anthropoid cousins is indirect testimony to its increased intelligence and mastery of its environment. It is manifest that we are in the presence here of a pre-human stock, neither chimpanzee nor gorilla, which possesses a series of differential characters not encountered hitherto in any anthropoid stock. This complex of characters exhibited is such that it cannot be interpreted as belonging to a form ancestral to any living anthropoid. For this reason, we may be equally confident that there can be no question here of a primitive anthropoid stock such as has been recovered from the Egyptian Fayüm. Fossil anthropoids, varieties of Dryopithecus, have been retrieved in many parts of Europe, Northern Africa, and Northern India, but the present specimen, despite its youth, cannot be confused with anthropoids having the dryopithecid dentition. Other fossil anthropoids from the Siwalik hills in India (Miocene and Pliocene) are known which, according to certain observers, may be ancestral to modern anthropoids and even to man.

Whether our present fossil is to be correlated with the discoveries made in India is not yet apparent; that question can only be solved by a careful comparison of the permanent molar teeth from both localities. It is obvious, meanwhile, that it represents a fossil group distinctly advanced beyond living anthropoids in those two dominantly human characters of facial and dental recession on one hand, and improved quality of the brain on the other. Unlike Pithecanthropus, it does not represent an ape-like man, a caricature of precocious hominid failure, but a creature well advanced beyond modern anthropoids in just those characters, facial and cerebral, which are to be anticipated in an extinct

它们已经具有了辨别事物外貌、触感以及声音的基础，这对于语言能力（能够清晰发音）的获得是一个重要的里程碑。

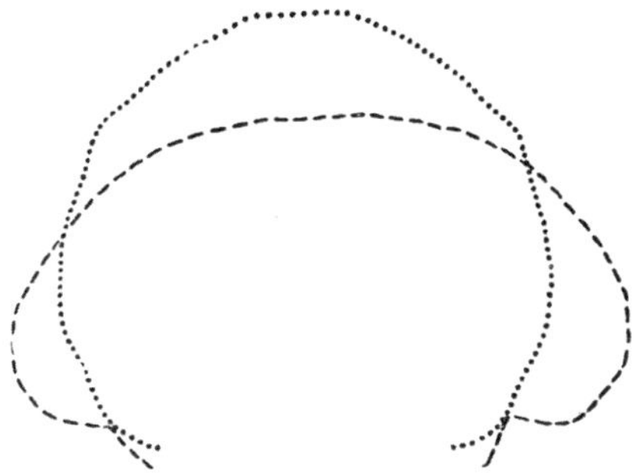

图 6. 南方古猿非洲种和大猩猩颅内模顶区最宽处的冠状面轮廓图

因此，这个未成熟的颅内模描绘的是一个具有超类人猿性质的脑，这与整个头骨解剖形态显示的超类人猿特征是一致的，并进一步证实了以此而来的各种推论。这个生物和它最近的现存类人猿兄弟们相隔了 2,000 英里的区域，这也间接证实了其智力的提高以及对环境的掌握。显然，摆在我们面前的是一个前人类类群，既不是黑猩猩也不是大猩猩，它已经拥有了一系列至今为止任何类人猿都不具有的特征。这些特征的复杂性表明，它不能被认为是任何现存类人猿的祖先。因此，我们同样可以确信，如同在埃及法尤姆所发现的，这里毫无疑问有一个原始类人猿类群的存在。在欧洲、南非、北印度的多个地区曾发现多种森林古猿的化石类人猿，然而我们的标本，尽管年轻，也不可能与具有森林古猿齿系的类人猿相混淆。根据其他人的观察，在印度（中新世和上新世）西瓦利克山脉发现的一些其他化石类人猿可能是现代类人猿甚至人类的祖先。

我们目前的化石是否与印度的发现相互关联还不清楚，这一问题只有通过仔细比较来自两地的恒臼齿才能解决。然而，很显然的是它所代表的化石类群毫无疑问超越了现存类人猿，主要表现在，一方面已经具有显著的人类特征的面部和齿系，另一方面脑质的提高。它不像猿人属代表的是与猿相像的人类，一种早熟的人类的失败类型，而是代表一种从面部及脑部特征上已经远远超过当代类人猿的生物，这些面部及脑部特征被认为可能是人类及其猿类祖先的已经灭绝的连接类型中应有的

link between man and his simian ancestor. At the same time, it is equally evident that a creature with anthropoid brain capacity, and lacking the distinctive, localised temporal expansions which appear to be concomitant with and necessary to articulate man, is no true man. It is therefore logically regarded as a man-like ape. I propose tentatively, then, that a new family of *Homo-simiadae* be created for the reception of the group of individuals which it represents, and that the first known species of the group be designated *Australopithecus africanus*, in commemoration, first, of the extreme southern and unexpected horizon of its discovery, and secondly, of the continent in which so many new and important discoveries connected with the early history of man have recently been made, thus vindicating the Darwinian claim that Africa would prove to be the cradle of mankind.

It will appear to many a remarkable fact that an ultra-simian and pre-human stock should be discovered, in the first place, at this extreme southern point in Africa, and, secondly, in Bechuanaland, for one does not associate with the present climatic conditions obtaining on the eastern fringe of the Kalahari desert an environment favourable to higher primate life. It is generally believed by geologists (*vide* A. W. Rogers, "Post-Cretaceous Climates of South Africa", *South African Journal of Science*, vol. XIX, 1922) that the climate has fluctuated within exceedingly narrow limits in this country since Cretaceous times. We must therefore conclude that it was only the enhanced cerebral powers possessed by this group which made their existence possible in this untoward environment.

In anticipating the discovery of the true links between the apes and man in tropical countries, there has been a tendency to overlook the fact that, in the luxuriant forests of the tropical belts, Nature was supplying with profligate and lavish hand an easy and sluggish solution, by adaptive specialisation, of the problem of existence in creatures so well equipped mentally as living anthropoids are. For the production of man a different apprenticeship was needed to sharpen the wits and quicken the higher manifestations of intellect—a more open veldt country where competition was keener between swiftness and stealth, and where adroitness of thinking and movement played a preponderating role in the preservation of the species. Darwin has said, "no country in the world abounds in a greater degree with dangerous beasts than Southern Africa", and, in my opinion, Southern Africa, by providing a vast open country with occasional wooded belts and a relative scarcity of water, together with a fierce and bitter mammalian competition, furnished a laboratory such as was essential to this penultimate phase of human evolution.

In Southern Africa, where climatic conditions appear to have fluctuated little since Cretaceous times, and where ample dolomitic formations have provided innumerable refuges during life, and burial-places after death, for our troglodytic forefathers, we may confidently anticipate many complementary discoveries concerning this period in our evolution.

In conclusion, I desire to place on record my indebtedness to Miss Salmons, Prof. Young, and Mr. Campbell, without whose aid the discovery would not have been made; to

特征。同时，同样可以确信，一个生物只具有与类人猿同样大小的脑量并且缺少局部的颞区扩张（而这些是成为具有语言能力的人所必需的），它还不是真正意义上的人。因此，从逻辑上讲它应当被称为像人的猿。这样我试探性地建议，创立一个名为人猿科的新科来接纳标本代表的生物类群，并且将这一类群第一个已知的种命名为南方古猿非洲种，以纪念：第一，其极南的发现地和出人意料的地层，第二，其所在的大陆。在这一大陆上近来有非常多的重要发现与人类的早期历史相关，这将为达尔文所主张的非洲是人类的摇篮这一提议提供依据。

鉴于卡拉哈里沙漠东缘现在的气候条件无法与适宜高等灵长类生存的环境条件相联系，将会发生的一个毋庸置疑的事实是，超级猿类和前人类类群的发现应当在，第一，非洲大陆的极南端，第二，贝专纳兰。地质学家普遍承认（参见罗杰斯的文章《后白垩纪南非的气候》，《南非科学杂志》，第 19 卷，1922 年），从白垩纪以来这片大陆的气候只是在一个非常小的范围内波动。因此我们必然能推出，只有这个大脑能力已经提高的类群才能在这种不利的环境下存在。

在预期位于热带地区的国家将会发现连接猿类与人类真正的“接环”时，存在一种忽略如下事实的倾向，那就是在热带繁茂的森林中，大自然用其多产而慷慨的手，通过适应性的特化，为具有类人猿智力程度的生物们的生存问题提供了一个简单但却迟缓的解决方案。为了人类的产生，需要一个不同的学徒期以磨砺它们的智力，提高它们的理解力。然而正是一片更为广阔的草原地带为之提供了必要的条件，在这里迅捷与隐秘之间的竞争更加尖锐，在这里敏捷的思维和运动在物种生存中扮演了极为重要的角色。达尔文说过，“世界上没有任何国家比南非拥有更多的危险野兽”，以我的观点，南非，由于拥有广阔的空旷地带和稀少的林带，水源相对匮乏，加上哺乳动物之间的残忍严酷的竞争，提供了对于人类进化史上这倒数第二个阶段来说至关重要的实验室。

南非，这里的气候从白垩纪以来波动就很小，而丰富的白云岩层则为我们的穴居人祖先提供了无数的生活居所和死亡墓地，我们可以自信地预期，还将会继续发现有关这一时期的更多的人类进化的补充证据。

最后，我要感谢萨蒙斯小姐、扬教授和坎贝尔先生，没有他们的帮助就没有这些发现；感谢莱恩·理查森先生提供的照片；感谢莱恩博士和我实验室的同事们的热

Mr. Len Richardson for providing the photographs; to Dr. Laing and my laboratory staff for their willing assistance; and particularly to Mr. H. Le Helloco, student demonstrator in the Anatomy Department, who has prepared the illustrations for this preliminary statement.

(**115**, 195-199; 1925)

Prof. Raymond A. Dart: University of the Witwatersrand, Johannesburg, South Africa.

情帮助；特别要感谢解剖学学院的学生管理员勒埃洛克先生为这个初步报告准备了插图。

（刘晓辉 翻译；赵凌霞 审稿）

The Fossil Anthropoid Ape from Taungs

Editor's Note

Raymond Dart's announcement of *Australopithecus africanus* in *Nature* the previous week, and his assertion that it was intermediate between apes and humans, was greeted with a chorus of very faint praise in this quartet of letters from the anthropological establishment. All welcomed the discovery, but preferred to consider Australopithecus as very definitely an ape whose human-like features could be attributed to the fact that the fossil was of a child, whose adult form could not yet be discerned. The tone of all letters was courteous—except for Smith Woodward's criticism of the "barbarous" merger of Latin and Greek to create "Australopithecus".

THE discovery of fossil remains of a "man ape" in South Africa raises many points of great interest for those who are studying the evolution of man and of man-like apes. No doubt when Prof. Dart publishes his full monograph of his discovery, he will settle many points which are now left open, but from the facts he has given us, and particularly from the accurate drawing of the endocranial cast and skull in profile, it is even now possible for an onlooker to assess the importance of his discovery. I found it easy to enlarge the profile drawing just mentioned to natural size and to compare it with corresponding drawings of the skulls of children and of young apes. When this is done, the peculiarities of Australopithecus become very manifest.

Prof. Dart regrets he has not access to literature which gives the data for gauging the age of young anthropoids. In the specimen he has discovered and described, the first permanent molar teeth are coming into use. Data which I collected 25 years ago show that these teeth reach this stage near the end of the 4th year, two years earlier than is the rule in man and two years later than is the rule in the higher monkeys. In evolution towards a human form there is a tendency to prolong the periods of growth. Man and the gorilla have approximately the same size of brain at birth; the rapid growth of man's brain continues to the end of the 4th year; in the gorilla rapid growth ceases soon after birth.

Prof. Dart recognises the many points of similarity which link Australopithecus to the great anthropoid apes—particularly to the chimpanzee and gorilla. Those who are familiar with the facial characters of the immature gorilla and of the chimpanzee will recognise a blend of the two in the face of Australopithecus, and yet in certain points it differs from both, particularly in the small size of its jaws.

In size of brain this new form is not human but anthropoid. In the 4th year a child has reached 81 percent of the total size of its brain; at the same period a young gorilla has obtained 85 percent of its full size, a chimpanzee 87 percent. From Prof. Dart's accurate diagrams one estimates the brain length to have been 118 mm—a dimension common in

汤恩发现的类人猿化石

编者按

一周前，雷蒙德·达特在《自然》上宣布了南方古猿非洲种的发现，他断言这是介于古猿与人类之间的中间类型。在下述4封信中，人类学研究领域的重要人物一致对此给出了明褒实贬的评价。所有人都表示欢迎这个发现，但同时也提出更倾向于认为南方古猿非常明显就是古猿，其与人类相像的特征可能是因为这块化石属于幼年的个体，其成年的形态还不能被辨认出来。史密斯·伍德沃德批评道，造出“南方古猿”这个名字属于“野蛮”拼合拉丁语和希腊语，除此之外，这几封信的语气都还算客气。

南非发现的“人猿”化石激起了那些正在研究人类和类人猿进化的人们的巨大兴趣。当达特教授发表关于其发现的详尽专著时，他无疑将解决现在尚待进一步讨论的许多问题，但是从他向我们展示的事实来看，尤其是从他对颅腔模型和头骨剖面精确的绘图来看，即使是现在，旁观者也可以掂量出其发现的重要性。我发现很容易将刚刚提到的剖面图放大到真实大小，也很容易将其与相应的人类的孩子和幼年猿类的头骨绘图进行比较。当这样做的时候，南方古猿的特性就会变得十分明显了。

达特教授遗憾的是自己没能使用那些提供鉴定幼年猿类年龄资料的文献。在由他发现并进行描述的这个标本中，第一恒臼齿正开始被使用。我在25年前搜集的数据表明这些牙齿达到这一阶段应该是接近4岁末的时候，比人类早了两年，比高等猴子则晚了两年。在向人类形式进化的过程中，存在发育期延长的趋势。人类和大猩猩在出生时具有大约相同大小的脑；人脑的快速生长一直持续到4岁末；而大猩猩的脑的快速生长在出生后不久就停止了。

达特教授识别出了将南方古猿和大型类人猿联系起来的许多相似点——尤其是将南方古猿与黑猩猩和大猩猩联系起来的相似之处。对未成年大猩猩和黑猩猩的面部特征熟悉的人会在南方古猿的脸上看到二者的混合体，然而有些方面南方古猿与这两者都不同，尤其是其颌骨较小。

在脑尺寸方面，这个新类型与人类不同，是类人猿式的。儿童的脑在4岁时达到了其脑总尺寸的81%；而同样年龄的幼年大猩猩已经达到了全部尺寸的85%，黑猩猩则达到了87%。根据达特教授精确的图表可以估计出脑的长度已经达到了

the brains of adult and also juvenile gorillas. The height of the brain above the ear-holes also corresponds in both Australopithecus and the gorilla—about 70 mm. But in width, as Prof. Dart has noted, the gorilla greatly exceeds the new anthropoid; in the gorilla the width of brain is usually about 100 mm; in Australopithecus the width is estimated at 84 mm. The average volume of the interior of gorillas' skulls (males and females) is 470 c.c., but occasional individuals run up to 620 c.c. One may safely infer that the volume of the brain in the juvenile Australopithecus described by Prof. Dart must be less than 450 c.c., and if we allow a 15 percent increase for the remaining stages of growth, the size of the adult brain will not exceed 520 c.c. At the utmost the volume of brain in this new anthropoid falls short of the gorilla maximum. Even if it be admitted, however, that Australopithecus is an anthropoid ape, it is a very remarkable one. It is a true long-headed or dolichocephalic anthropoid—the first so far known. In all living anthropoids the width of the brain is 82 percent or more of its length; they are round-brained or brachycephalic; but in Australopithecus the width is only 71 percent of the length. Here, then, we find amongst anthropoid apes, as among human races, a tendency to roundness of brain in some and to length in others. On this remarkable quality of Australopithecus Prof. Dart has laid due emphasis.

This side-to-side compression of the head taken in conjunction with the small size of jaws throw a side light on the essential features of Australopithecus. The jaws are considerably smaller than those of a chimpanzee of a corresponding age, and much smaller than those of a young gorilla. There is a tendency to preserve infantile characters, a tendency which has had much to do with the shaping of man from an anthropoid stage. The relatively high vault of the skull of Australopithecus and its narrow base may also be interpreted as infantile characters. It is not clearly enough recognised that the anthropoid and human skulls undergo remarkable growth changes leading to a great widening of the base and a lowering or flattening of the roof of the skull. In Australopithecus there is a tendency to preserve the foetal form.

When Prof. Dart produces his evidence in full he may convert those who, like myself, doubt the advisability of creating a new family for the reception of this new form. It may be that Australopithecus does turn out to be "intermediate between living anthropoids and man", but on the evidence now produced one is inclined to place Australopithecus in the same group or sub-family as the chimpanzee and gorilla. It is an allied genus. It seems to be near akin to both, differing from them in shape of head and brain and in a tendency to the retention of infantile characters. The geological evidence will help to settle its relationships. One must suppose we are dealing with fossil remains which have become embedded in the stalagmite of a filled-up cave or fissure of the limestone cliff.

May I, in conclusion, thank Prof. Dart for his full and clear description, and particularly for his accurate drawings. One wishes that discoverers of such precious relics would follow his example, and, in place of reproducing crude tracings and photographs, give the same kind of drawings as an engineer or an architect prepares when describing a new engine or a new building.

Arthur Keith

118毫米——这个尺寸在成年和幼年大猩猩的脑中也是很常见的。耳孔之上的脑高度在南方古猿和大猩猩中是相当的——大约70毫米。但是宽度方面，正如达特教授注明的那样，大猩猩远远超过了新发现的类人猿；大猩猩中，脑的宽度通常是100毫米左右；而南方古猿的脑的宽度估计在84毫米左右。大猩猩头骨（雄性和雌性）的平均内容量是470毫升，但是个别个体达到了620毫升。我们可以有把握地推测达特教授描述的幼年南方古猿的脑量肯定不足450毫升，如果我们容许其在剩余的生长阶段还有15%的增长空间的话，那么成年脑将不会超过520毫升。这种新型类人猿的最大脑量比大猩猩的最大脑量小。然而，即使承认南方古猿是一种类人猿，它们也是一种非常特别的类型。它们是一种真正的长头型或长颅型的类人猿——这是目前为止知道的第一种。现存的所有类人猿的脑宽度都是其长度的82%以上；它们的脑都是圆形的或短颅型的；而南方古猿的脑宽度只有其长度的71%。因此我们发现在类人猿中，就像各人种之间那样，有些具有圆头型趋势，而另一些则具有长头型趋势。达特教授突出强调了南方古猿的这一显著特征。

这种与颌骨较小有关的对头颅两侧的挤压从侧面为阐明南方古猿的本质特征提供了线索。其颌骨比相应年龄的黑猩猩小得多，比幼年大猩猩的小得更多。在进化过程中有着保留婴儿特征的趋势，这种趋势对于从类人猿阶段逐渐形成人类具有紧密关联。南方古猿相对较高的头骨顶盖以及狭窄的颅底也可以理解成是婴儿时期的特征。现在还不能十分清楚地认识到类人猿和人类头骨经历了导致颅底显著变宽以及颅顶变低或变平的显著的生长变化。南方古猿具有保留胎儿形式的趋势。

当达特教授将其证据悉数列出时，他可能转变了那些像我一样怀疑过创建一个新科来接纳这种新类型是否明智的人的观点。可能结果南方古猿确实是“介于现存的类人猿和人类之间的一种中间类型”，但是依据现在所列出的证据，人们倾向于将南方古猿放到与黑猩猩和大猩猩同样的群体或亚科中。南方古猿是一个同源的属，它似乎与黑猩猩和大猩猩都具有很近的亲缘关系，但是在头部和脑形状以及保留婴儿特征的趋势等方面有所不同。地质学证据将有助于解决其关系问题。必须想到我们正在研究的是那些埋藏在被填满了的山洞或者石灰岩悬崖裂缝的石笋中的化石。

最后，请允许我感谢达特教授给出的详实而清晰的描述，尤其是他那精确的绘图。人们希望这样一类珍贵化石的发现者会遵循他的榜样，像工程师或者建筑师在描述一种新发动机或新大楼时绘制的图画那样提供同样水平的绘图，而非只是复制粗糙的描图和照片。

阿瑟·基思

* * *

It is a great tribute to Prof. Dart's energy and insight to have discovered the only fossilised anthropoid ape so far obtained from Africa, excepting only the jaw of the diminutive Oligocene Propliopithecus from the Egyptian Fayum. Whether or not the interpretation of the wider significance he has claimed for the fossil should be corroborated in the light of further information and investigation, the fact remains that his discovery is of peculiar interest and importance.

The simian infant discovered by him is an unmistakable anthropoid ape that seems to be much on the same grade of development as the gorilla and the chimpanzee without being identical with either. So far Prof. Dart does not seem to have "developed" the specimen far enough to expose the crowns of the teeth and so obtain the kind of evidence which in the past has provided most of our information for the identification of the extinct anthropoids. Until this has been done and critical comparisons have been made with the remains of Dryopithecus and Sivapithecus, the two extinct anthropoids, that approach nearest to the line of man's ancestry, it would be rash to push the claim in support of the South African anthropoid's nearer kinship with man. Prof. Dart is probably justified in creating a new species and even a new genus for his interesting fossil: for if such wide divergences between the newly discovered anthropoid and the living African anthropoids are recognisable in an infant, probably not more than four years of age, the differences in the adults would surely be of a magnitude to warrant the institution of a generic distinction.

Many of the features cited by Prof. Dart as evidence of human affinities, especially the features of the jaw and teeth mentioned by him, are not unknown in the young of the giant anthropoids and even in the adult gibbon.

The most interesting, and perhaps significant, distinctive features are presented by the natural endocranial cast. They may possibly justify the claim that Australopithecus has really advanced a stage further in the direction of the human status than any other ape. But until Prof. Dart provides us with fuller information and full-size photographs revealing the details of the object, one is not justified in drawing any final conclusions as to the significance of the evidence.

The size of the brain affords very definite evidence that the fossil is an anthropoid on much the same plane as the gorilla and the chimpanzee. But while its brain is not so large as the big gorilla-cast used for comparison by Prof. Dart, it is obvious that it is bigger than a chimpanzee's brain and probably well above the average for the gorilla. But the fossil is an imperfectly developed child, whose brain would probably have increased in volume to the extent of a fifth had it attained the adult status. Hence it is probable the brain would have exceeded in bulk the biggest recorded cranial capacity for an anthropoid ape, about 650 c.c. As the most ancient and primitive human brain case, that of Pithecanthropus, is at least 900 c.c. in capacity, one might regard even a small advance on 650 c.c. as a definite approach to the human status. The most suggestive feature (in Prof. Dart's Fig. 5,

*　　*　　*

除了在埃及法尤姆发现的小型渐新世原上猿的唯一颌骨之外，达特教授的发现是迄今为止在非洲得到的唯一一个类人猿化石，他的精力和洞察力令人万分敬佩。无论他声明的这具化石的重要意义是否应该根据更多的信息与研究加以确认，他的发现具有特殊意义和重要性这一事实都不会改变。

达特教授发现的猿孩肯定是一只类人猿，该类人猿似乎处于与大猩猩和黑猩猩同样的发育阶段，但是与两者又都不相同。迄今为止，达特教授似乎并没有对该标本进行足够的"开发"以揭示其牙冠状况，所以得到的都是过去为我们鉴定已灭绝类人猿提供了绝大部分信息的那一类证据。除非完成了揭示牙冠状况的工作，并且对另两种与人类祖先世系最接近的已灭绝类人猿（森林古猿和西瓦古猿）的化石进行比较研究之后，我们才能有理由支持南非类人猿与人类的亲缘关系更近，否则这种说法就太轻率了。达特教授在为其感兴趣的化石创建一个新物种甚至一个新属方面可能是有道理的：因为，如果新发现的类人猿与现存的非洲类人猿在可能不超过 4 岁的婴儿期阶段就存在如此大的差异的话，那么它们的成年个体之间肯定差异更大，能确保形成属一级的区别。

达特教授引用了许多特征作为与人类具有亲缘关系的证据，尤其是他提到的颌骨和牙齿的特征，我们并不是不知道这些特征在巨型类人猿的幼年期是什么样子，甚至成年长臂猿的也很清楚。

最有趣的，可能也是最重要的独特特征是由天然的颅腔模型呈现出来的。这些颅腔模型可能可以证明如下说法的合理性，即南方古猿在向人类状态进化的方向上确实比其他猿类都迈进了更大一步。但是除非达特教授能够提供更充分的信息以及揭示我们研究对象形态细节的真实尺寸的照片，否则我们不能信服于对证据重要性所作的任何定论。

脑的大小提供了非常明确的证据，证明了该化石是与大猩猩和黑猩猩处于大致同一水平的一种类人猿。但是它的脑没有达特教授用来进行比较的大型大猩猩的模型大，不过它肯定比黑猩猩的脑大，并且可能远远超过了大猩猩的平均尺寸。但是该化石是一个尚未发育完全的孩子的，在其达到成年状态之前其脑量可能还有 1/5 的增长幅度。因此它的脑可能远远超过了现有大量记录的类人猿的约 650 毫升最大颅腔容量。作为最古老、最原始的人类脑壳，爪哇猿人的颅容量至少有 900 毫升，有人可能认为甚至还需要从 650 毫升经过小幅增长才能达到确定的接近于人类的状

p. 197) is the position of the sulcus lunatus and the extent of the parietal expansion that has pushed asunder the lunate and parallel sulci—a very characteristic human feature.

When fuller information regarding the brain is forthcoming—and no one is more competent than Prof. Dart to observe the evidence and interpret it—I for one shall be quite prepared to admit that an ape has been found the brain of which points the way to the emergence of the distinctive brain and mind of mankind. Africa will then have purveyed one more surprise—but only a real surprise to those who do not know their Charles Darwin. But what above all we want Prof. Dart to tell us is the geological evidence of age, the exact conditions under which the fossil was found, and the exact form of the teeth.

G. Elliot Smith

* * *

The new fossil from Taungs is of special interest as being the first-discovered skull of an extinct anthropoid ape, and Prof. Dart is to be congratulated on his lucid and suggestive preliminary description of the specimen. As usual, however, there are serious defects in the material for discussion, and before the published first impressions can be confirmed, more examples of the same skull are needed.

First, as Prof. Dart remarks, the fossil belongs to an immature individual with the milk-dentition, and, so far as can be judged from the photograph, I see nothing in the orbits, nasal bones, and canine teeth definitely nearer to the human condition than the corresponding parts of the skull of a modern young chimpanzee. The face seems to be relatively short, but the lower jaw of the Miocene Dryopithecus has already shown that this must have been one of the characters of the ancestral apes. The symphysis of the lower jaw may owe its shape and the absence of the "simian shelf" merely to immaturity; but it may be noted that a nearly similar symphysis has been described in an adult Dryopithecus, of which it may also be said that "the anterior symphyseal surface is scarcely less vertical than that of Heidelberg man" (see diagrams in *Quart. Journ. Geol. Soc.*, vol. 70, 1914, pp. 317, 319).

Secondly, the Taungs skull lacks the bones of the brain-case, so that the amount and direction of distortion of the specimen cannot be determined. I should therefore hesitate to attach much importance to rounding or flattening of any part of the brain-cast, and would even doubt whether the relative dimensions of the cast of the cerebellum can be relied on. Confirmatory evidence is needed of the reality of appearances in such a fossil.

In the absence of knowledge of the skulls of the fossil anthropoid apes represented by teeth and fragmentary jaws in the Tertiary formations of India, it is premature to express any opinion as to whether the direct ancestors of man are to be sought in Asia or in Africa. The new fossil from South Africa certainly has little bearing on the question.

态。最具提示性的特征（达特教授的图 5，第 197 页）是月状沟的位置和顶骨的扩展范围，后者将月状沟和平行沟推离开了——这是一种人类特有的特征。

当将来出现更全面的关于脑的信息时——没有人会比达特教授更应该在这个化石上观察证据并且给出解释——就我个人而言，将充分做好准备：承认已经发现一种猿类，它的脑指向人类特有的脑和智能出现的道路。那时非洲将提供又一个惊喜——但是只是对于那些不知道他们的查尔斯·达尔文为何许人的人来说，才是一个真正的惊喜。但是我们最希望达特教授告诉我们的是关于年代的地质学证据、发现化石的地点的准确状况以及牙齿的精确形式。

埃利奥特·史密斯

* * *

在汤恩发现的新化石特别有意思，因为这是第一次发现的一种已灭绝类人猿的头骨，在此祝贺达特教授对标本进行了清晰而具有提示性的初步描述。然而，与通常一样，在讨论部分还是存在严重的缺陷，所以在发表的第一印象能被确认之前，还需要更多同样头骨的例子。

首先，正如达特教授论述的，该化石属于一只具有乳牙齿系的未成年个体，就能从照片判断出来的信息而言，我看不到其具有比现代幼年黑猩猩头骨的相应部分明显更接近于人类状况的眼眶、鼻骨和犬齿。它的面部似乎相对较短，但是中新世森林古猿的下颌骨已经显示出这肯定是猿类祖先的特征之一。下颌联合部位的形状及没有“猿板”仅是因为该个体尚未成年；但是大家也许注意到一只成年森林古猿曾经被描述为具有一个很相似的下颌联合的情况，这也可以说是“下颌联合的前表面简直与海德堡人的一样垂直”（见《地质学会季刊》中的图，第 70 卷，1914 年，第 317、319 页）。

其次，汤恩头骨缺少脑壳骨骼，所以标本扭曲变形的程度和方向都无从确定。因此我对于认为脑模型任何部分的变圆和变得扁平有重要意义的观点都深表怀疑，甚至怀疑小脑模型的相对尺寸是否可信。对于这样一个化石，要想确定其真实的外观，还需要进一步证据来确认。

由于缺乏对在印度的第三纪地层发现的牙齿和颌骨断片所代表的类人猿头骨化石的了解，所以想要表达应该在亚洲还是非洲寻找人类直接祖先的任何观点还为时尚早。南非发现的这个新化石对于这一问题的解答毫无帮助。

Palaeontologists will await with interest Prof. Dart's detailed account of the new anthropoid, but cannot fail to regret that he has chosen for it so barbarous (Latin-Greek) a name as Australopithecus.

Arthur Smith Woodward

* * *

Prof. Dart's description of the fossil skull found at Taungs in Bechuanaland shows that this specimen possesses exceptional interest and importance. Should the claims made on its behalf prove good, then its discovery will in effect be comparable to those of the Pithecanthropus remains, of the Mauer mandible and the Piltdown fragments. In the following paragraphs I venture to make some comments based upon perusal of the article published in *Nature* of February 7.

First of all, the fact that the fragments came immediately under notice of so competent an anatomist as Prof. Dart establishes confidence in the thoroughness of the scrutiny to which they have been subjected. That the history of the specimen should be known precisely from the time of its release from the limestone matrix, provides another cause for satisfaction.

The specimen itself at once raises a number of questions, and, as Prof. Dart evidently realises, these fall into at least two categories. The first question arising out of the discovery is the status of the individual represented by these remains. But the answer to that question, and the presence of such a creature in South Africa, affect other problems. The latter include inquiry into the probable locality of origin of the simian and human types, and the search for evidence of dispersion from a centre, or along a line of successive migrations.

In dealing with the first problem, Prof. Dart has surveyed a considerable number of structural details, and he concludes that the specimen represents an extinct race of apes intermediate between living anthropoid apes and mankind. The specimen comprises the greater part of a skull with the lower jaw still in place (or nearly so). The number and characters of the teeth testify to the immaturity of the individual. The evidence on the last-mentioned point is quite definite, and interest thus comes to be centred in the status assigned to the specimen; namely, that of a form intermediate between the living anthropoid apes and man himself.

Prof. Dart places the specimen on the side of the living anthropoid apes in relation to the interval separating these from man. At the same time, it is claimed that this new form of ape is more man-like than any of the existing varieties of anthropoid apes; and so it comes about that the decision turns on the claims made for the superiority of the new ape to these other forms.

古生物学家会继续怀着兴趣等待达特教授对该新型类人猿作出更详细的说明，但是对于他为其选择了南方古猿这样一个如此野蛮的(拉丁–希腊语)名字不得不感到遗憾。

阿瑟·史密斯·伍德沃德

* * *

达特教授对发现于贝专纳兰的汤恩的头骨化石的描述表明该标本具有特别的影响和重要性。如果就这件化石发表的这个主张被证明是对的，那么实际上这件头骨的发现就堪比爪哇猿人化石、毛尔下颌骨化石和皮尔当颅骨破片化石的发现了。在接下来的段落中，我冒昧地根据自己对《自然》2 月 7 日发表的文章的精读进行评论。

首先，以上那些骨骼破片出现后很快就引起了像达特教授这样杰出的解剖学家的注意，这个事实使我们相信对这件标本的研究会是彻底的。应该由这件标本从石灰岩基质挖掘出来的时间开始精确地了解它的历史，这将提供另一个使人们对此发现感到满意的理由。

标本本身马上就引出了许多问题，正如达特教授明显意识到的，问题至少可以分为两类。从这个发现引出的第一个问题是这些化石所代表的个体的身份。但是这个问题的答案以及这样一种生物存在于南非影响了其他问题。后者包括对猿猴和人类诸多类型起源的可能地点的探究，以及对从中心扩散开来或沿着相继迁徙的路线向外扩散的证据的寻找。

在研究第一个问题时，达特教授调查了大量结构细节，他得出的结论是，标本代表的是一种已灭绝的猿类，介于现存的类人猿和人类之间的中间类型。该标本由头骨的大部分组成，该头骨上还连有仍然处于原位（或接近原位）的下颌骨。牙齿的数目和特征说明该个体是未成年的。最后提到的这个论点的证据是非常确定的，因此兴趣就集中在了这件化石应该属于什么身份；也就是，一种介于现存类人猿和人类自身的中间类型。

达特教授将该标本置于把现存的类人猿与人类分开的间隔当中的类人猿这一边。同时，他主张这种新类型的猿比任何现存种类的类人猿都更像人类；所以就决定转而主张这种新型猿比其他类型的猿更加优越。

The report shows that (as noted above) many structural details have been scrutinised, and that all accessible parts of the specimen have been examined. The observations relate not only to the external parts of the skull and lower jaw, but also to the endocranial parts exposed to view by the partial shattering of the brain-case. The claims advanced on behalf of the higher status of the specimen are based, therefore, upon a number and variety of such details. Should Prof. Dart succeed in justifying these claims, the status he proposes for the new ape-form should be conceded. Much will depend on the interpretation of the features exhibited by the surface of the brain, as also upon that of all the characters connected therewith; and since Prof. Dart is so well equipped for that aspect of the inquiry, his conclusions must needs carry special weight there. In regard to the brain and its characters, I find the tracing of the contour of an endocranial cast in a gorilla-skull shown in Fig. 6 rather surprisingly flattened, and almost suggestive of the influence of age.

Among the anatomical characters enumerated in the article, some appear to me to possess a higher value in evidence than others. As good points in favour of the claims, there may be cited, in addition to the cerebral features to which reference has just been made, the level of the lower border of the nasal bones in relation to the lower orbital margins, the (small) length of the nasal bones, the lack of brow-ridges (even though the first permanent tooth has appeared fully), the steeply-rising forehead, and the relatively short canine teeth.

On the other hand, I feel fairly certain that some of the other characters mentioned are related preponderantly to the youthfulness of the specimen. Fully to appreciate the latter, demands not only the handling of it, but also thorough survey of a collection of immature (anthropoid ape) crania. The development of the "shelf" at the back of the symphysis of the lower jaw may almost certainly be delayed in some individuals (gorillas). Even the level of the lower border of the nasal bones is subject to some variation, and in young gorillas before the first permanent tooth has emerged fully, that level may be (as in man) above the level of the orbital margin. Generally, the elimination and detachment of features influenced largely by the factor of age demand special attention.

If, however, the good points can be justified, then these characters of youth will not gravely affect the final decision.

However these discussions may end, the record remains of the occurrence of an anthropoid ape some two thousand miles to the south of the nearest region providing a record of their presence. So far as the illustrations allow one to judge, the new form resembles the gorilla rather than the chimpanzee, that is, an African, not an Asiatic form of anthropoid ape. In this respect the new ape does not introduce an obviously disturbing factor. Disturbance, and the recasting of disturbed views, might nevertheless be caused in two other directions. Thus, the determination of the geological antiquity of the embedding of the fossil remains might have such an effect, were the estimate such as to carry that event very far back in time. Again, a comparison of the new ape with the fossil forms from India (Siwaliks) remains to be made, and it may be productive of results bearing on the relation of the African and the Asiatic groups. In any case, opinion must needs conform to the situation created by this discovery.

该报告表明（正如上文所述），许多结构性细节都已经被仔细观察过了，标本所有可及的部分也都已经被查看过了。这些观察不仅涉及头骨和下颌骨的外部，还涉及脑壳一部分毁损后暴露出来的颅腔部分。因此为了表示该标本具有比较高级的身份而提出的主张是建立在许多种这类细节的基础上的。如果达特教授成功地证明这些说法是合理的，他建议的新型猿类的身份应当得到承认。更多信息将依赖于对脑表面所展示出的特征的解释，还依赖于对与此相关的所有特征的解释；而且由于达特教授对此方面的调查进行了非常充分的准备，所以他的结论肯定具有格外重要的分量。至于脑及其特征，我觉得图 6 所展示的大猩猩头骨的颅腔模型轮廓线的扁平程度堪称惊人，也暗示了年龄的影响。

在文章中列举的解剖学特征中，对于我来说，就证据价值而言，有些特征的价值比其他特征的更高。除了刚刚已经提到的大脑特征以外，可能被引用的作为支持这些说法的很好的证据还有与眼眶下缘相关的鼻骨下缘位置的水平、鼻骨的长度（短）、缺乏眉脊（尽管第一恒牙已经完全出现）、陡峭上升的前额以及相对短小的犬牙。

另一方面，我非常肯定地认为，提到的其他特征中有些肯定也是与标本的年轻性有关。为了充分理解后者，不仅需要对标本进行处理，还需要对收藏的未成年（类人猿）头盖骨进行彻底的调查。几乎可以肯定的是，下颌联合部位背面的“猿板”的发育在某些个体（大猩猩）中被延迟了。甚至鼻骨下缘的位置水平趋向于发生一定的变化，幼年大猩猩在第一颗恒牙完全出现之前，其位置水平可能（与人类一样）位于眶缘水平之上。通常来说，需要特别注意那些很大程度上受年龄因素影响而造成的特征的消失和分离。

然而，如果这些有利的证据可以被证实的话，那么这些幼年个体的特征将不会严重影响最终的决定。

无论这些讨论将会如何结束，这块化石记录了一只类人猿的存在，它距离现存最近的类人猿发现区域还要偏南 2,000 英里。就允许我们进行判断的现有说明而言，这种新类型与大猩猩的相像程度大于与黑猩猩的，也就是说，这是一种非洲类型的类人猿，而不是亚洲类型的。在这方面，这种新型猿并没有带来明显的干扰因素。干扰和对受到干扰的观点的重新认识可能来自另外两个方向。确定化石的埋藏发生在哪个地质学时期可能具有这样一种作用，这样的估计可能会将这一事件带回到非常远古的时期。再者，现在还没有将这种新型猿类与印度西瓦利克发现的化石类型进行比较，这种比较可能会得到关于非洲和亚洲群体之间关系的丰富结果。无论如何，意见必须与这一发现所产生的局面相符。

If in these notes there have been passed over those observations and reflections wherewith Prof. Dart has illustrated and supported his views, such omissions are not due to want of appreciation, but to lack of capacity and space for their adequate treatment.

W. L. H. Duckworth

(**115**, 234-236; 1925)

如果在这些说明中出现了忽略达特教授描述过并用来支持自己观点的观察和思考，那么这种忽略不是由于想索要赞赏，而是缺乏对它们进行充分处理的能力和空间。

迪克沃斯

（刘皓芳 翻译；吴新智 审稿）

Some Notes on the Taungs Skull

R. Broom

Editor's Note

While anthropologists in distant London debated the significance of Raymond Dart's new "Man-Ape", *Australopithecus africanus*, from South Africa, local palaeontologist Robert Broom went to see the actual specimen. This note looks at the geological setting, suggesting that the skull was of Pleistocene to Recent date; and also the cranial and dental anatomy, asserting the intermediate status of the creature. However, Broom presciently notes that if specimens of adults were to be discovered, "the light thrown on human evolution would be very great". A decade was to elapse before such finds came to light: and the discoverer would be Broom himself.

A few days ago I visited Johannesburg to have a look at the remarkable new skull discovered by Prof. Dart, and named by him *Australopithecus africanus*. Prof. Dart not only allowed me every facility for examining the skull, but also gave me with almost unexampled generosity full permission to publish any observations I made on it, and suggested further that I might send to *Nature* any notes that might amplify the account he had already given. As the skull is one of extreme importance, a full account with measurements and very detailed figures will in due course be published by Prof. Dart, but the world already realises the unique character of the discovery and is anxious for more immediate information.

From the cablegrams received in South Africa, it is manifest that the first demand is for further light on the geological age of the being, and unfortunately complete information on this point cannot now be given, and will possibly never be available. Though I have not myself visited the Taungs locality, I am fairly familiar with many similar deposits farther south along the Kaap escarpment. This escarpment runs for more than 150 miles along the west side of the Harts River and lower Vaal River valleys from a little south of Vryburg to 20 miles south of Douglas. The escarpment is formed for the most part of huge cliffs of dolomitic limestone of the Campbell Rand series, in most places some hundreds of feet thick. The wide valley has an interesting geological history. Originally it was carved out in Upper Carboniferous or Lower Permian times by the Dwyka glaciers. For millions of years it was steadily refilled by Dwyka, Ecca, and Beaufort beds until the whole valley was perhaps buried by more than 2,000 feet of Permian and Triassic shales. Then conditions changed and the valley was re-excavated, by denudation, until today we find it not unlike what it must have been when originally carved out by the Dwyka glaciers.

The dolomite escarpment forms the most striking feature of the landscape in this part of the world. All along the west of the Harts-Vaal valley lies the high dead-level Kaap

汤恩头骨的几点说明

布鲁姆

编者按

当伦敦的人类学家们正在为雷蒙德·达特在南非发现的新“人猿”（即南方古猿非洲种）的意义争论不休的时候，身在南非的古生物学家罗伯特·布鲁姆去看了真实的化石标本。这篇文章着眼于地质学背景，提出该头骨介于更新世到现代之间，同时也考虑了头盖骨和牙齿的解剖学特征，主张该生物处于中间的位置。不过，布鲁姆很有预见性地指出，如果能发现这种生物的成年个体，那么“其对于人类进化带来的启示将是非常重要的”。10 年之后迎来了这样的发现，而发现者正是布鲁姆本人。

几天前我访问了约翰内斯堡，参观了达特教授发现并将其命名为南方古猿非洲种的著名新头骨。达特教授不仅允许我使用查看该头骨的所有设备，而且完全许可我发表自己对该头骨进行的任何观察，这种慷慨几乎是史无前例的，他还建议我可以向《自然》投递可以扩充他发表过的报告内容的任何稿件。因为该头骨是极具重要性的标本之一，所以一份兼具测量尺寸和详细图像的完整报告将会由达特教授在适当的时间发表，但是世人已经意识到了这一发现的独特特征，并且急于得到更多的即时信息。

从南非收到的海底电报来看，很显然第一个要求就是希望对这种生物的地质学年代作更进一步的说明，不幸的是，对于这一点，目前还不能给出完整的信息，而且可能永远都无法得到。尽管我并没有亲自参观过汤恩遗址，但是我非常熟悉远在开普断崖沿线南部的许多相似的堆积物。这一断崖从弗雷堡稍南部沿着哈茨河和瓦尔河下游河谷西侧绵延 150 多英里直至道格拉斯南部 20 英里处。该断崖大部分由坎贝尔–兰德系列的巨大白云灰岩山崖组成，很多地方都有几百英尺厚。这里宽阔的河谷有一段很有趣的地质学史。最初它是在上石炭纪或早二叠纪时期由德怀卡冰川开拓出来的。数百万年间它不断地被德维卡、埃卡和博福特河床回填，直到整条河谷多半被 2,000 多英尺的二叠纪和三叠纪页岩掩埋掉为止。后来环境发生了改变，河谷受到剥蚀作用而被重新挖掘出来了，直到现在，我们发现这条河谷与最初被德维卡冰川开拓出来时的样子几乎一样。

白云石断崖在世界的这个部分形成了最具独特特征的景观。沿着哈茨–瓦尔河谷西部的所有部分都位于高而平坦的开普高原上，从 20 英里外眺望时，整个断崖看

plateau, and when viewed from 20 miles away the escarpment looks like a high black wall bounding the lower plain of the valley. Every five or ten miles along the black wall are to be seen large light-coloured patches which on examination prove to be great masses of calc-sinter formed by calcareous springs. These, of course, must have been formed after the dolomite cliffs had been denuded of their covering Dwyka shales, and may in some cases be of considerable age—perhaps even dating from moderately early Tertiary times. Other masses of this secondary limestone may be of comparatively recent date. In places the great masses of calc-sinter have been excavated by underground water and moderately large caves are formed.

At Taungs the mass of secondary limestone is some hundreds of feet thick and about 70 feet high where it is being worked. Already 250 feet have been quarried away. On the face about 50 feet below the top of the mass, an old cave is cut across which is filled up with sand partly cemented together with lime, and it is in this old cave that the skull of Australopithecus has been found. The only other bones that I have seen or heard of are skulls and bones of a baboon, a jaw of a hyrax, and remains of a tortoise. I have not seen the hyrax jaw, so cannot say if it belongs to one of the living species. The baboon has been examined by Dr. Haughton, who regards it as an extinct species and has named it *Papio capensis*. I have seen a number of imperfect skulls of this baboon, and while they belong to a different species from the living local *Papio porcarius*, the difference between them is not so very striking.

I think it can be safely asserted that the Taungs skull is thus not likely to be geologically of great antiquity—probably not older than Pleistocene, and perhaps even as recent as the *Homo rhodesiensis* skull. When later or other associated mammalian bones are discovered, it may be possible to give the age with greater definiteness. At present all we can say is that the skull is not likely to be older than what we regard as the human period. But the age of the specimen in no way interferes with its being a true "missing link", and the most important hitherto discovered.

Prof. Dart in his photographs has given the general features of the skull and the brain, but there are a number of important characters in the skull and dentition to which I should like to direct attention.

Though the parietals and occipital are almost completely lost from the brain cast, most of the sutures can be clearly made out, and are as I indicate in Fig. 1. The sutures in the temporal region can also be clearly seen. The suture between the temporal bone and the parietal is fairly horizontal as in the anthropoid apes, but in the upward development of the squamous portion we have a character which is human and not met with in the gorilla, the chimpanzee, the orang, or the gibbon.

起来就像一堵以河谷的下流平原为界的高大黑墙。沿着黑墙，每 5~10 英里就可以看到巨大的浅颜色的点缀，经检验证实它们是由石灰泉形成的大片钙华。当然，这些肯定是在白云石山崖被剥蚀掉了覆盖着的德维卡页岩之后才形成的，而且在有些情况下，它们可能有着相当长的年代——甚至可能一直追溯到第三纪的早期。这种次生石灰石的其他块体的年代可能比较晚近。有些地方，地下水挖掘了大片钙华，形成了中等的大型山洞。

在汤恩，次生石灰石有几百英尺厚，70 英尺高。采石场已经挖掉了 250 英尺。在大块石灰石顶部之下约 50 英尺处的表面上，挖掘中穿过了一个旧山洞，洞中填满了与石灰一起形成水泥的沙子，就在这个老山洞里发现了南方古猿头骨。我见过或听说过的骨骼只有狒狒的头骨和骨骼、岩狸的颌骨和龟的残骸。我没有见到岩狸的颌骨，所以不能说它是否属于现存物种之一。狒狒的遗骸已经由霍顿博士查看过了，他认为它是一种已灭绝的物种，并将其命名为狒狒开普种。我见过这只狒狒的许多不完整的头骨，尽管它们属于一种与现存的本地浅灰狒狒不同的物种，但二者之间的差异不是十分明显。

我认为可以有把握地说，从地质学角度而言，汤恩头骨不可能是非常古老的——可能不早于更新世，甚至可能与罗德西亚人头骨一样晚。在发现比较晚的或其他与之共生的哺乳动物骨骼时，就有可能给出更确定的年代了。现在我们能说的就是这个头骨不可能早于我们所认为的人类时期。但是该标本的年代绝不妨碍它作为一个真正的“缺失环节”和迄今为止发现的最重要的标本。

达特教授在他发表的照片中给出了头骨和脑的一般特征，但是我想关注的事情是，头骨和齿系还有许多重要特征。

尽管在脑的模型上几乎找不到顶骨和枕骨的影子，但是可以清楚地辨认出大部分骨缝，这些骨缝我在图 1 中都指明了。颞区的骨缝也可以清楚看到。颞骨和顶骨之间的骨缝与类人猿中的一样非常水平，但是它的鳞部向上发育，这是一个属于人类的特征，在大猩猩、黑猩猩、猩猩和长臂猿中都没有遇见过。

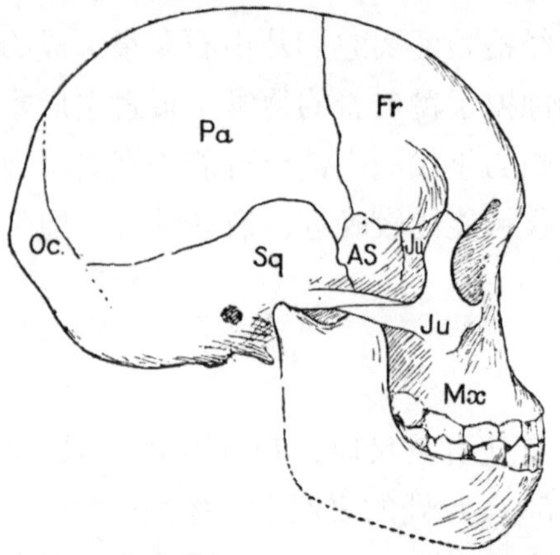

Fig. 1. Side view of skull of *Australopithecus africanus*, Dart. About $\frac{1}{3}$ natural size.

The arrangement of the sutures in the temporal region is also remarkably interesting. The upper part of the sphenoid articulates with both the parietal and the frontal. In the gorilla and chimpanzee in all the drawings I can find, the temporal bone meets the frontal and prevents the meeting of the sphenoid and the parietal. In the orang the condition varies, and I have in my possession a skull which has on the right side a spheno-parietal suture and on the left a fronto-temporal. In the baboon there is a large fronto-temporal suture, and in Cercopithecus a spheno-parietal suture. In the gibbon there is also a spheno-parietal suture. While the arrangement of the sutures in this region may not be of very great fundamental importance, it is interesting to note that Australopithecus agrees with man, the gibbon, and Cercopithecus, but differs from the gorilla, the chimpanzee, and the baboon.

The jugal or malar arch is interesting in that there is a long articulation between the jugal and squamosal. In this Australopithecus agrees rather with the anthropoids than with man.

On the face there are one or two striking characters, and of these perhaps the most important is the fusion of the premaxilla with the maxilla. On the palate the suture between these bones is seen almost as in the human child, the suture running out about two-thirds of the way towards the diastema between the second incisor and the canine. On the face there is no trace of any suture in the dental region, but on the left side of the nasal opening there is what is probably the upper part of the original premaxilla-maxillary suture. On the right side there is a faint indication of a suture just inside the nostril. In the chimpanzee the suture becomes obliterated in the dental region early, as apparently is the case in Australopithecus. In the orang and gorilla the suture remains distinct until a much later stage. In man, as is well known, all trace of the suture is obliterated from the face long before birth.

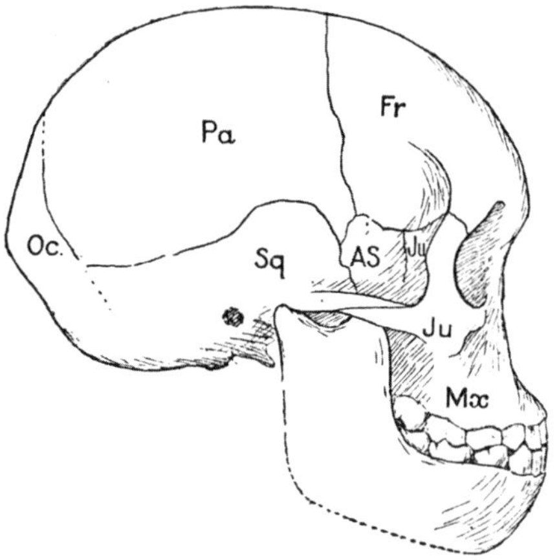

图 1. 南方古猿非洲种头骨的侧面观，达特。约相当于真实尺寸的 $\frac{1}{3}$。

颞区骨缝的分布也非常有意思。蝶骨的上部与顶骨和额骨都连接着。在我能找到的所有大猩猩和黑猩猩的图画中，颞骨都与额骨连接，而且颞骨阻止了蝶骨和顶骨相遇。猩猩则不是这种情况，我拥有的一个头骨的右侧有一条蝶顶缝，左侧有一条额颞缝。狒狒有一条大型的额颞缝，长尾猴有一条蝶顶缝。长臂猿也有一条蝶顶缝。尽管这一区域的骨缝的分布不具有重大的关键意义，但有趣的是，我们注意到南方古猿与人类、长臂猿和长尾猴都是一致的，而与大猩猩、黑猩猩和狒狒则是不同的。

颧弓的有趣之处在于，颧骨和颞鳞之间有一条长的关节。在这点上南方古猿与类人猿的一致性要多过与人类的一致性。

面部有一两点显著的特征，这些特征中最重要的可能是前颌骨和上颌骨之间的融合。可以看出硬腭的这些骨骼间的骨缝几乎与人类孩子的一样，骨缝向第二门齿和犬齿间的齿隙延伸约 2/3 的距离后消失。在面部，牙齿区域没有任何骨缝的痕迹，但是在鼻腔开口的左侧有一条骨缝可能是原先的前颌骨–上颌骨骨缝的上半部分。右侧在鼻孔里有一条微弱的骨缝迹象。黑猩猩这条骨缝早早地就在牙齿区域消失了，南方古猿中表面上也是这样。猩猩和大猩猩中这条骨缝直到很晚的阶段依然明显。众所周知，人类的所有骨缝的痕迹在出生之前就已经从脸上消失了。

Australopithecus agrees with man and the chimpanzee in having a single foramen for the superior maxillary nerve. In the orang, gibbon, and other apes there are usually two or more foramina. In the gorilla sometimes there is one foramen; sometimes two.

In the shortness of the nasal bones and the high position of the nasal opening the Taungs skull agrees more with the chimpanzee than with the gorilla.

The dentition is beautifully preserved, and the teeth have been cleared of matrix by Prof. Dart with the greatest care. Though, owing to the lower jaw being in position, a full view of the crowns of the teeth could only be obtained by detaching the lower jaw, a sufficiently satisfactory view can be obtained to give us practically all we require of the structure.

The whole deciduous denture is present in practically perfect condition. The incisors, which are small, have been much worn down by use, and most of the crowns of the median ones have been worn off. Prof. Dart has directed attention to the vertical position of the teeth, which is a human character and differs considerably from the conditions found in the chimpanzee and gorilla. The small size of the incisors is also a human character.

The relatively small size of the canine is a character in which Australopithecus agrees with both the chimpanzee and man, and lies practically between the two.

The deciduous molars agree more closely with those of man than with those of any of the apes.

The first permanent molars of both upper and lower jaws are perfectly preserved and singularly interesting.

The first molar of the upper jaw (Fig. 2) has four large cusps arranged as in man and the anthropoid apes.

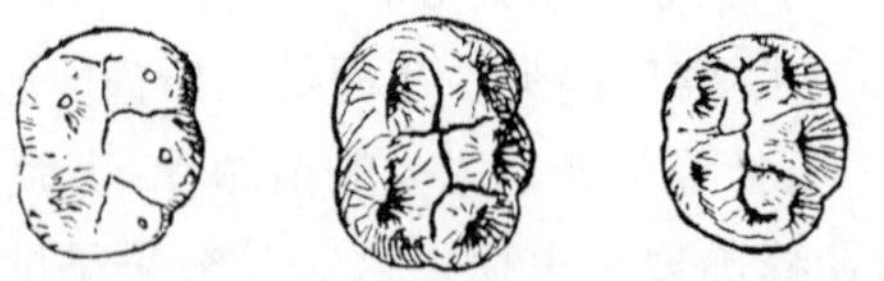

A B C

Fig. 2. First right upper molars: A, orang (after Röse); B, *Australopithecus africanus*, Dart, unworn; C, Bushman child, unworn. All natural size.

在上颌神经具有单一小孔这方面，南方古猿与人类和黑猩猩一致。而猩猩、长臂猿和其他猿类通常有两个或更多个小孔。大猩猩有时有一个小孔，有时有两个。

在鼻骨短小和鼻腔开口位置偏高这方面，汤恩头骨与黑猩猩的一致性比与大猩猩的高。

齿系的保存状况很好，而且达特教授已经十分慎重地清除掉了附着在牙齿上的基质。由于下颌骨仍保留在原位，所以要想对牙冠进行全面观察，只有将下颌骨分离下来才可以，这样就能得到一幅足以满足我们对全部结构的需要的视图。

全部乳牙的保存状况都相当完美。门齿小，由于使用而有了很大程度的磨损，大部分内侧门齿的牙冠都被磨损掉了。达特教授注意到牙齿的方位是垂直的，这是一种人类特征，与黑猩猩和大猩猩中观察到的情况差异很大。门齿尺寸较小也是一种人类特征。

相对小尺寸的犬齿是一种南方古猿与黑猩猩和人类都一致的特征，实际上南方古猿的犬齿尺寸介于这二者之间。

与所有猿类的乳臼齿相比，南方古猿的乳臼齿与人类的更加相符。

上颌骨和下颌骨的第一恒臼齿的保存状况都很好，它们都非常有意思。

上颌骨的第一臼齿（图 2）有 4 个大的牙尖，其排列情况与人类和类人猿的一样。

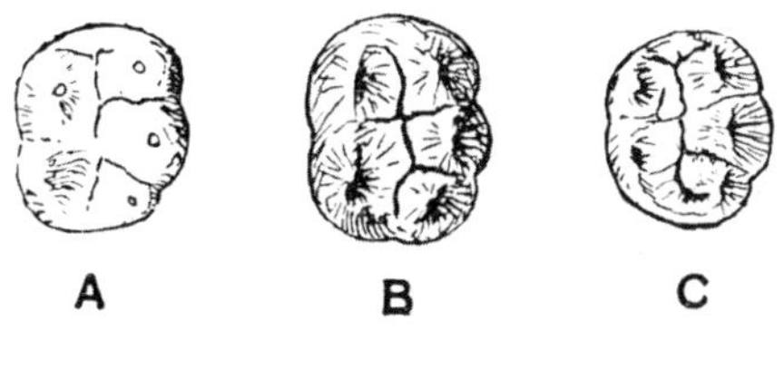

图 2. 第一右上臼齿：A，猩猩（依照罗斯）；B，南方古猿非洲种，达特，未磨损；C，儿童布希曼人，未磨损。全部都是真实尺寸。

The first lower molars (Fig. 3) has three well-developed sub-equal cusps on the outer side and two on the inner. Though in its great length and in the large development of the third outer cusp or hypoconulid the tooth differs considerably from the typical first lower molar of man, teeth of this pattern not infrequently occur in man. In general structure, however, the tooth more closely resembles that of the chimpanzee. It is interesting to compare this tooth with the corresponding tooth in Eoanthropus.

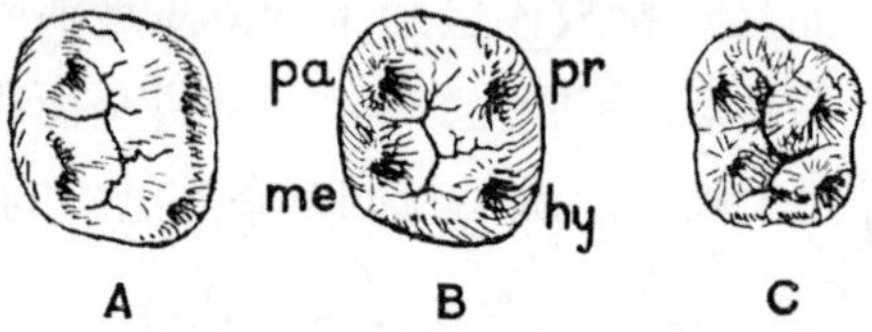

Fig. 3. First right lower molars: A, old chimpanzee, worn (after Miller); B, *Australopithecus africanus*, Dart; C, Bushman child. All natural size.

The arrangement of the furrows on the crown of the molar of Australopithecus is almost exactly similar to that in both the orang and the Bushman. In the chimpanzee and gorilla, there is usually a well-marked ridge passing from the protocone to the metacone, of which there is an indication in the Bushman tooth.

It will be seen that in *Australopithecus africanus* we have a large anthropoid ape resembling the chimpanzee in many characters, but approaching man in others. We can assert with considerable confidence that it could not have been a forest-living animal, and that almost certainly it lived among the rocks and on the plains, as does the baboon of today. Prof. Dart has shown that it must have walked more upright than the chimpanzee or gorilla, and it must thus have approached man more nearly than any other anthropoid hitherto discovered.

Eoanthropus has a human brain with still the chimpanzee jaw. In Australopithecus we have a being also with a chimpanzee-like jaw, but with a sub-human brain. We seem justified in concluding that in this new form discovered by Prof. Dart we have a connecting link between the higher apes and one of the lowest human types.

The accompanying table (Fig. 4) shows what I believe to be the relationships of Australopithecus. If an attempt be made to reconstruct the adult skull (Fig. 5), it is surprising how near it appears to come to *Pithecanthropus erectus*—differing only in the somewhat smaller brain, and less erect attitude.

While nearer to the anthropoid apes than man, it seems to be the forerunner of such a type as Eoanthropus, which may be regarded as the earliest human variety, the other probably branching off in different directions.

第一下臼齿（图 3）外侧有 3 个发育完好的几乎相等的牙尖，内侧有 2 个。尽管在牙尖的巨大长度以及第三外侧牙尖或下次小尖的发达程度上，这颗牙与典型的人类第一下臼齿非常不同，但是人类中这种形式的牙齿并不罕见。然而，就总体结构而言，这颗牙与黑猩猩的更像。将这颗牙与曙人相应的牙齿进行比较会得到很有趣的发现。

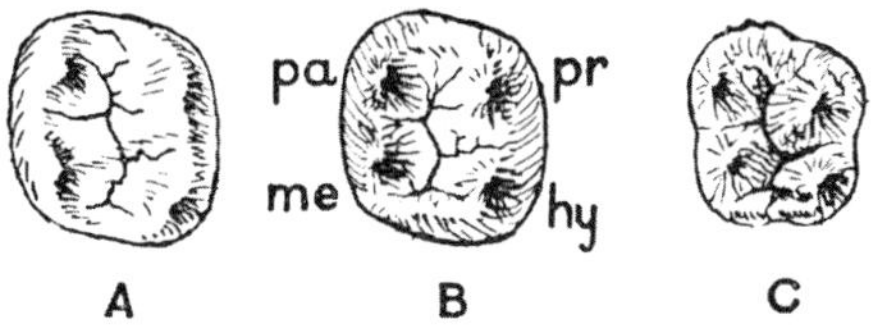

图 3. 第一右下臼齿：A，年老黑猩猩，有磨损 (依照米勒)；B，南方古猿非洲种，达特；C，儿童布希曼人。全部都是真实尺寸。

南方古猿臼齿牙冠上沟的排列几乎与猩猩和布希曼人的完全一样。黑猩猩和大猩猩中，通常有一条从上原尖到后尖的明显的脊，布希曼人牙齿中也有这种迹象。

我们将会看到，在南方古猿非洲种中出现了一只许多特征与黑猩猩相像、但其他方面又与人类相像的大型类人猿。我们可以充满信心地断言它不是一只生活在森林中的动物，并且几乎可以肯定它是生活在岩石间和平原上的动物，就像现在的狒狒一样。达特教授已经说明了，它行走的姿势肯定比黑猩猩和大猩猩的更加直立，因此它肯定比迄今发现的其他类人猿更接近于人类。

曙人具有人类的脑、黑猩猩的颌骨。南方古猿中我们也有一只黑猩猩式的颌骨和次人的脑。我们似乎有理由作出如下结论：这种由达特教授发现的新类型使得我们拥有了一个连接高等猿类和一种人类的最低级类型的环节。

附表（图 4）展示了我相信的南方古猿的亲缘关系图。如果试图对成年头骨进行复原（图 5），那么就会惊奇地发现它与爪哇猿人是非常接近的——只是在脑稍小、姿势欠直立方面有所不同。

尽管与人类相比它更接近于类人猿，但是似乎仍旧可以将其认为是像曙人这样的类型的先祖，并将其当作是最早期的人类物种，即可能是另一个与人类发生了分歧而向不同方向进化的物种。

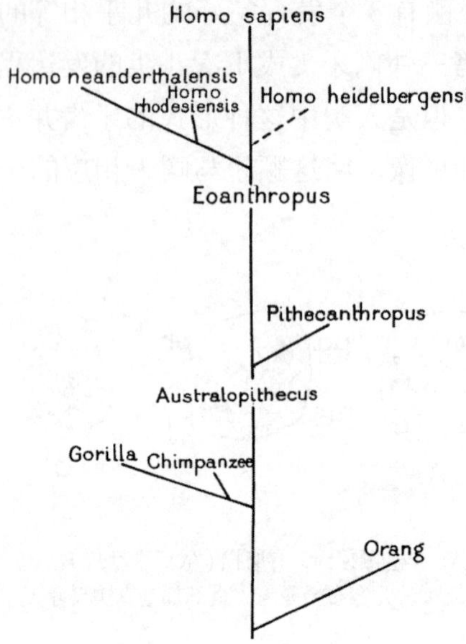

Fig. 4

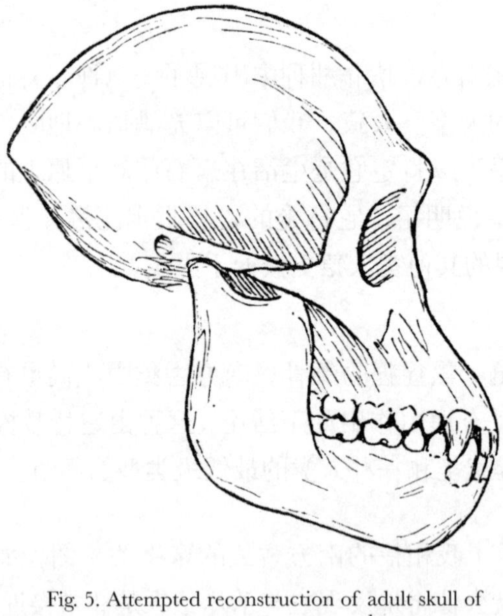

Fig. 5. Attempted reconstruction of adult skull of *Australopithecus africanus*, Dart. About $\frac{1}{3}$ natural size.

There seems considerable probability that adult specimens will yet be secured, and if the skeleton as well as the skull is preserved, the light thrown on human evolution will be very great.

(**115**, 569-571; 1925)

R. Broom: Douglas, South Africa.

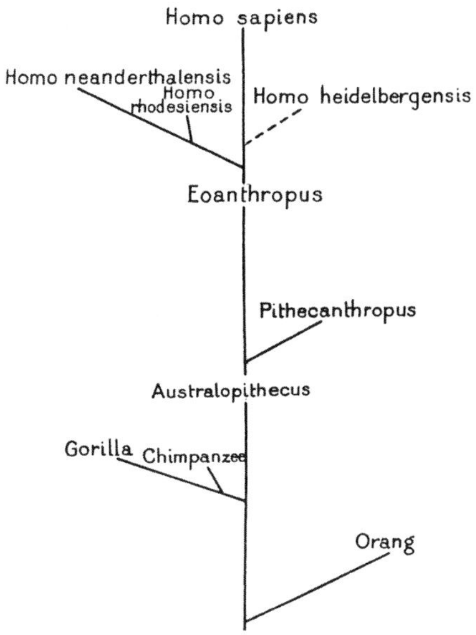

图 4

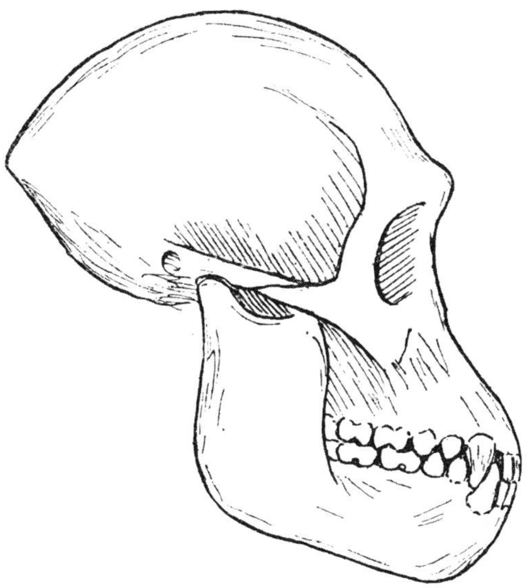

图 5. 成年南方古猿非洲种尝试性的复原头骨，达特。
约相当于真实尺寸的 $\frac{1}{3}$。

似乎有足够的可能性认为成年标本会使我们更安心，如果骨架和头骨都被保存下来的话，那么其对于人类进化带来的启示将是非常重要的。

（刘皓芳 翻译；吴新智 审稿）

The Taungs Skull

A. Keith

Editor's Note

Months after the discovery of *Australopithecus africanus*, London anthropologists remained frustrated that they could not see the specimen—not even a cast. This may explain the rage of this letter from the eminent Scottish anatomist and anthropologist Arthur Keith, that the closest he had got to the Taungs skull was to go to an exhibition and "peer at [the casts] in a glass case" along with ordinary members of the public. Keith's conviction that Australopithecus was a fossil ape rather than a transitional form between apes and humans hardened.

THE account which Prof. Dart published of the Taungs skull (*Nature*, Feb. 7, p. 195) left many of us in doubt as to the true status of the animal of which it had formed part, and we preferred, before coming to a decision, to await an examination of the fossil remains, or failing such an opportunity, to study exact casts of them. For some reason, which has not been made clear, students of fossil man have not been given an opportunity of purchasing these casts; if they wish to study them they must visit Wembley and peer at them in a glass case which has been given a place in the South African pavilion.

The chief point which awaited decision relates to the position which must be assigned in the animal kingdom to this newly discovered form of primate. Prof. Dart, in writing of it, has used the name of anthropoid ape; he has described it as representing "an extinct race of apes intermediate between living anthropoids and man"—which is tantamount to saying that at Taungs there has been discovered the form of being usually spoken of as the "missing link". That this is his real decision is evident from the fact that he speaks of it as "ultrasimian and prehuman" and proposes the creation of a new family for its reception.

An examination of the casts exhibited at Wembley will satisfy zoologists that this claim is preposterous. The skull is that of a young anthropoid ape—one which was in the fourth year of growth—a child—and showing so many points of affinity with the two living African anthropoids—the gorilla and chimpanzee—that there cannot be a moment's hesitation in placing the fossil form in this living group. At the most it represents a genus in the Gorilla-Chimpanzee group. It is true that it shows in the development of its jaws and face a refinement which is not met with in young gorillas and chimpanzees at a corresponding age. In these respects it does show human-like traits. It is true that it is markedly narrow-headed while the other African anthropoids are broad-headed—but we find the same kind of difference in human beings of closely allied races. Prof. Dart claimed that the brain showed certain definite human traits. This depends upon whether or not he had correctly identified the position of a certain fissure of the brain—the parallel fissure. In the show-case at Wembley a drawing is placed side by side with the "brain cast"; but when we examine the brain cast at the site where the fissure is shown on the drawing, we find only a broken surface where identification becomes a matter of guess-work.

汤恩头骨

基思

编者按

在达特发现南方古猿非洲种几个月之后，伦敦的人类学家们依旧很沮丧，因为他们看不到那个标本，甚至连模型也看不到。这大概可以解释苏格兰杰出解剖学家、人类学家阿瑟·基思在这封来信中表达的愤怒，他说自己距离那件汤恩头骨最近的时候就是随着普通观众参观头骨展览时“盯着看玻璃柜里的[模型]”。基思坚信，南方古猿更像是古猿的化石，而不是古猿与人类之间的过渡类型。

达特教授就汤恩头骨发表的报道（《自然》，2 月 7 日，第 195 页）中关于这种动物（这件化石就是它的一个部分）的真实身份给我们留下了许多可疑之处，在得到结论之前，我们更倾向于等待对化石进行查看，如果没有这样的机会，研究其精确的模型也行。没有人清楚是出于什么原因，研究人类化石的学者们还没有得到购买这些模型的机会；如果他们希望研究这些模型的话，他们就必须去温布利参观南非展览馆，然后盯着看玻璃柜里的模型。

有待确定的最主要的一点与将这种新发现的灵长类类型归属到动物界的哪个位置有关。达特教授写这部分的时候，使用了类人猿的名称，他将其描述为代表着“一种已经灭绝的猿类，介于现存的类人猿和人类之间的中间类型”——这种说法相当于是说汤恩发现了通常被称作是“缺失环节”的生物类型。显而易见这是他的真实判断，从他将其称为“超猿和前人类”的说法以及提议创建一个新科来接纳这种动物的事实就可以看出来。

对温布利陈列的模型进行观察将会使动物学家相信这种主张是荒谬可笑的。这具头骨是一只幼年类人猿的——该个体处于生长阶段的第 4 年——即一个孩子——它显示出了许多与两种现存的非洲类人猿——大猩猩和黑猩猩——的亲缘关系，于是将该化石类型列入这种现存群体中就变得不容置疑了。它最多代表了大猩猩–黑猩猩群体中的一个属。它确实表明颌骨和面部在发育过程中发生了细微的变化，这在相应年龄的幼年大猩猩和黑猩猩中都是不曾见到过的。这些方面它确实显示出了与人类相像的特点。它具有非常狭窄的头部也是真实的，而其他非洲类人猿都是阔头型的——但是我们发现具有密切亲缘关系的人类种族也存在同样的差异。达特教授声称该标本的脑显示了某些确定的人类特征。这依赖于他是否正确地辨认了某一大脑裂缝——平行裂缝的位置。在温布利的陈列柜里，一幅画与“脑模型”一起陈列在那里；但是当我们观察脑模型，查看图画上展示的裂缝所在位置时，我们发现只有一处裂开的表面，在这里，鉴定变成了一种猜测。

In every essential respect the Taungs skull is that of a young anthropoid ape, possessing a brain which, in point of size, is actually smaller than that of a gorilla of a corresponding age. Only in the lesser development of teeth, jaws, and bony structures connected with mastication can it claim a greater degree of humanity than the gorilla. Its first permanent molar teeth which have just cut are only slightly smaller than those of the gorilla, while the preparations which are being made in the face for the upper permanent canines show that these teeth were to be of the large anthropoid kind.

The other point on which we awaited information related to the geological age of the Taungs skull. Fortunately, Dr. Robert Broom (*Nature*, April 18, p. 569) has thrown a welcome light on this matter. The skull was blasted out of a cave which had become filled up by sand washed in from the Kalahari. The fossil baboons found in neighbouring caves differ in only minor structural details from baboons still living in South Africa. In Dr. Broom's opinion the Taungs skull is of recent geological date; it is not older than the Pleistocene; he thinks it probable that it may not be older than the fossil human skull found in a limestone cave at Broken Hills, Rhodesia. It is quite possible—nay, even probable—that the Taungs anthropoid and Rhodesian man were contemporaries. Students of man's evolution have sufficient evidence to justify them in supposing that the phylum of man had separated from that of anthropoid apes early in the Miocene period. The Taungs ape is much too late in the scale of time to have any place in man's ancestry.

In a large diagram, placed in the show-case at Wembley, Prof. Dart gives his final conception of the place occupied by the Taungs ape in the scale of man's evolution. He makes it the foundation stone of the human family tree. From the "African Ape Ancestors, typified by the Taungs Infant", Pithecanthropus, Piltdown man, Rhodesian man, and African races radiate off. A genealogist would make an identical mistake were he to claim a modern Sussex peasant as the ancestor of William the Conqueror.

In the show-case at Wembley plastic reconstructions are exhibited in order that visitors may form some conception of what the young Taungs Ape looked like in life. Although the skull is anthropoid it has been marked by a "make-up" into which there have been incorporated many human characters. It is true the ears are those of the chimpanzee, but the forehead is smooth and rounded, the hair of the scalp is sleek and parted; the bushy eyebrows are those of a man at fifty-five or sixty; the neck is fat, thick, and full—extending from chin to occiput. In modelling the nose, gorilla lines have been followed, whereas the nasal part of the skull imitates closely chimpanzee characters. The mouth is wide, with a smile at each corner.

Prof. Dart has made a discovery of great importance, and the last thing I want to do is to detract from it. He has shown that anthropoid apes had extended, during the Pleistocene period, right into South Africa—into a land where anthropoid apes could not gain a livelihood today. He has found an extinct relative of the chimpanzee and gorilla but one with more man-like features than are possessed by either of these. His discovery throws light on the history of anthropoid apes but not on that of man. Java-man (Pithecanthropus) still remains the only known link between man and ape, and this extinct type lies on the human side of the gap.

(**116**, 11; 1925)

汤恩头骨在各个关键方面，都显示出一只幼年类人猿的特点，它的脑在尺寸上确实比相应年龄的大猩猩的小。只在较欠发育的牙齿、颌骨和与咀嚼相关的骨质结构方面，可以将它称为一种比大猩猩更高等的似人动物。其刚刚萌出的第一恒臼齿只比大猩猩的略小，而面部针对上恒犬齿所作的准备则表明这些牙齿是属于大型类人猿那一类的。

我们等待的另外一个方面的资料与汤恩头骨的地质学年代有关。幸运的是，罗伯特·布鲁姆博士（《自然》，4月18日，第569页）已经对这个问题进行了阐述，并且他的说法被很多人接受。这个头骨是从一个填满了来自卡拉哈里沙漠的沙子的山洞里炸出来的。在旁边山洞发现的狒狒化石与南非现存的狒狒只在微小的结构细节上存在差异。布鲁姆博士认为，汤恩头骨的地质学年代较近；应该不早于更新世；他认为一种可能的情况是，这具头骨的年代并不比在罗德西亚布罗肯希尔山的石灰石山洞中发现的人类头骨化石的时期早。非常可能——不，几乎可以肯定的是——汤恩类人猿和罗德西亚人是同时代的。研究人类进化的学者们有足够的证据证实他们所假设的人类这一支系是在中新世早期从类人猿中分离出来的。汤恩猿在时间尺度上太晚了，所以根本不可能是人类的祖先。

在温布利的陈列柜里摆放的一张大图表中，达特教授给出了他对汤恩猿在人类进化上的位置的最终想法。他把汤恩猿当作奠定人类家族树的基石。从“汤恩婴儿代表的非洲猿祖先”开始，爪哇猿人、皮尔当人、罗德西亚人和非洲人种呈辐射状散出。如果一位系谱专家宣称一位现代萨塞克斯的农民是征服者威廉的祖先，那他就是犯了同样的错误。

在温布利的陈列柜里，陈列着复原像的造型以便参观者可以对幼年汤恩猿在生活状态下看起来是什么样子有一定的概念。尽管这具头骨是类人猿的，但是许多人类特征被“捏造”出来表现在这件复原像上。确实其耳朵是黑猩猩的，但是前额平滑而圆润，头皮上的头发光滑而疏散；浓密的眉毛与55~60岁的人类很像；脖子粗、厚而丰满——从下巴一直延伸到枕部。在制作鼻子的模型时，参照了大猩猩的线条，然而这具头骨的鼻子部分模仿了与黑猩猩非常接近的特征。嘴部宽阔，嘴角带笑。

达特教授取得了非常重要的发现，但是我想做的最后一件事就是贬低它。他向我们展示了类人猿在更新世时期曾经一直扩展到南非——这是一片现在的类人猿无法生存的地域。他发现了黑猩猩和大猩猩的一种已灭绝的亲属，但是它又具有比起二者来更像人类的特征。他的发现对于类人猿历史有一定的昭示作用，但是对人类历史并不尽然。爪哇猿人仍然是唯一已知的人和猿之间的环节，这种已灭绝的类型处于缺口中人类这一边。

（刘皓芳 翻译；吴新智 审稿）

The Taungs Skull

Editor's Note

Here Raymond Dart responds to Arthur Keith's assertive criticism of Dart's claim to have identified an intermediate form between man and ape, called *Australopithecus africanus*. Dart refutes Keith's charges robustly at every point except one, which relates to the question of whether a cast of the skull should be made available to other anthropologists.

IN *Nature* of July 4, 1925, p. 11, Sir Arthur Keith has attempted to show first that I called the Taungs skull a "missing link", and secondly, that it is not a "missing link".

As a matter of fact, although I undoubtedly regard the description as an adequate one, I have not used the term "missing link." On the other hand, Sir Arthur Keith in an article entitled "The New Missing Link" in the *British Medical Journal* (February 14, 1925) pointed out that "it is not only a missing link but a very complete and important one". After stating his views so definitely in February, it seems strange that, in July, he should state that "this claim is preposterous".

Despite this reversal of opinion, Sir Arthur tells us that the skull "does show human-like traits in the refinement of its jaws and face which is not met with in young gorillas and chimpanzees at a corresponding age." He appears to have overlooked the fact that in addition to these and other facts brought forward by myself, the temporal bone, sutures, and deciduous and permanent teeth (according to Dr. Robert Broom) also show human-like traits. Moreover, as Prof. Sollas has so ably shown, the whole profile of the skull is entirely different from that in living anthropoids, thus indirectly confirming my discovery that the brain inside the skull-dome which caused this profound difference was very different from the brains inside the skulls of modern apes.

The fact that Sir Arthur was unable to find the *parallel sulcus* depression in the replica cast sent to Wembley illustrates how unsatisfactory the study of the replica can be in the absence of the original.

With reference to the question of endocranial volume, I would state with Prof. Sollas that this "is a matter of only secondary importance". Nothing could exemplify this matter better than the condition of affairs in the Boskop race, where the endocranial volume was in the vicinity of 1,950 c.c. (The average European's endocranial volume is 1,400–1,500 c.c.) Indeed, the world's record in human endocranial volume (2,000 c.c.) was discovered in a "boskopoid" skull by Prof. Drennan in a dissecting room subject at Capetown this year. It is well known that the elephant and the whale have brains much larger than those of

汤恩头骨

编者按

雷蒙德·达特此前声称发现并鉴定了一种介于人与古猿之间，被命名为南方古猿非洲种的中间类型，阿瑟·基思对此提出了批评。在这里，雷蒙德·达特对阿瑟·基思的批评作出了回应。除了是否应该将这具头骨的模型对其他人类学家开放的问题，达特对基思的每一条意见都进行了反驳。

在 1925 年 7 月 4 日的《自然》第 11 页，阿瑟·基思爵士首先试图表明我把汤恩头骨称为一个“缺失的环节”，其次，试图表明汤恩头骨不是一个“缺失的环节”。

事实上，尽管我毫不怀疑地认为这种描述是适当的，但是我并没有使用过“缺失的环节”这个短语。另一方面，阿瑟·基思爵士在发表于《英国医学杂志》（1925 年 2 月 14 日）上的一篇题为《新的缺失环节》的文章里指出“它不仅是一个缺失的环节，而且是一个非常完整而重要的环节”。在他 2 月份如此确定地陈述了自己的观点之后，在 7 月他又说“这种说法是荒谬可笑的”，这似乎有点奇怪。

尽管有这样矛盾的观点，阿瑟爵士还是告诉我们这个头骨“在颌骨和面部的细节上确实显示出了与人类相像的特征，这种特征在相应年龄的幼年大猩猩和黑猩猩中都是不曾见到过的。”看起来他似乎忽略了一个事实，那就是除了我本人提出来的这些和其他事实之外，颞骨、骨缝、乳齿和恒牙（根据罗伯特·布鲁姆）也显示出了与人类相像的特征。另外，索拉斯教授也非常巧妙地说明了该头骨的整个剖面与现存的类人猿完全不同，因此间接地证实了我的发现，即引起这种深刻差异的位于头骨之内的脑与现代猿的头骨内的脑差异很大。

阿瑟爵士未能发现送到温布利的复制品模型上的**平行沟**凹陷，这个事实显示了在缺乏原型时只是对复制品进行研究会多么令人不满。

关于颅腔容量的问题，我已经与索拉斯教授共同声明过，即这“只是第二重要的事情”。除了颅腔容量接近 1,950 毫升（欧洲人的颅腔容量平均值为 1,400 ~ 1,500 毫升）的博斯科普人（南非石器时代中期的一个人种）的情况之外，可能没有什么更好的可以用来证明这个问题的例子了。实际上，人类颅腔容量的世界纪录（2,000 毫升）是德雷南教授今年在开普敦的一间解剖室里发现的一具“类博斯科普人的”头骨的容量。众所周知，大象和鲸的脑比人类的大得多，但是没有人会从这一点得出它们

human beings, but no one has inferred from that that their intelligence is greater. It is fairly certain that size of brain has some relation to size of body, as Dubois has shown. It is highly probable that the australopithecid man-apes were relatively small as compared with the gorilla. It is not the quantity so much as the quality of the brain that is significant.

Sir Arthur is harrowed unduly lest the skull *may* be Pleistocene. It is significant in this connexion that Dr. Broom, who first directed attention to this possibility (of which I was aware before my original paper was sent away), regarded it nevertheless as "the forerunner of such a type as Eoanthropus". It should not need explanation that the Taungs infant, being an infant, was ancestral to nothing, but the family that he typified are the nearest to the prehuman ancestral type that we have.

In view of these facts, there is little justification for the attempted witticism that in making the "African ancestors typified by the Taungs infant" the "foundation stone of the human family tree"—whatever that may be—I am making "a mistake identical with that of claiming a Sussex peasant as the ancestor of William the Conqueror". This is merely a case of mistaken identity on the part of Sir Arthur. I have but translated into everyday English the genealogical table suggested by Dr. Robert Broom (*Nature*, April 18, 1925), with which I agree almost entirely. I take it, however, as a mark of his personal favour that Sir Arthur should have attacked my utterance and spared Dr. Broom's.

Sir Arthur need have no qualms lest his remarks detract from the importance of the Taungs discovery—criticism generally enhances rather than detracts. Three decades ago Huxley refused to accept Pithecanthropus as a link. Today Sir Arthur Keith regards Pithecanthropus as the only known link. There is no record that Huxley first accepted it, then retracted it, but history sometimes repeats itself.

Raymond A. Dart

* * *

Prof. Dart is under a misapprehension in supposing that I have in any way or at any time altered my opinion regarding the fossil ape discovered at Taungs. From the description and illustrations given by him (*Nature*, Feb. 7, 1925, p. 195) the conclusion was forced on me that Australopithecus was a member of "the same group or sub-family as the chimpanzee and gorilla" (*Nature*, Feb. 14, 1925, p. 234). In the same issue of *Nature*, Prof. G. Elliot Smith expressed a similar opinion, describing Australopithecus as "an unmistakable anthropoid ape that seems to be much on the same grade of development as the gorilla and chimpanzee without being identical with either."

All the information which has come home since Prof. Dart made his original announcement in *Nature* has gone to support the close affinity of the Taungs ape to the gorilla and to the chimpanzee—it is a member of that group. Prof. Bolk, of Amsterdam,

的智慧也比人类高的推论。正如杜波伊斯指出的，脑的尺寸肯定与身体的尺寸有一定的关系。有一种很大的可能性是，与大猩猩相比，南方古猿属的人猿相对较小。显然脑的量不如质那样具有重大意义。

阿瑟爵士因为唯恐该头骨**可能**是更新世时期的而过度苦恼了。在这个关系上，值得注意的是，首先将注意力放到这种可能性（我意识到这种可能性是在我的初稿送出去之前）上的人是布鲁姆博士，他认为其不过是“像曙人一样类型的先祖”。需要解释的不是汤恩婴儿作为一个婴儿能不能作为任何生物的祖先，而是他所代表的科与我们拥有的人类之前的祖先型的亲缘关系是否是最近的。

鉴于这些事实，企图抓住将“汤恩婴儿代表的非洲祖先”当作“人类家族树的基石”——无论可能是什么——这些话，说我正在犯着“与宣称一位萨塞克斯农民是征服者威廉的祖先同样的错误”这种风凉话，几乎是没有道理的。这只不过是阿瑟爵士自己搞错罢了。我只是将罗伯特·布鲁姆建议的家族树（《自然》，1925 年 4 月 18 日）翻译成日常用的英语，对于该家族树，我几乎完全赞成。阿瑟爵士攻击我的意见，却没有批评布鲁姆博士的意见，我将这当作是他个人偏爱的一个标志。

阿瑟爵士不必担心他的言论是否会贬低汤恩发现的重要性——批评通常会提高影响而不会贬低影响。大约 30 年前，赫胥黎拒绝接受爪哇猿人作为一个环节而存在。今天阿瑟·基思爵士又认为爪哇猿人是唯一一个已知的环节。虽然没有关于赫胥黎首先接受它、而后又摒弃它的记载，但是历史有时是会重演的。

雷蒙德·达特

* * *

达特教授认为我随随便便改变对汤恩发现的猿化石的观点，他的这种想法是一种误解。从他给出的描述和说明（《自然》，1925 年 2 月 7 日，第 195 页）来看，他是将如下结论强加于我，即南方古猿是“与黑猩猩和大猩猩一样的群体或亚科”的成员之一（《自然》，1925 年 2 月 14 日，第 234 页）。在同期的《自然》中，埃利奥特·史密斯教授表达了相似的观点，将南方古猿描述为“肯定是一只类人猿，该类人猿似乎处于与大猩猩和黑猩猩同样的发育阶段，但是与两者又都不相同。”

自从达特教授在《自然》上最初公布结果之后，这里的所有信息都趋向于支持汤恩猿与大猩猩和黑猩猩存在亲密的亲缘关系——认为它是那一群体的成员之一。阿姆斯特丹的博尔克教授和俄亥俄州克利夫兰的温盖特·托德教授都注意到了一个事

and Prof. Wingate Todd, of Cleveland, Ohio, have directed attention to the fact that the skulls of occasional gorillas show the same kind of narrowing and lengthening as has been observed in that of the Taungs ape. Prof. Arthur Robinson has shown that there is a wide variation in the size of jaws of young chimpanzees of approximately the same age, the smaller of the jaws approaching in size and shape to the development seen in the Taungs ape. The dimensions of the erupting first permanent molar of the Taungs ape and the form of its cusps point to the same conclusion—that Australopithecus must be classified with the chimpanzee and gorilla. It is, therefore, "preposterous" that Prof. Dart should propose to create "a new family of Homo-simiadae for the reception of the group of individuals which it (Australopithecus) represents". It is preposterous because the group to which this fossil ape belongs has been known and named since the time of Sir Richard Owen.

The position which Prof. Dart assigns to the Taungs ape in the genealogical tree of man and ape has no foundation in fact. A large diagram in the exhibition in Wembley, prepared by Prof. Dart, informs visitors that the Taungs ape represents the ancestor of all forms of mankind, ancient and modern. Before making such a claim one would have expected that due inquiry would first be made as to whether or not the geological evidence can justify such a claim. From his letter one infers that Prof. Dart does not set much store by geological evidence. Yet it has been customary, and I think necessary, to take the time element into account in constructing pedigrees of every kind. Dr. Robert Broom and, later, Prof. Dart's colleague, Prof. R. B. Young, have reviewed the evidence relating to the geological antiquity of the Taungs fossil skull, and on data supplied by them one can be certain that early and true forms of men were already in existence before the ape's skull described by Prof. Dart was entombed in a cave at Taungs. To make a claim for the Taungs ape as a human ancestor is therefore "preposterous".

Finally, Prof. Dart reminds me that whales and elephants have massive brains and that many large-headed men and women show no outstanding mental ability. Still the fact remains that every human being whose brain fails to reach 850 grams in weight has been found to be an idiot. Size as well as convolutionary pattern of brain have to be taken into account in fixing the position of every fossil type of being that has any claim to be in the line of human evolution—the Taungs brain cast at Wembley possesses no feature which lifts it above an anthropoid status.

Arthur Keith

(**116**, 462-463; 1925)

Raymond A. Dart: University of the Witwatersrand, Johannesburg.
Arthur Keith: Royal College of Surgeons of England, Lincoln's Inn Fields, London, W.C., September 5.

实，即个别的大猩猩头骨显示出与在汤恩猿中观察到的同样形式的变窄、拉长的现象。阿瑟·鲁宾逊教授也表明大约同样年龄的幼年黑猩猩的颌骨大小存在很广泛的变异范围，其中稍小的颌骨就与汤恩猿中见到的大小和形状的发育程度接近。汤恩猿正在萌出的第一颗恒臼齿的尺寸和其牙尖的形式指向同样的结论——南方古猿肯定与黑猩猩和大猩猩属于同类。因此达特教授建议创建“一个名为人猿科的新科来接纳标本（南方古猿）代表的生物类群”的想法是“荒谬可笑的”。因为这例化石猿所属的群体是已知的，早在理查德·欧文先生时期就已经被命名过了，所以说他是荒谬可笑的。

达特教授给汤恩猿在人类和猿类家族树上指定的位置事实上没有任何基础。陈列在温布利的一幅达特教授制定的大图表告诉参观者，汤恩猿代表了古代和现代所有人类类型的祖先。一个人在提出这样的主张之前，应该会想到先通过适当的调查，看看地质学证据是否支持这种主张。从达特教授的信中可以推断出他并没有对地质学证据投入太多精力。但是通常习惯性的做法是，在构建每一类谱系时将时间因素考虑在内，我觉得这是必需的。罗伯特·布鲁姆博士审视过与汤恩头骨化石的地质学年代相关的证据，后来达特教授的同事扬教授也对此作过调查，根据他们所提供的数据，可以肯定早期的真正的人类类型早在达特教授描述的猿类头骨埋入汤恩洞穴之前就已经存在了。因此声称汤恩猿是人类祖先是“荒谬可笑的”。

最后，达特教授提醒我，鲸和大象具有巨大的脑，并且许多大脑袋的男人和女人并没有过人的智力。但是事实仍然是，研究发现每个脑重量不足 850 克的人都是白痴。在确定每种与人类进化世系有所关联的生物化石类型的地位时，脑的大小和脑回式样都是必须考虑的因素——温布利的汤恩脑模型不具备将其提升到类人猿以上地位的任何特征。

阿瑟·基思

（刘皓芳 翻译；吴新智 审稿）

Tertiary Man in Asia: the Chou Kou Tien Discovery*

D. Black

Editor's Note

The fossil-bearing cave site of Chou Kou Tien (Zhoukoudian in China) had been discovered in 1921, and soon yielded bones of various fossil mammals from horses to bats. This note is a secondary account of discoveries made to date that included, notably, two teeth attributable to "*Homo ? sp.*", the first human remains known to science in mainland Asia. The reporter, Canadian-born Davidson Black, Anatomy Professor at Peking Union Medical College, would go on to make his name at Chou Kou Tien with discoveries of other remains that he would call Sinanthropus, now regarded as *Homo erectus*. Black died in his office of heart problems in 1934, the remains of "Peking Man" close by. He was 49.

A rich fossiliferous deposit at Chou Kou Tien, 70 li [about 40 kilometres] to the southwest of Peking, was first discovered in the summer of 1921 by Dr. J. G. Andersson and later surveyed and partially excavated by Dr. O. Zdansky. A preliminary report on the site was published by Dr. Andersson in March 1923 (*Mem. Geol. Surv. China*, Ser. A, No. 5, pp. 83–89), followed in October of that year by a brief description of his survey by Dr. Zdansky (*Bull. Geo. Surv. China*, No. 5, pp. 83–89). The material recovered from the Chou Kou Tien cave deposit has been prepared in Prof. Wiman's laboratory in Upsala and afterwards studied there by Dr. Zdansky. As a result of this research, Dr. Andersson has now announced that in addition to the mammalian groups already known from this site, there have also been identified representatives of the Cheiroptera, one cynopithecid, and finally two specimens of extraordinary interest, namely, one premolar and one molar tooth of a species which cannot otherwise be named than *Homo ? sp.*

Judging from the presence of a true horse and the absence of Hipparion, Dr. Andersson in his preliminary report considered that the Chou Kou Tien fauna was possibly of Upper Pliocene age, an opinion also expressed by Dr. Zdansky. It is possible, however, in the light of recent research, that the horizon represented by this site may be of Lower Pleistocene age. Whether it be of late Tertiary or of early Quaternary age, the outstanding fact remains that, for the first time on the Asiatic continent north of the Himalayas, archaic hominid fossil material has been recovered, accompanied by complete and certain geological data. The actual presence of early man in eastern Asia is therefore now no longer a matter of conjecture.

---

* Announcement of the Chou Kou Tien discovery was first made by Dr. J. G. Andersson on the occasion of a joint scientific meeting of the Geological Society of China, the Peking Natural History Society and the Peking Union Medical College held in Peking on October 22, 1926, in honour of H. R. H. the Crown Prince of Sweden.

亚洲的第三纪人：周口店的发现*

步达生

编者按

1921 年人们在中国发现了藏有化石的周口店洞穴遗址，很快地，从中发掘出了从马到蝙蝠等多种哺乳动物的化石标本。这篇文章间接地介绍了之前在周口店遗址取得的各种发现，其中包括两件著名的来自“人属（种名不确定）”的牙骨，以及对于科学界来说亚洲大陆的首例人类化石。文章作者是出生于加拿大的戴维森·步达生，当时他是北京协和医学院的解剖学教授，后来他因为在周口店发现了一些其他的化石（他称之为中国猿人，现在被称为直立人）而闻名于世。1934 年，49 岁的步达生因心脏病死于办公室中，当时“北京人”的化石就在他身旁。

1921 年夏天，安特生博士在北京西南 70 里 [大约 40 公里] 的周口店首次发现了一套富含化石的沉积层，日贾尔斯基博士随后进行了调查及部分发掘工作。1923 年 3 月，有关这个遗址的初步报告由安特生博士发表（《中国地质专报》，A 辑，第 5 期，第 83~89 页）。同年 10 月，日贾尔斯基博士发表了他的初步调查结果（《中国地质学会会志》，第 5 期，第 83~89 页）。从周口店洞穴沉积中发现的材料在乌普萨拉的威曼教授的实验室进行修整，之后日贾尔斯基博士在那里对这些材料进行了研究。安特生博士发布了这项研究的一个结果：在这个化石点，除了已知的哺乳动物类群，还发现了一些翼手目的代表性物种，一个猕猴的标本，以及两个非常有趣的标本，即一个前臼齿和一个臼齿，对应的物种好像只能称为“人属（种名不确定）”。

根据存在真正的马以及没有发现三趾马来判断，安特生博士在他初期的报道中认为周口店动物群属于上新世早期，日贾尔斯基博士也表达了同样的观点。最近的研究表明，这一化石点代表的层位更可能属于上新世晚期。无论它属于第三纪晚期还是第四纪早期，最引人注目的事实是，这是第一次在喜马拉雅山北部的亚洲大陆发现原始人类的化石材料，并且具有完整的相关地质学资料。从此，关于早期人类事实上存在于东亚的说法不再只是一种猜测。

* 1926 年 10 月 22 日，为欢迎瑞典王储，中国地质学会、北京自然历史学会和北京协和医学院联合举办了科学会议，会上安特生博士首次宣布了周口店的发现。

While a complete description of these very important specimens may shortly be expected in *Palaeontologia Sinica*, the following brief notes may be of interest here. One of the teeth recovered is a right upper molar, probably the third, the relatively unworn crown of which presents characters appearing from the photographs to be essentially human. The posterior moiety of the crown is narrow and the roots appear to be fused. The other tooth is probably a lower anterior premolar, of which the crown only is preserved. The latter also is practically unworn, and appears in the photograph to be essentially bicuspid in character, a condition usually to be correlated with a reduction of the upper canine.

The Chou Kou Tien molar tooth, though unworn, would seem to resemble in general features the specimen purchased by Haberer in a Peking native drug shop and afterwards described in 1903 by Schlosser. The latter tooth was a left upper third molar having a very much worn crown, extensively fused lateral roots, and from the nature of its fossilisation considered by Schlosser to be in all probability Tertiary in age. It was provisionally designated as *Homo? Anthropoide?* It is of more than passing interest to recall that Schlosser, in concluding his description of the tooth, pointed out that future investigators might expect to find in China a new fossil anthropoid, Tertiary man or ancient Pleistocene man. The Chou Kou Tien discovery thus constitutes a striking confirmation of that prediction.

It is now evident that at the close of Tertiary or the beginning of Quaternary time man or a very closely related anthropoid actually did exist in eastern Asia. This knowledge is of fundamental importance in the field of prehistoric anthropology; for about this time also there lived in Java, Pithecanthropus; at Piltdown, Eoanthropus; and, but very shortly after, at Mauer, the man of Heidelberg. All these forms were thus practically contemporaneous with one another and occupied regions equally far removed respectively to the east, to the south-east, and to the west from the central Asiatic plateau which, it has been shown elsewhere, most probably coincides with their common dispersal centre. The Chou Kou Tien discovery therefore furnishes one more link in the already strong chain of evidence supporting the hypothesis of the central Asiatic origin of the Hominidae.

(**118**, 733-734; 1926)

Davidson Black: Department of Anatomy, Peking Union Medical College, Peking, China.

关于这些非常重要的标本的完整描述在《中国古生物志》上发表之前，以下这些简要的描述可能是有价值的。发现的牙齿中有一个是右上臼齿，可能是第三个，它的照片显示，牙冠几乎没有磨损，所显示出的基本上是人类的特征。其牙冠后半部分狭窄，根部融合。另一个牙齿可能是一个前部的下前臼齿，只有牙冠保存了下来。这一个也几乎没有磨损，它的照片显示出基本的二尖齿特征，这种情况通常与上犬齿退化有关。

周口店臼齿，尽管没有磨损，其总体特征却和哈贝雷尔在北京当地的一个药店买到的标本（1903 年，施洛瑟描述了这一标本的特征）似乎相似。这一来自药店的牙齿是一颗左上第三臼齿，其牙冠磨蚀严重，外侧两齿根大部分合并。从它石化的性质考虑，施洛瑟认为它很可能是第三纪的。它被临时归属于人属？或者猿属？回想起来，非常有趣的是，当施洛瑟总结他对那颗牙齿所作的描述时曾指出，将来的调查者将有希望在中国发现一种新的类似人的化石第三纪人或古老更新世人。周口店的发现是对那个预测的一个惹人注目的确认。

现在很明显了，在第三纪结束或第四纪开始之际，在东亚的确存在人类或者说一种与人类关系密切的类人猿。这在史前人类学领域是一个重要的基本发现。在大约同一个时期，爪哇存在爪哇直立猿人，皮尔当存在曙人，之后不久在毛尔地区出现海德堡人。所有这些生物类型都是属于同一时期的，并且分别在东部、东南部和西部各自占据了距中亚平原相同距离的区域，而其他的证据表明，中亚平原极有可能是它们共同的扩散中心。周口店的发现，在已有的有关人类中亚起源假说的有力证据链中又增添了一环。

（刘晓辉 翻译；徐星 审稿）

Thermal Agitation of Electricity in Conductors

J. B. Johnson

Editor's Note

In the early days of electronics and telephone communications, engineers aimed to reduce the noise in their circuits to a minimum. But there are fundamental limits, as physicist and engineer John Johnson of Bell Telephone Laboratories here reports. Johnson shows that ordinary electrical conductors exhibit spontaneous, thermally induced fluctuations of voltage, even when not being driven by external currents. This voltage noise depends directly on the temperature of the sample, leading Johnson to surmise that it must arise from continual agitation and excitation by random thermal energy. Thus, he concludes, the minimum voltage that can be usefully amplified in electronic circuits is limited by the very matter from which they are built.

ORDINARY electric conductors are sources of spontaneous fluctuations of voltage which can be measured with sufficiently sensitive instruments. This property of conductors appears to be the result of thermal agitation of the electric charges in the material of the conductor.

The effect has been observed and measured for various conductors, in the form of resistance units, by means of a vacuum tube amplifier terminated in a thermocouple. It manifests itself as a part of the phenomenon which is commonly called "tube noise". The part of the effect originating in the resistance gives rise to a mean square voltage fluctuation V^2 which is proportional to the value R of that resistance. The ratio V^2/R is independent of the nature or shape of the conductor, being the same for resistances of metal wire, graphite, thin metallic films, films of drawing ink, and strong or weak electrolytes. It does, however, depend on temperature and is proportional to the absolute temperature of the resistance. This dependence on temperature demonstrates that the component of the noise which is proportional to R comes from the conductor and not from the vacuum tube.

A similar phenomenon appears to have been observed and correctly interpreted in connexion with *a current sensitive* instrument, the string galvanometer (W. Einthoven, W. F. Einthoven, W. van der Horst, and H. Hirschfeld, *Physica*, 5, 358–360, No. 11/12, 1925). What is being measured in these cases is the effect upon the measuring device of continual shock excitation resulting from the random interchange of thermal energy and energy of electric potential or current in the conductor. Since the effect is the same for different conductors, it is evidently not dependent on the specific mechanism of conduction.

The amount and character of the observed noise depend upon the frequency-characteristic of the amplifier, as would be expected from experience with the small-shot

导体中电的热扰动

约翰逊

编者按

在电子通讯和电话通信的早期，工程师们的目标是将电路中的噪声降至最小。但贝尔电话实验室的物理学家、工程师约翰·约翰逊认为，噪声的减少是有基本极限的。约翰逊指出：即使在没有加载外部电流的情况下，导电体也会表现出由自发热扰动导致的电压波动。这种电压噪声与样品的温度直接相关，这使约翰逊联想到它应该是由随机热能的连续激励和扰动造成的。因此他得出以下结论：在电路中能够被有效放大的最小电压受到组成电路的材料的限制。

普通的导电体是电压自发涨落的来源，而这种波动电压可以用足够灵敏的仪器进行测量。导体的这种性质是导体材料中电荷热扰动的结果。

在各种导体中都已经观测到这类效应，并以电阻的形式，通过终端连接温差电偶的真空管放大器对此进行了测量。它通常作为所谓的“电子管噪声”现象的一部分而表现出来。源于电阻波动的这部分效应造成了电压均方值 V^2 的波动，此波动与电阻值 R 成正比。比率 V^2/R 与导体的性质和形状无关，对金属丝、石墨、金属薄膜、绘图墨水薄膜及强电解质或弱电解质来说，该比率都是相同的。然而，该比率却与温度有关，并与电阻的绝对温度成正比。这种温度依赖性表明，与 R 成比例的噪声分量来自导体，而不是来自真空管。

目前已经有人用**对电流灵敏**的仪器——弦线电流计观测到类似现象并正确地进行了解释（老艾因特霍芬、小艾因特霍芬、范德霍斯特以及赫希菲尔德，《物理学》，第 5 卷，第 358~360 页，第 11/12 期，1925 年）。在这些情况下测量到的是，导体中的热能与电势或电流能量的随机交换引起的持续冲击激发对测量装置的影响。因为该效应对不同导体是相同的，所以显然它与传导的特定机制无关。

正如从散粒效应的经验可以预期的那样，观测到的噪声的量及其特性与放大器的频率特性有关。在室温下，来自电阻的表观输入功率在 10^{-18} 瓦的量级。至少在声

effect. The apparent input power originating in the resistance is of the order 10^{-18} watt at room temperature. The corresponding output power is proportional to the area under the graph of *power amplification–frequency*, at least in the range of audio frequencies. The magnitude of the "initial noise", when the quietest tubes are used without input resistance, is about the same as that produced by a resistance of 5,000 ohms at room temperature in the input circuit. For the technique of amplification, therefore, the effect means that the limit to the smallness of voltage which can be usefully amplified is often set, not by the vacuum tube, but by the very matter of which electrical circuits are built.

(**119**, 50-51; 1927)

J. B. Johnson: Bell Telephone Laboratories, Inc., New York, N. Y., Nov. 17.

频范围内，相应的输出功率与**功率放大-频率**关系图中曲线下的面积成正比。当采用无输入电阻的最平稳的真空管时，“初始噪声”的量级约与在室温下输入回路中5,000欧姆的电阻产生的噪声量级相同。因此，对于放大技术而言，这种效应意味着可以被有效放大的电压的最小值与构成电路的材料有关，而不由真空管决定。

（沈乃澂 翻译；赵见高 审稿）

The Scattering of Electrons by a Single Crystal of Nickel

C. Davisson and L. H. Germer

Editor's Note

In his PhD thesis of 1924, Louis de Broglie hypothesized that electrons and other massive particles might behave as waves, just as photons reveal the particle-like aspects of otherwise wave-like electromagnetic radiation. This implied that material particles should exhibit wave-like phenomena such as interference or diffraction. Here Clinton Davisson and Lester Germer of the Bell Telephone Laboratories verified this prediction by measuring the diffraction of an electron beam from a nickel crystal. Using the known spacing of atomic planes in nickel, they were able to calculate the expected angles where diffracted beams should occur if the electrons were indeed acting as waves in the manner de Broglie suggested. The experimental results agreed with these predictions, and they won Davission the 1937 Nobel Prize in physics.

IN a series of experiments now in progress, we are directing a narrow beam of electrons normally against a target cut from a single crystal of nickel, and are measuring the intensity of scattering (number of electrons per unit solid angle with speeds near that of the bombarding electrons) in various directions in front of the target. The experimental arrangement is such that the intensity of scattering can be measured in any latitude from the equator (plane of the target) to within 20° of the pole (incident beam) and in any azimuth.

The face of the target is cut parallel to a set of {111}-planes of the crystal lattice, and etching by vaporisation has been employed to develop its surface into {111}-facets. The bombardment covers an area of about 2 mm^2 and is normal to these facets.

As viewed along the incident beam the arrangement of atoms in the crystal exhibits a threefold symmetry. Three {100}-normals equally spaced in azimuth emerge from the crystal in latitude 35°, and, midway in azimuth between these, three {111}-normals emerge in latitude 20°. It will be convenient to refer to the azimuth of any one of the {100}-normals as a {100}-azimuth, and to that of any one of the {111}-normals as a {111}-azimuth. A third set of azimuths must also be specified; this bisects the dihedral angle between adjacent {100}- and {111}-azimuths and includes a {110}-normal lying in the plane of the target. There are six such azimuths, and any one of these will be referred to as a {110}-azimuth. It follows from considerations of symmetry that if the intensity of scattering exhibits a dependence upon azimuth as we pass from a {100}-azimuth to the next adjacent {111}-azimuth (60°), the same dependence must be exhibited in the reverse

镍单晶对电子的散射

戴维森，革末

编者按

1924 年，路易斯·德布罗意在他的博士论文中猜测，电子以及其他一些有质量的粒子的运动方式可能与波类似，就像光子除了具有电磁辐射的波动行为以外，还显示出与此不同的粒子行为。这一假说意味着，物质粒子会表现出干涉或者衍射这样的波动特征。在这篇文章中，贝尔电话实验室的克林顿·戴维森和莱斯特·革末通过对从镍晶中发出的电子束的衍射效应进行测量证实了这一假说。如果正如德布罗意猜想的那样电子具有像波一样的性质，那么根据镍晶中已知的原子平面之间的间距，他们就能计算出应该出现衍射束的预期角度。实验结果与这些预测是相吻合的。戴维森因此获得了 1937 年的诺贝尔物理学奖。

现在正在进行的一系列实验中，我们用一窄束电子垂直轰击一块从镍单晶上切下来的靶子，并测量了靶前方不同方向上的电子散射强度（与每单位立体角中速度和轰击电子的速度相似的电子数量）。我们的实验装置可以测量从赤道（靶平面）到与轴（入射束）成 20° 范围内在任意纬度处和在任意方位角上的散射强度。

被切割下来的靶面平行于镍单晶点阵的一组 {111} 面，切下来之后用蒸气进行蚀刻，以使其表面形成一系列 {111} 小晶面。轰击的电子束打在约 2 平方毫米的范围内，并与以上的小晶面垂直。

沿着入射电子束的方向观察，我们发现原子在晶体中的排列呈现三重对称。从晶体出发，在纬度为 35° 处有三条相隔同样方位角的 {100} 法线，并且在这些方位角中间，有三条 {111} 法线出现在纬度 20° 处。为了方便起见，任何一条 {100} 法线的方位角都可以被看作是一个 {100} 方位角，任何一条 {111} 法线的方位角也可以被看作是一个 {111} 方位角。我们还必须指定第三组方位角；它将平分相邻的 {100} 方位角和 {111} 方位角之间的二面角，并且包含一条位于靶平面上的 {110} 法线。有 6 个这样的方位角，它们当中的任意一个都可以被看作是一个 {110} 方位角。考虑到对称性的要求，当我们从 {100} 方位角转到下一个相邻的 {111} 方位角 (60°) 时，如果散射强度的变化依赖于方位角的话，那么当我们从 60° 转到下一个相邻的

order as we continue on through 60° to the next following {100}-azimuth. Dependence on azimuth must be an even function of period $2\pi/3$.

In general, if bombarding potential and azimuth are fixed and exploration is made in latitude, nothing very striking is observed. The intensity of scattering increases continuously and regularly from zero in the plane of the target to highest value in co-latitude 20°, the limit of observations. If bombarding potential and co-latitude are fixed and exploration is made in azimuth, a variation in the intensity of scattering of the type to be expected is always observed, but in general this variation is slight, amounting in some cases to not more than a few percent of the average intensity. This is the nature of the scattering for bombarding potentials in the range from 15 volts to near 40 volts.

At 40 volts a slight hump appears near 60° in the co-latitude curve for azimuth-{111}. This hump develops rapidly with increasing voltage into a strong spur, at the same time moving slowly upward toward the incident beam. It attains a maximum intensity in co-latitude 50° for a bombarding potential of 54 volts, then decreases in intensity, and disappears in co-latitude 45° at about 66 volts. The growth and decay of this spur are traced in Fig. 1.

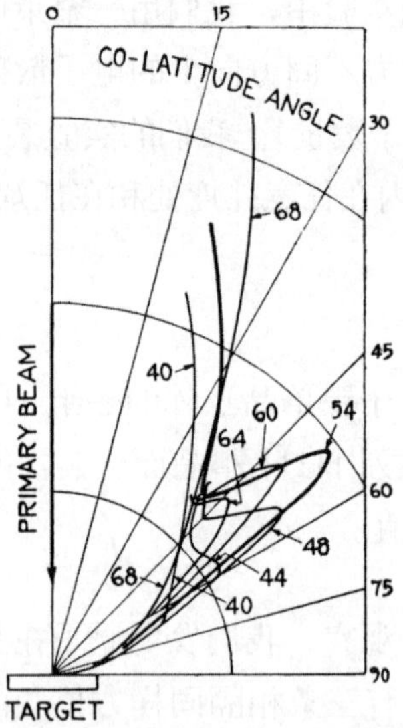

Fig. 1. Intensity of electron scattering vs. co-latitude angle for various bombarding voltages—azimuth-{111}-330°

A section in azimuth through this spur at its maximum (Fig. 2—Azimuth-330°) shows that it is sharp in azimuth as well as in latitude, and that it forms one of a set of three such spurs, as was to be expected. The width of these spurs both in latitude and in azimuth is almost completely accounted for by the low resolving power of the measuring device. *The spurs are due to beams of scattered electrons which are nearly if not quite as well defined as the primary*

{100} 方位角时，相同的依赖关系会按相反的次序再现。散射强度对方位角的依赖关系肯定是一个周期为 $2\pi/3$ 的偶函数。

一般来说，如果轰击电压和方位角固定，当沿着纬度测量的时候，不会观察到任何令人惊奇的现象。散射强度连续而有规律地从靶平面处的零增加到余纬 20° 时的最大值，余纬 20° 是我们的观测极限。如果轰击电压和余纬固定，而沿着方位角进行探测，则通常可以观察到预期之中的散射强度变化，但总的来说这种差别很小，在某些情况下总的变化量不会大于平均强度的百分之几。这就是散射在轰击电压从 15 伏特到接近 40 伏特之间变动时的特性。

当电压为 40 伏特时，在方位角为 {111} 的余纬曲线上接近 60° 处出现了一个小峰。随着电压的增加，这个峰迅速变大形成一个很强的尖峰，同时朝着入射电子束的方向缓慢向上提升。当轰击电压为 54 伏特，余纬为 50° 时，峰值达到最大，然后强度逐渐减小，在电压约为 66 伏特，余纬为 45° 时消失。图 1 描述了这个尖峰从增大到衰减的过程。

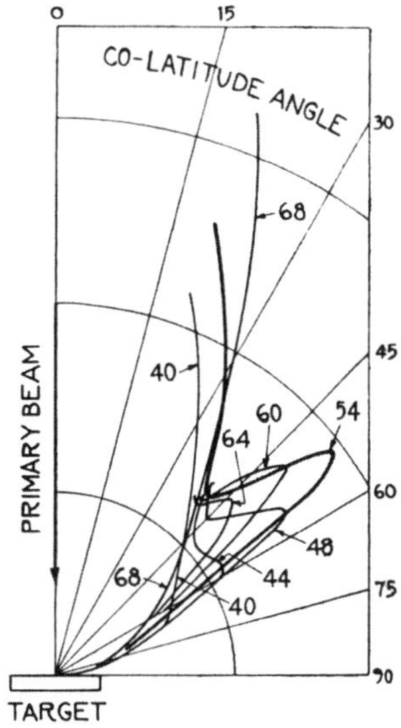

图 1. 不同轰击电压下电子散射强度随余纬角的变化——方位角{111}-330°

在最大值处穿过这个尖峰的方位角截面图（图 2 之方位角 330°）表明，它以方位角为变量的时候和以纬度为变量的时候一样尖锐，并且，如我们所料，它形成了 3 个尖峰组合中的一个。这些尖峰的宽度在随方位角变化和随纬度变化时都较大，几乎可以肯定地说这是由测量仪器的低分辨率造成的。**尖峰是由散射的电子束引起**

beam. The minor peaks occurring in the {100}-azimuth are sections of a similar set of spurs that attains its maximum development in co-latitude 44° for a bombarding potential of 65 volts.

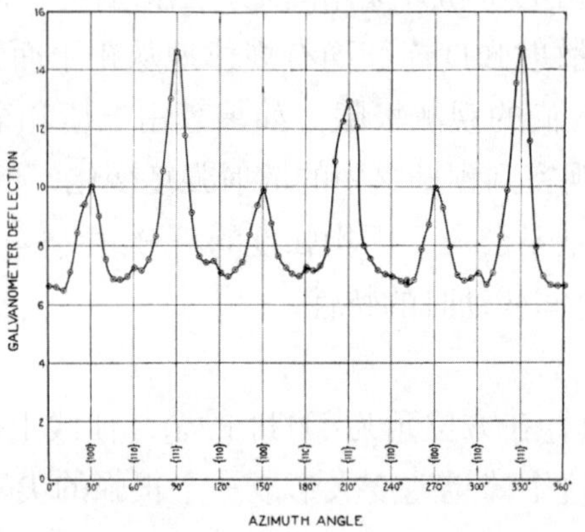

Fig. 2. Intensity of electron scattering vs. azimuth angle—54 volts, co-latitude 50°

Thirteen sets of beams similar to the one just described have been discovered in an exploration in the principal azimuths covering a voltage range from 15 volts to 200 volts. The data for these are set down on the left in Table I (columns 1–4). Small corrections have been applied to the observed co-latitude angles to allow for the variation with angle of the "background scattering", and for a small angular displacement of the normal to the facets from the incident beam.

Table I

	Electron Beams			X-ray Beams						
Azimuth	Bomb. Pot (volts)	Co-lat. θ	Intensity	Reflections	$\lambda\times10^8$ cm	Co-lat. θ	Co-lat. θ'	$v\times10^{-8}$ cm/sec	$n\lambda\times10^8$ cm	$n\left\{\frac{\lambda mv}{h}\right\}$
{111}	54	50°	0.5	{220}	2.03	70.5	52.7	4.36	1.65	0.99
	100	31	0.5	{331}	1.49	44.0	31.6	5.94	1.11	0.91
	174	21	0.9	{442}	1.13	31.6	22.4	7.84	0.77	0.83
	174	55	0.15	{440}	1.01	70.5	52.7	7.84	1.76	2(0.95)
{100}	65	44	0.5	{311}	1.84	59.0	43.2	4.79	1.49	0.98
	126	29	1.0	{422}	1.35	38.9	27.8	6.67	1.04	0.95
	190	20	1.0	{533}	1.04	28.8	20.4	8.19	0.74	0.83
	159	61	0.4	{511}	1.05	77.9	59.0	7.49	1.88	2(0.97)
{110}	138	59	0.07	{420}	1.22	78.5	59.5	6.98	1.06	1.02
	170	46	0.07	{531}	1.04	57.1	41.7	7.75	0.89	0.95
{111}	110	58	0.15	..	..	..	..	6.23	1.82	1.56
{100}	110	58	0.15	..	..	..	..	6.23	1.82	1.56
{110}	110	58	0.25	..	..	..	..	6.23	1.05	0.90

的，对这些散射电子束的界定即使不能像对原射线束的界定那样精确，至少也能相差不多。在 {100} 方位角上出现的次级峰是另一组类似的尖峰，它们在余纬 44°，轰击电压 65 伏特时达到最大值。

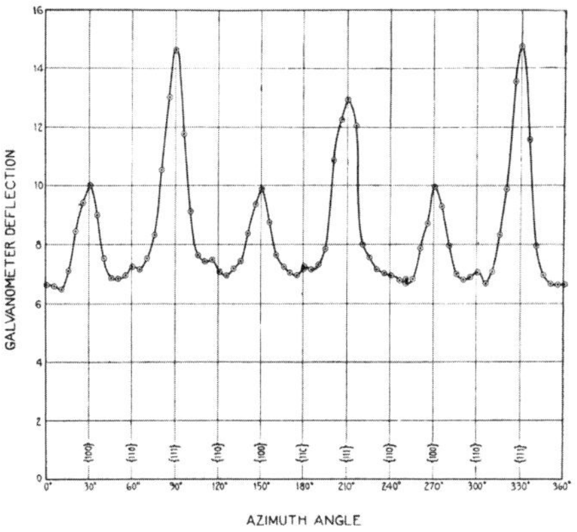

图 2. 电子散射强度随方位角的变化——54伏特，余纬50°

电压在从 15~200 伏特的范围内变动时，我们在主方位角方向发现了 13 组与上面所描述的电子束类似的射束。结果见表 I 左侧（第 1~4 列）。考虑到“背景散射”随角度的变化，以及晶面法线与入射电子束之间的角位移，我们对观察到的余纬角进行了微小的修正。

表 I

方位角	电子束			X 射线束						
	轰击电压（伏特）	余纬度 θ	强度	反射	$\lambda \times 10^8$厘米	余纬度 θ	余纬度 θ'	$v \times 10^{-8}$ 厘米/秒	$n\lambda \times 10^8$ 厘米	$n\left\{\frac{\lambda mv}{h}\right\}$
{111}	54	50°	0.5	{220}	2.03	70.5	52.7	4.36	1.65	0.99
	100	31	0.5	{331}	1.49	44.0	31.6	5.94	1.11	0.91
	174	21	0.9	{442}	1.13	31.6	22.4	7.84	0.77	0.83
	174	55	0.15	{440}	1.01	70.5	52.7	7.84	1.76	2(0.95)
{100}	65	44	0.5	{311}	1.84	59.0	43.2	4.79	1.49	0.98
	126	29	1.0	{422}	1.35	38.9	27.8	6.67	1.04	0.95
	90	20	1.0	{533}	1.04	28.8	20.4	8.19	0.74	0.83
	159	61	0.4	{511}	1.05	77.9	59.0	7.49	1.88	2(0.97)
{110}	138	59	0.07	{420}	1.22	78.5	59.5	6.98	1.06	1.02
	170	46	0.07	{531}	1.04	57.1	41.7	7.75	0.89	0.95
{111}	110	58	0.15	..	..	..	..	6.23	1.82	1.56
{100}	110	58	0.15	..	..	..	..	6.23	1.82	1.56
{110}	110	58	0.25	..	..	..	..	6.23	1.05	0.90

If the incident electron beam were replaced by a beam of monochromatic X-rays of adjustable wave-length, very similar phenomena would, of course, be observed. At particular values of wave-length, sets of three or of six diffraction beams would emerge from the incident side of the target. On the right in Table I (columns 5, 6 and 7) are set down data for the ten sets of X-ray beams of longest wave-length which would occur within the angular range of our observations. Each of these first ten occurs in one of our three principal azimuths.

Several points of correlation will be noted between the two sets of data. Two points of difference will also be noted; the co-latitude angles of the electron beams are not those of the X-ray beams, and the three electron beams listed at the end of the Table appear to have no X-ray analogues.

The first of these differences is systematic and may be summarised quantitatively in a simple manner. If the crystal were contracted in the direction of the incident beam by a factor 0.7, the X-ray beams would be shifted to the smaller co-latitude angles θ' (column 8), and would then agree in position fairly well with the observed electron beams—the average difference being 1.7°. Associated in this way there is a set of electron beams for each of the first ten sets of X-ray beams occurring in the range of observations, the electron beams for 110 volts alone being unaccounted for.

These results are highly suggestive, of course, of the ideas underlying the theory of wave mechanics, and we naturally inquire if the wave-length of the X-ray beam which we thus associate with a beam of electrons is in fact the h/mv of L. de Broglie. The comparison may be made, as it happens, without assuming a particular correspondence between X-ray and electron beams, and without use of the contraction factor. Quite independently of this factor, the wave-lengths of all possible X-ray beams satisfy the optical grating formula $n\lambda = d\sin\theta$, where d is the distance between lines or rows of atoms in the surface of the crystal—these lines being normal to the azimuth plane of the beam considered. For azimuths-{111} and-{100}, $d = 2.15\times10^{-8}$ cm and for azimuth-{110}, $d = 1.24\times10^{-8}$ cm. We apply this formula to the electron beams without regard to the conditions which determine their distribution in co-latitude angle. The correlation obtained by this procedure between wave-length and electron speed v is set down in the last three columns of Table I.

In considering the computed values of $n(\lambda mv/h)$, listed in the last column, we should perhaps disregard those for the 110-volt beams at the bottom of the Table, as we have had reason already to regard these beams as in some way anomalous. The values for the other beams do, indeed, show a strong bias toward small integers, quite in agreement with the type of phenomenon suggested by the theory of wave mechanics. These integers, one and two, occur just as predicted upon the basis of the correlation between electron beams and X-ray beams obtained by use of the contraction factor. The systematic character of the departures from integers may be significant. We believe, however, that this results from

如果入射电子束被一束波长可调的单色 X 射线取代，我们当然可以观察到非常类似的现象。当波长为一些特定值时，在靶的入射面一侧就会出现 3 条一组或 6 条一组的衍射光束。表 I 右侧（第 5、6、7 列）列出了在我们能观测到的角度范围内由波长最长的 10 组 X 射线束得到的数据。前 10 组中每组数据都来自我们先前设定的 3 个主方位角中的一个。

在以上表中关于电子束和 X 射线束的两部分数据中有几点关系值得我们注意。同时还要注意两个区别：电子束的余纬角不是 X 射线束的余纬角；对于表格底部列出的 3 行电子束数据，没有 X 射线束数据与之对应。

第一个区别是系统上的，可以用一种简单的方式定量地总结出来。如果晶体沿着入射束方向被压缩至 70%，即压缩因子为 0.7，那么 X 射线束将移至更小的余纬角 θ'（第 8 列），这样它的位置将与观测到的电子束的情况精确地吻合——平均差异为 1.7°。同理，对于我们能观察到的前 10 组 X 射线束中的每一个，都存在一组电子束与之对应，只有 110 伏特的电子束不能应用这种方法。

当然，这些结果完全证实了波动力学理论的基本观点，我们自然很想知道，与电子束相关联的 X 射线束的波长是否就是德布罗意所说的 h/mv。也许不用假设 X 射线束和电子束之间的特定关系，也不用使用压缩因子，我们就可以作出比较。所有的 X 射线束都满足光栅公式 $n\lambda = d\sin\theta$，这与压缩因子完全不相干，在这个公式中，d 是位于晶体表面的原子行或列的间距——这些行列都垂直于我们粒子束所在的方位角平面。对于方位角 {111} 和方位角 {100}，$d = 2.15 \times 10^{-8}$ 厘米；对于方位角 {110}，$d = 1.24 \times 10^{-8}$ 厘米。我们可以把这个公式应用于电子束中而无需考虑支配电子束沿余纬度分布情况的因素。用这种方式得到的波长和电子速度 v 之间的关系列于表 I 的最后 3 列中。

至于在最后一列中列出的计算值 $n(\lambda mv/h)$，我们可能应该不考虑表格底部 110 伏特的电子束数据，因为我们已经有理由把这些电子束看作是异常情况。对于其他电子束，$n(\lambda mv/h)$ 值趋向于一个比较小的整数，与波动力学理论预言的现象十分吻合。使用压缩因子对电子束和 X 射线束之间的差别进行校正之后，也得到了预想的结果，即整数 1 和 2。对整数的偏离可能主要是由系统特性造成的。我们认为误差来自入

imperfect alignment of the incident beam, or from other structural deficiencies in the apparatus. The greatest departures are for beams lying near the limit of our co-latitude range. The data for these are the least trustworthy.

(**119**, 558-560; 1927)

C. Davission, L. H. Germer: Bell Telephone Laboratories, Inc., New York, N.Y., Mar. 3.

射束不严格准直或者仪器的其他结构缺陷。最大的偏离出现在我们能观察到的余纬度极限附近，那里的数据是最不可靠的。

（王静 翻译；赵见高 审稿）

The Continuous Spectrum of β-Rays

C. D. Ellis and W. A. Wooster

Editor's Note

Since the discovery of radioactivity in 1895, three different mechanisms had been identified. In some cases an atom would disintegrate by shedding an α-particle (the nucleus of a helium atom), in others by emitting a γ-ray (a high-frequency X-ray). In both these mechanisms, the amount of energy lost in the disintegration was found to be the same for each kind of disintegrating atom. In the third mechanism of radioactive decay, however, in which the particle shed during the disintegration is an electron, the energy carried away by the particle is indeterminate, ranging from zero to a certain maximum characteristic of the atom concerned. This raises problems referred to in the letter below.

THE continuous spectrum of the β-rays arising from radio-active bodies is a matter of great importance in the study of their disintegration. Two opposite views have been held about the origin of this continuous spectrum. It has been suggested that, as in the α-ray case, the nucleus, at each disintegration, emits an electron having a fixed characteristic energy, and that this process is identical for different atoms of the same body. The continuous spectrum given by these disintegration electrons is then explained as being due to secondary effects, into the nature of which we need not enter here. The alternative theory supposes that the process of emission of the electron is not the same for different atoms, and that the continuous spectrum is a fundamental characteristic of the type of atom disintegrating. Discussion of these views has hitherto been concerned with the problem of whether or not certain specified secondary effects could produce the observed heterogenity, and although no satisfactory explanation has yet been given by the assumption of secondary effects, it was most important to clear up the problem by a direct method.

There is a ready means of distinguishing between the two views, since in one case a given quantity of energy would be emitted at each disintegration equal to or greater than the maximum energy observed in the electrons escaping from the atom, whereas in the second case the average energy per disintegration would be expected to equal the average energy of the particles emitted. If we were to measure the total energy given out by a known amount of material, as, for example, by enclosing it in a thick-walled calorimeter, then in the first case the heating effect should lead to an average energy per disintegration equal to or greater than the fastest electron emitted, no matter in what way this energy was afterwards split up by secondary effects. Since on the second hypothesis no secondary effects are presumed to be present, the heating effect should correspond simply to the average kinetic energy of the particles forming the continuous spectrum.

To avoid complications due to α-rays or to γ-rays from parent or successive atoms, we

β 射线的连续谱

埃里斯，伍斯特

编者按

自从 1895 年发现放射性以来，人们已经确认放射性存在三种不同的机制。在某些情况下，原子会通过放射出一个 α 粒子（氦原子的原子核）而衰变，而在另一些情况下，原子会通过放射 γ 射线（一种高频 X 射线）而衰变。在这两种机制中，每一种衰变原子在衰变过程中的能量损失都是相同的。然而在放射性衰变的第三种机制中，原子衰变时放射出的粒子是电子，这个电子携带的能量是不确定的，其范围是从零到相关原子的最大特征能量。这就引发了下文中提到的问题。

在物质衰变的研究中，放射性物质产生的 β 射线连续谱是一个非常重要的问题。关于这种连续谱的起源有两种相反的观点。一种观点认为，和放射出 α 射线的衰变一样，在每一次衰变中原子核都发射出一个具有固定的特征能量的电子，并且对同一物体的不同原子这个过程都是相同的。于是，这些衰变电子的连续谱就被归因为二次效应，而有关二次效应的本质我们不必在这里进行讨论。另外一种观点认为，发射电子的过程对不同原子是不同的，并且连续谱是衰变原子类型的基本特征。关于这两种观点的讨论至今停留在是否有某种具体的二次效应能够产生可观测的连续谱这一问题上，尽管二次效应的假设还没能给出一个令人满意的解释，但最重要的是，我们应该用一种直接的方法来解释以上的问题。

有一个简易的方法可以区分以上两种观点，因为从第一个观点来看，每次衰变中放出的特定能量应该等于或者大于观察到的从原子中逃逸出的电子的最大能量，而从第二种观点来看，每次衰变的平均能量应该等于发射的电子的平均能量。如果我们测量一定量的物质发射出的电子的总能量，例如将这些物质放在一个厚壁量热计中，那么按照第一种观点，由热效应就能得出每次衰变的平均能量，这个值应该是等于或大于发射出的最快的电子的能量，不管这个能量随后以什么方式被二次效应分解。而在第二种观点中，由于不存在二次效应，所以热效应应该简单地对应于形成连续谱的电子的平均动能。

为了避免衰变中母原子或后续原子产生的 α 射线或者 γ 射线的干扰，我们在

measured the heating effect in a thick-walled calorimeter of a known quantity of radium E. This measurement proved difficult because of the small rate of evolution of heat, but by taking special precautions it has been possible to show that the average energy emitted at each disintegration of radium E is 340,000 ± 30,000 volts. This result is a striking confirmation of the hypothesis that the continuous spectrum is emitted as such from the nucleus, since the average energy of the particles as determined by ionisation measurements over the whole spectrum gives a value about 390,000 volts, whereas if the energy emitted per disintegration were equal to that of the fastest β-rays, the corresponding value of the heating would be three times as large—in fact, 1,050,000 volts.

Many interesting points are raised by the question of how a nucleus, otherwise quantised, can emit electrons with velocities varying over a wide range, but consideration of these will be deferred until the publication of the full results.

(**119**, 563-564; 1927)

C. D. Ellis, W. A. Wooster: Cavendish Laboratory, Cambridge, Mar.23.

厚壁量热计中测量了一定量的镭 E 的热效应。由于放热速度很慢，这个测量很难进行，但是通过采用特别的措施，我们得到镭 E 每次衰变辐射出的平均能量是 340,000 ± 30,000 电子伏。这个结果对于连续谱来源于原子核中的发射这一假设是个有力的确证。因为通过电离测量整个能谱而得到的电子平均能量大约是 390,000 电子伏，如果每次衰变放出的能量等于最快的 β 射线的能量，那么相应的热效应的值就应该是当前测量值的 3 倍，即 1,050,000 电子伏。

一个量子化的原子核怎么能发射速度变化范围如此广泛的电子？这个问题引发了很多有趣的观点，不过还是等到详细的结果发表之后再来考虑这些。

（王锋 翻译；江丕栋 审稿）

A New Type of Secondary Radiation

C. V. Raman and K. S. Krishnan

Editor's Note

Physics in the early twentieth century was dominated by scientists in Europe and the United States. Yet a landmark discovery in quantum physics is reported here by two Indian physicists in Calcutta. As hitherto understood, light scattering from a stationary material object should preserve its frequency. But Chandrasekhara Venkata Raman and Kariamanickam Srinivasa Krishnan demonstrate that a small part of the scattered light can significantly change frequency. This "Raman effect" involves an exchange of energy between the scattered photons and the internal degrees of freedom of atoms or molecules. The effect is used today to probe molecular structure and motion, and the chemical nature of materials. For his discovery, Raman was awarded the 1930 Nobel Prize in physics.

IF we assume that the X-ray scattering of the "unmodified" type observed by Prof. Compton corresponds to the normal or average state of the atoms and molecules, while the "modified" scattering of altered wave-length corresponds to their fluctuations from that state, it would follow that we should expect also in the case of ordinary light two types of scattering, one determined by the normal optical properties of the atoms or molecules, and another representing the effect of their fluctuations from their normal state. It accordingly becomes necessary to test whether this is actually the case. The experiments we have made have confirmed this anticipation, and shown that in every case in which light is scattered by the molecules in dust-free liquids or gases, the diffuse radiation of the ordinary kind, having the same wave-length as the incident beam, is accompanied by a modified scattered radiation of degraded frequency.

The new type of light scattering discovered by us naturally requires very powerful illumination for its observation. In our experiments, a beam of sunlight was converged successively by a telescope objective of 18 cm aperture and 230 cm focal length, and by a second lens of 5 cm focal length. At the focus of the second lens was placed the scattering material, which is either a liquid (carefully purified by repeated distillation *in vacuo*) or its dust-free vapour. To detect the presence of a modified scattered radiation, the method of complementary light-filters was used. A blue-violet filter, when coupled with a yellow-green filter and placed in the incident light, completely extinguished the track of the light through the liquid or vapour. The reappearance of the track when the yellow filter is transferred to a place between it and the observer's eye is proof of the existence of a modified scattered radiation. Spectroscopic confirmation is also available.

Some sixty different common liquids have been examined in this way, and every one of them showed the effect in greater or less degree. That the effect is a true scattering and not a fluorescence is indicated in the first place by its feebleness in comparison with the

一种新型的二次辐射

拉曼，克里希南

编者按

20 世纪早期的物理学主要是由欧美科学家主导的，然而，一项在量子物理方面具有里程碑意义的发现却是由两名印度物理学家在加尔各答作出的。就当时人们所知，从一个静止实物散射出来的光线应该保持频率不变，这一点大家都可以理解，但钱德拉塞卡拉·文卡塔·拉曼和卡瑞马尼卡姆·斯里尼瓦桑·克里希南却用实验证实，有一小部分散射光频率变化很大。“拉曼效应”包含散射光子与原子或分子的内自由度之间能量的交换过程。现在利用这个效应可以检测分子结构和分子运动以及材料的化学性质。拉曼因发现了这个效应而获得了 1930 年的诺贝尔物理学奖。

如果我们假定，康普顿教授观察到的“不变”的 X 射线散射对应于原子和分子的正常态或平均态，而波长发生改变的“变”散射对应于原子和分子相对于正常态或平均态的涨落，那么我们就可以预测，普通光的散射应该也存在两种类型，一种取决于原子或分子的正常光学性质，另一种则代表了它们相对于正常态的涨落效应。因此有必要检验真实的情况是否确实如此。我们的实验证实了上述预测。实验表明，由任何一种无尘的液体或气体分子造成的光散射，都不仅包含了与入射光波长相同的正常漫射辐射，同时也伴随频率发生变化的变散射辐射。

要观察到我们发现的这种新型光散射，自然就需要非常强的光照。在我们的实验中，一束太阳光依次通过口径为 18 厘米、焦距为 230 厘米的望远镜物镜和一个焦距为 5 厘米的透镜而被会聚。在第二个透镜的焦点处放置散射材料，这些材料或者是在真空中反复蒸馏得到的非常纯净的液体，或者是无尘的蒸气。为了探测变散射辐射的存在，我们使用了互补的滤光片。当把一个蓝紫色滤光片和一个黄绿色滤光片一起放置在入射光处时，透过液体或蒸气的光路会完全消失。而当把入射光处的黄色滤光片移置到散射材料和观测者的眼睛之间时，透过散射材料的光路就会重新出现。这就证实了变散射辐射的存在。光谱分析的结果也证实了这一点。

我们用这种方法检测了六十多种不同的常见液体，所有结果中都或多或少地出现了这种效应。这是一种真正的散射效应而不是一种荧光现象，因为与普通的散射

ordinary scattering, and secondly by its polarisation, which is in many cases quite strong and comparable with the polarisation of the ordinary scattering. The investigation is naturally much more difficult in the case of gases and vapours, owing to the excessive feebleness of the effect. Nevertheless, when the vapour is of sufficient density, for example with ether or amylene, the modified scattering is readily demonstrable.

(**121**, 501-502; 1928)

C. V. Raman, K. S. Krishnan: 210 Bowbazar Street, Calcutta, India, Feb. 16.

相比它的强度非常微弱，而且在很多情况下它具有与普通的散射相当的非常强的偏振性。这种效应的强度非常微弱，因此要在气体和蒸气中开展这项研究自然是非常困难的。不过，当蒸气浓度足够大时，例如乙醚或戊烯的蒸气，还是很容易观察到变散射的。

（王锋 翻译；李芝芬 审稿）

Anomalous Groups in the Periodic System of Elements

E. Fermi

Editor's Note

Enrico Fermi was one of the outstanding physicists of the 1930s and 1940s. He is credited with the construction of the first atomic reactor at the University of Chicago in 1941. In the early 1930s he was more concerned with the structure of atoms, and this brief note is a summary of a paper afterwards published in the proceedings of the Accademia dei Lincei, an Italian scientific academy. The message Fermi wished to emphasise was that the first group of transition elements in the periodic table could be accounted for by theoretical calculations of the behaviour of electrons in atoms whose atomic weight exceeded 21. Fermi received a Nobel Prize in physics in 1938.

IN a paper which will shortly appear in the *Rend. Accad. Lincei*, I have calculated the distribution of the electrons in a heavy atom. The electrons were considered as forming an atmosphere of *completely degenerated* gas held in proximity to the nucleus by the attraction of the nuclear charge screened by the electrons. Formulae were given for the density of the electrons and the potential as functions of the distance r from the nucleus.

In continuation of the previous work, I have applied the same method to the study of the formation of anomalous groups in the periodic system of elements. From the density of the electrons and their velocity distribution, one can easily calculate how many electrons have a given angular momentum in their motion about the nucleus, that is, how many electrons have a given azimuthal quantum number k.

It is known, for example, that the formation of the group of the rare earths corresponds to the bounding of electrons in 4_4 orbits, that is, to the presence in the atom of electrons with $k = 4$. Now it follows from the theory that electrons with $k = 4$ exist in the normal state only for atoms with atomic number $z \geq 55$. This agrees well with the empirical result that the group of the rare earths begins at $z = 58$ (cerium).

Similarly, the bounding of 3_3 electrons with $k = 3$ corresponds to the anomaly of the first great period beginning at $z = 21$ (scandium); according to the theory, electrons with $k = 3$ should appear in the atom just at $z = 21$.

Further details will be published later.

(**121**, 502; 1928)

E. Fermi: Physical Institute of the University, Rome.

元素周期系中的反常族

费米

编者按

恩里科·费米是20世纪三四十年代最杰出的物理学家之一。他享誉于1941年在芝加哥大学建成了第一个原子反应堆。20世纪30年代初，他更关注于原子的结构，这篇短文是他随后发表在《林琴科学院院刊》（林琴科学院是意大利的一个科学院）上的一篇论文的摘要。在这篇短文中费米强调，对原子量超过21的原子中电子的行为的理论计算可以用来解释周期表中的第一组过渡元素。费米于1938年获得了诺贝尔物理学奖。

在不久将会发表于《林琴科学院院刊》的一篇短文中，我计算了重原子中电子的分布。电子被看作是一团**完全简并**的气体，它们被由电子屏蔽了的核电荷吸引而围绕在核附近。另外我也给出了计算电子密度和势能的公式，它们都是与核之间的距离 r 的函数。

作为此前工作的延续，我使用了同样的方法来研究元素周期系中反常族的形成。根据电子密度和它们的速率分布，可以很容易地计算出相对于核的运动角动量为某一指定值的电子的数目，也就是说，角量子数 k 为某一指定值的电子的数目。

例如，已经知道稀土族的形成对应于电子被束缚在 4_4 轨道上，也就是说，对应于原子内存在 $k = 4$ 的电子。而根据理论可以得出，只有在那些原子序数 $z \geq 55$ 的原子中 $k = 4$ 的电子才能以正常状态存在。这与稀土元素族从 $z = 58$（铈）开始这一经验结果吻合得很好。

类似的，$k = 3$ 的电子被束缚在 3_3 轨道上，对应于从 $z = 21$（钪）开始的第一个长周期的反常。而根据理论，$k = 3$ 的电子刚好应该出现在 $z = 21$ 的原子之中。

进一步的细节将随后发表。

（王耀杨 翻译；李芝芬 审稿）

Wave Mechanics and Radioactive Disintegration

R. W. Gurney and E. U. Condon

Editor's Note

The "wave mechanics" description of quantum theory arose from the work of Erwin Schrödinger in the 1920s, which built on the suggestion of Louis de Broglie in 1924 that matter can possess wave-like properties. Schrödinger's description of the behaviour of quantum particles was formulated purely in terms of waves (or wavefunctions) whose amplitude in different parts of space specified the probability of the particle being there. Here Ronald Gurney and Edward Condon perceptively states what this implies for the decay of radioactive atomic nuclei by emission of alpha particles. They points out that the escape of the alpha particle can be regarded simply in terms of overlap between wavefunctions inside and outside the nucleus. This is a form of quantum "tunnelling" through an energy barrier.

AFTER the exponential law in radioactive decay had been discovered in 1902, it soon became clear that the time of disintegration of an atom was independent of the previous history of the atom and depended solely on chance. Since a nuclear particle must be held in the nucleus by an attractive field, we must, in order to explain its ejection, arrange for a spontaneous change from an attractive to a repulsive field. It has hitherto been necessary to postulate some special arbitrary "instability" of the nucleus; but in the following note it is pointed out that disintegration is a natural consequence of the laws of quantum mechanics without any special hypothesis.

It is well known that the failure of classical mechanics in molecular events is due to the fact that the wave-length associated with the particles is not small compared with molecular dimensions. The wave-length associated with α-particles is some 10^5 smaller, but since the nuclear dimensions are smaller than atomic in about the same ratio, the applicability of the wave mechanics would seem to be ensured.

In the classical mechanics, the orbit of a moving particle is entirely confined to those parts of space for which its potential energy is less than its total energy. If a ball be moving in a valley of potential energy and have not enough energy to get over a mountain on one side of the valley, it must certainly stay in the valley for all time, unless it acquire the deficiency in energy somehow. But this is not so on the quantum mechanics. It will always have a small but finite chance of slipping through the mountain and escaping from the valley.

In the diagram (Fig.1), let *O* represent the centre of a nucleus, and let *ABCDEFG* represent a simplified one-dimensional plot of the potential energy. The parts *ABC* and *GHK* represent the Coulomb field of repulsion outside the nucleus, and the internal part *CDEFG* represents the attractive field which holds α-particles in their orbits. Let *DF* be an allowed orbit the energy of which, say 4 million volts, is given by the height of *DF* above

波动力学和放射性衰变

格尼，康登

编者按

20 世纪 20 年代，埃尔温·薛定谔在路易斯·德布罗意于 1924 年提出的物质可能具有波动性这个观点的基础上进行了一系列研究工作，开启了用“波动力学”描述量子理论的先河。薛定谔完全从波（或者说是波函数）的角度系统地描述了量子化粒子的行为，空间中不同位置上波的振幅代表了粒子在该处出现的概率。在这篇文章中，罗纳德·格尼和爱德华·康登敏锐地指出了薛定谔的描述对于理解放射性原子核发射出 α 粒子的衰变的意义。他们认为，α 粒子的逃逸可以简单地理解成核内外波函数的重叠。这正是一种穿透能量壁垒的量子“隧道效应”。

在 1902 年放射性衰变的指数律被发现后，人们很快认识到一个原子衰变的时间与这个原子以前的历史无关，而完全是由几率决定的。因为核子一定是被引力场束缚在原子核中的，所以为了解释核子的发射，我们必须假设一个从引力场到斥力场的自发转变。目前，有必要假设原子核具有某种特别的任意“不稳定性”，但在下文中，我们会指出衰变是量子力学原理的固有结果，并不需要任何特别的假设。

众所周知，经典力学在分子事件中失效是由于与分子尺度相比粒子的波长较大。α 粒子的波长约为分子尺度的 $1/10^5$，但因为原子核的尺度比原子尺度小差不多同样的比率，波动力学的适用性似乎是有保证的。

在经典力学中，一个运动粒子的轨道被严格限制在势能小于总能量的区域内。如果一个球在势阱中运动，并且其能量不足以使它翻过势垒，那么它将一直待在势阱中，除非它以某种方式获得足以翻越势垒的能量。但在量子力学中情况并非如此，球总是有一个很小但不为零的机会可以穿透势垒从势阱中逃逸出来。

如图 1 所示，设 O 表示一个原子核的中心，$ABCDEFG$ 表示一条简化了的一维势能曲线。其中，ABC 段和 GHK 段表示核外的库仑排斥场，中间的 $CDEFG$ 段表示将 α 粒子束缚在其轨道上的引力场。设 DF 是一个容许轨道，它的能量，比如说 400 万电子伏，由 OX 到 DF 的高度表示。我们可以近似地说，这个轨道对应的波函

OX. Approximately, we can say that with this orbit will be associated a wave-function which will die away exponentially from *D* to *B*. Again, corresponding to motion outside the nucleus along *BM*, there will be a wave-function which will die away exponentially from *B* to *D*. The fact that these two functions overlap in the region *BD* means that there is a small but finite probability that the particle in the orbit *DF* will escape from the nucleus along *BM*, acquiring kinetic energy equal to the height of *DFBM* above *OX*, say 4 million volts. This occurrence will be spontaneous and governed solely by chance.

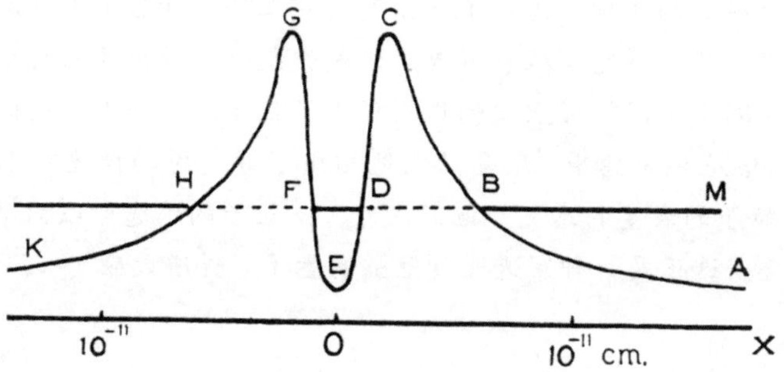

Fig. 1

The rate of disintegration, that is, the probability of escape, depends on the amount of overlapping of the wave-functions in the regions *DB* and *FH*, and this is extremely sensitive to the height to which the potential curve at *C* rises above *BDF*. By varying this height through a small range we can obtain all periods of radioactive decay from a fraction of a second, through the 10^9 years of uranium, to practical stability. (In considering the transmutation of a molecule into its isomer, Hund found a similar vast range of transformation periods, *Zeit. f. P.*, 43, 810; 1927) If the potential curves for the interaction of an α-particle with the various radioactive nuclei are similar, we can obtain a qualitative understanding of the Geiger-Nuttall relation between the rate of disintegration and the range of the emitted α-particles. For the α-particles of high energy the wave function for outside motion will overlap that for the inside motion more, and the rate of disintegration will be greater.

Besides obtaining a general idea of the mysterious instability of the nucleus, we can visualise in this way one of the most puzzling results of recent experimental work. An α-particle having the same range (2.7 cm) as those emitted by uranium should, if fired directly at the uranium nucleus, penetrate its structure; while faster α-particles should do so, even when not fired directly at the nucleus. It was therefore disconcerting when, on examining the scattering of fast α-particles fired at uranium, Rutherford and Chadwick (*Phil. Mag*, 50, 904; 1925) could find no indication of any departure from the inverse square laws. But from the model outlined above, this is what would be expected. For if the height of *BM* above *OX* represents the energy of the uranium α-particles, then a faster particle fired at the nucleus will simply run part way up the hill *ABC* and return without having encountered any change in the repulsive field or any nuclear particles (which are describing orbits within the region *GEC*).

数从 *D* 到 *B* 呈指数衰减。此外，相应于核外运动的 *BM* 段中，粒子运动的波函数从 *B* 到 *D* 也呈指数衰减。这两个波函数在 *BD* 区域交叠的事实意味着，存在一个很小但不为零的几率，使得在 *DF* 轨道上的粒子能够沿着 *BM* 逃逸出原子核，同时获得 *OX* 到 *DFBM* 的高度所表示的动能，比如说 400 万电子伏。这一事件是自发的并且只受几率的控制。

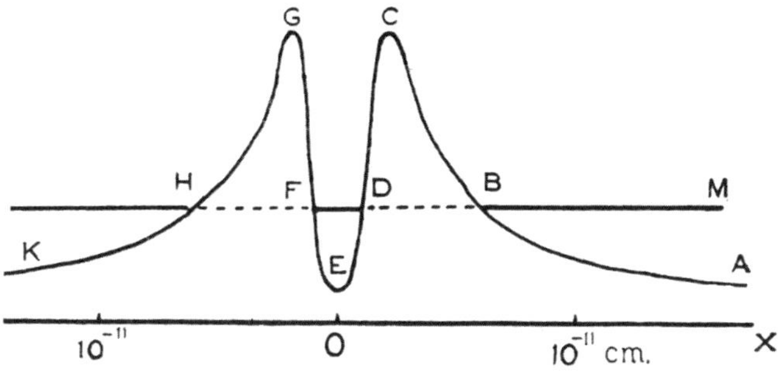

图 1

衰变速率，即逃逸几率，依赖于 *DB* 和 *FH* 区域中波函数的交叠量，并且对势能曲线中 *C* 点到 *BDF* 的高度十分敏感。通过在小范围内改变这个高度，我们可以获得所有放射性衰变元素的衰变周期，从几分之一秒、10^9 年（铀）到实质上是稳定的，各种情况都存在。（在考察分子的异构体之间的转化时，洪德发现了一个与此类似的转变周期的巨大范围，见《物理学杂志》，第 43 卷，第 810 页，1927 年）如果 α 粒子与各种放射性核相互作用的势能曲线是类似的，我们就能定性地理解衰变速率和所发射 α 粒子的射程之间的盖革–努塔耳关系。对于高能 α 粒子，核外运动波函数与核内运动波函数交叠得更多，从而衰变速率将会更大。

除了获得关于原子核神秘不稳定性的一般观点以外，以此方法我们还可以解释最近实验工作中最令人迷惑的一些结果中的一个。如果一个与铀发射出的 α 粒子射程相同（2.7 厘米）的 α 粒子直接射在铀核上，它将穿透铀核；而更快的 α 粒子即使不是直接射在铀核上，也会穿透铀核。因此，令人不解的是，在快 α 粒子射在铀上的散射实验中，卢瑟福和查德威克（《哲学杂志》，第 50 卷，第 904 页，1925 年）没能发现任何偏离库仑平方反比定律的迹象。但是从以上概述的模型来看，这应该是预期的结果。因为如果 *OX* 到 *BM* 的高度表示铀发射的 α 粒子的能量，那么射在核上的更快的粒子将仅仅沿着势垒 *ABC* 向上爬一段然后返回，其间没有经历排斥场的任何变化也没有遇到任何核子（核子在 *GEC* 区域内运动）。

The peculiar property of the wave mechanical equations which finds application here has also been applied to the theory of the emission of electrons from cold metals under the action of intense fields (Oppenheimer, *Proc. Nat. Acad. Sci.*, 14, 363; 1928; and Fowler and Nordheim, *Proc. Roy. Soc.*, A, 119, 173; 1928). Ordinarily, an atom does not lose its electrons because the attractive field of the atom remains attractive to all distances. But when an intense field is applied, then the attractive field is reversed in sign a short distance from the atom. This makes the resultant potential energy curve similar to that in the diagram, and so the atoms begin to shed their electrons.

Much has been written of the explosive violence with which the α-particle is hurled from its place in the nucleus. But from the process pictured above, one would rather say that the α-particle slips away almost unnoticed.

(**122**, 439; 1928)

Ronald W. Gurney, Edw. U. Condon: Palmer Physical Laboratory, Princeton University, July 30.

这里应用到的波动力学方程的特殊性质也曾被应用于强场作用下冷金属发射电子的理论中（奥本海默，《美国科学院院刊》，第 14 卷，第 363 页，1928 年；福勒和诺德海姆，《皇家学会学报》，A 辑，第 119 卷，第 173 页，1928 年）。一般情况下，原子不会失去它的电子，因为不论相隔多远的距离，原子的引力场都吸引着电子。然而，当施加一个强场时，原子附近一个短距离范围之内，原子引力场将被颠倒。这使得合成的势能曲线类似于图 1 中的曲线，因此原子开始发射它们的电子。

很多文献都提到了将 α 粒子从原子核内抛出的一种爆炸性的力量。然而，从以上描绘的过程来看，人们宁愿说 α 粒子几乎是神不知鬼不觉地溜出原子核的。

（王锋 翻译；李军刚 审稿）

Sterilisation as a Practical Eugenic Policy

E. W. MacBride

Editor's Note

Perhaps the most chilling aspect of this discussion of eugenics by zoologist Ernest William MacBride, reviewing of a book on eugenic policies in the United States, is its ignorance and prejudice several decades after Darwin's evolutionary theory sparked calls for controls on breeding in human society. MacBride does not even respect that theory: he was a notorious late advocate of Lamarckian inheritance of acquired characteristics, evident here in how he thinks people who acquire mental defects might pass them on to offspring. There is also an elision from sterilization of "mental defectives" to the control of fertility in people whom MacBride deems merely "stupid" or feckless—apparently, many of the poor. It is a stark reminder of how ideas about genetic heredity were confused and abused.

Sterilisation for Human Betterment: a Summary of Results of 6,000 Operations in California, 1909–1929. By E. S. Gosney and Dr. Paul Popenoe (A Publication of the Human Betterment Foundation.) pp. xviii+202. (New York: The Macmillan Co., 1929.) 8*s*. 6*d*. net.

THIS little book is a storehouse of information on the efforts which have been made in the United States to improve the human stock by sterilising the feeble-minded and the insane. It appears that although *more Americano* laws have been passed in about twenty states of the Union providing for the legal sterilisation of sexual perverts, and imbecile and insane patients in public institutions, these laws have been put into practical operation only in the State of California, so that in the book discussion is mainly concerned with the results obtained in that State.

The justification for these attempts to aid Nature in eliminating the unfit is set forth in the introduction. Amongst our unsentimental forefathers, no efforts were made to keep alive weakly and diseased children, and hence the race was propagated only from its most vigorous members; but nowadays, when unreflecting humanitarian sentiment is in fashion, all babies are kept alive so far as medical science can avail, and this science is paid for by levying tribute on the thrifty and self-supporting. The result is that this section of society limits its offspring, and future generations are likely to be recruited not from the fit but from the unfit.

How drastically and efficiently natural selection operated amongst the young in England during the eighteenth century may be gathered from figures given by Miss Buer in her book, "Health, Wealth, and Population in the Early Days of the Industrial Revolution". In 1730, out of all babies born in London, 74 percent died before they were five years of age; in 1750, 63 percent died; in 1770 the percentage was 50, and it did not sink to 30 until

作为一项实用优生学政策的绝育术

麦克布赖德

编者按

达尔文在进化论中曾提出应该对人类的生育有所控制，几十年后，动物学家欧内斯特·威廉·麦克布赖德仍然对此缺乏理解并存有偏见，他对一本关于美国优生政策的书所作的评论令人心寒。麦克布赖德甚至并不推崇达尔文的进化论。他作为拉马克获得性遗传假说的支持者之一而广为人知，从这篇文章中可以很明显地看出，他认为精神病患者可能会把他们的疾病基因传给后代。其实，不管是对“精神病患者”的绝育，还是对在麦克布赖德看来只是略显“愚钝”的无能之辈的生育控制，都是没有必要的，后者显然是指穷人。这篇文章是提醒人们关注遗传学的观点是如何被混淆和滥用的绝好实例。

《用于人类改良的绝育术：1909~1929 年加利福尼亚 6,000 例手术结果的概要》

戈斯尼和保罗·波普诺博士著（人类改良基金会的一部著作）

xviii + 202 页。（纽约：麦克米伦公司，1929 年。）8 先令 6 便士

这本小册子是一座知识宝库，蕴藏着美国通过使低能者和精神病患者绝育的方式来提高人口质量的过程中取得的成就。尽管美国联邦约 20 个州已经通过了**越来越多的美式**法律，为性异常者、智能低下者及精神病患者在公共机构进行绝育手术提供合法依据，但是看起来只有加利福尼亚州真正实施了这些法律，因此本书中主要是关于加利福尼亚州取得的成果的讨论。

为了帮助造物主减少不适于生存者而采取这些绝育措施的合理性，在本书的引言里已经有所陈述。我们那些无情的祖先，没有为使虚弱和得病的孩子能够活下来而进行过任何努力，因此种系都是从其最强壮的成员中繁衍而来的；但是如今，在这个浅薄的人道主义情感泛滥的时代，医学在其力所能及的范围内力求使所有婴儿都能活下来，并且通过向节俭的自食其力者征税来支付这些费用。这样的结果就是，社会上的这部分自食其力者限制自己的后代个数，未来的孩子们更可能是从那些不适者，而非适者中繁衍而来。

自然选择在 18 世纪英国的年轻人中所起的作用是多么明显有效，这可以从比埃小姐书中给出的数字推断出来，那本书的书名是《工业革命早期的健康、财富和人口》。1730 年伦敦出生的所有婴儿中有 74% 死于 5 岁之前，1750 年这一比例为 63%，1770 年为 50%，直到 1833 年这一比例才降到 30%。这一比例在该国的其他

1833. The percentage was probably even higher in other parts of the country. The help given by hospitals, and later by the State, to indigent mothers has all grown up in the last century, so that the argument that because we have maintained a vigorous, enterprising, fighting race in these islands for eight hundred years since the Norman Conquest, we shall continue to do so, is not one for which there is any sound basis.

It is, however, not practical politics to suggest a return to the old plan of *laissez-faire*. How then shall the elimination of the unfit be promoted? The authors of this book suggest "by legalised sterilisation". The method of sterilisation advocated is cutting the ducts (vasa deferentia in the male, and Fallopian tubes in the female) which convey the germ cells to the exterior. The authors point out that more than six thousand operations of this sort have been already performed in California, and that only seven failures are recorded (three in males, four in females). The operation does not interfere with sexual desire or the performance of the sexual act. The genital organs in man, as in Vertebrata generally, have two functions, namely: (1) to produce the germ cells; (2) to produce a hormone which diffuses through the system and maintains youth and vigour. In a man the spermatozoa forms a minor part of the sexual discharge, the main portion of which is constituted by the prostatic secretion, and some authorities hold that this secretion when absorbed by the female has an invigorating effect on the constitution. As to a woman, when it becomes necessary on account of tumours to remove the uterus, if a portion of one of the ovaries is preserved and sewn to the abdominal wall, this will prevent the premature onset of the menopause and maintain in the patient all the qualities of a young woman.

But are insanity and mental defect hereditary? Some British authorities hold that in many cases they are not. So far as insanity is concerned, however, there is general agreement, as our authors point out, that the condition known as "dementia praecox" is the result of an inborn weakened constitution, and that it is a mere question of time when it will manifest itself in the life of the unfortunate individual who has inherited this constitution. As to mental defect, the argument that it is sometimes not of hereditary origin, overlooks the consideration that all "mutations", of which mental defect is one, must ultimately have been produced by some external cause, and there is nothing to show that an "accidental" mental defective will not propagate mentally defective children. In any event, even if a defective should produce healthy children, such a person would make the worst possible parent to carry out the duty of caring for and training the children; and it is a little too much to ask the State to allow a defective to go on having children on the chance of some of them being normal, if the State has to support them all.

Our authors urge that sterilisation should not be regarded as a punishment but as a hygienic measure; that defectives confined in asylums might be allowed out on condition of their consenting to this operation. But whilst we agree that this argument is good so far as it goes, a little reflection will show that it only touches the fringe of the problem. The defectives most dangerous to society are those who are never confined in institutions at all! The high-grade defectives are just able to support themselves in the lowest paid and most unskilled occupations, and no civilised government would take the responsibility of

地方可能更高。上个世纪，医院给予这些贫困母亲的帮助（后来国家也给予了帮助）已经有所增加。因此有一种观点认为，诺曼征服以来的800年间，这些岛上的人们一直保持着一个精力充沛的、有进取心的、好战的民族应有的素质，所以应该继续保持这种做法，这一论点是缺乏可靠根据的。

然而，重新起用**自由放任**的旧政策并不切实可行。那么怎样才能促进不适于生存者数量的减少呢？本书的作者建议采取“合法绝育”的手段。这里提倡的绝育方法，是指切断向外输送生殖细胞的管道（男性是输精管，女性是输卵管）。作者指出加利福尼亚已经实施了六千多例此类手术，根据记录，只有7例失败（3例男性绝育术，4例女性绝育术）。该手术不会影响性欲和性行为的进行。与脊椎动物门其他成员一样，男性生殖器有两个功能，即：(1) 产生生殖细胞；(2) 产生散布于全身系统并维持机体青春与活力的激素。在男性分泌物中精子仅仅占很小部分，大部分是前列腺分泌液。一些权威人士认为，这些分泌液在被女性吸收后，会起到令女性的体质充满生气与活力的作用。对于女性，当她们因为肿瘤而必须摘除子宫时，如果保留一个卵巢的一部分，并将其与腹壁缝合在一起，就可以延缓更年期的到来并使病人保有年轻女人的所有特质。

但是精神病和心智缺陷是可遗传的吗？英国的一些权威人士认为，许多情况下它们并不遗传。然而就精神病而言，人们普遍认可的观点正如本书作者指出的那样，通常所说的“早发性痴呆”就是先天体质虚弱造成的，遗传了这一体质的不幸个体，发病只是时间问题。对于心智缺陷，有观点认为，它有时不是由于遗传。这一看法忽略了一点，那就是包括心智缺陷在内的所有“突变”最终都是由某一外因引起的，而没有任何证据表明“偶然的”心智缺陷者不会生育出心智缺陷的孩子。无论如何，即使一个有缺陷的个体可以生出健康的孩子，作为负有照顾和培养孩子的责任的父母来说，他们也是最不该成为父母的人；如果国家不得不供养所有有缺陷的个体的话，那么怀着“有缺陷的人也可能生出正常孩子”的投机心理而请求国家允许他们生育，这未免太过分了。

本书作者认为不应该将绝育视为一种惩罚，而应该看作一项卫生措施；对于那些被关在收容所的有缺陷的个体，在他们同意接受该手术的情况下，可以允许他们出去。就目前状况来看，我们承认这一想法很好，但是稍加考虑就会发现，这种做法只是刚刚触及这一问题的边缘而已。对社会最具危险性的有缺陷的个体是那些从来没有受到公共机构限制的人！高级的有缺陷的个体能够凭最低收入和最不需技能的工作来养活自己，并且任何文明社会的政府都不会承担约束他们的责任，所以

confining them, and so they go on propagating large families as stupid as themselves. As Mr. Lidbetter has shown*, it is from the ranks of just these classes that in the last hundred years the majority of paupers and criminals of London has been recruited.

It seems to us that in the last resort compulsory sterilisation will have to be inflicted as a penalty for the economic sin of producing more children than the parents can support. Whether a man has a large or a small family is—given a healthy wife—a matter of taste, so long as he provides for his own children; but when he comes to the State and demands that it—that is to say, his neighbours—should support these children, then the State can say, "Very well—we shall help you with the family which you have, but if after this you have any more children you shall be sterilised."

Before, however, such an alternative is presented to any citizen, he may justly claim that he should receive instruction from the State in the means of birth-control. It is obviously unfair that such knowledge should be denied to the poor whilst it is easily accessible to the rich. It is often said, and with justice, that the great objection to birth-control is that the wrong people practise it. But this knowledge once attained cannot be taken away; the middle classes possess it and cannot be prevented from putting it into practice. If, however, the knowledge and practice of birth-control were widely spread among the working-class, there would be created such a resentment against the reckless production of children that the movement to establish compulsory sterilisation of the unfit would prove irresistible.

(**125**, 40-42; 1930)

* "Pauperism and Heredity", by E. J. Lidbetter, *The Eugenics Review*, vol. 14, p. 152; 1923.

他们就会继续繁衍出和他们自己一样有缺陷的大家族。正如李德贝特先生所说*，最近的100年间，伦敦的大部分乞丐和罪犯正是从这类人中产生的。

对我们而言，强制性的绝育术似乎是对那些生育了很多孩子而无力供养的父母们所犯经济罪进行惩罚的最后手段。如果一个男人拥有一个健康的妻子，那么他选择拥有大家庭还是小家庭完全是个人喜好的问题，只要他能养得起自己的孩子；但是如果他要求国家（其实也就是他的左邻右舍们）帮助抚养他的孩子们，那么国家就可以说："没问题，我们会帮你养家，但是从现在起如果你再生孩子的话，那么你将会被绝育。"

不过，在向市民公布这种可供选择的方案之前，人们可能会理直气壮地声称自己有权利了解国家的生育控制政策。富人们可以很容易地获悉这些信息，而穷人们的知情权可能被剥夺了，这显然是非常不公平的。人们常常说，生育控制的最大问题在于是错误的人在实施政策，这么说也是公平的。但是，这些知识一旦为人们获得就不可能再拿走；中产阶级们知悉了这些，就不可避免会将其付诸实践。然而，假如节育的知识和实践在工人阶级中广泛传播，那么可能会出现一种对不计后果进行生育的怨恨，到那时对不适于生存者实施强制性绝育术的行动可能就无法遏止了。

（刘皓芳 翻译；刘京国 审稿）

*《贫困与遗传》，作者李德贝特，《优生学评论》，第14卷，第152页，1923年。

The "Wave Band" Theory of Wireless Transmission

A. Fleming

Editor's Note

Ambrose Fleming here takes issue with a way of understanding wave communications, for telephone or television. In both technologies, devices encode signals as amplitude modulations of a carrier wave. In terms of Fourier analysis, one can view the resulting wave as occupying a "band" of frequencies around the carrier frequency, and it had become common to consider how these bands should be apportioned, which bands were allowed and so forth. But Fleming argues that talk of bands obscures the role of the amplitude. Too large an amplitude could cause interference between different transmissions, much as speaking too loudly at the theatre can be disruptive. Focusing on amplitude rather than bands, Fleming suggests, will help avoid unnecessary restrictions on the new technologies.

IN scientific history we meet with many examples of scientific theories or explanations which have been widely adopted and employed, not because they can be proved to be true but because they provide a simple, easily grasped, plausible explanation of certain scientific phenomena. The majority of persons are not able to see their way through complicated phenomena and so thankfully adopt any short-cut to a supposed comprehension of them without objection.

Ease of comprehension is not, however, a primary quality of Nature, and it does not follow that because we can imagine a mechanism capable of explaining some natural phenomenon it is therefore accomplished in that way. There is a widely diffused belief in a certain theory of wireless telephonic transmission, and also of television, that for securing good effects it is necessary to restrict or include operations within a certain width of "wave band". But although this view has been very much adopted there is good reason to think that it is merely a kind of mathematical fiction and does not correspond to any reality in Nature.

Let us consider how it has arisen. We send out from all wireless telephone transmitters an electromagnetic radiation of a certain definite and constant frequency expressed in kilocycles. Thus 2LO London broadcasts on 842 kilocycles. This means that it sends out 842,000 electric vibrations or waves per second. Every broadcasting station has allotted to it a certain frequency of oscillation and it is not allowed to depart from it.

It is like a lighthouse which sends out rays of light of one pure colour or an organ which emits a single pure musical note. For most broadcasting stations this peculiar and individual frequency lies somewhere between a million and half a million per second, though for the long wave stations like Daventry it is so low as 193,000 or 193 kilocycles.

无线传输的“波带”理论

弗莱明

编者按

在本文中安布罗斯·弗莱明对应用于无线电话和电视的波通信有不同的理解。在这两项技术中，装置把信号编译成振幅调制的载波。在傅立叶分析中，我们可以认为得到的波占据了载波频率附近的一个频“带”。大家通常要考虑的是这些频带如何分配，哪些频带能够被允许等等。但弗莱明认为人们对频带的过分关注掩盖了振幅的作用。太大的振幅可能会导致不同传输过程之间的干扰，就像在戏院中大声讲话带来的麻烦一样。弗莱明指出，对振幅而不是频带的关注将会帮助我们避免在应用这项新技术时遇到不必要的麻烦或限制。

在科学史上，我们遇到的许多科学理论或解释被广泛接受并应用的情况，并不是因为它们能够被证明是真理，而是因为它们为某些科学现象提供了简单、易于被理解并且似乎合理的解释。大多数人并不能透过复杂现象发现真理，因此也就乐于不加质疑地接受某种能够便捷地解释复杂现象的假说。

然而，简单的理解并不是自然界的基本特征，也不能随即得到，即不会因为我们想象一种能够解释某些自然现象的机制，它就以那种方式来实现。在关于无线电话传输以及电视的某种理论中，存在一种广为流传的认识，即为了达到可靠的良好效果，必须限制在具有确定宽度的“波带”内操作。这种观点虽然已被广泛接受，但我们有理由认为，这只是一类数学虚构，并不能与自然界的任何现实相对应。

让我们考虑这是如何产生的。我们从所有的无线电话发射台发出一个具有确定的恒定频率（以千周表示）的电磁辐射。伦敦 2LO 电台按照这种方式以 842 千周的频率进行广播。这意味着它每秒内发出 842,000 次电子振荡或 842,000 个波。每个广播站已分配到一个确定的振荡频率，而且不允许偏离此频率。

这就像一座发出单色光的灯塔或一个发出单一音符的风琴。对于大多数广播站而言，这类特有的专用频率位于每秒 50 万~100 万之间，然而对于类似达文特里这样的长波站，它的频率低至 193,000，或 193 千周。

When we speak or sing or cause music to affect the microphone at a broadcasting studio the result is to cause the emitted vibrations, which are called the *carrier waves*, to fluctuate in height or wave amplitude, but does not alter the number of waves sent out per second. It is like altering the height or size of the waves on the surface of the sea without altering the distance from crest to crest which is called the wave-length.

Suppose the broadcasting station emits a carrier wave of frequency n and let $p = 2\pi n$. Then we may express the amplitude a of this wave at any time t by the function $a = A \sin pt$ where A is the maximum amplitude. If on this we impose a low frequency oscillation due to a musical note of frequency m and let $2\pi m = q$, then we can express the modulated vibration by the function

$$a = A \cos qt \sin pt$$

But by a well-known trigonometrical theorem this is equal to

$$\frac{A}{2}\left\{\sin(p+q)t + \sin(p-q)t\right\} ,$$

and thence may be supposed to be equivalent to the simultaneous emission of two carrier waves of frequency $n + m$ and $n - m$.

If the imposed note or acoustic vibration is very complex in form, then in virtue of Fourier's theorem it may be resolved into the sum of a number of simple harmonic terms of form $\cos qt$, and each of these may be considered to be equivalent to a pair of co-existent carrier waves. Hence the complex modulation of a single frequency carrier wave might be imitated by the emission of a whole spectrum or multitude of simultaneous carrier waves of frequencies ranging between the limits $n + N$ and $n - N$, where n is the fundamental carrier frequency and N is the maximum acoustic frequency occurring and $2N$ is the width of the wave band. This, however, is a purely mathematical analysis, and this band of multiple frequencies does not exist, but only a carrier wave of one single frequency which is modulated in amplitude regularly or irregularly.

If the sounds made to the microphone at the broadcasting station are very complex, such as those due to instrumental music or speech, then in virtue of this mathematical theorem the very irregular fluctuations in amplitude of the single carrier wave can be imitated if we suppose the station to send out simultaneously a vast number of carrier waves of various frequencies lying between certain limits called the "width of the wave band".

This, however, is merely a mathematical artifice similar to that employed when we resolve a single force or velocity in imagination into two or more component forces. Thus, if we consider a ball rolling down an inclined plane and desire to know how far it will roll in one second, we can resolve the single vertical gravitational force on the ball into two components, one along the plane and one perpendicular to it. But this is merely an ideal division for convenience of solution of the problem; the actual force is one single force acting vertically downwards. Similar reasoning is true with regard to wireless telephony. What happens, as a matter of fact, is that the carrier wave of one single constant

当我们在播音室里对着麦克风说话、唱歌或放出音乐，都会导致发射振荡的产生，这被称作**载波**，它会使波的高度或波幅出现起伏，但并不改变每秒钟发出的波数。这类似于只改变海面上波浪的高度或大小，而不改变被称作波长的从波峰到波峰的距离。

假定广播站发射一个频率为 n 的载波，并令 $p = 2\pi n$。我们通过函数 $a = A \sin pt$ 表示在任意时刻 t 这个波的振幅 a，其中 A 是最大振幅。如按这种方式，我们施加一个由频率为 m 的音符引起的低频振荡，并令 $2\pi m = q$，则我们可以通过下面的函数方程表示调制振荡

$$a = A \cos qt \sin pt$$

通过人们熟知的三角定理，上式等于

$$\frac{A}{2}\left\{\sin(p+q)t + \sin(p-q)t\right\},$$

因此可以认为其等效于同时发射频率为 $n + m$ 和 $n - m$ 的两个载波。

如果施加的音律或声音振荡在形式上很复杂，那么利用傅立叶定理，我们可将其分解为如 $\cos qt$ 形式的许多简谐项之和，可以等效地认为其中每一项都是一对共存的载波。因此单频载波的复杂调制可以用整个谱的发射或大量频率在 $n + N$ 到 $n - N$ 之间的同步载波来模拟，其中 n 是基本载波频率，而 N 是出现的最大声波频率，$2N$ 是波带的宽度。然而，这是纯数学分析，这个多频波带并不存在，而只存在单频的载波，可以对其振幅进行规则或不规则的调制。

如果向广播站里的麦克风发出的声音非常复杂，例如乐器发出的声音或讲话的声音，那么根据这个数学原理，我们可以假定广播站里同时发出大量频率被限定在“波带宽度”之内的载波来模拟单个振幅不规则波动的载波。

然而，这仅是一个数学技巧，类似于我们在假想中将单个力或速度分解为两个或更多个分量。因此，如果我们考虑一个沿倾斜平面滚下的球，并要求知道球在一秒钟内滚动多远，我们就可以将球受到的单一的垂直引力分解为两个分量，分别与平面平行和垂直。但这只是便于解决问题的一个理想的分解；实际的力仍只是作用方向垂直向下的单个力。对于无线电话也存在着相似的思考过程。事实上，发生的情况是，一个单一恒定频率的载波按照某种规则或不规则的规律在振幅上发生变

frequency suffers a variation in amplitude according to a certain regular or irregular law. There are no multiple wave-lengths or wave bands at all.

The receiver absorbs this radiation of fluctuating amplitude and causes the direct current through the loud speaker to vary in accordance with the fluctuations of amplitude of the carrier wave; the carrier wave vibrations being rectified by the detector valve.

The same thing takes place in the case of wireless transmission in television. The scanning spot passes over the object and the reflected light falls on the photoelectric cells and creates in them a direct current which varies exactly in proportion to the intensity of the reflected light. This photoelectric current is employed to modulate the amplitude of a carrier wave, and the neon lamp at the receiving end translates back these variations of carrier wave amplitude into variations in the cathode light of the neon tube.

There is neither in wireless telephony nor in television any question of various bands of wavelength. There is nothing but a carrier wave of one single frequency which experiences change of amplitude. The whole question at issue then is, what range in amplitude is admissible?

In the case of television it is usual for critics of present achievements to say that good or satisfactory television cannot be achieved within the limits of the nine kilocycle band allowed. But there is in reality no wave band involved at all. It is merely a question of what change in amplitude in a given carrier wave can be permitted without creating a nuisance.

It is something like the question: How loud can you whisper to your next neighbour at a concert or theatre without being considered to be a nuisance? People do whisper in this way, and provided not too loudly, it is passed over. But if anyone is so ill-mannered as to speak too loudly he is quickly called to order, or turned out.

It is, however, not an easy thing to define a limit to wave amplitudes. They are measured in microvolts per metre and are difficult to measure. But a wave-length is easy to define in kilocycles or in metres, and hence the method has been adopted of limiting emission to an imaginary band of wavelengths which, however, do not exist.

The definition is imperfect or elusive. It is something like the old-fashioned definition of metaphysics as "a blind man in a dark room groping for a black cat which isn't there". Similarly, the supposed wave band is not there. All that is there is a change, gradual or sudden, in the amplitude of the carrier wave. It is clear, then, that sooner or later we shall have to modify our code of wireless laws.

We have no reason for limiting the output of our broadcasting stations to some imaginary wave band of a certain width, say nine kilocycles or whatever may be the limiting width, but we have reason for limiting the range of amplitude of the carrier waves sent out.

化。根本不存在多重的波长或波带。

接收器吸收了这类振幅振荡变化的辐射，并使通过扬声器的直流按照载波振幅的振荡变化而变化；载波的振动通过检测器的电子管进行整流。

对于电视的无线传输来说也存在相同的情况。扫描点在物体上扫描，而反射光落在光电元件上，并在其中产生了与反射光强度精确成正比变化的直流。这个光电流用于载波振幅的调制，在接收端的氖灯会将这些载波振幅的变化转换回氖管的阴极光的变化。

在无线电话和电视中均不存在各种不同波带的问题。只是存在振幅被调制的单频载波。那么争论的全部问题就是，可以允许振幅在什么范围内变化呢？

对于电视而言，批评现有成果的评论者们通常会提出，在允许的 9 千周的限制范围内，性能良好的或令人满意的电视不可能实现。但实际上根本不存在所谓的波带。这仅仅是一个在不产生干扰的情况下可以允许给定的载波中振幅如何变化的问题。

这就类似于如下的问题：在音乐厅或剧场中你可以用多大的声音悄悄对邻座说话而不至于影响其他人呢？人们以这种方式耳语，并保持声音不是很大，这样可以不被注意。但是如果任何人非常不礼貌地用很大的声音说话，他很快就会被要求保持安静或被逐出会场。

然而，确定波幅的调制范围并不是件容易的事情。它们以每米微伏的数量级来进行计量，并且是很难被测定的。但波长很容易用千周或米来定义，因此现已采用的方法是将发射限定到虚构的波带范围内，然而，这个波带实际上并不存在。

这样的定义是不完善的，或者说是难以理解的。这有点像形而上学的老式定义，如“一个在暗室里摸索一只并不存在的黑猫的盲人”。类似地，我们假设的波带并不存在。存在的只是载波中振幅逐步或突然的变化。显然，总有一天我们将不得不修改我们的无线通信的编码规则。

我们没有理由将我们的广播电台的输出限制在某种确定宽度的假想波带内，比如说 9 千周或其他任何限制范围，但我们有理由限定输出载波振幅的范围。

Some easily applied method will have to be found of defining and measuring the maximum permissible amplitude of the carrier waves as affected by the microphone or other variational appliance. It may perhaps be thought that an unnecessary fuss is here being made on what may be regarded as simply a way of explaining things, but experience in other arts shows how invention may be greatly retarded by unessential official restrictions. Consider, for example, the manner in which mechanical traction was retarded in Great Britain for years by ridiculous regulations limiting the speed of such vehicles on highway roads. The only restrictions that should be imposed are those absolutely necessary in the interests of public safety or convenience, and all else tend to throttle and retard invention and progress.

(**125**, 92-93; 1930)

我们必将找到某些易于应用的方法，用来定义及测量被麦克风或其他有变化的设备影响的载波的最大允许振幅。也许有人会认为，这只是事物的一种解释方式，没有必要像我们这样小题大做，但其他技术中的经验表明，不必要的官方限定是如何使可能的重大发明被推迟的。考虑一下这样的事例，机械牵引在英国的发展多年受阻，正是由于制定了荒谬的法规限制公路上的这类车辆的速度而导致的。唯一应该加以限制的，是那些出于公共安全性或便利性的考虑绝对必要的方面，而其他所有方面的限制往往会扼杀并妨碍发明和进步。

（沈乃澂 翻译；李军刚 审稿）

Electrons and Protons

Editor's Note

Paul Dirac was one of the most creative physicists of the early twentieth century. In 1932 he was appointed Lucasian professor at Cambridge University, and his theory unifying quantum mechanics with Einstein's theory of special relativity won him the 1933 Nobel Prize in physics. Here *Nature* reports on one of the implications of this theory, as Dirac outlined in a paper in the *Proceedings of the Royal Society*. His equations predicted "electrons of negative energy", which meant, of positive charge. The report echoes Dirac's initial suspicion that these predicted positively charged particles would behave like protons, but they soon proved to be "positive electrons" or positrons, made of antimatter.

A theory of positive electricity has been put forward by Dr. P. A. M. Dirac in the January number of the *Proceedings of the Royal Society*. The relativity quantum theory of an electron leads to a wave equation which possesses solutions corresponding to negative energies—the energy of the electron of ordinary experiment being reckoned as positive—and although there are serious difficulties encountered in any immediate attempt to associate these negative states with protons, the existence of positive electricity can be predicted by a fairly direct line of argument. Since the stable states of an electron are those of lowest energy, all the electrons would tend to fall into the negative energy states—with emission of radiation—were it not for the Pauli exclusion principle, which prevents more than one electron from going to any one state. If, however, it is assumed that "there are so many electrons in the world that... all the states of negative energy are occupied except perhaps a few...", it may be supposed that the infinite number of electrons present in any volume will remain undetectable if uniformly distributed, and only the few "holes", or missing states of negative energy will be amenable to observation. The step is then made of regarding these "holes" as "*things of positive energy*" which are identified with the protons. A difficulty now arises in ordinary electromagnetic theory which apparently has to cope with the presence of negative electricity of infinite density; this is met by supposing that for ordinary purposes volume-charges must be measured by departures from a "normal state of electrification", which is "the one where every electronic state of negative energy and none of positive energy is occupied." The problem of the large mass of the proton, as compared with that of the electron, is not discussed in detail, but a possible line of attack is indicated. Dr. Dirac has included the minimum of mathematical analysis in this paper, which can be followed in all essential points by anyone acquainted with the principles of the quantum theory.

(**125**, 182; 1930)

电子和质子

编者按

保罗·狄拉克是20世纪初最有建树的物理学家之一。1932年，他被聘任为剑桥大学的卢卡斯教授。他因提出能将量子力学与爱因斯坦的狭义相对论统一起来的理论而获得了1933年的诺贝尔物理学奖。在发表于《皇家学会学报》的一篇论文中，狄拉克对他的理论所蕴含的一个推断进行了概述,《自然》的这篇文章报道的正是这些。狄拉克的方程预测出“具有负能量的电子”，即带正电荷的电子。狄拉克最初认为，这些被预测到的带正电荷的粒子的行为方式与质子类似，这篇报道再次复述了这一观点，但不久之后，这些粒子就被证明是组成反物质的“正电子”。

在1月的《皇家学会学报》上，狄拉克博士提出了正电子理论。他通过电子的相对论量子理论导出了一个波动方程，而这个波动方程包含相应于负能量的解（普通实验中电子的能量被认为是正的）。尽管将这种负能态与质子联系起来的尝试遇到了很多困难，但是，通过非常直接的论证可以预言正电子的存在。因为电子的稳定态是能量最低的态，所以伴随着发射辐射，所有电子都将落入负能态，但是这与泡利不相容原理不同，这一原理规定不可能有多于一个的电子处于任何一个相同的状态。然而，如果假设“世界上有非常多的电子以至于除了极少的负能态以外，几乎所有的负能态都被占据了”，那么可以推测，这些存在于负能态体系中的无数个电子将不会被探测到（若电子是均匀分布的），而仅有极少的“空穴”或遗漏的负能态可以很容易地被观察到。接下来他把这些“空穴”视为与质子相同的“**具有正能量的物质**”。密度无限大的负电的出现使普通电磁学理论遇到了困难。不过，当假定一般情况下必须通过“带电的正常状态”的偏离来测量体电荷时，便可以解决这个难题，所谓带电的正常状态是指“每一个具有负能量的电子态均被占据而没有一个正能态被占据的状态”。在这篇文章中，他并没有详细讨论质子质量比电子质量大这一问题，但是却提出了一种可能的解决思路。狄拉克博士的论文仅涉及少量数学分析，这就使任何一个学习过量子理论基本原理的人都能看懂这些数学分析的要点。

（王锋 翻译；李淼 审稿）

The Connexion of Mass with Luminosity for Stars

J. Larmor

Editor's Note

Here the physicist Joseph Larmor writes to *Nature* to discuss important new work of Edward Milne on the topic of stellar interiors, and the relationship of stellar structure to luminosity. Given that we know very little about matter at the extremely high densities likely in stellar interiors, Larmor notes, Milne's work is a laudable attempt to work from basic principles, such as those of thermodynamics. The results suggest that a star's surface properties must be strongly constrained by the physics deep inside the interior, thereby offering a possible explanation for the empirical fact that the luminosity of a star depends for the most part only on its mass.

VERY remarkable and fruitful correlations have in recent years been detected, mainly at Mount Wilson, between the magnitudes of stars and their spectroscopic characteristics. The interpretation that would naturally present itself is that magnitude can enter into relation with the radiative phenomena of the surface atmosphere only through the intensity of gravity at the surface, which when great flattens down a steady atmosphere far more than proportionately. But if, following Eddington's empirical relation, total radiation of a star is a function of its mass alone, there must be more than this involved; for the radius of the star persists in this relation when expressed in terms of intensities of surface radiation and of gravity, the former determining the temperature roughly by itself. Modern hypothesis, which treats confidently of an "electron gas" with an atomic weight, as Ramsay boldly and prematurely proposed long ago, and subject to the Maxwell-Boltzmann exponential energy formula for statistics of distribution, and to its consequences for the theory of dissociation of mixed gases in relation to pressure and temperature, has on the initiative mainly of Saha led to promising applications to stellar atmospheres, which are held to be of densities low enough at any rate not to forbid this mode of treatment.

It would seem then to be necessary to conclude that these empirical spectroscopic relations on the surface require that the stellar atmosphere must be dominated to some degree by the remote steady interior of the star. Accordingly, tentative theories of the internal constitution of the stars and their flux of radiation have been developed in much detail. With Eddington the stars are perfect gases right down to the centre, though the density may there be hundreds of times that of platinum, as has apparently been verified for the case of the companion of Sirius—the high density involving the view at one time not unfamiliar that two atoms can occupy the same space, if the picturesque conception of atoms "stripped" irrevocably to the bone is to be avoided; and the energy emitted as radiation would come from a dissociation or destruction of matter according to a law involving temperature. On the other hand, it is insisted on by Jeans that the necessary radioactivity for the very long evolutions that are contemplated must be of constant and

恒星的质量与发光度之间的关系

拉莫尔

编者按

物理学家约瑟夫·拉莫尔在《自然》上发表的这篇文章，论述了爱德华·米耳恩在恒星内部结构以及内部结构与发光度之间的关系这一问题上的新成果。拉莫尔指出：因为我们对恒星内部处于超高密度的物质状态所知甚少，所以米耳恩尝试用一些基本原理（如热力学定律）进行的研究是值得称道的。米耳恩的研究结果表明，恒星的表面性质与它的内部物理状态有很强的相关性，这为解释恒星的发光度在很大程度上只取决于恒星质量这一经验事实提供了可能。

近几年的天文观测发现，恒星的绝对星等与它们的光谱特征之间存在很多明显的关联，这些观测主要是在威尔逊山进行的。对于这一点很自然的解释是，恒星的绝对星等只有通过位于表面处的引力强度才能和表面大气的辐射现象联系起来，引力强度在稳定的大气中下降很快，远不是成比例的关系。但是如果根据爱丁顿的经验关系式，一个恒星的总辐射量只是其质量的函数，则这其中一定还有另外的影响因素；因为在表面辐射强度和引力强度的关系式中含有恒星半径这个量，而凭借表面辐射强度本身就能大致确定恒星的温度。主要由萨哈倡导的一个最新的假设已经成功地解决了恒星大气方面的许多问题，一般认为恒星大气的密度很低，绝对不会影响这种处理方式的应用。该假设大胆地采用了拉姆齐在很久以前提出的冒险且欠成熟的假设——给“电子气”赋予一个原子量，其统计分布遵从指数形式的麦克斯韦–玻尔兹曼能量公式，并由此得到了混合气体的离解与压强和温度有关的理论。

接下来我们有必要假设，若要得到恒星表面的这些经验性的光谱关系就要求深藏在下面的稳定的恒星内部结构在某些程度上支配着恒星的大气。根据这一点，人们已经将关于恒星内部结构和辐射通量的试验性理论发展到很精细的程度了。根据爱丁顿的观点，恒星从表面到中心都可以看作是理想气体，尽管中心区的密度也许能达到铂的几百倍。这一点可以在天狼星的伴星上得到很好的证明——为了避免出现原子被不可逆地“离解”到只剩下骨架的独特概念，人们曾经用两个原子共同占据同一个空间这个并不陌生的观点来解释高密度的存在；根据一个与温度有关的定律，以辐射形式放出的能量来自于物质的离解或毁灭过程。另一方面，琼斯坚持认为，在预期的漫长演化过程中，那些必要辐射的强度肯定是稳定而确切的，否则恒星将

absolute intensity, else the star would explode: and he has essayed to regard the star as "liquid" in his investigations, apparently, however, implying a very imperfect gas rather than a special phase with its surface of sharp transition. There are other theories of less statical type.

A determined effort to shed off all such special hypotheses has been published very recently by Milne (*Monthly Notices R.A.S.* for November, pp. 17–53), which accordingly invites close attention and scrutiny. The procedure is the natural one, to try to make continuity between the gases of the atmosphere subject to laws more or less already formulated, and a dense interior about which as little is to be assumed as can be helped. He holds that it suffices merely to consider laws of internal density that are in mechanical equilibrium radially under internal pressure P, of which the fraction $(1-\beta)P$ is pressure of the internal field of radiation. He does not find it necessary to consider how this field of radiation of pressure $(1-\beta)P$ is sustained against loss by outward flux: for if he can arrive at results in terms of surface values that are valid for all such equilibrated densities whether otherwise possible or not, they must hold good for the one that follows the actual law of distribution whatever it may be.

The essential feature, so far as a reader can extract the gist from the complication of formulas that seems to be inherent in these discussions, appears to be that the coefficient β, while increasing rapidly downward in the atmosphere in a manner which can be regarded as known, suddenly rises when a photospheric level is reached, altering with steep gradient until a nearly constant value of β is soon attained for the interior of the star: and the same must apply only in less degree to the density ρ. The condition of mere mechanical equilibrium of the interior is found to express the pressure at the interface between atmosphere and photosphere in terms of values at the centre and one quantity C arising from an integral along the radius involving the arbitrarily assumed law of density. The expression for the atmospheric pressure at the interface involves the same constants in such way that on equating the pressures on the two sides of the interface they divide out of the result and only C remains. This C is held, in the light unforeseen of comparison with facts, to be in some degree a characteristic constant for all the stars, and thus may be the new element beyond surface values, and without assuming anything about their interiors, that the law as formulated requires.

This seems to be right enough in a general way, were it not that the formula for C involves the gradient of density within the star close to the interface, and thus its value must be very substantially changed, in absence of some verification to the contrary, by a very slight radial displacement of the surface which is chosen for that interface. For inside the photosphere ρ is as θ^3, while P which is continuous across the interface is as $\theta^4\phi(\theta)$: so C^{-1} is as the value of $P^{-1}(d\rho/dr)^2$ in which the second factor is the internal gradient, at the surface. If this consideration be correct it would appear that it is not legitimate to connect the chromosphere with the interior across a sharp boundary surface, as if they were different phases of matter like a liquid and its vapour. This conclusion would involve that the formula itself for C cannot be well founded: and the reason can be assigned, that

会发生爆炸。他在研究中试图假设恒星是“液体”的，这显然意味着恒星是一种非常不理想的气体，而不是处于一种在表面处变化很突然的特殊状态。另外还有一些其他的弱静态型理论。

米尔恩在最近发表的一篇文章（《皇家天文学会月刊》，11 月，第 17~53 页）中，坚决地剔除了所有这些特殊的假说，因而引发了人们的密切关注和深入思考。他很自然地试图让大气层中的气体与致密的内部之间保持连续性，其中，大气层中气体遵守的定律已经在前面或多或少地提到过了，而致密的内层则几乎没有什么理论可以参照。他提出，如果仅考虑内压为 P 时沿半径方向处于力学平衡的内部密度的变化规律，内压的一部分 $(1-\beta)P$ 源自内部辐射场，则这种连续性是可能存在的。米尔恩认为，没有必要考虑这个压强为 $(1-\beta)P$ 的辐射场如何克服向外的通量损失来维持自身，因为不管是否存在其他可能性，只要他能够得到对所有平衡态密度都有效的表面值，这些结果都会满足一个真实的分布规律，无论其形式如何。

只要读者能够从以上讨论中似乎不可避免的复杂公式中抽取出要点，就会看到基本特征是，系数 β 的值以一种我们已经知道的方式在恒星大气层中由外向内迅速增长，当到达光球层时突然增加，然后以很陡的梯度发生变化，直到很快在恒星内部达到一个基本恒定的值；密度 ρ 的情况肯定也是如此，只是在程度上会弱一些。内部的力学平衡条件可以表示为大气层和光球层间分界面上的压强，这个压强是中心物理量和 C 值的函数，其中 C 是由任意假定的密度分布沿半径方向积分得到的。大气压强在界面处的表达式中也包含同样的常数，在化简结果以使界面两侧的压强相等之后，其他常数都消失了，结果中只有 C 保留了下来。C 在某种意义上可以被认为是所有恒星的一个特征常数，这是在与实际情况的对比中没有料到的。因此它可能是除表面值之外的新要素，上述原理需要这些要素而不必考虑它们的内涵。

从一般意义上说上述做法已经足够完美了，如果不是因为 C 的表达式中包含恒星内部靠近分界面处的密度梯度，则在没有相反证据的前提下，只要所选的分界面沿半径方向有微小的位移，C 的数值必然会发生很大的变化。如果把光球层内的 ρ 看作 θ^3，把在分界面处数值连续变化的 P 看作 $\theta^4\phi(\theta)$，则 C^{-1} 就等于 $P^{-1}(d\rho/dr)^2$，其中第二个因子是在表面处的内部梯度。如果上述结果是正确的，那么将色球层与相隔一个突变边界面的恒星内部看成一个整体就显得不太合理，它们就像同一种物质的液态和气态一样属于不同的相。由此可见关于 C 的表达式不可能是合理的，理

the transition from Milne's formula (21) to (22) is invalid because the interior gradient of ϕ at the interface is very large and cannot be neglected even when multiplied by θ. Apparently one can only assert that the mass of the star involves the value of dP/dr within the photosphere and other quantities relating to the centre of the star, and the luminosity involves the value of P outside it, while the pressure P is continuous across a transition but not dP/dr.

In any case, perhaps not much stress would be laid on the deduction. The formula is regarded probably by its author as essentially an empirical result. When the value of C had been adapted to two prominent stars, the sun and Capella, it turned out in his hands, as he relates, to his astonishment, that it was a universal constant the same for all stars, and if so perhaps not connected with their interior constitutions at all.

(**125**, 273-274; 1930)

Joseph Larmor: Cambridge, Jan. 18.

由是从米尔恩的公式（21）推不出公式（22），因为内部梯度 ϕ 在分界面处非常大，即便是乘以 θ 以后也不能被忽略。显然我们只能说恒星的质量包含在光球层内的 dP/dr 值以及其他与恒星中心相联系的量中，发光度包含光球层外的 P 值，而穿过交界处时保持连续的是压强 P 而不是 dP/dr。

不管怎么说，也许对推导过程没有给予足够的重视。作者可能认为这个公式基本上是依赖经验结果得到的。当 C 值被应用于太阳和五车二这两颗著名的恒星时，作者发现，出乎他的意料，C 是一个对于所有恒星都相同的普适常数。如果真是这样，那么也许 C 值与恒星的内部结构根本没有关联。

（史春晖 翻译；蒋世仰 审稿）

Discovery of a Trans-Neptunian Planet

A. C. D. Crommelin

Editor's Note

***Nature* reports that astronomers at the Lowell Observatory in Flagstaff Arizona have for seven weeks been observing an object which appears to be a planet in orbit beyond Neptune. Its behaviour agrees fairly closely with predictions based on anomalies in the orbit of Uranus, and its size, based on one visual observation, seems intermediate between that of the Earth and Uranus. The article notes that the gravitation of this object might also account for a deviation of several days in the arrival of Halley's comet in each of its last two returns. The planet, soon named Pluto, is today no longer considered to be a proper planet, but rather a "dwarf planet"—an asteroid-like object comprised of frozen material.**

ON the evening of Mar. 13 (an appropriate date, being the anniversary of the discovery of Uranus in 1781, and Mar. 14 being the birthday of the late Prof. Percival Lowell) a message was received from Prof. Harlow Shapley, director of Harvard Observatory, announcing that the astronomers at the Lowell Observatory, Flagstaff, Arizona, had been observing for seven weeks an object of the fifteenth magnitude the motion of which conformed with that of a planet outside Neptune, and agreed fairly closely with that of one of the hypothetical planets the elements of which had been inferred by the late Prof. Percival Lowell from a study of the small residuals between theory and observation in the positions of Uranus. That planet was better suited than Neptune for the study, since the latter had not been observed long enough to obtain the unperturbed elements.

Lowell's hypothetical planet had mean distance 43.0, eccentricity 0.202, longitude of perihelion 204°, mass 6.5 times that of the earth, period 282 years, longitude 84° at the date 1914–1915. Its position at the present time would be in the middle of Gemini, agreeing well with the observed place, which on Mar. 12 at 3h U. T. was 7 seconds of time west of δ Geminorum; the position of the star was R.A. 7h 15m 57.33s, north decl. 22° 6′ 52.2″, longitude 107.5°. This star is only 11′ south of the ecliptic, making it likely that the new planet has a small inclination. As regards the size of the body, the message states that it is intermediate between the earth and Uranus, implying perhaps a diameter of some 16,000 miles. A lower albedo than that of Neptune seems probable, to account for the faintness of the body. It appears from a New York telegram that at least one visual observation of the planet has been obtained, from which the estimate of size may have been deduced.

Mention should also be made of the predictions of Prof. W. H. Pickering; one of these, made in 1919 (*Harvard Annals*, vol. 61), gives the following elements; Epoch 1920; longitude 97.8°; distance 55.1, period 409 years; mean annual motion 0.880°; longitude of perihelion 280°; perihelion passage 1720, eccentricity 0.31; perihelion distance 38, mass twice earth's, present

发现海外行星

克罗姆林

编者按

据《自然》报道，天文学家在亚利桑那州弗拉格斯塔夫市的洛威尔天文台已经对一个看似是在海王星轨道外运行的行星进行了7周的观测。这颗行星的特征非常接近于人们根据天王星轨道的反常现象所作的预测，它的目测大小似乎介于地球和天王星之间。这篇文章指出，该天体的引力可能是使哈雷彗星最近两次回归地球的时间产生数天偏差的原因。这颗行星很快被命名为冥王星，现在它已不再被人们看作是一颗行星，而是一颗“矮行星”——一个由超低温物质组成的小行星。

3月13日晚（此日恰为1781年发现天王星的纪念日，而3月14日是已故天文学家珀西瓦尔·洛威尔的诞辰）从哈佛大学天文台台长哈洛·沙普利那里收到了一则消息，亚利桑那州弗拉格斯塔夫市洛威尔天文台的天文学家已经对一个星等为15等的天体进行了长达7周的观测，结果发现该天体的运动与海王星外的一颗行星同步，且与一颗假想行星的情况相当吻合。通过研究天王星的理论位置和观测位置之间的细小差别，已故的珀西瓦尔·洛威尔教授推断出了假想行星的根素。鉴于尚未对海王星观测足够长的时间来获得无扰动根素，相比之下这颗行星会更适合进行研究。

洛威尔假想行星的平均距离为43.0个天文单位，偏心率为0.202，近日点的黄经为204°，质量是地球质量的6.5倍，周期为282年，1914~1915年的黄经为84°。当前时刻该行星的位置应在双子座的中间，而在3月12日世界时3点观测到该星位于双子座δ星西部7s赤经的位置，因而观测与预言相吻合。该星的赤经是7h 15m 57.33s，赤纬是$+22°6'52.2''$，黄经是107.5°。该星位于黄道南部11′的位置，很可能是一颗有很小倾角的行星。消息中写道，该行星的大小介于地球和天王星之间，直径约为16,000英里。该星的反照率低于海王星，因而星体本身极为暗弱。从来自纽约的电报看，目前已经至少获得过一次关于该星的目视观测，从而可能已经推测出了该星的大小。

这里需要提及的是皮克林教授的预言，他在1919年的一个预言（《哈佛年鉴》，第61卷）中给出的根素如下：历元1920年；黄经97.8°；距离55.1个天文单位，周期409年；平均周年运动0.880°；近日点的黄经280°；1720年通过近日点，偏心

annual motion 0.489°. This prediction gives the longitude for 1930 as 103°, which is within five degrees of the truth; actually it was in longitude 108° at discovery. Prof. Pickering's later prediction is further from the truth, making the longitude about 131°.

Gaillot and Lau also made predictions; like the other computers they noted that there were two positions, about 180° apart, that would satisfy the residuals almost equally well. Taking the position nearest to the discovered body, Lau gave longitude 153°, distance 75, epoch 1900. Gaillot gave longitude 108°, distance 66, epoch 1900. The latter is not very far from the truth; with a circular orbit, the longitude in 1930 resulting from Gaillot's orbit would be 128°, some 20° too great. Gaillot performed the useful work of revising Le Verrier's theory of Uranus, thus giving more trustworthy residuals. Lowell pointed out that the residuals of Uranus that led to the discovery of Neptune amounted to 133″, while those available in the present research did not exceed 4.5″; yet even in the case of Neptune the elements of the true orbit differed widely from the predicted ones, though the direction of the disturbing body was given fairly well. He noted that in the present case it would be wholly unwarrantable to expect the precision of a rifle bullet; if that of a shot-gun is obtained, the computor has done his work well.

Another method of obtaining provisional distances of unknown planets is derived from periodic comets; the mean period of the comets of Neptune's family is 71 years; it is pointed out in the article on comets ("Encyc. Brit." 14th edition, vol. 6, p. 102) that there is a group of five comets the mean period of which is 137 years; as stated there, "This family gives some ground for suspecting the existence of an extra-Neptunian planet with period about 335 years, and distance 48.2 units." This seems to be in fair accord with the new discovery, but probably the distance is nearer 45 than 48. Comets also suggest another still more remote planet, with period about 1,000 years, a suggestion which has also been made by Prof. G. Forbes and by Prof. W. H. Pickering.

The question has been asked, "Does the new planet conform to Bode's law?" It is difficult to assign a definite meaning to this question, since Bode's law broke down badly in the case of Neptune; Neptune's predicted distance was 38.8, its actual distance 30.1. For Bode's law, each new distance ought to be almost double the preceding one; the constant term of the law becomes negligible when the distance is great. For the extension of the terms given by the law we might (1) ignore Neptune as an interloper and take the next distance as double that of Uranus, giving 38.5 units; (2) we might take the next distance as four times that of Uranus, which would give 77 units; or (3) we might take the next distance as double that of Neptune or 60 units; none of these values is good, but (1) is the nearest to what we suppose to be the distance. Probably the best course is to assume that after Uranus the law changes; each new distance is then 1.5 times the preceding one; on this assumption, the hypothetical planet with distance 100 and period 1,000 years would be the next but one after the Lowell planet.

The low albedo of the new planet might be explicable if its temperature were much lower than that of Neptune. Owing to its smaller size, it would have lost more of its primitive

率 0.31；近日点距离 38 个天文单位，质量是地球的 2 倍，目前的周年运动为 0.489°。该预言认为 1930 年行星的黄经为 103°，该值与真实值相差不到 5°；事实上，发现时它的黄经为 108°。皮克林教授后来的预言与真实值相差更远，他预言的黄经达到 131°。

加约和劳也作了一些预言；和其他天文学家一样，他们认为在远离 180° 的两个位置能够很好地满足残差。选取离被发现天体最近的位置，劳给出黄经 153°，距离 75 个天文单位，历元 1900 年。加约给出黄经 108°，距离 66 个天文单位，历元 1900 年。后者与真实值相差不远；对于圆形轨道，由加约的轨道得到 1930 年该星的黄经是 128°，比真实值大 20°。加约修正了勒威耶关于天王星的理论工作，给出了更可靠的残差。洛威尔指出导致发现海王星的天王星残差是 133″，然而在目前的研究中能够得到的残差小于 4.5″；然而即使在海王星的情况中，虽然精确地给出了扰动天体的方向，真实轨道根素也远远不同于预言结果。他还指出在目前的情况下要达到来复枪子弹那样的精度是完全不可能的；如果有更好的观测仪器，那么计算机可以将他的工作处理得更好。

另一种获得未知行星临时距离的方法来自周期彗星；海王星家族的彗星平均周期是 71 年；一篇关于彗星的文章（《不列颠百科全书》，第 14 版，第 6 卷，第 102 页）中指出一个由 5 颗彗星组成的团组的平均周期是 137 年；文章中说“这个家族提供了某种基础，可以假定存在周期大约为 335 年，距离为 48.2 个天文单位的海外行星。”这似乎与目前的新发现非常吻合，但是距离是接近 45 个天文单位而不是 48 个天文单位。福布斯和皮克林教授提出，彗星暗示了其他更远的、周期约为 1,000 年的行星的存在。

曾有人问到“新的行星满足波得定律吗？”由于海王星已经严重打破了波得定律，因而这个问题已不再有意义。海王星的预言距离是 38.8 个天文单位，实际距离是 30.1 个天文单位。根据波得定律，每一个新的天体距离都必须是前一个天体距离的 2 倍；当距离很远时可以忽略定律的常数项。对于定律的延展项，(1) 我们可以把海王星作为闯入者而忽略，下一个距离取天王星距离的 2 倍，得到 38.5 个天文单位；(2) 我们也可以把下一个距离取天王星距离的 4 倍，得到 77 个天文单位；(3) 或者我们还可以把下一个距离取成海王星的 2 倍即 60 个天文单位。这里的任何一个值都不太好，但是 (1) 的取值最接近我们估计的距离。那么在天王星之后，波得定律不再满足，这可能是最可取的方法；而每一个新的距离是前一个天体距离的 1.5 倍；基于这种假定，距离为 100 个天文单位、周期为 1,000 年的天体将是继洛威尔行星后的又一颗行星。

如果新行星的温度远低于海王星，那么就可以解释这类行星的低反照率。由于该行星很小，它将失去更多自身的原始热量而且只能吸收来自太阳能量的一半；从

heat, and would only receive half as much from the sun; hence its gases might be reduced to a liquid form, with great reduction of their volume. This would result in a relatively smaller disc than the one that might be inferred from its mass.

Some further particulars of the discovery are given by the New York correspondent of the *Times* in the issue for Mar. 15. Quoting an announcement which had been received there from the Lowell Observatory, it is stated that the planet was discovered on Jan. 21 on a plate taken with the Lawrence Lowell telescope; it has since been carefully followed, having been observed photographically by Mr. C. O. Lampland with the large Lowell reflector, and visually with the 24-inch refractor by various members of the staff. The observers estimate the distance of the planet from the sun as 45 units, which would give a period of 302 years, and mean annual motion of 1.2 degrees.

At discovery, the planet was about a week past opposition, and retrograding at the rate of about 1′ per day; this has now declined to 0.5′ per day, and the planet will be stationary in April. It should be possible to follow it until the middle of May, when the sun will interfere with observation until the autumn.

The details of the Lowell Observatory positions have not yet come to hand; when they do, it will be possible to derive sufficiently good elements to deduce ephemerides for preceding years. There are many plates that may contain images of the planet; those taken by the late Mr. Franklin Adams in his chart of the heavens, those taken of the region round Jupiter some twelve years ago for the positions of the outer satellites, and those taken at Königstuhl and elsewhere in the search for minor planets; these all show objects down to magnitude 15. If early images should be found, they will accelerate the determination of good elements of the new planet; in the case of Uranus, observations were found going back nearly a century before discovery, and in that of Neptune they went back fifty-one years. In the present case, forty years is the most that can be hoped for, and probably very few photographs showing objects of magnitude 15 are available before the beginning of this century.

One of the most difficult problems will be to find the mass of the new body; in Neptune's case, Lassell discovered the satellite a few months after the planet was found, and the mass was thus determined. It is to be feared, however, that the new planet would not have any satellite brighter than magnitude 21. Stars of this magnitude have been photographed with the 100-inch reflector at Mount Wilson, but it is doubtful whether it could be done within a few seconds of arc of a much brighter body. Failing the detection of a satellite, the mass can only be deduced from a rediscussion of the residuals of Uranus and Neptune; new tables of these planets will ultimately be called for, but that task must wait until the orbit of the new body is known fairly exactly.

The perturbations of Halley's comet will also require revision; at each of the last two returns, there has been a discordance of two or three days between the predicted and observed dates of perihelion passage; it will be interesting to see whether the introduction

而行星上的气体将会变成液态，体积也大大减小。这样它的盘面将小于由质量推算出的盘面的大小。

《泰晤士报》的纽约通讯员在 3 月 15 日的一期上刊载了发现新行星的更进一步的细节。文章引用了洛威尔天文台的一段宣告，说行星是在 1 月 21 日劳伦斯·洛威尔望远镜拍摄的一张底片中发现的；此后天文学家即对该星进行跟踪，兰普兰德先生用大洛威尔反射望远镜对该星进行了照相观测，其他的工作人员用口径 24 英寸的折射望远镜进行目视观测。观测者们推断该星与太阳的距离是 45 个天文单位，由此周期应当是 302 年，平均周年运动 1.2°。

发现之初，该行星刚通过对冲大约一星期，每天退行 1′；目前退行速度减小为每天 0.5′，并将在 4 月保持静止（留）。在 5 月中旬之前可以对该星进行跟踪观测，此后直到秋天来临之前这段时间内，太阳都会影响观测。

洛威尔天文台还未对观测结果进行详细处理；一旦进行处理，那么将会获得足够精确的轨道根素进而推导出前几年的天文历表。有很多底片可能包含这个行星的图像，如已故的富兰克林·亚当斯先生的天空星图，12 年前拍摄木星周围卫星的底片，以及那些在柯尼希斯施图尔山和其他地方拍摄的搜寻小行星的底片；这些底片都能够显示暗到 15 等的天体。如果可以找到更早期的图像将可以加速新行星轨道根素的确定；对于天王星，观测资料可以上溯到天王星发现前近一个世纪，而对于海王星则可以追溯到 51 年前。对于目前这颗行星至多追溯到 40 年前，本世纪之前可能很少有能够记录到 15 等星的照相底片。

最困难的一个问题是确定新发现天体的质量，对于海王星，在发现该行星几个月后拉塞尔就发现了它的卫星，从而确定了海王星的质量。然而，这颗新的行星的任何一个卫星的星等可能都将暗于 21 等。在威尔逊山用 100 英寸的反射式望远镜曾经拍摄过 21 星等的星，但是并不能确定使用该望远镜能否在几弧秒的范围内拍摄到这样的天体。由于不能探测到卫星，那么该行星的质量只能从对天王星和海王星的残差的再讨论中推导出来。而要进行这种再讨论推导就必须有关于新行星的新的位置数据表，而要得到这个表就必须有这个新行星的精确的轨道根素。

哈雷彗星的扰动也需要修正；在最近两次回归中，通过近日点日期的预言值与观测值之间都有两三天的差别；研究结果是否会由于引入新天体的扰动而有所改进

of the perturbations of the new body effects an improvement. The late Mr. S. A. Saunder made the suggestion at the time of the last apparition of the comet that an unknown planet might be the cause of the discordance, but it was not then possible to carry the suggestion further. The discovery of a new planet therefore opens a large field of work for mathematical astronomers. It will also appeal to students of cosmogony; Sir James Jeans, in an article in the *Observer* for Mar. 16, suggests that it may represent the extreme tip of the cigar-shaped filament thrown off from the sun by the passage of another star close to it. It would have been the first planet to cool down and solidify; he says, "As a consequence of this, it will probably prove to be unattended by satellites."

(**125**, 450-451; 1930)

将是一个有趣的课题。在彗星最近一次出现时，已故的桑德先生提出未知行星的存在将可能导致这种差异，但是并未对这个想法进行更进一步的研究。一颗新行星的发现则为数学天文学家提供了更广阔的工作空间。这也将吸引更多研究宇宙演化的学者投入其中；詹姆斯·金斯先生在 3 月 16 日《观察家报》的一篇文章中写道：这个新行星可能代表离太阳较近的另外一个恒星通过时太阳抛射出来的雪茄状纤维丝的顶端。它将是冷却凝固形成的第一颗行星；他说“作为这个现象的结果，将可能证明这颗新的行星没有卫星。”

（王宏彬 翻译；蒋世仰 审稿）

Lowell's Prediction of a Trans-Neptunian Planet

J. Jackson

Editor's Note

John Jackson of the UK Royal Observatory here discusses the possibility that the orbit of the newly discovered planet had been accurately predicted by the astronomer and polymath Percival Lowell. If true, Jackson suggests, then Lowell deserves high admiration, for the problem, while in principle similar to the prediction of Neptune, is in detail immeasurably more difficult. Jackson reviews Lowell's methods, in which, by hypothesizing a planet of particular mass and orbit, he could reduce the unexplained motion of Uranus by some 70%. However, the brightness of the observed planet is about ten times higher than predicted. Astronomers since have determined that Lowell's calculations were in error, and that Pluto was discovered principally through a painstaking empirical search of the sky.

THE reported discovery of a planet exterior to Neptune naturally arouses the interest of the general public. It will be of importance in theories concerning the genesis of the solar system as to how far it falls into line with the other planets as regards distance, mass, eccentricity and inclination of orbit, and presence or absence of satellites. Its physical appearance will be beyond observation. To those interested in dynamical astronomy, it may be of some interest to consider the data which led to its discovery and to make some comparison with the corresponding facts relating to Neptune.

If the planet which has been reported approximately follows the orbit predicted by Dr. Percival Lowell, the prediction and the discovery will demand the highest admiration which we can bestow. It is true that the problem as regards its general form is a repetition of that solved by Leverrier, Adams, and Galle more than eighty years ago; but its practical difficulty is of quite a different order of magnitude. In short, this discovery, if it turns out to be actually Lowell's predicted planet, was extremely difficult—while Neptune was in fact crying out to be found. Let us look at the actual data.

Uranus was discovered in 1781 by Herschel. Scrutiny of old records showed that it had been observed about a score of times dating back to 1690. The fact that Lemonnier observed it eight times within a month, including four consecutive days, without detecting its character, should be a lesson to anyone who makes observations without examining them. In 1820 Bouvard found that the old and the new observations could not be reconciled, and in constructing his tables boldly rejected the early observations, but the tables rapidly went from bad to worse; the residuals amounted to 20″ in 1830, 90″ in 1840, and to 120″ in 1844. Adams used in his first approximation data up to 1840, Leverrier data up to 1845. Now Uranus had passed Neptune in 1822. As the relative motion is about 2° a year, it means that for most of the time covered by the prediscovery observations the perturbations were very small, while from the fact that the difference

洛威尔对海外行星的预言

杰克逊

编者按

英国皇家天文台的约翰·杰克逊在此讨论了这颗新发现的行星的轨道已经被天文学家、博学者珀西瓦尔·洛威尔精确预测的可能性。在杰克逊看来，如果事实真的如此，那么洛威尔理应得到很高的荣誉，因为虽然计算的原则类似于海王星，但具体过程更加困难。杰克逊介绍了洛威尔的方法，洛威尔通过假设另一个具有一定质量和轨道的行星解释了天王星中 70% 的不明运动。可是，这颗行星的实际亮度比预测值高 10 倍。天文学家们后来认为洛威尔的计算存在错误，而冥王星的发现主要是通过有经验的观测者在太空中艰苦搜索得到的。

已报道的海王星外行星的发现引起了公众广泛的兴趣。该行星对于与太阳系的形成相关的理论来说将是至关重要的，诸如该星在线距离、质量、偏心率和轨道倾斜度，以及是否存在卫星等方面是否与其他行星的规律相符合的问题。该星的物理外貌是不能观测到的。如果对动力天文学感兴趣，可以研究一下发现该行星的数据并将它与海王星的相关资料进行比较。

如果报道的行星大致符合珀西瓦尔·洛威尔博士预言的轨道，那么这个预言和发现将是令人瞩目的。的确，就其一般形式而言，这个问题确实是 80 年前已经由勒威耶、亚当斯和伽勒解决了的问题的重现，但是实际研究中存在的困难在于数量级的差异。简而言之，如果这次发现的新行星的确是洛威尔预言的行星，那么这一发现是极其困难的，而海王星的发现其实是非常容易的。下面让我们看看实际的数据资料。

天王星是赫歇尔在 1781 年发现的。详细审查原始记录会发现，早在 1690 年就已经观测到了这颗星。勒莫尼耶一个月内 8 次观测天王星并且有 4 天是连续观测，但并未探测到它的特征，每个进行观测但未进行详细检测的观测者都应该引以为戒。1820 年布瓦尔发现新旧观测数据并不一致，他在建数据表的时候大胆舍弃了早期的观测资料，但是数据表的结果更糟了；1830 年残差达 20″，1840 年达 90″，到 1844 年则达到 120″。而亚当斯直到 1840 年使用的还是他的第一个近似数据，勒威耶则直到 1845 年还在用。1822 年天王星越过海王星。由于每年的相对运动只有 2°，那么在发现前观测所覆盖的大部分时间内，扰动是非常小的。然而考虑到行星

between the heliocentric distances is much smaller than expected from Bode's law, the perturbations at the time of conjunction were relatively large. Consequently the prediction of the longitude of the disturbing body was very easy, while the determination of the other elements were correspondingly difficult. The fact was that the simple hypothesis of the existence of an exterior planet with any sort of guess as to size and shape of orbit would suffice to predict the longitude. In other words, most of the residuals could be closely satisfied provided that substantially correct values of the longitude of the planet and its attractive force $m\left(\frac{1}{\Delta^2}+\frac{1}{r^2}\right)$ were used. Both Leverrier and Adams easily found values of these quantities, and Galle had no difficulty in detecting the planet.

We now turn to Lowell's "Memoir on a Trans-Neptunian Planet", published in 1915. The observational basis is the outstanding residuals in the motion of Uranus during two centuries, that is, rather more than two revolutions of that planet round the sun, of somewhat less than two revolutions relative to the predicted planet and of about one relative to Neptune. The following are the values of the observed residuals of Leverrier's and of Gaillot's theories taken from Lowell's memoir.

	Leverrier	Gaillot		Leverrier	Gaillot
1709	. .	+2.14″	1855	. .	−0.50″
1753	+5.52″	+4.45	1858	+0.50″	−0.20
1769	+4.77	+2.47	1861	. .	−0.36
1783	−3.30	−0.96	1864	+0.25	+0.18
1787	−5.12	−1.20	1867	. .	+1.20
1792	−3.50	+0.10	1870	−0.50	+1.32
1796	−1.88	−0.69	1873	. .	+0.75
1803	+0.40	−1.19	1876	−1.65	−0.50
1812	+2.00	−0.77	1879	. .	+0.58
1817	+0.50	−0.60	1882	−2.88	+0.52
1820	−0.75	−2.37	1885	. .	−0.17
1827	−2.10	+2.00	1888	−4.22	−0.85
1837	−1.10	−1.22	1891	. .	−1.11
1840	+0.63	+0.78	1894	−5.63	−0.50
1843	. .	+0.74	1897	. .	+0.35
1846	+0.38	−1.40	1900	−4.32	+1.00
1849	. .	−0.25	1903	−3.00	+0.65
1852	−1.17	−0.95	1907	. .	+0.25
			1910	. .	+1.10

The residuals show remarkable differences between the two theories, but Lowell deduced that the residuals exceeded their probable errors four or five times. The problem was to find from these residuals corrections to the elements of the orbit and to find the mass and the elements of the disturbing body. It might almost appear hopeless when we consider that the residuals must be affected by errors in the accepted masses of the known planets. There can be no doubt, however, that the masses adopted by Gaillot for Jupiter, Saturn, and Neptune are very accurate. Lowell's procedure was to adopt a value of the semimajor axis of the unknown body, and a complete series of values for its longitude, and then select the value of the longitude for which the sum of the squares of the residuals was

的日心距比由波得定律估算出的值小得多，那么在合发生时扰动将相对大一些。这样就可以较容易地确定扰动天体的黄经，然而确定该天体的其他根素就相对困难一些。实际上简单的外行星存在的假定，不管对轨道的大小和形状作何种猜测，都足以预言行星的黄经。换句话说，只要充分地利用行星黄经的正确值以及它的引力 $m\left(\frac{1}{\Delta^2}+\frac{1}{r^2}\right)$，那么大部分残差都可以得到很好的满足。勒威耶和亚当斯都很容易地找到了这些值，伽勒也很容易地探测到了行星。

下面我们看一下洛威尔 1915 年发表的《海外行星回忆录》。观测的主要内容是两个世纪中在那颗行星绕太阳旋转两圈多的过程中天王星运动中出现的残差，这种运动相对于预言中的行星不到两圈，相对于海王星大约为一圈。下面是来自洛威尔回忆录中相对于勒威耶理论和加约理论的观测残差值。

	勒威耶	加约
1709	. .	+2.14″
1753	+5.52″	+4.45
1769	+4.77	+2.47
1783	–3.30	–0.96
1787	–5.12	–1.20
1792	–3.50	+0.10
1796	–1.88	–0.69
1803	+0.40	–1.19
1812	+2.00	–0.77
1817	+0.50	–0.60
1820	–0.75	–2.37
1827	–2.10	+2.00
1837	–1.10	–1.22
1840	+0.63	+0.78
1843	. .	+0.74
1846	+0.38	–1.40
1849	. .	–0.25
1852	–1.17	–0.95
1855	. .	–0.50″
1858	+0.50″	–0.20
1861	. .	–0.36
1864	+0.25	+0.18
1867	. .	+1.20
1870	–0.50	+1.32
1873	. .	+0.75
1876	–1.65	–0.50
1879	. .	+0.58
1882	–2.88	+0.52
1885	. .	–0.17
1888	–4.22	–0.85
1891	. .	–1.11
1894	–5.63	–0.50
1897	. .	+0.35
1900	–4.32	+1.00
1903	–3.00	+0.65
1907	. .	+0.25
1910	. .	+1.10

这两种理论得到的残差值有着明显的差别，但是洛威尔推导发现残差与他们估测的误差相比超出 4~5 倍。现在的问题是，从这些残差的改正中找到轨道根素和扰动天体的质量及根素。当我们考虑到残差必然会受到已知行星质量误差的影响时，那么这些问题的解决就几乎是不太可能的了。毫无疑问，加约引用的木星、土星和海王星的质量都是非常精确的。洛威尔的做法是，引入未知天体的一个半长轴值以及该天体一系列完整的黄经值，然后选择残差平方和最小的黄经。用各种平均距离值（即未知行星的半长轴值）重复进行这一过程，直到变量的取值使得残差最小为

a minimum. The process was repeated with various values of the mean distance until values of the variables were found giving minimum residuals. The process was of course very laborious, but Lowell carried it through with great perseverance. The following extract from his final summary may be quoted: "By the most rigorous method, that of least squares throughout, taking the perturbative action through the first powers of the eccentricities, the outstanding squares of the residuals from 1750 to 1903 have been reduced 71 percent by the admission of an outside disturbing body."

The inclusion of further terms, of additional years and of the squares of the eccentricity, do not alter the results by any substantial amount. Lowell considered that the remaining irregularities could be explained by errors of observation. No trustworthy results could be found from the residuals in latitude so that the inclination of the orbit to the ecliptic could not be deduced, but Lowell considered that it might be of the order of 10°.

As the solution really depends on the difference of the attraction of the unknown planet on Uranus and on the sun, there are two possible solutions in which the longitudes differ by about 180°. The following elements are for the solution satisfying most nearly the position of the newly found body.

Heliocentric longitude on 1914, July	84.0°
Semimajor axis	43.0
Mass in terms of the sun's mass	1/50,000
Eccentricity	0.202
Longitude of perihelion	203.8°

This gives the longitude at the present time as about 104° compared with 107° of the new planet. The predicted magnitude was 12 to 13 or about ten times brighter than the observed; and a disc of more than 1″ was predicted. This is a rather serious discordance.

The smallness of the residuals indicated that the forces were small. The mass given above is only 0.4 of the mass of Neptune. At mean conjunction, the attraction of the predicted planet on Uranus would be only one-fifteenth of the attraction of Neptune in a similar position, and in addition it would last for a shorter time on account of the more rapid relative motion.

The discovery of a minor planet of the fifteenth magnitude is an everyday occurrence. The planet reveals itself by a decided motion relative to the stars in the course of taking a photograph. For a planet in the predicted orbit, the motion shown (mostly due to the earth's motion) would in the most favourable circumstances not be more than 2″ or 3″ an hour, and it would probably need a trail of at least 5″ for the planet to be detected. On the other hand, photographs taken on successive days would show decided motion, but the labour of finding the planet in a region containing many thousands of stars from separate photographs would be very great. Probably the Lowell observers have come across several minor planets before they were rewarded by the discovery of the very distant planet.

止。这无疑是一项非常艰苦的工作，但是洛威尔却坚持不懈地完成了。下面是从他的文章中引出的一句话，“利用最严密的方法，即最小二乘法，引入外界的一个扰动天体计算偏心率一次方的扰动行为，已经把 1750~1903 年天体明显的残差平方减少了 71%”。

进一步引入其他因素，加长年代以及使用偏心率的平方，并没有使结果有任何实质上的改变。洛威尔认为剩下的不规则性可以用观测误差来解释。从黄纬的残差中不能找到可信的结果，因而不能推导出轨道与黄道的倾角，但是洛威尔认为这个倾角可能是 10°。

由于问题的解直接取决于未知行星对天王星和太阳的引力差别，所以可能有两种使黄经相差 180° 的解。下面这些根素就是满足新发现天体条件的最近的解。

1914年7月太阳中心黄经	84.0°
半长轴	43.0
以太阳质量为单位表示的质量	1/50,000
偏心率	0.202
近日点黄经	203.8°

与这一新行星的当前黄经 107° 相比，据此根素得出的其当前黄经大约是 104°。预测星等为 12~13 等，或者说比观测到的星亮 10 倍；而且还预测到了一个大于 1″ 的盘。这里产生了很严重的矛盾。

残差小意味着力小。上面给出的质量仅是海王星质量的 0.4 倍。在平均会合期，预测行星给天王星的引力仅是同一位置海王星所施加引力的 1/15，并且由于较快的相对运动，这一状态持续的时间很短。

发现一个 15 等的小行星是件很常见的事情。在拍照过程中行星以一种相对于恒星确定的运动显露出来。对于预测轨道上的行星，一般在最好的环境下显现出的运动每小时不超过 2″ 或 3″（主要是由于地球的运动），对于可以被探测到的行星，至少需要 5″ 的踪迹。另一方面，连续几天的拍摄可以得到行星确定的运动，但是从分立的照片中在包含数千颗恒星的区域里找到这颗行星将是一件困难的工作。恐怕洛威尔天文台的观测者们在发现这颗遥远行星之前已经偶遇了很多小行星。

Astronomers all the world over will naturally look forward with great interest to see how nearly the newly discovered body moves in the orbit predicted by Lowell, and are anxiously waiting for further details of the observations.

(**125**, 451-452; 1930)

全世界的天文学家对新发现的这个天体的运动离洛威尔预测的轨迹有多近都非常感兴趣，而且充满好奇地期待着关于这一观测结果更详细的进展。

（王宏彬 翻译；蒋世仰 审稿）

Age of the Earth

A. P. Coleman

Editor's Note

This letter from Canadian geologist Arthur Coleman at the Royal Ontario Museum in Toronto argued that the Earth must be much older than then commonly believed. His starting point was his reading of a popular book by Sir James Jeans, a British astronomer and physicist. The modern estimate of the age of the Earth is about 4.6 billion years. Estimates of when life first appeared on the Earth are only a little smaller—perhaps 4.2 billion years.

I have just finished reading a most interesting book on "The Universe around Us", by Sir James Jeans. It opens up a complex and abstruse subject with admirable clearness, so that even a geologist possessed of very little mathematics can find his way through it without too much difficulty. The ease with which in this brilliant book millions of millions of stars are marshalled and their history outlined for millions of millions of years inspires no little awe and a large amount of envy in the breast of a plodding geologist who keeps to the solid earth. If the book contained only the inspiring visions of an astronomer in regard to the origin and the fate of the universe around us a geologist might refrain from comment; but at several points the history of the earth and its inhabitants is touched upon, giving him a right to a word of criticism.

Sir James in a page or two suggests that the earth began about 2,000,000,000 years ago as a globe of intensely hot gas, which gradually cooled down, becoming first a liquid, then plastic, and finally an outer crust solidified, "rocks and mountains forming a permanent record of the irregularities of its earlier plastic form." Life probably began on the earth about 300,000,000 years before the present.

A generation ago, when Lord Kelvin laid down the law as to geological time, this allowance would have seemed very liberal; but the discovery that the age of certain rocks can be determined by an analysis of the radioactive minerals they contain has completely changed our point of view. 2,000,000,000 years is decidedly too short a time for Pre-Cambrian history if the earth began in a gaseous form; and life existed far earlier than 300,000,000 years ago.

These points are easily proved by a brief study of the Grenville Series of Ontario and Quebec and of the Laurentian granite and gneiss which have erupted through it, since radioactive minerals are found in pegmatite dikes connected with the granite.

地球的年龄

科尔曼

编者按

这是一篇来自多伦多皇家安大略博物馆的加拿大地质学家科尔曼的快报文章，文中他认为地球的年龄远比大家公认的古老得多。他的出发点得益于英国天文学家、物理学家詹姆斯·金斯爵士的名著。目前估计的地球年龄为46亿年，而地球上首次出现生命的时间稍晚，可能为42亿年。

我刚读完一本十分有趣的书《我们周围的宇宙》，作者是詹姆斯·金斯爵士。这本书以清晰得令人惊叹的文字讲述了一个复杂而深奥的主题，即使是一位几乎没有数学基础的地质学家要读懂它也不会感到很困难。这部精妙绝伦的著作轻松自如地列举了数以百万计的恒星以及它们在数百万年中的演化史，让局限于固体地球的古板的地质学家感到既敬畏又十分羡慕。如果这本书只是一位天文学家对我们周围的宇宙的起源和结局的美好展望，那地质学家也就用不着说三道四了；但作者在某些方面又提到了地球的历史以及它的居民，所以地质学家还是有权提出批评意见的。

詹姆斯爵士用了一两页的篇幅说明地球诞生在约2,000,000,000年以前，开始是一个炽热的气团，随后逐渐冷却，先变成液体，然后是塑性体，最后外壳硬化，“岩石和山脉是地球早期塑性形态不规则性的永久记录。”地球上的生命可能始于距今约300,000,000年前。

30年前，当开尔文勋爵建立地质年代的理论时，这种定量看似非常自由，但是在发现某些岩石的年龄可以用其所含的放射性矿物测定后，我们的观点彻底改变了。如果地球是从气态开始演变的，那么地球上生命存在的时间将远远超过300,000,000年，而寒武纪之前的地球仅有2,000,000,000年的历史显然是太短了。

针对安大略和魁北克的格伦维尔岩系以及侵入其中的劳伦琴花岗岩和片麻岩的初步研究的结果可以很容易地证明以上观点，因为在混有花岗岩的伟晶岩脉处发现了放射性的矿物。

Ages of Pegmatite in Ontario and Quebec

Localities	Determined by T. L. Walker[1]	Determined by H. V. Ellsworth[2]
Parry Sound	1,274,000,000 years	1,179,000,000 years
Villeneuve	1,293,000,000 ,,	1,189,000,000 ,,
Cardiff Tp.	1,239,000,000 ,,	1,299,000,000 ,,
Butt Tp.		1,130,000,000 ,,

The localities are scattered over 200 miles from west to east, and it will be observed that the age of the pegmatites does not differ greatly at the various points and that the two analysts are not far apart in their determinations. We may conclude that the uraninites of the pegmatites were formed about 1,230,000,000 years ago.

The pegmatite dikes are the latest phases of the molten granite which heaved up the rocks of the Grenvile Series into domes forming important mountain chains which covered many thousand square miles, and the Grenville Series must have been solid rock long before this took place.

The Grenville rocks are now crystalline limestones, gneisses, and quartzites, but were originally ordinary limestones, shales, and sandstones, which were deposited as limy material, mud, and sand in a shallow sea.

The series is very thick according to Dr. Adams, who measured 17,824 feet in one section and 94,406 feet in another; and to the age shown by the radioactive minerals must be added many millions of years for the deposit of a great geological formation, its consolidation, and the thrusting up of the widespread Laurentian mountains.

But we are still far from the beginning. Before the Grenville sediments could be laid down the earth's crust must have been firm and solid to form shores and sea bottoms of a permanent kind, and must have been cool enough to allow rain to fall and rivers to bring down mud and sand into the sea. If the earth ever passed through a stage of heat and plasticity, that was completely over before the beginning of the Grenville sea; and the water was cool enough to allow algae or other lowly plants to thrive, since in places the sediments contain several percent of carbon, now in the form of graphite.

Life had already appeared in the sea.

The Grenville rocks have been chosen for this study because they are well known and are dated by analyses of radioactive minerals, but they are probably somewhat surpassed in age by the Keewatin, which is mainly volcanic, but with important amounts of sediments, and the Coutchiching, which is wholly sedimentary. No uraninites have yet been found in the granites which penetrate them.

安大略和魁北克伟晶岩的年龄

采样点	沃克的测定结果 [1]	埃尔斯沃思的测定结果 [2]
帕里湾	1,274,000,000 年	1,179,000,000 年
维伦纽夫	1,293,000,000 年	1,189,000,000 年
加的夫区	1,239,000,000 年	1,299,000,000 年
巴特区		1,130,000,000 年

样品采集区域分布在从西到东的 200 多英里范围内。我们可以看到不同地点的伟晶岩年龄没有太大差别，两个分析者的测定结果也相近，可以认为伟晶岩中的沥青铀矿形成于约 1,230,000,000 年前。

伟晶岩脉是熔融花岗岩结晶的最后阶段，熔融花岗岩岩浆拱起格伦维尔岩系的岩石，使其变形成穹隆状，从而形成了绵延数千平方英里的大山系，而格伦维尔岩系肯定在该过程发生之前早已是固化的岩石了。

格伦维尔岩系的岩石目前是结晶质石灰岩、片麻岩和石英岩，而原来曾是普通的石灰岩、页岩和砂岩，由浅海中含石灰的物质和泥沙沉积而成。

根据亚当斯博士的测定结果，该岩系很厚，在一个剖面的厚度为 17,824 英尺，另一个为 94,406 英尺。由于一个巨型地质层组的累积和固化过程以及广阔的劳伦琴山脉的推进需要很长时间，因此其地质年代比根据放射性矿物测定的年龄早数百万年。

然而，我们离出发点还很远。在格伦维尔沉积物形成之前，地壳必须足够坚硬致密，以形成永久性的海岸和海底。同时，地球一定已经足够冷，使雨水能够降落，河流能够把泥沙带到海里。如果地球曾经经历了炽热和塑性阶段，那么该阶段一定在格伦维尔海形成之前已经彻底结束；海水也冷得可以使海藻和其他低等植物旺盛生长，致使一些地方的沉积物中含有百分之几的碳，现在以石墨形式存在。

生命已经在海洋中出现。

本次研究之所以选择格伦维尔岩系，一是由于它的知名度，二是因为可以通过分析放射性矿物测量它的年代。但是，格伦维尔岩系可能稍晚于以火山岩为主但带有大量沉积物的基瓦丁岩系和完全由沉积岩组成的库奇金岩系。在侵入其中的花岗岩中至今没有发现沥青铀矿。

Taking all of these formations together, we have a known area of 1,000,000 square miles of cold and rigid rocks with well-established lands and seas in North America 1,300,000,000 or 1,500,000,000 years ago, with no suggestion of physical conditions fundamentally different from the present.

The oldest rocks in Brazil, South Africa, Australia, India, Scandinavia, and Scotland, judging from what I have seen of them, include similar sediments to those of the Grenville and Keewatin in Canada, though not on so broad a scale. In Holmes's interesting discussion of the "Age of the Earth", the Lower Pre-Cambrian of West Australia is stated to be 1,260,000,000 years old, which fits well with the age of the Grenville.

There were, then, solid land surfaces not too warm for lakes or seas to exist in all the continents in the earliest times known to the geologist; and there is, in fact, no geological evidence that the world ever was molten. If our globe passed through intensely hot gaseous, liquid, and plastic stages, the cooling had run its course completely many millions of years before the pegmatite veins penetrated the Grenville sediments; and the cold continents had undergone at least one major mountain-building revolution at an earlier time than 1,230,000,000 years ago.

Since then the earth has not been cooling down, but has kept its surface temperature surprisingly uniform, though with minor variations, including several ice ages. The carbon and limestone in the earliest rocks suggest lowly plant life in the waters from the very beginning of known geological history, and the Pre-Cambrian geologist is inclined to be a uniformitarian, and to ask the astronomer if the first quarter of the world's history was really so wild and turbulent as he describes it, when the later three-quarters were so temperate and uniform.

May not the earth have been built up of cold particles such as now reach us by the million from space, and may it not have escaped entirely the white hot stage of the nebular theory? Is it not possible that the hot gases cooled rapidly into innumerable solid particles which later came together to make the earth? The tiny scattered asteroids and meteorites suggest some such process; and this would provide the cold earth which the Pre-Cambrian geologist requires.

If the astronomers cannot provide a cold process of world building, they must allow the geologist a much longer time than 2,000,000,000 years to condense the hot cloud of gas into a solid world with continents and basins cool enough for the Grenville sea with its algae.

(**125**, 668-669; 1930)

A. P. Coleman: Royal Ontario Museum, Toronto.

References:

1. Ages of Some Canadian Pegmatites, *Contribs. to Can. Mineralogy*, Univ., Toronto, 1924.

2. Radioactive Minerals as Geological Age Indicators, *American Journal of Science*, No. 50, p. 127, etc.

如果综合考虑所有这些地层，那么在北美我们就拥有一块面积达 1,000,000 平方英里的区域，该区域由冷而坚硬的岩石构成，根据这些岩石，在 1,300,000,000 或 1,500,000,000 年前已确定存在的海陆，其物质环境与现在没有根本的不同。

根据我的判断，巴西、南非、澳大利亚、印度、斯堪的纳维亚和苏格兰最古老的岩石中含有和加拿大格伦维尔岩系和基瓦丁岩系相似的沉积物，尽管其规模稍小一些。霍姆斯在关于《地球的年龄》的有趣讨论中指出，西澳大利亚的前寒武纪在距今约 1,260,000,000 年前，这和格伦维尔岩系的年龄十分吻合。

在地质学家了解的最早地质历史时期，各个大陆的固态陆地表面不是太热，可以维持湖泊或海洋的存在。而且，事实上没有地质上的证据可以证明地球曾经处于熔融状态。如果我们的星球曾经经历过炽热气体、液体和塑性阶段，那么冷却过程的彻底完成要比伟晶岩脉侵入格伦维尔沉积岩早至少几百万年，而且冷却后的大陆至少要在 1,230,000,000 年前经历一次较大的造山运动。

从那以后，地球没有继续变冷，相反它的表面温度均衡得令人惊讶，尽管有包括几个冰川期在内的一些小的起伏。最早期的岩石中含有碳和石灰石说明生活在水中的低等植物出现于普遍认可的地质历史的开始阶段。同时，对于前寒武纪，地质学家倾向于均变论，他们会问天文学家：如果地球历史的前 1/4 真的如他们描述的那样荒凉和动荡，那为什么后 3/4 的历史如此平静和均匀？

难道地球不可以由与如今从宇宙大批飞向我们的冷物质类似的物质组成吗？难道它不可能完全避免星云历史上的白热阶段吗？难道炽热的气体不可能迅速冷却成无数构成地球的固体颗粒吗？那些散布于宇宙中的小行星和陨石表明确实存在这样的过程，这将支持地质学家认为寒武纪之前地球是冷的这个观点。

如果天文学家不能提供地球形成的冷却过程，他们必须允许地质学家认为至少需要 2,000,000,000 年时间才能使气团浓缩成固体世界，才能使大陆和盆地足够冷以便海藻可以生活在格伦维尔海中。

（金成伟 翻译；张忠杰 审稿）

Artificial Disintegration by α-Particles

J. Chadwick and G. Gamow

Editor's Note

In 1919 Ernest Rutherford had shown that alpha particles fired into nitrogen gas could create hydrogen, apparently because the particles kick a proton out of the nucleus while being themselves captured. Rutherford called the process "artificial disintegration" (popularly, splitting) of the atom. Here James Chadwick and George Gamow suggest that protons might also be produced without the alpha particle entering the nucleus. Gamow went on to propose that protons, with only half the charge of alpha particles, might approach the positively charged nucleus more easily. That idea stimulated work at Cambridge on a high-voltage device to accelerate protons to high energies, leading to the first induced splitting of a lithium nucleus by proton bombardment by John Cockroft and Ernest Walton.

IT is commonly assumed that the process of artificial disintegration of an atomic nucleus by collision of an α-particle is due to the penetration of the α-particle into the nuclear system; the α-particle is captured and a proton is emitted.

On general grounds it seems possible that another process may also occur, the ejection of a proton without the capture of the α-particle.

Consider a nucleus with a potential field of the type shown in Fig. 1, where the potential barrier for the α-particle is given by the full line and that for the proton by the dotted line. Let the stable level on which the proton exists in the nucleus be $-E_p^\circ$ and the level on which the α-particle remains after capture be $-E_\alpha^\circ$.

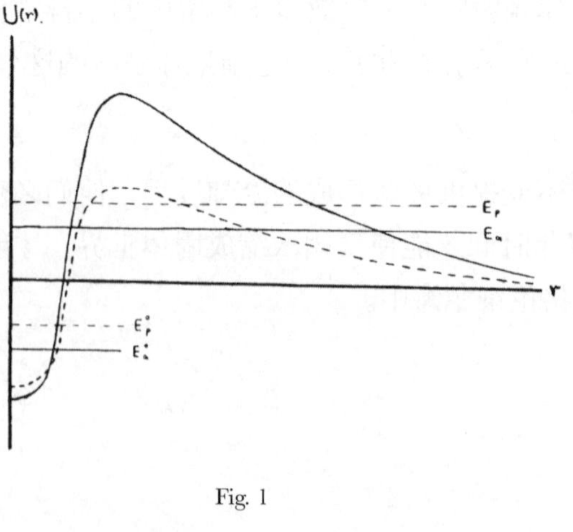

Fig. 1

α 粒子引发的人工衰变

查德威克，伽莫夫

编者按

1919 年，欧内斯特·卢瑟福发现 α 粒子射入氮气后可以产生氢，这显然是因为 α 粒子被俘获的同时把一个质子赶出了原子核。卢瑟福把这个过程称作原子的“人工衰变”（更通俗地说是裂变）。詹姆斯·查德威克和乔治·伽莫夫在本文中提出，在没有 α 粒子进入原子核的情况下我们也可以得到质子。伽莫夫还指出，质子所带电荷数只有 α 粒子的一半，它有可能更容易接近带正电的原子核。这种观点推动了剑桥大学的约翰·考克饶夫和欧内斯特·瓦耳顿的工作，他们用高压装置把质子加速到很高的能量，然后通过质子轰击首次实现了锂核的诱导裂变。

通常的假设是，一个 α 粒子与原子核发生碰撞的人工衰变过程，是由 α 粒子穿透到核系统内部导致的；在 α 粒子被俘获的同时，也会发射出一个质子。

按照通常的观点，另外一种过程也可能发生，即在没有俘获 α 粒子的情况下，释放出一个质子。

研究一个处于图 1 所示的势场条件下的核，其中实线代表 α 粒子的势垒，虚线代表质子的势垒。令原子核中质子的稳态能级为 $-E_p^{\circ}$，被俘获后的 α 粒子的能级为 $-E_{\alpha}^{\circ}$。

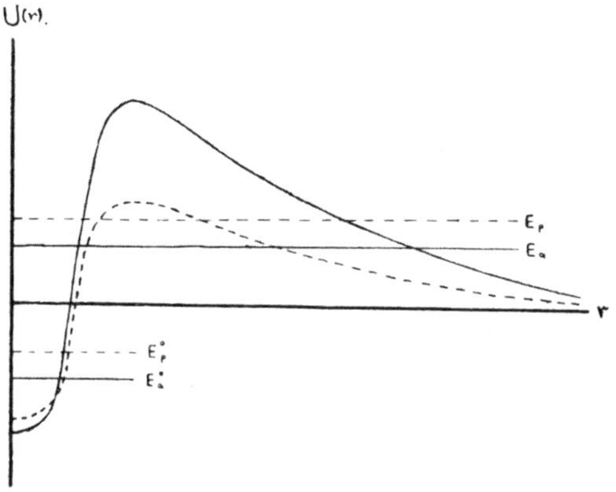

图 1

If an α-particle of kinetic energy E_α penetrates into this nucleus and is captured, the energy of the proton emitted in the disintegration will be $E_p = E_\alpha + E_\alpha^\circ - E_p^\circ$, neglecting the small kinetic energy of the recoiling nucleus. If the nucleus disintegrates without capture of the α-particle, the initial kinetic energy of the α-particle will be distributed between the emitted proton and the escaping α-particle (again neglecting the recoiling nucleus). The disintegration protons may have in this case any energy between $E_p = 0$ and $E_p = E_\alpha - E_p^\circ$.

Thus, if both these processes occur, the disintegration protons will consist of two groups: a continuous spectrum with a maximum energy less than that of the incident α-particles and a line spectrum with an energy greater or less than that of the original α-particles according as $E_\alpha^\circ > E_p^\circ$ or $E_\alpha^\circ < E_p^\circ$, but in either case considerably greater than the upper limit of the continuous spectrum (see Fig. 2).

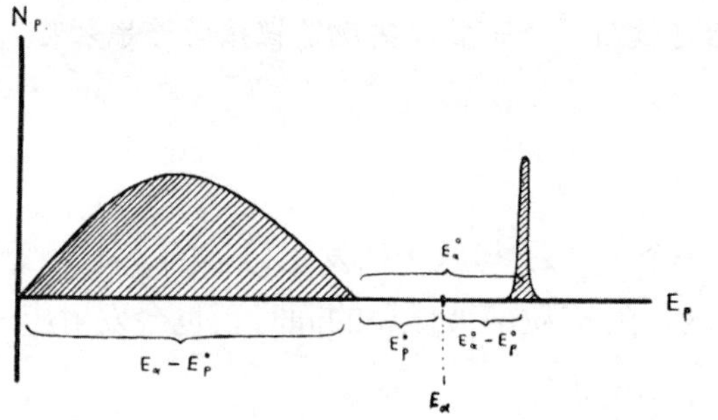

Fig. 2

In some experiments of one of us in collaboration with J. Constable and E. C. Pollard, the presence of these two groups of protons appears quite definitely in certain cases, for example, boron and aluminium. A full discussion of these and other cases of disintegration will be given elsewhere, but it may be noted that the existence of groups of protons has already been reported by Bothe and by Pose. In general the experimental results suggest that with incident α-particles of energy about 5×10^6 volts (α-particles of polonium) the process of non-capture is several times more frequent than the process of capture.

It is clear that, if our hypothesis is correct, accurate measurement of the upper limit of the continuous spectrum and of the line will allow us to estimate the values of the energy levels of the proton and α-particle in the nucleus. In the case of aluminium bombarded by the α-particles of polonium the protons in the continuous spectrum have a maximum range of 32 cm and those of the line spectrum a range of 64 cm. These measurements give the following approximate values for the energy levels:

$$E_p^\circ = 0.6 \times 10^6 \ e \text{ volts, and } E_\alpha^\circ = 2 \times 10^6 \ e \text{ volts.}$$

On the wave mechanics the probability of disintegration of both types is given by the square of the integral

$$W = \int f(r_{\alpha,p}) \cdot \psi_\alpha \cdot \psi_p \cdot \phi_\alpha \cdot \phi_p \cdot dV \cdot dV' \qquad (1)$$

如果动能为 E_α 的 α 粒子穿透到这个核的内部并被俘获，那么衰变中发射出的质子的能量是 $E_p = E_\alpha + E_\alpha^\circ - E_p^\circ$，这里忽略了此过程中反冲核的微小动能。如果在核衰变的过程中没有俘获 α 粒子，α 粒子的初始动能将在发射出的质子与逃逸的 α 粒子之间分配（再次忽略反冲核）。在这种情况下，衰变质子的能量可能是 $E_p = 0$ 与 $E_p = E_\alpha - E_p^\circ$ 之间的任意值。

因此，如果这两种过程都存在，那么衰变的质子将由两部分组成：最大能量小于入射 α 粒子能量的连续谱和能量大于或小于初始 α 粒子能量的线状谱，相当于 $E_\alpha^\circ > E_p^\circ$ 或 $E_\alpha^\circ < E_p^\circ$，但对于这两种情况中的任何一种，能量都比连续谱的上限大很多（见图 2）。

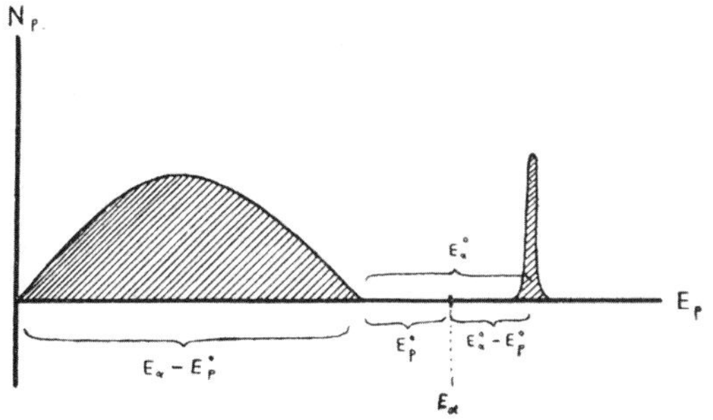

图 2

在我们中的一位与康斯特布尔和波拉德合作进行的一些实验中，在某些情况下这两组质子的存在会非常明确地表现出来，比如，硼和铝的情况。关于这些以及其他一些衰变的情况将会在别的地方给出全面的讨论，但可以注意到的是，博特和波泽已经报道了多组质子的存在。一般而言，实验结果显示，当入射 α 粒子（钋的 α 粒子）的能量约为 5×10^6 电子伏时，非俘获过程发生的频率比俘获过程高出几倍。

显然，如果我们的假设是正确的，那么对连续谱和线状谱上限的精确测量将使我们能对原子核中的质子和 α 粒子的能级数值进行估计。在用钋的 α 粒子轰击铝的情况下，在连续谱中质子的最大范围为 32 厘米，而在那些线状谱中的范围为 64 厘米。由这些测量得到的能级近似值如下：

$$E_p^\circ = 0.6 \times 10^6 \text{ 电子伏}，E_\alpha^\circ = 2 \times 10^6 \text{ 电子伏}。$$

根据波动力学，两种类型的衰变几率由积分的平方给出

$$W = \int f(r_{\alpha,p}) \cdot \psi_\alpha \cdot \psi_p \cdot \phi_\alpha \cdot \phi_p \cdot dV \cdot dV' \tag{1}$$

where $f(r_{\alpha,p})$ is the potential energy of an α-particle and a proton at the distance $r_{\alpha,p}$ apart, and the wave functions ψ_α, ψ_p represent the solutions for the α-particle and proton before and ϕ_α, ϕ_p after the disintegration. In calculating the integral (1) we must develop the incident plane wave of the α-particle into spherical harmonics corresponding to different azimuthal quantum numbers of the α-particle, and deal with each term separately.

In the case of capture of the α-particle the estimation of (1) can be carried out quite simply. It can be shown that the effect of the higher harmonics is very small, and that the disintegration is due almost entirely to the direct collisions. Thus we obtain for the probability of disintegration

$$W_1^2 = \frac{A}{v_\alpha^2} \cdot e^{-\frac{8\pi^2 e^2}{h} \cdot \frac{Z}{v_\alpha}} \cdot e^{-\frac{4\pi^2 e^2}{h} \cdot \frac{Z}{v_p}} \qquad (2)$$

where v_α and v_p are the velocities of the initial α-particle and the ejected proton respectively. Since only the first harmonic is important in disintegration of this type, it is to be expected that the protons will be distributed nearly uniformly in all directions.

When the α-particle is not captured the disintegrations will arise mainly from collisions in which the α-particle does not penetrate into the nucleus. For disintegration produced in this way the higher harmonics become of importance. The probability of disintegration can be roughly represented by the formula

$$W_2^2 = B \cdot e^{-\frac{8\pi^2 e^2}{h} \cdot Z\left(\frac{1}{v'_\alpha} - \frac{1}{v_\alpha}\right)} \cdot e^{-\frac{4\pi^2 e^2}{h} \cdot \frac{Z}{v_p}} \qquad (3)$$

where v'_α is the velocity of the α-particle after the collision, and B is a function of the angle of ejection of the proton. The protons of the continuous spectrum will not be emitted uniformly in all directions. According to the expression (3) the distribution with energy of the protons in the continuous spectrum will have a maximum value for an energy of ejection of about 0.3 of the upper limit, and will vanish for zero energy and at the upper limit.

More detailed accounts of the experimental results and of the theoretical calculations will be given shortly.

(**126**, 54-55; 1930)

J. Chadwick, G. Gamow: Cavendish Laboratory, Cambridge, June 18.

式中 $f(r_{\alpha,p})$ 是相距为 $r_{\alpha,p}$ 的一个 α 粒子和一个质子之间的势能，波函数 ψ_α 和 ψ_p 表示 α 粒子和质子在衰变前的解，ϕ_α 和 ϕ_p 表示它们在衰变后的解。在计算积分 (1) 时，我们必须将 α 粒子的入射平面波展开为对应于 α 粒子不同角量子数的球谐函数的形式，并对每一项分别处理。

在俘获 α 粒子的情况下，对 (1) 式进行估算非常简单。可以看到的是，高次谐波的效应很小，衰变几乎全部由直接碰撞产生。因此我们可以得到衰变几率

$$W_1^2 = \frac{A}{v_\alpha^{\ 2}} \cdot e^{-\frac{8\pi^2 e^2}{h} \cdot \frac{Z}{v_\alpha}} \cdot e^{-\frac{4\pi^2 e^2}{h} \cdot \frac{Z}{v_p}} \tag{2}$$

式中 v_α 和 v_p 分别是初始 α 粒子和发射出的质子的速度。因为在这类衰变中，只有一次谐波起主要作用，所以可以认为质子在所有方向上都是均匀分布的。

当 α 粒子未被俘获时，衰变将主要通过 α 粒子并不穿透到核内的碰撞产生。在由这种方式产生的衰变过程中，高次谐波开始起主要作用。衰变几率可以通过下述公式近似地表示

$$W_2^2 = B \cdot e^{-\frac{8\pi^2 e^2}{h} \cdot Z\left(\frac{1}{v'_\alpha} - \frac{1}{v_\alpha}\right)} \cdot e^{-\frac{4\pi^2 e^2}{h} \cdot \frac{Z}{v_p}} \tag{3}$$

式中 v'_α 是碰撞后 α 粒子的速度，B 是质子发射角度的函数。连续谱中的质子在各个方向上的发射并不均匀。根据表达式（3），连续谱中质子的能量分布，在约为上限的 0.3 处其发射能量达到最大值，而在上限处突然变为零。

不久之后，我们将给出关于实验结果和理论计算的更详细的解释。

（沈乃澂 翻译；江丕栋 审稿）

A New Theory of Magnetic Storms

S. Chapman and V. C. A. Ferraro

Editor's Note

In the 1920s and 1930s, radio communications became an important public issue with the creation of broadcasting systems and the use of radio waves for telephones. These systems were sometimes disrupted by an unknown process apparently linked to changes in the Earth's magnetism, called magnetic storms. Here Sydney Chapman and Vincente Ferraro at Imperial College, both later leading figures of "space physics", provide the basis of the explanation for magnetic storms. Such events, they say, are caused by interactions of the geomagnetic field with charged particles streaming from the Sun (the solar wind).

AN attempt to infer the course of events when a neutral ionised stream of particles from the sun is directed towards the earth has now led to results which we believe indicate how magnetic storms are produced. A full discussion of the phenomena involves the solution of numerous intricate mathematical problems, many of which have not yet been attacked in detail; but it seems possible to outline the main sequence of events.

The motion of a neutral ionised stream in the earth's magnetic field was investigated by one of us in 1923[1], and it was concluded that the stream would be scarcely deflected by the field, though some slight convergence would occur within about one earth-diameter from the earth's centre *O* (Fig. 1).

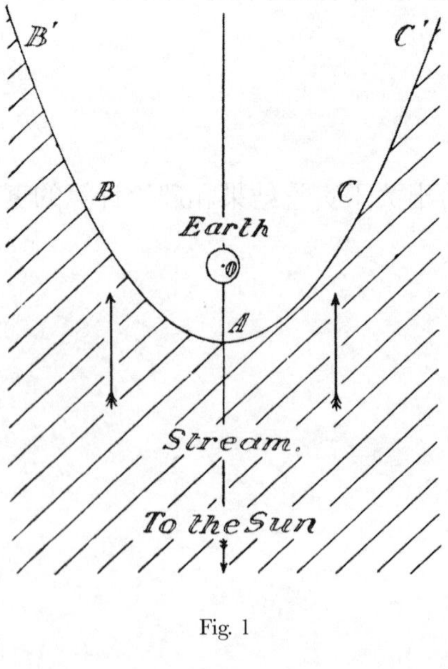

Fig. 1

一项关于磁暴的新理论

查普曼，费拉罗

编者按

20 世纪二三十年代，随着广播系统的发明和无线电波在电话上的应用，无线电通信成了一个重要的公共议题。这些无线电通信系统有时会被一种与地球磁场的变化有关的未知过程中断，这种未知过程叫做磁暴。在这篇文章中，帝国学院的辛迪·查普曼和文森特·费拉罗提出了磁暴现象的基本原理，他们两人后来都成了“空间物理学”领域的领军人物。他们认为，磁暴事件是由来自太阳的带电粒子流（太阳风）与地球磁场相互作用引起的。

我们尝试推断来自太阳的呈电中性的电离粒子流直接吹向地球时各事件的过程，如今有了结果，我们相信这一结果可以阐释磁暴是怎样产生的。完整地讨论这一过程需要求解大量复杂的数学问题，而且其中的许多问题尚未得到详细的研究。不过，概括出这些事件的大致顺序似乎还是可以做到的。

我们中的一位在 1923 年对呈电中性的电离粒子流在地球磁场中的运动作了研究 [1]，结果表明，虽然在距地心 O 约一个地球直径的范围内会发生一些轻微的会聚，但是这种流几乎不会因地磁作用而偏转（如图 1）。

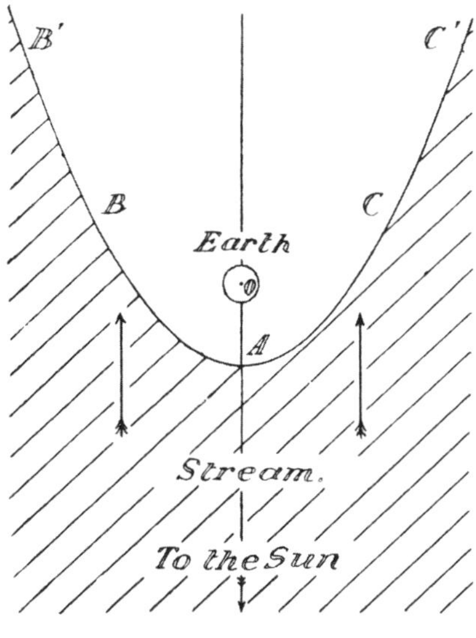

图 1

No indication as to how such a stream could produce magnetic storms and aurorae was obtained. It would seem that this failure was due to the assumption there made that the stream had enveloped the earth for a time long enough to enable a steady state to be set up, whereas it now appears that magnetic storms are essentially connected with the *approach* of the stream towards the earth. The important changes in the stream occur within a few earth radii from *O*, and beyond this distance the former conclusion that the stream travels almost without deflection remains valid.

The stream is in effect a highly conducting body, and as it enters the earth's field electric currents flowing parallel to its surface are induced in the surface layers, so that the interior of the stream is nearly shielded from the earth's field. Outside the stream the magnetic effect of the currents is roughly equivalent to that of an "image" magnetic doublet at a certain point inside the stream; in the equatorial plane the field between the earth and the stream is increased in intensity. It is as if the current layer, as it advances towards the earth, pushes forward and crowds together the earth's lines of force.

We identify this change with the observed increase in the earth's horizontal force during the first stage of a magnetic storm. Detailed examination shows that the magnitude of the effect, and its time scale, depend almost entirely on the kinetic energy of the stream per unit volume; if the velocity of the stream is of the order 1,000 km/s, the density requisite to explain the first phase of an average magnetic storm (taking account of the shielding effect of the Heaviside layer) is roughly of the order 10^{-22} gm/cc; this might be provided by about 1.5 calcium ions or 60 hydrogen ions per cc.

The magnetic energy of the field is increased during this phase at the expense of the kinetic energy of the stream; the retardation of the particles occurs in the current layer, which is continually increased in mass-density by the oncoming of particles from behind. The retardation is greatest at that part of the front of the stream (*A* in Fig. 1) which is moving along the direct line from the sun to the earth's centre *O*; on either side of *A* the stream will advance relatively to *A*, and the earth will become partly enclosed by the stream; the surface *B'BACC'* will continually close in, at a diminishing rate, upon the earth; whether it actually reaches the earth's atmosphere, in the equatorial plane, will depend on the density of the stream, and the length of time during which it is directed towards the earth (this is determined by the angular breadth of the stream viewed from the sun).

In the second (which is the main) phase of a magnetic storm the earth's horizontal force is decreased. We attribute this to the formation of a westerly current round the earth, due primarily to the flow of charges across the space "behind" the earth (viewed from the sun). Along the sides *BB'*, *CC'* of the enclosure there will be charged layers, *BB'* positive and *CC'* negative, due to the polarisation of the stream by the magnetic field. The charges in these layers will be subject to an outward electrostatic force, and the positive ions along *BB'* will cross over to *CC'*, partly guided by the earth's field. The electrons along *CC'* cannot flow along the reverse path because of the greater deflecting influence of the field upon them,

现在尚无法说明这样的流是如何形成磁暴和极光的。之所以无法说明，似乎是因为一项假设，即，认为粒子流包围地球的时间足够长，以至于建立起了一个平衡态。然而，在现在看来，磁暴在本质上与粒子流向地球**传输的过程**有关。在距离地心 O 几个地球半径的范围内，粒子流发生了重大变化。而在这个范围以外，前面的结论仍旧有效，即粒子流在运动过程中几乎没有发生偏转。

实际上，这种粒子流是一个良导体，当它进入地球磁场时，在其表面层会感应出平行于表面流动的电流，以至于粒子流内部几乎被地球磁场屏蔽。在粒子流外部，电流的磁效应则与粒子流内部某点的“镜像”磁偶极子的磁场大致相等；所以在赤道面上，地球与粒子流之间的磁场强度变大。当电流层向地球方向前进时，看起来就像是电流层推动地球磁力线向前并使之挤压在一起。

我们发现这一变化与磁暴初相时观测到的地球磁场水平分量的增加有关。进一步的研究表明，这种作用的数量级和持续时间几乎完全取决于单位体积粒子流的动能。假设粒子流速度的数量级是 1,000 km/s，那么一次中等强度磁暴的初相要求的粒子流密度应该大致在 10^{-22} gm/cc 这个数量级（考虑到亥维赛层的屏蔽效应），这大概相当于每立方厘米中有 1.5 个钙离子或 60 个氢离子。

在这一阶段，随着粒子流动能的消耗，磁场的磁能增大，电流层中的粒子出现阻滞现象，而其后相继而来的粒子使其质量密度不断增大。在沿太阳到地心 O 之间的直线上运动的粒子流的前端（图 1 中 A 点），阻滞现象最明显；A 两侧的粒子向 A 靠拢，进而把地球部分地包围起来。$B'BACC'$ 面将以持续减小的速度不断靠近地球；该面能否到达赤道面上空的大气层，取决于粒子流的密度和朝地球运动的时间(时间长短取决于粒子流以太阳为观测点的角宽度)。

在磁暴的第二阶段（即磁暴主相），地球磁场的水平分量开始减小。我们将这一现象归因于围绕地球的西向电流的形成，这种电流主要是由从地球“后面”（以太阳为观测点）穿过太空的电荷的流动引起的。由于磁场对带电粒子流的极化作用，沿封闭圈的 BB'、CC' 边将会形成电荷层，BB'边为正，CC'边为负。这些层上的电荷会受到一个向外的静电力。一定程度上由于地球磁场的作用，BB'边的正电离子将横渡到 CC'边。CC'边的电子不能向相反方向运动，因为地球磁场作用于其上的偏转力更

but negative charges from "above" and "below" the equatorial plane will travel along the earth's lines of force to neutralise the charge of the ions moving from *BB′* across the gap. The details of the process are not yet clear, but it appears likely that a westerly current can thus be set up round the earth. It can be shown that the current-ring, if formed, can persist in mechanical and electromagnetic equilibrium for some days after the cessation of the onward flow of particles from behind. The gradual dissipation of this ring current corresponds to the final phase of the storm.

One of the distinctive features of the theory here outlined is the distance from the earth within which the main electric currents flow, namely, a few earth radii; they are outside the earth's atmosphere (though secondary currents are induced therein), but they are much nearer the earth than the currents (in the equatorial plane) discussed by Birkeland, or the equatorial current proposed by Prof. Størmer and associated by him with the decrease of latitude of aurorae during magnetic storms.

We have not examined closely the extent to which the stream will cause inflow of ions and electrons into the earth's atmosphere in the polar regions, or how this inflow will give rise to the observed currents along the auroral zones; but it seems likely that present theories of the aurorae will need to be modified, because the particles of a neutral stream can approach much closer to the earth, in the equatorial plane, than the single charged particles hitherto considered. This must also have an important bearing on the theory of radio echoes, should it be proved that these are produced outside the earth's atmosphere.

(**126**, 129-130; 1930)

S. Chapman and V. C. A. Ferrapo: Imperial College of Science, South Kensington, London, S.W. 7, June 26.

Reference:

1. *Proc. Camb. Phil. Soc.*, 21, 577; 1923.

大，但来自赤道面“上侧”和“下侧”的负电荷会沿地球磁力线运动，以中和从 BB' 穿过间隙迁移过来的正电离子的电荷。尽管对这一过程的具体细节还不清楚，但可以看出这样就能形成围着地球的西向电流。我们发现，该环电流一旦形成，在没有后继粒子补充时仍能保持力学和电磁平衡状态并持续好几天。该环电流的逐渐消失对应于磁暴的末相。

本文所述理论的一个与众不同的特征是主电流的作用区域与地球之间的距离，即几个地球半径。这些区域在地球的大气层之外（虽然次级电流是在大气层内感应产生的），但它们离地球的距离远比伯克兰所说的电流（位于赤道面上）以及斯托末教授提出的赤道电流（斯托末教授还指出赤道电流与磁暴期间极光出现的纬度降低有关）近得多。

我们还没有仔细研究粒子流造成正电离子和电子进入地球两极地区大气层的广度，以及带电粒子进入大气层的过程如何形成了我们沿极光带观测到的电流。但看来现今关于极光的理论很可能需要修正，因为，与目前设想的单电荷粒子相比，在赤道面上，呈电中性的带电粒子流中的粒子可以到达距地球更近的位置。如果能够证明这些过程是在地球大气层之外发生的，那么本研究在无线电回波理论方面也一定会产生重大的影响。

（齐红艳 翻译；张忠杰 审稿）

Deep Sea Investigations by Submarine Observation Chamber

Editor's Note

C. William Beebe was a naturalist with little formal training who financed some of his later ornithological expeditions with best-selling accounts of earlier ones. He also pioneered deep-sea exploration, collaborating with the independently wealthy Otis Barton, who made the first true bathysphere in 1928, a steel sphere with thick quartz windows. Together they made record-breaking descents in the Atlantic. This report describes one of the first: a descent to 1,426 ft off Bermuda. Reaching such depths was unprecedented at the time, but Beebe had doubled that figure by 1934.

ON June 11, 1930, in lat. 32° 16′ N., long. 64° 39′ W., in the Atlantic Ocean off Bermuda, Dr. William Beebe, accompanied by Mr. Otis Barton, descended to a depth of 1,426 feet below the surface of the sea.[1]

This announcement marks a new era in the exploration of the sea. All previous diving records shrink into insignificance compared with this depth; it was with no wish for record-making achievements that the descent was undertaken, but a real explorer's desire to see the animals beneath the waters as they live and not at second-hand from the collections of deep sea nets.

The construction of the chamber was financed by Mr. Barton, and he and Dr. Beebe, working from the New York Zoological Society's Oceanographic Expedition's headquarters at Nonsuch Island, have now made several descents, of which three were to a depth of 800 feet and one to 1,426 feet. The chamber is a steel sphere 57.3 in. in outside diameter and 1.5 in. thick. Observations could be made through a 6 in. diameter port fitted with a quartz window. Outside the window was hung a bag of decayed fish and some baited hooks, and a strong electric searchlight could be used to illumine the surrounding water. Telephonic communication was maintained with the ship above and a supply of oxygen carried.

One of the most striking phenomena was the "blue brilliance of the watery light to the naked eye, long after every particle of colour had been drained from the spectrum." The visual degeneration of the spectrum was observed, in connexion with an intensity metre. In Dr. Beebe's own words[2]: "The red had gone completely a few feet down ...; orange had been absorbed at sixty feet below the surface and yellow at less than 400. At our depth (800 feet) lavender, too, was non-existent, together with the two opposite ends of the

基于水下观测室的深海调查

编者按

威廉·毕比是一位没有接受过太多正规训练的博物学家。他早期研究的报告非常畅销，这为他后来进行鸟类学考察积累了资金。他还与独立而又富有的奥蒂斯·巴顿一起在深海探测方面做了许多开创性的工作。巴顿于 1928 年制作出了第一个真正的深海球形潜水器，这是一种带有厚石英窗的钢球。他们一起打破了多项大西洋中潜海深度的记录。这篇文章报道的就是其中比较早的一项：在百慕大下潜达到 1,426 英尺。在当时能达到如此深度是史无前例的，而到 1934 年时毕比的下潜深度又翻了一倍。

1930 年 6 月 11 日，在北纬 32° 16′，西经 64° 39′，距离百慕大不远的大西洋中，威廉 · 毕比博士在奥蒂斯·巴顿先生的陪同下潜到了距海面 1,426 英尺的地方。[1]

这一通告标志着海洋探测的一个新时代的到来。与这一深度相比，之前所有的潜水记录都黯然失色了。进行这次下潜活动并不是为了创造纪录，而是基于一个真正的探险者的渴望，为了亲眼看见处于原生状态的水下动物，而不是靠从深海拖网捕捞获得的二手资料来观察这些动物。

观测室是由巴顿先生出资建造的，他和毕比博士均就职于位于楠萨奇岛的纽约动物学会的海洋探险队总部。目前他们已完成了多次下潜任务，其中 3 次潜到了 800 英尺，1 次潜到 1,426 英尺。该观测室是一个钢球，外径 57.3 英寸，厚 1.5 英寸。观察者可以通过一个直径为 6 英寸的石英窗进行观察。在窗外悬挂着一袋烂鱼和一些装有鱼饵的鱼钩，以及一个大功率的探照灯，可用于照亮周围的海水。水面上方的船只与观测室之间保持电话通信，并向观测室提供氧气。

最令人吃惊的现象之一是，“在所有的颜色都从光谱中消失很久以后，裸眼看到的是海水发出的蓝色光芒。”他们利用一个强度计来观测光谱中各种光的衰减。用毕比博士自己的话说就是[2]，“几英尺以下红光就完全消失了……；橙光在水面下 60 英尺处被吸收，而黄光在不到 400 英尺处被吸收。在我们所在的深度（800 英尺），淡紫色的光以及光谱两端的红外线和紫外线也都不复存在了，绿光依然存在，但

spectrum, infra-red and ultraviolet, while green still persisted, but greatly diluted. All that remained to our straining eyes were violet and blue, but blue such as no living man had ever seen."

It proved quite possible to observe pelagic animals drifting and swimming past the window, such as medusae, shrimps, and fish, and about a dozen true bathypelagic fish were identified. A very interesting result of these observations was the presence of certain species of fish and invertebrates in water layers well above the depth at which their occurrence is first indicated by net catches in the daytime.

Four descents have also been made in water up to 350 feet in depth along the shelving bottom of the Bermudian insular shelf as the vessel drifted seawards. Such exploration revealed a new fish fauna at these offshore depths, the recognisable shore fish also being of great size.

The observations will be continued another year, and it is to be hoped that this new weapon of marine research has come to stay and that similar submarine observation chambers may be built in time for a study of the floor of shallower seas and the habits of food fishes. Already shallow water diving has proved its scientific value. We shall await Dr. Beebe's and Mr. Barton's full reports with great interest.

(**126**, 220; 1930)

References:

1. *Science*, vol. 72, No. 1854, July 11, 1930, pp. 27-28. "A New Method of Deep Sea Observation First-hand". By Henry Fairfield Osborn.

2. *New York Times*, June 27, 1930.

也变淡了很多。我们睁大双眼只能看到紫光和蓝光，不过这种蓝是人类所不曾见过的。”

实践证明，极有可能会看到从窗前漂过或游过的浮游动物，如水母、虾类和鱼类，并且辨认出了约12种真正的深海鱼类。通过这些观察获得的一个非常有趣的结果是：在上层水体中发现的一些鱼类和无脊椎动物在这里也出现了，要知道它们在上层水体中的存在最早是通过白天用渔网捕捞而发现的。

当这艘船随海漂移时，沿着百慕大岛架的缓倾海底，在水中深达350英尺的地方又下潜过4次。这类探险活动在这些近海深度发现了一个新的鱼类区系，可识别的近海鱼类也达到了很大的数量。

明年水下观测还将继续进行，我们希望这一海洋研究的新装备能够得到普遍应用，并且能够及时建造出类似的水下观测室，用于研究浅海海底以及食用鱼类的生活习性。浅海潜水在科学上的价值已经得到了证明。让我们怀着极大的兴趣期待毕比博士和巴顿先生的详细报告吧。

（齐红艳 翻译；张泽渤 审稿）

Stellar Structure and the Origin of Stellar Energy

E. A. Milne

Editor's Note

Arthur Eddington had recently proposed a model for the structure of stellar interiors, which implied that the mass of a star determines its luminosity in a unique way. Here the English astronomer Edward Milne disputes Eddington's claim. Milne's alternative model predicts that stars should have an extremely dense core—the most likely setting, it was felt, for the processes giving stars their energy—surrounded by a gaseous body of lower density. The core temperatures in his theory could be as high as 10^{11} degrees, some 10,000 times higher than Eddington's theory suggests. Milne's theory made little use of the principles of quantum theory or emerging nuclear physics, but illustrated a growing interest in understanding stellar structure from first principles.

THE generally accepted theory of the internal conditions in stars, due to Sir A. S. Eddington, depends largely on a special solution of the fundamental equations, and according to this a definite calculable luminosity is associated with a given mass. If this were the only solution of the equations it would conflict, as I have repeatedly shown in recent papers, with the obvious physical considerations which show that we can build up a given mass in equilibrium so as to have an *arbitrary* luminosity (not too large) whatever the assumed physical properties of the material. I have recently noticed that the fundamental equations possess a whole family of solutions, corresponding to arbitrarily assigned luminosity for given mass. These solutions show immediately that Eddington's solution is a special solution and corresponds to an unstable distribution of mass. In the stable distributions the density and temperature tend to very high values as the centre is approached, theoretically becoming infinite if the classical gas laws held to unlimited compressibility.

The physical properties of the stable configurations can be described as follows. Suppose a star is built up according to Eddington's solution with his value of the rate of internal generation of energy. Let the rate of internal generation of energy diminish ever so slightly. Then the density distribution suffers a remarkable change. The mass suffers an intense concentration towards its centre, the external radius not necessarily being changed. The star tends to precipitate itself at its centre, to crystallise out so to speak, forming a core or nucleus of very dense material. The star tends to generate a kind of "white-dwarf" at its centre, surrounded of course by a gaseous distribution of more familiar type; the star is like a yolk in an egg. In this configuration the density and temperature are prevented from assuming infinite values by the failure of the classical gas laws, but they reach values incomparably higher than current estimates. For example, it seems probable (though the following estimates are subject to revision) that the central temperature exceeds 10^{11} degrees, in comparison with the current estimates of the order of 10^7 degrees; and the density may run up to the maximum density of which ionised matter is capable.

恒星的结构和恒星能量的起源

米尔恩

编者按

阿瑟·爱丁顿最近提出了一种恒星内部结构的模型，该模型认为恒星的质量唯一地决定了恒星的光度。在这篇文章中，英国天文学家爱德华·米尔恩对爱丁顿的观点提出了异议。米尔恩提出的另一种模型预言，恒星应该具有一个非常致密的核心，在这个核心中极有可能进行着为恒星提供能量的反应，而核心的周围被低密度的气体包围。按照米尔恩的理论，恒星核心的温度可能会高达 10^{11} 度，这比根据爱丁顿的理论估计出来的值高出 10,000 倍。米尔恩的理论几乎没有用到任何量子理论和新兴的核物理学的原理，不过这表明了人们从基本原则出发解释恒星结构的兴趣正在增长。

目前，由爱丁顿爵士提出的恒星内部结构理论被大家普遍接受，该理论主要基于基本方程的特殊解，由此计算出的恒星的光度与恒星质量密切相关。然而，正如我在最近几篇文章中反复强调的那样，如果这是方程的唯一解，那么就会与一些显而易见的物理学原理相矛盾，比如，一颗质量一定的处于平衡态的恒星，可以具有**任意的**光度（只要不是太大），而且与物质的物理特性无关。我最近发现，基本方程可以得出一组解，对于同一质量，会解出任意的光度。这组解显而易见地说明了，爱丁顿的解只是一个特解，并且对应着质量的不稳定分布。在稳定分布的情况下，中心区域附近的密度和温度会变得非常高；若经典的气体定律在无限压缩的条件下仍然适用，则密度和温度在理论上可以达到无穷大。

这种稳定结构的物理特性如下所述：假如一颗恒星是按照爱丁顿的解和他给出的内部产能率而构造的，则只要该产能率略微减小，密度分布就会发生明显的变化，质量向中心紧密聚集，而外部半径并不一定会相应地发生变化。随着这颗恒星中的物质逐渐向中心沉积，就会形成一个致密的核，甚至会达到结晶。恒星在其中心形成了一种“白矮星”，其周围分布着常见的气态物质，这时整颗恒星就像一个鸡蛋中的蛋黄。在这种结构中，由于经典气体定律的失效，密度与温度不再是无穷大，但仍然远高于现行的估计值。比如，中心温度很可能（虽然下面的估计值曾被修正）高于 10^{11} 度，而目前的估计值为 10^{7} 度；而中心密度可能达到了电离物质所能具有的极限密度。

The unstable density distribution of Eddington's model (curve *A*) and the stable density distribution of actual stars (curves *B*) are indicated roughly in Fig. 1, which is not drawn to scale. It may be mentioned that the instability is of a radically different kind from that discussed by Sir James Jeans. He concluded that perfect-gas stars of Eddington's model were *vibrationally* unstable. In my investigations, the instability of Eddington's model arises from any slight departure of the rate of generation of energy below the critical value found by Eddington. The perfect-gas distribution of my solutions is perfectly stable, but the density necessarily increases until degeneracy or imperfect compressibility takes control.

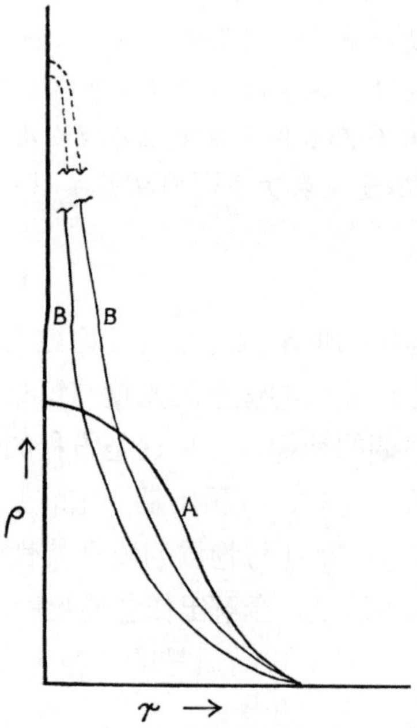

Fig. 1

The consequences amount to a complete revolution in our picture of the internal constitution of the stars. In the intensely hot, intensely dense nucleus, the temperatures and densities are high enough for the transformation of matter into radiation to take place with ease. It is to this nucleus that we must look for the origin of stellar energy, a nucleus the existence of which has previously been unsuspected. The difficulties previously felt as to stellar conditions being sufficiently drastic to permit the evolution of energy largely disappear. Many of the cherished results of current investigations of the interiors of stars must be abandoned; current estimates of central temperature, central density, the current theory of pulsating stars, the current view that high mass necessarily implies high radiation pressure, the supposed method of deducing opacity of stellar material from observed masses and luminosities, the supposed proof of the observed mass-luminosity correlation—all these require serious modification.

The new results are not a speculation. They are derived by taking the observed mass and luminosity of a star, and finding the restrictions these impose on the possible density

图 1 粗略地显示了爱丁顿模型中的非稳定密度分布（曲线 A），以及实际恒星的稳态密度分布（曲线 B），该示意图未标刻度。需要说明的是，这里的非稳定性完全不同于詹姆斯·金斯爵士的论述。詹姆斯·金斯认为，爱丁顿模型中的理想气体恒星处于非稳定的**振动**状态。而我的研究表明，爱丁顿模型的非稳定性源于产能率略低于爱丁顿给出的临界值。在我的解中，理想气体分布是完全稳定的，但是密度会不断增加，直到简并或非理想压缩起主导作用。

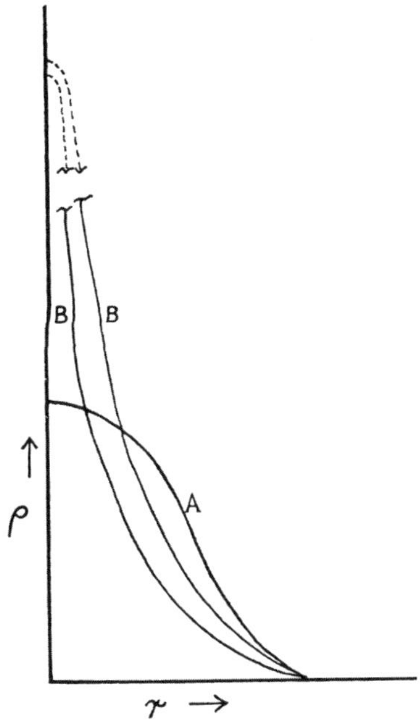

图 1

这些结果对于我们理解恒星内部结构具有革命性的意义。在极端炽热和致密的恒星核中，足够高的温度和压力可以轻易地使物质转化为辐射能。面对这样的核心，我们必须寻找恒星能量的来源，而核的存在以前并不为人所知。以前人们在考虑恒星内部必须存在极端条件才能完成能量演化时遇到的困难现在已经基本解决。然而，我们不得不放弃目前流行的许多关于恒星内部的珍贵研究结果：目前对于恒星中心温度、中心密度的估计，目前关于脉动星的理论，目前关于高质量必然导致高辐射压的观点，通过观测恒星的质量和光度得出恒星物质不透明度的方法，以及对观测到的质–光关系的证明——所有这些成果，都需要进行认真的修正。

这些新得到的结果不是猜测，而是通过测量恒星的质量和光度并给定一些限制条件后推导出来的，这些限制条件要求密度分布必须同恒星的质量和光度相容。通

distributions compatible with this mass and luminosity. By integrating the fundamental equations from the boundary inwards, we are inevitably led to high central temperatures and densities. So long as the classical gas laws persist, the solution is one of the family with a central singularity (infinities in ρ and T), and it is only the ultimate failure of the gas laws which rounds off the distribution with a finite though very large central ρ and T.

(**126**, 238; 1930)

E. A. Milne: Wadham College, Oxford, July 29.

过对基本方程从边界向内积分，我们必然会得到较高的中心温度和密度。只要经典气体定律仍然适用，就可以得到其中一个具有中心奇点的解（相应的密度 ρ 和温度 T 为无穷大）；而只有当气体定律最终不再适用时，才能够形成中心密度 ρ 和温度 T 虽然很高但不是无穷大的分布。

（金世超 翻译；何香涛 审稿）

Eugenic Sterilisation

Editor's Note

In the 1930s the potential benefits of eugenic sterilisation, first advocated by Darwin's cousin Francis Galton, were being considered in many European nations and in the United States. This editorial reacts to a proposal by the Eugenics Society of London for a legal change allowing sterilisation of the mentally impaired on their own consent, or that of a parent or legal guardian. The society had argued this would achieve a 17% reduction in mentally deficient individuals in one generation. The editorial counters that this figure is unreasonably optimistic, as it assumed permissions could be obtained for some 300,000 certifiable mental defectives in England and Wales. Yet like many scientists of the times, this editorial supports the principle of sterilisation.

SOME of the young people of Germany would have us believe that much of the time that can be spared from their more materially fruitful exploits is given over to singing a song which they call "Deutsche Jugend, heraus". Its language, borrowed from historical romanticism, permits, if it does not foster, a certain diversity of interpretation, and some lines with a frankly Christian significance may even be omitted at the discretion of the singer. Claim to popularity is thus made more catholic.

Wollt Ihr ein neues bauen
mit Händen stark und rein,
in gläubigem Vertrauen
lasst dies die Losung sein:
Den Feind in eigner Mitte
gefällt in ernstem Strauss....

Moralists, it is easy to see, may use these lines to assist them in focusing attention upon that enemy in their midst distinguished as the *beam*, while the nationalist may recognise more immediately its particular referability to the communistic *mote*.

We are assured, however, that the resiliency of this *credo* unites rather than divides, and such demonstrations as we have enjoyed tend to reinforce the assurance audibly. But we cannot help wondering what will happen in the world when the youth of one country or another not only present accessible enemies in their patriotic songs but also define them with scientific precision.

The real enemies of mankind are made, yearly, more and more accessible to attack by science, and if it were not for the protective screens, intangible and often fantastic, thrown up by the unscientific for whom nakedness, even the enemy's, still seems to possess terrible powers, mankind might subjugate very speedily its worst foes. But if, as Sir Walter Fletcher

优生绝育

编者按

20 世纪 30 年代，美国和很多欧洲国家都开始认识到，由达尔文表弟弗朗西斯·高尔顿最先提出的优生绝育将对社会产生积极的影响。这篇评论回应了伦敦优生学会的一项倡议，该倡议要求修改法律，以便在取得本人或父母中的一方或法定监护人同意的条件下，允许对心智缺陷者实施绝育。优生学会认为，这样做将使下一代中心智缺陷者的数量减少 17%。这篇评论认为这个数字过于乐观，因为它假设了英格兰和威尔士的 300,000 名确诊的心智缺陷者都会同意进行绝育。不过和当时的许多科学家一样，这篇评论的作者也是支持绝育原则的。

富有成效的物质文明建设使德国的年轻人有了更多的闲暇时间，但他们中的一些人把大部分多出来的时间都花费在吟唱一首被他们称为《德国青年》的歌曲上了。其带有历史浪漫主义色彩的歌词给整首歌赋予了多种解释，甚至是一些本来并不包含的意思。歌词中有些颇具虔诚基督教意义的部分可能会被歌手酌情删掉，从而可以更广泛地向大众普及。

你愿意建立一个新世界吗？
用强壮的双手建立一个纯净的世界，
坚定这一信念
并使它成为口号：
让处于中间阶层的敌人
就像虔诚的施特劳斯信徒一样……

很容易就能想到，道德家们可能会借用这些歌词来帮助他们将注意力集中在那些混杂于人民中间并以**国家栋梁**而著称的敌人，而民族主义者们通过这些歌词可能会更直接地意识到，相对于共产主义的**瑕疵**而言他们的主张具有的特殊借鉴意义。

尽管如此，我们还是确信这一**信条**的弹性有利于团结而非分裂，而且这种我们一直很喜欢的表述方式显然很有利于增强其可信性。但我们不禁要问，当一个国家或另一个国家的年轻人不仅在自己的爱国歌曲里提到触手可及的敌人，而且用科学精确的语言定义这些敌人的时候，这个世界将会发生什么呢？

一年一年过去，人类真正的敌人变得越来越容易被科学击倒，如果不是因为无知者甚至是反对者们非科学地抛出的一些无形的、通常是幻想出来的保护屏障看上去似乎拥有可怕力量的话，人类可能很快就能征服他们最大的敌人了。但是正如沃尔特·弗莱彻爵士最近指出的，如果对于一种纯粹的疾病（例如癌症），只有在破

has lately pointed out, a mere ailment, like cancer, has only been made accessible to scientific study through the lifting of foolish and superstitious taboos, how can we expect the direr social maladies to be approached courageously? A protective hedge of errors and superstitions hems them in on every side, so rank and poisonous that it seems that even science is infected and intimidated while it attacks.

How else is it possible to explain the demand just put forward* by a committee of the Eugenics Society for permissive legislation which would take a whole generation to achieve a reduction in the incidence of mental defect not of a hundred, not of fifty, not of twenty-five, but, problematically, of seventeen percent? Between our people and the realisation of this slender benefit stands "an ambiguity of the law" which the Society proposes to remove. A person may, with consent, be sterilised in the interests of his *own* health. In the interest of the public health, present or future, he may not be sterilised. By a curious legal inversion, the "willing mind" of the individual cannot take away the offence against the public even should he be prepared to save it from all possibility of contamination by his own progeny. The offence consists in a "maim" which deprives the individual, or so it may be contended, of martial courage, and the State of a vessel, however unsuitable otherwise, for this same virtue. To contentions of this sort, surely the monosyllabic genius of Mr. H. G. Wells's latest novel has supplied the only effective answer.

To meet the practical situation, the committee proposes a Bill legalising eugenic sterilisation. This would authorise the mental deficiency authority or superintendent of an institution to sterilise a mental defective, subject to the consent of the parent or guardian and of the Board of Control, and of the spouse if the defective is married. In the case of defectives deemed capable of giving consent, sterilisation would not be performed otherwise than with this consent. It would authorise the voluntary sterilisation of a person about to be discharged from a mental hospital for the insane as recovered, again with the added consent of Board and of parent, guardian, or spouse; and it would legalise voluntary sterilisation for the sole purpose of preventing the transmission of hereditary defect seriously impairing physical or mental health or efficiency.

Five members of the committee and another contributed to the *Lancet* for July 19 a letter defending this policy. The defence combats the assertion that if every certifiable mental defective had been sterilised twenty or thirty years ago it would have made little appreciable difference to the number of mental defectives existing today. It repeats a sentence of the committee's report urging that "if all the defectives in the community could be prevented from having children the effect would be even on the most unfavourable genetic assumptions with regard to defectiveness, to reduce the incidence of mental defect by as much as 17 percent in one generation".

---

* Committee for Legalising Eugenic Sterilisation. Eugenics Society, London, 1930.

除愚蠢迷信的忌讳后才能对其进行科学研究的话，那么我们又怎么能期望研究人员会大胆地去研究更加可怕的社会弊病呢？错误和迷信的保护罩将各种问题和弊病团团包围，当这些讨厌而又恶毒的保护罩发起进攻时，看起来似乎连科学都受到了影响和威胁。

优生学会的一个委员会最近提出,希望一项提案能获得立法通过*。他们认为提案的措施将会使下一代中心智缺陷的发生率下降，这种降低不是减少 100 个、50 个或者 25 个缺陷个体，而是使缺陷的发生率整体上降低 17%。对他们的此项要求，还有什么别的可能的解释吗？在这一微薄利益的实现和我们的民众意识之间，还存在着“法律上的模糊地带”，这正是该学会主张消除的。为了**自身**健康，一个人可能会同意做绝育手术。但如果是为了公众健康，那么无论是现在还是将来，他可能都不会去做绝育手术的。通过一种奇妙的立法转换,个人的“意愿”就不能再对公众利益有所冒犯，甚至他必须做好准备以免因为自己后代可能的缺陷而使公众利益受到损害。这里所说的冒犯包含在一种“伤害”中，这种伤害可能存在争议，它剥夺了个人的战斗勇气，剥夺了国家的命脉，然而同样的特点并不适合于其他情况。对于这种争论，单音节天才威尔斯先生的最后一部小说肯定已经提供了唯一有效的答案。

为了符合实际情况，委员会提出了一份使优生绝育合法化的议案。这将在心智缺陷者的父母或监护人和管理委员会及缺陷者的配偶（如果该缺陷者已婚）同意的情况下，赋予心智缺陷相关的权威机构或者机构管理人对心智缺陷者进行绝育的权利。如果缺陷者被认为具有作出决定的能力，则只有在本人同意的情况下才能对其进行绝育。自愿同意进行绝育的患者在康复后可以允许其离开精神病院，当然这需要来自委员会以及父母、监护人或配偶的同意；这将使把阻止遗传缺陷发生传递以防止其对生理或精神健康或功效造成严重损害作为唯一目的的自愿绝育合法化。

该委员会的 5 位委员和另外一人于 7 月 19 日向《柳叶刀》投稿捍卫这项政策。他们在辩护中反对了如下的断言：即使在二三十年前就对每一个确认具有心智缺陷的人进行了绝育，那也几乎不会对现在的心智缺陷者的数量带来多么显著的影响。文中重复了一个委员会报告中的句子，极力主张“如果能够使社会上所有有缺陷的个体不生育孩子的话，那么，即使是在最坏的遗传假设下，其影响也将使下一代中心智缺陷的发生率降低 17% 左右”。

* 优生绝育合法化委员会。优生学会，伦敦，1930年。

Obviously a 17 percent reduction in the incidence of mental deficiency is more desirable than a 17 percent increase. But do the committee's proposals ensure this reduction? Clearly, no. The words quoted promise at least that reduction if the fertility of *all* living mental defectives is prevented. The committee's proposals, with their emphasis upon the voluntary principle, by no means ensure that the 300,000 certifiable mental defectives in England and Wales would be sterilised. Who must consent? (1) The patient, if he is capable. (2) The parent or guardian. (3) The spouse if the patient is married. (4) The Board of Control. The calculation, it is true, is based on two assumptions "highly unfavourable" to the effectiveness of the proposals—that the genetic factor responsible for defectiveness (primary amentia) would be much "carried" and would only rarely produce manifest defectives, and that defectiveness is uniformly distributed throughout the community. (The fertility of defectives also is assumed to be that of the average of the population.)

How unfavourable, on the other hand, are the chances of permission? Nothing is gained by attempts to write off opposing assumptions. A figure is a figure, right or wrong.

Again, is a 17 percent reduction all that eugenic science can promise? Disregarding altogether those so-intelligent defectives who will strive to serve the country by seeking this minor mutilation, is it the institutional class that constitutes the chief danger to society? Prof. MacBride (*Nature*, Jan. 11, 1930, p. 40) says emphatically that this but touches the fringe of the problem. "The defectives most dangerous to society are those who are never confined in institutions at all! The high-grade defectives are just able to support themselves in the lowest paid and most unskilled occupations, and no civilised government would take the responsibility of confining them, and so they go on propagating large families as stupid as themselves." His idea of penal sterilisation, a punishment "for the economic sin of producing more children than the parents can support", is one which becomes more and more difficult to apply as more and more ways are devised by the State for screening the individual from biological estimation.

Is there not a real danger that the advocates of such legislation as here may mistake the assent of the political machine for victory? If assent were gained, would it not be much more accurately determined as the hall-mark of failure? It is not the assent of the State, but the initiative and creative power of the State, that is needed to secure essential progress, and that will not exist until our legislators of all parties or of any party derive their inspiration from the cultivation of natural knowledge.

(**126**, 301-302; 1930)

很显然，心智缺陷的发生率下降 17% 比增加 17% 更有利。但是该委员会的提案是否可以保证这一减少量的实现呢？当然不能。此处引用的文字认为，如果使**所有**活着的心智缺陷者不生育的话，那么至少可以保证实现这一减少量。按照委员会所强调的自愿原则，他们的提议根本不能保证英格兰和威尔士的 300,000 例确定具有心智缺陷的个体都被绝育。谁必须同意呢？（1）病人，如果他有自主能力的话。(2)父母或监护人。(3)配偶，如果病人已婚的话。(4)管理委员会。该计算是无误的，但它建立在两个"非常不支持"该提案有效性的假设之上。一个假设是：导致缺陷（先天痴愚）的遗传因子会在大量的后代中"被携带"，但仅仅在很少情况下才会产生明显的缺陷；另一个假设是：这种缺陷在整个社会中是均匀分布的。(缺陷者的生育能力也被假定为相当于整个人群的平均生育能力)。

另一方面，允许缺陷个体生育的话情况会有多么不利呢？通过试图取消对立的假设并没有取得任何成果。无论正误，数字就是数字。

再者，17% 的减少是否是优生学能够承诺的全部效果？如果完全不考虑那些非常聪明并尽力为国家作贡献的缺陷个体来判断最小损害的话，那么是不是公共机构就成了对社会的主要威胁呢？麦克布赖德教授（《自然》,1930 年 1 月 11 日,第 40 页）强调说这仅仅触碰到问题的边缘。"对社会最具危险性的有缺陷的个体是那些从来没有受到公共机构限制的人！高级的有缺陷的个体能够凭最低收入和最不需技能的工作来养活自己，并且任何文明社会的政府都不会承担约束他们的责任，所以他们就会继续繁衍出和他们自己一样有缺陷的大家族。"当国家设计出越来越多的方法通过生物评估来筛选缺陷个体时，他那将绝育术视为"对那些生育了很多孩子而无力供养的父母们所犯经济罪"的惩罚措施的观点就变得越来越难应用于实际了。

像这里提到的支持如此立法的倡导者们可能会将政治机构的同意误解为自己取得了胜利，这难道不是真正的危险吗？一旦得到同意，那么更准确地说，这更应该被视作失败的标志。因为这不是国家的同意，而是国家自发的创造性力量的同意，这种力量是用来保障最基本的发展的，除非所有政党的立法者从自然知识的熏陶中获得启发，否则这种力量将不会存在。

（刘皓芳 翻译；刘京国 审稿）

Fine Structure of α-Rays

G. Gamow

Editor's Note

George Gamow was a Russian scientist who left the Soviet Union in 1932 and worked at several Western European universities until moving to the United States before the Second World War. He made a powerful impression by his versatility as a scientist, his capacity to write clearly for the general public and his engagement in public causes such as advocacy of building nuclear weapons in the United States. This brief letter offers an explanation of why γ-rays emitted by radioactive atoms may have a variety of energies.

IT is usually assumed that the long range α-particles observed in C'-products of radioactive series correspond to different quantum levels of the α-particle in the nucleus. If after the preceding β-disintegration the nucleus is left in an excited state with the α-particle on one of the levels of higher energy, one of the two following processes can take place: either the α-particle will cross the potential barrier surrounding the nucleus and will fly away with the total energy of the excited level (long range α-particle), or it will fall down to the lowest level, emitting the rest of its energy in the form of electromagnetic radiation (γ-rays), and will later fly away as an ordinary α-particle of the element in question. Thus there must exist a correspondence between the different long range α-particles and the γ-rays of the preceding radioactive body. If p is the relative number of nuclei in the excited state, λ the corresponding decay constant, and θ the probability of transition of the nucleus from the excited state to one of the states of lower energy with emission of energy (in form of γ-quanta or an electron from the electronic shells of the atom), the relative number of long range α-particles must be $N = p\frac{\lambda}{\theta}$. Knowing the number of α-particles in each long range group and calculating, from the wave mechanical theory of radioactive disintegration, the corresponding values of λ, we can estimate for each group the value θ/p, giving a lower limit for the probability of γ-emission. For example, for thorium-C' possessing besides the ordinary α-particles also two groups of long range α-particles, we have for transition probabilities from two excited states to the normal state $\theta_1 < 0.4 \times 10^{12}$ sec^{-1} and $\theta_2 < 2 \times 10^{12}$ sec^{-1}, which is the right order of magnitude for the emission of light quanta of these energies. With decreasing energy λ decreases much more rapidly (exponentially) than θ, so that the number of long range α-particles from the lower excited levels will be very small. (From this point of view we can also easily understand why the long range α-particles were observed only for C'-products for which the energy of normal α-particles is already much greater than for any other known radioactive element.)

α 射线的精细结构

伽莫夫

编者按

乔治·伽莫夫是一位俄国科学家，他于 1932 年离开苏联，在几所西欧的大学里工作。第二次世界大战爆发前夕，他去了美国。他是一位多才多艺的科学家，为公众撰写的普及读物非常清晰易懂，他在美国投身于公众事业，支持原子武器，这些都给人们留下了深刻的印象。在这篇简短的快报文章中，他解释了为什么放射性原子发射出的 γ 射线可能具有不同的能量。

我们通常假设，C' 放射系列产物中观测到的长程 α 粒子对应于原子核中 α 粒子的不同量子能级。如果在 β 衰变之后，原子核处于 α 粒子占据某个更高能级的激发态，那么就可能发生下面两个过程中的一个：或者是 α 粒子穿过原子核周围的势垒，携带激发态的所有能量而逃逸（长程 α 粒子）；或者是 α 粒子降至最低能级，将剩余能量以电磁辐射（γ 射线）的形式发射出去，然后再以普通 α 粒子的形式逃逸出该原子核。这样，在长程 α 粒子和之前的放射体放出的 γ 射线之间就应该存在某种关联。如果 p 是处于激发态的原子核的相对数量，λ 是相应的衰减常数，θ 是原子核从激发态跃迁到某个低能态并辐射出能量（以 γ 量子或者是从该原子电子壳层中发射出的电子的形式）的几率，那么长程 α 粒子的相对数量就是$N = p\dfrac{\lambda}{\theta}$。我们已经知道了每一个长程组内的 α 粒子数量，并且可以通过辐射衰变的波动力学理论计算出相应的 λ 值，那么我们就可以估算出每一组的 θ/p 值，得到一个 γ 辐射发生几率的下限。比如，在钍 C' 的衰变中，除释放普通的 α 粒子之外，还有两组长程 α 粒子，对于从这两个激发态跃迁到正常态的几率，我们知道：$\theta_1 < 0.4 \times 10^{12}$/ 秒，$\theta_2 < 2 \times 10^{12}$/ 秒，这个数量级对于辐射这些能量的光量子来说是合适的。当能量减小时，λ（呈指数减小）比 θ 减小的速度快很多，因此来自较低激发态的长程 α 粒子数量将会非常少。（这样看来，就不难理解为什么长程 α 粒子只有在 C' 的衰变产物中才会出现了，C' 过程产物中的正常 α 粒子的能量远远高于其他已知的任何放射性元素产生的能量。）

A difficulty arises with the recent experiments of S. Rosenblum (*C. R.*, p. 1,549; 1929; p. 1,124; 1930), who found that the α-rays of thorium-*C* consist of five different groups lying very close together. The energy differences and intensities of the different groups relative to the strongest one (α_0) are, according to Rosenblum:

$$E\alpha_1 - E\alpha_0 = +40.6 \text{ kv} \quad I\alpha_1 = 0.3$$
$$E\alpha_2 - E\alpha_0 = -287 \text{ ,,} \quad I\alpha_2 = 0.03$$
$$E\alpha_3 - E\alpha_0 = -442 \text{ ,,} \quad I\alpha_3 = 0.02$$
$$E\alpha_4 - E\alpha_0 = -421 \text{ ,,} \quad I\alpha_4 = 0.005$$

If we suppose that these groups are due to α-particles escaping from different excited quantum levels in the nucleus, we meet with very serious difficulties. The decay constant λ for the energy of thorium-*C* fine structure particles is very small ($\lambda \sim 10^{-2}$ sec^{-1}), and in order to explain the relatively great number of particles in different groups we must assume also very small transition probabilities. We must assume that thorium-*C* nucleus can stay in an excited state without emission of energy for a period of half an hour!

We can, however, obtain the explanation of these groups by assuming that we have here a process quite different from the emission of long range α-particles. Suppose that two (or more) α-particles stay on the normal level of the thorium-*C* nucleus. It can happen that after one of the α-particles has escaped the nucleus will remain in an excited state with the other particle on a certain level of higher energy. (In this case the energy of the escaping α-particle will be smaller than the normal level and obviously will not correspond to any quantum level inside the nucleus.) From the excited state the nucleus (thorium-*C''* now) can afterwards jump down to the normal level, emitting the energy difference in form of a γ-quantum.

Thus the relative number of different groups will not depend on the probability of γ-emission but only on the transition integral:

$$W = \int f(r_{1,2})\, \psi E_0(\alpha_1)\, \psi E_0(\alpha_2)\, \psi E_n(\alpha_1)\, \psi E\alpha_n(\alpha_2)\, dv_1\, dv_2$$

where $f(r)$ is the interaction energy of two α-particles at a distance r apart, ψE_0 and ψE_n the eigenfunctions of an α-particle in the normal and n^{th} excited states, and ψE_a the eigenfunction of an escaping α-particle with the energy: $E_{\alpha_n} = E_0 - (E_n - E_0)$.

According to this scheme, the γ-rays corresponding to different fine structure groups of thorium-*C* must be observed as γ-rays of thorium-*C* (ejecting electrons from *K*, *L*, *M*, … shells of the thorium-C''-atom) and not as the rays of thorium-*B*, as we would expect in the case of long range particle explanation. The level scheme of the thorium-*C''*-nucleus as given by fine structure energies is represented in Fig. 1.

罗森布拉姆在最近的实验中遇到了一些困难（《法国科学院院刊》，1929 年第 1,549 页，1930 年第 1,124 页），他发现钍 C 放射的 α 射线是由非常接近的 5 个不同的组组成的。罗森布拉姆给出了其余各组相对于最强的那一组（α_0）的能量差和强度：

$$E\alpha_1 - E\alpha_0 = +40.6 \text{ 千电子伏} \quad I\alpha_1 = 0.3$$
$$E\alpha_2 - E\alpha_0 = -287 \text{ 千电子伏} \quad I\alpha_2 = 0.03$$
$$E\alpha_3 - E\alpha_0 = -442 \text{ 千电子伏} \quad I\alpha_3 = 0.02$$
$$E\alpha_4 - E\alpha_0 = -421 \text{ 千电子伏} \quad I\alpha_4 = 0.005$$

如果我们假设这些组分是由于逃逸的 α 粒子曾处于原子核内的不同激发量子能级而形成的，那么我们将遇到很大的麻烦。钍 C 精细结构粒子的能量衰减常数 λ 非常小（λ 约为 10^{-2}/ 秒），而且为了解释各组中何以有相对那么大数量的粒子，我们必须同时假设跃迁几率非常小。我们必须假设钍 C 原子核可以停留在激发态而不辐射能量长达半个小时！

然而，如果我们设想一个完全不同于发射长程 α 粒子的过程，就可以解释这 5 组 α 粒子了。假设有两个（或者更多的）α 粒子处于钍 C 原子核的正常能级上。其中一个 α 粒子逃逸到原子核外后，原子核有可能还保持在激发态，因为剩下的 α 粒子可能处于某个能量较高的能级上。（在这种情况下，逃逸 α 粒子的能量低于正常能级，而且它显然与原子核内的任何量子能级都不相等。）随后原子核（这里是钍 C''）可以从激发态跃迁到正常态，放出一个 γ 量子以释放两态之间的能量差。

这样，不同组的相对数量将与 γ 辐射的几率无关，而只与跃迁积分相关：

$$W = \int f(r_{1,2})\,\psi E_0(\alpha_1)\,\psi E_0(\alpha_2)\,\psi E_n(\alpha_1)\,\psi E\alpha_n(\alpha_2)\,dv_1\,dv_2$$

式中，$f(r)$ 是两个 α 粒子在相距为 r 时的相互作用能，ψE_0 和 ψE_n 分别是 α 粒子在正常态和第 n 个激发态的本征函数，ψE_a 是能量为 $E_{\alpha n} = E_0 - (E_n - E_0)$ 的逃逸 α 粒子的本征函数。

按照这种解释，对应于钍 C 不同精细结构组分的 γ 射线应该被看作是钍 C 的 γ 射线（发射出的电子来自于钍 C'' 原子的 K，L，M 等壳层），而不能被看作是钍 B 的 γ 射线，正如我们在解释长程粒子时预期的那样。图 1 中所示的钍 C'' 原子核能级图画出了能量的精细结构。

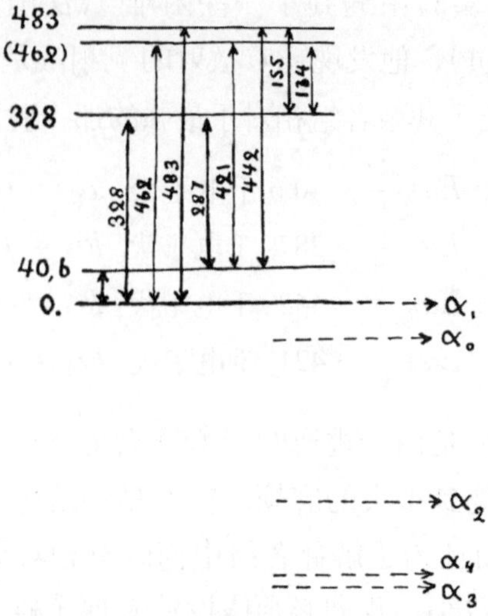

Fig. 1

In the observed γ-ray spectra of thorium-$C + C''$ (Black, *Proc. Roy. Soc.*, pp. 109–166; 1925) we can find lines with the energies: 40.8; 163.3; 279.4; 345.8; 439.0; 478.8; 144.6 kv fitting nicely with the energy differences in Fig. 1.

Thus we see that the fine structure group of highest energy corresponds to the normal level of the nucleus, while the other groups are due to the ordinary α-particles which have lost part of their energy, leaving the nucleus in an excited state.

I am glad to express my thanks to Dr. R. Peierls and Dr. L. Rosenfeld for the opportunity to work here.

(**126**, 397; 1930)

G. Gamow: Piz da Daint, Switzerland, July 25.

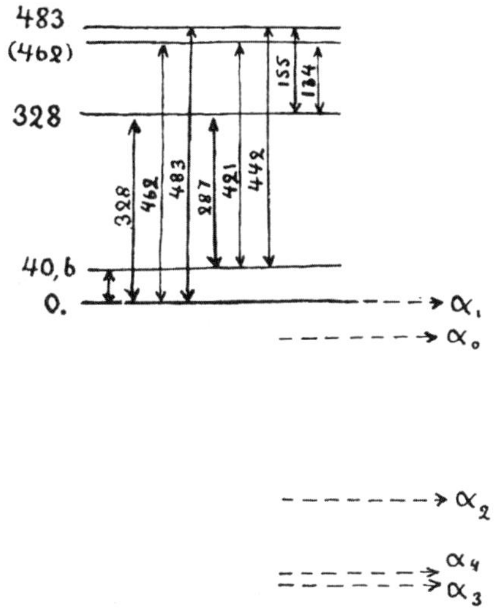

图 1

在实测的钍 C 和钍 C'' 的 γ 射线能谱中（布莱克，《皇家学会学报》，1925 年，第 109~166 页），我们可以找到对应于以下能量的谱线：40.8，163.3，279.4，345.8，439.0，478.8，144.6 千电子伏，这与图 1 中的能级差吻合得很好。

我们可以看到，能量最高的精细结构组分对应于原子核的正常态，而其他组分则是由损失掉部分能量的普通 α 粒子离开处于激发态的原子核造成的。

我非常感谢佩尔斯博士和罗森菲尔德博士给我提供在这里工作的机会。

（王静 翻译；江丕栋 审稿）

Eugenic Sterilisation

J. S. Huxley

Editor's Note

Eugenics—the attempted elimination of "bad genes" in a population—was widely held to be important for maintaining a healthy society for long after Charles Darwin published his evolutionary theory. It was advocated by Darwin's cousin Francis Galton, and Darwin himself assented. So did Julian Huxley, grandson of Darwin's staunch advocate Thomas Henry Huxley, who served in the British Eugenics Society until the 1960s. Here he writes to defend the society's recommendations of enforced sterilization of "mental defectives" against a criticism in a *Nature* editorial. Tellingly, that criticism was of the proposed mechanism, not the principle—*Nature* fully supported eugenic arguments in the 1930s. Only later did they become seen as not just morally but scientifically flawed.

AS a member of the Committee of the Eugenics Society for Legalising Eugenic Sterilisation, I should like to be allowed to say a few words concerning the leading article in *Nature* of Aug. 30 on our proposals. It is stated there: "Is there not a real danger that the advocates of such legislation as here may mistake the assent of the political machine for victory? If assent were gained, would it not be much more accurately determined as the hall-mark of failure? It is not the assent of the State, but the initiative and creative power of the State, that is needed to secure essential progress...."

With the last sentence I entirely agree; but I fail to perceive how a step in the right direction can be regarded as the hall-mark of failure—unless, indeed, the Committee should be so stupid as to believe that the taking of this one step had brought us to our final goal, which is certainly not the case. The article opens with references to the difficulties in the way of progress which are created by timid and ignorant public opinion, and continues, "if, as Sir Walter Fletcher has lately pointed out, a mere ailment, like cancer, has only been made accessible to scientific study through the lifting of foolish and superstitious taboos, how can we expect the direr social maladies to be approached courageously?" I think I can speak for the Committee in saying that we realise to the full the extent of these intangible difficulties, and that it is precisely for that reason that we have concentrated on a small but tangible and urgent beginning. Somehow or other the public has to be made race-conscious, has to be imbued with the eugenic idea as a basic political and ethical ideal. We believe that a campaign of the kind we have launched, directing attention to a gross racial defect, will be the best possible way of turning their thoughts in the desired direction.

优生绝育

赫胥黎

编者按

在查尔斯·达尔文提出进化论很长时间以后，优生学这种试图消除人类“不良基因”的学说被大家公认为是确保社会成员健康的重要方法。优生学是由达尔文的表弟弗朗西斯·高尔顿倡导的，并且得到了达尔文本人的支持。本文的作者朱利安·赫胥黎也是优生学的支持者，他是托马斯·亨利·赫胥黎的孙子。20世纪60年代以前，他一直在英国优生学会工作。他写这篇文章的目的是，为该学会倡导的“精神病患者”应该被强制进行绝育作辩护，以反驳《自然》上持反对意见的一篇评论。事实上，那篇评论批驳的是提案中所说的运作机制，而不是批驳基本原则。早在20世纪30年代，《自然》就完全赞成优生学的观点。只是后来这些观点变得不仅不人道，而且出现了科学上的谬误。

作为优生学会优生绝育合法化委员会的一名成员，我想就《自然》8月30日发表的那篇针对我们的提议的重要文章说几句。文中说到：“像这里提到的支持如此立法的倡导者们可能会将政治机构的同意误解为自己取得了胜利，这难道不是真正的危险吗？一旦得到同意，那么更准确地说，这更应该被视作失败的标志。因为这不是国家的同意，而只是国家自发的创造性力量的同意，这种力量是用来保障最基本的发展的……”

我完全同意最后一句话，但是我却不理解方向正确的措施怎么会被当作失败的标志呢？除非委员会愚蠢到相信仅仅通过实施这一措施就可以实现我们的终极目标，但事实上他们肯定没有这么认为。那篇文章一开篇就提出，怯懦而又无知的公众意识对优生绝育的实施造成了困难，紧接着写道“正如沃尔特·弗莱彻爵士最近指出的，如果对于一种纯粹的疾病（例如癌症），只有在破除愚蠢迷信的忌讳后才能对其进行科学研究的话，那么我们又怎么能期望研究人员会大胆地去研究更加可怕的社会弊病呢？”我想我可以代表委员会说，我们已经认识到了所有这些无形的困难，也正是因为这个原因，我们才集中精力从一个比较小但很明确很紧迫的问题入手。不管通过什么方法，都必须使公众具有种族意识，必须让他们把优生思想当作一项基本的政治道德理想。我们相信，我们发起的这种引导公众关注整个种族缺陷的运动，是把公众思想扭转到我们预期方向上的最可能有效的途径。

Comment is also made on the fact that the prevention of reproduction by all defectives would only lower the incidence of mental defect by about 17 percent in one generation. The article fails to remind readers that the process is cumulative, and also does not point to any other way in which it could be reduced more rapidly. Finally, the most relevant fact of all is omitted, namely, that one of the greatest obstacles to securing assent to the sterilisation of defectives has been and is the widespread belief that, since two normal persons may have a defective child, therefore preventing defectives from reproducing will have no effect on the proportion of defectives in later generations. Dr. R. A. Fisher has gone carefully into the matter, and has shown that, even when the most unfavourable assumptions are made, prevention of reproduction by all defectives would result in a reduction of some 17 percent—which to me at least seems considerable, as it would mean that there would be above 50,000 less defectives in Great Britain after the lapse of the, biologically speaking, trivial span of one generation.

I am glad that *Nature* has directed attention to the gravity of the problem, and look forward with interest to further discussion of the problem in its columns.

(**126**, 503; 1930)

J. S. Huxley: King's College, London, W. C. 2.

该文章还评论了如下事实：一代人中所有的缺陷个体都不生育，也将只能使下一代中心智缺陷的发生率减少大约17%。该文章并没有提醒读者注意这一过程是累积的，也没有提出任何别的能使心智缺陷的发生率减少得更加迅速的方法。最后，该文章还遗漏了与减少心智缺陷的发生率关系最密切的事实，即，确保对缺陷个体实施绝育得到人们赞同的最大障碍之一，曾经是并且现在依然是广为流传的一种看法，那就是即使两个正常人也可能会生出一个有缺陷的孩子，因此阻止缺陷个体的生育这一做法对于后代中缺陷个体的出现比例不会有任何影响。费希尔博士已经对这一问题进行了仔细深入的研究，结果表明即使是在最坏的假设下，通过阻止所有缺陷个体的生育也会使下一代中的缺陷个体减少17%左右，这个数字至少对我来说算是相当可观了，因为从生物学意义上来说，这意味着仅仅是经过一代人的生育之后，英国心智缺陷者的数量的减少就会超过50,000。

我很高兴《自然》能够关注这一问题的重要性，我也热切期待其专栏里能够出现对此问题的深入讨论。

（刘皓芳 翻译；刘京国 审稿）

The Proton*

P. A. M. Dirac

Editor's Note

Here English physicist Paul Dirac describes his prediction, reported earlier in another journal, that electrons may have "negative energy" and consequently positive charge. This prediction followed from Dirac's "relativistic" quantum theory, which made quantum mechanics compatible with special relativity. The "positive electrons" appear as "holes" in a pervasive electron sea. Dirac suspects they behave as protons, but there are problems with that: protons have very different masses from electrons, and the electrons and holes were predicted to annihilate. Dirac alludes to an idea of J. Robert Oppenheimer that a positive electron may in fact be a different particle with the mass of an electron. And so it later proved: they were positrons, anti-matter versions of electrons.

MATTER is made up of atoms, each consisting of a number of electrons moving round a central nucleus. It is likely that the nuclei are not simple particles, but are themselves made up of electrons, together with hydrogen nuclei, or protons as they are called, bound very strongly together. There would thus be only two kinds of simple particles out of which all matter is built, the electrons, each carrying a charge $-e$, and the protons, each carrying a charge $+e$.

It should be mentioned here that there is a difficulty in this point of view provided by the nitrogen atom. One can infer from the charge and mass of the nitrogen nucleus that it should consist of 14 protons and 7 electrons, but it appears to have properties inconsistent with its being composed of an odd number of simple particles. However, very little is really known about nuclei, and the opinion is generally held by physicists that some way of evading this difficulty will be found and that all nuclei will ultimately be shown to be made up of electrons and protons.

It has always been the dream of philosophers to have all matter built up from one fundamental kind of particle, so that it is not altogether satisfactory to have two in our theory, the electron and the proton. There are, however, reasons for believing that the electron and proton are really not independent, but are just two manifestations of one elementary kind of particle. This connexion between the electron and proton is, in fact, rather forced upon us by general considerations about the symmetry between positive and negative electric charge, which symmetry prevents us from building up a theory of the negatively charged electrons without bringing in also the positively charged protons. Let us examine how this comes about.

---

* Based on a paper read before Section A (Mathematical and Physical Science) of the British Association at Bristol on Sept. 8.

质　子*

狄拉克

编者按

这篇文章报道了英国物理学家保罗·狄拉克就电子可能具有“负能量”从而带有正电荷这一预测所作的论述，该预测在更早些时候已发表在其他期刊上。狄拉克的“相对论性的”量子理论使量子力学与狭义相对论得以相容，前述的预测正是这一理论的结果。“带正电荷的电子”就像是无处不在的电子海中的“空穴”。狄拉克猜测，这些“带正电荷的电子”的行为方式与质子类似，不过这一猜测存在一些问题：质子与电子在质量上的差别非常大，而且据预测电子和这些空穴相遇会湮灭。在这里，狄拉克也提到了罗伯特·奥本海默的观点，即带正电荷的电子可能就是与电子的质量相同的另一种粒子。后来人们证明确实如此：这些粒子就是正电子，电子的反物质形式。

物质是由原子构成的，每一个原子是由若干个围绕中心原子核转动的电子组成的。原子核很可能不是基本粒子，而是由电子和氢原子核（或者所谓质子）紧密束缚在一起构成的。这样所有的物质都只由这两种基本粒子构成，其中每一个电子带电荷 $-e$，每一个质子带电荷 $+e$。

这里需要指出的是，氮原子的存在给这个观点提出了一个难题。由氮原子核的电荷和质量，我们可以推断出氮原子核是由 14 个质子和 7 个电子组成的，但是氮原子核表现出来的性质似乎与它是由奇数个基本粒子构成这一点不符。然而，关于原子核，我们知之甚少，而且物理学家们普遍认为将来总会有办法克服这个困难，并且最终将会证明所有的原子核都是由电子和质子构成的。

哲学家总是梦想所有的物质都是由一种基本粒子构成的，所以我们的理论——包含两种基本粒子（电子和质子）——并不能使所有人都满意。然而人们有理由相信电子和质子并不是毫无关系的，它们只是一种基本粒子的两种表现形式。而事实上，电子和质子之间的联系在某种程度上是关于正负电荷之间对称性的一般认识强加给我们的，这种对称性使我们不能构建一套只包含带负电的电子，而不包含带正电的质子的理论。下面让我们看看为什么会是这样。

* 基于 9 月 8 日在布里斯托尔向英国科学促进会的 A 分部（数学和物理科学）宣读的一篇论文。

The energy W of a particle in free space is determined in terms of its momentum p according to relativity theory by the equation

$$W^2/c^2 - p^2 - m^2c^2 = 0,$$

where m is the rest-mass of the particle and c is the velocity of light. This equation can easily be generalised to apply to a charged particle moving in an electromagnetic field and can be used as a Hamiltonian to give the equations of motion of the particle, and thus its possible tracks in space-time.

Now the above equation is quadratic in W, allowing of both positive and negative values for W. Thus for some of the tracks in space-time the energy W will have positive values and for the others negative values. Of course a particle with negative energy (kinetic energy is referred to throughout) has no physical meaning. Such a particle would have less energy the faster it is moving and one would have to put energy into it to bring it to rest, quite contrary to anything that has ever been observed.

The usual way of getting over this difficulty is to say that the tracks for which W is negative do not correspond to anything real in Nature and are to be simply ignored. This is permissible only provided that for every track W is either always positive or always negative, so that one can tell definitely which tracks are to be ignored. This condition is fulfilled in the classical theory, where W must vary continuously, since W can never be numerically less than mc^2 and is thus precluded from changing from a positive to a negative value. In the quantum theory, however, discontinuous variations in a dynamical variable such as W are permissible, and detailed calculation shows that W certainly will make transitions from positive to negative values. We can now no longer ignore the states corresponding to a negative energy and it becomes imperative to find some physical meaning for them.

We can deal with these states mathematically, in spite of their being physically nonsense. We find that an electron with negative energy moves in an electromagnetic field in the same way as an ordinary electron with positive energy would move if its charge were reversed in sign, so as to be $+e$ instead of $-e$. This immediately suggests a connexion between negative-energy electrons and protons. One might be tempted at first sight to say that a negative-energy electron *is* a proton, but this, of course, will not do, since protons certainly do not have negative kinetic energy. We must therefore establish the connexion on a different basis.

For this purpose we must take into consideration another property of electrons, namely, the fact that they satisfy the exclusion principle of Pauli. According to this principle, it is impossible for two electrons ever to be in the same quantum state. Now the quantum theory allows only a finite number of states for an electron in a given volume (if we put a restriction on the energy), so that if only one electron can go in each state, there is room for only a finite number of electrons in the given volume. We thus get the idea of a *saturated* distribution of electrons.

根据相对论，自由空间中粒子的能量 W 由它的动量 p 决定，即

$$W^2/c^2 - p^2 - m^2c^2 = 0\ ,$$

其中 m 是粒子的静止质量，c 是光速。这个方程可以很容易地推广到带电粒子在电磁场中运动的情况，并且可以被用作哈密顿量，给出带电粒子的运动方程，从而得到带电粒子在时空中可能的径迹。

上面的方程中 W 项是二次的，所以 W 既可能是正的，也可能是负的。因此对时空中的一些径迹而言能量 W 是正的，而对其他一些则是负的。当然粒子具有负能量（动能总是会涉及）是没有物理意义的。这样的粒子运动得越快，它的能量就越小，我们不得不给它能量使它静止，然而这与我们观察到的所有现象都是截然不同的。

通常克服这个困难的办法是认为具有负 W 的径迹不对应于任何真实的自然现象，而只需要简单地把它忽略掉。不过这只有在每一条径迹的 W 值恒正或者恒负的前提下才成立，因为只有这样我们才可以明确地判断哪一条径迹应该被忽略。这个条件在 W 连续变化的经典理论中是满足的，因为 W 在数值上不能小于 mc^2，所以排除了 W 从正值变化到负值的可能性。然而，在量子理论中，像 W 这样的动力学量可以不是连续变化的，并且详细的计算表明 W 确实可以从正值变化到负值。因此我们不能再忽略负能量对应的状态，而必须为它们寻找某种物理意义。

我们可以从数学上处理这些状态，而先不去管它们是否具有物理意义。我们发现，如果普通电子的电荷符号发生翻转，即从 $-e$ 变为 $+e$，那么一个具有负能量的电子在电磁场中的运动方式和一个普通的具有正能量的正电子一样。这就意味着负能电子和质子之间存在某种联系。乍一看这种情况，人们可能会说负能电子**就是**质子，但是这无疑是不成立的，因为质子的动能不可能是负的。因此，我们必须在另外的基础上构建它们的联系。

为此我们必须考虑电子的另外一个特性，即它们满足泡利不相容原理。根据这一原理，两个电子永远不可能处于同一个量子态。因为量子理论在给定的空间内只允许有限数目的电子态（如果我们给能量一个限制），所以如果每一个态只允许一个电子占据，那么在给定的空间内只能容纳有限数目的电子。这样我们就会得到电子**饱和**分布的概念。

Let us now make the assumption that almost all the states of negative energy for an electron are occupied, and thus the whole negative-energy domain is almost saturated with electrons. There will be a few unoccupied negative-energy states, which will be like holes in the otherwise saturated distribution. How would one of these holes appear to our observations? In the first place, to make the hole disappear, which we can do by filling it up with a negative-energy electron, we must put into it a negative amount of energy. Thus to the hole itself must be ascribed a positive energy. Again, the motion of the hole in an electromagnetic field will be the same as the motion of the electron that would fill up the hole, and this, as we have seen, is just the motion of an ordinary particle with a charge $+e$. These two facts make it reasonable to assert that *the hole is a proton.*

In this way we see the proper role to be played by the negative-energy states. There is an almost saturated distribution of negative-energy electrons extending over the whole of space, but owing to its uniformity and regularity it is not directly perceptible to us. Only the small departures from perfect uniformity, brought about through some of the negative-energy states being unoccupied, are perceptible, and these appear to us like particles of positive energy and positive charge and are what we call protons.

This theory of the proton involves certain difficulties, which will now be discussed. The theory postulates the existence everywhere of an infinite number of negative-energy electrons per unit volume, and thus an infinite density of electric charge. According to Maxwell's equations, this would give rise to an infinite electric field. We can easily avoid this difficulty by a re-interpretation of Maxwell's equations. A perfect vacuum is now to be considered as a region in which all the states of negative energy and none of those of positive energy are occupied. The electron distribution in such a region must be assumed to produce no field, and only the departures from this vacuum distribution can produce a field according to Maxwell's equations. Thus, in the equation for the electric field E

$$\operatorname{div} E = -4\pi\rho,$$

the electric density ρ must consist of a charge $-e$ for each state of positive energy that is occupied, together with a charge $+e$ for each state of negative energy that is unoccupied. This gives complete agreement with the usual ideas of the production of electric fields by electrons and protons.

A second difficulty is concerned with the possible transitions of an electron from a state of positive energy to one of negative energy, which transitions were the original cause of our having to give a physical meaning to the negative-energy states. These transitions are very much restricted when nearly all the negative-energy states are occupied, since an electron in a positive-energy state can then drop only into one of the unoccupied negative-energy states. Such a transition process would result in the simultaneous disappearance of an ordinary positive-energy electron and a hole, and would thus be interpreted as an electron and proton annihilating one another, their energy being emitted in the form of electromagnetic radiation.

现在我们假设电子的负能态几乎都被占据了，因此整个负能区域电子几乎是饱和的。有一些没被占据的负能态，它们就像饱和分布的负能态电子海中的一些空穴。在我们看来这些空穴是什么样的呢？首先，为了使这些空穴消失，我们需要填充一个负能量的电子，即放入一个负能量。这样空穴本身必须具有正能量。其次，空穴在电磁场中的运动方式和填充空穴的电子的运动方式一样，就像我们之前看到的那样，就是一个带 $+e$ 电荷的普通粒子的运动。这两个事实使我们有理由断言——**空穴就是质子。**

这样我们就看到了负能态起到的作用。整个空间中负能电子几乎处于饱和分布，但由于它们表现出来的均匀性和规律性，因而不能直接被我们觉察到。只有在完美的均匀性上出现一些小的偏离，即有一些负能态没被占据，才能被我们觉察到。这些偏离在我们看来就是些具有正能量和带正电荷的粒子，也就是我们所谓的质子。

这个关于质子的理论存在一些问题，下面我们就来讨论这些问题。这个理论假设空间每处单位体积内都存在无限数目的负能电子，这样电荷密度就是无穷大了。根据麦克斯韦方程，这将会产生一个无限大的电场。不过我们可以很容易地通过重新解释麦克斯韦方程来克服这个困难。理想的真空被认为是所有的负能态都被占据，而所有的正能态都没被占据的空间。我们认为在这样的空间中，电子分布不会产生任何场，只有当电子分布偏离真空分布时才会产生根据麦克斯韦方程得到的场。因此，在电场 E 的方程中

$$\operatorname{div} E = -4\pi\rho\ ,$$

电荷密度 ρ 是由每一个被占据的正能态上的电荷 $-e$ 和每一个没被占据的负能态上的电荷 $+e$ 组成的。这和通常电子和质子产生电场的观点是完全一致的。

第二个困难涉及电子可能存在从正能态向负能态的跃迁，这种跃迁是我们必须赋予负能态以物理意义的最初原因。当几乎所有的负能态都被占据时，这种跃迁是非常受限制的，因为处于正能态的电子只能落入没被占据的负能态。这样的跃迁过程导致一个普通的正能态电子和一个空穴同时消失，所以可以解释成一个电子和一个质子的互相湮灭，它们的能量以电磁辐射的形式发射出来。

There appears to be no reason why such processes should not actually occur somewhere in the world. They would be consistent with all the general laws of Nature, in particular with the law of conservation of electric charge. But they would have to occur only very seldom under ordinary conditions, as they have never been observed in the laboratory. The frequency of occurrence of these processes according to theory has been calculated independently by several investigators, with neglect of the interaction between the electron and proton (that is, the Coulomb force between them). The calculations give a result much too large to be true. In fact, the order of magnitude is altogether wrong. The explanation of this discrepancy is not yet known. Possibly the neglect of the interaction is not justifiable, but it is difficult to see how it could cause such a very big error.

Another unsolved difficulty, perhaps connected with the previous one, is that of the masses. The theory, when one neglects interaction, requires the electron and proton to have the same mass, while experiment shows the mass ratio to be about 1,840. Perhaps when one takes interaction into account the theoretical masses will differ, but it is again difficult to see how one could get the large difference required by experiment.

An idea has recently been put forward by Oppenheimer (*Phys. Rev.*, vol. 35, p. 562) which does get over these difficulties, but only at the expense of the unitary theory of the nature of electrons and protons. Oppenheimer supposes that all, and not merely nearly all, of the states of negative energy are occupied, so that a positive-energy electron can never make a transition to a negative-energy state. There being now no holes which we can call protons, we must assume that protons are really independent particles. The proton will now itself have negative-energy states, which we must again assume to be all occupied. The independence of the electron and proton according to this view allows us to give them any masses we please, and further, there will be no mutual annihilation of electrons and protons.

At present it is too early to decide what the ultimate theory of the proton will be. One would like, if possible, to preserve the connexion between the proton and electron, in spite of the difficulties it leads to, as it accounts in a very satisfactory way for the fact that the electron and proton have charges equal in magnitude and opposite in sign. Further advances in the theory of quantum electrodynamics will have to be made before one can deal accurately with the interaction and see whether it will settle the difficulties, or whether, perhaps, a new idea can be introduced which will answer this purpose.

(**126**, 605-606; 1930)

看起来没有什么理由可以说明为什么这样的过程不能在现实世界的某处发生。它们会遵守自然界所有的一般规律，特别是电荷守恒定律。但是它们在普通条件下必然很少发生，因为即使在实验室中它们也还没有被观察到。一些研究者已经独立地计算出这些过程发生的理论几率，计算中忽略了电子和质子之间的相互作用（即它们之间的库仑力）。计算给出的结果太大了，肯定是不正确的。事实上，结果的数量级都是完全错误的。现在还不知道为什么会出现这样的差异。可能忽略相互作用是不合理的，但是仍然很难理解为什么会导致这么大的错误。

另外一个没有解决的困难就是质量问题，这可能和前一个困难有关。如果忽略相互作用，这个理论就要求电子和质子具有相同的质量，然而实验表明它们的质量比约为 1,840。也许考虑相互作用后理论上的质量会有所不同，但还是很难理解怎样才能得到实验要求的那么大的质量差。

奥本海默最近提出的一个观点（《物理学评论》，第 35 卷，第 562 页）的确可以解决这个困难，但是它却牺牲了关于电子和质子本质的统一理论。奥本海默假定所有的（不仅仅是几乎所有的）负能态都被占据了，因此正能电子不能跃迁到负能态。这里没有我们可以称之为质子的空穴，所以我们必须假定质子是真正独立的粒子。这样，质子本身也有自己的负能态，而且我们必须假设它们也被完全占据了。根据这个观点，电子和质子的独立性允许我们随心所欲地给它们的质量赋值，而且它们之间也不会相互湮灭。

目前断定质子的最终理论还为时尚早。如果可能的话，人们愿意保留电子和质子之间的这种关系，而不管它带来的困难，因为它非常圆满地解释了这个事实——电子和质子携带的电荷大小相等，而符号相反。量子电动力学需要进一步的发展，人们才可以准确地计算相互作用，才可以知道我们的理论是否可以解决这些困难，或者是否会出现新的可以回答这个问题的观点。

（王锋 翻译；李淼 审稿）

The Problem of Epigenesis

E. W. MacBride

Editor's Note

Ernest William MacBride is perhaps in retrospect not the ideal author for this discourse on embryogenesis, being a late supporter of the Lamarckian view of inheritance. Yet he was considered an expert on embryology, and here he anticipates some of the key themes in modern biology: how is the development of an embryo related to its evolutionary heritage (the topic now dubbed "evo-devo"), and how does the organism, "considered as a machine", function? The former question had received much attention in Germany (MacBride's article is a review of three recent German books), especially in Ernst Haeckel's notion that embryogenesis recapitulates evolutionary history. The main issue debated here, however, is the origin of the force that organizes an undifferentiated egg into a structured body, and how that depends on the emerging concept of genes.

(1) *Grundriss der Entwicklungsmechanik.* Von Prof. Dr. Bernhard Dürken. Pp. vii + 208. (Berlin: Gebrüder Borntraeger, 1929.) 12.50 gold marks.

(2) *Die Determination der Primitiventwicklung: eine zusammenfassende Darstellung der Ergebnisse über das Determinationsgeschehen in den ersten Entwicklungsstadien der Tiere.* Von Prof. Dr. Waldemar Schleip. Pp. xii + 914. (Leipzig: Akademische Verlagsgesellschaft m.b.H., 1929.) 85 gold marks.

(3) *Experimentelle Zoologie: eine Zusammenfassung der durch Versuche ermittelten Gesetzmässigkeiten tierischer Formen und Verrichtungen.* Von Prof. Dr. Hans Przibram. Band 6 : *Zoonomie; eine Zusammenfassung der durch Versuche ermittelten Gesetzmässigkeiten tierischer Formbildung* (*Experimentelle, theoretische und literarische Übersicht bis einschliesslich 1928*). Von Prof. Dr. Hans Przibram. Pp. viii + 431 + 16 Tafeln. (Leipzig und Wien: Franz Deuticke, 1929.) 40 gold marks.

THE question of epigenesis may be justly said to constitute one of the two root problems of zoology. For if we think it out there are two main things to be discovered about an animal, namely: (1) How does it fulfil its functions?—in a word, considered as a machine, how does it work? and (2) How does it come into being?—that is, how did it develop and grow? A subsidiary question to the last is: If there be such a thing as evolution, how and why did the powers of growth change from generation to generation? For, as the late Dr. Bateson reminded us so long ago as 1894, the conception of evolution as the remoulding of the adult structures of an animal as we could alter the features of a wax doll by melting the wax and remodelling it, is an entire illusion, since the members of the parent species and of that to which it gives rise both begin as tiny formless germs and what is changed is *the powers of growth.* Now when we begin to analyse growth, we can either directly observe its successive phases—and this is the scope of descriptive embryology; or by operating on the germ by chemical and physical agencies we can seek to discover the

关于渐成论的问题

麦克布赖德

编者按

回想起来，由欧内斯特·威廉·麦克布赖德来作这个胚胎发生学方面的报告可能不是很理想，因为他是拉马克遗传学说的晚期支持者之一。不过，他毕竟是胚胎学方面的专家，而且在这篇报告中他也预见到了现代生物学的一些关键性课题：胚胎的发育过程与其种系的进化过程有什么关系（这一研究课题现在被称为“进化发育生物学”）？“被视为机器”的有机体如何发挥功能？实际上，在此之前德国科学家就对前一个问题给予了极大的关注（麦克布赖德的这篇文章就是对最近相关的3本德国科学家的著作所作的评论），特别是恩斯特·海克尔提出了胚胎发生重演了进化过程的观点。不过，这篇文章讨论的主要问题是，由未分化的受精卵形成有结构的机体的原始动力是什么，以及这一动力与人们最近提出的基因这一概念有何关系。

(1)《发育机制概论》。伯恩哈德·迪肯博士，教授。vii + 208页。（柏林：施普林格兄弟，1929年）。12.50金马克。

(2)《原始发育的决定性：关于动物早期发育中决定因素的结果总结》。瓦尔德马·施莱普博士，教授。xii + 914页。（莱比锡：学术出版有限责任集团公司，1929年）。85金马克。

(3)《实验动物学：关于实验动物挑选的法律规范和步骤概要》。汉斯博士，教授。第6卷：动物学；有关实验动物的法律的总结概述（理论和文学概述，以及试验，1928年）。汉斯·普西布兰博士，教授，xiii + 431页 + 16表格。（莱比锡和维也纳：弗朗茨·多伊蒂克，1929年）。40金马克。

公正地说，渐成论的问题可能是动物学的两大根本问题之一。因为我们在研究一个动物时会思考两个主要问题：(1)它是怎样实现它的功能的？简言之，如果把动物看作一台机器的话，它是如何工作的？(2)它是怎样产生的？即，它是如何发育和生长的？伴随第二个问题又会出现另一个问题：如果真的存在进化的话，那么生长的动力是如何一代代发生变化的，以及为什么会发生变化呢？正如已故的贝特森博士早在1894年就提醒我们的，把进化看成是对成体动物结构的重塑，就如同我们通过熔化蜡并进行重建来改变蜡人的造型特点一样，这是完全错误的概念，因为亲代物种的成员以及他们产生的子代都是从无定形的微小生殖细胞开始生长而来的，而**生长的动力**在变化。现在当我们开始分析生长的时候，我们可以直接观察到其连续进行的各个阶段——这属于描述胚胎学的范畴，我们也可以通过我们能够找到的化学和物理手段对生殖细胞进行实验来观察每一个可见因素在成年个体的生长

part which each visible element plays in the upbuilding of the adult individual—and this is the object of experimental embryology.

How this science has grown since its first beginnings with His in 1874 ("Unser Körperform und die physiologische Problem ihrer Entstehung") is witnessed by the three splendid works which are the subject of this review. Each of the three is worthy of unstinted praise: though we may differ from the authors in some of the conclusions reached by them, yet in each case the collection and setting forth of the matter is worthy of our sincere admiration. We hope that too long a time may not elapse before all are translated into English.

As an introduction to the subject Dürken's manual is to be preferred, because it is concise, well illustrated, and includes only typical cases which serve to exemplify the main principles of the subject, so that a beginner can get a good grasp of these principles without being overwhelmed by too much detail. Schleip's large and well-illustrated volume attempts to give a more or less complete account of the present state of our knowledge of the subject, and it will for a long time constitute a classic work of reference. Przibram's work—thorough and excellent as all his work is—is even more ambitious in its scope than that of Schleip, for it includes not only the facts of experimental embryology in the narrower sense, but also a considerable amount of the results of Mendelian experiments. It is, however, extremely condensed and, not being adequately illustrated, somewhat difficult to follow: it seems to us that its chief value will reside in its being a manual in which references to all the important papers on the subject can be easily looked up.

It must be obvious to the reader that, within the limits of the longest review for which space can be found in *Nature*, it would be impossible to refer to a tithe of the new matter contained in these volumes, and so we must limit ourselves to a discussion of the main problems involved and to the attitude of the three authors towards them. In fairness, however, it should be added that this new matter is almost entirely confined to an elaboration of subjects dealt with by the older authors such as Roux, Hertwig, Driesch, Herbst, Boveri, Conklin, and Wilson, and does not consist to consist to any considerable extent of discoveries in newer fields. The number of animals the eggs of which can conveniently be handled and which are tolerant of experiments is limited, and the same familiar figures crop up in successive text-books of experimental embryology. After all, as Driesch has wisely remarked, the biological experimenter cannot produce life at will—he must wait until he finds it, and he is therefore in the same position as a physicist would be if he could only study fire when he found it in the crater of a volcano.

When we approach the analysis of the development of the egg, the first question we encounter is whether the organs of the adult exist in the egg preformed in miniature and development consists essentially in an unfolding and growing bigger of these rudiments, or whether the egg is at first undifferentiated material which from unknown causes afterwards becomes more and more complicated and development is consequently an "epigenesis". This problem is *the* problem of experimental embryology; in varied forms it reappears in every experiment on development which has been made.

过程中都起着什么作用——这是实验胚胎学的目标。

这里即将评论的3部出色的著作描述了这一领域从1874年西斯首次发表著作(《我们的身体形态、构造和生理问题》)以来的发展过程。这3部作品中的任何一部都是值得高度赞扬的，尽管我们可能在某些结论上和作者有不同观点，但是每部作品中对事件的收集及详尽阐述都值得我们致以由衷的敬意。我们希望不久之后，所有这些著作都可以被翻译成英语。

迪肯的这本指南是该学科的首选入门书籍，因为该书简洁明了、插图丰富，只包含解释该学科主要原理的经典事例，所以初学者可以很好地掌握这些原理而不会被铺天盖地的细节吓倒。施莱普的那本插图丰富的大部头试图尽可能全面地向我们阐明该学科的知识现状，该书将在很长时间内成为这一领域的经典参考著作。普西布兰的这部著作和他的其他所有著作一样全面而出色，就其视野来说比施莱普的那本更具远见，因为在这本书里，不仅包括狭义实验胚胎学的内容，也包括相当一部分孟德尔实验的结果。不过，这本书极度浓缩，描述不够详细，因而想要读懂会有些难度。对于我们而言，它的主要价值似乎在于，这是一本可以从中很容易地找到该学科所有重要文章引用的参考文献的手册。

读者们都知道，《自然》上发表的评论是有字数限制的，受此所限，即使只是想谈论这些书中包含的一小部分新内容也是不可能的，因此我们只好仅限于讨论涉及的主要问题以及3位作者对这些问题的看法。不过，应该补充说明的是，公平地讲，这些新内容几乎完全局限于对老一辈作者所讨论主题的细化，这些作者包括鲁、赫特维希、德里施、赫布斯特、博韦里、康克林和威尔逊，而并没有涉及较新领域的任何重要发现。能够方便地对其卵子进行操作并保证卵子可以耐受实验条件的受试动物的数量是有限的，那些熟悉的图片后来连续出现在实验胚胎学的教科书中。毕竟，正如德里施曾经明智地提出过的，生物学实验者不能随意制造生命——他只有在找到受试的生命体后才能对其进行实验，因此从这个角度来看，生物学实验者的处境就如同只有在火山爆发时发现了火之后才能对火进行研究的物理学家的处境。

当我们对卵的发育进行分析时，我们面临的第一个问题就是：是不是卵子中就存在预先成形了的成体器官的缩微版，而发育过程实质上是这些器官雏形逐渐展开显露并变大的过程？又或者卵子最初只是未分化的物质，后来由于未知的原因而变得越来越复杂，因而发育是一个“渐成的”过程？这正是实验胚胎学要解决的问题。在研究发育的实验中，这一问题以各种各样的形式重复出现。

The answer to this question given by the earlier experiments of Driesch was that some eggs, such as those of starfish and sea-urchins, consist of undifferentiated material; but others, like those of Ctenophores, show a specialisation into parts destined to form particular organs of the adult. The experiments of Wilson, Conklin, and Crampton proved that the eggs of Annelida and Mollusca belong also to this latter category. To eggs of the first kind Driesch gave the name of "equipotential systems", since when the egg had divided into eight cells any one of these was capable of forming a tiny larva perfect in all details, and, moreover, when the egg had developed into a hollow sphere or blastula, any considerable piece of this blastula would round itself off and form a perfect blastula of reduced size, which would give rise to a correspondingly reduced larva. On these results, which were a complete surprise to him, Driesch founded his theory of vitalism, arguing that if the organism were to be regarded as a physico-chemical machine, such things could not happen, for no conceivable machine could be divided into parts, each of which would function as a similar machine of reduced size. He inferred that there must be in every egg a non-material force or "entelechy" which was capable of controlling the physical and chemical changes taking place in the germ, so as to direct them towards a definite end. This power of direction was named by Driesch "regulation". This revolutionary idea of Driesch, transcending the bounds of materialistic explanation, evoked the fiercest opposition amongst those biologists by whom life was regarded as nothing more than complicated chemistry. Yet the arguments of Driesch have never been successfully met. The utmost that can be urged against them is the assertion that, although we cannot explain life by physics and chemistry now, some day in the distant future, when we have made further discoveries, we may possibly be able to do so.

Of the authors reviewed in this article, Dürken is inclined to favour Driesch whilst Schleip and Przibram oppose him, but the alternative explanations of the two latter authors when examined in detail resolve themselves into saying the same things that Driesch said, in different phrases. All three authors agree in showing that between equipotential and specialised eggs every conceivable grade of intermediate exists, and that even the eggs of *Echinus* itself are not quite so equipotential as Driesch imagined. Schleip quotes the work of Hörstatius as proving that when the upper half of a blastula is cut off, though it will round itself off so as to form a reduced blastula, yet this will never form endoderm or proceed any further in development. The vegetative half, however, when severed will produce a completely viable gastrula. By a triumph of manipulative skill, Hörstatius succeeded in separating the vegetative pole of a blastula and grafting it in various positions on another blastula in which an appropriate defect had been produced. He thus proved that in all cases development begins in the graft, and that this graft can change cells that would otherwise produce ectoderm into endoderm, in other words, act as an "organiser" of development.

Driesch attributed specialisation in eggs to a "premature stiffening of the cytoplasm" which prevented the "entelechy" from moulding the fragment of the egg into a reduced whole. Przibram in other language comes to exactly the same conclusion. He says that the formation of definite organs is in all cases due to a *solidifying* of a portion of the

德里施早期的实验对于这一问题给出的答案是，像海星和海胆这一类的生物的卵是由未分化的物质组成的；而其他生物，如栉水母类，它们的卵则特化为几个不同的部分，每一个部分都特定发育为成体的特定器官。威尔逊、康克林和克兰普顿的实验证明，环节动物和软体动物的卵也属于第二类。对于第一类卵，德里施称它们是“等潜能系统”，因为当这类卵分裂成 8 个细胞时，其中任何一个都具有成长为所有细节部分都完整的小幼虫的能力。此外，当卵发育成中空的球体或囊胚时，该囊胚的任何一部分有一定大小的片段都可以完善自身，并形成一个体积相对较小的完好的囊胚，相应地这一囊胚可以产生一个体积较小的幼虫。这些结果使德里施感到无比惊讶，基于此，他建立了自己的活力论。该学说认为，如果把生物当作一台物理化学机器的话，这些情况就不会发生，因为想象不出任何机器可以在被分成几部分后，各个部分仍然能够像一台只是尺寸有所减小的类似机器一样正常运转。他推断每个卵中一定都存在一种非物质的驱动力或者“生机”，它具有控制生殖细胞中发生的物理和化学变化的能力，因而可以指导这些变化朝着特定的方向发展。德里施将这种指导能力称为“调控”。德里施这一革命性的想法超越了唯物论解释的范畴，激起了那些认为生命仅仅是一些复杂化学变化的生物学家们最强烈的反对。然而没有任何人在与德里施的观点的交战中取胜。这些极力反对的观点中，分量最重的也只不过是如下的论断：尽管我们现在不能用物理化学变化来解释生命，但是在遥远未来的某一天，当我们取得了进一步发现的时候，我们可能就有能力对其进行解释了。

在本文所评论的这些作者中，迪肯倾向于支持德里施的观点，而施莱普和普西布兰则反对德里施的观点，但是经过仔细推敲后可以看到，后两位作者提出的另外的解释其实与德里施的观点是一样的，只是说法不同而已。所有这 3 位作者都同意，每一种可以想象到的介于等潜能的卵与特化的卵之间的中间状态都是存在的，甚至连海胆本身的卵也并不像德里施想象的那样处于完全等潜能的状态。施莱普引用了赫斯塔提乌斯的工作，以此为证据来证明，当囊胚的上半部分被分离下来时，尽管它可以完善自己而形成一个体积减小了的囊胚，然而却并不能形成内胚层，也不会继续发育。植物极的那一半在被切下来后则可以生成一个完全能存活下去的原肠胚。由于操作技术上的巨大突破，赫斯塔提乌斯成功地分离了囊胚的植物极并将其植入到另一囊胚（此囊胚事先已经进行了相应的切除）的多个位置上。他由此证明了：在所有这些例子中，发育都是从植入物上开始的，这种植入物可以改变细胞，使本来会产生外胚层的细胞转而形成内胚层，换言之，植入物扮演了发育的“组织原”的角色。

德里施将卵的特化归因于“细胞质的过早硬化”，这阻止了“生机”将卵的片段塑造成尺寸有所减小的完整卵。普西布兰用不同的语言书写了几乎完全相同的结论。他说，在所有情况下特定器官的形成都是由部分细胞质的**硬化**引起的，从而形成了

cytoplasm, forming what he calls an "apoplasm" which, if we understand him right, he does not regard as fully alive. In proportion as "apoplasms" are deposited the potentialities of the germ are successively limited, and the reason why the higher animals approximate in their working to mechanisms is the large number of "apoplasms" included in their make-up. Only fluid cytoplasm is completely living and possesses all the potentialities of the race, and Przibram is driven to conclude that these potentialities, so far as embodied at all, must be contained in the molecules of the cytoplasm, and that, therefore, these molecules constitute the real entelechy. Schleip similarly concludes that there must be an ultra-microscopic structure in the cytoplasm which, like a crystal, tends to assume a definite form and to complete itself when a fragment is severed.

In making these admissions, however, it seems to us that both Schleip and Przibram deliver themselves into the hands of Driesch. For in the crystallisation of an inorganic substance from a solution, the crystal assumes a definite form because its molecules have definite corresponding shapes, as Sir William Bragg has taught us. But what kind of structure, whether molecular or super-molecular, are we to envisage in cytoplasm? When the limb of a young newt is cut off and the stump proceeds to regenerate a new limb, are the molecules in the stump in the form of infinitesimal fingers and toes? Moreover, when the stump is cut at different levels and only the missing piece is regenerated, are we to assume that at each level in the limb before amputation the molecules are miniatures of the part distal to them? If we are able to swallow these fantastic assumptions, what are we to say of the experiment recorded by Dürken in which the tail bud of one newt embryo was grafted into the body of another near its forelimb and developed into a new limb? Presumably the cytoplasm of the tail bud was "organised" so as to produce the tissues of an adult tail. How then was this organisation so completely changed as to produce a limb instead? No wonder that Dürken says that in cases like this, physical and chemical explanations leave us completely in the lurch, and we must have recourse to the conception of the "biological field", an influence not in the living matter itself, but in the space, presumably the ether, around it.

Schleip seeks to disprove Driesch's theory by pointing out that the supposititious entelechy sometimes does foolish things, as in the case of the eggs of Nematoda subjected to centrifugal force each of which produces two partial embryos instead of one whole one. But in this objection lurks the childish conception that the entelechy, if it exists, must be the embodiment of Divine Wisdom. the entelechy is not all-seeing—it is a rudimentary "striving" which reacts to its immediate environment, in this case the "apoplasm" or ball of dead matter ejected from the egg by centrifugal force.

The term "organiser" we owe, of course, to Spemann, who wisely abstains from giving any chemical explanation of it. In the course of his marvellous experiments on the newt, Spemann showed that a piece of the dorsal lip of the blastopore of one newt gastrula grafted on the flank of another would change the fate of all the cells in its neighbourhood and force them to develop into a supplementary nerve-cord and underlying notochord. The reviewer might humbly plead that exactly the same conception was reached by him

他称为“质外体”的结构。如果我们没有理解错的话，德里施并不认为“质外体”完全具有生命的结构。根据所形成的“质外体”的多少，生殖细胞的潜能也会成比例地受到相应的限制。高等动物具有相似的生长机理正是由于它们的组成物质中包含大量的“质外体”。只有流动态的细胞质具有完全生命力并拥有该种系的全部潜能。普西布兰认为，所有个体发育中都包含的这些潜能一定存在于细胞质的分子中，因此这些分子构成了真正的生机。与此相似，施莱普认为细胞质中一定存在一种超微观结构，这种结构像晶体一样倾向于形成一种固定的形式，并且当其中一部分被切下来时，它可以再完善自己。

不过，如果认可这些说法，那在我们看来施莱普和普西布兰似乎都成了德里施的支持者。一种无机物从溶液中结晶析出时，正如威廉·布拉格爵士告诉我们的，析出的晶体会呈现出一定的形状，因为这种物质的分子具有相应的确定形状。但是我们应该弄清楚的是，细胞质中的结构到底是什么样子的？构成细胞质的是分子还是超分子？当蝾螈幼体的四肢被切除后，其残肢可以继续再生出新的四肢，那么残肢的分子是不是以极小的手指和脚趾的形式存在？此外，当上述残肢被不同程度地切除时，只有缺失的部分能够再生出来，那么我们是否可以认为，在以不同程度切除四肢之前，四肢末端的微缩版就已经存在于四肢的分子之中？如果我们轻信这些荒谬的设想的话，那么对于迪肯记录的现象，即，在实验中将一只蝾螈胚胎的尾芽移植到另一只蝾螈前肢附近的躯体上，结果发育出一只新的前肢，我们又如何解释呢？大概是尾芽的细胞质被“组织化”而产生了成体尾巴的组织。那么这一组织是如何完全改变发育方向而产生前肢的呢？难怪迪肯说，像这种情况，物理和化学解释根本无能为力，我们必须依靠“生物学领域”的概念，这种影响力不在生命物质上，而可能是在其周围的空间中，有可能是以太。

施莱普不同意德里施的理论。他指出，生机假设有时是很愚蠢的，例如当线虫类的卵受到离心力作用时，每个卵并不是形成一个完整的胚胎，而是产生两个不完整的胚胎。但是这一反对观点中潜藏着一个幼稚的概念，即，如果生机真的存在的话，那么它一定是神性智慧的化身。生机并不是全能的——它是对周围环境作出反应的最基本的“努力”，在上述例子中，它就是“质外体”或者被离心力驱逐出卵的无机球状物。

当然，我们认为“组织原”一词的发明者是施佩曼，他很明智地拒绝给这个词赋予任何化学解释。在其令人称奇的蝾螈实验中，他向我们展示了，将一只蝾螈的原肠胚中的胚孔背唇片断移植到另一只蝾螈的原肠胚的侧面，就会改变附近所有细胞的命运，促使它们发育成一条辅助的神经索和深层脊索。本人在此谦恭地为自己辩护，早在1918年我就提出来了同样的概念，并且发表在一篇题为《具有两套水管

and published in a paper which appeared in 1918 entitled "The artificial production of Echinoderm larvae with two water-vascular systems and also of larvae devoid of a water-vascular system" (*Proc. Roy. Soc.*, B, vol. 90). In this paper he showed that when under the stimulus to hypertonic sea-water a second hydrocoele bud was produced in the pluteus, it completely altered the fate of all the tissues near it. It unfortunately did not occur to him to invent the term "organiser."

Of what nature is the influence emitted from the "organiser"? Here again all physical and chemical analogies fail to help us. If the influence were merely a physical or chemical force it would *combine* with the growth-forces of the organised tissue, and what we should observe would be the *resultant* of the two forces. The complete domination of one part by another is not a physical but a vital phenomenon and an instance of Driesch's "regulation".

It would be a fair conclusion to draw from all that has been discovered in the field of embryology to say that in broad outline there are three stages in development, namely: (1) Division of the egg into cells—that is, segmentation; (2) differentiation of these cells so as to form the three primary layers—ectoderm, endoderm, and mesoderm; (3) The action of portions of one layer on the neighbouring parts of other layers so as to form definite organs—that is, the action of organisers.

The ultimate question, however, whence the original organisation of the cytoplasm of the egg is derived, must now be faced. The only answer possible is the nucleus. It is true that, as we have seen, many eggs when ready for fertilisation have an already differentiated cytoplasm. But the cytoplasm of these eggs *when young* is undifferentiated, and during ripening their nuclei are engaged in emissions into the cytoplasm. In particular the nucleolus has been repeatedly observed to become broken into fragments which pass through the nuclear membrane and become dissolved in the cytoplasm. If we take such a specialised egg as that of the Nematode *Ascaris*, Boveri has shown that if it is subjected to centrifugal force *when young*, large portions of the cytoplasm can be shorn away and yet the reduced egg will give rise to a typical embryo. To this conclusion Schleip and Przibram also consent. But it seems to us that a further conclusion follows which they have not clearly envisaged. When differentiation of the cells of the blastula takes place, this must be due to further emissions from the nuclei. But the nuclei in these early stages of development are all alike, and by means of pressure experiments, these nuclei, as Hertwig has put it, may be juggled about like a heap of marbles without altering the result. Moreover, so far as can be judged by the most minute cytological examination, they remain unchanged in their essential make-up throughout the whole of development. So we reach the conception of an *intermittent action of the nuclei on the cytoplasm* giving rise to successive differentiations, that is, stages of development; and as it is by means of these stages that development is directed towards a definite end, if there be an entelechy, we may conclude that the mode of its action is by nuclear emissions. These emissions are the physical correlates of what Uexküll in his "Theoretische Biologie" (1927) calls the "Impulse" to development and the distinguishing of which, he avers, constitutes the utmost limit to which biological analysis can go.

系统的棘皮动物及其缺少水管系统的幼虫的人工培育》(《皇家学会学报》,B 辑,第 90 卷)的论文中。我在该论文中提到,当受到高渗海水的刺激时,长腕幼虫就可以长出第二个水系腔芽体,它的出现会完全改变附近所有组织的命运。但是很遗憾,并不是我发明了“组织原”这个词。

那么“组织原”产生的影响的本质是什么呢?对于这个问题的解释,物理的和化学的推理再一次无能为力。如果这一影响仅仅是一种物理的或化学的力量,那么它就会与有序组织的生长力**结合**起来,我们观察到的就应该是这两种力量作用的**合成**。一部分相对于另一部分来说成为完全主导,这不属于物理现象而是生命现象,这是德里施的“调控”的一个实例。

根据胚胎学领域现已观察到的所有现象,可以很客观地得到如下结论,概括地说,发育可以分为 3 个阶段,即:(1)卵分裂为细胞——即卵裂;(2)这些细胞分化形成 3 个主要的胚层——外胚层、内胚层和中胚层;(3)一个胚层的某些部分作用于邻近的属于其他胚层的部分从而形成特定的器官——即组织原的作用。

然而,最根本的问题是,卵细胞质的最初组成物质来源于何处?这是我们现在必须面对的问题。唯一可能的答案是细胞核。正如我们看到的,事实上许多将要受精的卵子都拥有已分化了的细胞质。但是这些细胞质在卵子**未成熟时**并没有分化,而是在卵子逐渐成熟的过程中,它们的细胞核向细胞质中释放了物质。特别是,在许多研究中都重复观察到核仁分裂成碎片,然后这些碎片穿过核膜,融合到细胞质中。就拿线虫类蛔虫的卵子这样一个特化的卵子来说,博韦里的实验已向我们表明,如果这些卵子**未成熟时**受到离心力的作用,那么大部分细胞质都会丢掉,但是减小了的卵子仍能发育成一个典型的胚胎。施莱普和普西布兰也赞成这一结论。不过我们觉得似乎可以进一步得到他们还没有想清楚的某些结论。囊胚细胞的分化一定是由于细胞核又释放出了某些物质。但是处于这些早期发育阶段的细胞核都是很相似的。赫特维希通过压力实验发现,这些细胞核可能就像被拨弄的一堆大理石一样并没有发生任何变化。此外,从大多数细微的细胞学检查结果可以判断,在整个发育过程中,这些细胞核的基本组成都没有发生变化。所以我们想到,**细胞核对细胞质的间歇性作用**引起了细胞的连续分化,即各个发育阶段。发育过程正是通过这些阶段逐渐走向一个确定的结果,所以如果存在生机的话,我们可以断定它是通过细胞核释放出的物质来发挥作用的。这些释放物就是于克斯屈尔在其《理论生物学》(1927 年)中称为发育“推动力”的相关物质。于克斯屈尔断言,对这些释放物的区分是生物分析所能达到的最大极限。

Comparative embryology, however, can go further, and Schleip rightly insists that experimental embryology ought to be comparative. These embryonic stages are soon discovered to be merely smudged and simplified forms of larval stages which in allied forms lead a free life in the open, seeking their own food and combating their own enemies. These larval forms in turn are seen to be nothing but modified and simplified editions of adult forms in the past history of the race. Therefore, in the last resort, development is found to be due to the successive coming to the surface of a series of racial memories, and the entelechy might be defined as a "bundle" of such memories.

The so-called Mendelian "genes", however, constitute a problem for the embryologist; for the conception of the hereditary make-up which they induce in the minds of geneticists is totally at variance with that which the embryologist draws from the study of development. Schleip and Przibram struggle valiantly to reconcile the two conceptions and fail. Dürken alone boldly questions the validity of the whole conception of the genes and points out how much it is purely arbitrary and theoretical. If the results of a crossing experiment agree with expectation based on the ordinary Mendelian rules, then it proves the reality of genes; if the results do not agree, the geneticist denies that it disproves them, because he immediately postulates the action of an undiscovered "gene" which complicates the result. The real answer to the conundrum was given by Johannsen, when, in his latest publication, deploring the damage and confusion of thought caused by the invention of the word "gene", he states that it represents a mere superficial disturbance of the chromosomes and gives no insight into the real nature of heredity. Even Przibram points out that X-rays will produce "unzählige" mutations, and that there is no correlation between the rays and the nature of the mutation. With these remarks we thoroughly agree.

(**126**, 639-643;1930)

不过，比较胚胎学还可以走得更远，施莱普始终坚持实验胚胎学应该引入比较的方法。人们很快发现这些胚胎发育的阶段仅仅是幼体阶段被混杂和简化后的形式，各个幼体阶段的联合作用最终形成了可以在野外生存的自由生命，它们可以自己寻找食物并与敌人战斗。反过来，这些幼虫形式又仅仅被看作是该种系在过去的发育史中成年形式的修饰简化版本。因此，最终我们会得出发育是一系列种族记忆的逐次苏醒，而生机可以被定义成这些记忆中的“一束”。

但是所谓的孟德尔式“基因”给胚胎学家带来了一个难题，因为遗传学家们推导出的遗传组成的概念与胚胎学家从发育研究中得出的完全不同。施莱普和普西布兰大胆地尝试去调和这两种概念，但最终失败了。迪肯独自勇敢地质疑了基因整个概念的有效性，并且指出这一概念是非常主观和理论性的。如果有交叉实验结果与基于普通孟德尔法则预期的结果一致的话，那么就能证实基因的真实性；如果不一致，那么遗传学家就不会承认该结果能成为他们理论的反证，因为他可以马上假定出一个作用是使该结果变得更加复杂的尚未被发现的“基因”。约翰森给出了这一谜底的真正答案，在他的最后一部著作中，他谴责了“基因”一词的发明给人们造成的思维混乱和困惑，他指出“基因”这个词只不过是给染色体的概念带来了浅薄的干扰，而对于洞察遗传的真正本质并无帮助。甚至普西布兰也指出，X 射线能产生“无数”突变，而射线和突变的本质之间没有任何相关性。我们完全同意这些观点。

（刘皓芳 翻译；刘京国 审稿）

Natural Selection Intensity as a Function of Mortality Rate

J. B. S. Haldane

Editor's Note

J. B. S. Haldane was a prime mover behind the "modern synthesis" in evolutionary biology. This fused Darwin's ideas about selection and Mendel's insights into how traits pass from parents to offspring into a mathematical description of the genetic makeup of populations and how it changes. At the time, some believed these ideas to be antithetical to one another. *Nature* previously published a commentary on Haldane's note by E. W. MacBride, who concluded that he was "convinced that Mendelism has nothing to do with evolution." Here Haldane gives a specific example of his advocacy of mathematical analysis in evolutionary biology, and its superiority over verbal arguments, disproving the notion that selection is limited to the stage of highest mortality.

IN *Nature* of May 31, Prof. Salisbury points out that most of the mortality among higher plants occurs at the seedling stage, and concludes that natural selection is mainly confined to this stage. I believe, however, that this apparently obvious conclusion is fallacious, for the following reason:

Consider two pure lines A and B originally present in equal numbers, and with a common measurable character, normally distributed according to Gauss's law in each group. Let the standard deviations of the character be equal in each group, but its mean value in group A slightly larger than that in group B. Johanssen's beans furnish examples of this type of distribution. Now let selection act so as to kill off all individuals in which the character falls below a certain value. I think that this type of artificial selection furnishes a fair parallel to natural selection, in which chance commonly plays a larger part than heritable differences. Let x be the proportion of individuals eliminated to survivors, and $1+y$ the proportion of A to B among the survivors, so that x measures the intensity of competition, y that of selection.

Then when x is small y is roughly proportional to it. Thus when x increases from 10^{-4} to 10^{-1}, y increases 200 times. But when x is large y becomes proportional to $\sqrt{\log x}$. In consequence y only increases 9 times when x increases from 1 to 10^{12}, and is only doubled when x increases from 1 to 1,800. In other words, when more than 50 percent of the population is eliminated by natural selection, the additional number eliminated makes little difference to the intensity of selection. The theory, which I hope to publish shortly, has been extended to cover cases where the standard deviations differ, and also where populations consist of many genotypes. In general y changes its sign with x, but when x is large y never increases more rapidly than $\log x$.

自然选择强度与死亡率的关系

霍尔丹

编者按

在进化生物学领域，霍尔丹是“现代综合论”的先驱。现代综合论融合了达尔文的自然选择学说和孟德尔关于性状如何从亲代传递到子代的观点，并用数学方法描述了人类的基因构成和基因如何变化。当时，一些人认为达尔文的理论和孟德尔的观点是相互对立的。《自然》早先曾发表了麦克布赖德对霍尔丹的一篇论文的评论，麦克布赖德在评论中声称，他“对孟德尔的遗传学说对于进化毫无意义这一点深信不疑”。在这篇文章中，霍尔丹为了反驳自然选择只局限于最高死亡率阶段的观点，他给出了一个具体的例子。这个例子表明，他支持在进化生物学中采用数学分析的方法，并认为这种方法比文字说明更好。

在 5 月 31 日的《自然》中，索尔兹伯里教授指出高等植物中绝大多数的死亡发生在幼苗时期，由此他断定自然选择主要发生在这个阶段。我认为，这个看似很显然的结论是不正确的，主要原因如下：

假定纯系 A 和纯系 B 在初始状态时具有相同的个体数并都具有某一相同的可测性状，并且这一可测性状在每组群体中都服从高斯分布。假定每组群体中性状分布具有相同的标准差，而 A 系群体的均值比 B 系群体的均值略高。约翰森豆正是能够满足这种分布类型的实例。现在让选择起作用，杀死那些性状低于某一给定值的所有个体。我认为这样的人为选择是非常类似于自然选择的，在自然选择中通常是偶然事件所起的作用比遗传差异更大。用 x 表示死亡个体与存活个体的比例，$1+y$ 表示纯系 A 的存活个体与纯系 B 的存活个体的比例。那么 x 表示的就是竞争的强度，而 y 则表示选择的强度。

这样，当 x 较小时 y 与 x 大致是成比例的。当 x 从 10^{-4} 增加到 10^{-1} 时，y 增加了 200 倍。但是当 x 较大时，y 则变成与 $\sqrt{\log x}$ 成比例。其结果是当 x 从 1 增加到 10^{12} 时，y 仅增加了 9 倍，当 x 从 1 增加到 1,800 时，y 仅翻了一番。换言之，当超过 50% 的个体在自然选择中被淘汰时，再淘汰更多个体对应的自然选择强度的改变是很小的。这一理论已经扩展到纯系间性状分布的标准差不同的情况，以及群体中包含多种基因型的情况。我希望能很快发表这一理论。总体上讲，y 的数值随 x 的改变而改变，但是当 x 较大时 y 增加的速度不会比 $\log x$ 更快。

Careful mathematical analysis seems to disclose the extraordinary subtlety of the natural selection principle, and merely verbal arguments concerning it are likely to conceal serious fallacies.

(**126**, 883; 1930)

J. B. S. Haldane: John Innes Horticultural Institution, Merton Park, London, S.W. 19, Nov. 1.

看起来，细致的数学分析才能够解析自然选择原理的精细之处，而文字说明中很可能隐藏着严重的谬误。

（刘晓辉 翻译；刘京国 审稿）

The Ether and Relativity

J. H. Jeans

Editor's Note

Here the English physicist James Hopwood Jeans responds to a letter from Oliver Lodge criticizing Jeans' recent claim that the laws of the universe would only be penetrated by the use of mathematics. Jeans affirms his belief that "No one except a mathematician need ever hope fully to understand those branches of science which try to unravel the fundamental nature of the universe—the theory of relativity, the theory of quanta and the wave mechanics." Lodge suggested that the universe might ultimately turn out to have been created or designed on aesthetic, rather than mathematical lines. If so, one might expect artists, not mathematicians, to be best suited to fundamental science. But Jeans notices no such aptitude in his artist friends.

I obviously must not ask for space to discuss all the points raised in Sir Oliver Lodge's interesting letter in *Nature* of Nov. 22, and so will attempt no reply to those parts of it which run counter to the ordinarily accepted theory of relativity. For I am sure nothing I could say would change his views here. But I am naturally distressed at his thinking I have quoted him with a "kind of unfairness", and should be much more so, had I not an absolutely clear conscience and, as I think, the facts on my side.

In the part of my book to which Sir Oliver objects most, I explained how the hard facts of experiment left no room for the old material ether of the nineteenth century. (Sir Oliver explains in *Nature* that he, too, has abandoned this old material ether.) I then quoted Sir Oliver's own words to the effect that many people prefer to call the ether "space", and his sentence, "The term used does not matter much."

I took these last words to mean, not merely that the ether by any other name would smell as sweet to Sir Oliver, but also that he thought that "space" was really a very suitable name for the new ether. He now explains he was willing to call the ether "space", "for the sake of peace and agreement". If I had thought it was only *qua* pacifist and not *qua* scientist that he was willing to call the ether "space", I naturally would not have quoted him as I did, and will, of course, if he wishes, delete the quotation from future editions of my book. But I did not know his reasons at the time, and so cannot feel that I acted unfairly in quoting his own words verbatim from an Encyclopaedia article.

Against this, I seem to find Sir Oliver attributing things to me that, to the best of my belief, I did not say at all, as, for example, that a mathematician alone can hope to understand the universe. My own words were (p. 128):

以太与相对论

金斯

编者按

英国物理学家詹姆斯·霍普伍德·金斯写这篇文章是为了回应奥利弗·洛奇的一篇快报，奥利弗·洛奇在该快报文章中批评了金斯最近提出的观点，即宇宙中的定律只有用数学方法才能解释清楚。金斯坚持认为“除了数学家以外，没有人能完全理解那些试图揭示宇宙基本性质的科学分支——相对论、量子理论和波动力学。”洛奇认为宇宙的创造和设计最终有可能是按照美学原则而非数学原理进行的。如果事实真的如此，那么艺术家应该比数学家更适合研究基础科学。但金斯在他的艺术界朋友中没有发现这种特别的倾向。

显然，我没有必要要求一个很大的版面来讨论奥利弗·洛奇爵士发表在 11 月 22 日《自然》上的那篇引人注意的快报中提到的全部要点，因而也不会试图回复该快报文章中那些与人们普遍接受的相对论不相符合的内容。因为我知道无论我在这里说什么也不能改变他的观点。但是，如果他认为我是以“一种不公正的方式”引述他的话，我自然会感到不安，而且，如果我不曾拥有一个绝对清晰的意识并认为事实一定站在我这一边的信念，我将会感到更加不安。

在我的书中，奥利弗爵士最不赞同的部分是我关于铁一般的实验事实如何使得 19 世纪的旧的物质性以太再无容身之地的解释。（奥利弗爵士在《自然》上曾解释说他也已经抛弃了这种旧的物质性以太。）接着我引述了奥利弗爵士本人针对很多人更愿意称以太为“空间”这一现象所说的话，他的原话是，“用什么样的术语关系不大。”

我引用最后这句话是想说明，不仅以任何其他方式命名的以太一样合乎奥利弗爵士的胃口，而且他认为“空间”对于新的以太来说确实是一个很合适的名称。他现在解释说自己乐于称以太为“空间”，“为的是息事宁人和意见统一”。如果我以前认为他只是作为和平主义者而不是作为科学家才乐于称以太为“空间”，自然就不会像我之前那样引述他的话，当然，如果他希望的话，我将在我那本书的新版中删去那段引文。但当时我并不知道他是这样想的，所以我不知道我从一篇百科全书的文章中一字不差地引述他的原话是不公正的做法。

与此相反，我可以非常负责任地说，我发现奥利弗爵士把我根本没有说过的话强加在我身上，比如，单凭数学家就能理解宇宙。我的原话是这样的（第 128 页）：

"No one except a mathematician need ever hope fully to understand those branches of science which try to unravel the fundamental nature of the universe—the theory of relativity, the theory of quanta and the wave-mechanics."

This I stick to, having had much experience of trying to explain these branches of science to non-mathematicians. In the same way, if the material universe had been created or designed on aesthetic lines—a possibility which others have contemplated besides Sir Oliver Lodge—then artists ought to be specially apt at these fundamental branches of science. I have noticed no such special aptitude on the part of my artist friends. Incidentally, I think this answers the question propounded in the News and Views columns of *Nature* of Nov. 8, which was, in brief:—If the universe were fundamentally aesthetic, how could an aesthetic description of it possibly be given by the methods of physics? Surely the answer is that if the objective universe were fundamentally aesthetic in its design, physics (defined as the science which explores the fundamental nature of the objective universe) would be very different from what it actually is; it would be a *milieu* for artistic emotion and not for mathematical symbols. Of course, we may come to this yet, but if so, modern physics would seem rather to have lost the scent.

However, I am glad to be able to agree with much that Sir Oliver writes, including the quotations from Einstein which he seems to bring up as heavy artillery to give me the final *coup de grace*:—"In this sense, therefore, there exists an ether", and so on. On this I would comment that nothing in science seems to exist any more in the good old-fashioned sense—that is, without qualifications; and modern physics always answers the question, "To be or not to be?" by some hesitation compromise, ambiguity, or evasion. All this, to my mind, gives strong support to my main thesis.

(**126**, 877; 1930)

J. H. Jeans: Cleveland Lodge, Dorking, Nov. 23.

“除了数学家以外，没有人能完全理解那些试图揭示宇宙基本性质的科学分支——相对论、量子理论和波动力学。”

我仍然坚持这一点，因为在试图向非数学家解释上述科学分支方面我已经积累了许多经验。同样地，如果物质性宇宙的创造和设计是按美学原则进行的——除了奥利弗·洛奇爵士之外还有其他人也曾考虑过这种可能性，那么艺术家应该特别容易理解这些基础的科学分支。迄今为止，在我的艺术家朋友中我并没有发现这种特别的倾向。顺便提一句，我认为这解答了 11 月 8 日《自然》的“新闻与视点”栏目中提出的问题，简单地说就是：如果宇宙基本上是美学的，那么物理学方法又怎么能给出一个关于宇宙的美学描述呢？答案只能是这样的，如果客观的宇宙基本上是依美学观点设计的，那么物理学（定义为研究客观宇宙基本性质的科学）将会与它实际的样子大为不同；这种依美学观点设计的**宇宙环境**将适合于艺术情感而非数学符号。当然，我们可以这样做，但是果真如此的话，现代物理学似乎会迷失方向。

不过，我很高兴在很多方面与奥利弗爵士意见一致，包括他引用的爱因斯坦的话——“所以就这个意义而言，是有某种以太存在”等，看起来他要把这些当作重炮给我最终的**致命一击**。对此我的评论是，在科学中没有什么东西能以旧有的形式安然存在，这是绝对的；现代物理学经常会以某种迟疑不决的折中、模棱两可或者遁词来回答“存在还是不存在？”的问题。在我看来，所有这些都强有力地支持了我的主要观点。

（王耀杨 翻译；张元仲 审稿）

The X-Ray Interpretation of the Structure and Elastic Properties of Hair Keratin

W. T. Astbury and H. J. Woods

Editor's Note

William Astbury was an X-ray crystallographer based in a university department supported by funds from the wool and leather industries of the county of Yorkshire in northern England. He was one of the first to use X-ray analysis for the study of the structure of complicated polymer molecules; during his career in Leeds, he also studied the structure of deoxyribonucleic acid (DNA) and even arrived at the correct spacing between successive units in the polymer. His biggest handicap was that he made enemies easily. Here he describes the X-ray analysis of a fibrous "structural" protein, keratin, the main component of hair. Such proteins were Astbury's forte.

RECENT experiments,[1] carried out for the most part on human hair and various types of sheep's wool, have shown that animal hairs can give rise to two X-ray "fibre photographs" according as the hairs are unstretched or stretched, and that the change from one photograph to the other corresponds to a reversible transformation between two forms of the keratin complex. Hair rapidly recovers its original length on wetting after removal of the stretching force, and either of the two possible photographs may be produced at will an indefinite number of times. Both are typical "fibre photographs" in the sense that they arise from crystallites or pseudo-crystallites of which the average length along the fibre axis is much larger than the average thickness, and which are almost certainly built up in a rather imperfect manner of molecular chains—what Meyer and Mark[2] have called *Hauptvalenzketten*—running roughly parallel to the fibre axis.

Hair photographs are much poorer in reflections than are those of vegetable fibres, but it is clear that the α-keratin, that is, the unstretched form, is characterised by a very marked periodicity of 5.15 Å along the fibre axis and two chief side-spacings of 9.8 Å and 27 Å (? mean value), respectively; while the β-keratin, the stretched form, shows a strong periodicity of 3.4 Å along the fibre axis in combination with side-spacings of 9.8 Å and 4.65 Å, of which the latter is at least a second-order reflection. The β-form becomes apparent in the photographs at extensions of about 25 percent and continues to increase, while the α-form fades, up to the breaking extension in cold water, which is rarely above 70 percent. Under the action of steam, hair may be stretched perhaps still another 30 percent, but no other fundamentally new X-ray photograph is produced. The question is thus immediately raised as to what is the significance of a crystallographically measurable transformation interpolated between two regions of similar extent where no change of a comparable order, so far as X-ray photographs show, can be detected.

X 射线衍射法解析毛发角蛋白的结构与弹性

阿斯特伯里，伍兹

编者按

威廉·阿斯特伯里是一位在大学工作的 X 射线晶体学家，他所在的系受到了北英格兰约克郡羊毛和皮革业多家企业提供的研究基金的支持。他是利用 X 射线解析复杂多聚体分子结构的先驱之一。在利兹工作期间，他还研究了脱氧核糖核酸 (DNA) 的结构，甚至得出了这一聚合物中相邻单体分子之间的正确距离。对他来说最不利的是他太容易树敌。在这篇文章中，他描述了对角蛋白这种纤维状“构造的”蛋白质的 X 射线衍射分析。角蛋白是毛发的主要组成部分。阿斯特伯里非常擅长于研究这种蛋白。

最近的一些主要针对人类毛发和各种类型羊毛的实验[1]表明，动物毛发在拉伸状态和非拉伸状态下可以产生两种不同的 X 射线“纤维衍射图”，而且从一种衍射图向另一种衍射图的转变对应于角蛋白复合体的两种形式之间的可逆转变。在撤去外界拉力后，浸湿的毛发会很快恢复到原来的长度，因此可以无限次地重复得到这两种衍射图中的任意一种。这两种衍射图都来自于晶体或伪晶体，从这一意义上来说，这两种衍射图都是典型的“纤维衍射图”。在相应的晶体和伪晶体中，沿中心轴方向的平均长度比平均厚度大得多，这些晶体和伪晶体极有可能是由一种非常不完整的、几乎平行于中心轴的分子链（被迈尔和马克[2]称为**主分子链**）搭建起来的。

与植物纤维的衍射图相比，毛发衍射图中的反射线非常少，但很清楚的是，非拉伸状态的 α 角蛋白明显表现出了沿中心轴 5.15 Å 的周期性以及宽度分别为 9.8 Å 和 27 Å（均值 ?）的两个主要的侧向间隔。处于伸展状态的 β 角蛋白则表现出沿中心轴 3.4 Å 的很强的周期性以及宽度分别为 9.8 Å 和 4.65 Å 的两个侧向间隔，其中后一个侧向间隔至少是二次反射。当毛发在冷水中被拉伸 25% 时，衍射图中开始出现 β 类型，随着拉伸幅度的增大 β 类型越来越多，与此同时，α 类型逐渐消失，直到毛发被拉断（在冷水中毛发的拉伸幅度很少超过 70%）。在蒸汽的作用下，毛发也许还能再被拉伸 30%，但即便这样也不会出现本质上全新的 X 射线衍射图。这样立刻就提出了一个问题，就 X 射线图而言，在拉伸程度接近、没有发生相对次序改变的两个区域之间，晶体学上可测量的相互转变的意义是什么呢？

The elastic properties of hair present a complex problem in molecular mechanics which up to the present has resisted all efforts at a satisfactory explanation, either qualitative or quantitative. Space forbids a detailed discussion here of the almost bewildering series of changes that have been observed, and we shall merely state what now, after a close examination of the X-ray and general physical and chemical data, appear to be the most fundamental.

(1) Hair in cold water may be stretched about twice as far, and hair in steam about three times as far, as hair which is perfectly dry. (2) On the average, hair may be stretched (in steam) to about twice its original length without rupture. (3) By suitable treatment with steam the discontinuities in the load/extension curve may be permanently smoothed out, the original zero is lost, so that the hair may be even contracted by as much as one-third of its original length, and elasticity of form may be demonstrated *in cold water* over a range of extensions from −30 percent to +100 percent. (4) The elastic behaviour in steam is complicated by "temporary setting" of the elastic chain and ultimately by a "permanent setting" of that part which gives rise to the fibre photograph. (5) That part of the elastic chain which is revealed by X-rays acts *in series* with the preceding and subsequent changes.

On the basis of these properties and the X-ray data, it is now possible to put forward a "skeleton" of the keratin complex which gives a quantitative interpretation of the fundamentals, and may later lead to a correct solution of the details. The skeleton model is shown in Fig. 1. It is simply a peptide chain folded into a series of hexagons, with the precise nature of the side links as yet undetermined. Its most important features may be summarised as follows:—(1) It explains why the main periodicity (5.15 Å) in unstretched hair corresponds so closely with that which has already been observed in cellulose, chitin, etc., in which the hexagonal glucose residues are linked together by oxygens. (2) When once the side links are freed, it permits an extension from 5.15 Å to a simple zigzag chain

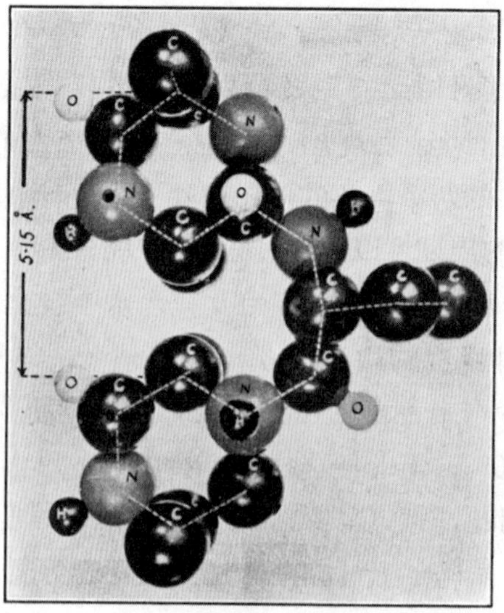

Fig. 1

毛发弹性的分子机制是个复杂的问题，无论是定性的还是定量的，至今为止还没有得到任何令人满意的结论。篇幅所限，我们这里将不再详细讨论曾观测到的纷繁复杂的一系列变化，只介绍对 X 射线结果和基本的物理和化学数据进行详细考察后得到的一些看起来最重要的结论。

(1) 和完全干燥的毛发相比，在冷水中毛发可以被拉伸到大约原长的两倍，而在蒸汽中可以被拉伸到大约原长的三倍。(2) 平均而言，毛发（在蒸汽中）可以被拉伸到原长的两倍而不断裂。(3) 适当的蒸汽处理可以永久地消除毛发载荷–伸长曲线的不连续性，原来的零点就没有了，这样毛发甚至可以收缩到其原长的 2/3，而**在冷水中**其所表现的弹性介于收缩 30% 到伸长 100% 之间。(4) 在蒸汽中毛发的弹性行为由于存在弹性链的“临时形态”而变得复杂，但最后变成“固定形态”，从而产生纤维衍射图。(5) X 射线衍射图显示，部分弹性链**连续地**进行先前和随后的变化。

在这些特性以及 X 射线衍射数据的基础上，我们现在可以提出一种角蛋白复合体的“骨架”，这可以定量地解释一些基本问题，将来或许还能引导对细节问题的正确解答。骨架模型如图 1 所示。该骨架模型只是简单地展示了折叠成几个六边形的一条多肽链，其中侧链的准确性质目前还没有被测定。该模型最重要的特征可以总结如下：(1) 它解释了为什么非拉伸状态下毛发中的主要周期 (5.15 Å) 与在纤维素、几丁质等物质中观察到的周期十分相近，这是因为在这些物质中六边形

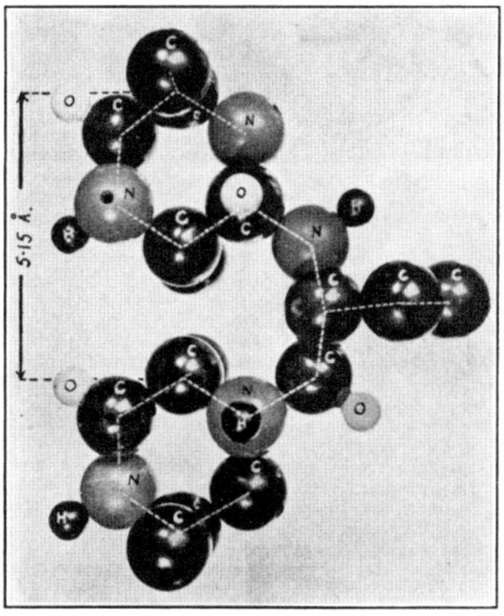

图 1

of length 3 × 3.4 Å, that is, 98 percent, and also allows for possible contraction below the original length, without altering the inter-atomic distances and the angles between the bonds. (3) It explains why natural silk does not show the long-range elasticity of hair, since it is for the most part already in the extended state,[3] with a chief periodicity of 3.5 Å. We may now hope to understand why it is that the photographs of β-hair and silk are so much alike. (4) It gives a first picture of the "lubricating action" of water and steam on the chain, since X-rays show that the direction of attack is perpendicular to the hexagons and that this spacing remains unchanged on stretching. Furthermore, it now seems clear that the new spacing, 4.65 Å, is related to the old by the equation $27/(3\times4.65) = 3\times3.4/5.15$ (very nearly), that is, the transformation elongation takes place directly at the expense of the larger of the two side-spacings. In the particular arrangement of the hexagons shown in the model, the side chains occur in pairs on each face, and it may well be that the action of water is the opening-up of an internal anhydride between such adjacent side chains. (5) The chain being built up of a succession of ring systems stabilised and linked together in some way by side chains of the various amino-acids, we have here an explanation of the well-known resistance of the keratins to solvents and enzyme action. In addition, each hexagon is effectively a diketo-piperazine ring, an interesting point in view of the evidence which has been brought forward by Abderhalden and Komm[4] that such groups pre-exist in the protein molecule. It may also throw light on the stimulating researches of Troensegaard.[5] (6) There are three principal ways of constructing the model, according to which group lies at the apex of a hexagon. It thus affords an explanation of the apportioning of a transformation involving a 100 percent elongation into three approximately equal regions which may be opened up in turn under the influence of water and temperature and other reagents. The modification shown in the model must be ascribed to the crystalline phase, since it would, alone of the three, be expected to give rise to a strong reflection at 5.15 Å, as in the α-photograph.

A detailed account of the above work will be published shortly.

(**126**, 913-914; 1930)

W. T. Astbury, H. J. Woods: Textile Physics Laboratory, The University, Leeds, Nov. 15.

References:

1. W. T. Astbury, *J. Soc. Chem. Ind.*, **49**, 441; 1930.

2. Meyer and Mark, "Der Aufbau der hochpolymeren organischen Naturstoffe".

3. Meyer and Mark, *Berichte*, **61**, 1932; 1928.

4. Abderhalden and Komm, *Z. Physiol. Chem.*, **139**, 181; 1924.

5. Troensegaard, *Z. physiol. Chem.*, **127**, 137; 1923.

的葡萄糖残基可以通过氧原子相互连接起来。(2) 当侧向的连接解开后，模型中长为 5.15 Å 的主链单元就会伸展成 3×3.4 Å 的简单“之”字形长链，也就是伸展了 98%，同样也可能会因为收缩而使长链短于初始长度，但不论伸展还是收缩，原子间距和键–键之间的角度都不会发生改变。(3) 它解释了为什么自然丝的伸缩性没有毛发那么大，因为大多数情况下自然状态的丝已经处于主周期为 3.5 Å 的伸展状态。[3] 这样我们就能够理解为什么 β 型毛发的衍射图与丝蛋白的衍射图十分相似。(4) 这一模型首次给出了水和蒸汽对主链的“润滑作用”的图片。X 射线衍射结果显示，水分子攻击主链的方向是垂直于六边形平面的，而且在主链拉伸过程中六边形中各原子之间的空间间隔没有发生改变。另外，比较明确的是，拉伸后新的侧向间隔 (4.65 Å) 与原来的侧向间距之间的关系符合方程 $27/(3\times4.65)=3\times3.4/5.15$ (非常近似)，这就是说，主链的伸长是以两个侧向间隔的减小为直接代价的。在模型中显示的六边形的某些特殊排列中，在六边形平面任何一边的侧链总是成对出现，水分子的作用很可能就是打开这些相邻侧链内部的酐键。(5) 由连续的环形系统构成的主链是通过各种不同的氨基酸的侧链以某种方式保持稳定并连接在一起的，这样我们就可以解释众所周知的角蛋白对溶剂作用和酶作用的抗性。另外，每个六边形都是一个有效的二酮哌嗪环，这一点很有趣，考虑到阿布德哈尔登和科姆 [4] 曾提出证据说这种基团结构预先就存在于蛋白分子中，那么这个有趣的特点也许会对特森加德的研究工作有所启示。[5] (6) 根据位于六边形顶点上的基团的不同，有 3 种主要的模型构建方法。这样就可以把长链 100%的伸长分配到 3 个近似相等的区域上。在水、温度或其他试剂的影响下，这些区域会依次伸展开。对模型的这些修正要归因于晶相结构的特点，因为可以预期在 3 种结构中只有晶相结构会给出如 α 衍射图所示的在 5.15 Å 处出现的强反射。

关于以上工作的详细说明将在不久之后发表。

(高如丽 翻译；刘京国 审稿)

Embryology and Evolution

Editor's Note

In 1930 Irish-born zoologist Ernest William MacBride had exposed his Lamarckian leanings in a review of several books on experimental embryology, arguing that the genes-eye view of the world was "totally at variance with that which the embryologist draws from the study of development." Here, embryologist G. L. Purser conjures up a nice analogy that he thinks illuminates the role that genes do (and do not) play in embryonic development or "epigenesis". This letter contains some prescient reflections about the mutual interdependence of genes and their immediate environment. However, Purser goes too far in claiming a new proof for the inheritance of acquired characteristics when he suggests "that the environment is in some way responsible for the appearance of the gene".

I have read with much interest Prof. MacBride's review entitled "The Problem of Epigenesis", and I should like to make a few remarks upon what he says at the end. First of all, I wonder if the following analogy will help him, as it has helped me, to reconcile the conceptions of the geneticist with those of the embryologist. In a modern motor works the cars, so I understand, move along a track past a series of workmen, each of whom has one particular job to do, which is related to what has already been done and also to what is going to be done afterwards. Now if we imagine that all the parts and materials which are going to make up the finished car represent the substances in the developing embryo and that the workmen are the genes, we have an analogy which can be carried surprisingly far. Not only will it give us a picture of normal development, but we can see, by altering one of the parts, how a variation may occur; by altering a workman, how "sports" may arise; and, by adding a new workman with a new job, how progressive evolution may take place.

There is no need for me to occupy space in working the analogy out, for anyone can do it for himself: what is more important is to point out where the analogy fails. A motor-car is adapted for life on the road, and, until it is completed, it has, for all practical purposes, no environment at all comparable with that which bears upon an embryo throughout its development. So whereas a feature of a car is simply due to the action of the workman on the materials, a feature of an animal is the result of the combined action of the genes and of the environment upon the materials of the embryo. Genes without the appropriate materials can produce nothing; genes with the appropriate materials can only produce a partially developed structure; but genes with the appropriate materials and environment can produce the fully developed functional character. Hence it is that in the development of the frog, for example, the gill-clefts, etc., are full developed, whereas in the Amniota, with the radical change in the environment of the early stages, such structures are only

胚胎学与进化

编者按

1930年，出生于爱尔兰的动物学家欧内斯特·威廉·麦克布赖德在评论几本关于实验胚胎学的著作时，表示了自己对拉马克学说的认同。他认为从基因的角度看到的世界“与胚胎学家从发育研究中得出的完全不同。”在这里，胚胎学家珀泽想出了一个很好的类比，他认为这样就可以解释基因在胚胎发育或“渐成论”中是否起作用。这篇文章在基因与周围环境的相互依赖关系这个问题上提出了一些很有远见的观点。但是珀泽在阐述获得性遗传假说的新证据时说，“环境在某种程度上决定了基因的表现形式”，这就过于偏激了。

我怀着极大的兴趣读完了麦克布赖德教授那篇名为《关于渐成论的问题》的书评，看完后我想对他的书评结尾处的内容发表一下我的看法。首先，我想知道如下的类比是否会像其帮助过我一样，也能够帮助他化解遗传学家的观念与胚胎学家的观念之间的分歧。我是这样理解的，现代汽车工业中，小汽车是沿着一条由一系列工人组成的生产线移动的，生产线上的每个人都有自己特定的工作要做，每个人的工作都与已经完成的工作以及之后将要进行的工作有关。这时如果我们想象组成成品车辆的所有零件和材料代表正在发育的胚胎里的物质，工人代表基因，那么我们就有了一个蕴含着深远意义的类比。这一类比不仅可以给我们描绘出一幅正常发育的图像，而且从中我们可以看到：变异是如何通过改变众多零件中的一个而实现的；“变种”是如何通过更换一名工人而产生的；以及渐进演化是如何通过增加一名从事新工作的新工人而发生的。

我没有必要浪费版面在这里推演这一类比，因为任何人都可以独自完成，但更重要的是指出这种类比在哪些地方不适用。一辆汽车是适于在路面上奔跑的，并且在生产出来后它就已经适于各种实用目的，而在胚胎生长发育的整个过程中都没有任何环境可以与此相类比。因此汽车的特征仅仅取决于工人对原材料进行的处理，而动物的特征则是基因和环境对胚胎物质共同作用的结果。如果没有适当的胚胎物质，那么基因什么也产生不了；如果只有适当的胚胎物质，那么基因只能产生部分发育的结构；只有既有适当的胚胎物质，又有环境时，基因才能产生充分发育的有功能的结构。因此，如腮裂等结构，在青蛙的发育过程中是充分发育的，在羊膜动物中则因为在发育早期遭遇到剧烈的环境变化而只是部分发育的。引用麦克布赖德

partially developed and the stages, to quote Prof. MacBride, are smudged.

Looked at from this point of view, two other conclusions of great importance are unavoidable. The first is that the recapitulation of an ancestral stage of the evolution of an animal, as distinct from the repetition of an ancestral character, will only occur when the early stage of development is passed in the same environment as that of the ancestor, which environment is different from that of the present-day adult. Only under such conditions will the genes responsible for the adult ancestral characters give rise to them all together without any great admixture of other features; though it must always be borne in mind that such stages in the life history, being larvae, may evolve on their own account and, therefore, may have features which the ancestor never had. In parenthesis, I should just like to add here that, so far as I know, a larva has never been properly defined: such a definition would be "A free-living stage in an animal's life history which fends for itself and possesses certain characters which it has to lose before it can become a young adult": the possession of *positive* characters distinguishes a larva, not its lack of adult ones.

The other conclusion is reached thus. The appearance of a functional feature is dependent, as we have seen, upon the interaction of three things: the materials of the embryo, the genes, and the environment. Now the facts of Mendelian inheritance give clear evidence that there need be no change in the materials of an embryo for a new gene to modify the form, so, in discussing the origin of a new feature, there is no need to consider a change in the materials as one of the essential factors. The fortuitous appearance of a gene without the appropriate environment would produce a partially developed character, but, in actual experience, we do not find features in a partially developed condition which *have never been functional* at any period in the history of the race. So the genes must, in actual fact, only arise after the suitable environment is present; and the only conclusion to be drawn from that is that there is a causal relation between the two; that is, that the environment is in some way responsible for the appearance of the gene, which is surely nothing more or less than the basis of a new proof of the inheritance of acquired characters.

G. L. Purser

* * *

I have read with interest Mr. Purser's thoughtful letter on the subject of my review. If he will substitute the term "race-memory" for "gene", we shall not be far apart. But the gene of the Mendelian stands out as something that is never functional. "No one," said the late Sir Archdall Reid, "ever heard of a useful gene." When one takes into consideration the fact that the Mendelian genes in *Drosophila* have been shown to increase in their damaging effect on the viability of the organism in proportion to the structural change which they involve, and when further it is discovered that genes can be artificially produced by irradiating insect eggs with X-rays—a process which kills most of the eggs—one is driven

教授的话说，这些阶段都是遗留下的痕迹。

从这个观点来看，将会不可避免地得出另外两个非常重要的结论。第一个是，只有当胚胎在发育早期阶段所处的环境与该种生物的祖先所处环境一样时，才会出现对其祖先进化过程的重演。这种环境与现在的成体所处的环境是不同的，这种重演与重现祖先特征是完全不同的概念。只有在这种情况下，那些决定祖先成体特征的基因才完全产生效果，而不与其他特征发生大规模的混合；但是我们必须时刻牢记，生命过程中的这些阶段（即幼虫）都是自行进化的，因此它们可能具有祖先从不具有的特征。另外，我想在此补充说明的是，据我所知，幼虫这一概念从未被恰当地定义过。如下描述可能是一个比较恰当的定义，“在动物生命史中的一个自由生活的阶段，在此阶段中它们自己照料自己并具有一定的特征，但一旦它们成长为年轻的成体，这些特征就会消失。”幼虫拥有**初级的**特征而成虫没有，根据这一点能够将它们区别开来。

第二个结论是按照如下所述得出的。正如我们看到的一样，功能性特征的出现依赖于 3 个条件的相互作用：胚胎物质、基因和环境。现在，孟德尔的遗传结果已经给出了明确的证据，表明一个新基因形式发生改变时胚胎物质并没有发生改变，所以在讨论新特征的起源时，就没有必要将胚胎物质的变化作为基本因素之一来考虑。偶尔会出现某个没有适当环境的基因，这时该基因会产生部分发育的结构，然而实际上在种系史的任何阶段我们都没发现部分发育的**无功能**结构的出现。因此，事实上基因肯定是在适当环境出现之后才产生的。从这一点，我们可以得到的唯一结论是：基因与环境之间存在因果关系，即，从某种意义来说，环境在某种程度上决定了基因的表现形式，这正好可以作为获得性性状遗传的新证据。

珀泽

* * *

我已经饶有兴趣地读完了珀泽先生就我的评论所写的颇具思想性的快报。如果他用“种族记忆”一词来代替“基因”，那么我们的分歧并不大。但是，孟德尔式基因正是因其根本不具有功能而格外引人注目。已故的阿奇德尔 · 里德爵士说：“没有人曾听说过一个有用的基因。”考虑到果蝇的孟德尔式基因对该物种生存能力的损伤有所增加，且损伤增加的程度与相关基因发生的结构变化成比例这一事实，再加上人们又发现可以通过 X 射线辐射昆虫卵子来人为地产生基因（这一辐射过程能够杀

to the conclusion that a gene is germ damage of which the outward manifestation is a mutation. The only effect that natural selection would have on such aberrations would be to wipe them out. In my opinion, mutations and adaptations have nothing to do with one another and only adaptations are recapitulated in ontogeny.

E. W. MacBride

(**126**, 918-919; 1930)

G.L. Purser: The University, Aberdeen, Oct. 29.

死大部分卵子)，我们就会被引向如下结论：基因是外在表现为突变的配子损伤。自然选择对这种异常的唯一反应就是清除它们。在我看来，突变和适应没有任何关系，只有适应性能够在个体发育中重演。

麦克布赖德

（刘皓芳 翻译；刘京国 审稿）

Unit of Atomic Weight

F. W. Aston

Editor's Note

Francis Aston, working at the Cavendish Laboratory of the University of Cambridge, had devised an instrument for measuring the atomic masses of individual atoms, now called the mass spectrometer. This led him earlier to postulate the notion of isotopes, which have identical chemical properties but different masses. Having used the device to identify the isotopes of more than 80 different chemical elements, Aston here advocates the need for a new standard of atomic mass, to replace the practice then current of referring all masses to that of oxygen—for this element, having several isotopes, is not an appropriate reference point.

THE discovery of the complexity of oxygen clearly necessitates a reconsideration of the scale on which we express the weights of atoms. Owing to the occurrence of O^{17} and O^{18}, now generally accepted, it follows that the mean atomic weight of this element, the present chemical standard, is slightly greater than the weight of its main constituent O^{16}. The most recent estimate of the divergence is 1.25 parts per 10,000.

This quantity, even apart from its smallness, is not of much significance to chemists, for the experience of the last twelve years has shown that complex elements do not vary appreciably in their isotopic constitution in natural processes or in ordinary chemical operations. Physics, on the other hand, is concerned with the weights of the individual atoms, and by the methods of the mass-spectrograph and the analysis of band spectra it is already possible to compare some of these with an accuracy of 1 in 10,000. Furthermore, the theoretical considerations of the structure of nuclei demand an accuracy of 1 in 100,000, which there is reasonable hope of attaining in the near future. The chemical unit is clearly unsuitable, and it seems highly desirable that a proper unit for expressing these quantities should be decided upon.

The proton, the neutral hydrogen atom, one-quarter of the neutral helium atom, one-sixteenth of the neutral oxygen atom 16, and several other possible units have been suggested. None of these is quite free from objection. It is desirable that this matter should be given attention, so that when a suitable opportunity occurs for a general discussion of the subject, each point of view may be afforded its proper weight in arriving at a conclusion.

(**126**, 953; 1930)

F. W. Aston: Trinity College, Cambridge, Dec. 4.

原子量的单位

阿斯顿

编者按

在剑桥大学卡文迪什实验室工作的弗朗西斯·阿斯顿设计了一种用来测量单个原子的原子质量的仪器，现在我们称之为质谱仪。这使得他更早地提出了同位素（化学性质完全相同但质量不同）的概念。利用该仪器鉴定了超过 80 种化学元素的同位素以后，在这篇文章中阿斯顿主张，有必要采用一种新的原子质量的标准，以取代当时测量所有其他原子质量时利用氧作为标准的做法——因为氧有多种同位素，不适合作为原子质量的参考标准。

氧元素具有多种同位素，这一发现无疑使我们必须重新考虑用以表述原子量的标度。目前在化学上是以氧元素的平均原子质量作为原子量的标准，但由于现在同位素 O^{17} 和 O^{18} 的发现已经得到了普遍的认可，所以氧元素的平均原子质量略大于氧元素中的主要组成部分 O^{16} 的原子质量。最新的估计表明这一差别是 0.125‰。

这个差值对于化学家来说没有太大的意义，更别说它还非常微小，因为最近 12 年的研究经验已经表明，对于同位素形式复杂多样的元素来说，在自然过程或者化学操作中其各种同位素的相对丰度并不发生明显改变。但是，物理学研究要考虑单个原子的质量，通过质谱仪和谱带分析的方法目前已经可以以万分之一的精度来分辨某些原子质量的差别。此外，对原子核结构的理论研究需要十万分之一的精度，在不久的将来对原子质量的测定有望能够达到这一精度。因此，目前化学上采用的原子量的单位显然是不合适的，对于物理学研究来说，似乎迫切需要确定一个能够描述这些量的合适的原子量单位。

质子的质量，中性氢原子的质量，氦原子质量的 1/4，中性氧同位素 O^{16} 原子质量的 1/16，以及其他几种可能的原子量单位都被提出来了。但所有这些无一例外都遭到了一些反对。这个问题应该受到关注，一旦出现一个对这一主题进行广泛讨论的合适时机，那么为了得出结论每一种观点都可以适当地发挥作用。

（王锋 翻译；李芝芬 审稿）

Embryology and Evolution

J. B. S. Haldane

Editor's Note

Renowned biologist J. B. S. Haldane was one of the key figures in the development of population genetics, a field underpinned by a Mendelian view of inheritance. Here Haldane has no time for the views of the neo-Lamarckian zoologist Ernest William MacBride, one of a dwindling number of scientists still prepared to dismiss the notion of a gene. With characteristic rhetorical flair, Haldane ridicules MacBride for his outdated views. Publicly aired disagreements like these played an important part in turning the scientific community towards a modern evolutionary synthesis, the idea that a Mendelian mechanism of inheritance could result in the sort of gradual natural selection that Charles Darwin envisaged.

FOUR of Prof. MacBride's statements, in *Nature* of Dec. 6, call for comment. "...no one has ever seen 'genes' in a chromosome." Genes cannot generally be seen, because in most organisms they are too small. In *Drosophila* more than 100, probably more than 1,000, are contained in a chromosome about 1 μ in length. They are therefore invisible for exactly the same reasons as molecules. But the evidence for their existence is, to many minds, as cogent. Where the chromosomes are larger, as in monocotyledons, competent microscopists—for example, Belling, in *Nature* of Jan. 11, 1930—claim to have seen genes. In a case where I (among others) postulated the absence of a gene in certain races of *Matthiola*, my friend Mr. Philp has since detected the absence of a trabant, which is normally present, from a certain chromosome. I shall be glad to show this visible gene to Prof. MacBride.

"...if Prof. Gates were a zoologist instead of being a botanist, he would know that the assumption that 'genes' have anything to do with evolution leads to results...that can only be described as farcical." I should like to direct Prof. MacBride's attention to the droll fact that in a good many interspecific crosses various characters behave in a Mendelian manner, that is, are due to genes. This is so, for example, with the coat colour of *Cavia rufescens*, which, on crossing with the domestic guinea-pig, behaves as a recessive to the normal coat colour, but a dominant to the black. Hence there has been a change in a gene concerned in its production during the course of evolution. Scores of similar cases could be cited.

"All known chemical actions are inhibited by the accumulation of the products of the reaction. An 'autocatalytic' reaction, in which the products of the reaction accelerated it, must surely be a vitalistic one!" Autocatalytic reactions are common both in ordinary physical chemistry and in that of enzymes. Thus the acid produced by the hydrolysis of an

胚胎学与进化

霍尔丹

编者按

在以孟德尔遗传学说为基础的群体遗传学的发展中，著名生物学家霍尔丹算是一个重要的人物。在这篇文章中，霍尔丹没有理会支持新拉马克主义的动物学家欧内斯特·威廉·麦克布赖德（他仍然拒绝接受基因的概念，持这种观点的科学家已经越来越少了。）的观点，而是以他特有的文字才能嘲弄了麦克布赖德的过时思想。在公开场合进行这样的争论，对于学术界最终转向现代进化综合论起到了重要的作用。在现代进化综合论中，可以由孟德尔的遗传机制推出查尔斯·达尔文设想的自然选择学说。

我要对麦克布赖德教授发表于12月6日的《自然》上的4项陈述稍作评论。他说，“……没有人看见过染色体上的‘基因’。”确实，基因一般是看不见的，因为大多数生物的基因都太小了。果蝇的一条长约1μ的染色体上可能就有一百多个甚至一千多个基因。因此，就像分子是不可见的一样，基因也是不可见的。不过许多人都认为用来证明基因确实存在的证据是令人信服的。另外，一些有能力的显微镜专家曾公布过他们在具有较大染色体的单子叶植物中看到了基因，例如贝林就在1930年1月11日的《自然》上宣称看到了基因。我（和其他人一起）曾经在一项研究中推测紫罗兰属的某些种是没有基因的，后来我的朋友菲尔普先生检测到了特定染色体上随体的缺失，而在正常情况下这一染色体上是存在随体的。我很乐意向麦克布赖德教授展示这个可见的基因。

“……如果盖茨教授不是植物学家而是动物学家的话，他将了解‘基因’与进化具有某些关联，这样的假设产生的结果……只能用滑稽可笑来形容。”我期望麦克布赖德教授能注意到下面这个很有趣的事情：即在许多种间杂交的实例中，很多性状都表现出孟德尔式遗传的特点，也就是说它们是由基因决定的。例如，有一种野生巴西豚鼠，当它与家养豚鼠杂交时，其毛色对于正常毛色表现为隐性，但对于黑色却表现为显性。因此，在进化过程中，与产生该性状相关的基因一定发生了某种变化。与此类似的例子还有很多。

“所有已知的化学反应都会被反应产物的积累所抑制。而在‘自催化’反应中，反应产物却可以加快反应的进行，因而这种反应显然就是活力论的一个实例！”其实，自催化反应在普通物理化学和酶化学中都是很常见的。酯水解产生的酸可以促

ester may accelerate its further hydrolysis. As an example of an enzyme action, which for quite simple physico-chemical reasons proceeds with increasing velocity up to 75 percent completion, I would refer Prof. MacBride to Table 7 of Bamann and Schmeller's[1] paper on liver lipase.

In view of such facts, Prof. MacBride's statement that "The term 'autocatalysis' is a piece of bluff invented by the late Prof. Loeb to cover up a hole in the argument in his book" would seem to be a wholly unfounded attack on a great man who can no longer defend himself. If Prof. MacBride would acquaint himself with the facts of chemistry and genetics, he might be somewhat more careful in his criticism of those who attempt to analyse the phenomena of life. He might also cease to ask the question propounded by him in *Nature* of Oct. 25, "whether the organs of the adult exist in the egg preformed in miniature and development consists essentially in an unfolding and growing bigger of these rudiments, or whether the egg is at first undifferentiated material which from unknown causes afterwards becomes more and more complicated and development is consequently an 'epigenesis'." The formation of bone in the embryo chick was shown by Fell and Robison[2] to be due to the action of the enzyme phosphatase, which is neither a miniature bone nor an unknown cause. But so long as he does not take cognisance of recent developments in science, Prof. MacBride will no doubt remain a convinced vitalist.

(**126**, 956; 1930)

J. B. S. Haldane: Biochemical Laboratory, Cambridge University, Dec. 8.

References:

1. *Zeit. Physiol. Chem.*, **188**, p. 167.

2. *Biochem. Jour.*, **23**, p. 766.

进酯的进一步水解。酶反应中也有因为非常简单的物理化学原因而使得反应速率提高 75% 的例子，我想请麦克布赖德教授看一看巴曼和施梅勒关于肝脂酶的那篇文章[1]中的表 7。

麦克布赖德教授还认为，"'自催化'是已故的洛布教授为掩盖其书中观点的漏洞而发明的欺骗性词语。"而从前述的事实来看，麦克布赖德教授的这一观点似乎完全是对一位不可能再为自己辩解的伟人的毫无根据的攻击。如果麦克布赖德教授了解化学和遗传学事实的话，那他在批评那些试图分析生命现象的人时可能会更加谨慎些。他可能也就不会提出自己在 10 月 25 日的《自然》上提到的问题："是不是卵子中就存在预先成形了的成体器官的缩微版，而发育过程实质上是这些器官雏形逐渐展开显露并变大的过程？又或者卵子起初只是未分化的物质，后来由于未知的原因而变得越来越复杂，因而发育是一个'渐成的'过程？"费尔和罗比森的研究[2]显示，鸡胚中骨骼的形成是由于磷酸酶的作用，因而既不存在微型骨骼，也不是由于未知的原因。但是，在麦克布赖德教授认识到科学领域的最新进展之前，他无疑将依旧是一个固执的活力论者。

（刘皓芳 翻译；刘京国 审稿）

Appendix: Index by Subject
附录：学科分类目录

Physics
物理学

Chemistry
化学

Biology
生物学

Astronomy
天文学

Geoscience
地球科学

Engineering & Technology
工程技术

Others
其他